ENCYCLOPEDIA OF GLOBAL WARMING & CLIMATE CHANGE

SECOND EDITION

3

ENCYCLOPEDIA OF GLOBAL WARMING & CLIMATE CHANGE

SECOND EDITION

GENERAL EDITOR

S. GEORGE PHILANDER

Princeton University

SAGE reference

Los Angeles | London | New Delhi
Singapore | Washington DC

Los Angeles | London | New Delhi
Singapore | Washington DC

FOR INFORMATION:

SAGE Publications, Inc.
2455 Teller Road
Thousand Oaks, California 91320
E-mail: order@sagepub.com

SAGE Publications India Pvt. Ltd.
B 1/I 1 Mohan Cooperative Industrial Area
Mathura Road, New Delhi 110 044
India

SAGE Publications Ltd.
1 Oliver's Yard
55 City Road
London EC1Y 1SP
United Kingdom

SAGE Publications Asia-Pacific Pte. Ltd.
3 Church Street
#10-04 Samsung Hub
Singapore 049483

Vice President and Publisher: Rolf A. Janke
Senior Editor: Jim Brace-Thompson
Project Editor: Tracy Buyan
Cover Designer: Gail Buschman
Editorial Assistant: Michele Thompson
Reference Systems Manager: Leticia Gutierrez
Reference Systems Coordinator: Laura Notton
Marketing Managers: Kristi Ward, Ben Krasney

Golson Media
President and Editor: J. Geoffrey Golson
Director, Author Management: Susan Moskowitz
Production Director: Mary Jo Scibetta
Layout and Copy Editor: Stephanie Larson
Proofreaders: Mary Le Rouge, Rebecca Kuzins
Indexer: J S Editorial

Printed in the United States of America

Library of Congress Cataloging-in-Publication Data

Encyclopedia of global warming and climate change / S. George Philander, general editor. – 2nd ed.
p. cm.
Includes bibliographical references and index.
ISBN 978-1-4129-9261-9 (cloth)
1. Global warming–Encyclopedias. 2. Climatic changes–Encyclopedias. I. Philander, S. George.
QC981.8.G56E47 2012
363.738'7403–dc23

2012002545

12 13 14 15 16 10 9 8 7 6 5 4 3 2 1

Contents

Volume 3

List of Articles

D

E

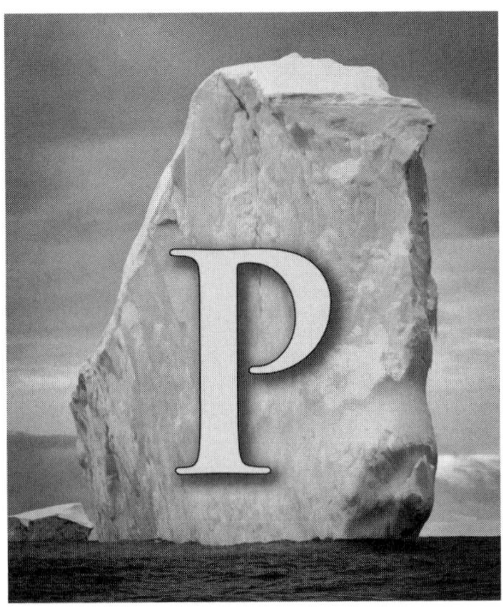

Pacific Ocean

The Pacific Ocean—named the "peaceful sea" by Ferdinand Magellan, a Portuguese explorer leading a Spanish expedition—is the largest ocean in the world, covering 65.3 million sq. mi. (169.2 million sq. km), encompassing 32 percent of the total surface of the Earth and holding 46 percent of the Earth's water. Altogether, there are 25,000 islands in the Pacific, the vast majority south of the equator, which bisects the ocean.

Rising Sea Levels

Global warming and climate change pose many real threats to the Pacific Ocean. The major focus of much attention around the world has been on the rising water levels, which is likely to inundate many of the low-lying Pacific Islands. Independent countries such as Fiji, Kiribati, the Federated States of Micronesia, Nauru, Palau, Samoa, and Tuvalu risk losing the vast majority of their land if the rising world temperature continues to raise the water level of the ocean. Atolls in French Polynesia and in Wallis and Futuna are also under threat. In addition to those places, all the countries in the Pacific have an increased risk of flooding, which could lead to permanent soil loss, as well as an increased risk of the prevalence of insect-borne diseases such as malaria and dengue fever as mosquitoes find further breeding grounds. The rising sea levels also threaten mangrove swamps in many areas, including off the northeastern coast of Australia and in many Pacific Islands, with 13 percent of the world's mangrove swamps at risk of being lost.

For this reason, many of the countries in the Pacific have been at the forefront of urging countries around the word to embrace the Kyoto Protocol and limit carbon dioxide emissions. The Republic of Nauru, the country with the highest per capita rate of carbon dioxide emissions in the Pacific, went as far as adding a long addenda to the Kyoto Protocol, arguing that it did not feel that the protocol went far enough. Two U.S. territories in the Pacific, Guam and American Samoa, have considerable carbon dioxide emissions. The Solomon Islands, Papua New Guinea, and Vanuatu have, respectively, the lowest rates of carbon dioxide emissions in the Pacific, at rates similar to that of many African countries.

Changes to Marine Life

Other problems in the Pacific Ocean regarding global warming focus on the marine flora and fauna. The most dramatic effect has been the bleaching of coral reefs around the Pacific, with studies by the International Ocean Institute of the University of the South Pacific in Fiji conducting surveys of coral reefs in the southwest Pacific as part of the International Coral Reef Initiative. In

Reefs in Fiji, in the Pacific Ocean. Global warming and climate change many cause a rise in sea level, threatening lowland areas with the increased risk of flooding, soil loss, and human disease from insect-borne diseases such as malaria.

many cases, the damage to coral reefs has come from overpopulation, and through overexploitation through tourism, but even many reefs located in remote parts of the Pacific have experienced bleaching, showing that the damage can be ascribed as much to global warming as to other problems.

As well as coral reefs, there have been significant changes to marine life, especially the fish in the Pacific. The most dramatic changes have been the reduction in the diversity of fish shoals, as well as the decline in the number of fish, the latter probably as much from overfishing as from global warming. However, there still remain large numbers of tuna fish and also some cluepoids in the central part of the Pacific Ocean, as well as sardines and jack mackerel along the coast of Chile, anchovy off the coast of Ecuador and Peru, mackerel and Saury off the Pacific coasts of Mexico and the United States, and sardine and salmon off the Pacific coast of Canada.

Some 4 percent of the ozone in the Earth's stratosphere is lost each decade, and a hole has appeared over Antarctica, leading to a higher risk of skin cancer from ultraviolet light in places such as Australia, New Zealand, Chile, and southern Argentina. Although there has been a great focus on their effect on humans, ultraviolet rays have also been linked to a reduction in the plankton population in the southern part of the Pacific Ocean. The removal of much of the plankton has major effects on the food chain throughout the Pacific, especially on the whale population, which has been growing following a moratorium on commercial whaling in 1986, although Japan continues whaling for ostensibly "scientific" reasons.

One last major area of problems in the Pacific Ocean through global warming and climate change has been changes in the ocean currents, which have been caused by the rise in the temperature of the water. Although few Polynesians travel long distances in traditional canoes, as they did about 1,000 years ago during the populating of many of the islands, the currents are very important, not just for shipping, but also for the movement of marine life such as shoals of fish. The warmer temperature and changes in the current have had major effects on the spawning process of some fish species, and this may be responsible for a decline in the population of certain fish.

Although there is a serious worry about global warming and its effects on the Pacific Ocean, one report in 1997 by scientists from the Lamont-Doherty Earth Observatory at Columbia University claims that the vast size of the Pacific Ocean has led to the dissipation of many of the effects of global warming and climate change, and might account for the fact that the world's temperature has only risen half the level of that in some projections.

Justin Corfield
Geelong Grammar School

See Also: Alliance of Small Island States; Floods; Marine Mammals; Oceanic Changes; Sea Level, Rising.

Further Readings

Kawahata, Hodaka and Yoshio Awaya. *Global Climate Change and Response of Carbon Cycle in the Equatorial Pacific and Indian Oceans and Adjacent Land Masses*. Oxford: Elsevier, 2006.

Morrison, John, Paul Geraghty, and Linda Crowl, eds. *Science of Pacific Island Peoples*. Suva, Fiji: Institute of Pacific Studies, 1994.

Shibuya, Eric. "Roaring Mice Against the Tide: The South Pacific Islands and Agenda-Building on Global Warming." *Pacific Affairs*, v.69/4 (1996–97).

von Storch, Hans and Ann Smallegauge. *The Phase of the 30- to 60-Day Oscillation and the Genesis of Tropical Cyclones in the Western Pacific.* Hambourg: Max-Planck-Institut für Meteorologie, 1991.

Wilkinson, Clive, ed. *Status of Coral Reefs of the World.* Cape Ferguson: Australian Institute of Marine Science, 2000.

Pakistan

Pakistan is a south Asian country sharing borders with India, Iran, Afghanistan, and China. It has 649 mi. (1,046 km) of coastline on the Arabian Sea and is the 36th-largest country by area at 307,373 sq. mi. (796,095 sq. km), about twice the size of California. As of July 2011, it was the sixth most populous country in the world with a population of 187.3 million. Pakistan is mostly hot, dry desert with a temperate climate in the northwest and arctic in the north. The land is a flat plain in the east with the Balochistan plateau in the west and mountains in the north and northwest, and the Indus River runs the entire length of the country. About a quarter of the land is arable, and 43 percent of the workforce is engaged in agriculture. Per-capita gross domestic product (GDP) in 2010 was $2,500, ranking 179th in the world, with a highly uneven distribution of income (Gini index of 41) and 24 percent of the population living below the poverty line. The fertility rate of 3.17 children per woman (54th highest in the world) and birth rate of 24.81 per 1,000 population are somewhat tempered by a negative migration rate (meaning more people leave the country than enter it) of minus 2.17, resulting in a population growth rate of 1.57 percent.

Pakistan has proven oil reserves of 0.29 billion barrels (49th highest in the world) and 31 billion cu. ft. of proven natural gas reserves (25th highest in the world). Oil production in 2008 was 61.57 thousand barrels per day (ranked 58th in the world), much less than what was necessary to meet consumption of 386 thousand barrels per day (34th in the world). Refinery capacity in 2008 was 22.2 thousand barrels per day. Natural gas production in 2009 was 1,356 billion cu. ft. (22nd in the world), as was natural gas consumption. Pakistan is a net importer of coal, producing 4.12 million short tons of coal in 2008 (35th highest in the world), while consuming 9.23 million short tons. In 2009, Pakistan generated 90.8 billion kilowatt hours (kWh) of electricity (32nd highest in the world), consuming 72.2 billion kWh. Installed electrical generating capacity in 2008 was 19.41 gigawatts. Carbon dioxide (CO_2) emissions from consumption of fossil fuels in 2008 amounted to 139.75 million metric tons, 34th in the world.

Although Pakistan's per-capita emissions of greenhouse gases (GHGs) falls below the global average (true of most developing countries), it has already been seriously affected by global warming. Pakistan is prone to many types of natural disasters, including floods, drought, cyclones, earthquakes, and heavy rainfall, and an estimated 40 percent of Pakistani citizens are highly vulnerable to natural disaster. This has increased in recent years as many parts of the country have experienced an increase in extreme climatic events (heavier rainfall, more intense cyclones, more pronounced droughts, and increased flooding, depending on the area). Additionally, in many parts of the country, rainfall has become increasingly erratic, making it difficult for farmers to predict the crop season.

Summers have become hotter and winters warmer, and traditional measures to deal with water shortages are no longer sufficient to deal with current conditions. Agricultural productivity has been lower in many areas and the fish and prawn catch has been reduced. Groundwater has become more brackish and the land has become more saline in coastal areas because of intrusion of seawater, while water shortages have led to a reduction in both quality and quantity of rangeland available for livestock.

The most extreme weather event to date that has been tied to global warming is the flooding caused by unusually intense monsoon rains in July and August 2010, which resulted in flooding over about 20 percent of Pakistan's territory. The United Nations and many climatologists attributed this event to global warming, specifically to higher temperatures in the Atlantic Ocean. An estimated 1,500–2,000 people died during the flooding, while millions more were affected by loss of property and livelihood; damage to infrastructure was estimated at over $1 billion.

Sarah Boslaugh
Kennesaw State University

See Also: Agriculture; Developing Countries; Drought; Floods; Sea Level, Rising.

Further Readings

Abbas, Zehar. "Climate Change, Poverty, and Environmental Crisis in the Disaster-Prone Areas of Pakistan." Oxfam GB (2009). http://www.oxfam.org.uk/resources/policy/climate_change/climate-change-poverty-pakistan.html (Accessed July 2011).

Gronewaold, Nathanial. "Is the Flooding in Pakistan a Climate Change Disaster?" *Scientific American* (Aug. 18, 2010). http://www.scientificamerican.com/article.cfm?id=is-the-flooding-in-pakist (Accessed July 2011).

U.S. Energy Information Administration. "Country Information: Pakistan" (June 30, 2010). http://www.eia.gov/countries/country-data.cfm?fips=PK (Accessed July 2011).

Palau

The Republic of Palau (sometimes spelled Belau, as it is pronounced by natives) is an island nation in the Pacific Ocean, 500 mi. (804 km) east of the Philippines and 2,000 mi. (3,218 km) south of Japan. Formerly controlled by the Spanish, Germans, Japanese, and United States, the islands were passed to the trusteeship of the United Nations (UN) in 1947. Palau declined to join the federated Micronesian state created in 1978 and instead drafted a constitution and formed the Republic in 1981, eventually being released from UN trusteeship in 1994. It is one of the world's youngest sovereign nations, and one of the smallest, with a 2010 population of 20,879.

The republic's most populous islands are Koror (inhabited by two-thirds of the population), Babeldaob, and Peleliu, all along the same barrier reef, and Angaur to the south. The islands lie outside the typhoon zone and are not often subject to extreme weather. The climate is tropical year-round, with little variation in temperature from month to month, remaining close to the annual mean of 82 degrees F (27 degrees C). Humidity is nearly always high, and rainfall is concentrated in the summer and early autumn months.

Palau is a prosperous country, and its per-capita carbon dioxide (CO_2) emissions are high, although declining after public- and private-sector efforts. Air conditioning is a significant contributor to emissions and energy use, and automobile use and ownership is prevalent. In 2005, Palau joined the regional initiative called the Micronesia challenge, aimed at conserving 30 percent of near-shore coastal waters and 20 percent of forestland by 2020. Palau is also a signatory to the UN Framework Convention on Climate Change signed in Rio de Janeiro in 1992 and the Kyoto Protocol signed in 1999, which entered into force in 2005.

The public water supply is an ongoing concern, in part because of the poor quality of solid-waste disposal facilities in Koror. Rising sea levels are of great concern, threatening the drinking water supply, agriculture, and many residences, as well as the potential loss of some of the low-lying islands.

Bill Kte'pi
Independent Scholar

See Also: Australia; Kyoto Protocol; Micronesia; Philippines.

Further Readings

Black, Richard. "Palau Pioneers Shark Sanctuary." BBC News. http://usproxy.bbc.com/2/hi/science/nature/8272508.stm (Accessed July 2011).

Hong Kong Observatory. "Climatological Information for Palau Islands." http://www.weather.gov.hk/wxinfo/climat/world/eng/australia/pacific/palau_islands_e.htm (Accessed July 2011).

Paleoclimates

Paleoclimates are past climates. The term *paleo* is, however, normally reserved for those time periods that are prehistorical, although usage of the term to describe climate in the first millennium C.E. and earlier is not uncommon. Like reading a historical text, paleoclimates give glimpses into the varied past of Earth's climate system, and understanding of paleoclimates may well be critical to understanding how Earth's climate will change in the future.

Unfortunately, understanding of this critical climate history does not come easily. Unlike present climate that is carefully recorded and catalogued by various observational and physical measurements, paleoclimate records must be extracted from Earth's geological history; in particular, the archives of sedimentary rocks and large ice sheets. These geological archives do not contain precise measurements of climatic variables such as temperature and precipitation, but rather record environmental responses to ambient climate conditions. As such, the climate indicators extracted from the geological record are known as proxy data.

Proxy Data

An example of the use of proxy data would be conducting a North American winter snow survey by searching North American garages for snow shovels in the summer. In regions where snow shovels are common in garages, a researcher might reasonably interpret cold winter temperatures and the presence of snow. The shovels, though not actually snow, are a proxy for the occurrence of winter snow. In regions without snow shovels, winter temperatures are either warmer with no snow, or snow is removed in a different way. Because proxy data may be definitive in one sense, but ambiguous in another, using them to decipher the nature of paleoclimates requires a suite of data types, or a multiproxy approach. Fortunately, paleoclimatologists have proven extremely innovative in their development of climate proxies, which range from preserved vegetation in ancient pack rat middens (nests) to the relative abundances of different stable oxygen isotopes (notably ^{16}O and ^{18}O) in the shells of small, ocean-dwelling plankton known as foraminifera.

Some of the oldest proxy data are those related to preserved vegetation. That vegetation may be actual fossils (permineralized plant material or traces of plant material) or simply desiccated (mummified) plant remains, some of which can be tens of millions of years old in cold climates. Vegetation is used as a climate indicator in two primary ways. The first is based on the observation that certain types of plants inhabit certain types of climates. Alders, for example, prefer wetter climates, while Ponderosa pines thrive in drier conditions. As such, if a plant fossil is found in a given location and there is reasonable certainty that the plant actually grew there, climate in that location can be interpreted based on the known preferences of the identified vegetation.

This is relatively straightforward when the preserved plant material is from a known, living species such as an alder or Ponderosa pine. It becomes increasingly difficult further back in geological time when many plants are unknown. In these instances, the climate tolerances of the extinct plant's nearest living relative are used to interpret climate. The other common use of preserved vegetation as a climate indicator is through a statistical process known as leaf margin analysis. Statistical studies of large, modern datasets have indicated that the shape of a leaf's margin (smooth or toothed, elongated or rounded) has some correlation to the climatic conditions that the plant prefers. By performing the same analyses on ancient leaves, scientists can compare ancient leaf samples to the modern climate interpretation and make some assessment of paleoclimatic conditions.

The Petrified Forest in northeastern Arizona has one of the best fossilized records of the Late Triassic period in the world. Fossils found here show that the forest was once a tropical region, filled with towering trees and species of plants and reptiles.

While analyses of preserved vegetation continue to provide significant insight into paleoclimates, more recently, scientists have developed a suite of chemically based climate proxies that make use of the lengthy climate records stored in the sediments of the ocean floor. The most common of these chemical proxies is based on the stable isotopes of oxygen (in particular ^{16}O and ^{18}O, which are the most abundant). Because different isotopes of a given element have different masses, they are relatively easier or harder to evaporate and, inversely, relatively harder or easier to precipitate. This means that water vapor will always have more ^{16}O than the liquid it evaporated from, and is therefore lighter. Liquid precipitation will always have more ^{18}O than the vapor it condensed from, and is therefore heavier. These processes can be used to decipher past climate.

For icehouse climates with large, continent-spanning ice sheets, the relative volume of water stored in large glaciers largely influences the oxygen isotopic concentration of global seawater. Because lighter isotopes evaporate preferentially, precipitation is almost always lighter than the global ocean. If precipitation falls as snow and is then trapped as glacial ice, the overall ocean isotopic composition will get heavier as more and more of the light isotope is locked up in growing ice sheets. If climate warms and ice sheets begin to melt, the lighter isotopes are returned to the ocean and the ocean isotopic composition becomes lighter. Thus, changes in the oxygen isotopic composition of ocean water can be tied to growing and shrinking ice sheets and, by extension, cooling and warming climate.

The primary source of information on past oceanic oxygen isotopic composition is the shells of ocean dwelling microorganisms, in particular, foraminifera. Foraminifera create their shells out of calcium carbonate ($CaCO_3$). The oxygen incorporated into the foraminifera shells comes from the ocean water, and therefore reflects the ocean's oxygen isotopic composition. The exact isotopic ratio of the shell, however, is also influenced by the ambient temperature and to a lesser and largely negligible extent salinity, as well as the organic processes of the foraminifera. Empirically derived equations relate the oxygen isotopic composition of calcium carbonate shells to the ocean temperature and oxygen isotopic composition. If any two of those values are known, the third can be calculated. While the isotopic composition of ancient foraminifera shells can be measured in the laboratory, the overall oceanic oxygen isotopic composition and temperature in the past must either be estimated or derived from other proxies.

Climate Extremes

This panoply of proxy climate indicators records a startling array of paleoclimates in Earth's history. Climatic conditions on Earth have ranged from extreme icehouse conditions with potentially the entire planet covered in glaciers (a paleoclimate known as snowball Earth) to extreme hothouse conditions, with atmospheric carbon dioxide concentrations as much as 20 times higher than those at present, and tropical forests extending nearly pole to pole. Earth's climate has also apparently resided everywhere inbetween these extremes and at times moved rapidly from one to another.

Whether it is because the current climate falls relatively in the middle of the climate spectrum, or because extremes are more likely to be preserved in the geological record, or because understanding of extremes may provide the greatest insight into the climate system as a whole, the extreme paleoclimate events are the most studied. In the realm of extreme warmth, there were the hothouse climates of the Cenomanian/Turonian boundary (90 million years ago) and the Early Eocene Climatic Optimum (52 million years ago), or nearer-term warm climates like the Miocene Climatic Optimum (14 million years ago), the mid-Pliocene warm period (3.5 million years ago) or the Altithermal of the middle Holocene (5,000 years ago).

At the other end of the spectrum lies the extreme cooling of snowball Earth (630 million years ago), the rapid inception of large Antarctic ice sheets (35 million years ago), or the peak glaciation of the last glacial maximum (18,000 years ago). Equally fascinating, though even more difficult to quantify, are transient or abrupt climate changes such as Pleistocene Heinrich and Dansgaard-Oeschger events where circum North Atlantic temperature changed by as much as 9 degrees F (5 degrees C) in 30 or 40 years, or the Initial Eocene Thermal Maximum, when temperatures in the Arctic Ocean reached 73 degrees F (23 degrees C) for 50,000 to 100,000 years. Climate extremes such as these in Earth's history and incomplete explanations for them help show that the current climate samples a very finite portion of Earth's climatic possibilities, and if scientists wish to have a solid understanding of what the climate of the future may hold, paleoclimates must first be understood.

Jacob O. Sewall
Virginia Polytechnic Institute and State University

See Also: Cenozoic Era; Climap Project; Climatic Data, Proxy Records; Earth's Climate History; Ice Ages; Mesozoic Era; Paleozoic Era; Snowball Earth.

Further Readings

Bradley, R. S. *Paleoclimatology: Reconstructing Climates of the Quaternary*. San Diego, CA: Academic Press, 1999.
Crowley, T. J. and G. R. North. *Paleoclimatology*. Oxford: Oxford University Press, 1995.

Paleozoic Era

The Paleozoic era is the earliest of three geologic eras of the Phanerozoic eon. This era spanned from roughly 542 million years ago to roughly 251 million years ago. The Paleozoic era is subdivided into six geologic periods: the Cambrian, Ordovician, Silurian, Devonian, Carboniferous, and Permian. The Paleozoic covers the time from the first appearance of abundant, hard-shelled fossils to the time when the continents were beginning to be dominated by large reptiles and modern plants. The oldest geological period was classically set at the first appearance of creatures known as trilobites and archeocyathids. The youngest geological period marks a major extinction event 300 million years ago, known as the Permian extinction.

At the start of the era, all life was confined to bacteria, algae, sponges, and a variety of enigmatic forms known collectively as the Ediacaran fauna. The Cambrian explosion resulted in an exponential increase of lifeforms. There is some evidence that simple life may already have invaded the land at the start of the Paleozoic, but substantial plants and animals did not take to the land until the Silurian and did not thrive until the Devonian. Although primitive vertebrates are known near the start of the Paleozoic, invertebrates were the dominant lifeforms until the mid-Paleozoic. Fish populations exploded in the Devonian. During the late Paleozoic, great forests of primitive plants thrived on land, forming the great coal beds of Europe and eastern North America. By the end of the era, the first large, sophisticated reptiles and the first modern plants had developed.

The Paleozoic era began shortly after the breakup of a supercontinent called Pannotia and at the end of a global ice age. During the early Paleozoic, the Earth's landmass was broken up into a number of relatively small continents. Toward the end of the era, the continents gathered together into a supercontinent called Pangaea, which included most of the Earth's land area.

The early Cambrian climate was probably moderate at first, becoming warmer over the course of the Cambrian, as the second-greatest sustained sea-level rise in the Phanerozoic got underway. Gondwana moved south with considerable speed. By the Ordovician period, most of West Gondwana (Africa and South America) lay directly over the South Pole. The early Paleozoic climate was also strongly zonal. The climate became warmer, but the continental shelf marine environment became steadily colder. The early Paleozoic ended rather abruptly with the short, but apparently severe, late Ordovician ice age. This cold spell caused the second-greatest mass extinction of Phanerozoic time. The middle Paleozoic was a time of considerable stability. Sea levels had dropped coincident with the ice age, but slowly recovered over the course of the Silurian and Devonian.

The slow merger of Baltica and Laurentia and the northward movement of bits and pieces of Gondwana created numerous new regions of relatively warm, shallow seafloor. The far southern continental margins of Antarctica and West Gondwana became increasingly less barren. The Devonian period (410 to 360 million years ago) resulted in diversifiaction of life on the land, including the first terrestrial vertebrates, amphibians, and forests of trees. In the waters, fish continued their diversification with the rise of the lobe-finned and ray-finned fish. The Devonian ended with a series of turnover pulses that killed off much of middle Paleozoic vertebrate life, without noticeably reducing species diversity overall. Global cooling tied to Gondwanan glaciation has been proposed as the cause of the Devonian extinction, as it was also suspected of causing the terminal Ordovician extinction. Rocks in parts of Gondwana suggest a glacial event. The forms of marine life most affected by the extinction were the warm-water to tropical species.

The late Paleozoic consisted of the Carboniferous period (360 to 286 million years ago), also known as the Mississippian period. The period began with a spike in atmospheric oxygen, while carbon dioxide plummeted. This destabilized the climate and led to multiple ice age events during the Carboniferous. The supercontinent of Pangaea was assembled during this time, causing the uplift of seafloor as continental land masses collided to build the Appalachian and other mountains. This created huge arid inland areas subject to temperature extremes. The Permian period spanned the time interval from 286 to 245 million years ago. During the Permian, the assembly of Pangaea was completed and a whole host of new groups of organisms evolved.

The Permian ended in the greatest of the mass extinctions, where over 90 percent of all species were extinguished. With the assembly of Pangaea and resulting mountain building, many of the shallow seas retreated from the continents. The Permian saw the spread of conifers and cycads, two groups that would dominate the floras of the world until the Cretaceous period with the rise of the flowering plants. The end of the Permian, also the end of the Paleozoic era, was marked by the greatest extinction of the Phanerozoic eon. During the Permian extinction event, over 95 percent of marine species went extinct, while 70 percent of terrestrial taxonomic families suffered the same fate.

The fusulinid foraminiferans went completely extinct, as did the trilobites. The majority of extinctions seem to have occurred at low paleolatitudes, possibly suggesting some event involving the ocean. The exact cause of the terminal Permian extinction remains unknown; however, many theories have been hypothesized. Regardless, this event proved to be a massive and severe crisis for life. Many groups of organisms went extinct at that time. Surviving groups diversified during the Triassic period, and gradually, a more modern world developed.

Fernando Herrera
University of California, San Diego

See Also: Climatic Data, Proxy Records; Earth's Climate History; Ice Ages.

Further Readings

Montañez, I. P. "CO_2-Forced Climate and Vegetation Instability During Late Paleozoic Deglaciation." *Science*, v.315 (2007).

Palaeos. "Paleozoic Era." http://palaeos.com/paleozoic/index.html (Accessed March 2012).

University of California Museum of Paleontology. http://www.ucmp.berkeley.edu (Accessed March 2012).

Panama

Located in Central America, the Republic of Panama has a land area of 29,157 sq. mi. (78,200 sq. km), a population of just over 3.4 million (May 2010 census), and a population density of 115.3 people per sq. mi. (44.5 per sq. km). In spite of being in the tropics, only 7 percent of the land in the country is arable, the second lowest percentage in Central America, with 20 percent used for meadows and pasture, and 44 percent forestland. The level of carbon dioxide (CO_2) emissions in Panama was 1.3 metric tons per capita in 1990, rising to 2.3 metric tons per person in 2001, and falling slightly to 1.9 metric tons per person by 2003. It rose to 2.2 metric tons per person by 2007, leaving Panama at 121st in per-capita carbon emissions by country. Most of these emissions come from liquid fuels, which make up 89 percent of all CO_2 emissions from the country, with cement manufacturing contributing 6 percent and solid fuels (coal and charcoal) contributing another 3 percent. In 2009, there were 167 cars per 1,000 people in the country, ranking Panama 61st in the world, the highest of any country in Central America.

The Caribbean coast of Panama has long had problems with hurricanes, but the rising water temperatures in both the Caribbean Sea and the Pacific Ocean have led to increased worry over flooding and have caused some bleaching of coral reefs in the Archipelago de Bocas del Toro off the northwest coast. Although some 30 percent of the country has been set aside for conservation, the deforestation of many areas used for pasture, especially for cattle, has led to soil erosion, which has also contributed to the destruction of mangrove swamps. There have also been effects on wildlife in the pristine cloud forest on the Quetzal Trail around the Parque Nacional Volcán Barú, and there are concerns that flooding could lead to a spread of insect-borne diseases such as malaria and dengue fever. These were prevalent until the early 20th century, when there were major moves to prevent infection, and during the 1950s there were major antimalaria campaigns to drain breeding grounds of mosquitoes. More important are the likely effects that rising temperatures will have on water resources in some parts of the country.

The Panamanian government of Guillermo Endara took part in the United Nations Framework Convention on Climate Change (UNFCCC) signed in Rio de Janeiro in May 1992, and the government of Ernesto Pérez Balladares signed the Kyoto Protocol to the UNFCCC on June 8, 1998. The Kyoto Protocol was ratified on March 5, 1999, and came into force on February 16, 2005.

Justin Corfield
Geelong Grammar School
Robin S. Corfield
Independent Scholar

See Also: Diseases; Floods; Hurricanes and Typhoons.

Further Readings

EarthTrends. "Panama: Climate and Atmosphere." http://earthtrends.wri.org/text/climate-atmosphere/country-profile-141.html (Accessed October 2011).

Espinosa, Daly, Abril Mendez, Irina Madrid, and Raul Rivera. "Assessment of Climate Change Impacts on the Water Resources of Panama: The Case of the La Villa, Chiriqui, and Chagres River Basins." *Climate Research*, v.9 (1997).

Pan American Health Organization. "Dengue Fever in Costa Rica and Panama." *Epidemiological Bulletin*, v.15/2 (1994).

Pan American Health Organization. "Dengue in Central America: The Epidemics of 2000." *Epidemiological Bulletin*, v.21/4 (2000).

Rainforest Foundation. "Indigenous Climate Change Strategy in Panama." http://www.rainforestfoundation.org/indigenous-climate-change-strategy-panama (Accessed July 2011).

Papua New Guinea

Papua New Guinea, which became an independent country in 1975, consists of the eastern half of the island of New Guinea as well as numerous smaller islands in the South Pacific. The climate is tropical and the land is mostly mountainous, with less than 1 percent arable. Two-thirds of Papua New Guinea's export earnings come from mineral extraction, while most of the population works in subsistence agriculture. The per-capita gross domestic product (GDP) in 2010 was $2,500, with a Gini index of 50.9 (18th highest in the world), indicating a high level of income inequality.

Papua New Guinea has 0.09 billion barrels of proven oil reserves (65th in the world) and 8 billion cu. ft. of proven natural gas reserves (40th in the world). Exploitation of both resources began in the early 1990s, and in 2009, the country produced 35,050 barrels of oil per day and 5 billion cu. ft. of natural gas. Coal is no longer in use, although it was used in Papua New Guinea in the 1980s and 1990s. Net generation of electricity has risen fairly steadily over the past three decades, from 1.189 billion kilowatt hours (kWh) in 1980 to 2.99 billion kWh in 2008. Carbon dioxide (CO_2) emissions from consumption of fossil fuels was 4.55 million metric tons in 2008 (ranked 127th in the world). As of 2010, almost 40 percent of the electricity in the country was generated by hydroelectric power and 9.1 percent by geothermal sources. However, the country remains a net energy importer, and over 90 percent of residents have no access to electricity. Papua New Guinea has high potential for further development of geothermal energy (which could theoretically meet all its energy needs), as well as solar and hydroelectric energy, but lack of funds and technical expertise remain barriers, as do the extreme cultural diversity of the country and its rugged terrain.

The climate in Papua New Guinea has warmed about 0.54 degree F (0.3 degree C) each decade since the 1970s, one of the fastest warming rates in the world. Mapping of glaciers on the nation's highest peak, Mount Jaya, indicate that they have retreated about 984 ft. (300 m) since the 1970s. In June 2008, residents of Carteret Island, an atoll in the Autonomous Region of Bougainville, requested assistance to be relocated to higher ground, causing them to be dubbed the "world's first climate change refugees" by IRIN, a news service of the United Nations. A sea-level rise of 3.9 in. (10 cm) in the previous 20 years destroyed much of the island's food supply because of the incursion of sea water into the land, and many other atolls are expected to be similarly effected if the sea level continues to rise.

Papua New Guinea has the third-largest tropical rainforest in the world, but widespread logging destroyed about 1.4 percent of the country's forest annually between 1972 and 2002, primarily for logging and land clearance. Estimates are that by 2021, if deforestation continues at this pace, 83 percent of accessible forest and 53 percent of total forest would be seriously degraded or destroyed. Destruction of forest also leads to loss of habitat for many species of plants and animals, of particular concern since scientists believe that the interior of Papua New Guinea is home to many species that are not yet known to science.

Malaria remains a significant health threat in Papua New Guinea, with about 700 deaths from the disease in 2007 (a number the World Health Organization considers a gross undercount). This problem is expected to increase with global warming, particularly in the highlands regions. According to a study by the Papua New Guinea Institute of Medical Research, 65 percent of the country's population is currently at risk from endemic malaria, but given the expected increase in temperatures over the next century, 95 percent of the country's population will be at risk by 2100.

Sarah Boslaugh
Kennesaw State University

See Also: Deforestation; Developing Countries; Sea Level, Rising.

Further Readings

Renewable Energy and Energy Efficiency Partnership. "Policy Database Details: Independent State of Papua New Guinea" (2010). http://www.reeep.org/index.php?id=9353&text=policy&special=viewitem&cid=71 (Accessed July 2011).

United Nations Office for the Coordination of Humanitarian Affairs. "Papua New Guinea: The World's First Climate Change 'Refugees.'" *IRIN Humanitarian News and Analysis* (June 8, 2008). irinnews.org/Report.aspx?Reportid=78630 (Accessed July 2011).

U.S. Energy Information Administration. "Country Information: Papua New Guinea" (June 30, 2010). http://www.eia.gov/countries/country-data.cfm?fips=PP (Accessed July 2011).

Paraguay

A landlocked country in South America surrounding the Paraguay River and surrounded in turn by Argentina, Brazil, and Bolivia, Paraguay has been independent from Spain since 1811. It has the fastest-growing economy on the continent. From 1970 to 2009, economic growth averaged 7.2 percent, reaching 9 percent in 2010, despite the worldwide financial crisis. Many of its rural areas are still quite poor; a large part of the population works as subsistence farmers. Only 4 percent of the employed population works for a company with more than 50 employees. The country's export earnings come from agricultural crops and consumer goods imported from elsewhere and re-exported, usually illegally.

Although 55 percent of Paraguay's land is used to graze livestock, only 6 percent of the land is arable. The Chaco Desert, comprising much of the western half of the country, is used for ranching, but has soil too poor to be used for other agricultural purposes. Even the far more fertile eastern area has little to offer. Deforestation in an attempt to create more farmland also leads to soil erosion and the destruction of ecosystems.

Lack of development and low standards of living have kept the country from being a significant contributor to greenhouse gas emissions, but automobile usage has been steadily increasing since the 1970s, while reliance on public transport has been decreasing. One of the country's major railway networks basically closed down in the 1970s, with its only remaining service operating for tourists, increasing the use of cars and trucks for long-distance transport and freight.

The country has adopted environmental controls domestically, and is a signatory to the United Nations Framework Convention on Climate Change (UNFCCC) signed in Rio de Janeiro in 1992 and the Kyoto Protocol in 1999. Most of its electricity—over 90 percent—comes from the Itaipu dam, which began generating power in 1982, and was supplying electricity to most of the country by 2000. The dam also provides 20 percent of the electricity used in Brazil. The Yacyreta dam, opened in 1994, was intended to supply a significant amount of electricity to Argentina, but because of management problems, was only able to operate at 60 percent capacity until 2011.

Bill Kte'pi
Independent Scholar

See Also: Bolivia; Brazil; Uruguay.

Further Readings

Galindo-Leal, Carlos, et al. *The Atlantic Forest of South America: Biodiversity Status, Threats, and Outlook*. Washington, DC: Island Press, 2003.

Moscov, Stephen. *Paraguay: An Annotated Bibliography*. Buffalo: State University of New York at Buffalo, 1972.

Penguins

From the tropics to Antarctica, penguins depend on predictable regions of high ocean productivity where their prey aggregate. Penguins are sentinels of the marine environment locally and regionally, but also at the global level, as more than half of the species are on the International Union for Conservation of Nature's Red List of species that are in trouble. There are between 16 and 19 species of penguins, all generally restricted to the Southern Hemisphere, with the greatest species diversity found in New Zealand. Changes in precipitation, sea ice, ocean temperature and productivity, and prey distributions associated with global climate warming are affecting penguins and changing their distribution and abundance. BirdLife International includes climate change among the threats for eight of 11 species of penguins listed as endangered or vulnerable.

Signals of a Changing Climate

The global climate signal is strongest in the Antarctic Peninsula, where air temperature has increased, glaciers have retreated, ice shelves have collapsed, and krill, small shrimp-like creatures that many Antarctic birds and marine mammals feed upon, have declined because of ocean warming. Adélie and emperor penguins, the most ice-associated and southerly of the penguin species, are suffering more reproductive failures because of increases in rain and snow, early breakup of ice, and blocking of colony access by icebergs.

Emperor penguins, as seen in the 2005 documentary film *March of the Penguins*, breed almost exclusively on fast ice (sea ice held in place along the shore by the bottom or capes). Early breakup of fast ice results in reproductive failure, as the ice must remain intact for nine months for emperor penguins to successfully raise chicks. Even without complete loss of ice, breeding colonies disappear when the ice season becomes too short. The exact number of colonies is unknown, but 42 colonies are known to be occupied. The colony at Dion Island, in the western Antarctic Peninsula, has already disappeared, probably because of warming of the peninsula.

Gentoo penguins, more northerly and less ice tolerant, have extended their range farther south and appear to have benefited from the warming in the Antarctic Peninsula. However, population trends are hard to establish because of large year-to-year fluctuations in their breeding population. Chinstrap penguins, long thought to benefit from less sea ice, are declining throughout their range because of reductions in their main prey of krill, which is linked to climate warming. King penguins, the second-largest species, harvested during the whaling era for their oil, have increased in number and have expanded their range northward.

Many of the temperate species of penguins are in decline because of human perturbations, including harvest, accidental capture by fisheries, petroleum

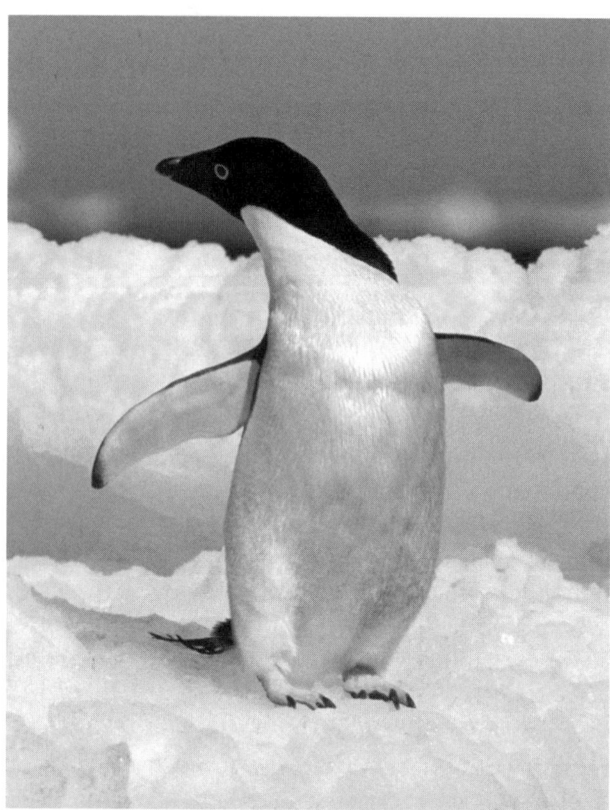

Climate change is most affecting Adélie and emperor penguins, the southerly species most closely associated with ice, through reproductive failures because of increases in rain and snow, early breakup of ice, and blocking of colony access by icebergs.

pollution, breeding habitat destruction, and climate variation that changes the distribution and abundance of their prey. Temperate penguins also suffer from increases in rainfall or air temperatures, as young chicks cannot regulate their body temperatures when their down is wet. Increased frequency of El Niño events associated with global warming reduced Galapagos penguins to about half of what they were in the 1970s. During El Niño events, the water warms, ocean productivity declines, and penguins quit breeding. Under the strongest and longest events, penguins die. Peruvian penguins declined after the 1972 El Niño event and have never recovered, in large part because the anchovy fishery, the second-largest fishery in the world, harvests much of their prey to make fishmeal.

Patagonian penguins, the most common of the temperate penguins, have declined by 22 percent since 1987 at their largest breeding colony at Punta Tombo, Argentina. They swim farther north during incubation than they did a decade ago, likely a reflection of shifts in their prey in response to climate change and reductions in prey abundance because of commercial fishing. Penguins raise fewer chicks when they have to swim farther from the colony to forage. African penguins declined from about one million pairs in the 1920s to about 25,000 pairs in 2009. They have not recovered from large oil spills in the 20th century because humans are harvesting their prey and because of changes in prey distributions linked to climate. Some African penguins have shifted their breeding locations to follow their prey, and the new colonies are not in protected reserves. Penguins are usually faithful to breeding colonies, but African penguins that do not move are no longer breeding near sources of food.

Petroleum Dangers

Human use of petroleum affects penguins indirectly through global warming and increased climate variation, and directly through pollution. Temperate penguins are particularly at risk from oil pollution because they breed and winter in areas where petroleum is extracted and transported, and thousands of penguins get covered with petroleum each year. In 2011, possibly half of the northern rockhopper penguins in the world were oiled in one accident on Nightingale Island when a cargo ship ran aground. Authorities estimated that 88 percent of the penguins that were rescued died. Illegal discharge of petroleum kills Patagonian penguins from Brazil to Argentina, and African penguins frequently encounter petroleum around the Cape of Good Hope.

Ocean productivity will likely continue to decline as the climate warms, which is a problem for penguins, as they need areas of high productivity to survive. As global sentinels, penguins are showing that global climate change is creating new challenges for their populations.

P. Dee Boersma
Ginger A. Rebstock
University of Washington

See Also: Benguela Current; Charismatic Megafauna; Detection of Climate Changes; El Niño and La Niña; Marine Mammals; Oceanic Changes; Phytoplankton; Polar Bears; Sea Ice; Zooplankton.

Further Readings

BirdLife International. http://www.birdlife.org (Accessed May 2011).

Williams, T. D. *The Penguins: Spheniscidae*. Oxford: Oxford University Press, 1995.

Pennsylvania

In recent years, a heightened focus on the possible effects of and solutions to global warming has made the issue very prominent. Because of Pennsylvania's interior location in the northeast, it is in prime position to experience many negative effects associated with global warming.

Effects of Global Warming in Pennsylvania

By 2100, average summer temperatures in Pennsylvania could increase between 7 and 9 degrees F (3.8 and 5 degrees C). This temperature change could cause extreme cases of precipitation, with some parts of the state experiencing up to a 50 percent increase in rainfall, while other areas face drought conditions.

In Philadelphia, heat-related deaths during a typical summer could increase by 90 percent by 2050, up from about 130 deaths per summer to more than 240. Currently, "red alert" air-quality days happen about two days every summer in Pittsburgh. By the middle of the century, this could climb up to five days. Ozone levels in the city are already above the Environmental Protection Agency's (EPA) healthy standard at least 10 days out of the year. Global warming could cause this number to reach 22 days in the near future, meaning more cases of respiratory diseases such as asthma. Loss of wildlife and habitat are also possible threats caused by global warming, which could mean a loss of tourism dollars.

Over the last century, the average temperature in Harrisburg, Pennsylvania, has increased 1.2 degrees F (.67 degrees C) and precipitation has increased by up to 20 percent in many parts of the state. Over the next century, climate in Pennsylvania is expected to change even more. Precipitation is estimated to increase by about 10 percent in spring, by 20 percent in winter and summer, and by as much as 50 percent in fall. The amount of precipitation on extreme wet or snowy days is also likely to increase, which would cause an increase in extremely hot days in summer because of the general warming trend. Although it is not clear how severe storms would change, an increase in the frequency and intensity of summer thunderstorms is possible. Higher temperatures and increased frequency of heat waves may increase the number of heat-related deaths and the incidence of heat-related illnesses. Pennsylvania, with its irregular, intense heat waves could be especially susceptible. Similar but smaller increases have been projected for Pittsburgh, with a possible 50 percent increase in heat-related deaths, or from about 40 to 60. However, winter-related deaths in Philadelphia could drop from 85 to about 35 per winter if winter temperatures increase.

The future complications that global warming could have on Pennsylvania's major cities will also be felt in the large rural areas that make up a generous portion of the state. Pennsylvania's farming and agriculture industries are vital to the state's economy as well as the outdoors tourism market; both would be greatly debilitated by the effects of global warming. In 2001, more than 4.5 million people spent nearly $3 billion on hunting, fishing, and wildlife viewing in Pennsylvania, which in turn supported over 56,000 jobs in the state.

The extraction of natural gas from the Marcellus shale formation is another issue connected to global warming and climate change in the northeastern United States, and Pennsylvania is most affected. While natural gas burns cleaner than other fuel sources like coal and oil, the negative effects of extracting the gas is argued to outweigh its benefits. Hydraulic fracturing, also known as "fracking," is the process of extracting natural gas from shale and results in the dispersal of greenhouse gases into the environment. Another cause for concern is that the majority of the gas expelled during production is methane, which has a greater potential for influencing global warming than carbon dioxide.

This is particularly threatening to Pennsylvania because the majority of the Marcellus shale formation is located underneath the state, affecting almost every region, with the exception of the southeast. The large potential profits associated with this drilling is alluring to many landowners, and lack of regulation on hydraulic fracturing makes the payout of millions of dollars even more tempting, regardless of the threat to the state's land, water, and air. Climate change may also increase ground-level ozone levels. For example, high temperatures, strong sunlight, and stable air masses tend to increase urban levels of ozone, a major component of smog. If a warmed climate causes increased use of air conditioners, air-pollutant emissions from power plants will also increase. A preliminary modeling study of the midwest, which included the area around Pittsburgh, found that a warming of 4 degrees F (2.2 degrees C), with no other change in weather or emissions, could increase concentrations of ozone by as much as 8 percent. Currently, ground-level ozone concentrations exceed national ozone health standards in several areas throughout the state. Ground-level ozone has been shown to heighten respiratory illnesses such as asthma as well as other complications. In addition, ambient ozone reduces crop yields and is harmful to ecosystems.

Warming and other climate changes may cause an increase in disease-carrying insects and thus the potential for the spread of diseases such as malaria and dengue ("break bone") fever. Mosquitoes flourish in many areas around Pennsylvania. Some can carry malaria, while others can carry encephalitis, which can be lethal or cause neurological damage. Incidents of Lyme disease, which is carried by ticks, have also increased in the northeast. If conditions become warmer and wetter, mosquito and tick populations could increase in Pennsylvania, increasing the risk of these types of diseases.

Pennsylvania's valuable water resources would also be affected by changes in precipitation, temperature, humidity, wind, and sunshine. Changes in stream flow tend to coincide with changes in precipitation. Water resources in drier climates are more sensitive to climate changes. Because evaporation often increases with the onset of a warmer climate, the result could be lower river flow and lower lake levels, especially in the summer. If this happens, groundwater will consequently be reduced. In addition, a rise in precipitation could lead to increased flooding. Pennsylvania's Susquehanna River drains much of the eastern two-thirds of the state, and the Allegheny and the upper Ohio rivers drain most of the western third. A warmer climate would lead to earlier spring snowmelt and could result in higher stream flows in winter and spring and lower stream flows in summer and fall. However, changes in rainfall could also have significant effects on stream flow and runoff. This alerts many Pennsylvanians because some of the most intense flooding on record in the United States has occurred in Pennsylvania.

Pennsylvania and other states across the nation will continue to be affected by increased temperatures because of global warming. The long-term affects may prove to be more dramatic to the environment unless pollution from power plants and passenger vehicles is drastically reduced.

Arthur Matthew Holst
Widener University

See Also: Floods; Natural Gas; Pollution, Air; Rainfall Patterns.

Further Readings

Benn, Douglas I. and David J. A. Evans. *Glaciers and Glaciations*. Hoboken, NJ: Wiley, 1998.

Clean Air Council. http://www.cleanair.org/press Room/press%20 release%20-%20globalwarming dec03.htm (Accessed March 2012).

National Wildlife Federation. "Global Warming and Pennsylvania." cf.nwf.org/globalwarming/pdfs/pennsylvania.pdf (Accessed July 2011).

Oerlemans, Johannes. *Glaciers and Climate Change*. Lisse, Netherlands: A. A. Balkema, 2001.

Penn Environment. "Global Warming Reports." http://www.pennenvironment.org/reports/global-warming/globalwarming-reports/the-cost (Accessed March 2012).

U.S. Environmental Protection Agency. "Climate Change and Pennsylvania." http://64.233.167.104/search?q=cache: RevlHDfNv5oJ:yosemite.epa.gov/OAR/globalwarming.nsf/Unique KeyLookup/SHSU5BVMDY/%24File/pa impct .pdf (Accessed March 2012).

Perfluorocarbons

The perfluorocarbons (PFCs) are a group of chemically related greenhouse gases (GHGs) covered by the Kyoto Protocol. Although emissions of PFCs are low compared to many other pollutants, they are of great concern because they are extremely powerful GHGs with very long atmospheric lifetimes. Furthermore, the release of manufactured PFCs is on the rise because of increasing aluminum and semiconductor-chip manufacturing. Annual releases of perfluoromethane (PFM), the most abundant PFC, are the global warming equivalent of about 70 megatons of carbon dioxide (CO_2), roughly one two-hundredth of the amount of CO_2 released annually.

In the context of climate change, the most important PFCs are PFM and perfluoroethane (PFE). Also of interest in a wider environmental context are the oxygenated PFCs: perfluorooctanoic acid (PFOA) and its salt, perfluorooctane sulfonate (PFOS). These compounds, which are highly soluble in water, are found in both the ocean environment and living tissues, but rarely in the atmosphere. Since 1980, the atmospheric concentration of PFM has risen by around 30 percent, despite reductions in emissions per ton from the aluminum industry, and is thought to have risen by about 70 percent since 1960. The atmospheric concentration of PFE has doubled from its concentration in 1980, and is believed to be more than 10 times higher than its 1960 value.

PFC molecules are strong absorbers of infrared radiation and are therefore powerful GHGs. PFM is a much more powerful GHG than CO_2 as measured by its global warming potential. Although the atmospheric concentration of PFM is around 100,000 times lower than CO_2, the radiative forcing because of this atmospheric loading of PFM is as much as one five-hundredth of the radiative forcing because of CO_2. (Radiative forcing is a measure of the global warming effect of a chemical at a given atmospheric concentration.) The global warming potentials of PFM and PFE are 7,390 and 12,200, respectively (see Table 1).

PFCs are extremely environmentally stable; they are only very slowly destroyed by the action of sunlight and oxygen. The main pathway for removal of PFCs from the environment is through a high-temperature combustion process, such as when air is taken into vehicle engines or power-station furnaces. This environmental stability arises from PFC molecules' chemical structure. PFCs are related to simple hydrocarbons by replacement of all hydrogen atoms by fluorine. For example, the simplest hydrocarbon is methane (CH_4). The corresponding PFC is perfluoromethane (CF_4). The carbon–fluorine chemical

Table 1 Properties of common perfluorocarbons

Name	Chemical formula	Boiling point/ degrees C	Atmospheric lifetime/years[a]	Atmospheric concentration/pptv[a,b]	GWP[a]
Perfluoromethane (PFM) PFC-14	CF_4	-128	50,000	74 ± 1.6	7,390
Perfluoroethane (PFE) PFC-116	C_2F_6	-78	10,000	2.9 ± 0.025	12,200
Perfluoropropane PFC-218	C_3F_8	-37	2,600		8,830
Perfluorocyclobutane PFC-318	C_4F_8	-7	3,200		10,300
Perfluorobutane PFC-3-1-10	C_4F_{10}	-2	2,600		8,860
Perfluoropentane PFC-4-1-12	C_5F_{12}	30	4,100		9,160
Perfluorohexane PFC-5-1-14	C_6F_{14}	56	3,200		9,300

a. Intergovernmental Panel on Climate Change, *Fourth Assessment Report.*
b. parts per trillion by volume.

bond is tremendously robust with respect to normal mechanisms by which the atmosphere cleans itself. Consequently, the atmospheric lifetimes of PFM and PFE are from 50,000 to 10,000 years, respectively (see Table 1).

Natural PFM emissions from soils give rise to a background "clean air" concentration of about 40 parts per trillion by volume (pptv). Its concentration has been increasing throughout the latter half of the 20th century to its current value of about 75 pptv because of industrial activity. Both PFM and PFE are produced as a byproduct of the electrochemical extraction of aluminum from its ores.

Over the past 20 years, the global aluminum industry has significantly improved its performance: currently, an average of 400 grams of PFM and about 40 grams of PFE are released per ton of aluminum produced, down by almost two-thirds since the 1980s. However, this per-ton emissions reduction has been somewhat offset by increases in aluminum production volumes. The semiconductor-chip manufacturing industry is a major source of PFE and a secondary source of PFM. Consequently, the increase in chip manufacture is a major influence on the growing emissions of PFE. The oxygenated perfluorocarbons PFOS and PFOA are considered harmful to human health. In fact, the U.S. Environmental Protection Agency regards PFOA as a "likely carcinogen"; it has been proven to be carcinogenic to rodents, as well as to cause immune and reproductive system damage. PFOA and PFOS are released to the environment from the manufacture and use of nonstick materials, fabric protectors, and firefighting foams. Because of their high water solubility and extremely long environmental lifetime, they are found in low concentrations in the blood of humans worldwide and in many animals, including U.S. dolphins, Chinese pandas, and Arctic polar bears.

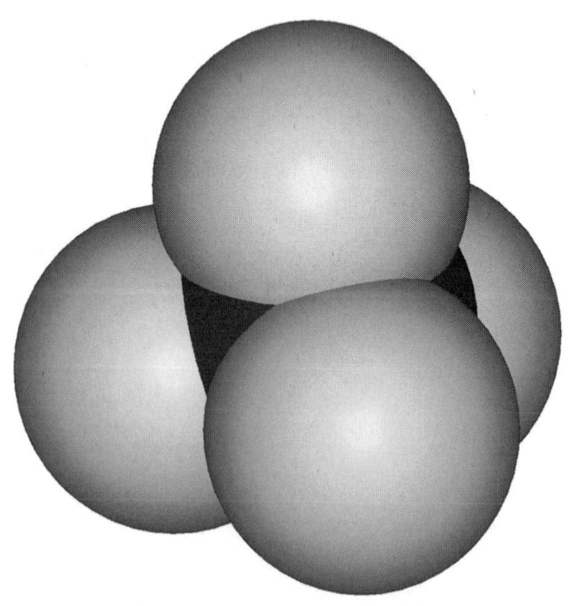

A tetrafluoromethane molecule, the most abundant and simplest atmospheric perfluorocarbon. It is also the most potent and stable greenhouse gas, with an atmospheric lifespan of 50,000 years. However, it exists in relatively small amounts in the atmosphere.

Christopher J. Ennis
University of Teesside

See Also: Greenhouse Gas Emissions; Hydrofluorocarbons; Intergovernmental Panel on Climate Change; Kyoto Protocol; Radiative Feedbacks.

Further Readings

Intergovernmental Panel on Climate Change. http://www.ipcc.ch (Accessed March 2012.

Khalil, M. A. K., et al. "Atmospheric Perfluorocarbons." *Environmental Science and Technology*, v.37 (2003).

Test America. "Determination of Perfluorocarbons and Related Compounds by LC/MS/MS." http://www.testamericainc.com/pfoa (Accessed October 2011).

Peru

Located in the western part of South America, Peru is the third-largest country in the region with a land area of 494,209 sq. mi. (1,279,996 sq. km) and a population of approximately 30 million people. It has 28 of the 32 climates of the world and comprises more than 70 percent of the world's tropical glaciers. The glaciers in the Andes Mountains drain primarily into the Amazon rainforest, but also serve as the main source of water for the arid areas in the coastal region close to the Pacific Ocean. Approximately 60 percent of the Peruvian territory is covered by the Amazon rainforest, often referred to as the "lungs of the world." This accounts for approximately 70 million hectares of forests, which are threatened by subsistence farmers that practice slash-and-burn techniques, illegal loggers that seek a limited number of species and/or disregard sustainable practices, and illegal miners mining for gold and other valuable minerals. Peru is one of the top 10 most biodiverse countries in the world, and the rainforest is considered one of the world's major biodiversity hotspots, with thousands of species still not identified. These natural resources are now also threatened by climate change.

Peru contributes less than 0.5 percent of the world's greenhouse gas emissions. Close to 50 percent of these emissions is a consequence of land-use change, which is primarily a consequence of migratory agricultural practices in the Amazon rainforest. Per-capita emissions were 2.5 tons of carbon dioxide (CO_2) equivalent (CO_2e) in 2000 and 4.7 CO_2e when including nonproductive activities. Despite Peru's low contribution to climate change, several reports observe that the country is one of the most vulnerable in the world to its effects.

Notable Effects of Climate Change

The ecological, economic, and human effects in Peru have been widely reported. For example, commercial fish catches decrease dramatically during El Niño events, which are expected to increase in frequency and magnitude with warmer temperatures. Effects on biodiversity because of the alteration of ecosystems, decreases in agriculture outputs, impacts on human populations via the spread of vector-borne diseases such as malaria and dengue fever, and the melting of tropical glaciers are just some of the main problems facing the country.

Several studies have documented the dramatic retreat of tropical glaciers during the last decades, with some estimates suggesting a rate of over 20 percent in the last 30–35 years. The change in river flows is having an increasing and dramatic effect on water accessibility (some estimates suggest a 12 percent decrease in coastal areas, where 60 percent of the population lives) and could potentially impact hydroenergy production. According to a report by the Andean Community (CAN), the cost of these climate change related impacts in Peru could account for approximately 4.5 percent of the country's gross domestic product by 2025.

In order to deal with these issues, the Peruvian government became a signatory to the United Nations Framework on Climate Change (UNFCC) in 1992 and ratified the Kyoto Protocol in 2002. Additionally, the government created the National Committee of Climate Change in 1993, which was not successful. In 2003, the government reactivated and restructured the committee, making the National Strategy of Climate Change official. More recently, the government created the Ministry of the Environment, which has been focusing on climate change as one of its central lines of action. This effort resulted in the development of the Mitigation and Adaptation Action Plan Against Climate Change in late 2010.

In terms of adaptation measures, the Ministry of the Environment has recently begun to develop projects and studies to better understand the impacts of climate change. Until 2010, close to $34 million had been invested in adaptation measures. However, only some regional governments have developed their own adaptation management plans.

In regard to mitigation actions, the country has developed a number of Clean Development Mechanism (CDM) projects, and has strongly advocated for the implementation of projects under the United Nations Reducing Emissions from Deforestation and Forest Degradation in Developing Countries (REDD) program.

Bruno Takahashi
State University of New York

See Also: El Niño and La Niña; Forests; Glaciers, Retreating; Reducing Emissions from Deforestation and Forest Degradation (UN-REDD).

Further Readings

Adger, W. N., N. Brooks, G. Bentham, M. Agnew, and S. Eriksen. "New Indicators of Vulnerability and Adaptive Capacity." Norwich, UK: Tyndall Centre for Climate Change Research, 2004.

Andean Community. *Climate Change Has No Frontiers*. Lima, Peru: Andean Community, 2008.

Andean Community. *The End of the Snowy Peaks? Glaciers and Climate Change in the Andean Community.* Lima, Peru: Andean Community, 2007.
Ministry of the Environment. *Second National Communication from Peru to the UNFCC.* Lima, Peru: Ministry of the Environment, 2010.

Peruvian Current

Originating in the frigid waters off the coast of Antarctica, the Peruvian Current moves north along the western coast of South America. When it reaches the continental shelf along South America, the current rises, carrying cold water with it to the surface of the Pacific Ocean. The prevailing winds of the South Pacific and Earth's rotation cause the Peruvian Current to rotate; the Coriolis force causes the current to rotate clockwise. The Peruvian Current extends 125 mi. (201 km) west from the coast of South America. As the current moves north through the coasts of Chile, Peru, and Ecuador, it splits into two masses where Cabo Blanco, Peru, meets the Gulf of Guyanquil. The main current turns west into the Pacific Ocean, while the remnant of the current moves along the coast of Ecuador. At that point, the second branch of the Peruvian Current also moves west, rejoining the main current near the Galapagos Islands.

Limiting Rainfall

The Peruvian Current is also known as the Humboldt Current after its discoverer, German scientist Alexander von Humboldt. The Peruvian Current affects Peru year round and moderates the climate of Chile in spring and summer, when it displaces a subtropical center of high pressure. Ordinarily, the coast of Chile would warm in spring and summer, but the onset of the Peruvian Current diminishes temperatures and forestalls any rain. The air that accompanies the current is dry, keeping the coast arid. Some weather stations along the Chilean coast have never recorded rainfall; others areas receive considerably less than 1 in. (2.5 cm) of rain per year. The northern coast of Peru is dry from May to November, and receives light rain between December and April. Even though some areas of the coast are humid, rain does not fall. The arid coastline supports few plants, and so sunlight either is absorbed by the land or radiates back into space. Rainfall along the southern coast of Ecuador totals 12 in. (30 cm) per year, though in the north, where the Peruvian Current weakens, rainfall increases tenfold. Some regions receive as many as 197 in. (500 cm) of rain per year.

With Peru located near the equator, one might expect warm temperatures, but the Peruvian Current keeps the coast of Peru at 75 degrees F (24 degrees C). Lima varies from 70 degrees F (21 degrees C) in January and 50 degrees F (10 degrees C) in June. Areas inland from the current often record temperatures of 90 degrees F (32 degrees C). Periodically, El Niño disrupts the Peruvian Current, bringing warm water from the tropical Pacific to the western coast of South America. Temperatures along the coast rise and rain falls on some parts of the coast.

As it cools the western coast of South America, the Peruvian Current creates a climate of unremitting dryness. Temperatures are moderate, but rainfall is scant. Seabirds inhabit the western coast of South America but humans have only colonized the region in small numbers. The deserts of Chile are especially forbidding. Without rain, the land ceases to sustain plant life. In contrast to the sterility of the desert, life abounds in the ocean. The Peruvian Current carries plankton to the surface of the ocean, and fish feed on it in large numbers. Seabirds in turn feed on the fish. Despite creating an arid climate, the Peruvian Current teems with life.

Christopher Cumo
Independent Scholar

See Also: Benguela Current; Climate; Coriolis Force; El Niño and La Niña; Somali Current; Western Boundary Currents.

Further Readings

Philander, S. George H. *Our Affair With El Niño: How We Transformed an Enchanting Peruvian Current Into a Global Climate Hazard.* Princeton, NJ: Princeton University Press, 2004.
Ya Kondratyev, Kirill and Vladimir F. Krapivin. *Global Environmental Change: Modeling and Monitoring.* New York: Springer, 2002.

Philippines

The Philippines is an archipelago composed of 7,107 islands located in southeast Asia in the western Pacific Ocean. Its tropical climate has endowed it with biodiversity that is considered some of the richest in the world. At 94 million people, the country is the 12th most populous country in the world.

Including emissions from land use and forestry, the Philippines is still a relatively minor contributor to greenhouse gas (GHG) emissions, and represented a minute 0.51 percent of the world's total in 2000. However, climate change issues affect all facets of national development. Response to climate change in the Philippines began as early as the 1990s through a series of legal and policy initiatives.

The effects of climate change have already been manifest in the country in the form of temperature spikes and variable weather. Hot days and hot nights tend to be more frequent. Extreme weather events such as typhoons, floods, and landslides have also recurred. The country's vulnerability to climate change is considered high, with impacts adversely affecting agriculture and fisheries, two of the country's major industries.

With increasing frequency and intensity of heat waves, floods, droughts, and typhoons, agricultural and coastal marine ecosystem output and productivity will be altered. Moreover, water availability and quality will be reduced. The incidence of climate-sensitive infectious diseases will also trend upward. The poor are especially at risk from these impacts, as many of them live in naturally hazard-prone areas and are dependent on natural resources for their livelihoods.

Aside from being located at the western rim of the Pacific Ring of Fire, the Philippines also lie along a typhoon belt in the Pacific. This makes the country more geographically vulnerable to the adverse impacts of climate change. With a coastline length of approximately 20,132 mi. (32,400 km), the Philippine coast is one of the longest in the world. More than 60 percent of the coastal population are dependent on marine resources for their livelihoods; thus, impacts to coastal communities would be more pronounced, especially in terms of significant sea-level rise, coral bleaching, and fish kills.

The Philippines is one of the world's most natural disaster-prone countries because of a combination of high incidence of typhoons, floods, droughts, volcanic eruptions, earthquakes, and landslides. It is estimated that 81.3 percent of the population occupying 50.3 percent of the country's total area are vulnerable to these natural disturbances, which primarily affect their economic lives. Of these natural hazards, typhoons claim the most lives and cause the most damage to property.

The impacts of climate change to stream flow and groundwater surcharge brought about by long-spell droughts has already affected water quality and availability in the country, especially in urban centers. Metro Manila had already experienced the rationing of its potable water supply. In addition, climate change is also expected to impact health conditions related to nutrition, growth and development, and the rise of mosquito-related diseases such as dengue fever.

The primary expected impacts of climate change to the Philippines, however, will come as a result of climate variability—changes in precipitation patterns and increase in temperature. These have close correlation to vulnerabilities that are linked to poverty and environmental degradation, since it is the poor who are located in disaster-prone and environmentally fragile areas. It could be felt even more through interrelated effects on agriculture, soil and land quality, and forest cover by means of soil degradation, flooding, drought, and the low volume of irrigation water. The decline in agricultural production and productivity would threaten the country's food security. Because of increased irrigation demands and low crop yields, groups involved in rice and corn production would be the most affected sector.

The Philippines have worked to adapt to these impacts through a series of responses, ranging from addressing vulnerabilities of specific sectors to focusing on disaster-prone settlements, high-risk population centers, and food-production areas. In terms of mitigation, the Philippines have been promoting a climate-friendly energy supply mix, policy incentives for renewable energy, and diverse energy system interventions through energy efficiency and energy generation.

Laurence Laurencio Delina
Independent Scholar

See Also: Abrupt Climate Changes; Adaptation; Climate Change, Effects of; Drought; Global Warming, Impacts of; Land Use; Land Use, Land-Use Change, and Forestry; Preparedness; Rainfall Patterns; Vulnerability.

Further Readings

Allen, K. "Community-Based Disaster Preparedness and Climate Adaptation: Local Capacity-Building in the Philippines." *Disasters*, v.30/1 (2006).

Buan, R. D., et. al. "Vulnerability of Rice and Corn in the Philippines." *Water, Air and Soil Pollution*, v.92/1–2 (1996).

Government of the Philippines. *Philippines First National Communication on Climate Change*. Manila: Government of the Philippines, 1999.

Jose, A. and N. Cruz. "Climate Change Impacts and Responses in the Philippines Coastal Sector." *Climate Research*, v.12 (1999).

Pulhin, F and R. Lasco. "Climate Change and Biodiversity in the Philippines: Potential Impacts and Adaptation Strategies." In *Moving Forward: Southeast Asian Perspectives on Climate Change and Biodiversity*, edited by P. Sajise, et al. Singapore: Institute of Southeast Asian Studies, 2010.

Phillips, Norman

Norman Phillips is a theoretical meteorologist who pioneered the use of numerical methods for the prediction of weather and climate changes. His influential studies led to the first computer models of weather and climate, as well as to an understanding of the general circulation of the atmosphere, including the transports of heat and moisture that determine the Earth's climate. His 1955 model is generally regarded as a groundbreaking device that helped to win scientific skepticism in reproducing the patterns of wind and pressure of the entire atmosphere within a computer model.

Phillips received his B.S. from the University of Chicago in 1947, and his Ph.D. in 1951. He was the first to show, with a simple general circulation model, that weather prediction with numerical models was possible. The advent of numerical weather predictions in the 1950s also marked the transformation of weather forecasting from a highly individualistic effort to a cooperative task in which teams of experts developed complex computer programs. With the first digital computer in the 1950s, scientists tried to represent the complexity of the atmosphere and its circulation in numerical equations. Nineteenth- and early 20th-century mathematicians such as Vilhelm Bjerknes and Lewis Fry Richardson had failed to come up with adequate mathematical models. Through the 1950s, some leading meteorologists tried to replace Bjerknes's and Richardson's numerical approach with methods based on mathematical functions, working with simplified forms of the physics equations that described the entire global atmosphere. They succeeded in getting only partial mathematical models. These reproduced some features of atmospheric layers, but they could not show the features of the general circulation persuasively.

Their suggested solutions contained instabilities because they could not account for eddies and other crucial features. Discouraged by such failures, scientists began to think that the real atmosphere was too complex to be described by a few lines of mathematics. The comment of such a leading climatologist as Bert Bolin is revealing of this skepticism. In 1952, Bolin argued that there was very little hope for the possibility of deducing a theory for the general circulation of the atmosphere from the complete hydrodynamic and thermodynamic equations. Yet, computers opened up new possibilities in the field, although the first digital specimens were extremely slow and often broke down.

Jule Charney was the first to devise a two-dimensional weather simulation. Dividing North America into a grid of cells, the computer started with real weather data for a particular day and then solved all the equations, working out how the air should respond to the differences in conditions between each pair of adjacent cells. It then stepped forward using a three-hour step and computed all the cells again. The system was slow to operate and it had imperfections, but its completion paved the way for more research. Norman Phillips sought to address the problems in Charney's model. The challenge for meteorologists now became the computation of the unchanging average of the weather given a set of unchanging conditions such as the physics of air and sunlight and the geography of mountains and oceans. This was a "boundary problem." A parallel problem that they had to face was that of the "initial value," where the operation of calculating how the system evolves from a particular set of conditions found at one moment becomes less accurate as the prediction moves forward in time.

Phillips was inspired by "dishpan" experiments carried out in Chicago, where patterns resembling weather had been modeled in a rotating pan of water that was heated at the edge. For Phillips, this showed that "at least the gross features of the general circulation of the atmosphere can be predicted without having to specify the heating and cooling in great detail." Phillips argued that if such an elementary laboratory system could effectively model a hemisphere of the atmosphere, a more advanced tool such as a computer should also be able to do it. Although certainly more advanced than a dishpan, Phillips's computer was still quite primitive. Thus, his model had to be extremely simple. By mid-1955, Phillips had devised improved equations for a two-layer atmosphere. To avoid mathematical difficulties, his grid covered not a hemisphere but a cylinder 17 cells high and 16 in. in circumference. The calculations allowed the representation of a plausible jet stream and the evolution of a realistic-looking weather disturbance over a period of a month.

This settled an old controversy over what procedures set up the pattern of circulation. The simulation-based approach became the generally accepted method to devise circulation models. For the first time, scientists could visualize how giant eddies spinning through the atmosphere played a key role in moving energy from place to place. Phillips's model was quickly hailed as a "classic experiment," the first true general circulation model (GCM). Phillips used only six basic equations (PDEs), which have been since described as the "primitive equations." They are generally conceived of as the physical basis of climatology. These equations represent the well-known physics of hydrodynamics. The model was able to reproduce the global flow patterns of the real atmosphere. Phillips was awarded the Benjamin Franklin Award in 2003.

Luca Prono
Independent Scholar

See Also: Atmospheric General Circulation Models; Bolin, Bert; Computer Models; Richardson, Lewis Fry.

Further Readings

Lorenz, Edward N. *The Global Circulation of the Atmosphere*. Princeton, NJ: Princeton University Press, 2007.
Weart, Spencer. *The Discovery of Global Warming*. Cambridge, MA: Harvard University Press, 2004.

Phytoplankton

Phytoplankton are a group of free floating, microscopic organisms predominantly classified as algae. Over 4,000 species of phytoplankton have been identified and this list is rapidly growing. Phytoplankton are mostly single cellular organisms and all are autotrophic (i.e., they contain photosynthetic pigments). These pigments allow phytoplankton to use the sun's energy to convert CO_2 and inorganic nutrients, through photosynthesis, into biological molecules such as proteins and carbohydrates. This process of creating new biological molecules is called primary production. Phytoplankton are the only primary producers in the open oceans, and thus form the basis of the food chain in over 70 percent of the world's surface area.

Specific aspects of primary production are often considered more carefully. The net primary production (NPP) refers to the amount of organic carbon available after respiration has been subtracted from the total amount of photosynthesis. It is an important term because this is the amount of carbon that is available to the rest of the foodweb, and is the upper limit on respiration. The net community production (NCP) is the difference between net primary production and heterotrophic respiration. It is measured by gross changes in oxygen or biomass in a specific time. The new production is the fraction of primary production driven by newly available nitrogen. This is principally the nitrate and nitrite that becomes available when deep ocean waters are brought up into the euphotic zone, but could include sources from the atmosphere or river inputs.

The global population of phytoplankton is not evenly distributed throughout the world's oceans.

This distribution is the result of growth being limited by the availability of nutrients. High concentrations of phytoplankton are seen in upwelling areas, such as the Benguela Current off the southwest coast of Africa.

Phytoplankton and Climate Change

CO_2 does not limit phytoplankton growth. CO_2 is not thought to limit phytoplankton growth in any region of the ocean. Primary production, in general, is limited by the availability of inorganic nutrients. Therefore, it is not expected that increased atmospheric concentrations of CO_2 will have a significant impact on the phytoplankton population in the oceans.

The "biological pump" locks away CO_2 The biological pump is the mechanism by which anthropogenic CO_2 is taken from the atmosphere and stored in the deep ocean, which transfers carbon from the surface to the deep ocean, across the barrier of the permanent thermocline. The pump is powered by phytoplankton fixing carbon and sinks to the deep ocean. This removes CO_2 from the surface ocean, causing more CO_2 to be drawn from the air to maintain equilibrium. Carbon is locked away from the atmosphere as a result of being taken below the permanent thermocline. This contrasts with primary productivity on land, which builds plants (such as trees) and in turn animals, locking CO_2 away from the atmosphere.

Although increasing levels of CO_2 do not have a fertilizing effect on the oceans, the longer growing season in temperate and polar regions is expected to lead to an increase in primary productivity.

Carl Palmer
University of Cape Town

See Also: Benguela Current; Biogeochemical Feedbacks; Carbon Sequestration; Marine Mammals; Zooplankton.

Further Readings

Bopp, L., et al. "Will Marine Dimethylsulfide Emissions Amplify or Alleviate Global Warming?" *Canadian Journal of Fisheries and Aquatic Sciences*, v. 61 (2004).

Lalli, C. M. and Timothy R Parsons. *Biological Oceanography*. 2nd ed. Oxford: Oxford University Press, 1997.

Plants

The ancient Greeks were the first to identify a relationship between climate and plants. Following this insight, other naturalists recognized that fossilized plants revealed the climate in prehistory. Plants colonized the land 410 million years ago, shaping the climate as they spread throughout the globe. By absorbing carbon dioxide, plants have the potential to cool the climate. By releasing water vapor into the atmosphere, plants have the potential to warm the climate by trapping heat, or to cool the climate by forming clouds. Rooted to the ground, plants must either adapt to the climate, or die. Sudden climate changes threaten some species with extinction, whereas hardier species survive. From equator to pole, climate determines what plants grow at given latitudes.

Relationship Between Climate and Plants

Theophrastus, a pupil of Aristotle and the founder of botany, may have been the first to ponder the relationship between plants and climate. He understood that each species of plant is adapted to a particular climate, and that in a foreign climate, a plant will not thrive and may not survive. Plants are thus an indicator of climate. The mangrove, for example, is an indictor of a climate wet enough to form swamps. With the work of Theophrastus, the promising synthesis of botany and the study of climate was in its infancy, but the Romans did not bring this synthesis to maturity. The Romans were a practical people, with no interest in the theoretical relationship between plants and climate. The Middle Ages were no better for the study of the relationship between plants and climate. The emphasis on theology undercut any progress in science.

In 1876, Norwegian botanist Axel Blytt revived Theophrastus's notion that plants are an indicator of climate. Working on prehistoric climates, Blytt identified fossils of trees that no longer grew in Denmark. From this observation, he posited that the climate in Denmark had once been suitable for these species of trees, but was no longer. The climate was not therefore a static, unchanging entity. Rather, the climate has changed over time. Working in a similar vein, Swedish geologist Ernst Von Post identified fossils of *Alnus*, a genus of tree, in the strata of rocks throughout Europe.

Alnus is adapted to a warm wet climate, and Von Post tracked the tree fossils as they migrated from southern to northern Europe at the end of the Cenozoic ice age. As the glaciers retreated, *Alnus* rooted itself in the warm wetlands that followed the ice age.

Plants and Climate Change

Fossilized pollen can likewise pinpoint changes in climate. The evergreen red beech is adapted to warm locales. The abundance of its pollen 8,000 years ago implies that this time correlates with the maximum temperature since the end of the Cenozoic ice age. On the other hand, the abundance of pollen from *Phyllocladus*, a shrub that is adapted to the cold, at 26,000 years ago and again at 20,000 years ago, indicates these years as the coldest during the ice age.

Photosynthetic algae evolved in the ocean as early as four billion years ago. From this beginning, the first plants colonized land 410 million years ago, before animals. By 360 million years ago, the beginning of the Carboniferous period, plants had spread throughout the planet, forming lush forests. The temperature and amount of carbon dioxide in the atmosphere were both higher than today. So lush was the growth of plants that when they died, they formed one layer upon another. Under heat and pressure, these many layers of plants formed the vast deposits of coal, natural gas, and petroleum that humans are now extracting. The immensity of these deposits underscores the massive growth of plants during the Carboniferous period.

The fact that plants absorb carbon dioxide during photosynthesis has an important consequence for the climate. Carbon dioxide correlates with temperature. A high concentration of carbon dioxide correlates with high temperatures, whereas a low concentration of the gas correlates with low temperatures. This relationship holds true because carbon dioxide is a greenhouse gas: it traps sunlight that reflects from Earth, preventing light, in the form of infrared radiation, from returning to space. In trapping sunlight, carbon dioxide traps heat, thereby increasing the temperature of the atmosphere. Plants lower the concentration of carbon dioxide by absorbing it during photosynthesis. When plants absorb carbon dioxide faster than Earth produces it through volcanism, the concentration of carbon dioxide diminishes and temperatures decline. In this context, plants may have contributed to the onset of the ice ages by absorbing carbon dioxide.

The absorption of carbon dioxide is cyclical. In spring and summer, when plants grow vigorously, they absorb large amounts of carbon dioxide. In autumn, however, plant growth slows and in winter it stops. The decay of dead plants in autumn returns the carbon dioxide that they had absorbed while alive to the atmosphere. The Northern and Southern Hemispheres contribute to the carbon dioxide cycle in opposite fashion, for when plants are growing vigorously in the Northern Hemisphere, they are dead and decaying in the Southern Hemisphere, and vice versa.

In another sense, the carbon dioxide cycle spans eons of time. In the Carboniferous period, lush forests absorbed prodigious amounts of carbon dioxide, storing it in their tissues as sugars. Upon the death of plants, their conversion to coal, natural gas, and petroleum locked up these vast amounts of carbon dioxide. Since the Industrial Revolution, humans have burned these fossil fuels for energy, and in the process have liberated carbon dioxide, returning to the atmosphere the carbon dioxide that plants had absorbed during the Carboniferous period. The liberation of this carbon dioxide has increased global temperatures. Humans are burning fossil fuels constantly, leading climatologists to predict that humans will, by 2080, double the amount of carbon dioxide in the atmosphere, further increasing temperatures and perhaps leading to the flooding of coastal cities.

Plants absorb carbon dioxide through their stomata, pores on the leaves. At the same time, plants shed water through their stomata. Plants transpire as water vapor more than 90 percent of the water that they absorb through their roots. Through transpiration, plants change the climate, though not in a straightforward way. On the one hand, water vapor is a greenhouse gas, absorbing as heat the sunlight reflected from Earth.

The warmth of the Carboniferous period, a time of abundant plant growth, was because of the greenhouse effect. Along with carbon dioxide, the water vapor transpired by plants contributed to a warm climate. On the other hand, the water vapor that plants transpire forms clouds, which reflect 30 percent of sunlight back into space before it can heat the Earth. In their role in forming clouds, plants cool the climate. One scientist predicted that a doubling of carbon dioxide concentration might cool, rather than heat Earth, because plants will hasten the formation of clouds by transpiring water vapor.

Climate Changes and the Greenhouse Effect

The product of a long evolutionary history, plants are adapted to the climate in a way that humans are not. Humans fashion their material culture to suit the climate, or, when climate worsens, they migrate to a more hospitable locale. Plants, rooted to the ground, must adapt or die. Migration is not an option, though seed dispersal is a kind of intergenerational migration. Through dispersal, seeds travel roughly one kilometer per year, a rate too slow to adapt a plant to rapid climate change. As the climate deteriorates, some plant species die out, to be replaced by hardier species. In Neolithic and Bronze Age Denmark, for example, the climate was warmer and wetter than it is today.

The climate was warm and wet enough to sustain the growth of oak trees. In the Iron Age, however, temperatures dropped to current figures. Unable to cope with the decline in temperature and moisture, oak trees died out and were replaced by grasses and heather. In New Hampshire, the tree *Pinus strobus* was for centuries the most numerous tree in the region's deciduous forests. *Pinus* came through the fire of 1665 unscathed, but was unable to cope with climatic catastrophe. In 1921, a tornado, and in 1938, a hurricane, swept through the forests, felling large trees of several species. The catastrophes wiped out *Pinus*, ending centuries of its dominance, and leaving other species to reconstitute the forests.

These sudden changes in climate and flora are dramatic and easy to quantify. No less dramatic has been the effect of hydrofluorocarbons on climate and plants. Humans have released large quantities of hydrofluorocarbons into the atmosphere, where they have thinned the ozone layer. Consequently UV-B radiation, a type of ultraviolet light, penetrates the ozone layer in greater amounts than in the past. UV-B radiation damages half of all plant species. Damaged plants grow small leaves and short shoots and photosynthesize at a slow rate. These effects are magnified by the fact that several crops are among the plants sensitive to UV-B radiation. One study concluded that a 25 percent reduction in the ozone layer would halve soybean yields.

The tropics support lush vegetation, with more than 80 in. of rain per year, temperatures nearly uniform and warm year-round, and abundant sunshine. Trees have thin bark, for they don't need insulation against the cold or protection against water loss. Trees grow in layers, with those in the innermost layers able to survive without exposure to direct sunlight. Because sunlight does not penetrate to the rainforest floor, little vegetation grows along the ground. At higher latitudes north and south of the tropics, rainfall diminishes and temperatures vary year round. These climates have seasons, with vigorous plant growth in spring and summer and dormancy in autumn and winter.

To cope with a diminution in rain, plants in temperate climates have evolved small leaves to minimize the loss of water through transpiration. In areas that have a dry season, trees evolved the shedding of leaves to stop transpiration and, in other species of trees, the growth of needles rather than leaves to minimize transpiration. With their needles, the evergreens and conifers are adapted to short summers because they can carry out photosynthesis as soon as temperatures warm, whereas deciduous trees must regrow their leaves before they can photosynthesize. Evergreens and conifers grow where winter is cold enough to freeze the ground. Once the ground freezes, roots cannot absorb water, making winter a period of drought and favoring trees that can minimize transpiration in response to frost-induced drought.

In contrast to the rainforest, sunlight penetrates to the forest floor in temperate forests, permitting the growth of plants, often grasses, in abundance along the ground. As the dry season lengthens and rainfall diminishes still further, forest gives way to grassland. Grasses need less water than trees. Trees grow alongside grasses on the African savanna, but few trees grow on the Russian steppe. There, grasses are the dominant flora. At high latitudes, temperatures fall below 40 degrees F (4 degrees C) for six to nine months per year. Summers are brief, with temperatures above 50 degrees F (10 degrees C). Rainfall ranges between 10 and 40 in. (25 and 102 cm) per year, with between 15 and 24 in. (38 and 61 cm) typical. This climate favors

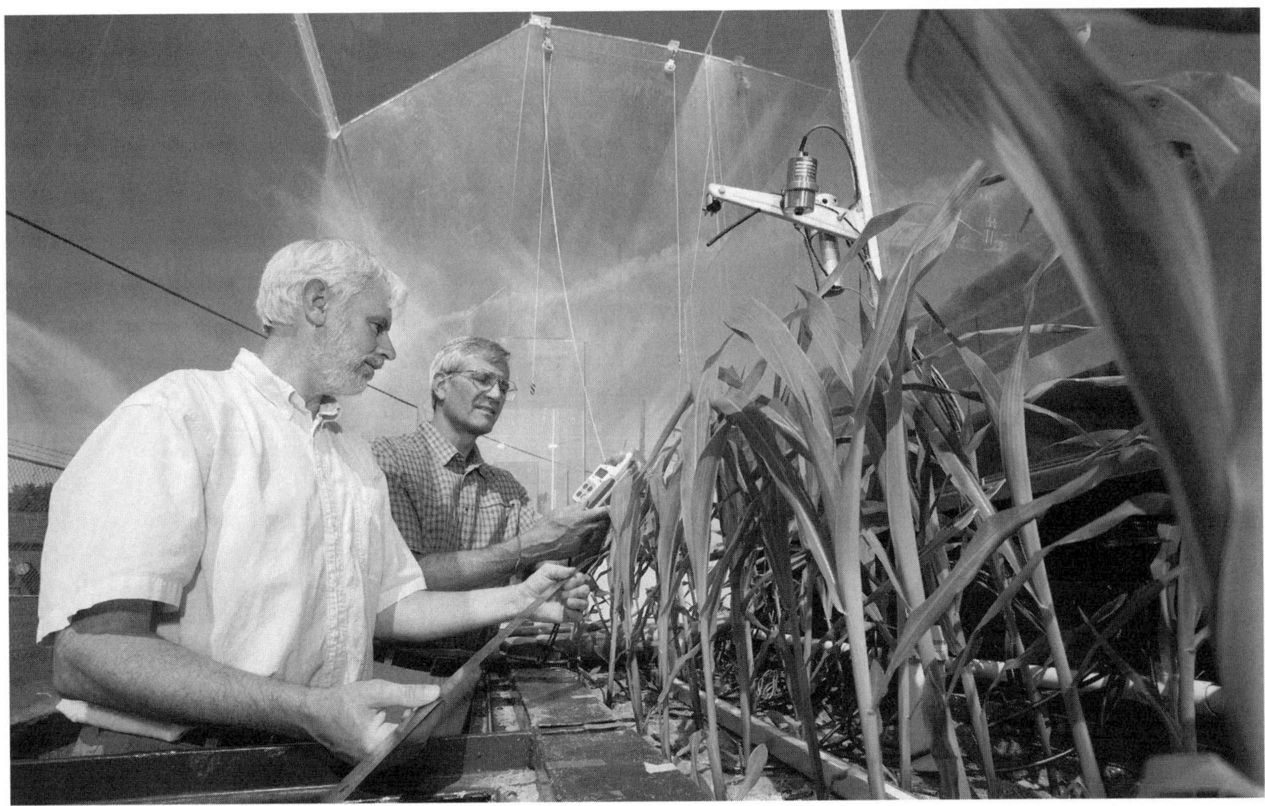

USDA soil scientists Dennis Timlin (left) and plant physiologist Richard Sicher measure chlorophyll concentrations in corn leaves exposed to elevated carbon dioxide. The absorption of carbon dioxide by plants during photosynthesis has important consequences for the climate. As a greenhouse gas, carbon dioxide traps sunlight and heat, increasing atmospheric temperatures. When plants absorb carbon dioxide faster than Earth produces it through volcanism, the concentration of carbon dioxide diminishes, cooling the planet.

the growth of coniferous forests. In addition to their needles, conifers have thick bark to retard water loss and to protect against the cold. Temperature separates coniferous forest from tundra. Where summer temperatures exceed 50 degrees F (10 degrees C), conifers predominate, but wherever summer temperatures fall below 50 degrees F (10 degrees C), tundra results. Grasses and sedges are the tundra flora. During the 50 or 60 days of summer, the sun melts a thin strip of ground. Free from the grip of ice and benefiting from the nearly continuous sunlight of summers in high latitudes, plants grow vigorously and then are dormant for the long, bitter winter.

Effects of Climate Change on Plants

The increase in temperatures that is the likely outcome of the greenhouse effect will affect plants. By one estimate, a 2 or 3 degree F (1.7 degrees C) increase in temperature will raise crop yields in the temperate zone, though an increase above 3 degrees F (1.1 to 1.7 degrees C) will decrease yields. Any temperature increase will likely reduce yields in the tropics, where crops are already at their maximum heat tolerance.

The climate of the future is sure to affect plants. Despite predictions to the contrary, the increase in carbon dioxide will likely increase temperatures. One study suggests that a doubling of carbon dioxide in the atmosphere will triple the growth rate of plants and trees. Forests will grow more densely and will extend their range to higher latitudes. Plants will grow more vigorously on marginal land. Another study indicates that a doubling of carbon dioxide will shift temperate forests 310 to 621 m (500 to 1,000 km) north in the Northern Hemisphere, and south in the Southern Hemisphere. The concentration of carbon dioxide is likely to double by 2080, but trees are not likely to migrate so far so fast. Global warming therefore endangers temperate forests. As the climate warms, trees will advance north and south, taking over ground that had been tundra. As temperatures rise, dead plants will decay more rapidly, liberating still more carbon dioxide into the atmosphere.

The cutting down of forests will harm the plants that survive. The amount of rainfall will decrease as forests are cut down. With fewer forests, the rate of transpiration will diminish. Whereas forests absorb sunlight, bare ground reflects sunlight back into space. A 15 percent decline in rainfall would replace the forests of South America with grassland. A 30 percent decrease in rainfall would replace the forests of Zaire with grasses. A 70 percent decrease in rainfall would make the Amazon basin a desert.

The climate of the future may imperil many plant species, but as a kingdom, plants are resilient. They have survived the ice ages and the predation of herbivores in warm climates. Humans are fortunate that plants are so adaptable, for with their agriculture, humans are dependent on plants. Life would not exist without the diversity of plants. The most numerous form of terrestrial life during the Carboniferous period, plants occupy every biome. Even in deserts, their seeds lie dormant, awaiting the infrequent rains. Plants have adapted to every climate, from the tropics to frigid tundra. Even bodies of water are home to plants. The survival of plants depends on their ability to adapt to the climate of the future. The survival of the rest of the biota depends on the success of plants.

Christopher Cumo
Independent Scholar

See Also: Botanical Gardens; Carbon Sequestration; Carbon Sinks; Cenozoic Era; Climate; Cretaceous Period; Forests; Greenhouse Effect; Greenhouse Gas Emissions.

Further Readings

Institute for Biospheric Research. *The Greening of Plant Earth: The Effects of Carbon Dioxide on the Biosphere.* Washington, DC: Western Fuels Association, 1991.

Money, D. C. *Climate, Soils and Vegetation.* London: University Tutorial Press, 1965.

Morison, J .I. L. and M. D. Morecroft, eds. *Plant Growth and Climate Change.* Oxford: Blackwell, 2006.

Rozema, Jelte, Rien Aerts, and Hans Cornelissen, eds. *Plants and Climate Change.* New York: Springer, 2006.

Woodward, F. I. *Climate and Plant Distribution.* Cambridge: Cambridge University Press, 1987.

Pleistocene Epoch

The increasing frequency and intensity of glacial-interglacial cycles toward the end of the Pliocene (1.806–5.332 million years ago) set the stage for the Pleistocene epoch (11,800 years ago–1.806 million years ago), which is the final phase of the Quaternary period. Some argue that the lower Pleistocene boundary may be set too late because the general trend toward significant cooling and glaciation had begun in the mid–late Pliocene (2.75 million years ago). Hence, the term *Plio-Pleistocene* may be used to delineate this transitional phase between the two epochs.

Marked by Glacial Phases

Strong glacial-interglacial phases are the key climatic features that characterize the Pleistocene epoch and have shaped much of the modern landscape. Glacial stages may be referred to as ice ages, and are used to describe a period of extensive ice sheet presence in the polar, high latitude continental, and alpine regions. Glacial phases are synonymous with reduced global temperatures. Quaternary glacial-interglacial cycles occurred with a 41,000-year periodicity, starting in the late Pliocene (2.75 million years ago) to mid-Pleistocene (1.11 million years ago), followed by a 100,000-year cycle in the mid- to late Pleistocene. The most intensely studied glacial stage during the Pleistocene is the last glacial maximum (21,000 years ago).

Marine fossil material and isotopic proxies were used to simulate sea surface temperatures, sea ice, continental ice sheets, and albedo during the last glacial maximum, with results indicating that high latitudes in the northern hemisphere cooled by 7–11 degrees F (4–6 degrees C), while simulated sea temperatures increased by 2–5 degrees F (1–3 degrees C) in the Pacific and Indian oceans. Most recent evidence suggests that with the exception of Central America and the Indo-Pacific, the climate was much drier than today, because of the combination of reduced evaporation, greater coverage of land surfaces by ice sheets, and wind anomalies.

Glaciation was most extensive in the Northern Hemisphere, with 2–2.5 mi.- (3–4 km-) thick ice sheets covering Canada and parts of the northern United States, Greenland, northern Europe, Russia, and, perhaps to a lesser extent, the Tibetan Plateau. In the Southern Hemisphere, the glaciation of Antarctica that began in the Pliocene continued through to the last glacial maximum, the Andes were highly glaciated, the Patagonian Ice Sheet covered much of southern Chile, and small glaciers formed in Africa, the Middle East, and southeast Asia, where simultaneously deserts were expanding. Sea levels may have been up to 426.5 ft. (130 m) lower than today. The hydrologic and geological consequences of the last glacial maximum and other glacial stages are still evident, particularly at the higher latitudes of the Northern Hemisphere, where the abundance of fresh water is effectively the result of glacial retreat and runoff. Remnants of Pleistocene glaciers also remain in high-altitude tropical localities such as on Mount Kilimanjaro and the Peruvian Andes, but these glaciers are quickly retreating.

Vostok Ice and the Link to Greenhouse Gases

The causes of the Pliocene-Pleistocene glacial-interglacial cyclicity are largely attributed to climate forcing caused by variations in the Earth's orbital parameters (Milankovitch cycles), but the sequence of events is difficult to establish. However, there is strong evidence that greenhouse gas levels fell at the start of glacials and rose during the interglacial retreat of the ice sheets. So far, eight glacial cycles have been identified from cores in Antarctica dating back to 740,000 years ago, but currently, it is the ice core dating back 420,000 years ago that provides the clearest perspective on the link between greenhouse gases and sea surface temperatures over the last four glacial-interglacial cycles. CO_2 concentrations fell between 180 and 200 ppm during the coldest glacial periods, and 280 and 300 ppm during full interglacials, while methane concentrations were approximately 350 ppb during glacials, and roughly twice that amount during interglacials. Current thinking is that Pleistocene changes in greenhouse gas levels were probably caused by disturbance to the sources of these gases, of which the oceanic and terrestrial sources were most significant.

During the last glacial maximum, the presence of large ice sheets over the high latitudes of the Northern Hemisphere significantly reduced the amount of exposed vegetation and combined with low atmospheric CO_2 and other regional climatic changes, creating biomes and vegetation assemblages that have no modern analogue. The Laurentide Ice Sheet completely covered Canada and the northern United States, with taiga, desert, and grassland ecotones dominating the mid-latitudes. At this time, woodland and shrub communities were also present, but highly fragmented. An exception to this is the Canadian and Alaskan Pacific coasts, where the continuity of woody flora remains largely unchanged from the last glacial maximum.

Substantial winter cooling reduced the global extent of tropics and subtropics and caused local extinctions, but equable areas may have acted as regional refuges for species that otherwise would have disappeared. The expansion of more arid ecosystems is well documented from pollen data showing that grasslands and shrub ecotones spread into previously tropical areas such as the Amazonian basin, equatorial Africa, and southern Asia. The persistence of rainforests in central North America and Indonesia during the last glacial maximum can be attributed to the consistently high rainfall in these regions. By contrast, over half of central Australia was desert, with tropical grasslands lying in the north, and scrub-woodland vegetation dominating the eastern and western regions.

Substantial evidence exists to support the hypotheses that Pleistocene fauna was dually affected by the climatic oscillations of the early–mid Quaternary, and the hunting activity of ancestral humans. The disappearance of species that had evolved in colder climates, such as the woolly mammoth, woolly rhinoceros, and musk ox, is most consistent with the appearance of humans in North America. By contrast, the Pleistocene extinction of Eurasian megafauna was likely because of climate. The disappearance of African and South American mammals is unresolved, but current evidence points to the arrival of humans as a key factor.

Jarmila Pittermann
University of California, Santa Cruz

See Also: Earth's Climate History; Glaciology; Ice Ages; Pliocene Epoch; Quaternary Period.

Further Readings

Alverson, K. D., R. S. Bradley, and T. F. Pederson. *Paleoclimate, Global Change and the Future*. New York: Springer-Verlag, 2003.

Barnosky, A. D., et al. "Assessing the Causes of Late Pleistocene Extinctions on the Continents." *Science*, v.306 (2004).

Van Couvering, J. A. *The Pleistocene Boundary and the Beginning of the Quaternary*. Cambridge: Cambridge University Press, 1997.

Pliocene Epoch

The Pliocene epoch is the uppermost subdivision of the Tertiary period (65.5 to 2.588 million years ago), and represents a geological stage from about 1.806 to 5.332 million years ago. Although the Pliocene was generally warmer than the present, this epoch is characterized by pronounced climatic oscillations that ultimately led to the characteristic cooling of the late Quaternary glacial-interglacial cycles. Pliocene climate data are inferred from oxygen isotope, dust, microfossil, and in some cases pollen data from cores collected under the flag of the Ocean Drilling Program (ODP), as well as terrestrial deposits. These records have allowed climatologists to refine the absolute chronology of the Pliocene epoch, and provide a continuous climatic record of global ice volume, sea surface temperatures, aridity, and terrestrial vegetation patterns.

The first Pliocene cooling event is documented at 4.5 million years ago, and was followed by variable, but persistent reductions in temperature after 3.6 million years ago. A brief period of warmth followed until 3.5 million years ago, at which time a second cooling event took place. A well-characterized mid-Pliocene warm period dates to approximately 3.3 to 3.15 million years ago, and is followed by the return to progressive cooling that culminated in the arrival of early Northern Hemisphere glacial-interglacial cycles about 2.75 million years ago. Significant growth of ice sheets did not begin in Greenland and North America until approximately 3 million years ago, following the formation of the Isthmus of Panama. Many agree that this final Pliocene cooling period set the stage for strongly developed glacial events of the Pleistocene (1.8 million to 11,550 years ago) and thus represents a climatic stage that is most relevant to the climates of late Tertiary and early Quaternary.

Model for Future Climate Change

The contemporary significance of the mid-Pliocene warm period lies in its utility as a model for future scenarios of global warming. This is because continental distributions and climate-indicative plant taxa are thought to have been very similar to today. Members of the Goddard Institute for Space Studies (GISS) and the PRISM (Pliocene Research, Interpretations and Synoptic Mapping) group have exploited these paleofeatures in their efforts to model global Pliocene climate and vegetation distributions. Average mid-Pliocene global sea levels are modeled at 33 to 82 ft. (10 to 25 m) higher than today, because of reduced Greenland and Antarctic ice cover, while sea surface temperatures were approximately 6.5 degrees F (3.6 degrees C) warmer than at present day. Mid-Pliocene climate simulations generally indicate increased surface air temperatures, particularly during the winter, and increased annual rainfall, evaporation, and soil moisture. Pollen records from land-based cores are less chronologically accurate, but consistent with a 7–18 degrees F (4–10 degree C) warmer Northern Hemisphere climate, coupled with higher continental moisture levels. This is especially evident in high latitude regions such as the Arctic.

The PRISM group has used fossil and pollen data to document vegetation patterns across the globe during the mid-Pliocene warm period. Their work indicates extensive conifer and mixed forests in the mid-Pliocene Arctic, and generally more northerly distributions of the mixed deciduous forests of eastern North America. Interior North America was likely moister and warmer than today, with evidence of lakes in southeastern California, Arizona, and Utah. Northern Europe was warmer and wetter, with a greater abundance of swamps and wetlands. Little information exists about Central and South America, but the limited numbers of pollen studies are consistent with GISS climate models suggesting a warmer, wetter climate, with a greater abundance of steppe and prairie vegetation. The Australian mid-Pliocene warm period is poorly documented, but it is thought to be wetter than today, with broader distributions of forest flora. Regions of Antarctica were significantly warmer than today, so increased exposure of soils supported the presence of mixed beech forests.

The cause of the mid-Pliocene warming is uncertain, but some combination of CO_2 increase and change in ocean heat transport may have been responsible. Carbon isotopic data from deep-sea microfossils, coupled with GISS climate models, support the increased strength of thermohaline circulation during the mid-Pliocene, particularly with respect to North Atlantic deep water production. However, simulations where CO_2 is the single variable show that the proposed, realistic patterns of mid-Pliocene oceanic heat transports would only have been possible at CO_2 levels greater than 1,200 ppm. There is no evidence supporting such elevated CO_2 excursions, but some workers suggest that even the predicted minor increases up to 380 ppm, in combination with altered ocean heat transport, may have been enough to catalyze mid-Pliocene warming.

Early Pliocene fauna was transitional, favoring grazers over browsers, as grasslands and savannas expanded in central North America and Africa, thereby replacing woodlands and their associated fauna. Charismatic Pliocene fauna included mammoths, mastodons, camels, and hippopotamus in the mid–Northern Hemisphere latitudes, while large turtles and marsupials were found in the Southern Hemisphere. Pliocene high-Arctic fauna was primarily Eurasian, characterized by now extinct species of beavers, badgers, deer, and caniids, the presence of which is consistent with mixed-evergreen forest vegetation. The Pliocene deposits of eastern North America revealed mostly Eurasian fauna, most notably new species red panda.

Pliocene Africa, prior to 2.8 million years ago, was wetter than today, as evidenced by deposits of mangrove swamps and tropical forests, which retreated southward as desertification intensified. The western Sahara Desert likely formed 2.8 million years ago. The Pliocene is a particularly important time for the evolution and diversification of hominids. The aridity-humidity cycles that were related to the late Pliocene glacials-interglacials in the Northern Hemisphere in the Pliocene climate of Africa may have shaped hominid evolution by creating cyclic opportunities for species extinction and innovation.

Jarmila Pittermann
University of California, Santa Cruz

See Also: Earth's Climate History; Global Warming; Pleistocene Epoch; Quaternary Period; Tertiary Period.

Further Readings

Chandler, M. A., D. Rind, and R. S. Thompson. "Joint Investigations of the Middle Pliocene Climate II: GISS GCM Northern Hemisphere Results." *Palaeogeography, Palaeoclimatology and Palaeoecology*, v.9 (1994).

Vrba, E. S., et al. *Paleoclimate and Evolution, With Emphasis on Human Origins*. New Haven, CT: Yale University Press, 1995.

Wrenn, J. H., J.-P. Suc, and S. A. G. Leroy. *The Pliocene: Time of Change*. Dallas, TX: AASPF, 1999.

Poland

Located in eastern Europe, Poland has a land area of 120,696.41 sq. mi. (312,685 sq. km), a population of nearly 38.2 million (2010 est.), and a population density of 319.9 people per sq. mi. (120 per sq. km). Some 47 percent of the country is arable, with a further 13 percent used as pastures and meadows, and 29 percent covered in forestland. Farming is still the mainstay of much of the economy, helped by the fertile soil.

Regarding electricity generation in Poland, 98.1 percent comes from fossil fuels, mainly coal that is mined in many parts of the country and is comparatively cheap, and only 1.5 percent comes from hydropower. As a result, even though Poland is less industrialized than many other European countries, it has a high per-capita rate of carbon dioxide (CO_2) emissions. The rate was 9.1 metric tons in 1990, falling slowly to 8 metric tons by 2004, and rising to 8.3 metric tons by 2008. About 57 percent of all CO_2 emissions in the country come from the production of electricity, with 17 percent from manufacturing and construction, 11 percent from transportation, and 11 percent for residential purposes. The reliance on coal has meant that 76 percent of Poland's CO_2 emissions have originated from solid fuels, with 15 percent from liquid fuels, and 7 percent from gaseous fuels.

Since the end of communism in 1989, Poland has experienced a burgeoning private automobile ownership, with 382 vehicles per 1,000 people, ranking the country 33rd in terms of vehicle ownership. Only three former communist European countries have a higher level of vehicle ownership per 1,000 people: Lithuania (453), Estonia (410), and the Czech Republic (399). This heavy use of automobiles, a reflection of the growing wealth in the country, is in spite of an effective freight railroad network and a high-speed rail network—the only one in east-central Europe—that links Warsaw, Krakow, and Katowice. Part of the problem with rail freight in Poland is that the country has a different track gauge than that of Lithuania, Belarus, the Russian Federation, and Ukraine. There is an effective bus network and over 30 Polish cities still use trams.

Rising Temperatures and Climate Mitigation

The rising average temperatures in Poland as a result of global warming and climate change have caused hot summers in Lesser Poland, a region in the south of the country. Poland has been actively involved in various schemes to introduce carbon trading and has even managed to reduce its own emissions rate, although it has expressed determination to cut back further. As a result, Poland has attempted to follow a project developed by the Global Environment Facility, by which Mexico and Norway managed to reduce their power use through widespread introduction of compact fluorescent lamps in two major cities. In the case of Poland, this would also involve the conversion of coal-fired boilers to use gas. The main problem with this approach has been the political power of the coal-mining areas, which has hindered many attempts to reduce the dependence on coal.

The Polish government took part in the United Nations Framework Convention on Climate Change (UNFCCC) signed in Rio de Janeiro in May 1992. It signed the Kyoto Protocol to the UNFCCC on July 15, 1998, committing to a 3 percent reduction prior to ratification, which took place on December 13, 2002. It entered into force on February 16, 2005. However, in some Polish political circles, there were attempts to delay additional moves to combat climate change within the European Union. In spite of this, a UN Conference on climate change was held in the Polish city of Poznan on December 1–12, 2008. In the following year, Poland made known its opposition at a meeting of finance ministers at Luxembourg in June 2009, ahead of the climate change conference at Copenhagen that took place on December 7–18, 2009.

Justin Corfield
Geelong Grammar School

See Also: Coal; European Union; Global Warming.

Further Readings

EarthTrends. "Poland: Climate and Atmosphere." http://earthtrends.wri.org/text/climate-atmosphere/country-profile-146.html (Accessed October 2011).

Hicks, Barbara. *Environmental Politics in Poland: A Social Movement Between Regime and Opposition.* New York: Columbia University Press, 1996.

Hughes, Gordon and Julia Bucknall. *Poland: Complying With E.U. Environmental Legislation.* Washington, DC: World Bank, 2000.

O'Riordan, Tim and Jill Jäger, eds. *Politics of Climate Change: A European Perspective.* London: Routledge, 1996.

Rankin, Jennifer. "Poland Blocking Climate Change Deal." *European Voice* (June 9, 2009). http://www.europeanvoice.com/article/2009/06/poland-blocking-climate-change-deal/65117.aspx (Accessed July 2009).

Yamin, Farhana, ed. *Climate Change and Carbon Markets: A Handbook of Emissions Reduction Mechanisms.* London: Earthscan, 2005.

Polar Bears

The polar bear (Order Carnivora, family Ursidae) is the largest bear species and is thought to have evolved from brown bears, *Ursus arctos*, approximately 1 million years ago. There are 19 recognized populations distributed in Canada, the United States (Alaska), Norway (Svalbard Islands), Denmark (Greenland), and Russia. The current estimated worldwide population is 20,000–25,000.

Polar bear territories can cover tens of thousands of sq. km. They live solitarily, but often congregate around food sources. Their diet consists primarily of ringed seals, but includes other seals, walruses, and beluga whales. They obtain the majority of their nutritional intake in the spring and summer from seal pups, which can be as much as 50 percent fat. Polar bears can swim for 60 km without resting, at speeds up to 10 km per hour. They possess several adaptations for a semiaquatic existence, including partially webbed front paws, eyes adapted to see underwater, and a thick fat layer of 3 to 5 in. (8 to 12 cm) that provides buoyancy and insulation in water.

Polar bears are also highly adapted to the Arctic climate. Arctic temperatures can drop to minus 49 degrees F (minus 45 degrees C) for days or weeks, but polar bears can withstand this because of their thick fat layer, a dense undercoat of fur with longer guard hairs, and black skin, which absorbs heat from sunlight. Other Arctic adaptations include white-appearing fur for camouflage (the hairs are actually colorless and hollow); small ears and tail, which reduce heat loss; large furry feet, which act like snowshoes; and a digestive system that is very effective at absorbing and storing fat.

Females become sexually mature at 4 to 5 years of age. Males may not mate successfully until they are 8 to 10 years old. Mating occurs from April to June and each male may mate with more than one female. The females have induced ovulation; mating multiple times causes the release of an egg. The implantation of the blastula (the fertilized egg after several cell divisions) is delayed until September/October, and in November/December the female excavates a den in the snow. She eventually gives birth in December or January. One to three cubs are born (two-thirds of cubs are twins) and they are nursed in the den until March or April, when they emerge. Cubs are weaned at the age of two or three years. During this period, they learn important hunting skills from their mother. Occasionally, cubs are attacked and killed by males, so mothers are fiercely protective. Females breed after they wean their current cubs, but will not reproduce at all if conditions are unfavorable. Cub mortality rates are high.

The most common cause of death for subadult bears is starvation, as they do not yet have a territory and must compete with larger bears, pushing many into marginal habitats. In adulthood, mortality drops sharply (to less than 5 percent annually); the most common natural cause of death in adults is attacks by other bears. Human activities and impacts that pose the greatest risk to polar bears include hunting, pollution, industrial development, and climate change.

Human Effects on Polar Bears

Humans kill polar bears for aboriginal subsistence, sport, and defense of human life and property. In some areas, monitoring of polar bear kills is effective (in Norway and the United States), but in other areas, such as Russia and Greenland, there is little reliable information. Concerns have recently been expressed about the threat posed by trophy hunting, which is currently allowed only in Canada. Quotas are often based on poor population data. Until 2008, approximately 80 trophies were imported into the United States each year from Canada.

Arctic marine mammals can accumulate high concentrations of pollutants in their blubber. As top predators, polar bears accumulate the highest concentrations, and in some areas there is concern about pollution effects on the bears' health. Polar bears with abnormal genitalia and other defects have been recorded, and there have been suggestions that these are caused by exposure to certain contaminants. Development in the Arctic is also an issue. The Arctic is rich in natural resources, especially oil and gas. Exploration for and extraction of these resources involve construction that could potentially reduce habitat, produce pollution, or cause disturbances.

Because of their exclusively Arctic habitat and their charismatic nature, polar bears have become the "poster child" for the impacts of climate change. The primary concern is the loss of sea ice, which would remove essential habitat both for the bears and their marine mammal prey. Greater numbers of drowned polar bears have been reported, presumably because of overexertion during swimming between the increasing areas of open water between ice floes. The loss of ice cover could also disrupt migration routes, reduce the ability to access prey and mates, and increase distances that animals have to travel to find food, exacting an energetic cost. Many females also construct their birthing dens on ice. In the past 20 years, the proportion of dens located on sea ice has halved.

Habitat loss and prey reduction may force bears closer to human habitation to find food. Sightings of animals near Arctic towns and villages are occurring with greater frequency. This increases the likelihood of negative human/polar bear interactions. In addition, researchers have reported a significant decrease in polar bear body condition over the past 20 years, and longer periods of ice-free seas have caused reduced litter sizes and decreased female and calf survival rates.

Climate change may also cause chronic overheating in the highly cold-adapted bear. Finally, rising temperatures are causing a shift in the distributions of other bear species; for example, brown bear populations are shifting further north into the range of the polar bear and could compete, or hybridize, with polar bears.

Conservation of Polar Bears

In 1965, the "polar bear" nations met and agreed that each country should take whatever steps were necessary to conserve the species. It was determined that cubs, and females accompanied by cubs, should be protected throughout the year; that each nation should, to the best of its ability, conduct research on polar bears within its territory; and that each nation should exchange information on polar bears freely. This eventually led to the signing of the International Agreement on the Conservation of Polar Bears in May 1973 in Oslo, Norway. This agreement is currently the major international polar bear treaty.

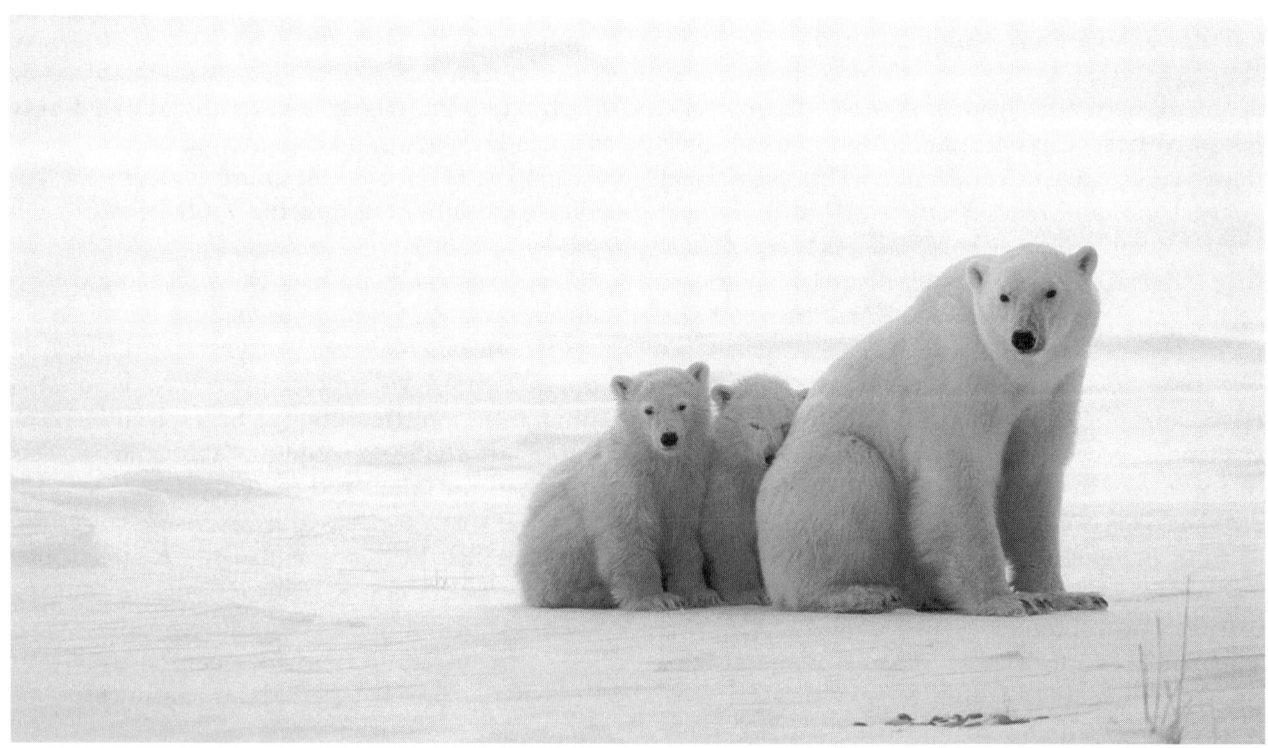

Because of their exclusively Arctic habitat and their charismatic nature, polar bears have become the "poster child" for climate change and global warming awareness. A reduction of sea ice would result in the loss of polar bear habitat. It would also affect their access to food and to mates, limit the availability of den sites, and disrupt their migration routes. Many drowned polar bears have been reported, possibly the result of larger areas of open water between ice floes.

In 1982, largely because of declines resulting from hunting pressure, the International Union for Conservation of Nature (IUCN) listed the polar bear as "vulnerable." This rating was reduced to "conservation dependent" in 1996. However, in May 2006, polar bears were relisted as vulnerable because of a predicted population reduction of more than 30 percent within three generations (45 years) and a decline in area of occupancy, extent of occurrence, and habitat quality resulting from climate change. In their 2005 meeting, the IUCN Polar Bear Specialist Group conducted a review of the world's 19 polar bear populations and determined that five populations were in decline, five were stable, and two were increasing. However, at their 2009 meeting, the IUCN Polar Bear Specialist Group determined that the situation was worse, with eight populations in decline, three stable, and only one population increasing (this population was seriously depleted before the hunting quota was reduced). The rest of the populations are data deficient.

In the United States, polar bears are managed by the Fish and Wildlife Service and are protected by the Marine Mammal Protection Act (MMPA), which prohibits acts or attempts of harassment, hunting, capture, or killing. There are exemptions for Alaska Native subsistence hunting, as well as scientific research and "incidental harassment" from activities such as oil and gas exploration. However, a controversial amendment in 1994 permitted the import of sport-hunted polar bear trophies into the United States from some Canadian populations. In 2010, the U.S. government proposed listing polar bears under Appendix I of the Convention on the International Trade in Endangered Species of Wild Fauna and Flora, which would have ended the commercial trade of polar bear products internationally and limited movement of trophies. However, this proposal was defeated, largely because of an IUCN analysis of the proposal that stated that listing the polar bear on Appendix I was not warranted—an analysis that has received subsequent criticism for ignoring the science that showed greater likelihoods of polar bear declines.

In February 2005, the Center for Biological Diversity and Greenpeace USA petitioned the U.S. government to list polar bears under the U.S. Endangered Species Act (ESA). In January 2008, a proposed rule to list the bear as threatened (that is, likely to become endangered within the foreseeable future throughout all or a significant portion of its range) was published, and the species was listed as such on May 15, 2008. The import of trophies from Canada became prohibited, as ESA-listed species are automatically considered "depleted" under the MMPA, and depleted species cannot be imported.

The listing was primarily the result of scientific analyses of climate change data and polar bear distribution conducted by the U.S. government, which projected a two-thirds reduction in population within the next 40 years. This listing obligates the U.S. government to reduce anthropogenic impacts on polar bears and devise a plan to aid their recovery. A recent analysis by Steven Amstrup and colleagues calculated that ice loss correlated to greenhouse gas (GHG) emissions was the biggest factor in determining the likelihood of polar bear extinction—and that GHG emissions control (coupled with the best management scenario of other impacts) minimizes the extinction risk of this iconic polar species.

Edward Christien Michael Parsons
B. J. Milmoe
George Mason University
Naomi A. Rose
Humane Society International

See Also: Charismatic Megafauna; Climate Change, Effects of; Detection of Climate Changes; Marine Mammals; Penguins; Sea Ice.

Further Readings

Amstrup, Steven, et al. "Greenhouse Gas Mitigation Can Reduce Sea-Ice Loss and Increase Polar Bear Persistence." *Nature*, v.468/7326 (2010).

Derocher, A. E., et al. "Polar Bears in a Warming Climate." *Integrative and Comparative Biology*, v.44, 2004.

Durner, G. M., et al. "Predicting 21st-Century Polar Bear Habitat Distribution From Global Climate Models." *Ecological Monographs*, v.79 (2009).

Fischbach, A. S., et al. "Landward and Eastward Shift of Alaskan Polar Bear Denning Associated with Recent Sea Ice Changes." *Polar Biology*, v.30 (2007).

Monnett, Charles and J. S. Gleason. "Observations of Mortality Associated With Extended Open-Water

Swimming by Polar Bears in the Alaskan Beaufort Sea." *Polar Biology*, v.29 (2006).

Obbard, M. E., et al. "Temporal Trends in Body Condition of Southern Hudson Bay Polar Bears." *Climate Change Research Information Note*, v.3 (2006).

Parsons, E. C. M. and L. A. Cornick. "Sweeping Scientific Data Under a Polar Bear Skin Rug: The IUCN and the Proposed Listing of Polar Bears Under CITES Appendix I." *Marine Policy*, v.35 (2011).

Regehr, E. V., et al. "Survival and Breeding of Polar Bears in the Southern Beaufort Sea in Relation to Sea Ice." *Journal of Animal Ecology*, v.79 (2010).

Rosing–Asvid, Aqqalu. "The Influence of Climate Variability on Polar Bear (*Ursus maritimus*) and Ringed Seal (*Pusa hispida*) Population Dynamics." *Canadian Journal of Zoology*, v.84/3 (2006).

Stirling, Ian and A. E. Derocher. "Possible Impacts of Climatic Warming on Polar Bears." *Arctic*, v.46/3, 1993.

Stirling, Ian, Nicholas J. Lunn, and John Iacozza. "Long Term Trends in the Population Ecology of Polar Bears in Western Hudson Bay in Relation to Climatic Change." *Arctic*, v.52/3 (1999).

Stirling, Ian and C. L. Parkinson. "Possible Effects of Climate Warming on Selected Populations of Polar Bears (*Ursus maritimus*) in the Canadian Arctic." *Arctic*, v.59/3 (2006).

Policy, International

Calls for concerted international action on climate change date back more than two decades. Early cooperation on climate policy largely occurred under the auspices of the United Nations (UN), resulting in the two major international treaties, the United Nations Framework Convention on Climate Change (UNFCCC) and the Kyoto Protocol. Yet, multilateral engagement within this regime has often proven tedious, and recent efforts to flesh out commitments and extend these beyond 2012 have been marred by diplomatic stalemate. Other venues such as the Group of Eight (G8), Group of 20 (G20), and Major Economies Forum (MEF) have increased the pace of their activities, yet the same divisions marring cooperation under the UN have also prevented such alternative channels from achieving a significant breakthrough. Going forward, international climate policy is likely to become more fragmented and driven by bilateral or regional cooperation.

Defining Principles and Institutions

In 1988, the UN General Assembly declared global warming a "common concern of mankind," paving the way for formal negotiations under the auspices of the UN. In the same year, a newly established scientific body, the Intergovernmental Panel on Climate Change (IPCC), was mandated with assessing the actual threats posed by climate change, becoming the most authoritative source of scientific advice on global warming and providing much of the factual background for diplomatic negotiations on an international response. Such negotiations were launched in 1990 with the establishment of an Intergovernmental Negotiating Committee (INC) and ultimately resulted in the adoption of the UNFCCC in 1992. A milestone in early climate cooperation, the UNFCCC entered into force on March 21, 1994, and has since been ratified by 194 parties, affording it one of the broadest memberships of any international agreement.

Following a pattern found in other multilateral environmental agreements (MEAs), the UNFCCC establishes a sophisticated framework of institutions and procedures, deferring the adoption of more detailed obligations to subsequent protocols or amendments. Rather than calling for quantified emissions reductions, the UNFCCC declares its "ultimate objective" to be "stabilization of greenhouse gas concentrations in the atmosphere at a level that would prevent dangerous anthropogenic interference with the climate system." Scientific uncertainty and political dissent have, to date, prevented international consensus on a threshold for "dangerous anthropogenic interference," although the international community has since backed the objective of limiting the increase in global average temperatures to 3.6 degrees F (2 degrees C) above preindustrial levels.

A set of principles guides the achievement of the foregoing stabilization objective, including the principle of common but differentiated responsibilities. Introducing strong considerations of equity and distributional justice, this principle acknowledges the different contributions of industrialized and developing countries to global warming. It also gives consideration to the uneven distribution of its impacts, which are likely to be most severe in developing countries, where poverty, a weak infrastructure, and a degraded natural resource base all lead to high vulnerability and lessen the capacity for adaptation. Accordingly, the international community has agreed to confront climate change on a differentiated basis, assigning different levels of commitment to different states.

All parties are required to establish national programs outlining mitigation and adaptation measures; cooperate in research, education, and the development of clean technologies; and compile and publish national greenhouse gas (GHG) inventories. A number of additional commitments apply solely to the industrialized countries listed in Annexes I and II to the convention, including a duty to provide "new and additional financial resources" to cover the costs of compliance by developing countries and assist particularly vulnerable countries, and a quantified—but not legally binding—aim of returning GHG emissions to 1990 levels by 2000. With few exceptions, industrialized parties to the UNFCCC failed to achieve this objective.

Another important function of the UNFCCC is the creation of an institutional framework to monitor implementation of its provisions, channel information and cooperation, and promote the negotiation of further commitments. An annual Conference of the Parties (COP) is vested with the authority to review operation of the convention and to "make, within its mandate, the decisions necessary to promote" its effective implementation. Recurrent monitoring and administration tasks are carried out by its secretariat located in Bonn, Germany, which is endowed with substantial financial and personnel resources; more detailed aspects of implementation and scientific and technical advice are addressed by two subsidiary bodies.

Quantifying Commitments

Given the need for unanimous consent prior to adoption, the broad participation of the UNFCCC also translated into substantive commitments that were largely programmatic in nature, with more specific obligations deferred to a subsequent instrument. To this end, parties to the UNFCCC convened in Berlin, Germany, in 1995 for the first session of the COP to adopt a series of decisions elaborating on the climate regime and further commitments. Charged with reviewing the adequacy of the commitments of developed countries, the COP adopted a decision—later known as the Berlin Mandate—opening a new round of negotiations on "a protocol or another legal instrument" with the aim of setting quantified emissions limitation and reduction objectives (QELROs).

Negotiations continued with the aim of presenting a draft protocol to the third COP, which was to meet in Kyoto, Japan, in 1997. A highly contentious negotiation process followed, with different coalitions of states with countervailing interests pitting against each other. Reconciling the various positions only succeeded after a marathon of consultations and reluctant concessions from each side, setting the tone for future climate summits. The unanimously adopted outcome, the Kyoto Protocol, marked the birth of a sophisticated regime built on quantitative limitation and reduction commitments for developed countries, as well as a set of highly innovative market instruments—the "flexible mechanisms"—to meet these obligations.

Rather than amending the parent convention, the Kyoto Protocol is a separate instrument under international law that requires ratification to enter into force. A Plan of Action adopted at the fourth Conference of the Parties (COP 4) in Buenos Aires was meant to finalize the text of the Kyoto Protocol, paving the way for ratification by its signatories. Various setbacks, however, notably during the sixth Conference of the Parties (COP 6) held at The Hague, the Netherlands, in 2000 and coupled with a rejection of the Kyoto Protocol by the United States in 2001, threatened to derail the multilateral climate process. Despite this diplomatic stalemate, COP 6 resumed in Bonn, Germany, in 2001 and culminated in the adoption of the Bonn Agreement, a political arrangement on core elements of the Buenos Aires Plan of Action.

With some of the most contentious issues thereby resolved, the seventh Conference of the Parties (COP 7) in Marrakesh, Morocco, in 2001 was in a position to settle remaining technical issues with a set of detailed rules, procedures, and guidelines known as the Marrakesh Accords. Central features of the climate regime had thus been put in place. Still outstanding, however, were a sufficient number of ratifications to prompt the entry into force of the Kyoto Protocol. Given the withdrawal of the United States, only ratification by Russia would ensure that the threshold specified in the protocol was met. After much hesitation and political bargaining, the Russian government finally submitted its ratification instrument to the UNFCCC Secretariat on November 18, 2004, securing the entrance into force of the Kyoto Protocol on February 16, 2005.

As an international treaty with only 28 provisions, the Kyoto Protocol relies on decisions by the Meeting of the Parties for elaboration and specification. Sustaining the differentiation of commitments for industrialized and developing countries, it sets forth a number of substantive requirements for developed countries, including a detailed list of policies and measures (PAMs) these shall adopt, and defines general obligations for all parties to the protocol. Other provisions contain guidance on financial aspects; an assignment of institutional roles to the bodies established under the parent convention, including designation of a supreme body; the "Conference of the Parties serving as the meeting of the Parties"; and rules on compliance procedures and the settlement of disputes. It is rounded off by several provisions relating to its amendment, entry into force, voting, and withdrawal. The annexes list GHGs and sectors covered by the protocol, as well as quantified emissions limitations and reduction commitments for specified industrialized countries.

These legally binding commitments are at the core of the Kyoto Protocol, imposed on the same developed nations—Annex I Parties—listed in the parent convention, the UNFCCC, and requiring said nations to individually or jointly ensure that their anthropogenic GHG emissions do not exceed individually specified amounts. As the Kyoto Protocol goes on to state, the objective of these quantified commitments is to reduce "overall emissions of such gases by at least 5 percent below 1990 levels in the commitment period 2008 to 2012." In its annexes, the Kyoto Protocol goes on to define absolute limit values for a basket of GHGs, expressed in a percentage of base year emissions, to be met over the first commitment period from 2008 to 2012. Ranging from a reduction of 8 percent for the European Union to an increase of 10 percent for Iceland, these commitments are calculated against a 1990 baseline, although countries in transition to a market economy have the option of selecting an alternative base year. Sequestration from certain land use, land-use change and forestry activities (LULUCF) may be counted toward compliance with the mitigation commitments.

Carbon Markets

Aside from these policies and measures, parties may choose to meet mitigation commitments with a set of three, market-based flexible mechanisms. All three mechanisms involve cooperation between parties and are based on the notion of tradable carbon units that may be applied toward compliance with the mitigation commitments. Quantified emissions objectives are hence a vital condition for the carbon market because they define the units that can subsequently be traded. Largely adopted in response to pressure from certain parties in the negotiations, such as the United States, these mechanisms aim to harness differences in abatement cost between countries. Because atmospheric levels of GHGs will decline regardless of where reductions occur, such heterogeneity can help lower the overall costs of mitigation measures. By providing an ongoing incentive to reduce emissions, moreover, the flexible mechanisms also promise to encourage innovation, while helping channel foreign investment and sustainable technologies to developing countries.

One mechanism, international emissions trading (IET), creates an international market for tradable emissions allowances between parties, whereas the other two—Joint Implementation (JI) and the Clean Development Mechanism (CDM)—allow for generation of offset credits through mitigation projects in developed and developing countries. Use of the flexible mechanisms is conditional on a sophisticated set of rules and methodologies. Most importantly, states must be parties to the Kyoto Protocol. Developed countries, moreover, are required to calculate their assigned amounts pursuant to specified accounting modalities and establish a national system for the estimation of anthropogenic GHG emissions by sources and removals by sinks. In order to ensure "the accurate accounting of the issuance, holding, transfer, acquisition, cancellation, and retirement" of carbon units, these parties are also required to establish a national registry—essentially a standardized electronic database with different types of accounts—and designate an organization serving as its administrator. Annual submission of accurate inventories, finally, is the "backbone" of the eligibility criteria and subject to a strict "threshold of failure" specified in a separate decision. Parties failing to meet these criteria may be subject to several sanctions, including exclusion from the use of the flexible mechanisms. While the Marrakesh Accords contain no quantitative limits on the use of the flexible mechanisms to meet commitments, parties are required to provide information demonstrating that use of the mechanisms is "supplemental to domestic action," and domestic policies and measures must constitute "a significant element" of efforts to meet commitments.

Altogether, the flexible mechanisms have contributed to the emergence of a sizeable market for carbon, outgrowing several traditional commodity markets and resulting in the creation of an independent services sector comprising project developers, brokers, verifiers, and other services required for operation of the mechanisms. Of the three mechanisms, the CDM in particular has been considered a success, seeing more than half a billion Certified Emissions Reductions (CERs) issued since its launch and resulting in net financial flows of several billion U.S. dollars to mitigation projects in developing countries. It has also been subject to criticism, however, for a perceived lack of environmental integrity, cumbersome approval procedures, and the concentration of mitigation projects in a small number of emerging economies. Calls for reform seek to address these shortcomings, yet engagement in the international carbon market has subsided significantly as the end of the first commitment period of the Kyoto Protocol approaches. Meanwhile, domestic and regional markets such as the European Union Emissions Trading Scheme (EU ETS) have quickly surpassed the international market in size and scope, foreshadowing a broader trend in international climate cooperation.

Fragmentation and Stagnation

Because the quantified emissions limitations and reduction objectives (QELROs) for developed countries specified in the Kyoto Protocol expire in 2012, its governing body quickly adopted a mandate to negotiate new commitments by its parties. Significantly, this mandate had to account for the divergent membership of the UNFCCC and the Kyoto Protocol, forcing the negotiations to proceed on two separate, yet overlapping tracks, with distinct bodies and procedures. Also, the difficulties experienced in its negotiation and ensuing ratification prompted the emergence of new channels for international engagement on climate change, including several regional and bilateral initiatives, which further increased the complexity of international climate cooperation.

By December 2007, discussions under the UNFCCC and Kyoto Protocol had progressed sufficiently to adopt a more sophisticated mandate, the Bali Roadmap, which called for a focused process to conclude two years later. Yet when leaders from around the world converged in Copenhagen in December 2009, the parallel negotiation processes had failed to narrow down potential options sufficiently to allow for passage of an international agreement in the tradition of the UNFCCC or the Kyoto Protocol. Instead, in an atmosphere of tension and mistrust, a group of heads of state and government elaborated a new document that was sufficiently vague and limited in scope to meet with general approval. Given the absence of alternative options, a majority of states agreed to "take note" of the ensuing Copenhagen Accord, with several parties censuring its lack of ambition and the undemocratic process in which it had been adopted.

Although parties soon resumed the negotiations following the Copenhagen climate summit, faith in the UNFCCC regime had been severely shaken. Only when measured against significantly lowered expectations can the climate summit held in Cancún one year later be considered a success. In effect, the central decisions adopted at this meeting—collectively referred to as the Cancún Agreements—were largely limited to enshrining the broad terms of the Copenhagen Accord in the more formal terms of official decisions. New institutional arrangements such as the Technology Mechanism and the Green Climate Fund were rendered operational, but more divisive issues, including the legal form of a future climate agreement and the extension of commitments under the Kyoto Protocol, were consciously deferred to later meetings. Negotiations in the wake of the Cancún summit showed that these questions once again threatened to unravel diplomatic progress.

In many ways, the Copenhagen summit marked an important turning point in the practice of multilateral climate cooperation. Although fissures had been visible at many stages during the earlier negotiations, international engagement on the issue became so overloaded with a multiplicity of reciprocally contingent, technically complex, and politically controversial issues that progress was no longer defined in terms of resolving the climate challenge, but merely in keeping the multilateral process alive. At least in part, this development followed from the difficulties faced in achieving universal agreement on binding international commitments, giving new momentum to an earlier debate about the merits of alternative venues of climate cooperation. Accordingly, a number of existing forums such as the G8 or the G20 have expanded the scope of their activities to include climate change, while new high-level or technical dialogues have been created with the MEF, the Cartagena Group, the now defunct Asia–Pacific Partnership on Clean Development and Climate (APP), or the International Partnership for Mitigation and MRV (Measuring, Reporting, and Verficiation).

At least some of these venues, such as the MEF, were originally launched to rival the negotiations under the UNFCCC, yet they now primarily serve a complementary function, facilitating a constructive exchange on contentious issues outside the context of formal negotiations. None of these venues has an official negotiating mandate, thus they are largely limited to adopting political recommendations or declarations of intent. Likewise, these bodies lack the financial resources and staff of the UNFCCC Secretariat to promote implementation of climate policies. As such, they cannot substitute the international climate regime established under the auspices of the UN. Given the diplomatic obstacles encountered there, it is therefore likely that future international climate policy will be increasingly driven by voluntary, informal, and bilateral or regional initiatives, resulting in a far more fragmented and multifaceted policy landscape. Whether this "bottom-up" approach to climate cooperation proves more successful at addressing the challenges of mitigation and adaptation than the multilateralism of earlier phases has yet to be seen.

Michael Mehling
Ecologic Institute

See Also: Asia–Pacific Partnership on Clean Development and Climate; G8/G20; Intergovernmental Panel on Climate Change; Kyoto Protocol; United Nations Framework Convention on Climate Change.

Further Readings

Aldy, J. E., et al., eds. *Architectures for Agreement: Addressing Global Climate Change in the Post-Kyoto World*. Cambridge: Cambridge University Press, 2007.

Biermann, F. "Beyond the Intergovernmental Regime: Recent Trends in Global Carbon Governance." *Current Opinion in Environmental Sustainability*, v.2 (2010).

Dubash, N. K., et al. "Beyond Copenhagen: Next Steps." *Climate Policy*, v.10 (2010).

Okereke, C., et al. "Conceptualizing Climate Governance Beyond the International Regime." *Global Environmental Politics*, v.9 (2009).

Yamin, F., et al. *The International Climate Change Regime: A Guide to Rules, Institutions, and Procedures*. Cambridge: Cambridge University Press, 2004.

Pollution, Air

Air pollution is defined as the addition of contaminants to the atmosphere that impact the air quality. Sources may be anthropogenic or natural. Contaminants render the air less pure and may result in harm to humans, animals, plants, water, soils, and the built environment.

These pollutants have also been implicated in climate change, including alterations to temperature and visibility, at various temporal and spatial scales. They may alter ambient global and local climates in various ways. The addition of greenhouse gases (GHGs) into the atmosphere increases temperatures. Particulate pollutants render the air less safe to breathe. Certain pollutants may impact visibility by the creation of haze or smog.

Air Contaminants and Their Consequences

Many contaminants are the products of human activities. These include industrial activity, wood combustion for cooking, and fossil fuel usage for heating, cooling, and transportation. Air pollution is often thought of as a human-caused problem. However, not all sources of climate-affecting air pollution are anthropogenic; natural sources also exist. Volcanoes, for example, can spew ejecta (ash and sulfurous gases) into the atmosphere, affecting both local and global climates. The 1980 eruption of Mount St. Helens in the state of Washington resulted in a reduction of solar radiation at the surface in the vicinity of the volcano. This effect was relatively short. The 1991 eruption of Mount Pinatubo in the Philippines, on the other hand, reduced the Earth's average temperature by approximately 0.9 degree F (0.5 degree C).

Volcanic eruptions that release large amounts of sulfurous gases into the stratosphere are related to short-term global cooling, generally on the order of no more than a few years. Sulfur combines with water vapor to produce sulfuric acid; this in turn scatters incoming short-wave radiation back to space. However, the effect of these eruptions may be tempered if an El Niño climate disturbance is present. Other sources of natural air pollution are wetlands, forest fires, and biological decay. The water-saturated environments of wetlands promote the formation and release of methane (CH_4), one of the most significant GHGs. Wildland fires add particulates, carbon monoxide (CO), and nitrogen oxides (NOx). Biological decomposition adds CO_2 and CH_4 to the atmosphere. The addition of GHGs leads to the increase in global temperatures referred to as global warming.

Atmospheric particulate matter (PM)—whether solid or liquid—varies in density, size, mass, and residence time. Particulates can be either primary or secondary. Primary particulates are sent directly into the atmosphere. Examples of these include soot, volcanic ash, salts, and pollen. Secondary particulates are the products of chemical processes between gases, aerosols, and other components of the atmosphere. Sulfuric acid and nitric acid are examples of secondary particulates. Particulates exist in a range of sizes; they may be anywhere from less than 0.01 micrometers (μm) to several hundreds of μm in size. The size of PM helps determine its effect on the scattering of light. While heavier particles quickly settle out, lighter particles can remain suspended in the atmosphere for some time, affecting ambient climate. The PM affects the scattering of short-wave radiation and precipitation. For the latter, particulates actually serve the useful function of condensation nuclei (particles onto which water condenses) or ice nuclei (particles promoting the formation of ice crystals); both condensation and ice nuclei are required for precipitation. However, the presence of PM in the atmosphere also has negative health consequences. One ramification is the negative impact on the human respiratory system. Particles smaller than 10 μm are easily inhaled and may build up in human lungs. Those sized 2.5 μm (or less) can be even more damaging. These small-sized particles may lodge even more deeply into the lungs, affecting breathing in compromised individuals.

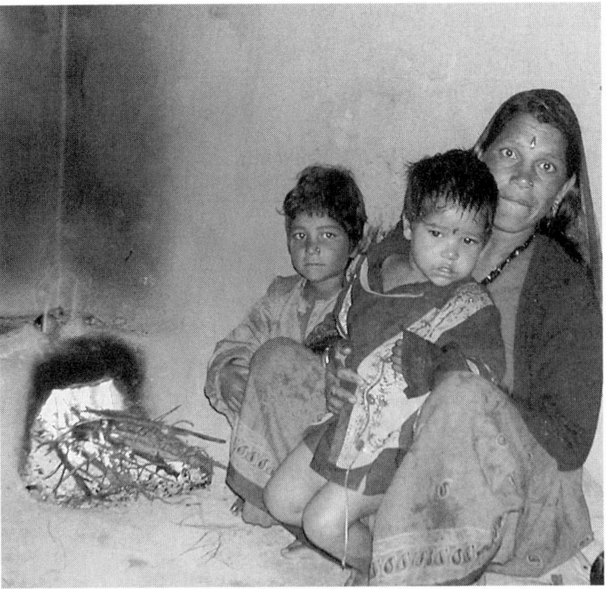

In many developing countries, families such as this one in India rely on dung or wood to fuel cooking stoves. Annually, nearly 2 million people worldwide die prematurely from illness attributable to indoor air pollution from household solid fuel use.

Smog and Acid Rain

Smog, a term derived from the words *smoke* and *fog* and first used in 1905, describes a form of pollution associated with industrial activity and vehicular emissions. Reduced visibilities and negative health impacts are the hallmarks of smog. Smog was originally associated with coal combustion, when the burning of coal would produce smoke containing sulfur dioxide (SO_2). The Great Smog of 1952, which occurred in London, was one of the United Kingdom's worst environmental disasters. Up to 4,000 deaths were attributed to this event, which was precipitated by a combination of meteorological factors and the burning of greater than usual amounts of low-quality sulfurous coal.

Presently, much smog is produced by large amounts of emissions from motor vehicles. This is termed *photochemical smog*, which describes the reactions between sunlight and vehicular exhaust. Solar radiation reacts with NO_x and volatile organic compounds (primary pollutants that are largely anthropogenic in origin). These reactions leave the lower troposphere with an unhealthy "soup" of secondary pollutants, including tropospheric ozone and peroxyalkyl nitrate (PAN). Photochemical smog negatively affects humans and plants; humans suffer from an exacerbation of respiratory ailments, while plant cells become damaged. Photochemical smog, which is typically associated with densely populated urban areas, can be magnified by meteorological and physiographic conditions.

Anticyclones (high-pressure systems) promote smog episodes since they are accompanied by clear skies and light (or calm) winds; therefore, the ability of the atmosphere to advect the pollution away is greatly reduced. Anticyclones are associated with stable atmospheres (from temperature inversions) that keep pollutants near the ground. These inversions act as a lid. Additionally, some cities are geographically situated in settings that enhance the formation of smog. Mexico City's location in a basin, for example, means that smog is trapped. Similarly, the mountains surrounding Salt Lake City, Utah, confine pollutants. Los Angeles, California, a city synonymous with smog, is located in a basin and surrounded by the San Bernardino Mountains. In these cases, anticyclones magnify the pollution effects.

Another consequence of air pollution, the Asian brown cloud, has been implicated in regional climate changes, particularly in preventing solar radiation from reaching the surface. The Asian brown cloud is a 1.86 mi. (3 km) thick mass of pollutants found over south Asia and the Indian Ocean; it appears as a brown-colored layer. Its presence typically occurs during the winter (wet) monsoon season. The culprit is soot (black carbon), which absorbs solar radiation. Sources of soot include wood fires (from cooking and agriculture), dung fires, industrial activity, and vehicular exhaust. The presence of the cloud impacts local and regional radiation balances, as well as visibility. Melting of Himalayan glaciers has also been partially attributed to the presence of the Asian brown cloud, as the cloud has been implicated in the compounding of atmospheric warming. The carbon settles out onto the glaciers, exacerbating melting. Additional climate impacts of the Asian brown cloud include reduced precipitation in some regions, including Pakistan and western China.

One of the earliest publicized climate change issues related to contaminated air was acid rain (or acid deposition). The pH of normal rain is 5.6, rendering it slightly acidic. Water reacts with atmospheric carbon dioxide (CO_2) to form carbonic acid. When pH values fall below that of normal rain, the result is referred to as acid deposition.

Acid rain derives primarily from industrial activities and power generation. Most commonly, NO_x and SO_2 are converted into nitric acid and sulfuric acid. When these are precipitated out onto the Earth's surface, they have damaging effects. Given global-scale winds and the ability of the atmosphere to transport contaminants across great distances, acid rain may have disastrous effects in locations far from the source of the initial pollution. For example, acid precipitation in Sweden has been linked to industrial activity in Germany.

Acid deposition has been shown to deteriorate building surfaces, escalating the breakdown associated with normal weathering. Additionally, damage to vegetation has resulted from acidic precipitation. *Waldsterben* (forest death) has been famously documented in Germany's Black Forest region; this has been linked to industrial pollutants that form acid rain. Furthermore, acid precipitation runs off into streams and rivers, which can upset aquatic ecology and harm wildlife. In Sweden, for example, acid deposition runoff has resulted in fish deaths occurring in thousands of lakes. These damaging effects are exacerbated in water bodies with an underlying geology comprised of neutral or acidic rock.

Ozone Layer

Another climate change issue related to pollution that has received much public attention is the decrease in stratospheric ozone (O_3). Anthropogenically derived atmospheric contaminants, most notably chlorofluorocarbons (CFCs), have resulted in damage to the protective layer of ozone found in the stratosphere. Ozone prevents much of the ultraviolet (UV) radiation from reaching the troposphere and the Earth's surface, thus shielding humans from its harmful effects. Most of the atmosphere's oxygen is in its diatomic form (O_2), a more stable form than ozone; O_3, unlike O_2, reacts readily with other compounds. The stratosphere houses 97 percent of the atmosphere's ozone, with the remainder found in the troposphere, where it acts as a pollutant and is dangerous to inhale. Loss of stratospheric ozone has resulted in an ozone "hole" above Antarctica, with a smaller one appearing in the Arctic. The hole is a seasonal thinning of the ozone layer. The loss of this climatic protection against harmful UV radiation has been linked to photo-aging, immune system imbalances, and increased cases of skin cancers.

CFCs and other compounds found in refrigerants and solvents, halons (fire extinguishers), and methyl bromide (pesticides) can wreak havoc if they migrate to the stratosphere. They give up chlorine and bromine atoms, which are highly efficient at destroying ozone. A single chlorine atom can obliterate over 100,000 ozone molecules, and a bromine atom is 50 times as destructive as chlorine. The time required for these compounds to travel from the troposphere to the stratosphere can take years, but once there, they remain for a long time. The actual residence times of CFCs and similar chemicals vary, but they can stay in the stratosphere for over 100 years.

The destructiveness of these chemicals to the ozone layer prompted the international community to outlaw their use. The Montreal Protocol on Substances That Deplete the Ozone Layer, signed by 196 countries, is a treaty constructed to protect the ozone layer from further damage. Entered into force in 1989, its aims are to eliminate the production and consumption of ozone-damaging substances. Since this agreement went into effect, the amounts of many of those substances in the stratosphere have declined, in some cases rather markedly. Given the destructive impact on the atmosphere's ability to block UV radiation resulting from a thinned ozone layer, this treaty can be considered a successful model of international cooperation.

Unlike its counterpart in the stratosphere, tropospheric ozone is regarded as a pollutant and a GHG, since it contributes to the overall absorption of outgoing long-wave radiation. Although it still comprises a relatively small amount of total GHGs, concentrations of ozone near the Earth's surface have increased markedly since 1900, largely because of fossil fuel consumption, particularly gasoline-powered vehicles and coal-fired power plants. Tropospheric ozone is seasonal; higher concentrations are typically found in warmer months. Additionally, levels begin to rise in the morning, reaching a peak in the afternoon. Levels decrease after the sun goes down, because of the role that solar radiation plays in ozone formation. In the lower troposphere, nitrogen dioxide (NO_2) from fossil fuel combustion reacts with short-wave radiation, forming nitric oxide (NO) and atomic oxygen (O). Oxygen subsequently forms ozone by combining with the rather abundant O_2. Since solar radiation plays a major role in the formation of tropospheric ozone, the pollutant is a bigger problem during the high sun season (May through September, in the Northern Hemisphere). Too much tropospheric ozone can affect the pulmonary system of humans, becoming especially dangerous to those suffering from lung disease. Because of this, governmental agencies such as the National Weather Service in the United States may issue warnings or alerts when ground-level ozone concentrations are high.

Pollution–Climate Feedbacks

Feedbacks between air pollution and ambient climates occur in both directions. Certain pollutants such as ozone and particulates affect climate by altering the radiation balance, either by disrupting the amount of short-wave radiation reaching the Earth or altering the amount of long-wave radiation that leaves the Earth. A changing climate system also influences air quality. The exact nature of the modifications depends on the given contaminant. For example, particulate pollution may decrease if precipitation increases, as these pollutants flush out of the atmosphere. Additionally, should the number of warm weather, high-pressure systems grow, more days with potentially dangerous concentrations of tropospheric ozone could potentially occur.

General circulation models (GCMs), which simulate the climate based on its physics, incorporate atmospheric chemistry with global circulation. This permits the examination of the interactions between the climate system and air pollution. Transport of pollutants is also investigated, since the effects of pollution on climate may well occur far from the origins of those contaminants. However, the magnitudes of those impacts are not well known and need to be studied further. One issue involves spatial scale; GCMs, though becoming more finely resolved, may still be too coarse to capture some sources of pollution. Additionally, local factors such as meteorology and topography also exist at scales too small to be incorporated explicitly into GCMs.

Because the atmosphere is dynamic and air constantly circulates at global, regional, and local scales, air pollution rarely remains at the point or location of origin. Even when contaminants don't travel far, over time they become less concentrated as they disperse. Long-range atmospheric transport is a transboundary issue, affecting places far removed from the source of pollution. Arctic haze, for example, is a seasonal decrease in visibility that appears either grayish or reddish-brown (depending on viewing angle). The effect is strongest in winter, when the lack of precipitation precludes removal of pollutants. This high-latitude phenomenon has been traced largely to industrial emissions and coal combustion from mid-latitude Europe and Asia. Arctic regions are highly sensitive to increased temperatures and arctic haze has been implicated in winter warming in polar latitudes.

Because of the potential danger to humans, plants, and animals, many governmental agencies have passed laws to regulate emissions to the atmosphere; many have also undertaken air-quality monitoring to assess the status and impact of air pollution on the ambient environment. The U.S. Clean Air Act and its subsequent amendments, for example, established standards and requirements designed to control contaminants in the atmosphere. Other countries, including New Zealand, Canada, and Germany, have passed similar legislation to assess and enforce air quality. To evaluate the quality of air, agencies monitor the atmosphere; this may be done at federal, regional, or local levels, often to gather different measurements.

Petra A. Zimmermann
Ball State University

See Also: Aerosols; Chlorofluorcarbons; Clean Air Act, U.S.; Climate Models; Pollution, Land; Pollution, Water.

Further Readings

Godish, Thad. *Air Quality.* 4th ed. Boca Raton, FL: Lewis Publishers, 2004.

Kidd, J. S. and Renee A. Kidd. *Air Pollution Problems and Solutions.* New York: Chelsea House, 2006.

Solomon, S., D. Qin, M. Manning, Z. Chen, M. Marquis, K. B. Averyt, M. Tignor, and H. L. Miller, eds. *Climate Change 2007: The Physical Science Basis. Contribution of Working Group I to the Fourth Assessment Report of the Intergovernmental Panel on Climate Change.* Cambridge: Cambridge University Press, 2007.

Pollution, Land

Land pollution is the degradation of the land surface through misuse of the soil by poor agricultural practices, mineral exploitation, industrial waste dumping, and indiscriminate disposal of urban wastes. It includes visible waste and litter, as well as pollution of the soil. The contamination of land usually results from its commercial and industrial uses or from the spillage and dumping of waste, including landfill. These activities leave behind levels of trace metals, hydrocarbons, and other compounds on the land, which have the potential to cause harm to people or the environment. The main human contributors to land pollution are landfills. About half of the waste is disposed of in landfills. The gradual decomposition of landfill wastes over several decades also generates new environmental problems in the form of air pollutants. Trace organic gases are emitted from landfills, along with significant amounts of methane and carbon (IV) oxide, both of which are greenhouse gases. Garbage and other forms of waste arising from homes, municipalities, industries, and agricultural practices are the major sources of pollution on the land environment. The indiscriminate discharge of these wastes creates a filthy environment.

Unlike contaminated air and water, which directly affect human health, pollution of the land from the dumping or burial of solid wastes affects people less directly. The primary environmental concern is that a waste material in the soil may migrate into surface water or groundwater where it can be ingested and harm living organisms. Soil pollution is mainly because of chemicals in pesticides. Soil erosion and degradation are some of the problems facing the state of the land. The world is losing 24 billion tons of topsoil every year. Globally, a minimum of 15 million acres of prime agricultural land is lost to overuse and mismanagement every year. Desertification is threatening about one-third of the world's land surface.

Types of Waste

Litter is waste material dumped in public places such as streets, parks, picnic areas, bus stops, and near shops. The accumulation of waste threatens the health of people in residential areas. Waste decays, encourages household pests, and turns urban areas into unsightly, dirty, and unhealthy places. The following measures can be used to control land pollution. Antilitter campaigns can educate people against littering, organic waste can be dumped in places far from residential areas, and inorganic materials such as metals, glass, paper, and plastic can be reclaimed and recycled. One of the main factors influencing fast generation of municipal sewage and garbage and agricultural, commercial, and industrial wastes is population growth. The world human population has increased tremendously, and there has been phenomenal urban growth because of the migration of rural-area dwellers to urban areas. The larger the population, the larger the wastes generated, and the greater the pollution. Pollution becomes even more pronounced when the population is crowded into a smaller space.

The sources of domestic wastes are garbage, rubbish, and ashes. Municipal wastes emanate from bulky wastes, street refuse, and dead animals. Municipal solid wastes are wastes collected by private or public authorities from domestic, commercial, and industrial sources. No two wastes are the same. The wastes generated within a municipality vary widely depending on the community and its level of commercial venture. The data on waste will depend on the level of sophistication of the waste management operation. Domestic waste from a house will vary from week to week and from season to season. Waste varies from socioeconomic groups and from country to country. In most cases, the number of refuse dumps decreases with increasing distance from the city center. Other factors that influence the distribution of solid waste dumps in cities are distance from main markets, positions of residential houses, commercial and industrial centers, and topographic characteristics of the city that determine accessibility by vehicles.

Commercial wastes are traceable to markets, stores, and shops; while industrial wastes are from factories, power plants, and treatment plants. Commercial, domestic, agricultural, and industrial activities generate vast amounts of wastes, which include paper, food, metals, glass, wood, plastics, and dust. Effluents from domestic and industrial sources are also potential land pollutants. Many commercial houses and industries, especially in developing countries, do not have an organized method of disposing of their wastes. They are dumped indiscriminately, thus constituting a menace, and if they are toxic or in any way harmful, they become hazardous to the health of the public. Spillage of oil on land is a source of pollution. Land can also be polluted by the introduction of pesticides. Acid deposition also changes the integrity of the land. Contamination of land gives rise to impairment of the quality of groundwater, and impoverishment of soil to the extent of not supporting plant and animal life.

Land pollution occurs in many forms. Some of the sources of land pollution are agricultural, commercial, industrial, military, and from the general public. About half of the waste is disposed of in landfills, which leaves behind trace metals, hydrocarbons, and other compounds. Decades of waste decomposition generates air pollutants such as trace organic gases and large amounts of the greenhouse gases methane and carbon (IV) oxide.

Land pollution leads to the uptake of pollutants by plants, thereby introducing the pollutant to the food chain.

Garbage or trash is a component of municipal solid waste, which includes all of the wastes commonly generated in residences, commercial buildings, and institutional buildings. Municipal solid wastes consist of such things as paper, packaging, plastics, food wastes, glass, wood, and discarded appliances. Similar kinds of wastes generated by industrial facilities also are part of municipal solid wastes. The additional wastes generated by manufacturing processes, construction activities, mining and drilling operations, agriculture, and electric power production are referred to as industrial wastes. The environmental threats posed by municipal and industrial wastes are varied. Though defined as nonhazardous wastes, many of these wastes are capable of harming human health and the viability of other living species. They contain discarded hazardous wastes like batteries, paints, solvents, and waste motor oil, items that add trace metals and organic compounds to the inventory of potential contaminants in soil.

Environmental pollution by industrial wastes has become a threat to the continued existence of plants, animals, and humans. Industrial pollution contains traces of quantities of raw materials, intermediate products, final products, coproducts, and byproducts, and of any ancillary or processing chemicals used. They include detergents, solvents, cyanide, trace metals, mineral and organic acids, nitrogenous substances, fats, salts, bleaching agents, dyes, pigments, phenolic compounds, tanning agents, sulfide, and ammonia. Many of these substances are toxic. Because of the larger volumes of waste materials, landfills are the preferred method of waste disposal. The pollutants arising from a particular industry are different from those arising from another industry. The waste generated differs from industry to industry. The level of pollution arising from the industry depends on the nature and magnitude of its wastes.

Pollution by trace metals occurs largely from industries, trade wastes, agricultural wastes, and automobile exhausts. These wastes are large in magnitude and varied in types. They include large quantities of raw materials, byproducts, coproducts, and final products. Mining is a major area where metal pollution occurs. Apart from natural occurrence such as erosion, metal pollution on land is a direct result of anthropogenic activities. The dumping of old or damaged vehicles on land occurs especially in developing countries. Also, the dumping of obsolete or dangerous military wastes on sites is another source of pollution. Apart from trace metals, the wastes contain organic materials, biological and chemical warfare explosives, pesticides, solid objects, and other materials peculiar to military operations. Trace metals in soil can also enter the food chain via uptake by plants and vegetation that are subsequently consumed by animals and humans, with deleterious consequences. Land disposal sites serve as breeding grounds for disease-carrying organisms.

Pollution from agricultural practices is because of animal wastes, materials eroded from farmlands, plant nutrients, vegetation, inorganic salts, and minerals resulting from irrigation and pesticides that farmers use on their farms to increase agricultural yield and fight pests and weeds. Agricultural wastes are made up of unwanted parts of crops during harvesting season. Examples are maize sheaves and cobs, maize stalks of guinea corn, millet and rice and their chaffs, yam vines, cassava stems, and yam and cassava peelings. Studies have shown that groundwater can be contaminated through seepage by leachate arising from solid wastes dumped on the ground. Land application of wastes is the most economical, practical, and environmentally sustainable method for managing agricultural wastes, especially animal wastes. Application of agricultural wastes to the land recycles valuable nutrients and organic matter into the system from which they originated. Land application can also be an effective component of management strategies for other organic wastes like food processing wastes.

Radioactive wastes are peculiar and dangerous. Their harmful effects on living organisms are induced by radiation, rather than by chemical mechanisms. They also remain dangerous for several years. Radioactive wastes are products of usage of nuclear energy. An example is the mining of uranium ore and its processing into nuclear fuel, which is used for electric power production. Power plants may also be radioactive. The environmental impacts of nuclear waste vary with the nature and form of the waste material. The most dangerous of these include the spent fuel from nuclear reactors, as well as the radioactive liquids and solids produced from any reprocessing of spent fuel. This high-level waste is characterized by the intensity of its radioactivity and long half-life. Death from exposure to intense radiation can occur, depending on the intensity and duration of the exposure. Human exposure can occur through inhalation of radioactive substances and ingestion of food containing radioactive materials.

Ways to Minimize Waste

The best way to avoid the environmental problems of solid waste disposal is to desist from generating wastes in the first place. Pollution prevention programs aimed at this objective have become widespread. Recycling and reuse of materials are ways to avoid waste generation. At the residential level, recycling programs for newspapers, glass, and metal containers have been implemented. However, some municipal programs have been criticized for increasing environmental emissions of air pollutants from fuel combustion.

The ultimate land disposal methods used for municipal solid wastes are land filling, land farming, and deep well injection. Land filling of solid wastes involves the controlled disposal of solid wastes on or in the upper layer of the Earth's mantle, which has been excavated to a depth of about 13 ft. (4 m). When solid wastes are placed in sanitary landfills, biological, chemical, and physical processes occur. Biological decay of organic materials occurs by either aerobic or anaerobic processes, resulting in the evolution of gases or liquids. The chemical oxidation of waste materials occurs; dissolving and leaching of organic and inorganic materials by water and leachate moving through the fill also occur.

Land filling in moist climates produces large quantities of leachate that are toxic and of high organic strength and require treatment in wastewater plants. Land filling in dry climates produces localized air pollution problems. There is also movement of dissolved material by concentration gradients and osmosis. Initially, the organic material in the landfill undergoes aerobic decomposition because of some oxygen amount obtained in air trapped in the landfill. Within a few days, the oxygen content is exhausted and long-term decomposition occurs under aerobic conditions. The anaerobic conversion of organic compounds occurs in the transformation of high molecular weight compounds catalyzed by enzymes in soil bacteria into compounds suitable for use as a source. However, landfill sites cause soil and groundwater contamination if not properly operated. Additional environmental problems with landfills are odors, litter, scavengers, and rat infestation.

Solid wastes are wastes from human and animal activities. In the domestic environment, solid wastes include paper, plastics, food wastes, and ash. Improper management of solid wastes has direct adverse effects on health. Solid wastes may contain human pathogens, animal pathogens, and soil pathogens. Inadequate storage of such wastes provides a breeding ground for vermin, flies, and cockroaches, which may act as passive vectors in disease transmission. The pathogens that can cause fecal-related diseases are viruses, bacteria, protozoa, and helminths. As proper waste management involves recycling, reuse, transformation, and disposal, it is relevant to know the physical, chemical, energy, and biological properties of wastes. The physical properties that are relevant include density, moisture content, particle size distribution, field capacity, hydraulic conductivity, and shear strength. Chemical analyses required are proximate analysis, ultimate analysis, and energy content analysis. The important elements in waste energy transformation are carbon, hydrogen, oxygen, nitrogen, and sulfur.

Only 2 percent of waste is actually recycled. Solid waste recycling implies recovery of a component of waste for use in a manner different from its initial function. Recycling consists of recovering from waste the matter of which a product was made and reintroducing it into the production cycle for reproduction of the same item. Composting after decomposition by aerobic bacteria readily recycles garbage, grass, and organic matter. Composting may be defined as the decomposition of moist, solid, organic matter by the use of aerobic microorganisms under controlled conditions. The end product of the decomposition is a sanitary, nuisance-free, humus-like material that can be used as soil conditioner and as partial replacement for fertilizer. In a typical operation, municipal wastes are presorted to remove noncombustible materials and those that might have salvage value such as paper, cardboard, rags, metals, and glass. Refuse is then shredded and stacked in long piles where it degrades to humus, much as it would in soil. Usually, the decomposed material contains less than 1 percent of each of the three primary fertilizer nutrients. The final step is grinding and bagging for ultimate sale as soil conditioner.

Plants die because of land pollution. Crops are affected, because they do not mature and grow well. There are three ways that people pollute the land: littering, improper garbage disposal, and dumping of chemical fluids. It is not uncommon to see people throw trash on the road while in the car. Every day, people are polluting the land. Because of pollution, people do not only affect the cleanliness of the land, but also destroy the beauty and increase avenues of contracting diseases. These negative tendencies have effects on tourism potentials of nations as tourists are turned off. Tourists won't like to take risks in an unsafe environment because of pollution. Mosquitoes live in littered empty cans. Thus, the threat of mosquito bite is imminent in a polluted land. A greater proportion of land pollution is instigated and carried out by man. Governments of nations should be alive to their responsibilities of providing a safe and secured world environment to their people.

Akan Bassey Williams
Covenant University

See Also: Conservation; Diseases; Health; Methane Cycle; Nuclear Power; Pollution, Air; Pollution, Water; Poverty and Climate Change.

Further Readings

Ademoroti, C. M. A. *Environmental Chemistry and Toxicology*. Ibadan, Nigeria: Foludex Press, 1996.

Rubin, E. S. and C. I. Davidson. *Introduction to Engineering and the Environment*. New York: McGraw-Hill, 2001.

Vesilind, P. Aarne. *Introduction to Environmental Engineering*. Stamford, CT: Cenage Learning, 2009.

Pollution, Water

According to the National Pollutant Discharge Elimination System (NPDES), "Water pollution degrades surface waters making them unsafe for drinking, fishing, swimming, and other activities." Worldwide, it is a challenge to obtain quality drinkable water from bodies of water such as streams, lakes, oceans, or groundwater/aquifers. Except for rainwater or water from deep aquifers, most of the water sources in the world are polluted—unsuitable for human or animal consumption. Even rainwater in many locations is polluted because of the presence of greenhouse gases (GHGs) such as sulfur dioxide and oxides in the atmosphere.

Sources of Water Pollutants

When pollutants are discharged to bodies of water directly or indirectly, it causes water pollution. According to the Environmental Protection Agency (EPA), water pollutants come from two sources: point and nonpoint. Direct discharge of pollutants to bodies of water is mainly from point sources such as municipal storm sewer systems, sewage treatment plants, factories, ships, and construction sites. Examples of direct pollutants include food-processing wastes (fats and grease), personal hygiene and cosmetic products, shipwreck debris, silt from eroded soil in construction sites, and other trashes. Nonpoint sources are indirect sources of pollution and are more diffuse, as they do not originate from a single discrete source. Agricultural runoff; runoff coming from other land uses such as urban, forest, and mining activities; and paved roads are nonpoint source pollutants. Crop fertilizers, manures, pesticides, and insecticides that are washed away with runoff into streams and other bodies of water cause water pollution. Animal or bird (wild or domestic) excreta over pastures, forests, or other land uses also wash away into streams as nonpoint source pollution. Human and animal excreta discharged directly into bodies of water are major point source pollutants.

Specific contaminants that cause water pollution include a wide spectrum of chemicals and pathogens. Both organic and inorganic chemicals are water pollution contaminants. Common chemical contaminants include detergents; disinfection byproducts; petroleum byproducts (gasoline, diesel, jet fuels, kerosene, other fuel oil, lubricants, and fuel-combustion byproducts); industrial solvents; chlorinated solvents; sulfur dioxide and other chemical wastes from industrial sources; heavy metals from motor vehicles via urban storm-water runoff; and nutrients (e.g.,nitrogen, phosphorous, and potassium) from fertilizers.

Increased pathogen count in water also causes water pollution. Fecal coliform bacteria are the common indicator of water pollution. They are generally present in human, animal, or bird excreta. Other pathogens found in polluted water are Burkholderia pseudomallei, Cryptosporidium parvum, Giardia lamblia, salmonella, norovirus and other viruses, and other parasitic worms.

Effects of Pollutants in Water

Other than making water unsuitable for human or animal consumption, these water pollutants cause unfavorable conditions for aquatic plants and animals. Physical or sensory changes such as elevated temperature and discoloration occur in the water. Elevated water temperature decreases the amount of dissolved oxygen (DO) in water, thus limiting the supply of oxygen to aquatic animals. One of the sources of water temperature elevation is urban runoff. Increased nutrient levels cause algal bloom in surface water and decreases sunlight interaction. Excess nutrients from algal death and decomposition decreases the amount of DO in water, causing death of aquatic animals. Because of these water pollutants, the pH level in water changes, harming aquatic life. Global warming is believed to cause the acidification (lowering of pH) of ocean water because of the uptake of anthropogenic carbon dioxide (CO_2) from the atmosphere. Table 1 provides a list of

Table 1 Aquatic chemistry reference sheet

Temperature (degrees C)	
<13	Suitable for cold-water species such as trout, salmon, mayflies, caddisflies, and stoneflies
13-20	Suitable for salmon, mayflies, caddisflies, stoneflies, and beetles
20-25	Suitable for most other fish, invertebrates, and warm-water species
>25	Lethal to trout, salmon, many aquatic insects, and most cold-water species
pH	
0-4.0	Aquatic life is severely stressed
4.0-4.5	Few fish or invertebrates can survive
4.5-6.5	Acid-tolerant invertebrates can survive
6.5-8.5	Suitable for most aquatic animals
6.5-13.0	Suitable for most aquatic plants
5.0-9.0	Suitable for human consumption
DO (ppm or mg/L)	
0-3	Few organisms can survive
3-4	Only a few fish or invertebrates can survive
4-7	Most non-trout and warm-water fish can survive
5	EPA suggested lower limit for maintenance of healthy aquatic biota
>7	Necessary for trout, salmon, and many invertebrates
Alkalinity (mg/L $CaCO_3$)	
20	EPA suggested lower limit for maintenance of healthy aquatic biota
<25	Poorly buffered
25-75	Moderately buffered
>75	Highly buffered (Highly buffered streams are better able to support acidity-sensitive organisms)
Phosphate (ppm or mg/L)	
0.005-0.05	Typical of undisturbed forest streams
<0.5	Suitable for human consumption
0.5-0.1	May increase aquatic plant growth
>0.1	Likely to cause algal bloom
1.0	Approximate ideal upper limit wastewater treatment plant effluent
Nitrate (ppm or mg/L)	
0.1	Typical of undisturbed forest streams
0.1-1.0	May increase aquatic plant growth
>1.0	Likely to cause algal bloom
<10	Suitable for human consumption
<90	No direct effect on fish
Chloride (ppm or mg/L)	
<12	Typical of undisturbed forest streams
0-200	Suitable for most fish and invertebrates
200-1000	Suitability varies by species
<500	Suitable for human consumption
>1000	Most aquatic invertebrates are negatively affected

Source: Carlsen, William S. *Watershed Dynamics, Teachers Edition.* Arlington VA: National Science Teachers Association, 2004.

appropriate nutrient and other aquatic parameter reference values for aquatic plants, animals, and human consumption suitability. Most of the regulated discharge should abide by these limits.

Debris (plastic and other harmful materials) transported to oceans or lakes are consumed by aquatic birds or animals, in many cases resulting in their deaths as the trash obstructs their digestive pathways. Groundwater pollution occurs from pollutants discharged to groundwater by industries or contaminated water infiltrating the deeper soil horizon. When polluted groundwater is used for crop irrigation, it harms the quality of the food produced. Contaminated groundwater also causes health hazards when used for drinking. Polluted surface water can be treated more easily than polluted groundwater because groundwater moves undetected in aquifers for great distances.

Measurement and Detection

Water pollution is measured through three broad categories of testing: physical, chemical, and biological. Water samples are collected from different locations of bodies of water and tested in the laboratory with specialized analytical methods. With the advancement in instrumentation, in situ analyses are also feasible and being conducted, even by novice stakeholders. Common physical tests include tests for temperature, total suspended solids (TSS), or sediment concentration in water. General chemical tests include tests for pH; alkalinity; DO; biochemical oxygen demand (BOD); chemical oxygen demand (COD); nutrients such as nitrogen (nitrate, nitrite, ammonia); phosphorous (orthophosphate, phosphate); metals like copper, zinc, lead, mercury, and arsenic; oil and grease; petroleum products; and different type of pesticides. Common biological tests of water quality include the presence and health analysis of biological indicators like plants, invertebrates, and other microbial organisms.

The U.S. Geological Survey (USGS) monitors the water quality of most of the country's major bodies of water by setting up real-time monitoring stations at different spatial locations. Other than the stream discharge measurements, these gauging stations record most of the physical and chemical characteristics of water. These measurements are available on the USGS National Water Information System (NWIS) Web interface.

Since the 1960s, various water quality and quantity monitoring models have been developed to quantify the amount of pollutants coming from nonpoint sources. The Universal Soil Loss Equation (USLE) model was the first attempt to quantify the soil erosion rate (eroded soil transported to bodies of water) from watersheds with the use of a rainfall erosivity index, a soil erodibility factor, slope length and gradient factors, and vegetation characteristics. Successive models developed for this purpose and their year of development include Chemicals, Runoff, and Erosion from Agricultural Management Systems model (CREAMS, 1970s); Groundwater Loading Effects of Agricultural Management Systems (GLEAMS, 1980s); Environment Policy Integrated Climate (EPIC, 1980s); Water Erosion Prediction Project (WEPP, 1980s); Agricultural Non-Point Source Pollution (AGNPS, 1980s); Simulator for Water Resources in Rural Basins (SWRRB, 1980s); Hydrologic Simulation Program–FORTRAN (HSPF, 1990s); Storm Water Management Model (SWMM, 1990s); Better Assessment Science Integrating Point and Non-Point Sources (BASINS, 1990s); and the Soil and Water Assessment Tool (SWAT, 1990s). These are all free models developed by federal, academia, and industry collaboration. All models quantify different water-quality parameters from different sources of pollutants using physical ground parameters such as soil, vegetation, and slope.

Virginia Tech has developed an online Bacteria Source Load Calculator (BSLC) that can calculate bacteria loadings from point sources such as wild and domestic animals, human excreta delivery through open pipes or septic tanks, and a few nonpoint sources like forest and pasture. According to the BSLC user manual, this software is designed to simplify the complex and time-consuming work involved in determining bacterial loadings with automated processes involved in many of the characterization steps. It provides a high level of consistency in data development and processing and is considered a useful tool for developing total maximum daily load (TMDL) studies, including the allocation scenario development process. Researchers S. S. Panda and R. Randal published a study in 2009 in which they developed an automated geospatial model to estimate the bacteria load from nonpoint sources such as urban, forest, agriculture, and water sources. The model can quickly provide the amount of fecal coliform bacteria loadings from individual land-use pixels using the land-use land-cover raster of the watershed and the runoff raster developed from the watershed precipitation data.

With the advancement of geospatial technology, stream health can be determined on a rating scale from the land-uses surrounding the stream, along with some associated parameters like presence of roads, point source, and wetlands close to the stream. The Watershed Habitat Evaluation And Biotic Integrity Protocol (WHEBIP), developed by Dr. Reuben Goforth of Cornell University, helps scientists determine the stream integrity using the following 12 parameters on a grading score: (1) land-use along the stream, (2) average width of riparian belt, (3) riparian canopy continuity, (4) presence of wetlands, (5) active agriculture, (6) forest or brush, (7) upstream riparian vegetation, (8) upstream forest or brush, (9) watershed land gradient, (10) presence of point source pollution, (11) presence of roads, and (12) conservation activity. Panda and K. Dalton have developed semiautomated geospatial models that use geospatial data layers for all 12 parameters to determine the stream integrity on a segment basis in a watershed. This model helps in determining the quality of the stream in an inexpensive and time-efficient way.

Water Pollution Control Mechanisms

Pollutants from point sources can be regulated, but it is quite a difficult problem to regulate nonpoint source pollutants. Although it was previously difficult to identify the sources of pollution from nonpoint sources, the available models make it easier to identify the loading sources, on spatial basis as well. Once the sources and locations of the water pollutants are identified, it becomes easier to control the pollution.

Under the Clean Water Act (CWA), the NPDES permit program controls water pollution in the United States by regulating point sources. Since 1977, the amended CWA was enacted, making it unlawful to discharge any pollutant from a point source into navigable waters without a permit. The EPA works with state and local authorities to monitor pollution levels in the country's bodies of water and to provide status and trend information on a representative variety of ecosystems. The EPA imposes the following rule:

> Under section 303(d) of the CWA, states, territories, and authorized tribes are required to develop lists of impaired waters. These are waters that are too polluted or otherwise degraded to meet the water quality standards set by states, territories, or authorized tribes. The law requires that these jurisdictions establish priority rankings for waters on the lists and develop TMDLs for these waters. A TMDL is a calculation of the maximum amount of a pollutant that a water body can receive and still safely meet water quality standards.

The 303(d) Listed Impaired Waters program provides impaired water data and impaired water features in the country, showing the spatial extent of river segments, lakes, and estuaries that cross the mandated water-quality parameter limits. The EPA's Assessment and TMDL Tracking and Implementation System (ATTAINS) contains the listed water identification, which can be linked to the source feature identification field of the bodies of water file available with 303(d) data.

Conclusion

In the context of World Water Day on March 2010, UN Secretary General Ban Ki-moon stated that clean water has become scarce and will become even scarcer with the onset of climate change. He added that every year, more people die from unsafe drinking water than any form of violence or natural disaster.

Nearly 900 million people worldwide are still without safe drinking water. Therefore, it is everyone's job to reduce water pollution. With the advancement of technology, it is easier to pinpoint the source of pollutant loadings, even on a spatial basis. Therefore, it will be easier to take preventive measures to keep the world's water supplies safe.

Sudhanshu Sekhar Panda
Gainesville State College

See Also: Agriculture; Diseases; Health; Oceanic Changes; Pollution, Air; Pollution, Land.

Further Readings

Benham, B. "*Bacteria Source Load Calculator V3.0–Users Manual.*" Blacksburg, VA: Virginia Tech, 2011. http://www.tmdl.bse.vt.edu/uploads/File/pub_db_files/BSLC_v3_UsersManual.pdf (Accessed October 2011).

Caldeira, K. and M. E. Wickett. "Anthropogenic Carbon and Ocean pH." *Nature*, v.425/6956 (2003).

Carlsen, W. and N. Trautmann. *Watershed Dynamics, Teachers Edition*. Arlington, VA: NSTA, 2004.

Cordy, G. E. "A Primer on Water Quality." http://pubs.usgs.gov/fs/fs-027-01/pdf/FS-027-01.pdf (Accessed October 2011).

Environmental Protection Agency. "Water Pollution." http://www.epa.gov/ebtpages/watewaterpollution.html (Accessed October 2011).

Panda, S. S. and K. Dalton. "Development of an Automated Watershed Stream Health Determination Model Using the 12-Point Physical Watershed Parameters." National Water Conference, Hilton Head, SC (February 21–25, 2010).

Panda, S. S. and R. Randall. "Geospatial Model Development for Watershed Based Fecal Coliform Estimation and Comparison With Virginia Tech's Bacteria Loading Calculator." National Water Conference, St. Louis, MO (February 8–12, 2009).

Schueler, T. R. and H. K. Holland. "Microbes in Urban Watersheds: Concentrations, Sources, and Pathways." In *The Practice of Watershed Protection*, edited by T. Scheuler and H. K. Holland. Ellicott City, MD: Center for Watershed Protection, 2000.

Virginia Tech. "Bacteria Source Load Calculator (BSLC)." http://www.tmdl.bse.vt.edu/outreach/C85 (Accessed October 2011).

Population

The term *population* refers to a group of individuals that belong to the same species and live in the same geographical area. Normally, the term is used to refer to a collection of human beings and derives from the Latin words *populare* and *populatio*. Population Council demographer Geoffrey McNicoll states that originally, the verb *populare* was commonly used to mean to lay waste, and the noun *population* had the meaning of plundering or despoliation. He also affirms that both meanings entered into English, but become obsolete by the 1700s.

Demography is the branch of science that statistically studies human populations, focusing on its size, structure, and dynamics in space and time and identifying and measuring the responses to environmental, economic, and social changes in terms of births, migrations, aging, and deaths. It can be divided between formal demography, which studies the measurement of demographic processes; and population studies, which is more focused on the analysis of the relationships between economic, social, cultural, and biological processes influencing a population.

Human activities can be considered as one of the primary drivers for almost all types of environmental degradation and change, especially in the present day. Additionally, the population size directly and indirectly affects the scale and form of human activities and, consequently, their environmental impacts. Global environmental changes are occurring in ways fundamentally different than in any other time in history. Experts and scientists are showing that virtually all ecosystems on the planet have been significantly transformed through human activities, and that about 60 percent of the Earth's ecosystems have been degraded or used unsustainably. These changes have been especially rapid in the past 50 years or so, and are expected to continue into the projected future.

Although changes in these large-scale processes are complex and difficult to understand, assess, and attribute cause or responsibility, the core problems seem to point back to the growing scale of human activities. In this sense, there are links between population and environmental change, and they are dependent on the socioeconomic development pathway of a society. In simple terms, the human population is growing, consuming natural resources at unprecedented rates, and the planet is demonstrating the first effects.

Population Growth

Population growth can be defined as the balance between the number of people who enter a population (through births and migration) and the number of people who leave the same population (through deaths and migration) during the same period of time. If the number of people entering is greater than the number of people leaving, the population growth is positive and the size of the population tends to become larger along the timeline. If the number of people leaving the population is fewer than the number of people entering, the population growth is negative and the population size tends to become smaller along the timeline. Population growth will be zero if the number of incoming people equals the number of exiting people and the population size tends to remain stable along the timeline.

Beyond this oversimplified definition, population growth is an issue that mobilizes public opinion and discussion anywhere in the world, likely because it is one of the most pervasive demographic issues and is something reasonably easy to define and understand. Population growth has been discussed since ancient times by many scholars who have dedicated their time and energy to try to understand the consequences of population growth for the economy, community, environment, and natural resources.

Population Growth and the Environment

For the first time in human history, society is using many of the planet's resources faster than they can regenerate. The impacts are becoming increasingly obvious with the vulnerability of freshwater resources, the mounting number of plant and animal species becoming endangered or extinct, the acidification of the oceans, deforestation, and global climate change. In the last decades, environmental conditions have failed to improve in most sectors. A summary of specific environmental sectors includes the following:

- *Public health:* Unclean water along with poor sanitation kills over 12 million people each year, most in developing countries. Air pollution kills nearly 3 million more.
- *Food supply:* In 64 of the 105 developing countries studied by the United Nations Food and Agriculture Organization (FAO), the population is growing faster than food supplies. Population pressures have degraded some 2 billion hectares of arable land, an area the size of Canada and the United States.
- *Coastal zones:* Half of all coastal ecosystems are pressured by high population densities and urban development. A tide of pollution is rising in the world's seas. Ocean fisheries are overexploited and fish catches are decreasing in many places.
- *Forests:* Nearly half of the world's original forest cover has been lost, and each year another 16 million hectares are cut or burned. Forests provide over $400 billion to the world economy annually and are vital to maintaining healthy ecosystems. Yet, current demand for forest products may exceed the limit of sustainable consumption by 25 percent.
- *Biodiversity:* The Earth's biodiversity is crucial to the continued vitality of agriculture, medicine and, perhaps, even to life on the planet itself. Yet human activities are pushing many thousands of plant and animal species into extinction. Two of every three species is estimated to be in decline.
- *Global climate change:* The surface of the planet is warming because of natural and anthropogenic forces, mainly greenhouse gas emissions derived from fossil fuel burning. If global temperatures rise as projected, sea levels will rise by several meters, causing widespread flooding. Global warming could also cause droughts and disrupt agriculture in some places of the world.

During the 19th and 20th centuries, the consumption of natural resources increased in a manner not previously observed, primarily in developed countries during the process of industrialization and economic development. After World War II, these production and consumption patterns spread rapidly to the rest the world. This period has been marked by the generation of unprecedented technological development that led to a greater capacity of humans to influence and impact the environment. It is possible that the impact of population over the environment depends on a series of features and also varies according to the selected aspect of the environment and the impact that is analyzed.

For conceptualizing and measuring the impacts of population growth on natural resources and the environment, Paul R. Ehrlich and John Holdren developed a classic accounting equation that has been widely used since the 1970s. It states that human impact on the environment (I) equals the product of population (P), affluence (A), and technology (T): I = P x A x T, or simply I = PAT. Human impact on the environment (I) can be described as how species, ecosystems, and natural resources are affected or impacted by humans (such as air pollution or water consumption), while (P) represents the population size or total number of people. Population always acts in combination with the other IPAT factors; (A) represents the affluence (or the consumption) of the environment. It is normally correlated with income, or how much each person consumes in terms of resources such as water, energy, passenger miles, space and resources used for housing. The waste generated through resource consumption is part of this equation. Finally, (T) represents the technology, representing how a resource is used and how much waste and pollution is created by the production and consumption of the resource. Sometimes, it improves environmental impact (such as through efficient coal-burning power plants).

Using the IPAT equation, it can be determined that population growth causes a multiplicative effect on the environment. In other words, if the size of a population doubles over a certain period of time while all other conditions remain the same, the impact of this population on the environment will double over the same period of time. However, the equation—and all models derived from it—considers that technological advances and improvements of a determined analyzed system can lead to lesser impact.

Technology and Other Solutions

The belief in the power of technology to reduce the impact of human population and its activities on the environment provides the conceptual basis for most contemporary public and corporate policies regarding environmental impacts. Thus, it stimulates significant investments in research, development, and technological innovation in areas such as alternative cleaner technologies, administrative and institutional changes in the organization of production, and the generation of significant changes in people's lifestyles and decision-making processes. However, technology alone does not appear to be enough. The need for a global commitment in terms of resource conservation to be able to meet the challenge of sustainable development is increasingly acknowledged. Practicing sustainable development requires a number of social, political, and cultural measures to change human behavior toward cleaner agriculture, less industrial pollution, and more effective natural-resource management, among many other required changes.

With the stabilization of the world's population, worries about a "population bomb" because of uncontrolled population growth have reduced with the years, although the population will still be growing until the second half of the 21st century. It is also crucial to consider that the current population level has never been experienced before by the planet, as well as the per-capita consumption that continues to rise with unsustainable economic development. Compared to earlier debates about population growth, the picture today seems to be much more complex and demands much more political engagement than has been seen in the past.

Leonardo Freire de Mello
University of Valedo Paraíba

See Also: Anthropocene; Anthropogenic Forcing; Carbon Footprint; Climate Justice; Dangerous Anthropogenic Interference; Economics, Impact From Climate Change; Gender and Climate Change; Indigenous Communities; Maximum Sustainable Yield; Oil, Consumption of; Poverty and Climate Change; Refugees, Environmental and Climate; Small Farmers; Vulnerability.

Further Readings

Ehrlich, P. R. and J. P. Holdren. "Impact of Population Growth." *Science, New Series*, v.171/3977 (March 26, 1971).

Ehrlich, P. R. *The Population Bomb*. New York: Ballantine Books, 1968.

LeGrand, T. K. "World Population Growth and the Environment in Human Resources Management." In *Encyclopedia of Life Support Systems (EOLSS)*, edited by Michael J. Marquardt. Oxford: EOLSS, 2006.

Lutz, W., ed. *The Future Population of the World. What Can We Assume Today*? London: Earthscan, 1996.

Lutz, W. and R. Qiang. "Determinants of Human Population Growth." *Philosophical Transactions of the Royal Society of London*, v.357/1425 (September 29, 2002).

Lutz, W., W. C. Sanderson, and S. Scherbov, eds. *The End of the World Population Growth in the 21st Century: New Challenges for Human Capital Formation and Sustainable Development*. London: Earthscan, 2004.

O'Neill, Brian C., F. Landis MacKellar, and Wolfgang Lutz. *Population and Climate Change*. Cambridge: Cambridge University Press, 2001.

Searle, R. *Population Growth, Resource Consumption, and the Environment: Seeking a Common Vision for a Troubled World*. Victoria, BC, Canada: Wilfrid Laurier University Press, 1995.

Portugal

One of the oldest states in Europe, Portugal traces its modern history to the mid-12th century and its present-day borders to the mid-13th century. Slightly smaller than Indiana, it borders the North Atlantic Ocean. The nation's 520 mi. (836 km) coastline makes it particularly vulnerable to soil erosion and rising sea levels. Since the 1970s, temperatures have been increasing at double the world rate, making the risk of forest fires and deforestation a concern. All of these factors make Portugal sensitive to the need to adapt to climate change and address the problem of global warming. Hydropower accounted for nearly a third of the country's electricity generation in 2001, and the use of renewable energy sources has steadily increased since. Portugal entered the European Union (EU) in 1986 and is an Annex I Party to the Kyoto Protocol. Membership in both groups means that the country is committed to reducing greenhouse gas (GHG) emissions and cutting overall energy consumption.

According to a 2006 report from Quercus, a Portuguese conservation group, 67 percent of Portugal's coastline is at risk of erosion. Nearly 30 percent has already been affected, with some areas receding as much as 29 ft. (9 m) in a single year. Attributed to rising sea levels and changes in the direction and power of waves, the erosion could lead to the permanent inundation of both dry areas and wetland in low-lying areas. With 60 percent of the country's population living in the coastal zone and with tourism accounting for 5 percent of gross domestic product, the growing threat of coastal erosion could mean not only the loss of ecosystems but also major losses to an economy already troubled by debt and a double-digit unemployment rate.

Forest Fires

Battling forest fires seems to be an annual summer activity in Portugal. Increased temperatures and reduced rainfall have increased fire risks in a country where nearly 40 percent of the land is forest cover. Between 1980 and 2009, almost 38 percent burned. The worst fire season in 30 years occurred in 2003, when 430,000 hectares of forest and scrubland were destroyed at a cost of €1.2 million, 117 homes, and 20 human lives. The fires of 2009 released 1 million tons of carbon dioxide (CO_2) into the atmosphere. In 2008, Portugal's forests absorbed 4.42 million tons of CO_2, but experts say that the nation's forested areas have lost 3 percent of their carbon-fixing capacity, a loss that could take decades to recover, given the time it takes for trees to reach their full absorption capacity.

Excluding forestry and land use, Portugal's GHG emissions have increased about 2 percent per year since 1990. Total emissions for 2008 were about 5.5 percent above the Kyoto Protocol target for the period from 2008 to 2012. Adding to concerns about emissions levels has been Portugal's dependence upon imported energy, mostly fossil fuels. In 2003, the country imported 85 percent of the energy it consumed, well above the EU average. Two years later, Portugal began an ambitious effort to reduce both GHG emissions and its dependence on imported fossil fuels by using the country's resources to establish renewable energy projects. In 2006, Portugal approved the building of the world's largest solar plant, launched a wind-power project that could supply electricity for 750,000 homes, and made plans for the world's first commercial wave farm. By 2010, nearly 45 percent of Portugal's electricity came from renewable sources, an increase of close to 30 percent in

5 years. The current plan is for the renewable energy program to allow the country to close at least two conventional power plants and reduce the operation of others by 2014.

In 2011, Portugal became the first country in Europe to reach average automotive emissions below 130 g. per km, the European target for 2015. The future looks even more promising. Portugal opened the world's first nationwide electric-vehicle charging network in 2011. MOBI.E has 50 charging stations in place, with plans to reach 1,300 stations in 25 municipalities throughout the country. With 80 percent of the nation's drivers traveling less than 100 mi. (160 km), Portugal is an ideal choice for the experiment. Portugal has also made the purchase of electric vehicles more attractive through tax incentives. The country's long-term target is 750,000 electric cars using these stations. Prime Minister José Socrates is setting an example by using a Nissan Leaf for his official travels around Lisbon. Portugal is on track to reach its goal of using domestically produced renewable energy, including large-scale hydropower, for 60 percent of its electricity and 31 percent of its total energy needs by 2020.

Wylene Rholetter
Independent Scholar

See Also: Angola; Automobiles; Brazil; Deforestation; Forests; Guinea-Bissau; Malta; Renewable Energy, Wind; Spain.

Further Readings

European Environment Agency. "Climate Change Mitigation—National Responses (Portugal)." http://www.eea.europa.eu/soer/countries/pt/climate-change-mitigation-national-responses-portugal (Accessed July 2011).

Organization for Economic Cooperation and Development. "OECD Stat Extracts: Country Statistical Profile 2010, Portugal." http://stats.oecd.org/Index.aspx?DatasetCode=CSP2010 (Accessed July 2011).

Rosenthal, Elizabeth. "Portugal Gives Itself a Clean-Energy Makeover." *New York Times* (August 9, 2010). http://www.nytimes.com/2010/08/10/science/earth/10portugal.html?_r=1&pagewanted=all (Accessed July 2011).

Poverty and Climate Change

Poverty and climate change are inextricably linked; it is likely that the greatest effects of climate change will be on rural and urban populations that have the least access to the world's resources. This is because poor populations are already extremely vulnerable to environmental risks and hazards and have limited resources at their disposal to allow them to cope with the impacts of climate change. Moreover, poor populations in rural areas are highly dependent on natural resources that are affected by climate variability. Currently, there are 1.4 billion people living on less than $1.25 per day and close to 1 billion people suffering from hunger. Although there has been substantial progress in reducing poverty in the last few decades because of developments in East Asia, the burgeoning impacts of climate change might reverse this positive trend.

Among others effects, climate change is expected to further reduce access to drinking water, negatively affect the health of poor people, threaten food security, and result in loss of property and livelihoods in many regions and countries in Africa, Asia, and Latin America. Poverty in these areas might worsen, thus further compromising the capacity of poor people to adapt to the burgeoning effects of climate change. International efforts at reducing poverty are then at risk of becoming significantly hampered by climate change. Given that more affluent populations will have better access to resources necessary to cope with environmental risk, climate change can exacerbate the already salient income inequality in the world. As more affluent populations continue to generate a great deal of the world's greenhouse gas (GHG) emissions, the poor will endure much of the burdens of climate change. It is then critical that the relationship between poverty and climate change be understood and acknowledged and that climate change interventions be integrated with development and poverty-alleviation endeavors.

Poor in Rural Areas

At least 70 percent of the world's extremely poor are found in the rural areas of the developing world, a large proportion of which are women and children. It is projected that even in the next few decades, at least 50 percent of the world's poor will continue to live in rural areas. The greatest number of poor rural people is found in sub-Saharan Africa and south Asia, regions that may be impacted greatly by climate change. The poor living in these regions suffer from hunger, malnutrition, and diseases.

While the livelihoods of impoverished populations are diversified across regions and countries, most are highly dependent on agriculture and natural resources. Over 80 percent of rural households farm, with the poorest households relying mostly on farming and agricultural labor. There is a risk that increases in temperature, climate variability, and the occurrence of extreme climatic events will significantly limit agricultural productivity and the availability and abundance of natural resources.

Climate change could potentially worsen the occurrence of hunger and induce migration through direct negative effects on crop and livestock production. Dramatic changes in temperature, precipitation, climatic variability, and sea level will affect crop and livestock productivity in many developing-country regions. Water stress and salinization of soils because of sea-level rise will result in the degradation of soils and the loss of productivity of agricultural lands. For example, in 2003, P. G. Jones and P. K. Thornton projected a 10 percent decrease in aggregate yields of maize in Africa and Latin America by 2055. This decline in maize yields may encumber approximately 40 million impoverished livestock keepers in Latin America and 130 million in sub-Saharan Africa. Given that access to water and productive land is important for reducing poverty in rural areas, the impacts of climate change will probably inhibit efforts to alleviate rural poverty.

Biodiversity and poverty in the context of climate change are also closely related; changes to ecosystems brought about by climate variability influence poor people's ability to exploit natural resources and cope with climate change impacts. In many parts of developing countries, the rural poor rely heavily on a plethora of wild food sources as food and fodder, medicinal plants to maintain health, and forest products as fuel and construction material. Those living in coastal areas, for instance, rely on diverse marine resources as sources of food and income. Moreover, many people use biodiversity, including genetic diversity, as an adaptation strategy. In India, for example, some farmers practicing traditional knowledge access and develop new varieties in order to cope with warmer temperatures, pest infestations, and diseases. Conserved biodiverse areas also offer ecosystem services that can be important for adapting to climate change impacts, such as hydrologic regulation and protection for coastal communities.

Health effects of climate change, particularly as manifested in the spread of infectious diseases, are also of paramount concern in poor rural communities where access to proper healthcare and nutrition are severely limited. There is evidence showing that vector-borne diseases, such as malaria and dengue fever, increase with climate change-related temperature rise. Areas particularly affected are the highlands of developing countries where poor rural populations reside. In the highland Debre Zeit sector of central Ethiopia, for example, it has been documented that malaria prevalence increased with regional warming trends from 1968 to 1993. Using climate-based dengue models, it has been shown that in Central America and southeast Asian countries, in particular Honduras, Nicaragua, and Thailand, annual variations in dengue correlate with climate fluctuations.

Conflict in impoverished rural regions is also a serious issue relating to climate change. The confluence of food insecurity, loss of livelihoods, and increase in vulnerability to infectious diseases may result in various forms of adaptation that will generate new and exacerbate existing conflicts. Declining access to land and natural resources leads to loss of livelihoods and food insecurity, forcing affected populations to migrate to and encroach on already-occupied productive lands. This act of encroachment leads to land-related disputes and conflicts. Violent forms of conflicts erupt when poor people resort to aggressive means to obtain access to land and resources and provide for their immediate needs. The risk of people joining armed groups also increases as affected populations perceive their vulnerability to climate change as a consequence of the ineffectiveness of governments. Moreover, because the impacts of climate change can potentially exacerbate income inequality, class-based conflicts may also arise. Conflicts, in turn, deepen poverty as resources that are supposed to be used for providing basic needs are instead diverted in attempts to resolve the conflicts.

Poor in Urban Areas

The poor in urban areas may also be significantly affected by climate change. Although climate change impacts may affect entire populations in urban areas, at high risk are the urban poor who do not have adequate access to proper shelter, nutrition, and healthcare. It is estimated that there are at least 900 million urban dwellers living in low- and middle-income countries. In many cities, about 30–60 percent of the urban population live in informal settlements, which are located in areas with inadequate provision for water, sanitation, and drainage. Moreover, many of these informal settlements are in areas of high environmental risks and hazards, such as in drains, underneath bridges, along riverbanks, surrounding towering waste dumps, and near or at steep, unstable slopes. The number of those suffering from urban poverty is expected to rise in the future as more cities become overwhelmed with population growth and rural-to-urban migration. This type of migration may be induced by climate change impacts that force rural populations to find alternative sources of livelihood beyond agriculture and natural-resource exploitation.

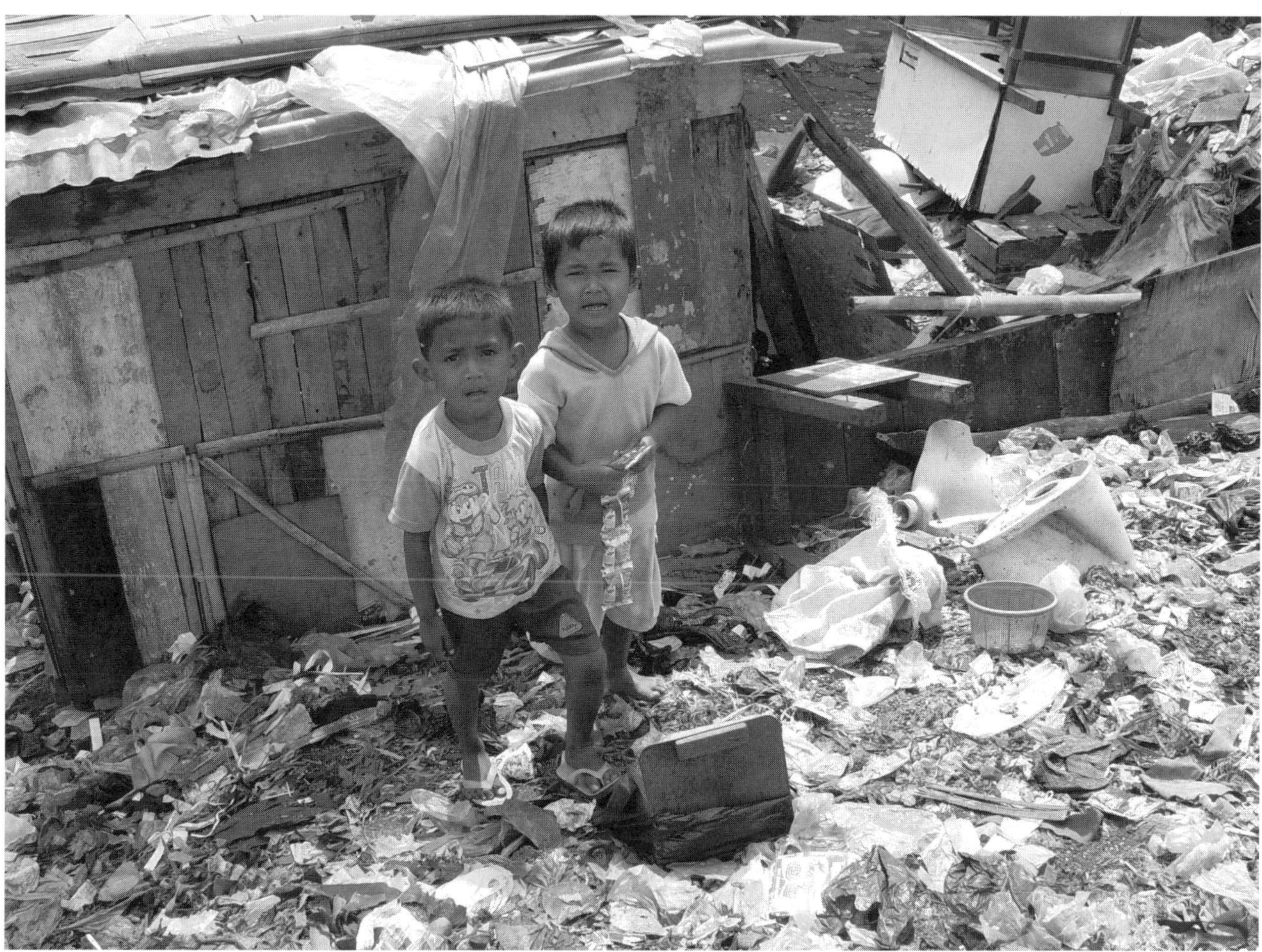

Slum life in Jakarta, Indonesia, which is prone to natural disasters like flooding and volcanic eruptions. The increase in natural disasters in this country, as in many other developing or underdeveloped nations, directly multiplies the number of people living in poverty. Increasing vulnerability to the impacts of climate change therefore exacerbates the cycle of poverty. In the United Nations Human Development Report for 2007, it is cited that climate change threatens human development because it undermines efforts to reduce extreme poverty,

Given that the urban poor are already living in areas of high risks and hazards, extreme weather events such as increased frequency of heavy rains and storms may cause displacements, damage of household assets, injuries, and deaths. The urban poor living in and around floodplains, unstable slopes, and water bodies are at risk of inundation by floods and landslides. In Asia alone, particularly in very large cities like Mumbai, Jakarta, Dhaka, and Manila, thousands of poor people perished and livelihoods and homes were lost in serious floods that occurred between 2000 and 2010. In the Latin American region, nearly 30,000 people, predominantly the poor, were killed in the flash floods and landslides that overwhelmed Caracas, Venezuela, in 1999. These are just some of the numerous examples of how the urban poor are harmed by extreme weather-related events.

Climate change may also cause indirect effects to the urban poor in the form of reduced access to freshwater supply and food as prices for these commodities increase, along with decreases in crop productivity and freshwater supply. It is expected that many cities and their water catchments will receive less precipitation, thus exacerbating the already salient water stress and scarcity issues in these urban areas. Since most urban areas are dependent on food produced in rural areas, the effects of climate change on agriculture may induce food scarcity and encumber the urban poor.

There are two main challenges that inhibit city and municipal governments to adapt and protect the populations highly vulnerable to climate change and other environmental hazards. First, is the inherent limitation of urban governments, particularly in underdeveloped and developing countries, to acquire the resources needed for adaptation. Second, is the adverse relationship between urban governments and the low-income cohort of the population; in most cases, the urban poor are not seen as critical elements of the city economy and are often not given adequate attention in urban development planning.

Poverty Reduction and Climate Change

It is expected that climate change can impede efforts to poverty reduction and threaten to undo decades of development efforts. In the United Nations Human Development Report for 2007, climate change has been identified as a serious threat to human development because of its propensity to undermine international efforts to reduce extreme poverty, such as the Millennium Development Goals that aim to significantly reduce poverty and hunger by 2015. Therefore, stabilizing GHG emissions and bolstering adaptive capacities to limit the impacts of climate change are essential in the overall effort to alleviate global poverty. Climate change efforts should then be integrated with poverty alleviation and development plans.

There is now a call among developmental and environmental agencies to ensure that climate issues are integrated and mainstreamed into general development and poverty-reduction strategies. The integration would necessitate examination of existing programs that reduce current vulnerabilities of the poor, improvement of institutional processes to address these vulnerabilities, and avoidance of development activities that undermine the poor to cope with climate change impacts.

Furthermore, there are emerging opportunities in regard to climate change mitigation programs that can potentially improve the livelihoods of the rural sector. For instance, reforestation and the production of sustainable forms of biofuels, if implemented well, can conserve natural resources and revitalize rural development.

Marvin Joseph Fonacier Montefrio
State University of New York
College of Environmental Science and Forestry

See Also: Adaptation; Agriculture; Climate Change, Effects; Global Warming, Impacts of; Livelihoods; Small Farmers; Vulnerability.

Further Readings

Arnell, N. W. "Climate Change and Global Water Resources: SRES Emissions and Socio-Economic Scenarios." *Global Environmental Change*, v.14 (2004).

Barnett, J. and W. N. Adger. "Climate Change, Human Security, and Violent Conflict." *Political Geography*, v.26 (2007).

Costello, A., et al. "Managing the Health Effects of Climate Change." *The Lancet*, v.373 (2009).

Hardoy, J. and G. Pandiella. "Urban Poverty and Vulnerability to Climate Change in Latin America." *Environment and Urbanization*, v.21 (2009).
International Fund for Agricultural Development. *Rural Poverty Report 2011. New Realities, New Challenges: New Opportunities for Tomorrow's Generation.* Rome, Italy: IFAD, 2011.
Jones, P. G. and P. K. Thornton. "The Potential Impacts of Climate Change on Maize Production in Africa and Latin America in 2055." *Global Environmental Change*, v.13 (2003).
Patz, J. A., et al. "Impact of Regional Climate Change on Human Health." *Nature*, v.438 (2005).
Reid, H. and K. Swiderska. *Biodiversity, Climate Change, and Poverty: Exploring the Links.* London: IIED, 2008.
Satterthwaite, D., et al. *Adapting to Climate Change in Urban Areas: The Possibilities and Constraints in Low- and Middle-Income Nations.* London: IIED, 2007.
United Nations Development Programme. *Human Development Report 2007/2008. Fighting Climate Change: Human Solidarity in a Divided World.* New York: Palgrave Macmillan, 2008.

Precambrian Eon

The Precambrian eon, or Supereon, refers to the geological time comprising the eons that came before the Phanerozoic eon. This time spans from the formation of Earth around 4.5 billion years ago to the evolution of abundant macroscopic hard-shelled animals, which marked the beginning of the Cambrian era, the first period of the first era of the Phanerozoic eon. The Precambrian eon encompasses 86 percent of the Earth's history, however, very little is known about this time period. In fact, the few fossil discoveries from this period were recently made in the late 20th century. Precambrian time can be further divided into three large eons, the Hadean, Archean, and Proterozoic eons.

The Precambrian's oldest eon, the Hadean (4.5 to 3.9 billion years ago), predates most of the geologic record. During the Hadean, the solar system formed out of gas and dust, the sun began to emit light and heat, and Earth took shape. Meteors and other galactic debris showered the planet over the first half-billion years, making it entirely uninhabitable. Planet Earth was very hot during its initial formation. As the Earth began to cool and its mass increased, its gravitational field strengthened. This attracted meteorites and other debris, which continued to bombard the planet for at least another 500 million years, producing enough energy and heat to vaporize any water or melt any rock that may have been present. Iron continued to sink to form the Earth's core, while silicon, magnesium, and aluminum gradually rose toward the surface. Gases released from magma inside the Earth escaped through cracks in the surface and began to collect in the early atmosphere. The likely presence of methane and ammonia among the gases made for conditions that would be highly toxic to life as we know it. Because there was little to no free oxygen, no protective ozone layers existed and damaging ultraviolet rays showered the Earth at full strength.

As the meteorite bombardment finally slowed, Earth was able to cool, and its surface hardened as a crust, rocks, and continental plates began to form. Water began to condense in the atmosphere, resulting in torrential rainfall. After several hundred million years of falling rain, great oceans were formed. By about 3.9 billion years ago, Earth's environment had been transformed from a highly unstable state into a more hospitable place. This marked the beginning of the Archean eon (3.9 to 2.5 billion years ago). It was early in the Archaean eon that life first appeared on Earth.

Glaciations and Cold Climate

The climate of the late Precambrian time, the Proterozoic eon (2.5 billion years ago to 543 million years ago) was typically cold, with glaciations spreading over much of the Earth. One of the most important events of the Proterozoic was the gathering of oxygen in the Earth's atmosphere. Although oxygen was undoubtedly released by photosynthesis well back in Archean times, it could not build up to any significant degree until chemical sinks of unoxidized sulfur and iron had been filled. The first advanced single-celled and multicellular life roughly coincides with oxygen accumulation. It was also during this period that the first symbiotic relationships between mitochondria and chloroplasts and their hosts evolved. At this time, the continents were bunched up into a single supercontinent known as Rodinia. It broke up starting around 750 million years ago, and as continental fragments reached the north and south poles they likely contributed to the great ice ages.

In the latest Proterozoic era, a new supercontinent called Pannotia came together. A number of glacial periods have been identified going as far back as the Huronian epoch, roughly 2,200 million years ago. The best studied is the Sturtian-Varangian glaciation, around 600 million years ago, which may have brought glacial conditions all the way to the equator, resulting in a "snowball Earth." This theory states that the continents and oceans were covered in ice approximately 600 million years ago. The Earth may have remained in this frozen form, but it was rescued by the release of volcanic gases. While the Earth was in a deep freeze, chemical cycles were halted; as a result, carbon dioxide accumulated in the atmosphere, causing an extreme greenhouse effect. After 10 million years of deep freeze, the Earth thawed in only a few hundred years. These dramatic events may have caused the explosion of life-forms seen in Cambrian fossils.

At this point, at the start of the Cambrian period, Earth had taken on its current form in the life-filled oceans and oxygenated atmosphere. Coevolution of biosphere and lithosphere over billions of years led to this point. Anaerobes and oxygen-breathers had evolved complementary chemical cycles, and biogenic carbonates entered the plate-tectonic cycle of the crust and upper mantle with new efficiency.

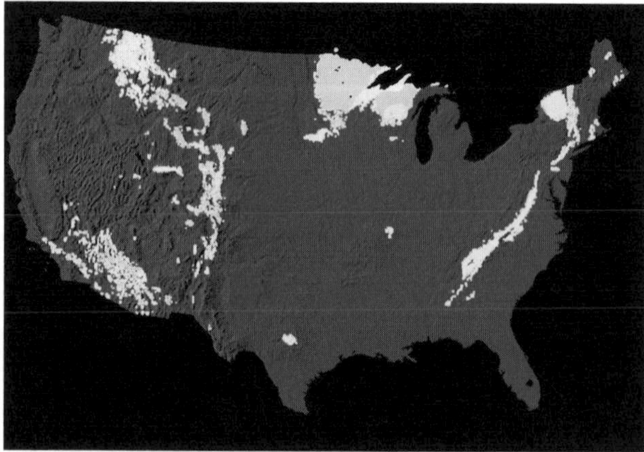

Highlighted in light areas are Precambrian rocks, formed between 560 million and 2.6 billion years ago. The Precambrian Era spans from the formation of Earth around 4.5 billion years ago to the evolution of abundant macroscopic hard-shelled animals.

Fernando Herrera
University of California, San Diego

See Also: Climate Change, Effects of; Climate Cycles; Snowball Earth.

Further Readings

Walker, J. C. "Precambrian Evolution of the Climate System." *Global Planet Change*, v.82 (1990).

Zahnle, K. "A Constant Daylength During the Precambrian Era?" *Precambrian Research*, v.37 (1987).

Precautionary Principle

Precaution, which is the idea that it is better to be safe than sorry when there are severe or irreversible consequences, has been a very important notion in environmental and public health policy. Precaution has been advocated in several issues, ranging from climate change to genetic engineering to the phase-out of persistent organic pollutants. The invocation of precaution has been particularly controversial when there are significant business interests at stake. The problem with simply asserting precaution whenever a technology, policy, or action involves possible negative outcomes is that it often poses significant challenges in evaluating public versus private tradeoffs.

Precaution is often invoked when outcomes are uncertain. The notion of uncertainty is used to characterize how well future events or scientific truths can be predicted or known. It is used in both social and natural science disciplines from mathematics to philosophy to risk assessment to public policy. If probability is a measure of likelihood, then uncertainty is a measure of how well the probability is known. Uncertainty can be classified into known and unknown probabilities. Events with known probabilities are referred to as events with statistical uncertainties. Events with unknown probabilities are often called events with true uncertainty.

Uncertainty in the context of the environment primarily refers to scientific uncertainty. Here, science generates truths through the testing of hypotheses. Often, however, the affirmation of hypotheses involves a certain degree of uncertainty because of the method or research design. Scientists often use the benchmark of 95 percent certainty when deciding whether or not cause and effect have been correctly identified. Scientists often report confidence limits based on research design and sampling error in their studies to account for uncertainty.

The precautionary principle is often invoked under uncertain circumstances, particularly when the consequences are irreversible or permanent. This differs from the choice that scientists make when deciding what to do under conditions of uncertainty. Typically, scientists are interested in avoiding false negatives because science is epistemologically conservative. Scientists do not want to suggest something as truth when in fact it may not be. In public or environmental policy, however, because the consequences are not epistemological but are ethical, there is desire to avoid false positives and be ethically conservative.

In public and environmental policy, it is important to understand how to make decisions in the absence of perfect information. Knowing the degree of uncertainty is particularly important when questions about risk arise. Risk assessment, a policy approach that deals with uncertainty, is widely used by the Environmental Protection Agency (EPA), but mainly focuses on known probabilities. Because of difficulties with codifying the precautionary principle into policy, the EPA has yet to include true uncertainty in environmental policy.

Classes of Scientific Uncertainty

Kristin Schrader–Frechette describes four classes of scientific uncertainty dealt with by scientists and policymakers: framing uncertainty, modeling uncertainty, statistical uncertainty, and decision–theoretic uncertainty. In framing uncertainty, scientists often use a two-value frame to accept or reject a hypothesis. Frechette argues that in public policy, it is more appropriate to adopt a three-value frame that creates a category to deal with situations where significant uncertainty and serious consequences suggest adopting the precautionary principle. Modeling uncertainties involve those in the prediction of future scenarios. These are highly speculative, despite claims to be verified and validated models.

In public and environmental policy, statistical uncertainty should highlight the difference between epistemological and ethical consequences. When faced with decision–theoretic uncertainty, scientists are forced to distinguish between using expected value rules and the minimax rule. The former argues that decisions should be based on the expected value, while the latter seeks to prevent the worst-case scenario. More recently, Bayesian statistics has been used to help evaluate data under conditions of uncertainty by updating the probabilities as new data come to view. A Bayesian approach involves the introduction of prior knowledge into statistical models.

There are many environmental policy debates where questions about uncertainty are raised. In debates about genetic engineering, uncertainty about the prediction of how transgenic organisms will behave in the environment, or uncertainty about how markets will react to the adoption of transgenic organisms, is cited as a reason to invoke the precautionary principle. In debates about nuclear waste disposal at Yucca Mountain, uncertainty about how the storage facility will perform in the long term is cited as a reason to question the suitability of this nuclear waste repository.

In debates about global climate change, scientists typically agree that there is significant uncertainty in the projection of future climate change models. Those wishing to discredit climate change science often highlight uncertainty.

Dustin R. Mulvaney
University of California, Berkeley

See Also: Climatic Data, Nature of the Data; Environmental Protection Agency, U.S.; Measurement and Assessment; Mitigation; Preparedness; Public Awareness.

Further Readings

Bodansky, Daniel. "Scientific Uncertainty and the Precautionary Principle." *Environment*, v.33/7 (1991).

Lemons, John, ed. *Scientific Uncertainty and Environmental Problem Solving*. Oxford: Blackwell Science Press, 1996.
MacFarlane, Allison and R. C. Ewing. *Uncertainty Underground: Yucca Mountain and the Nation's High-Level Nuclear Waste*. Cambridge, MA: MIT Press, 2006.

Precipitation

Precipitation is the primary factor that controls the hydrologic cycle. It takes different forms, such as rain, snow, hail, sleet, drizzle, dew, and fog. It supplies most of the freshwater on the Earth. Most precipitation starts from space as snow, as the upper space is cooler. If the temperature of the surface closer to the ground is below 32 degrees F (0 degrees C), then the precipitation falls on the ground in the form of snow. If the ground and closer surface temperature is above 32 degrees F (0 degrees C), the precipitation takes the form of rain. When the air at the ground is below freezing, the raindrops can freeze while hitting the ground, and that is known as freezing rain. When a dust particle in the atmosphere attracts a moisture drop, hail is formed. Drizzle consists of very small raindrops, 1/1000 of a normal raindrop size. Sleet is a type of precipitation between rain and snow, but very distinct from hail. Dew is another form of precipitation that can be seen in the early morning on colder days. Water vapor in the atmosphere condenses on the surface of exposed objects at a greater rate than that at which it can evaporate, developing dew. Fog as such is not precipitation, but is considered one because of its low-altitude occurrence. This consists of a cloud in contact with the ground, and produces water droplets when intercepted with vegetation or other exposed objects. If the precipitation evaporates before reaching the ground, it is then known as virga.

How Precipitation Occurs

Dynamic and adiabatic cooling causes precipitation. Because of this process, condensation of water vapor occurs and then falls to the Earth as rain, snow, or other forms of precipitation. Vertical air motion is the leading factor of all rainfall from clouds after condensation. Rising air in the tropics and midlatitudes (40 to 60 degrees N and S latitudes) causes more precipitation and descending air patterns in the subtropics (20 to 30 degrees N and S latitudes) and in the poles causes less. Precipitation is classified into different categories based on the conditions that generate vertical air motion: convective, orographic, and cyclonic.

Convective precipitation is mostly seen in the tropics. Heating up of the air at ground level, then moving upward and mixing with the water vapor in the atmosphere, with dynamic cooling in space, causes precipitation to fall on Earth. This is called convective precipitation. Orographic precipitation occurs because of the interception of moisture-laden air or clouds by mountain ranges. Cyclonic precipitation is caused by movement of moist air masses from high-pressure regions to low-pressure regions. In the hydrologic cycle, the total volume of precipitation onto land is measured at 42,471 sq. mi. (110,000 sq. km), while the total volume of precipitation on the ocean surface is 176,834 sq. mi. (458,000 sq. km).

Precipitation has greater ecological, geographical, and regional impact because of its characteristics, such as relative amount, seasonal timing, and most importantly the size and intensity. Low-intensity and well-distributed seasonal precipitation is good for agriculture. High-intensity, long-duration precipitation creates more problems than good. Precipitation is governed primarily by atmospheric water vapor, but its variation depends upon other climatic factors such as temperature, wind, and atmospheric pressure at different locations and season. The amount of water vapor is very high in the atmosphere closer to water bodies. Therefore, coastal areas always have heavier precipitation (high-intensity, long-duration) than inland areas. Thunderstorms, hurricanes, typhoons, cyclones, blizzards, and hailstorms are high-intensity precipitations with damaging ability.

It is essential to understand the precipitation distribution process and its temporal and spatial variation for water resources planning and management for the betterment of the society. Precipitation distribution mechanisms include interception by vegetation, filling in depression storage, infiltration to soil and ground water, surface detention, and overland or surface flow or runoff. Low-intensity, short-duration storms are good for land and vegetation because most of the water either stays in depression storage or infiltrates to soil and groundwater for groundwater recharge. However, high-intensity, long-duration storms are detrimental because most of the precipitation (rainfall) is wasted as runoff. The excessive runoff erodes soil, and creates floods, landslides, and other damaging effects. Blizzards are examples of such high-intensity, long-duration precipitation. A heavy amount of snowfall accompanied by high wind creates problems. Many high-intensity, long-duration storms are currently taking place as a result of global warming.

Effects of Global Warming on Precipitation

Global warming is the effect of increasing atmospheric concentrations of greenhouse gases. The near surface of the Earth has warmed by nearly 1 degree F (0.6 degree C) during the 20th century. It may continue in this century, warming the globe further, so that a warmer ocean surface would result. There would be an increased evaporation rate and subsequent increase in the other components of the hydrologic cycle, like water vapor in the atmosphere and consequential higher precipitation amounts. Computer simulation models found that global warming by 7.2 degrees F (4 degrees C) is expected to increase global precipitation by about 10 percent, as well as increase rainfall intensity. Scientists using models found that the upper tropospheric water vapor amount will increase by 15 percent with each degree of atmospheric temperature rise. The global water vapor amount will increase by 7 percent with each degree of atmospheric temperature rise.

Another major downside of global warming is less snowfall throughout the world. As the surface temperature is rising, raindrops, in many parts of the Earth, cannot take the shape of snow before reaching the ground. Snow is better for the land than rain because it helps in water conservation. Rainfall becomes runoff to oceans and is a waste if not conserved artificially through dams, earthen embankments, and soil and water conservation structures, whereas snow remains on the ground and melts slowly to release water for agriculture and other consumption. Slow melting of snow also facilitates more accumulation of soil moisture and ground water recharging. Agriculture in the northern United States and southern Canada depends upon soil moisture conserved by snowpack on the agricultural land. With less snowfall, because of an increase in surface temperature, there may be less snow deposition on agricultural land, and agriculture would suffer.

The irony of global warming is the occurrence of more rain resulting in less water. Because of warmer temperatures in the spring, snowpacks in the mountains are melting unseasonably and at a rapid rate. Many perennial rivers in the world are experiencing shortages of flowing water in summer months and groundwater recharging has lessened. The water from a quick snowmelt cannot be arrested in reservoirs because of lack of space. Therefore, the systems failing to hold the entire season of runoff would face challenges to meet the water demand for agriculture and other purposes. Increased quick melting of snowpacks (because of a rise in surface temperature) in the northwestern United States and India causes spring and summer floods.

It is estimated that there could be a 15 to 30 percent reduction of water available for human consumption from California's Sierra Nevada mountains. Agriculture in the Canadian prairies could be the worst affected, because of spring water runoff. Sufficient water may not be available for subsequent crop seasons. Many rivers in the world will experience water shortages in dry seasons, because during dry months most of the perennial rivers get water from snow and glacier melting. Glacial retreats are clearly visible in the poles, Greenland, and Antarctica. This occurs because of an insufficient supply of water to glaciers from precipitation as compared to the loss of water from melting and sublimation.

Global warming would cause more precipitation on the Earth, but would create more detrimental effects to agriculture, ecology, and, above all, society. The wetter areas of Earth would become wetter, and the drier areas would become even drier.

Sudhanshu Sekhar Panda
Gainesville State College

See Also: Climate Change, Effects of; Global Warming; Rain; Rainfall Patterns; Weather.

Further Readings

American Geophysical Union. *Water Vapor in the Climate System*. Washington, DC: American Geophysical Union, 1995.

Britt, R. R. "The Irony of Global Warming: More Rain, Less Water." *Live Science* (November 16, 2005).

Dunne, Thomas and L. B. Leopold. *Water in Environmental Planning*. New York: W. H. Freeman, 1978.

Groisman, P., et al. "Changes in the Probability of Heavy Precipitation: Important Indicators of Climatic Change." *Climatic Change*, v.42 (1999).

Huang, J. and H. M. van den Dool. "Monthly Precipitation–Temperature Relations and Temperature Prediction Over the United States." *Journal of Climate*, v.6 (1993).

Hulme, M. "Estimating Global Changes in Precipitation." *Weather*, v.50 (1995).

U.S. Geologic Survey. "Retreats of Glaciers in Glacier National Park." http://nrmsc.usgs.gov/research/glacier_retreat.htm (Accessed March 2012).

Preparedness

Preparedness for global warming and climate change requires multilevel planning at the international, national, local, and individual levels to deal with the direct effects of climate changes in temperature, precipitation, wind, storm patterns, and sea level, as well as the indirect strain on world resources leading to migration, famine, and conflicts.

While mitigation strategies like reducing greenhouse gas emissions are a start, they do not take precedence over readiness to respond to natural disaster emergencies associated with the impact of climate change (e.g., intense storms, flooding, wildfires, and public health) or the necessity of dealing with future environmental pressures, the decreasing longevity of infrastructure with roads, bridges, waterworks, buildings, and facilities requiring earlier replacement, repair or modifications to remain safe for use. Adaptive measures in the form of physiological, social, and cultural measures will allow people to live more safely throughout the world.

Assessment of Impact

In order to prepare for climate change, assessments must be completed and available for decision-making processes. On the international level, the International Governmental Panel on Climate Change encourages collaboration among scientists, allied professionals, and policymakers, providing a forum for the collection of research material and posing questions to spur planning. The take-away message is evaluation to determine global policy toward a common goal, while addressing implications and increasing cooperation between governments.

On the national and local levels, any assessment of the extent of impact climate changes (higher temperatures, rising sea levels, changing weather patterns) will have on human health, ecosystem diversity and productivity, agricultural production, water supply, sanitation, and infrastructure must factor in the associated costs and benefits in developing sustainability plans, preparation for emergency situations, and adaptive measures.

Each area is different, with varying susceptibilities, policies, institutions, and social/cultural structures. National and local assessments, to be most effective, must evaluate the capacity to handle emergency measures associated with natural disasters and identifying susceptible areas of the environment and resources including agriculture, fisheries, forestry, fauna, human health, water supply and sanitation, infrastructure and construction, land use in hazard-prone areas (flood plain, islands), and disaster management.

Challenges to assessment for impact include factors not related to climate change that will impact the areas being considered at risk because of climate change include the dynamics of society and economy (demographic trends, agricultural management, improving and new technologies, cultural preferences, opportunities for employment, availability, and changes in transportation). Complex dynamics in human relationships with the environment, self-reliance, growing population, and increased urbanization may produce very different impacts between urban and rural areas.

Impacts to Consider

When making assessments, a variety of elements need to be considered that cross boundaries of social structures and practices, as well as environmental conditions. Severe weather destruction has highlighted areas that lack preparedness. The current global situation for preparedness is still lacking. Severe weather causes great devastation. Even in a highly developed country, as seen in the aftermath of the 2005 Hurricane Katrina hitting New Orleans and other areas on the U.S. Gulf Coast, problems were experienced with infrastructure, the failure of levees to hold back storm surge water, and mobilizing emergency supplies.

Planning for impacts from rising global temperatures must take into account a wide range of factors, including the impact on human health, comfort, lifestyle, food production, economic activity, and residential and migration patterns, including tourism.

Housing or shelter must be sufficient to meet the needs of an emergency situation, for migration from an area no longer suitable for living, or to meet expectations for increased tourism to climates that so far have not been inundated by travelers. In addition, supportive measures must be able to supply water, sanitation, communication, energy, transport, and industry; and social and cultural services including health services, education, police protection, recreational services, parks, and museums for areas with an increased population.

With rising global temperatures, weather patterns could alter in frequency and seasonality of precipitation, weather-related events could increase, including droughts, floods, severe tropical storms with associated storm surge flooding,

Hurricane Fran (1996), shown in a weather satellite image, caused about $5 billion in damages in North Carolina alone. Computer simulation models with a global warming increase of 7.2 degrees F (4 degrees C) forecast an increase in global precipitation by about 10 percent. Rainfall intensity is also predicted to increase. If weather patterns change as predicted, coastal areas need to prepare for effective evacuations to deal with heavy rains, flooding, and more severe hurricanes.

and wildfires from increased temperatures and drought conditions. With a rise in evaporation and precipitation rates, water availability and quality could be affected, with lower groundwater levels, decreased surface area and water levels of many lakes or inland waterways, as well as altering natural habitats. Potential impacts range from loss of property, effects on housing and street/road conditions, effects on construction materials, stress on sewage systems, and potential overflow from excess storm water and drainage failures.

Rising sea levels could displace residents of delta regions (Nile, Ganges, Yangtze, Mississippi) from homes and livelihoods. Island nations could become uninhabitable. Coastal erosion on gradually sloping coasts by encroaching water could affect densely populated cities (New York, London) and important seaports.

Cooler climates may see an increase in agriculture and warmer climates may see a decrease in agriculture. Feeding the world's population means an inherent dependence on agricultural production, which is highly sensitive to climate change. Land degradation may produce either abandonment of the land or require changes in cultivation practices to improve yields and restoration of soil. Possible changes in fertileness of the land could increase or decrease food production capability for the agricultural country and as an export to other world regions.

Health impacts include temperature stress from either extreme heat or cold, especially among high-risk groups (children, the elderly, and those with already compromised health); air pollution exposure with increased incidence of respiratory disease; chemical pollution; water quality or water shortages from precipitation changes, flooding, or in some areas already a lack of safe water; and vector-borne diseases; lack of physical or economic access to health services or insufficient capacity of health services. In the event of disaster refugees or migration, sanitary facilities and housing could become quickly overburdened, enhancing the spread of communicable diseases.

Climate change could modify supplies and consumption patterns. These impacts would vary by region based on cost of various types of food and fiber. Changing availability of resources might lead to changed diets, production patterns, and employment levels. Major impacts could be felt by the energy, transport, and industrial sectors, with increasing need balanced against dwindling supply.

Impacts on physical and social environment would also vary by region and could include loss of housing (from wildfires, flooding, mud slides); loss of living resources (water, energy supplies, food, or employment); loss of social and cultural resources (cultural properties, neighborhood or community networks); decline in living standards (conditions caused by mandatory evacuation, contamination of water supply); total loss of livelihood following land degradation (erosion of top-soil, over-cultivation, or deforestation); or a major natural disaster like flooding or drought. In some areas, physical and social environments could improve. Communication technologies (cell phones, computers, and fax machines) could have a positive impact, including increased potential for decentralization of the population by enabling many professional and technical people to perform work in homes far removed from major metropolitan areas.

Possible Solutions

The Hyogo Framework for Action signed in January 2005 by 168 countries details steps to reduce the impact of natural hazards on populations. The result has been 40 countries adjusting national policies to give priority to disaster risk reduction.

Examples of disaster planning paying off in action include Jamaica in 2004. A community disaster response team issued early warning by megaphone, and used risk maps and equipment assembled by the Red Cross for successful evacuation of all area residents. Hurricane awareness is taught in schools in Cuba, along with practice drills. The Citizens' Disaster Response Center in the Philippines helps to create disaster management plans and provides emergency response. These examples indicate effectiveness and indicate the need for global, regional, national, and local early warning systems to alert populations of impending disasters.

While the previous examples are for disasters, the same ideas can be used to implement climate change preparedness, of which natural disasters are one element. One caveat for planning is the necessity to include feedback loops to allow for changes as new information becomes available or conditions are modified.

The challenge is the predictive factor (strategies planning far into the future, 10 years, 25 years, 50 years, or 100 years) in the presence of current needs of access to clean water management, production of energy allowing for cost-effectiveness and to allow developing countries to utilize cheap fossil fuels available to them, and supplying enough food for populations with limited agriculture because of already stressed conditions. To meet food supply needs far into the future, research to develop new varieties of wheat, corn, and soybeans for resistance to drought and heat and still produce good yields and strategies for future irrigation sources should begin now, building new wells and reservoirs instead of waiting until water shortages persist. Centralized stockpiles of grain could provide for increases in food needs or to supply areas with crop devastation.

With changing weather patterns, coastal areas should prepare for and be able to demonstrate effective evacuation procedures to deal with rising sea levels and more severe hurricanes, as well as to assess the risks of new construction in low-lying areas. Global, regional, national, and local early warning systems to alert populations of impending disasters should be developed, implemented, and tested. Water supplies, both in rural areas and in municipal water infrastructure, should be set up for equitable availability and pricing. Some areas have no access to freshwater and must use desalinization systems to make clean water from seawater.

Environmental policy should strengthen institutional and legislative environmental framework at national and regional levels, including environmental authorities, and incorporating new laws, as well as implementing environmental control standards for air quality, including reduced emissions of greenhouse gases, reducing ground level ozone and particulates from stationary and transport sources, and more tightly enforcing pollution and land use regulations.

Emphasis should be placed on nature and biodiversity, such as management of protected areas and a national biodiversity strategy and provide for better management of all natural resources, including forests, fishing, soils, water, air, wildlife. Altered habitats will cause animal migration. Barriers standing in their way could be removed by setting up migration corridors to connect natural areas and allow them to migrate safely. In the event that these measures aren't enough, wildlife managers may have to capture and move certain species.

Reducing health impacts from climate change can be acted upon early by preventing the onset of disease from environmental disturbances, in an otherwise unaffected population (such as supply bed nets to all members of a population at risk of exposure to encroaching malaria, taking precautions against mosquito bites to prevent West Nile virus, early weather watch warning systems). Surveillance systems could be improved in sensitive geographic areas with the potential for epidemics under certain climatic conditions, including those bordering areas of current distribution of vector-borne diseases (plague from prairie dogs, hanta virus from rats, malaria from mosquitoes). Vaccination programs could be intensified and pesticides used for vector control. Drugs for prophylaxis and treatment could be stockpiled.

In certain areas of the world, irrespective of climate change, breakdowns in public health measures have been responsible for many recent outbreaks of disease. In those areas, climate change would add to the health burden; current and future health problems related to the environment share similar causes of economic factors and access issues. All areas must have sufficient trained staff to handle healthcare issues, as well as technology for diagnosing and treatment and medications or medical interventions required.

Construction practices and advancing technology make possible new building techniques, including floating houses to weather storm surge and flooding in hazard areas. Current building material (including roads) may not be able to withstand future climate change conditions and will require replacement or restoration. Public reconstruction to ensure safe bridges and other infrastructure is necessary. In some areas, drainage improvements will need to be made for water and sewage.

Adaptation refers to actions taken to lessen the impacts of the anticipated changes in climate. The ultimate goal of adaptation interventions is the reduction, with the least cost, of disruptions in living standards, of suffering from diseases, injuries, or disabilities, and of destruction of habitats, ecosystems, and species (both plant and animal).

Raising Awareness

To make any plan for preparedness work, the information collected in assessments must reach all relevant stakeholders in the process of building and reaching a consensus on adaptation strategy. Providing environmental information to the public is essential to any preparedness method; if people don't know what to do in the event of an emergency or steps to take to prevent or adapt to changing conditions, the plan will fail. Awareness campaigns directed at the general public and to those persons who are directly affected should be made available through all forms of media and should be easily understood.

On the individual level, certain precautions can be taken to prepare for natural disasters, and some associated climate change, including extreme heat or winter weather, flooding, hurricanes, landslides and mudslides, tornadoes, tsunamis, and wildfires. Taking preparedness actions helps people deal with disasters in an effective manner. Having emergency supplies of water, food, and first aid prepared and easily available allows an individual or families to either evacuate quickly or shelter in place as appropriate. Having a plan of action for where to go, how to reconnect with family members, and dealing with finances lessens the stress involved with disasters.

Lyn Michaud
Independent Scholar

See Also: Diseases; Floods; Health; Hurricanes and Typhoons; Intergovernmental Panel on Climate Change; Mitigation; Public Awareness; World Health Organization.

Further Readings

Sitarz, Daniel ed. *Agenda 21: The Earth Summit Strategy to Save our Planet*. Boulder, CO: EarthPress, 1993.

World Health Organization. "Climate Change and Stratospheric Ozone Depletion. Early Effects on Our Health in Europe" (2000). http://www.euro.who.int/en/what-we-publish/abstracts/climate-change-and-stratospheric-ozone-depletion.-early-effects-on-our-health-in-europe (Accessed March 2012).

Worldwatch Institute. *Vital Signs 2006–2007*. New York: W. W. Norton & Co., 2006.

Princeton University

Princeton University, chartered in 1746 and located in Princeton, New Jersey, is a world-renowned, private research university with 1,100 faculty members, 5,000 undergraduate students, and 2,500 graduate students. The university provides undergraduate and graduate degrees in the humanities, social sciences, natural sciences, and engineering. On a per-student basis, Princeton has the largest university endowment in the world. Princeton has stated a commitment to climate change research and academic freedom.

Princeton Partnerships and Studies

The Carbon Mitigation Initiative (CMI) is a partnership between Princeton University and British Petroleum (BP) that formed in 2000. Additional funding comes from Ford Motor Company. The original grant was for \$1.5 million, which increased the following year to \$2.2 million annually. The additional funding will allow researchers to expand their efforts to understand capture facilities and the use of captured carbon dioxide (CO_2) in production of electricity, hydrogen, and synthetic fuels. Initially a 10-year project, BP added an additional five years to extend the CMI's lifespan to 2015. CMI involves over 60 researchers in national and international discussions; these researchers come from departments such as civil and mechanical engineering, geosciences, ecology, and evolutionary biology.

Also involved are the Energy Group, the Woodrow Wilson Center for Public and International Affairs, and international collaborators. CMI includes the Capture Group, which explores technology for capturing greenhouse gas emissions and researches alternative fuels. The Storage Group performs modeling, lab experiments, and field studies of the feasibility of injecting CO_2 underground permanently. The Science Group is collecting oceanic, atmospheric, land, and ice-core data to study carbon sinks and impacts of emissions on climate. Policy and Integration puts the research together, synthesizes it, and deals with the policy ramifications of the findings. One innovative and unconventional area of research is direct capture of CO_2 from the atmosphere.

CMI's major contribution has been the idea of stabilization wedges, a new paradigm in the effort to reduce 200 billion tons of carbon in the next 50 years. CMI divides the 200 tons into wedges like a pie. The wedges are the eight, 25 billion-ton elements that must be used to achieve the target—energy efficiency, capture and storage, conversion of coal to natural gas, wind, solar, biomass, nuclear, and natural sinks. CMI provides informational material and a wedge musical video to the public, as well as high school teaching tools such as the Wedge Game.

The Princeton Environmental Institute (PEI) is an interdisciplinary center opened in 1994. Scholarship, outreach, and education are the mission of the 90 faculty members from over 25 disciplines who explore technical, scientific, policy, and human aspects of environmental issues. Differing from the other Princeton centers researching the environment, PEI takes a cross-discipline approach to complex problems such as energy and climate, biogeochemical cycles, disease, biodiversity, and climate change.

PEI provides certification at the undergraduate level and graduate work in conjunction with the Woodrow Wilson School's Science, Technology, and Environmental Policy (STEP) program, providing an environmental policy component to graduate degrees. In 2007, the PEI, Woodrow Wilson School, and Engineering and Applied Science School jointly established the Grand Challenges Program to promote integrated research and teaching on the world's most difficult environmental issues, including energy and climate, infectious disease, and sustainable development. PEI and the university's center for human values offer the Seibel Energy Grand Challenges Lecture Series, which in 2009 focused on the ethics of climate change.

The Princeton Institute for International and Regional Studies (PIIRS) awarded its first grant after its creation in early 2011 to an interdisciplinary group of 16 faculty from nine departments and three programs/projects proposing the topic, "Communicating Uncertainty: Science, Institutions and Ethics in the Politics of Global Climate Change." With the study, Princeton stood to become a leader in a relatively new field. The study was funded at $750,000, and was to last three years from the fall of 2011 to the fall of 2014. Its goals were to work from the perspectives of the social, political, and natural sciences, with uncertainty arising from climate change and other international environmental problems. Robert Keohane of the Woodrow Wilson School of Public and International Affairs noted that there was a long-standing difficulty in translating the science of environmental concerns to lay terms, and the interdisciplinary group was to clarify that the issues had already been resolved and delay would only exacerbate the problems.

However, the debate does not yet appear to be resolved, not even within the Princeton community. Princeton University is home to Dr. Will Happer, who in 2009 warned Congress that it cannot stop climate change by law. An award-winning physicist and author of over 200 peer-reviewed scientific papers, Happer contends that there is a "climate change cult" consisting of ignorant zealots whose ideas are ludicrous, but also extremely dangerous to the poor people of the world. Happer argues that carbon dioxide (CO_2) is not a pollutant, but an essential that is in shortage, and life will be better with more of it. He and other scientists wrote an open letter to Congress in 2009 stating that the Earth has been cooling for the past decade. He is also one of 54 physicists who have requested the American Physical Society to revise its stance on global warming because the records show nothing remarkable about temperatures in the 20th and 21st centuries.

John H. Barnhill
Independent Scholar

See Also: BP; BP Deepwater Horizon Oil Spill; Carbon Footprint; Carbon Sinks; Emissions, Trading; Global Warming Debate.

Further Readings

BP. "Princeton University–Carbon Mitigation Initiative." http://www.bp.com/sectiongenericarticle.do?categoryId=9033341&contentId=7061346 (Accessed July 2011).

Brusca, Raymond. "Professor Denies Global Warming Theory." *Daily Princetonian* (January 12, 2009).

http://www.dailyprincetonian.com/2009/01/12/22506 (Accessed October 2011).
Carbon Mitigation Initiative. "About the Carbon Mitigation Initiative." http://cmi.princeton.edu/about (Accessed July 2011).
Carbon Mitigation Initiative. "Stabilization Wedges." http://cmi.princeton.edu/wedges (Accessed July 2011).
Holt, V. Sarada. "Princeton Institute to Support New Project on Politics of Global Climate Change." *Princeton Patch* (May 18, 2011). http://princeton.patch.com/announcements/princeton-institute-to-support-new-project-on-politics-of-global-climate-change (Accessed July 2011).
Princeton Environmental Institute. "About the Princeton Environmental Institute" (June 22, 2011). http://www.princeton.edu/pei/about (Accessed July 2011).
Princeton Environmental Institute. "Ethics and Climate Change: Siebel Energy Grand Challenges Lecture Series Spring 2009." http://www.princeton.edu/pei/ethics-climate (Accessed July 2011).
Princeton Institute for International and Regional Studies. "PIIRS Research Communities." https://www.princeton.edu/piirs/research/research-communities (Accessed July 2011).

Public Awareness

Public awareness in the United States of the issue of global warming increased from about one-third in the early 1980s to near 100 percent 25 years later. By 2007, climate change was featured in the media almost daily. Awareness does not necessarily imply acceptance; although polls indicate that over half of Americans consider climate change to be real, there remains widespread public uncertainty about the degree to which human activities are involved, and to what extent CO_2 emissions need to be curtailed. There also remain widespread misconceptions about the meaning of global warming, and likely effects.

Public acceptance of human-induced climate change as a real phenomenon has lagged well behind the scientific consensus. In the mid-1970s, the popular media widely reported that the Earth was cooling and may be entering the next glacial interval, accelerated by light reflected off atmospheric particulates from pollution. The reports were based on the ideas of several scientists, espoused primarily outside peer-reviewed literature. By the late 1970s, scientific consensus emerged from early generation global climate models that the warming influence of greenhouse gases was stronger than the cooling influence of particulates and insolation change. Scientific evidence that the climate was warming first received major coverage in a 1981 front-page article in the *New York Times*. Considerable advances in scientific understanding of current and past climate change occurred in the 1980s; this received enhanced public recognition with the 1988 congressional testimony by climatologists that coincided with a record-hot summer.

As calls for government controls to reduce greenhouse gases increased, climate change discussions and media coverage of it grew politicized. In the 1990s, media, in efforts to offer "balanced" reporting, covered a small number of climate change skeptics in roughly equal proportion to the scientific consensus that climate is warming, which had grown to close to 100 percent in peer-reviewed scientific literature. The public was thereby given the impression that a considerable scientific controversy still existed. Debate about U.S. participation in the international Kyoto Accord in late 1997 and again in 2001 further increased politicization of the issue.

Several events in the mid-2000s swung U.S. public opinion from simple awareness of the issue to greater acceptance that global warming was happening. During this time, skeptics also changed stances from whether climate change was happening to whether humans were causing observed changes. The severe hurricane season of 2005 (in particular, Hurricanes Katrina and Rita) centered U.S. public attention on potential domestic human and financial costs of climate change. Al Gore's 2006 documentary, *An Inconvenient Truth*, one of the most watched documentaries of all time, stimulated a groundswell of activity to further increase awareness, though to some degree maintaining the politicization of the issue. Several very warm years globally during the 2000s also helped give climate change greater reality to a broader geographic segment of U.S. citizens accustomed to hearing about warming in other areas of the world.

Limiting Factors

There are several factors that influence a person's understanding on the issue of climate change:

- *Scale*: It is difficult for most people to grasp scales of space the size of the Earth and its atmosphere; scales of time that include analyzing data from thousands of years in the past and up to decades or centuries into the future; and scales of human influence that involve billions of people, each contributing some quantity of CO_2 to the atmosphere.
- *Complexity*: Climate is a complex system that is difficult even for experts to understand in total; in global warming, some places cool, which is why *climate change* is now the preferred term; spatial and temporal variability in weather systems means that even places that are warming on average may occasionally be unusually cold; in some places, climate change is manifested by precipitation change.
- *Models and uncertainty*: Computer models of climate are complex sets of mathematical equations that are "black boxes" for most; it can be confusing that different results occur for different researchers' models, as is evaluating the probability of events decades in the future.
- *Personal reality*: In many places, climate change may not be readily evident from casual observations; changes at the poles may seem personally irrelevant; warming may sound attractive in cooler climates.
- *Personal beliefs*: Spiritual or philosophical beliefs may indicate that the Earth does not change, that humans need not concern themselves with personal influence on the Earth, or that near-term spiritual events on Earth will render such changes meaningless.

Education

Many efforts have developed in recent years to go beyond increasing public awareness to educating the public on the nature of climate change and what can be done about it. The former is especially the purview of science education, and the latter of personal and governmental action. Few K-12 school curricula have climate change as a major topic, though building block concepts such as greenhouse warming may be found in Earth and environmental courses at the high school level. Professional development for teachers at the state and national level may be needed in large numbers in the coming decades. Nonformal educational groups such as Scouts and 4-H are alternative settings for introducing climate education to youth. Major research partnerships such as Global Learning and Observations to Benefit the Environment (GLOBE) provide opportunities to collect data for scientific research that is relevant to climate change. Many colleges and universities are developing campus sustainability models and associated student action groups and courses.

Some organizations of informal education, such as museums and science centers, are in a good position to present public climate change education to a wide range of age groups. The International Polar Year (2007–09) provided an opportunity to present outreach associated with research on polar processes, including polar warming. Many grassroots groups have started in towns and cities across the United States to provide information on climate change to local citizens, particularly on how to take action to reduce CO_2 emissions. Among the most influential means of informal education remain radio and television documentaries on the topic.

Robert M. Ross
Warren D. Allmon
Paleontological Research Institution

See Also: Education; Gore, Albert, Jr.; *An Inconvenient Truth*; Media, Internet; Media, TV; United States.

Further Readings

Gore, Al, Jr. *An Inconvenient Truth*. New York: Rodale, 2006.

Stamm, K. R., Fiona Clark, and P. R. Eblacas. "Mass Communication and Public Understanding of Environmental Problems: The Case of Global Warming." *Public Understanding of Science*, v.9 (2000).

Weart, Spencer. "The Public and Climate Change." American Institute of Physics. http://www.aip.org/history/climate/Public.htm (Accessed March 2012).

Qatar

Located on a peninsula bordering the Persian Gulf and Saudi Arabia, the State of Qatar, an Arab emirate, is one of the Persian Gulf's wealthiest states. Despite its meager size of 4,416 sq. mi. (11,586 sq. km) and small population of 848,000 (2011 est.), Qatar's economy has the world's highest growth rate and ranks first in the world with a gross domestic product per capita of $179,000 (651,000 Qatar riyals). Less than 5 percent of Qataris live outside urban areas, and the country has the lowest unemployment rate in the world.

Qatar has a large sector of industries involved in the extraction and processing of the country's vast oil and gas reserves. The economy has benefitted from having the third-largest reserves of natural gas, producing 76.98 billion cubic meters (cu. m), as well as the world's 12th-largest proven fossil fuel reserves, ranking it 22nd in terms of global exports and producing an average of 142,000 barrels per day. Qatar also ranks first in the world in per-capita energy consumption. An arid desert climate and general lack of freshwater resources has meant that Qatar must rely on limited groundwater resources and large saltwater desalinization plants to provide its water supply, which is an energy-intensive process. Qataris use 381 percent of their water resources, which means that about 180 million cu. m of desalinated water is produced to meet the annual demand of 444 million cu. m.

While having major petroleum and natural gas deposits has greatly benefited Qatar's economy, it has also contributed to Qatar's status as one the largest emitters of carbon dioxide (CO_2). Although a relatively modest emitter of greenhouse gases (GHGs) in absolute terms, Qatar ranks first globally in CO_2 emissions per capita, reaching 48.8 metric tons of CO_2 equivalent in 2007, with national CO_2 emissions increasing nearly 300 percent from 1990 levels, mostly due to rapid industrial growth. About 70 percent of national GHG emissions are accounted for by the oil and gas industry alone. The electricity and water sector account for almost half of all energy-related GHG emissions, and manufacturing and construction contribute another 29.1 percent. Both are expected to increase in the next decade if no immediate action is taken to mitigate a growing trend in rising emissions levels.

Because of industry efforts to mitigate emissions, total national GHG emissions from the oil and gas industry were reduced from 45 percent in 2001 to less than 20 percent in 2006. With the registration of the first United National Framework Convention on Climate Change (UNFCCC) Clean Development Mechanism project in the Gulf region in 2007, which was carried out by Qatar Petroleum under the Kyoto Protocol, the hope is that other similar projects will be carried out in the future. In the *Qatar National Vision 2030* report, the State of Qatar acknowledged the pressing need to address the impacts of climate change, recognizing the multifaceted nature of the issue and viewing it as integral to the larger challenge of creating sustainable development. Qatar is also the first country in the region to join the World Bank's Global Gas Flaring Reduction Partnership (GGFR), and the Qatari government continues to actively participate in the UNFCCC negotiations, taking an active role in the emerging opportunities provided by the transition toward a post-carbon society.

Cary Yungmee Hendrickson
University of Rome

See Also: Carbon Dioxide; Clean Development Mechanism; United Nations Framework Convention on Climate Change; World Bank.

Further Readings

Al-Mulla, A., et al. "Climate Change and Human Development in Qatar: Issues, Challenges and Opportunities." In *Qatar National Vision 2030*. New York: United Nations Development Programme, 2009.

Food and Agriculture Organization of the United Nations. *Aquastat Profile: Qatar.* Rome: FAO, 2008.

United Nations Development Programme. *Qatar's Second Human Development Report: Advancing Sustainable Development.* New York: UNDP, 2009.

United Nations Development Programme Regional Bureau for Arab States. "Arab Statistics: Qatar." http://www.arabstats.org (Accessed July 2011).

World Resources Institute. "World Resources Climate Analysis Indicators Tool (CAIT): Qatar." http://cait.wri.org (Accessed July 2011).

Quaternary Period

The Quaternary period, the most recent geologic interval, represents the last 1.8 million years of time. Its most striking feature is that the Earth had cold polar regions, which led to periodic development of continental glaciers. The prolonged ice ages, comprising the main part of the Quaternary interval, ended 10,000 years ago, when the continental glaciers had melted. Compared to the many cold periods (glacial intervals) of the Quaternary, the last 10,000 years of the present interglacial has been comparatively warm.

The changes in climate over the past 150 years have shown a varied history. The world endured a historic cool period during the late 1800s (the Little Ice Age), followed by a warm period of the 1930s (the Dust Bowl years), and since about 2000, the climate has been more variable, reaching extremes in warmth at high latitudes. Reports from Vikings show that 1,000 years ago, the climate of Greenland and Iceland was warmer than today (the Medieval warming). While these historic shifts in temperature are relatively moderate, much larger changes in temperature on Earth have occurred in the geologic past.

Proxy Data

Using the changing oxygen isotope ratios from ice cores and marine sediment cores, scientists have discovered global changes, recorded synchronously over a wide range of latitudes. One long climate proxy record was taken in the Antarctic, the Vostok ice core. From it, the inferred temperature over a long interval is based on the temperature-sensitive ratio of oxygen isotopes, ^{18}O to ^{16}O. In the Vostok ice core, isotopes of oxygen have been used to develop Earth temperature histories extending over 400,000 years. Trapped gas bubbles record the history of atmospheric CO_2 concentrations for this period (data from the National Climatic Data Center, Asheville, North Carolina). Because the isotope ^{18}O is heavier than ^{16}O, the proportions of each vary depending on the climate region. In alpine areas such as in the Alps of Switzerland and in the polar regions, ^{18}O is more abundant, while in the lowlands of the middle and low latitudes, ^{16}O prevails in the atmosphere. By calibrating these, scientists can use the proportions of oxygen isotopes taken from ancient sediments or ice cores as an index of average annual temperature.

The changing proportions of isotopes can be matched and dated with a geomagnetic signal of polar reversals (the Earth's poles changed their magnetic signals), a known time scale based on

A glacier in New Zealand. The Quaternary period records a series of extreme climate changes in Earth's history. Plants and animals that had dispersed to lower latitudes and in areas of southerly environments during glacials dispersed northward again during the postglacial warming. The prolonged ice ages, comprising the main part of the Quaternary interval, ended 10,000 years ago. The continental glaciers melted during a comparatively warm cold period, but some have remained.

isotopically dated magnetic signals that are worldwide. Another source of long climate records are the deep-sea sediment cores from which oxygen isotopes can be extracted from calcareous plankton (foraminifera). These data carry the paleoclimatic records back through more than 60 million years through the Tertiary period and the time of the last dinosaurs.

The ice ages, which comprised the main part the last million years, was a globally cold period with increasingly variable swings of climate. The glacial pattern continued over long intervals, with only a few comparatively short warm or interglacial periods. The last major glaciation came in two parts; in the United States, these are called the early Wisconsin (80,000–28,000 years before present, oxygen isotope stage 4) and the Full Glacial or late Wisconsin (23,000–15,000 years ago, oxygen isotope stage 2); between these a somewhat warmer middle Wisconsin period occurred 28,000–23,000 years ago. The maximum of the last major glaciation occurred about 18,000–15,000 years ago (called the Full Glacial). After 15,000 years ago, a global warming began, and continental ice sheets melted by about 10,000 years ago. The period after 10,000 years ago, or postglacial, is called the Holocene or Recent period representing the present interglacial.

The period of the last glaciation (Full Glacial) brought continental ice down to the middle latitudes in both hemispheres. In Europe, ice from the Scandinavian highlands spread southward over the Netherlands and mountain glaciers covered the Alps. Ice stood over parts of northern Siberia, Greenland, and parts of Alaska. Permafrost ice developed underground in Siberia, northern Canada, and Alaska as much as 300 ft. (91 m) thick. Equivalent ice expansions occurred in the Southern Hemisphere.

As temperatures warmed between 15,000 and 9,000 years ago, the average annual temperatures at mid-latitudes increased by about 11 degrees F (6 degrees C). At Lamont National Observatory, scientists estimated that the difference in solar insolation between the Full Glacial and the mid-postglacial was about 8 percent.

An overall trend within the Quaternary is apparent. In the first half, the amount of variance from year to year was fairly low, while in the last half of the Quaternary, variance became more and more extreme. Looking back over records of the past million years, geologists estimate that there were at least 40 cold or glacial intervals, and these were of varying length and were not regular in occurrence. In the last 400,000 years of the ice ages, proxy data indicate that there were four extreme cool periods (glacials) interspersed with five variable warm periods (interglacials). The last major interglacial (spanning the interval of about 80,000–122,000 years before present) was similar in warmth to the present interglacial climate, but the previous interglacials were definitely cooler than the present one.

Causes

The main climatic changes of the Quaternary are linked to the orbital position of Earth in relation to the sun. Astronomer Milutin Milankovitch proposed the orbital theory that is established by a variety of evidence, starting with tree ring variation. The precession of the equinox, obliquity of the Earth's orbit around the sun, and eccentricity of the Earth's orbit all contribute to the level of insolation received by Earth and are therefore the cause of geologic shifts in climate. Before the Quaternary was the warm Tertiary period that brought tropical conditions to the mid-latitudes, and extensive temperate forests grew in the Northern Hemisphere. The north and south polar areas were at least 36 degrees F (20 degrees C) warmer than today, perhaps around 41 degrees F (5 degrees C) average annual temperature. Orbital causes for this part of Earth's history have not been specified, but probably were important factors.

Impacts

Global changes in climate forced major changes in plant and animal distributions, especially in the high and mid-latitudes. During the Full Glacial, remains of trees that now grow in the northern boreal forest were found in Tennessee where their fossils were associated with deciduous hardwoods and even cypress-swamp types, creating strange mixtures of genera. Plants and animals that had dispersed to lower latitudes and in areas of southerly environments during glacials dispersed northward again during the postglacial warming. Thus, the Quaternary period records some of the most extreme climate changes known in Earth's history.

Estella B. Leopold
University of Washington

See Also: Ice Ages; Little Ice Age; Milankovitch, Milutin; Orbital Parameters, Eccentricity; Orbital Parameters, Obliquity; Orbital Parameters, Precession.

Further Readings

Delcourt, H. R. and P. A. Delcourt. *Quaternary Ecology: A Paleoecological Perspective*. London: Chapman & Hall, 1998.

Graham, Alan. *Late Cretaceous and Cenozoic History of North American Vegetation*. Oxford: Oxford University Press, 1999.

Moore, P. D., W. G. Chaloner, and P. Scott. *Global Environmental Change*. Oxford: Blackwell Science, 1996.

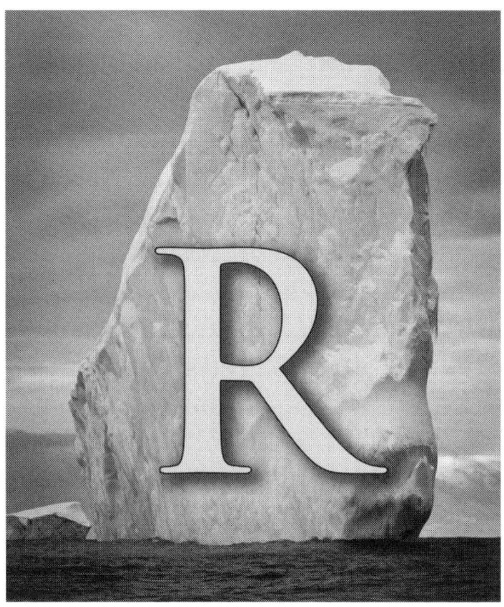

Radiation, Absorption

Radiation is energy transmitted by electromagnetic waves. Electromagnetic waves travel at the speed of light (when passing through a vacuum) and have a characteristic wavelength, λ, which is inversely proportional to their frequency, ν, by

$$\lambda\,(m) = c\ (m\ s\text{-}1)\ /\ \nu\ (s\text{-}1),$$

where c is the speed of light. Electromagnetic radiation is conceptualized in contemporary theory both as a wave and as a stream of particles called photons (this dual approach is referred to as wave-particle duality). The energy of any photon, E, of radiation is inversely proportional to the wavelength by

$$E = h\,\nu,$$

where h is Planck's constant. This relationship allows scientists to order electromagnetic waves from high energy/short wavelength (e.g., x-rays), to low energy/long wavelength (e.g., radio waves). The resulting progression is referred to as the electromagnetic spectrum (Figure 1). The visible region of the electromagnetic spectrum is bound by infrared (IR) radiation on the lower energy side of the visible region (around 1 µm to 1 mm in wavelength), and by UV radiation (UV) on the higher energy side (from 400 nm to 1 nm). Microwave radiation is slightly lower in energy than IR, with a wavelength of around 1 cm.

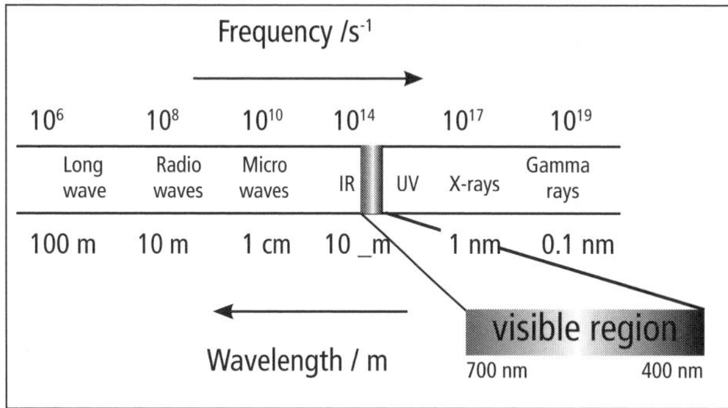

Figure 1 The electromagnetic spectrum

All objects both emit and absorb radiation. Although all objects emit radiation at all wavelengths, the frequency of maximum emission, λmax, is proportional to the temperature of the object by Wien's law

$$\lambda max = \alpha\ /\ T,$$

where α is a constant equal to 2897 µm K. This implies that hotter objects emit higher energy radiation, as would be expected from everyday experiences. From Wien's law, the surface temperature of the sun can be calculated based on its emission peak at ~0.5 µm (green light) to be around 5800 K. The average temperature of the Earth's surface is around 18 degrees C (290 K), which corresponds to a peak emission at around 10 μm, in the infrared to microwave region.

There are three basic modes of motion: translational (movement through space), rotational, and vibrational. These are important, because along with electronic energy, they are the ways in which gas molecules can store energy. Quantum theory dictates that energy levels are discrete, not continuous; this implies that molecules will only absorb discrete frequencies of radiation that correspond to the gap between a high and lower energy state. UV radiation corresponds to the gap in energy between electronic energy levels in a molecule. When a molecule absorbs UV radiation, it may be promoted to an electronically excited state. In the general, this will make the bonds holding the atoms together weaker and may help facilitate reactions or the breakup of molecules. For example, the reactions that complete the Chapman mechanism in the stratosphere:

(i) O2 + hv -> O + O (λ< 240 nm)
(ii) O+O2 + M -> O3 + M
(iii) O3 + hv -> O2 + O* (λ < 320 nm)
(iv) O* + M -> O + M

The Chapman mechanism is chemically a null cycle, but it is important in the context of life, as it prevents most of the high-energy radiation below 320 nm from reaching the Earth's surface. IR and microwave radiation, being lower in energy than UV, correspond to the gaps between rotational and vibrational energy levels, respectively. Quantum theory dictates that molecules interact with IR/microwave radiation only when two conditions (selection rules) are met:

- The energy of the radiation corresponds to the energy gap between two of the discrete energy levels in the molecule
- The resulting motion causes a change in the dipole moment (electron distribution) of the molecule

This implies that atmospheric components that are symmetrical molecules, such as N_2 and O_2 that have an even electron distribution and cannot be rotated, bent, or stretched in such as way as to create one, do not absorb microwave and IR radiation. Molecules possessing a dipole, which can be altered by bending and stretching, absorb radiation. Atmospheric components that fill this requirement include CO_2, CH_4, H_2O, and CFC's, for example, the asymmetric stretch of CO_2. These gases are collectively known as greenhouse gases. Radiation from the sun is received at the Earth's surface, mainly in the UV and visible region, with other frequencies cut out by the atmosphere and electromagnetic field.

Carl Palmer
University of Cape Town

See Also: Chemistry; Radiation, Infrared; Radiation, Long Wave; Radiation, Short Wave; Radiation, Ultraviolet.

Further Readings

Henrikson, Thormod and David H. Maillie. *Radiation and Health*. Oxfordshire, UK: Taylor and Francis, 2002.
Mahan, James R. *Radiation Heat Transfer.* Hoboken, NJ: Wiley & Sons, 2002.

Radiation, Infrared

Infrared radiation is the part of the electromagnetic spectrum popularly (but not entirely accurately) conceptualized as heat. The IR region covers wavelengths that span nearly three orders of magnitude; it is conventional therefore to break this down into further subgroups (Table 1).

The main application of IR radiation in relation to climate change is in the fields of meteorology and climatology. Satellite measurements of IR radiation received from the Earth can be used to derive cloud types and heights; these are used for weather forecasting, but knowledge of the number and type of clouds present is useful in calculating the Earth's radiative budget. It is possible to use the IR radiation returned to space from the Earth to measure land and sea surface temperature, both of which are important parameters in calculating the Earth's heat budget and in monitoring global climate

Table 1 Conventional breakdown of IR radiation

Name	Wavelength	Comments
Near-infrared (NIR)	0.75–1.4 μm	Absorbed by water and commonly used in fiber-optic technology.
Short wave-length infrared (SWIR)	1.4–3 μm	Strongly absorbed by water and used in long-range telecommunications.
Mid wave-length infrared (MWIR)	3–8 μm	Used in heat-seeking missile technology.
Long wave infrared / Far IR (LWIR)	8–1000 μm	Not absorbed by water, therefore used for thermal-image sensors.

change. Satellite measurements of IR are also employed to determine cloud height and rates of convection. IR radiation also finds a variety of industrial, military, other scientific, and domestic applications, for example:

- *Night vision*: Night vision devices use a photon muliplier to amplify the signal from the available ambient light which is then augmented with the IR radiation.
- *Thermal imaging*: Noncontact, nondestructive technique that generates a false-color thermal image of a subject, finding use in a wide range of industries.
- *Heating*: IR lamps can be used for heating; examples include for frozen aircraft wings and for patio heaters.
- *Spectroscopy*: IR spectroscopy (also called rotational spectroscopy) is a technique used by chemists for the identification of molecules and for elution of chemical structure.

Carl Palmer
University of Cape Town

See Also: Radiation, Absorption; Radiation, Microwave; Radiation, Ultraviolet.

Further Readings

Guelachvili, Guy and K. Ramamohan Rao. *Handbook of Infrared Standards*. Oxford: Elsevier Science and Technology, 1986.

Henrikson, Thormod and David H. Maillie. *Radiation and Health*. Oxfordshire, UK: Taylor and Francis, 2002.

Radiation, Long Wave

Long wave radiation is the part of the electromagnetic spectrum emitted at spectral wavelengths generally greater than 1 micrometer (μm). Types of long wave radiation include infrared, microwave, and radio waves. Emittance of radiation is a function of temperature, and objects giving off long wave radiation are colder than those radiating at short wavelengths. For example, the sun (approximately 5800 K) radiates primarily in the short wave part of the spectrum (especially visible light from 0.4 to 0.7 micrometer), whereas the Earth (approximately 290 K) emits radiation at much larger wavelengths. Climatologically, long wave radiation generally refers to radiation emitted by the Earth-atmosphere system (also called terrestrial radiation), largely at wavelengths of 5–15 μm. Long wave radiation emitted by the Earth's surface and atmosphere falls primarily within the thermal infrared ("below the red") region of the electromagnetic spectrum. It can be sensed through the sensation of heat.

Counterradiation

In the Earth-atmosphere system, short wave radiation from the sun is absorbed and converted to long wave radiation. Various components of the Earth-atmosphere system absorb the incoming short wave radiation; among those are the Earth's surface, gas and dust molecules in the atmosphere, and clouds. Long wave radiation is then reradiated from those components, after which it is referred to as outgoing long wave radiation or counterradiation. Counterradiation may be reabsorbed (and reradiated) by those very same components that initially absorbed short wave radiation. This process is behind the greenhouse effect.

Globally, the Earth's atmosphere is relatively transparent to radiation between 8 and 15 μm. This atmospheric window allows much long wave radiation to be lost to space. However, the window may be closed locally by the presence of large amounts of water vapor or clouds. Additionally, increasing amounts of greenhouse gases can also potentially close this window. Thus, the role of long wave radiation in the greenhouse effect is fundamental. In the absence of an atmosphere containing long wave–absorbing greenhouse gases (e.g., water vapor, or carbon dioxide), the Earth's average temperature would be approximately 0 degree F, (minus 18 degrees C or 255 K). However, because of the efficiency with which greenhouse gases reabsorb counterradiation, the Earth's average temperature is 59 degrees F (15 degrees C, or 288 K).

Water vapor and carbon dioxide are the most abundant greenhouse gases by volume. They are particularly effective at absorbing counterradiation. The amount of water vapor in the atmosphere is a direct response to temperature. Other long wave–absorbing gases can be produced by human activity. Carbon dioxide and certain other trace gases (such as methane, nitrous oxide, and chlorofluorocarbons) can potentially upset the long wave exchanges involved in the Earth's energy balance. Such deviations can lead to changes in the average global temperature. Potential consequences of an escalating amount of these atmospheric gases include a growing proportion of outgoing long wave radiation being "trapped" in the atmosphere, leading to an increase in the Earth's temperature.

Clouds serve as very effective absorbers of outgoing long wave radiation. This, in turn, affects surface and near-surface temperatures. The presence of clouds in a nighttime sky results in warmer temperatures than a cloudless one. In the absence of incoming short wave radiation, only long wave exchanges occur at night. With clouds trapping much of the outgoing long wave radiation, substantial amounts are redirected back toward the surface. This increases the near-surface temperatures. Similarly, the lack of clouds leads to greater amounts of long wave radiation escaping the Earth-atmosphere system to space.

Petra A. Zimmermann
Ball State University

See Also: Albedo; Radiation, Microwave; Radiation, Short Wave; Radiative Feedbacks.

Further Readings

Ahrens, C. D. *Meteorology Today.* Belmont, CA: Thomson Brooks/Cole, 2007.

Oke, T. R. *Boundary Layer Climates.* London: Routledge, 1987.

Robinson, P. J. and Ann Henderson-Sellers. *Contemporary Climatology.* Upper Saddle River, NJ: Prentice Hall, 1999.

Radiation, Microwave

The existence of microwaves was first postulated by James Clerk Maxwell in 1864 and confirmed by the experiments of Heinrich Hertz some 20 years later. Microwaves are subdivided into categories as listed in Table 1.

Microwaves are most well known in popular culture for their role in heating food (in a microwave oven), and for their use in mobile telecommunications such as mobile phones and wire-

Table 1 Microwave frequency bands as defined by the radio association of Great Britain

Designation	Wavelength *
L band	15 cm – 30 cm
S band	8 cm – 15 cm
C band	3.75 cm – 8 cm
X band	2.50 cm – 3.76 cm
Ku band	1.7 cm – 2.50 cm
K band	1.1cm – 1.7 cm
Ka band	0.75 cm – 1.1 cm
Q band	0.6 cm – 1 cm
U band	0.5 cm – 0.75 cm
V band	0.4 cm – 0.6 cm
E band	0.33 cm – 0.5 cm
W band	0.3 cm – 0.4 cm
F band	0.2 cm – 0.33 cm
D band	0.27 cm – 0.18 cm

* The definition is originally in units of GHz and has been converted here to cm for comparison to the other measurements. This has led to some rounding.

less networking. Microwave ovens work by passing S-band radiation though food, which excites water, sugar, and fat molecules. These molecules in turn reradiate in the infrared to heat the food. With the rapid expansion of mobile phone technology, many studies have been done to look at the effects of microwave radiation on human health. The studies show mixed results and the general conclusion is that only doses high enough to heat up tissue are likely to have negative impacts.

Carl Palmer
University of Cape Town

See Also: Radiation, Absorption; Radiation, Ultraviolet; Technology.

Further Readings

Henrikson, Thormod and David H. Maillie. *Radiation and Health*. Oxfordshire, UK: Taylor and Francis, 2002.

Mahan, James R. *Radiation Heat Transfer: A Statistical Approach*. Hoboken, NJ: Wiley and Sons, 2002.

Willert-Porada, Monika. *Advances in Microwave and Radio Frequency Processing: 8th International Conference on Microwave and High-Frequency Heating*. New York: Springer-Verlag, 2006.

Radiation, Short Wave

Radiation traveling in waves shorter than 1 micrometer (μm) is characterized as short wave, and includes gamma rays, x-rays, ultraviolet light, and visible light. Climatologically, short wave radiation commonly refers to the incoming radiation from the sun. There is an inverse relationship between the temperature of an object and the wavelengths at which it primarily emits. Because the sun is a hot object (approximately 5800 K), it emits radiation at short wavelengths. Since shorter wavelengths carry more energy than longer ones, they are more intense.

Most of the short wave radiation emitted by the sun is in the visible region of the electromagnetic spectrum, which spans from 0.4 μm (violet) to 0.7 μm (red). The sun's wavelength of maximum emission is found at 0.5 μm.

The amount of emitted solar radiation that reaches the Earth decreases inversely with the square of the distance between the Earth and sun, by the inverse square law. The total of short wave radiation that reaches the top of the Earth-atmosphere system is called the solar constant. Although the amount varies slightly throughout the year, because of the elliptical nature of the Earth's orbit around the sun, the solar constant averages around 1370 W m^{-2}.

Because the eccentricity of the Earth's orbit varies between nearly 0 to 5 percent (with a periodicity of 110,000 years), the amount of short wave radiation received increases or decreases over time from present values. The current orbit is nearly circular, leading to little seasonal variation of incoming short wave radiation. However, a more highly elliptical orbit would render a difference of up to 30 percent between aphelion (maximum Earth-sun distance) and perihelion (minimum Earth-sun distance). Incoming solar radiation can take several avenues once it enters the atmosphere. Over a year, 30 percent of total short wave radiation is reflected back to space, either by gas molecules or other particles in the atmosphere, by clouds, or by the Earth's surface; this is the Earth's albedo (the proportion of radiation that is reflected from a surface). The atmosphere and clouds absorb an additional 20 percent of short wave radiation. Approximately 50 percent strikes the surface,

Optically thick clouds reflect more short wave radiation back to space than the darker surface would without the cloud. Therefore, less solar energy is available to heat the surface and atmosphere, which tends to cool the Earth's climate.

where it is absorbed. These values may vary locally and at shorter time scales.

Some of the short wave radiation reaching the Earth's surface arrives directly, unimpeded by clouds or atmospheric constituents; this is direct radiation. Some is scattered about the atmosphere and arrives at the surface indirectly. This is diffuse radiation. Diffuse radiation is a product of scattering, which occurs when short wave radiation strikes small particles in the atmosphere, including gas molecules. Upon impact, the radiation is scattered omnidirectionally. Some short wave radiation will be scattered toward the surface. Energy received at the Earth's surface is the total amount of direct and diffuse radiation.

The presence and type of clouds in the atmosphere reduces the amount of short wave radiation reaching the Earth's surface. Thin clouds, such as cirrus, have lower albedos than thick ones, such as cumulus. Once short wave solar radiation reaches the surface and is absorbed, it is converted to long wave forms of radiation.

Short wave radiation impacting the Earth also includes ultraviolet light. These wavelengths are shorter than visible radiation, so ultraviolet light carries more energy than visible light. Ultraviolet radiation is classed into three categories: UV-A (0.32–0.40 μm), UV-B (0.29–0.32 μm), and UV-C (0.20–0.29 μm). Excessive exposure to UV-A and UV-B radiation has been linked to skin cancers and skin damage. UV-C radiation is largely absorbed by stratospheric ozone. Absorption of ultraviolet radiation breaks ozone down into atomic and molecular oxygen. The presence of stratospheric ozone helps protect the Earth's surface from the damaging effects of this form of short wave radiation.

Petra A. Zimmermann
Ball State University

See Also: Radiation, Ultraviolet; Sunlight.

Further Readings

Ahrens, C. D. *Meteorology Today*. Belmont, CA: Thomson Brooks/Cole, 2007.
Hartmann, D. L. *Global Physical Climatology*. Maryland Heights, MO: Academic Press, 1994.
Stull, R. B. *Meteorology for Scientists and Engineers*. Monterey, CA: Brooks/Cole, 1999.

Radiation, Ultraviolet

Ultraviolet radiation was discovered as a result of the observation that silver salts darken on exposure to sunlight. In 1801, German physicist Johann Wilhelm Ritter first observed that invisible electromagnetic radiation was responsible for this darkening. These rays eventually became known collectively as UV, so named as this radiation is immediately beyond violet in the electromagnetic spectrum. This implies that UV is more energetic than visible light. Conventionally, UV radiation is broken down into further subdivisions as shown in Table 1.

In humans, UV radiation is important for health. UV-B has health benefits, as it is responsible for the production of vitamin D. A deficiency of vitamin D is thought to lead to a range of cancers and also to osteomalacia (the adult equivalent of rickets), with symptoms ranging from painful bones to brittleness and fractures.

Exposure to UV radiation also causes the skin to release a pigment (melanin), giving the skin a darker color that is regarded as healthy in most Western cultures. Melanin provides some protection against the more harmful effects of UV exposure. However, exposure to UV radiation also has a range of negative health effects. The most widely publicized of these is the link between exposure

Table 1 Conventional subdivisions of UV radiation

Name	Wavelength	Comments
1. Near	400 nm – 200 nm	Referred to as 'Blacklight'
UV-A	400 nm – 320 nm	Strongly absorbed by O_3
UV-B	320 nm – 280 nm	Strongly absorbed by O_3
UV-C	290 nm – 200 nm	Strongly absorbed by O_2
2. Far	200 nm – 31 nm	Strongly absorbed by O_2
3. Extreme	31 nm – 1 nm	

to UV radiation and skin cancer. UV radiation is strongly absorbed by DNA, the cellular molecule responsible for the transfer of hereditary information. The absorption of UV radiation causes chemical bonds in the DNA to be broken and reformed in the wrong order. This can lead to mutations and cancerous growths. UV radiation is also harmful to the eyes, leading to short-term uncomfortable conditions such as arc eye or to more serious conditions such as cataracts. There may also be a link between excessive UV exposure and poor immune response. UV radiation's range of industrial and domestic applications include the following:

- *Astronomy*: Many hot objects in the universe emit large amounts of UV radiation and are therefore better observed in the UV region. However, as the atmosphere absorbs a lot of this UV, these observations are generally only made from space.
- *Spectrophotometry*: A widely used technique in analytical chemistry to determine chemical structure.
- *Analyzing minerals*: Many minerals glow characteristic colors under a UV lamp, aiding identification.
- *Sterilization of surface and drinking water*: UV radiation is effective at killing pathogens and is used to sterilize critical workspaces (such as biochemistry labs) as an alternative to chlorination. Methods to sterilize water based on using UV from sunlight may provide a carbon neutral solution to domestic water treatment.

Carl Palmer
University of Cape Town

See Also: Chemistry; Geography; Health; Radiation, Absorption; Radiation, Infrared; Pollution, Water; Sunlight.

Further Readings

Grant, W. B. "An Estimate of Premature Cancer Mortality in the U.S. Due to Inadequate Doses of Solar Ultraviolet-B Radiation." *Cancer*, v.94/6 (2002).

Henrikson, Thormod and David H. Maillie. *Radiation and Health*. Oxfordshire: Taylor and Francis, 2002.

Radiative Feedbacks

While most of the feedbacks at play in the Earth's ecosystem are subject to the closed system of matter, radiative feedbacks are those that deal with the open system in which the sun's energy is transferred to the Earth and then absorbed or reflected back, to varying degrees. Because the amount of absorbed or reflected solar energy is the principal component in the planet's temperature (the other factors then determine what happens to that heat, but the solar energy provides the initial quantity), these feedbacks are key to understanding and modeling global climate. Feedbacks are relationships found in complex systems, in which the output (or result) of the system is returned to the input. For instance, as summers get hotter, people run their air conditioners longer, releasing more chlorofluorocarbons, accelerating global warming and causing hotter summers. That is a simple example of a positive feedback often used in schools. The radiative feedbacks in question involve principally clouds, ice, and water vapor.

A critical feedback in global warming is the albedo-ice feedback. Though the melting of the polar ice caps is often associated with rising sea levels, it is only one of the important effects, and simply the most vivid to illustrate. The more complicated effect is in the change it enacts on how much solar energy is retained by the Earth. Ice is far more reflective than land or liquid water, and as the polar ice melts, the area of the surface that it occupied is replaced by one of those two things. As a result, more sunlight is absorbed by the surface instead of being reflected, which then warms the poles further, melting more ice.

Clouds are also critical. Clouds act as a sort of imperfect barrier, absorbing a limited amount of heat moving in either direction, heat emanating from the Earth into space, and sunlight shining on the Earth. Models disagree on whether their overall feedback effect on global warming will prove positive or negative. On the one hand, the evaporative feedback in general accelerates global warming: as the temperature increases, the capacity of the air to hold moisture increases exponentially, so as water evaporates into the warm air, it is able to stay there. Since water vapor is a greenhouse gas, as it accumulates, it causes the temperature to get warmer and warmer. On the other hand, sufficiently great cloud cover shields the Earth from solar radiation, and cloudy days aren't as hot. High-enough clouds can reflect sunlight back down on the Earth, however, more than balancing out the sunlight they have blocked. Water vapor also retains heat better than the atmosphere does; a wetter atmosphere loses heat more slowly. As more and more water evaporates, more and more heat from the sun can be retained, which makes for a warmer Earth and greater evaporation.

It is possible that the effects of positive radiative feedbacks on global warming have been muted until recently by global dimming, the gradual reduction of the Earth's irradiance (emitted radiation, including heat and light) in the second half of the 20th century, which has reversed in the 21st century. Global dimming is probably caused by the coalescing of water vapor around anthropogenic particles in the air, the product of pollution and other industrial activities. The resulting water droplets form a little differently than they would otherwise, and make more highly reflective clouds. The Clean Air Act and other antipollution efforts are presumed to be responsible for ending global dimming. The apparent recent increase in global warming may be because that dimming had been reducing global warming's effects by limiting the amount of sunlight entering the system.

Bill Kte'pi
Independent Scholar

See Also: Albedo; Atmospheric Absorption of Solar Radiation; Biogeochemical Feedbacks; Climate Feedback; Cloud Feedback; Dynamical Feedbacks; Evaporation Feedbacks; Ice Albedo Feedback; Radiation, Absorption; Sunlight.

Further Readings

Buesseler, K. O., et al. "Revisiting Carbon Flux Through the Ocean's Twilight Zone." *Science*, v.316/5824 (2007).

Kenneth, J. P., et al. *Methane Hydrates in Quaternary Climate Change: The Clathrate Gun Hypothesis.* Washington, DC: American Geophysical Union, 2003.

Meehl, G. A., et al. "How Much More Global Warming and Sea Level Rise?" *Science*, v.307/5716 (2005).

Torn, M. and J. Harte. "Missing Feedbacks, Asymmetric Uncertainties, and the Underestimation of Future Warming." *Geophysical Research Letters*, v.33/10 (2006).

Rain

Rain is liquid precipitation. It is caused when the water vapor in the atmosphere condenses into raindrops, either because of a drop in temperature, or an increase in the amount of water vapor in the body of air. Rainfall is often associated with specific geographical locations, where moving bodies of air meet other colder bodies of air, rise, and then cool. This happens in most coastal areas and with small islands. The areas of the world that receive the most rainfall include Mount Waiaeleale in Hawai'i and Cherrapunji in Meghalaya, while deserts (areas of especially low precipitation such as rain) tend to be located in the interior of large land masses.

Raindrops tend to be spherical and reach around 4 mm in diameter. Larger drops tend to break up while falling; smaller raindrops, those with a diameter of 0.5 mm or less, are referred to as drizzle. The density of raindrops in the atmosphere ranges from around 100–1,000 per cubic meter (cu. m), with higher densities recorded for drizzle. The rate at which rain falls, measured in millimeters per hour, determines whether it is classified as light rain (less than 2.5 mm per hour), moderate rain (2.5–7.5 mm per hour), and heavy rain (7.6 mm per hour).

A combination of geographic and climatological variables determines the type and extent of rainfall at any particular point in the world; most areas have a fairly predictable amount and extent of rain, which tends to be seasonal in nature. Across most of Europe, South America, central Africa, and the eastern part of North America, rainfall reaches an excess of 500 mm per year. The amount is lower in most of Asia; it is lower than 250 mm in a broad strip of land reaching from Mongolia in the east to Arabia in the west. This lower level is also observed in central Australia, intermontane North America, southwest Africa, and the central western coast of South America. In contrast, annual rainfall can exceed 2,500 mm in parts of Myanmar, the west coast and Assam regions of

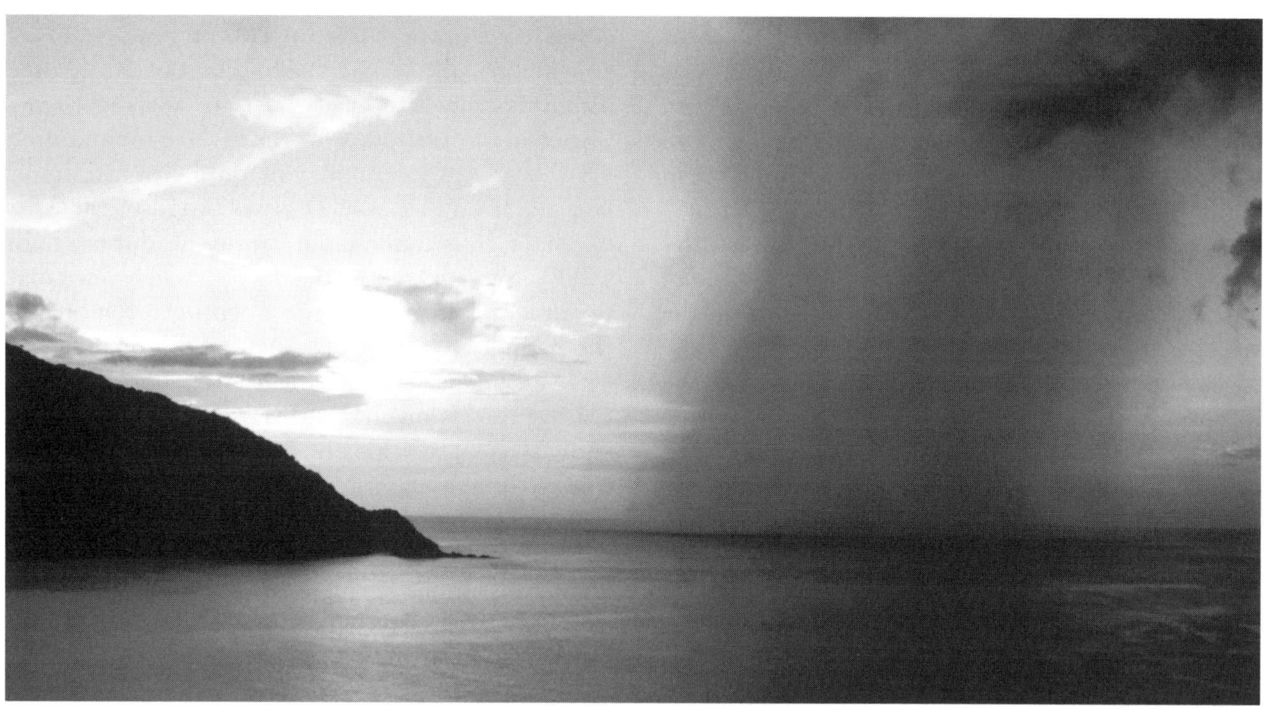

A patch of rain moving through Tobago in the Caribbean. It is yet unclear whether individual episodes of unusual weather are the result of climate change. However, up to early 2011, extreme weather incidents trended upward, with unusual rainfall incidents among these episodes—pointing to a pattern of climate change. Increased amounts of rain in some areas are countered by decreased levels in other areas, which is beginning to have a significant impact on agricultural productivity and may lead to water shortages in some areas.

India, and the windward slopes of coastal western North America. In addition to the total level of rainfall, differences exist in the persistence of rain and its intensity. Much of the tropical world is part of a region that is subject to the monsoon, a wind system that brings tropical rainstorms in a largely predictable season of several months, outside of which there is minimal precipitation. In parts of western Europe, on the other hand, rain is possible throughout the year and is evenly distributed among the different seasons. In desert regions, including the Sahara, for example, there may be no rain at all for several years before a sudden and unpredicted storm briefly rages. Irrespective of how and when it falls, rain is essential to the livelihoods of most of the world's population, because of its contribution to water supplies and agriculture. Significant changes to known rainfall patterns would have potentially devastating consequences for the majority of humanity and the other living creatures of the world.

In regions that are susceptible to cyclones, much of this rain can fall in a very short and potentially dangerous period of time. This type of excessive rainfall can lead to flooding and contribute to landslides. Flooding can threaten lives and destroy property, including crops, that lead to subsequent hunger and the outbreak of waterborne diseases. However, in some cases, regular and controllable flooding, such as historical flooding of the River Nile, is an important part of the agricultural system through spreading fertile land in an area of otherwise limited agricultural productivity.

Global Climate Change and Rainfall Patterns

The impact of global climate change on rainfall and its long-term consequences tend to have been more intensively studied in the Northern Hemisphere, at least in part because the majority of the world's developed economies and their researchers and universities are located there. However, research in tropical and Southern Hemisphere locations has also indicated that changes are taking place in these regions. In general, it is difficult to determine which effects are more important over the long-term and which represent challenges that can be overcome. A study of rainfall changes in Madagascar, for example, found that variations in precipitation and cyclone activity can have an impact on the fecundity of lemurs and make them vulnerable to population reduction. Because so many of the processes of nature involve complex interactions between a wide range of plants and animals, it is very difficult to accurately predict what an impact on one species will have on the wider ecosystem it inhabits. This kind of empirical work is very important in seeking to make the link between the results predicted by global change models and the results found at the local or regional level.

It is also difficult to determine whether individual episodes of unusual weather are the result of climate change. However, throughout the first half of 2011, in common with most recent years, the number of extreme weather incidents continued to increase and unusual rainfall incidents were among these episodes. Over the course of time, however, those incidents are leading to a pattern of definite change. In the case of rain, increased amounts in some areas are countered by decreased levels in other areas, which is beginning to have a significant impact on agricultural productivity. The United Nations estimates that, under current trends, by 2020 there will be between 75 and 250 million people in Africa facing water shortages because of changes in rainfall, and rain-fed agricultural yields could fall by as much as 50 percent in some countries. Those most seriously affected will be the poorest and the most vulnerable, as well as those living in areas without effective or compassionate government agencies. The famine caused by drought in east Africa in 2011 is most severely affecting Somalia and leading to hundreds of deaths per day because of the failed nature of the government and the effects of years of fighting and poverty. Somali citizens have been suffering more than their neighbors in Ethiopia and Eritrea, which have been equally affected by the drought, but at least have access to some national and international sources of aid.

Artificial Rain

Since the beginning of the 20th century, various agencies have made systematic attempts to artificially alter patterns of precipitation, particularly of rainfall. Manipulating the level of rain is considered particularly important because of its role in agriculture and in avoiding drought. Some attempts have also been made for the short-term dispersal of rainclouds for the sake of maintaining good weather for important public occasions. It was reported in *China Daily* in August 2008, for example, that Chinese authorities fired more than 1,000 cloud-dispersal rockets to try to ensure that no rain would disturb the opening ceremony of the Beijing Olympics.

The evidence for the efficacy of artificially stimulating rain is mixed, and statistical tests tend to show that large-scale efforts are generally ineffective, although short-term and short-range movement of rainfall may be possible. However, under current levels of technology, it is unlikely that artificial stimulation will be of significant use in mitigating the impact of climate change on rainfall patterns.

John Walsh
Shinawatra University

See Also: Agriculture; Clouds, Cumulus; Cyclones; Drought; Floods; Geoengineering; Hurricanes and Typhoons; Livelihoods; Plants; Precipitation; Vulnerability.

Further Readings

Biles, Peter. "Horn of Africa Drought: Why Is Somalia Worst Affected?" BBC News Online (July 14, 2011). http://www.bbc.co.uk/news/world-africa-14143562. (Accessed July 2011).

China Daily. "Beijing Disperses Rain to Dry Olympic Night" (August 9, 2008). http://www.chinadaily.com.cn/olympics/2008-08/09/content_6919493.htm (Accessed July 2011).

Dunham, Amy E., Elizabeth M. Erhart, and Patricia C. Wright. "Global Climate Cycles and Cyclones: Consequences for Rainfall Patterns and Lemur Reproduction in Southeastern Madagascar." *Global Change Biology*, v.17/1 (January 2011).

United Nations. "The Science." http://www.un.org/wcm/content/site/climatechange/pages/gateway/the-science (Accessed July 2011).

Von Storch, Hans, Eduardo Zorita, and Ulrich Cubasch. "Downscaling of Global Climate Change Estimates to Regional Scales: An Application to Iberian Rainfall in Wintertime." *Journal of Climate*, v.6 (June 1993).

Rainfall Patterns

A rainfall pattern describes the distribution of rain geographically, temporally, and seasonally. The tropics receive more rainfall than deserts on a more regular basis. Cooler places like the poles receive no rainfall as the moisture is converted to snow before it falls to the ground. Rainfall occurs more often during particular times of the year, creating rainy seasons. In other seasons, rainfall is scant. Worldwide, rain-fed agriculture is planned based on rainfall's natural pattern. Water storage, irrigation networks, and urban water-supply systems are designed according to the average annual rainfall. A significant amount of rain on a continuous basis for a long time increases the possibility of flood and subsequent disaster to infrastructure. No or little rainfall for a longer period (years) in an inhabited area could lead to drought and famine.

Changing rainfall patterns are a consequence of global warming, which causes a change in ocean thermohaline circulation. The world's agriculture, especially third-world agriculture, depends upon seasonal rainfall patterns. Recent erratic changes in rainfall patterns lead to low agriculture production, thus creating food insecurity for ever-increasing world populations. Untimely floods, droughts, and famine are the consequences of these unassuming patterns.

Global Warming and Rainfall Patterns

Global warming leads to a near-term collapse of ocean thermohaline circulation, a global ocean circulation pattern that distributes water and heat both vertically through the water column and horizontally across the globe. Because of this collapse, warm surface waters move from the tropics to the North Atlantic and extra-warm water surfaces in the Pacific Ocean surrounding the equator. Thus, western Europe, some parts of Asia, and many parts of the Americas become warmer than average and some parts of Europe become rapidly cooler. El Niño and La Niña are examples of this phenomenon. The latest deviant trend is generating dramatic weather impacts such as rapid cooling in some parts of the world and greatly diminished rainfall in agricultural and urban areas. The United Nations Educational, Scientific and Cultural Organization and other studies found that changes in rainfall pattern can be attributed to shifts in the global wind pattern. These shifts are because of changes in ocean surface temperatures. The effect of human activity on surface vegetation is also causing rainfall pattern variation. Widespread deforestation in parts of Africa and Asia is causing scarce rainfall and subsequent drought.

Global warming affecting rainfall pattern variability is a commonly accepted phenomenon among the world's scientific community. More precipitation is occurring in northern Europe, Canada, and northern Russia, but less in swaths of sub-Saharan Africa, southern India, and southeast Asia. A Canadian research team found in 75 years (1925–99) of rainfall data analysis through 14 powerful computer models that the Northern Hemisphere's midlatitude (a region 40–70 degrees north) received increased precipitation, which corroborates with the change in thermohaline circulation. The models also showed that in contrast, the Northern Hemisphere's tropics and subtropics (a region between the equator and 30 degrees N) became drier, while the Southern Hemisphere's similar region became wetter. This study was conducted for rainfall patterns over land.

Time and again, researchers worldwide have proven that natural pattern rainfall is good for plant growth, while variable rainfall patterns lead to lower amounts of water in the subsurface level of soil (in the upper 30 cm). Variable rainfall patterns also cause plant diversity in particular areas, indicating that weeds grow rapidly with variable rainfall. The significance of these changes is evident from recent large-scale, worldwide failure of crops, rangelands, and water-supply systems. Mass starvation in the Sahel region of Africa is stark proof of this.

Some argue that changes in rainfall patterns are unfounded because of a lack of instrumental records for a long duration. However, studies using indirect methods have proven that global warming, in fact, is causing serious variability in rainfall patterns. Tree-ring analysis for determining rainfall amounts in previous years (hundreds of years back) is one such study proving that rainfall pattern variability has been extensive in recent years.

If this trend continues, environmental managers will be pressed to make new decisions about the management of water and land. They will need to accurately understand the interannual variability of rainfall and a possibility of runs of dry and wet years, which may cause important changes in runoff, sedimentation, soil erosion, and communities of vegetation and animals, and effect the viability of large water-resource developments. Variability in rainfall patterns would also cause mass human migration.

Sudhanshu Sekhar Panda
Gainesville State College

See Also: Agriculture; Floods; Precipitation; Rain.

Further Readings

ABC Science. "Climate Change Already Affecting Rainfall" (July 24, 2007). http://www.abc.net.au/science/articles/2007/07/24/1986545.htm (Accessed April 2011).

Dunne, Thomas and Luna B. Leopold. *Water in Environmental Planning*. New York: W. H. Freeman, 1978.

Dybas, Cheryl. "Increase in Rainfall Variability Related to Global Climate Change." *Earth Observatory, NASA* (December 12, 2002).

United Nations Educational, Scientific and Cultural Organization. "Changes in Climate." *Arid Zone Research*, v.20 (1963).

Viessman, Warren, Jr. and Gary L. Lewis. *Introduction to Hydrology*. 5th ed. New York: Prentice Hall, 2003.

Reducing Emissions from Deforestation and Forest Degradation (UN-REDD)

The United Nations Collaborative Programme on Reducing Emissions from Deforestation and Forest Degradation in Developing Countries (REDD) is an initiative currently being negotiated within the United Nations Framework Convention on Climate Change (UNFCCC). Although no final agreements have yet been reached on REDD, there are dozens of "REDD Readiness" projects already on the ground, which provide illustrative examples of the potential successes and structural problems of the program.

According to its backers, REDD and REDD+ hold out the enticing prospect of mitigating climate change, conserving threatened biodiversity, and bringing much-needed development finance to poor indigenous peoples and local forest-dwelling communities, while simultaneously offering significant profits to investors.

According to the Global Canopy Program, "The idea behind REDD is simple: countries that are willing and able to reduce emissions from deforestation should be financially compensated for doing so." However, Australian scientist Peter Wood argues, "there are a number of fundamental issues that remain unresolved that hang in the balance, including environmental, social, and governance safeguards, monitoring reporting and verification of safeguards, and the inclusion of logging in natural forests."

Likewise, civil society groups, particularly those representing constituencies of climate justice, indigenous peoples, youth, and women, warn that REDD and REDD+ will benefit timber, oil, and gas companies; create perverse incentives to increase deforestation; and exacerbate already-existing toxic hotspots in developed countries.

In addition, they argue that even before it formally exists, just the idea and promise of REDD has already created the conditions for a global land grab, and that REDD-readiness projects have already displaced indigenous and forest-dependent communities from their ancestral lands; severely curtailed their abilities to practice traditional customs on those lands; failed to meet minimal requirements for free, prior, and informed consent; and in some cases failed to reduce overall deforestation.

Background

According to the UNFCCC, the initiative Reducing Emissions from Deforestation in Developing Countries and Approaches to Stimulate Action was first introduced into the Conference of the Parties (COP) agenda at its 11th session in Montreal (December 2005).

Parties to the UNFCCC process assert that greenhouse gas (GHG) emissions from deforestation in developing countries is a major factor in contributing to climate change and, therefore, there is need to take action to reduce such emissions. In 2007, the COP adopted a decision on Reducing Emissions from Deforestation in Developing Countries: Approaches to Stimulate Action (Decision 2/CP.13). Parties are also encouraged to apply the IPCC Good Practice Guidance for Land Use, Land-Use Change, and Forestry for estimating and reporting of emissions and removals.

The UNFCCC differentiates between REDD and REDD+ by saying that REDD is an effort to create a financial value for the carbon stored in forests, while REDD+ goes beyond deforestation and forest degradation and includes the role of conservation, sustainable management of forests, and enhancement of forest carbon stocks. The UN-REDD program claims that REDD+ will fulfill the requirements of full engagement and respect for the rights of indigenous peoples and other forest-dependent communities.

While various aspects of REDD and REDD+ policy are debated within the UN, at least three major kinds of REDD and REDD+ demonstration activities have arisen between governments, with other subnational governmental players and actors also developing policies and demonstration projects.

REDD became a major flashpoint of civil society actions and debate inside and outside the UNFCCC 16th Conference of the Parties (COP 16) in 2011, in Cancún, Mexico. The same result may occur at the 2012 17th Conference of the Parties (COP 17) in Durban, South Africa.

Key Pro-REDD Programs and Players

The UN-REDD Program, launched in September 2008 to assist developing countries in preparing and implementing national REDD+ strategies, is a collaboration of the Food and Agriculture Organization of the United Nations (FAO), the UN Development Program (UNDP), and the UN Environment Program (UNEP).

The program currently has over 35 partner countries spanning Africa, Asia-Pacific, and Latin America, of which at least 13 are receiving support for National Program activities. To date, the UN-REDD Program's Policy Board has approved a total of over $55.4 million for its nine initial pilot countries and four new countries (Cambodia, Ecuador, the Philippines, and Solomon Islands). Norway is the UN-REDD Program's largest donor.

REDD projects are being piloted in many countries under the auspices of the UN-REDD Program, the World Bank Forest Carbon Partnership Facility, the U.S. Agency for International Development, and other global bodies. There are also bilateral pilot programs such as the Kalimantan Forests and Climate Project (between Indonesia and Australia) and some voluntary market programs that involve some conversation nongovernmental organizations (NGOs) such as the World Wildlife Fund and Conservation International.

In addition, journalist Jeff Conant reports in the July/August 2011 issue of *Z Magazine* that at least one subnational government-level REDD readiness initiative was unveiled at COP 16 in 2010 between California and the state of Chiapas, Mexico.

Critiques of REDD Projects and Policies

According to Friends of the Earth International, Carbon Trade Watch, REDD-Monitor, the Indigenous Environmental Network, the Global Justice Ecology Project, and hundreds of other civil society organizations representing environmental, indigenous peoples, women, and youth constituencies: Although REDD may benefit some communities and biodiversity in certain specific areas, overall it is emerging as a mechanism that has the potential to exacerbate inequality, reaping profits for corporate and other large investors while bringing considerably fewer benefits—or even serious disadvantages—to indigenous peoples and other forest-dependent communities. In addition, if governments focus on REDD in isolation, it could become a dangerous and ineffective distraction from the business of implementing real and effective policies for climate change mitigation and adaptation.

An emerging literature in political science, sociology, and international-development studies also shows the emergence of a global land grab in relationship to global land governance questions, with REDD and other market-based initiatives such as payment for ecosystem services (PES) as some of the key drivers and uncertainty in the land tenure of indigenous peoples or other national minority groups a major factor in vulnerability to displacement.

Civil society organizations such as Friends of the Earth International (2010) conclude the following:

> Large transnational corporations, especially those involved in the energy sector or energy-intensive industries, are rapidly honing in on REDD because it offers them—perhaps more than any other participant—a true 'win-win' opportunity. Through REDD, these corporations recast themselves as climate change champions even as they continue, or even expand, operations to extract fossil fuels and other pollution-intensive activities. At the same time, they stand to profit from REDD at the level of hundreds of millions of dollars.

In many countries, there is also ongoing uncertainty about land tenure and carbon rights, and in some areas it seems that REDD is muddying these waters even further. Case studies from Ecuador and Chiapas also demonstrate that areas with conflict or uncertainty over indigenous peoples' land tenure, in conjunction with government or state-sponsored interests in income generation, provide ample opportunity for these projects to exacerbate human rights abuses. Jeff Conant documented that the community of Amador Hernandez in the Lacandon region of Chiapas had their health services cut off in early 2011, most likely in preparation for "REDD Readiness." REDD is being championed as a source of revenue, both by the Chiapas state government and the Mexican national government.

Previous studies demonstrated the eviction of Ogiek peoples from their lands in Kenya in anticipation of a REDD-readiness project, and research conducted or reported by Rebecca Sommer and Chris Lang have demonstrated conservation NGOs or individual "carbon cowboys" attempting to get indigenous communities to sign over traditional lands and rights without meeting the substantive standards of free, prior, and informed consent in Brazil and Papua New Guinea.

There is also an emerging debate about whether REDD can really work at the project level. A 2011 consultation process with southeast Asian groups demonstrated that in at least one pilot project, community members have yet to receive any revenues, although the project has been ongoing for about five years. A similar experience was expressed by members of a Tanzanian NGO at a U.S. government-sponsored side event at COP 16. Also, most studies demonstrate that in and near pilot projects, deforestation has not actually decreased, such as in the case of the Kuna and Emberá territories in the Darien region of Panama.

Current research on carbon markets demonstrate that the current trading price of carbon is below that of potential profits from deforestation or replacement by monocrop forest plantations of eucalyptus, acacia, or oil palm. That is, given that the largest culprits of global deforestation are larger timber and forestry companies, plantation forestry, industrial agro- or bio-fuels and exploration for fossil fuel extraction, existing market prices are insufficient to prevent deforestation by market-driven actors. This reality, if it continues, would undermine the market feasibility of any carbon trading or offset mechanism, including REDD.

In their 2010 publication, *REDD: The Realities in Black and White*, Friends of the Earth International offers the following alternative to a market-based solution to deforestation:

> If governments are to succeed in mitigating climate change by addressing deforestation, they must agree to an equitable mechanism that actually aims to stop deforestation. This will require reducing demand for agricultural and timber products, and addressing other underlying causes of deforestation. Such a mechanism should reward those that have already conserved their forests. It should build on the experiences of Indigenous Peoples and communities around the world, who already know how to manage and benefit from forests sustainably.

Diana Pei Wu
Antioch University, Los Angeles

See Also: Afforestation; Carbon Markets; Carbon Offsets; Carbon Sinks; Deforestation; Developing Countries; Forests; Land Use, Land-Use Change, and Forestry; United Nations Framework Convention on Climate Change.

Further Readings

Angelsen, Arild, Sandra Brown, Cyril Loisel, Leo Peskett, Charlotte Streck, and Daniel Zarin. *Reducing Emissions From Deforestation and*

Forest Degradation (REDD): An Options Assessment Report. Oslo, Norway: Meridian Institute and REDD-OAR, 2009.

Cabello, Joanna and Tamra Gilbertson, eds. *No REDD! A Reader.* Barcelona, Spain: Carbon Trade Watch and Indigenous Environmental Network, 2010.

Carbon Trade Watch. "Key Arguments Against Reducing Emissions from Deforestation and Degradation (REDD+)" (June 2011). http://www.carbontradewatch.org/publications/key-arguments-against-reducing-emissions-from-deforestation-and-degradation.html (Accessed July 2011).

Carbon Trade Watch. "Some Key REDD+ Players." (June 17, 2011). http://www.carbontradewatch.org/articles/some-key-redd-players.html (Accessed July 2011).

Conant, Jeff and Orin Langelle. "Turning the Lacandon Jungle to the Carbon Market." *Z Magazine* (July/August 2011).

Filippini, Ana. *REDD and Gender Impacts.* Montevideo, Uruguay: World Rainforest Movement, 2010.

Focus on the Global South. *REDD in South East Asia: A Political Economy Perspective* (2011). Conference proceedings. http://www.focusweb.org/content/downloads-redd-polical-economy-perspective-conference-presentations (Accessed July 2011).

Friends of the Earth International. *REDD: The Realities in Black and White.* Amsterdam: Friends of the Earth International, 2010.

Gender CC Women for Climate Justice. "Gender in the Climate Money Grail." *Outreach.* http://www.stakeholderforum.org/sf/outreach/index.php/day6-item3 (Accessed July 2011).

Global Canopy Programme. *The Little REDD Book.* Oxford: Global Canopy Programme, 2008.

United Nations Framework Convention on Climate Change. "Demonstration Activities." http://unfccc.int/methods_science/redd/demonstration_activities/items/4536.php (Accessed July 2011).

United Nations Framework Convention on Climate Change. "REDD Web Platform." http://unfccc.int/methods_science/redd/items/4531.php (Accessed July 2011).

UN-REDD Program. http://www.un-redd.org (Accessed July 2011).

Wood, Peter. "REDD+: Reducing the Risk." *Outreach* (2010). http://www.stakeholderforum.org/sf/outreach/index.php/day6-item1 (Accessed July 2011).

Wu, Diana P., Aurora Conley, and Ana Filippini. "Women and REDD." *Outreach* (2010). http://www.stakeholderforum.org/sf/outreach/index.php/day6-item8 (Accessed July 2011).

Refugees, Environmental and Climate

In the early 2000s, the issue of environmental refugees emerged as a pressing issue. Most refugees flee from natural disasters such as the tsunami in Asia in 2004 or as a result of the impacts of global climate change such as sea-level rise. As the executive director of the United Nations Environment Programme (UNEP) noted in 1989, "as many as 50 million people could become environmental refugees" if the world does not support sustainable development. Since then, many studies have considered this topic, with Norman Myers (one of the leading thinkers in this field) estimating that environmental refugees will soon become the largest group of involuntary migrants.

Classifying Environmental Refugees

One of the difficulties in managing the issue of environmental refugees is their classification. Who should be classified as an environmental refugee? The 1951 Convention Relating to the Status of Refugees classifies a refugee as a person who, "owing to a well-founded fear of being persecuted for reasons of race, religion, nationality, membership of a particular social group, or political opinion, is outside the country of his nationality, and is unable to or, owing to such fear, is unwilling to avail himself of the protection of that country." Refugees who are outside of their country for environmental reasons do not strictly fit within this category. The first definition of environmental refugees came from UNEP researcher Essam El-Hinnawi in 1985: "environmental refugees are those people who have been forced to leave their traditional habitat, temporarily or permanently, because of a marked environmental disruption (natural and/or triggered by people) that jeopardized their existence and/or seriously affected the quality of their life [sic].” In this definition, the term *environmental disruption* means any “physical, chemical, and/or biological changes in the ecosystem (or resource base) that render it, temporarily or permanently, unsuitable to support human life.” The important point is that environmental refugees are different from environmental migrants in that they do not have any choice about their situation; it is an involuntary condition.

Today, there are at least 25 million environmental refugees. This is a huge number, considering that there are 22 million people in the world that classify as refugees under the traditional definition. The ratio of environmental refugees to the general population is 1 person to every 225 worldwide. A further 900 million people may also become environmental refugees because they live in marginal environments, or are driven into marginal environments, for political, economic, social, cultural, legal, and institutional reasons.

There are a number of instances that might cause someone to become an environmental refugee. First, natural disasters such as floods, cyclones, earthquakes, tsunamis, or any other major event can make the lived environment temporarily or permanently uninhabitable. One example is the eruptions of the Soufriere Hills Volcano on the Caribbean Island of Montserrat from 1995 through 1998.

As a result of these eruptions, 7,000 residents were forced to evacuate. Second, individuals or whole groups of people might become environmental refugees because of the appropriation of habitat or land by external parties, thus dispossessing and permanently displacing people. For example, the building of the Three Gorges Dam in China has displaced over 1 million people. Third, people may become environmental refu-

Refugees wait in line to register and receive their initial bundle of supplies at the Dagahaley refugee camp, in Dadaab, Kenya, on August 8, 2011. It is the world’s largest refugee camp in Kenya, where thousands of exhausted and starving refugees seek food, water, and medical care after fleeing their drought and famine-stricken lands in southern Somalia. There are an estimated 25 million environmental refugees around the world today, a ratio of one person to every 225 people in the general population.

gees as a result of an ongoing deterioration of their land and seas. Desertification and sea-level rise are good examples of the type of activity that ultimately creates environmental refugees in their homelands. Movement from one area to another occurs as families and settlements find it harder to sustain livelihoods. This is the group that most often finds it difficult to get support, as it is hardest for this group to be recognized as having refugee status.

Problem and Response

There will be many examples of this situation across the world because of climate change. For example, water shortages caused by climate change will cause huge reductions in agriculture and therefore people's way of life. Tropical forests are estimated to lose another 40 to 50 percent of their cover because of climate change, which will dispossess many millions of people living within and dependent upon it for survival. Up to 500 million people could be experiencing absolute shortages in fuelwood supply as a result of climate change impacts. Other (and preliminary) estimates highlight that the total number of people at risk from sea-level rise could be 26 million in Bangladesh, 12 million in Egypt, 73 million in China, 20 million in India, and 31 million elsewhere, for an aggregate total of 162 million.

At least another 50 million people are at risk through increased drought, desertification, and related climate disruptions. In 2010, there was a drastic increase in migrants into southern Europe, caused primarily by food shortages caused by climate change. Environmental degradation is still a major problem in 32 countries in Africa, while one-third of 1 billion people suffer from water scarcity. Africans count for almost 50 percent of the worlds' displaced persons.

Many responses are needed for this problem. One is to ensure that environmental refugees have a formal classification within the 1951 Convention Relating to the Status of Refugees. Another is to address the root causes of the environmental problems motivating relocation of people across the world. Promoting policies and achieving sustainable development programs is a good first step to help achieve this goal. Nations must also develop policy to address what types of support structures and frameworks they are going to implement to acknowledge and deal with their environmental refugees. Implementing specific projects in countries most affected is another pathway nations could take to support each other on this issue, as are the relief of foreign debt and the granting of foreign aid that will help ameliorate this problem. While the issue of how to deal with refugees is always politically contentious, the plight of environmental refugees merits special attention from policymakers and governments because any country—no matter how rich or poor—may be vulnerable and need support in the future.

Melissa Jane Nursey-Bray
University of Adelaide

See Also: Drought; Economics, Impact From Climate Change; Gender and Climate Change; Indigenous Communities; Livelihoods; Population; Poverty and Climate Change; Sea Level, Rising; Tsunamis; United Nations Environment Programme.

Further Readings

Bates, Diane. "Environmental Refugees? Classifying Human Migrations Caused by Environmental Change." *Population and Environment*, v.23/5 (May 2002).

Bell, Derek. "Environmental Refugees: What Refugees: What Rights? Which Duties?" *Res Publica*, v.10 (2004).

Dyson, Tim. "On Development, Demography, and Climate Change: The End of the World as We Know It?" *Population and Environment*, v.27/2 (November 2005).

El-Hinnawi, E. *Environmental Refugees*. Nairobi, Kenya: United Nations Environmental Programme, 1985.

Friends of the Earth International. *The Citizens Guide to Climate Refugees*. Amsterdam: FOE, 2007.

Hunter, Lori. "Migration and Environmental Hazards." *Population and Environment*, v.26/4 (March 2005).

Lou, Y. "Immigration Policy Adjusted in Three Gorges." *Beijing Review*, v.43 (2000).

Myers, Norman. "Environmental Refugees." *Population and Environment*, v.19/2 (November 1997).

Ramlogan, R. "Environmental Refugees: A Review." *Environmental Conservation*, v.23 (1996).

Tolba, M. "Our Biological Heritage Under Siege." *Bioscience*, v.39 (1989).

Religion

In recent years, religious actors have become more visible in response to global environmental change and other environmental concerns. This has taken a variety of forms, with some religious groups and beliefs being barriers, and other religious forms strongly advocating the need to deal with climate change. Many climate change activists are motivated out of a deep spiritual or religious sense. Even in places where institutionalized religious practice has declined, personalized "spirituality" may remain high and historic religious institutions often still have influence.

The world's major religious traditions are increasingly concerned with environmental issues, often through rethinking traditional practices and doctrines in light of evidence of environmental decline. Climate change in particular is challenging the world's religions on practical and cosmological fronts. On the practical front, religious institutions are often at the forefront of providing material services and are involved in people's everyday lives. However, the sheer scale of global environmental change and the proposal that humans have become so powerful as to alter the planet challenges traditional religious understandings of the capabilities and role of humankind, the divine, and the rest of the created order.

Defining Religion

Defining religion is difficult, as such definitions either exclude some of what is commonly understood as religious, or include things that are not usually labeled as such. Definitions that center on a supernatural content exclude such faiths as Confucianism. Definitions that center on faith—that is, beliefs based on empirically unprovable (often meaning "unscientific") tenets—fail to eliminate philosophies like Marxism or secular humanism, which also rest on belief statements about the nature of reality, the human place in the cosmos, and value assertions. Indigenous religious people usually do not have a codified doctrine. Other spiritualities do not have an institutional structure. Finally, the controversies surrounding climate change have led to it being called a religion, sometimes by those who assert that climate change "believers" are not being rational because their "faith" is founded on scientific tenets that are not absolutely certain.

These difficulties in defining religion demonstrate that sharp dichotomies such as between faith and reason, science and religion, or secular and religious are not tenable. Early social scientists focused on religion's societal function, such as providing shared values, producing social cohesion, and ritualizing life stages. Religion is also understood as a system of meaning that can provide answers to central questions about behavior, social order, and purpose. Meaning systems may be enacted through social practices such as prayer, ritual, good works, or fidelity to the church, mosque, temple, or tribe. European history shows a decline in organized religion, although it still has cultural salience. Such secularization is not so evident in much of the rest of the world, including countries like the United States.

Most descriptions of religion encompass a number of integrated dimensions such as sacred narratives or texts, ethics, ritual, doctrine, collective and personal practices, and social institutions. Just as religion encompasses more than just beliefs or faith, it is also more than just values or ethics. Religion makes epistemic and ontological claims about the nature of reality. For example, belief that a divine force is in control of the cosmos, or that humankind is not so powerful as God or nature, will affect what people believe about climate change. Or, emphasis on a creator may be seen by those with nature-focused faiths as "otherworldly." Thus, some conflicts over the issue may reflect differences in value orientations, while other conflicts may point to fundamental differences in understanding the world and the human place in it.

Intersection of Religion and Climate Change

Some religions have been accused of being hostile or ambivalent to ecology (in particular, Christianity). Evidence does not show this to be particularly true, especially when religious complexity is taken into account. In practice, this means that an Ahmadiyyan Muslim at a mosque where environment is discussed (e.g., a study of the Koran on "pollution" or a community-wide promotion of energy efficiency) will likely have different perspectives than an Ahmadiyyan, where such a social milieu is not an influence. Thus, determining the interaction of religion/spirituality and personal or collective practice is difficult. This also demonstrates that religion operates at different scales, such as the personal; collective (as in local mosques, churches, temples, and faith communities); denomination (as in the Presbyterian Church or Shin Buddhism); and transnationally and cross-temporally (as in historic faith traditions).

Climate change is as contested in religious settings as in society at large and must be placed in the context of religious traditions' broader teachings on the environment, human purpose, social justice, and relations with the divine. Drawing on their faith traditions and holy texts, many religious institutional bodies have developed statements about climate change. Relatively little systematic research has been conducted on what religious groups at any scale are actually doing, particularly at the level of faith communities. Existing research on the interaction of religion and climate change has been dominated by survey research on the correlates of individual faith beliefs and climate change beliefs. Given the complexity of lived religion, critics have suggested that methodological expansion would be fruitful.

Much of the research and media attention has been devoted to the contrarian perspectives of some American evangelical Christians. As the political and media focus of climate change has become polarized, the correlation between evangelical Christianity and the rejection of anthropogenic climate change has strengthened, likely indicating political ideology, more than religious fidelity. American evangelical organizing culminated in the formal launch of the Evangelical Climate Initiative (ECI) in 2006, with the support of the National Evangelical Alliance (NAE). As political polarization became more pronounced, other high-profile evangelical leaders began opposing the theory of anthropogenic climate change. The Cornwall Alliance produced *An Evangelical Declaration on Climate Change* in 2009, modeled after and refuting the ECI position.

Attention to evangelical Christians obscures other Christian action, and if evangelicalism is taken as par exemplar of Christianity as a whole, may even reify notions that climate change science is not congruent with Christian faith. Surveys show that African American evangelicals have higher levels of belief in climate change than white evangelicals. Many other Protestant denominations have issued official statements calling on governments to address climate change. The U.S. Roman Catholic bishops released a major statement in 2001 and several reports since then. Some of the numerous denominational, interdenominational, and interfaith climate change campaigns are oriented around the moral duty for stewardship or "climate justice" because of the unfairness of impacts, such as on small island states, and economic disparity between developed and developing countries. Even the Cornwall Alliance declaration expresses its opposition to the theory of anthropogenic climate change as concern for the global poor. Specifically, it declares that costly mitigation efforts result in reduced resources for disadvantaged peoples.

The issue of climate change is less contested in faith communities elsewhere in the world. Sir John Houghton, lead editor of the first three Intergovernmental Panel on Climate Change (IPCC) reports, is also an evangelical Christian. Christian development agencies like World Vision have identified climate change as an impediment to development assistance for the world's poor. Climate justice has become a prevalent framework based on principles such as equity, compassion, and charity. The Roman Catholic Church under both Pope John Paul II and Benedict XVI has been a prominent player. Besides papal speeches and reference to climate change in other documents, the Vatican became the world's first carbon-neutral state in 2007 through conservation, solar energy, and offset projects.

Websites that list official religious statements on climate change (such as the Forum on Religion and Ecology), Web searches, and searches for popular press books from different religious traditions suggest that Christian responses are more prevalent than other religious traditions. Caution must follow this observation, as apparent differences may be the result of organizational factors, population numbers, or other social characteristics. Muslim activists in developed countries comment that in a post-9/11 world, faith-based organizing on environmental issues has been difficult.

The issue of climate change has provided the opportunity for interfaith dialogue and collaborative action. Reports from the 2009 World Parliament of Religions indicate that fully 15 percent of the program was directly devoted to the environment, and most of that was associated with climate change. Interfaith coalitions have been active in the climate justice movement. For instance, the World Council of Churches organized a climate justice session at the United Nations Conference of the Parties (COP 16) in combination with the Asian Muslim Action Network and a Roman Catholic international development agency. Coincident with the annual meetings of the G8 leaders, world religious leaders meet to discuss progress toward the Millennium Development Goals. Recent meetings have been used to declare that climate change causes conditions under which the development goals cannot be met.

Environmental and Religious Collaboration

Many environment, development, and justice nongovernmental organizations have sought to collaborate with faith communities. The United Kingdom–based Alliance on Religion and Conservation (ARC), which works with 11 faiths worldwide, claims that these faiths encompass 85 percent of the world's population, run half the world's schools, and own 7 percent of the habitable surface of the planet. The original faith participants of Buddhism, Christianity, Hinduism, Islam, and Judaism have been joined by Baha'i, Daoism, Shintoism, Zoroastrianism, Jain, and Sikh faiths.

One example of ARC's work was an initiative supported by the government of Kuwait and the United Nations Development Programme that involved representatives from 14 countries to develop the Muslim Seven Year Action Plan to Deal with Global Climate Change. The British Council organized an interfaith gathering in Nigeria in 2010 with Muslim, Christian, and indigenous African spiritual leaders. In their final report, the leaders recognized the impact of climate change on their constituents and made several commitments indicative of the reach of faith institutions. They agreed to strengthen the communication of environmental ethics consistent with their faith tradition and scriptures and to share best practices.

However, despite the many projects and statements by religious institutions, systematic evaluation on the extent to which faith groups are actively engaging in such practices would help determine the actual impact of religion on climate change mitigation or adaptation.

Finally, action about climate change is at least partly associated with ideas about nature, weather, metaphysics, and human capacity. Some of the reasons for resistance to the theory of anthropogenic climate change are worldview assumptions, such as doubt about human capacity to change an entire planet's livability, the purported fragility and resilience of nature, or whether a creator would allow creation to be severely damaged. The dominant modern worldview can also be seen as a religious-like faith in progress. For instance, geoengineering can be seen as a form of climate change salvation through technology. Religions often challenge such exuberant presumptions. Major religious traditions are adapting to environmental concerns by "greening" their doctrines and institutions. At the same time, some observers believe that other "dark green" religious faiths are developing strength.

From a religious perspective, climate change can be a moral and spiritual issue. The impact of religious institutions extends beyond their adherents and affects attitudes, public policy, social capital, education, and human rights—all of which are part of the climate change conundrum. However, while there are plenty of lofty ideals and institutional statements, the effective bridge to religious practices and the operations of religious institutions must still be demonstrated.

Randolph Haluza-DeLay
King's University College

See Also: Climate Justice; Global Warming, Attribution of; Global Warming Debate; Movements, Environmental; Needs and Wants; Poverty and Climate Change; Social Ecology.

Further Readings

Forum on Religion and Ecology. "Climate Change Statements from World Religions." http://fore.research.yale.edu/climate-change/statements-from-world-religions (Accessed July 2011).

Gerten, Dieter and Sigurd Bergmann, eds. *Religion in Environmental and Climate Change*. London: Continuum Books, 2012.
Haluza–DeLay, Randolph, Andrew Szasz, and Robin Globus, eds. "World Religions and Global Climate Change." *Journal for the Study of Religion, Nature, and Culture* (2012).
Hulme, Mike. "Why We Disagree About Climate Change: Understanding Controversy, Inaction, and Opportunity." Cambridge: Cambridge University Press, 2009.
Kerber, Guillermo and Martin Robra. "Climate Change." *Ecumenical Review*, v.62/2 (2010).
Millais, Corin, ed. *Common Belief: Australia's Faith Communities on Climate Change*. Sydney: The Climate Institute, 2006.
Northcott, Michael S. *A Moral Climate: The Ethics of Global Warming*. Maryknoll, NY: Orbis Books, 2007.
Reuter, Thomas. "Faith in the Future: Climate Change at the World Parliament Of Religions, Melbourne 2009." *Australian Journal of Anthropology*, v.22/2 (2011).

Renewable Energy, Bioenergy

Renewable bioenergy is derived from diverse sources and in multiple forms: gas, electricity, and liquid fuel. The energy is released by combustion of organic matter that contains carbon compounds. Significant questions surround the production of bioenergy, including the level of greenhouse gas (GHG) emissions released by bioenergy facilities and the displacement of other economic products by the expansion of bioenergy sources. Although any carbon fuel, including coal, petroleum, and natural gas, is organic; bioenergy refers to the alternative, renewable sources that are not as harmful to the atmosphere and are not depleted through use. Most discussions of bioenergy also anticipate production that does not require mining or drilling.

Bioenergy sources include (1) anaerobic digestion of many materials, such as animal and human waste; (2) biomass processing, as in the pyrolysis of wood chips; (3) algae production of oils or chemicals, including fuels; (4) chemical conversion of carbohydrates, such as sugar from corn or sugarcane into ethanol; and (5) direct combustion of such fuels as cosmetics oils or food waste.

The most common product of the decomposition of organic substances is the gas methane (CH_4). Methane is the most important component of natural gas from wells and provides the majority of the heating and much of the electric power generation in industrial countries. Methane in natural form is trapped in many locations, including peat bogs, ocean sediment, and coal. For the most part, renewable or alternative-energy policies refer to the diverse ways to produce methane under controlled conditions from natural materials, rather than from drilling wells.

A very important concern in regard to global warming is the potential for uncontrolled release of methane from these natural reservoirs, such as the Siberian peat bogs, directly into the atmosphere. A large increase in methane in the atmosphere would stimulate further global warming, perhaps inducing a cycle of warming–release–warming. The most significant alternative or bioenergy objective is to control the combustion of methane into less harmful carbon dioxide (CO_2) and water.

Biogas

Biogas production generally refers to the production of methane from natural feedstocks. Methane is the second-largest global warming gas in the atmosphere, after CO_2. It has been calculated by the International Panel on Climate Change (IPCC) to have a global warming potential (GWP) of 21 times the GWP of CO_2. Methane (CH_4) combusts as follows:

$$CH_4 + 2O_2 \rightarrow CO_2 + 2H_2O.$$

Thus, methane projects that combust the biogas emit substantially less harmful exhaust gas. Many such applications also produce renewable energy credits in the United States and other countries, and carbon credits for the cap-and-trade markets.

Global projects under the Kyoto Protocol such as the Clean Development Mechanism (CDM) projects have opened up awareness of methane-producing sources from dairy and hog farms to sewage plants to coal seams in undeveloped regions. As projects around the world have expanded, North America has gradually joined the industry by planning, building, and investing in biogas plants.

Thousands of methane sources are distributed throughout many countries. It is recognized that wind and solar generation have advantages over huge, traditional, central coal-fired plants, in part because they can be independent for local power or connected to a new, more expansive, power system. Just as small farms and ranches have relied on wind generators for generations to power pumps, illuminate work areas, and other uses, the 21st-century power system is likely to incorporate large applications of the same useful, economical source of energy. If methane gas supplies, or methane-powered generation at dairy farms, landfills, and wastewater plants are considered, the notion of a sustainable, effective, and clean distribution is conceivable.

Algae Oil

In recent years, another promising source of bioenergy from algae has been developing. The technology is proven. Currently, industrial-scale plants are being tested. The product is an oil resembling petroleum, which can be refined into diesel fuel, petrochemicals, and pharmaceuticals. Algae oil burns with less CO_2 release than standard carbon fuels. Bioenergetic processes in which algae ingest CO_2 and convert it to a carbon-rich fuel similar to petroleum offer an emerging industry that can—for the first time—substantially reduce the burning of coal and oil, thereby reducing the emission of GHGs.

Algae fuel production will also reduce the environmental effect of coal mining and drilling for petroleum. A promising technology currently being implemented at a pilot plant in West Virginia collects CO_2 from coal-fired power plants and provides it to algae liquid-fuel facilities. This symbiotic combination of traditional power generation and innovative liquid-fuel production is likely to be the next stage of "clean coal" production on a long path to the replacement of coal as a primary energy source.

At this point, algae also offers the only promising source of vehicle fuel in large quantities as a substitute for imported petroleum. Algae fuel production could soon be scaled up to relieve the economic burdens on petroleum-dependent countries such as Japan, India, Korea, China, and the United States.

Ethanol and Corn Power

In the United States, the ethanol fuel industry is by far the largest investment in bioenergy that has been strongly encouraged, subsidized through tax incentives and grants, and stimulated by federal government policies. Large acreages of cornfields have been planted to produce the corn for ethanol plants. Federal, state, and even local regulations have been adopted requiring ethanol–gasoline mixes, especially during warmer months. E-15 ethanol fuel (15 percent ethanol) is now a common product in many parts of the United States.

Ethanol from carbohydrate processing is both cleaner than gasoline and sustainable. Many countries have initiated ethanol production for both economic and environmental gains. For many, reducing the dependence on petroleum and the combustion of petroleum products will be a major goal.

Ethanol production, especially in the United States, has become controversial and an object of conflicting environmental and economic interests. Taking large areas of corn production out of the food-supply chain and devoting it to ethanol production has been identified as a contributor to rising food and animal feed costs, a significant burden on poor populations in many countries. The expansion of ethanol production from food crops in developing countries will very likely increase conflict between competing interests.

A second controversy surrounding corn ethanol relates to energy balance—the result when the fuel consumption of the vehicles required to plant and harvest corn, transport corn to refineries, and transport ethanol to the distributors, as well as the electricity consumption on a large scale at the refineries, is calculated into the cost and environmental demands. One study from the University of California, Berkeley, estimated that corn ethanol produced in the United States actually increased energy consumption with its attendant environmental degradation. This was vigorously disputed by other researchers.

Brazil has become the world's leader in ethanol fuel production by processing sugarcane, a more efficient source for carbohydrates than corn. The raw product, called bagasse, is a major product that fuels energy plants, automobiles, public transportation, and even aviation, and constitutes a significant part of Brazil's national GHG control and economic-development policies.

Attention has recently been directed to increasing the productivity of such ethanol energy sources as switchgrass and other fibrous plants. This production is constrained by current technology, which does not break down the energy locked in the cellulose fiber of the plants. As research and experimentation make progress on breaking down cellulose, cellulosic ethanol production could become the industry standard with significant improvements in the economics and environmental feasibility of expanded ethanol production.

Biomass Energy

The categories of bioenergy are not exclusive, and terminology is flexible. Biomass can be applied to a number of different processes for producing either gas or fuel energy. However, the most common use of the term *biomass energy* relates to producing electricity or ethanol from natural materials commonly available in nature or human activities. A good example is the conversion of wood waste to a fuel for power generation.

The technology of pyrolysis is most commonly used for breaking down hard material such as wood into the hydrocarbon necessary for efficient combustion. The challenge in processing wood waste is to break apart the molecules trapped in the form of wood so they can be recombined into a hydrocarbon fuel. Wood releases flammable gas, mostly methane, when it is heated. With the application of enough heat in an oxygen environment, wood will burn at the surface, the heat gradually releasing more gas, until the wood is largely consumed. The ash remaining is composed of unburned carbon and other waste. Pyrolysis works in an environment without oxygen, but the heat releases methane, which can then be combusted in a boiler or turbine.

Carbonless flight? An artist's concept of a blended wing body aircraft that could become a prototype by 2020. NASA plans to continue working on its years-long quest to develop zero net carbon aviation into mid-century. Developed in stages, the prototype, based on the standard Boeing 737 design, would use aviation biofuel that would take in carbon while being grown, then emit carbon when the fuel is burned. The trick, says NASA, is to balance the intake and output so that the net effect on the atmosphere is zero.

A current application of pyrolysis connects the gasification process to gas-fired engines or turbines, which can achieve higher electrical efficiencies than biomass combustion alone. Fuel cells increase the power output efficiency to the range of 25–50 percent in current applications.

Pyrolysis production of the gas fuel also reduces the air pollution from CO_2 and nitrogen. This is especially effective in reducing pollution if the fuel is used in fuel cells, rather than in gas-fired engines or gas turbines.

Bioenergy in the Market

The future growth of bioenergy in its various forms as renewable energy depends on the viability of new market conditions. If prices for traditional fuels rise as they have in recent years and investment in bioenergy projects continues to grow, prices should converge. Interest in using bioenergy must also expand among major market influences, such as large commercial office centers, factories, research laboratories, military installations, government capitals and institutions, universities, public schools, and electric trains.

Investment in bioenergy has been impressive, but skewed heavily to subsidized technology such as solar panels and ethanol. The opportunity for distributed generation through the private sector with smaller wind-generator facilities and methane producers is not widely recognized. Entrepreneurial efforts for distributed generation as a base for renewable energy production needs to expand.

Investors could do a better job of evaluating the costs of relatively available and efficient technologies such as biogas conditioning that employs pressure-shift adsorption or solar arrays combined with wind generation that can generate under varying conditions.

Bioenergy can be expected, perhaps even predicted, to become an increasingly familiar and vital component of energy production in every economy.

Alan B. Reed
University of New Mexico

See Also: Agriculture; Georgia (U.S. State); Indiana; Malaysia; Minnesota; Natural Gas; Plants; Renewable Energy, Biomass; Renewable Energy Policy Project; University of Washington.

Further Readings

Dolan, Kerry A. "Procter & Gamble Partners With Bio-Based Chemicals Startup ZeaChem." *Forbes* (June 1, 2011). http://blogs.forbes.com/kerryadolan/2011/06/01/procter-gamble-partners-with-bio-based-chemicals-startup-zeachem (Accessed September 2011).

Farrell, Alexander E., Richard J. Plevin, Brian T. Turner, Andrew D. Jones, Michael O'Hare, and Daniel M. Kammen. "Ethanol Can Contribute to Energy and Environmental Goals." *Science*, v.311 (January 27, 2006).

U.S. Environmental Protection Agency. "Landfill Methane Outreach Program." http://www.epa.gov/lmop (Accessed September 2011).

Renewable Energy, Biomass

Biomass energy is produced by extracting volatile gases, primarily methane, from organic materials such as wood, switch grass, corn stalks, or municipal solid waste. The variety of methods and levels of efficiency for extracting the gas are extensive. Biomass energy production, either for direct combustion of the gas or electric power generation, has been utilized widely for decades, but recently, large-scale production of biomass energy as a substitute or enhancement for traditional carbon fuel sources has received a great deal of investment and experimentation. Traditionally, a winter fireplace produces biomass energy from wood. What the new biomass energy industry adds is mass production of energy for use in the economy as well as residential comfort. Gas from biomass sources can be conditioned (cleaned) and injected into existing or new gas lines. Its more frequent use is in combined heat and power (CHP) installations, particularly at landfill sites, for generating electricity. These more recent applications fill a double purpose, in the same vein as the much more recognized solar and wind generation, by both producing energy and reducing greenhouse gas (GHG) emissions into the atmosphere.

Methane, the primary biomass energy source, is rated by the International Panel on Climate

Change (IPCC) as 21 times as warming in the atmosphere as carbon dioxide (CO_2), the most abundant GHG. When a biomass energy project can extract and combust methane, the potential warming affect is significantly lower than if the methane escapes into the air—even though burning methane produces CO_2 along with water:

$$CH_4 + 2O_2 = CO_2 + 2H_2O$$

Feedstocks and Production Technology

Biomass developments must pair feedstock with technology. Applying the wrong technology or process to a biomass source will be unproductive and waste money. Some technology is still being tested, but several are proven and in common use.

The most common, reliable, and profitable technology for extracting methane energy is anaerobic (without oxygen) digestion. Retention of a liquid slurry of organic biomass in a storage container for a period, usually, 21–30 days, will allow naturally occurring bacteria to digest the hydrocarbons in the material and produce methane. The bacteria work best in a moderately warm environment, around 98–100 degrees F (36–37 degrees C), and are self-generating. With care, digesters will produce a sustainable, predictable stream of gas for years.

There are many types of organic feedstocks for digesters. The content of the feedstock determines the quantity and quality of biogas production. For centuries, small farms in central Asia have deposited manure from their animals into pits and captured the methane for heating and cooking. Today, thousands of operational digesters produce commercial quantities of gas from dairy and hog manure, waste from farm operations, and even commercial fields of convertible feedstock such as grasses and fibrous plants.

Wastewater (sewage) treatment plants in developed countries employ digesters for reducing pollutants in water before returning it to the environment. The digesters flare excess digester gas that is not used for heating the digesters. This flare gas has become a credible source of bioenergy and can either generate electricity for the power grid or be conditioned for injection to the gas grid.

Digesters are attractive technology because they produce a purified dry cake of nitrogen-rich fertilizer, as well as the remaining liquid fertilizer after the biogenesis process is completed.

European enterprises have built pilot plants using dry biomass technology in which large quantities of industrial, commercial, and municipal solid waste can be ground into fine particles and processed to produce the proper moisture content for combustion in boilers or gasifiers. Often, the best use for such dry biomass is in the form of pellets for ease of storage and shipment.

Dry biomass has some of the advantages in efficiency and reduced pollution of biogas, but it is still in its infancy for large production for the energy market. It is an especially attractive power source in densely populated urban areas because it is reduced to a much smaller quantity of ash and debris that can be disposed of in a landfill.

Much effort is being invested in improving pyrolysis of organic feedstock. Pyrolysis occurs when organic materials are heated in the range of about 662–932 degrees F (350–500 degrees C) in a closed container that prevents oxidation. At high temperatures and depending on the fuel, light oils can be produced, as well as biochar, a soil enhancement much sought after in the Amazon region; charcoal; activated charcoal for filters; and other valuable products.

Pyrolysis has the advantage of eliminating infectious (pathogenic) organisms in the biomass, as does anaerobic digestion, producing a clean gas for use in power generation. However, the multiple stages in the process of producing biogas through pyrolysis have so far been uneconomical. Very large pyrolytic plants are in use for enhancing the power production of coal-fired generators, and in northern Europe, pyrolytic plants provide power from wood. Currently, the capital investment required for large, complex engineered plants for pyrolysis reduces a number of economical pyrolysis applications.

Biomass feedstocks (plants, organic wastes, wood, and liquid waste) and biomass energy technology (anaerobic digestion, dry biomass processing, and pyrolysis) offer potential for extensive enhancement and even restructuring of the energy systems of the industrial world. This energy source, which is close at hand, available without great transportation infrastructure costs, and produces less GHG, is attractive for investment.

Biomass energy can provide the energy expansion needed in the future through distributed generation and gas production far beyond the limited power and gas grids of the 20th century.

Alan B. Reed
University of New Mexico

See Also: Alternative Energy, Overview; Carbon, Black; Carbon Capture and Storage; Carbon Dioxide Equivelant; Carbon Sequestration; Carbon Sinks; Deforestation; Forests; Land Use, Land-Use Change, and Forestry; Plants; Pollution, Air; Soil Organic Carbon; Zooplankton.

Further Readings

Habmigern. "Pyrolysis of Biomass." http://www.habmigern2003.info/biogas/Pyrolysis.htm (Accessed September 2011).

Northwest CHP Application Center and Carolyn J. Roos. "Biomass Drying and Dewatering for Clean Heat and Power" (September 2008). http://www.chpcenternw.org/NwChpDocs/BiomassDryingAndDewateringForCleanHeatAndPower.pdf (Accessed September 2011).

U.S. Environmental Protection Agency. "Biomass CHP." http://www.epa.gov/chp/basic/renewable.html (Accessed September 2011).

U.S. Environmental Protection Agency. "Biomass Combined Heat and Power Catalog of Technologies." (September 2007). http://www.epa.gov/chp/documents/biomass_chp_catalog.pdf (Accessed September 2011).

Renewable Energy, Geothermal

Geothermal energy is a renewable resource that is generated by natural processes within the Earth. Since prehistoric times, humans have enjoyed the recreational and perceived therapeutic properties of surface geysers and thermal hot springs. It has also been used to produce electricity on an industrial scale for more than 100 years. Geothermal energy provides a viable, sustainable, low-carbon alternative to fossil fuels. Furthermore, the potential global energy production capacity of the Earth's exploitable geothermal resources far exceeds both current and predicted future primary energy demand. However, in 2010, geothermal energy production comprised less than 1 percent of total global primary energy consumption.

The concept of geothermal energy raises some important questions: What is geothermal energy and under what conditions can it be exploited? What are some direct and indirect strategies for harnessing geothermal energy? What are some of the benefits and limitations of geothermal energy? And finally, what are the prospects for future development of this abundant and valuable energy source?

Geothermal Energy Basics

Over 99 percent of the Earth's mass is extremely hot—above 1,830 degrees F (1,000 degrees C). In general, temperature increases with depth. The core of the planet, which is about 2,174 mi. (3,500 km) in radius, is composed of a solid iron inner core and a superheated liquid outer core that generates a magnetic field. The inner and outer cores are enveloped by another region called the mantle, which is about 1,800 mi. (2,900 km) thick. The mantle, in turn, is enclosed by an extremely thin and largely impervious outer crust (12–40 mi. or 20–65 km thick in continental areas, less in ocean regions), which in relative terms is analogous to the skin of an apple. In most places, the outer crust blocks a significant amount of heat from rising to the surface. However, there are some locations, such as a space between tectonic plates or active volcano systems, where the outer crust is unusually thin, permeable, or cracked, and heat can rise close enough to the planet's exterior to allow human exploitation. Commercially exploitable resources are generally 1.8 mi. (3,000 m) or less below the Earth's surface.

Besides heat and close proximity to the Earth's surface, exploitable geothermal systems also require water and permeability. Hot water or steam is sometimes trapped in permeable rock formations, forming geothermal reservoirs below layers of impermeable crust. In some cases, these reservoirs can be accessed via geothermal wells; in other situations, the hot water or steam may seep to the surface naturally. It is also possible to artificially introduce water into a field of fractured hot rock where it is heated and then extracted by a production well.

Exploitable Geothermal Energy

Geothermal energy can be exploited either directly or indirectly. Direct applications involve harnessing heat directly from its source. For thousands of years, hot springs have been used for bathing and recreation. Ancient Rome's bathhouses are a well-known example. Today, Iceland's Blue Lagoon draws tourists from around the world. Space heating is another common direct application. For built environments located near an active geothermal heat source, hot water can be piped directly into structures. Indoor temperature is regulated via radiators or similar technologies. These same piping systems can also be used to supply hot water for cooking, bathing, or industrial processes. Direct heating is also widely used in agriculture and aquaculture.

Another space heating technology that does not require close proximity to a geothermal reservoir is the geothermal heat pump. These devices access geothermal heat from shallow ground or groundwater sources (where temperatures are relatively constant) via a local piping system. Heat pumps can also be used to draw away heat and recirculate cooler air during the summer.

According to the International Geothermal Association (IGA), estimated global exploitation of direct geothermal energy has increased substantially, from 112,441 terajoules/year in 1995 to 438,071 terajoules/year in 2010. The largest national producers of direct geothermal energy include China, the United States, Sweden, Turkey, Norway, and Iceland. The main indirect use of geothermal energy is electricity production. Typically, geothermal plants transfer steam and/or water (depending on the temperature) from hot groundwater reservoirs via a production well to electricity generating turbines. After passing the turbines, the steam/water is cooled and may either be vented via cooling towers or reinjected back into the geothermal reservoir. Some newer plants allow both direct and indirect exploitation. One example of this type of multiple-use facility is the recently constructed Hellisheidi plant in Iceland. The plant provides both electricity and space heating via a district heating system to local consumers.

Geothermal power generation is the fourth-largest source of electricity in the world behind coal, hydropower, and natural gas. A few countries, including the Philippines, El Salvador, Nicaragua, and Iceland, derive 15 percent or more of total electricity production from geothermal sources. The United States is by far the largest producer of geothermal electricity, followed by the Philippines and Indonesia. According to the IGA, between 1990 and 2010, total installed generating capacity almost doubled from 5,832 megawatts electric (MWe) to 10,716 MWe.

Benefits and Limitations

Exploitation of geothermal energy has grown substantially in recent decades. This virtually limitless sustainable power source produces far fewer greenhouse gases or other harmful emissions than fossil fuel alternatives. However, global geothermal energy still contributes less than 1 percent of total primary energy production. Why has geothermal energy not been exploited more widely? A number of geological, economic, environmental, and political factors continue to hold back further exploitation. At present, most commercially exploitable resources are widely dispersed. For instance, electric generation generally requires access to near-surface, high-temperature reservoirs (generally above 300 degrees F, or 150 degrees C). However, even already-tapped geothermal reservoirs are exploited far below their potential. Geothermal heat pumps are not affected by this geological limitation.

A variety of economic barriers inhibit geothermal development. The biggest barrier is the availably of lower-cost energy sources. Geothermal energy competes directly with coal and natural gas. Fossil fuel energy still remains relatively inexpensive in relation to other alternatives. Also, geothermal plant construction costs are generally higher than fossil fuel equivalents. However, operation and maintenance costs are far lower, which makes geothermal an attractive investment over time. For instance, operation of the Hellisheidi plant in Iceland requires fewer than 20 employees, and customers pay around 2¢ per kilowatt hour (kWh), which is less than one-fifth of the average cost in the United States (2010 est.). Geothermal producers also directly compete with solar and wind energy providers.

Geothermal energy production is not free from negative environmental externalities. Extracted water and steam typically include toxic elements such as hydrogen sulfide, arsenic, mercury, boric acid, and other trace compounds that may be hazardous to human health. However, these emissions can be almost completely abated through the use of scrubbers or reinjection. Geothermal energy producers also discharge carbon dioxide (CO_2), although these emissions are much lower than that of natural gas plants. In rare cases, high-pressure water reinjection may cause localized seismic events, though the jury is still out regarding the strength of this relationship. An additional concern is the possible contamination of potable groundwater via reinjection. This potential problem can be abated by the construction of reinjection sites far away from known groundwater sources.

Finally, there are political barriers that inhibit the development of geothermal energy. Fossil fuel interests have every reason to use their political clout to limit government research and investment in geothermal development. Also, public awareness of the benefits of geothermal energy is relatively low, which further inhibits both public- and private-sector support for this clean, sustainable, and virtually limitless resource.

Conclusion

What are the future prospects for geothermal energy development? While geothermal energy use is still relatively low in relation to other energy sources, it is growing quickly, especially in fossil fuel–poor countries such as the Philippines and Iceland. Fossil fuels will become scarcer and more expensive over time. At some point, environmental externalities associated with global climate change may become so pronounced that governments around the world will be compelled to put a high price on carbon emissions, which will further enhance the attractiveness of geothermal energy and quicken the transition to a low-carbon economy.

Jonathan Harrington
Troy University

See Also: Alternative Energy, Overview; Climate Change, Effects of; Energy, Climate; Iceland; Indonesia; Natural Gas; Philippines; Sustainability; United States; Volcanism.

Further Readings

Anonymous. "Forward to the Steam Age?" *Futurist*, v.45/1 (2011).

Barbier, E. "Geothermal Energy Technology and Current Status: An Overview." *Renewable and Sustainable Energy Reviews*, v.6/1–2 (2002).

DiPippo, R. "Geology of Geothermal Regions." *Geothermal Power Plant*. Oxford: Butterworth–Heinemann, 2008.

Gallup, D. "Production Engineering in Geothermal Technology: A Review." *Geothermics*, v.38/3 (2009).

Hammons, T. J. "Geothermal Power Generation Worldwide: Global Perspective, Technology, Field Experience, and Research and Development." *Electric Power Components and Systems*, v.32/5 (2004).

International Geothermal Association. "Installed Generating Capacity (2010)." http://www.geothermal-energy.org (Accessed June 2011).

Younis, M., et al. "Ground Source Heat Pump Systems: Current Status." *International Journal of Environmental Studies*, v.67/3 (2010).

Renewable Energy, Overview

According to the International Energy Agency, renewable energy is energy that is derived from natural processes that are replenished constantly. In its various forms, it is derived directly or indirectly from the sun, or from heat generated deep within the Earth. Included in the definition is energy generated from solar, wind, biomass, geothermal, hydropower and ocean resources, and biofuels and hydrogen derived from renewable resources. Renewable energy sources from natural energy sources such as sun, wind, waves, and tides have developed rapidly in recent years and have been used widely around the world as an alternative-energy system.

Renewable energy is considered one of the viable options to meet the challenge of achieving sustainable development and conserving natural resources that have been depleted because of the rapid growth in population, urbanization, and fossil fuel consumption while addressing the issues of energy security. Renewable energy has been widely considered an indispensable basis of sustainable energy systems, as electricity generation from renewable sources helps in reducing greenhouse gas (GHG) emissions, contributing to sustainable development and addressing climate change over the past decades as compared to electricity from conventional fossil fuels. Renewable-electricity generation capacity, including large hydropower, reached an estimated 1,140 gigawatt (GW) worldwide in 2008, with a total share of about 18 percent—of which 15 percent was from hydroelectricity and 3 percent from new renewables such as small hydropower, modern biomass, wind, geothermal, and biofuels. According to Renewable Energy Network 21 (REN 21), about 19 percent of global final energy consumption came from renewables in 2008, with 13 percent from traditional biomass used for heating, 3.2 percent from hydroelectricity, and a small percentage—about 2.7 percent—from the new renewables.

Apart from the environmental benefits, renewable energy has also been playing an important role in alleviating poverty and enhancing the development of remote regions through offering opportunities for work and social advancement of underemployed and disadvantaged segments of the population. However, their adoption has been mainly driven by impending environmental and energy security considerations arising from the use of fossil fuel–based energy (coal, oil, and gas) and the fact that fossil-based energy sources are finite.

Renewable Energy and Developing Countries

A majority of the estimated 2 billion people without access to adequate, affordable, clean energy sources live in developing countries, raising the need for alternative energy for developing countries. The predictions of REN 21 show that dependence on oil imports in developing countries like China is likely to increase from 22 to 77 percent by 2020. India's dependence is predicted to rise from 87 to 92 percent, east Asia's from 54 to 81 percent, and the rest of south Asia's from 87 to 96 percent by 2020, demonstrating the growing concern of energy security and the need for renewable energy in developing countries. Apart from the concern over climate change mitigation and emissions reduction targets of at least 5.2 percent below that of 1990 by 2020 for developed countries (termed as Annex I countries) per the Kyoto Protocol, the transfer of renewable energy to developing countries (termed as Non-Annex I countries) is of primary interest to many organizations in developed countries.

The geography of renewable energy is thus changing. Wind power existed in just a handful of countries in the 1990s, but now exists in over 82 countries, with manufacturing leadership shifting from Europe to Asian developing countries like China, India, and South Korea. Many

Table 1 Top five countries for existing capacity of renewable energy by 2009 show that the developing countries are leap-frogging in the development of RES

Top Five Countries	1	2	3	4	5
Annual amounts for 2009					
Renewables power capacity (including only small hydropower)	China	United States	Germany	Spain	India
Renewables power capacity (including all hydropower)	China	United States	Canada	Brazil	Japan
Wind power	United States	China	Germany	Spain	India
Biomass power	United States	Brazil	Germany	China	Sweden
Geothermal power	United States	Philippines	Indonesia	Mexico	Italy
Solar PV (grid connected)	Germany	Spain	Japan	United States	Italy
Solar hot water/heat	China	Turkey	Germany	Japan	Greece

Source: REN21, 2010.

The Bonneville Lock and Dam is part of a hydropower and dam system on the Columbia River between Oregon and Washington. Hydropower makes up about 15 percent of renewable energy's share of electricity generation capacity.

recent trends also reflect the significance of developing countries in renewable energy; China leads in several indicators of market growth, and India is fifth worldwide, with overall developing countries now making up over half of all countries with policy targets (45 out of 85 worldwide) and making up half of all countries with some type of renewable energy promotion policy (42 out of 83 worldwide). Renewable energy is well suited to many developing countries, but whether it is possible for third-world countries to "leapfrog" over the dirty, inefficient, intermediate stage of industrialization remains to be seen.

Renewables and Socioeconomic Development

Apart from providing alternative energy sources, renewable energy contributes to socioeconomic development providing jobs in communities. One of the forces propelling renewable energy development is the potential to create new industries and generate millions of new jobs. Jobs from renewables now number in the hundreds of thousands in several countries. Globally, there are an estimated three million direct jobs in renewable energy industries, with additional indirect jobs well beyond this figure. For example, it was reported that about 43 jobs had recently been created in a group of renewable energy projects in a small Maphephetheni community in KwaZulu Natal in South Africa. China and Brazil account for a large share of global total employment, having strong roles in solar hot water and biofuels industries. Jobs are expected to grow apace with industry and market growth.

Renewable energy also has a major impact on improving social-service amenities like schools and hospitals in developing countries by providing facilities in remote areas where obtaining energy through a grid is impossible. For example, the 1,000 solar photovoltaic (PV) schools project in the Northern Province and eastern Cape Province in South Africa had a significant impact on the education of schoolchildren in those rural areas, with noted benefits for their respective communities through adult education and communal use such as community functions or entertainment using solar-powered audio-visual equipment. Another example in the health sector is the Solar Energy for Rural Health Facilities of the Health Sector Recovery Programme of the Government of Mozambique, which in 1997 consisted of solar PV electrification of approximately 250 rural health facilities in all 10 provinces of Mozambique.

Social Acceptance of Renewable Energy

Opposition to technology is not new; there are fears that new technologies will result in major social or environmental dislocations, which have been proven in many cases. Social acceptance is a prerequisite for the adoption of new technologies, and the same is applicable to renewable energy; if the local community does not accept the technology, there will be no demand for its services. For example, if it is culturally or socially unacceptable in a community for women to cook in the middle of the day instead of in the evening, then solar cookers will not be easily accepted in that community. Opposition to renewable energy can also occur because such technologies ignite social conflicts that have little to do with the systems.

The landscape issue for renewable energy is one of the most debated, especially in the case of wind-energy projects. For example, rural residents sometimes want renewable energy for their use as a vehicle for economic development and resent what seems like an intrusion by urban residents' intent on preserving the countryside for its scenic and recreational value. In general, landscape values are central and rooted in strong sociocultural traditions, which also explains the resistance by communities to the siting of controversial land-use facilities. Another major issue affecting community perceptions is the gap in the technical knowhow of renewable energy technologies in these communities. This is because of the low information levels about the operation and the lack of skilled technicians and technology standards. The Patsari cooking stove program in rural Mexico, raising awareness of technology, health, and environmental benefits (especially to women and children) of the renewable energy technology of modified biomass, is a good example of triggering a successful implementation of renewable energy programs through creating awareness and training local people.

Another key problem facing renewable energy technologies is the attempt to establish them in an institutional, market, and industrial context base on the existing type of conventional-energy technology. The involvement and participation of a community in renewable energy projects creates a relationship and synergy between the people who live in the community and the people who set up the technology, facilitating the effective implementation of these projects. Although technical and financial support is often necessary for successful deployment of renewable energy technologies, social acceptance becomes of prime importance for the systems to remain sustainable in the long run after deployment.

Risks and Barriers for Renewable Energy

There can be many risks and barriers for implementing renewable energy relating to policy implementation, financing, and technology:

Environmental or social risks: Some projects, such as hydropower, impact the environment and society—for instance, because of civil works and resettlement. This means that environmental and social impact assessments are necessary for financial support, especially in regard to financial institutions. The cost of the assessments may be high. In addition, the response from the community may be negative, for instance, as a result of political influence.

Lack of information or data risks: Market failure because of inadequate assessment of project viability because of unavailable or inadequate data is a serious problem for renewable energy financing and implementation.

Financial risks: Renewable-energy technologies, which are new in most countries, come with the baggage of inexperience. This leads to risks involving the capital structure of the project and the ability to generate cash flows sufficient to fund planned investments, operations, and maintenance expenditures; service debt; and provide reasonable returns to the sponsor.

Technology barriers: Conventional fossil fuel energy projects generally employ mature technologies that have been proven through years of successful commercial application. Conversely, renewable energy projects often employ recent technologies that risk-averse insurers and financiers penalize with prohibitive premiums and terms.

High cost barriers: The lack of mature technologies is a barrier and leads to importing technology from Western countries, which leads to high costs. For the components to be manufactured locally, technology providers must open manufacturing plants locally or research and development results must rise to the level of commercial technologies.

Policy barriers: In most countries, renewable energy is not in the forefront of central or local government concerns. This leads to lack of robust policies to support renewable energy development. Rather than developing a policy framework to invite renewable energy developers, many local governments wait for developers to initiate projects and then consider policy development. This forms a vicious cycle, as neither is ready to take the first step. It poses a serious risk to renewable energy development, as without the support of a proper policy framework, it is impossible to proceed further. At the onset, government intervention is necessary to unlock and scale-up investment in renewable energy technologies, but the key stakeholders also need to cooperate.

Lessons Learned From Renewable Energy

Although the bulk of research on renewable energy over past decades has been on technological development, research on sociocultural issues has slowly been gaining momentum. Although the "not in my backyard" attitude has often been identified as the cause for community resistance to renewable energy implementation, other factors, such as financial benefits, the possibility of stakeholder participation, ownership of the technology, how well local communities are informed about the technology, previous uses of the chosen site, the quality of communication with communities, and public participation in the planning process of siting the technology are important issues in the successful implementation of renewable energy technologies and projects. There is a broad range of issues, with a broad range of perspectives on renewable energy among the research community, although most of them end on the issues of cultural identity, insufficient information, and inequitable distribution of benefits. An example of equitable distribution might be a wind-energy project, with the "winners" as the project-hosting landlords who receive an annual income from the installer and the "losers" as neighboring landlords who suffer the effects of the turbines, but enjoy no revenue.

Despite billions of dollars invested in research and development, procurement, tax incentives, tax credits, subsidies, standards, and financial assistance, the impediments to renewable energy technologies remain largely social and cultural, and there is a need for policymakers and governments to increase their efforts in understanding the challenges faced through community resistance to these technologies. One of the fundamental barriers for renewable energy technologies is that the systems are not appropriate to the local context and demands, or are not adapted to the local environment. Adoption and effective implementation of renewable energy will not be realized until it is seen as an economic-development issue. For example, it can be linked with another project that improves development, depending on the community in which the project is installed—such as a solar-powered fish drier, or a coconut drier for making soap.

Komalirani Yenneti
University of Birmingham

See Also: Alternative Energy, Overview; Climate Change, Effects of; Greenhouse Gas Emissions; International Energy Agency; International Renewable Energy Agency; Renewable Energy, Bioenergy; Renewable Energy, Geothermal; Renewable Energy, Solar; Renewable Energy, Wind.

Further Readings

Akella, A. K., et al. "Social, Economical, and Environmental Impacts of Renewable Energy Systems." *Renewable Energy*, v.34 (2009).

Australian Agency for International Development. *Power for the People: Renewable Energy in Developing Countries*. Canberra: Commonwealth of Australia, 2000.

Devine–Wright, Patrick. "Reconsidering Public Acceptance of Renewable Energy Technologies: A Critical Review." In *Delivering a Low Carbon Electricity System: Technologies, Economics and Policy*, edited by Michael Grubb, Tooraj Jamasb, and Michael G. Pollitt. Cambridge: Cambridge University Press, 2008.

Laumanns, Ulrich, Danyel Reiche, and Mischa Bechberger. "Renewable Energy Markets in Developing Countries: Providing Green Power for Sustainable Development." In *Green Power Markets: Support Schemes, Case Studies, and Perspectives*, edited by Lutz Mez. Essex, UK: Multi-Science Publishing, 2007.

Painuly, J. P. "Barriers to Renewable Energy Penetration: A Framework for Analysis." *Renewable Energy*, v.24 (2001).

REN 21. *Renewables 2010: Global Status Report*. Paris: GTZ-GmBH, 2010.

Wilkins, Gill and the Royal Institute of International Affairs Sustainable Development Programme, eds. *Technology Transfer for Renewable Energy: Overcoming Barriers in Developing Countries*. London: Earthscan, 2002.

Wüstenhagen, Rolf, Maarten Wolsink, and Mary Jean Burera. "Social Acceptance of Renewable Energy Innovation: An Introduction to the Concept." *Energy Policy*, v.35 (2007).

Renewable Energy, Solar

Solar energy is a renewable energy source in the form of radiant light from the sun. The term *renewable solar energy* refers to a broad range of technologies and techniques for capturing and utilizing this energy. In one hour, approximately 440 exajoules (EJ) of energy from the sun reach the Earth's outer atmosphere. This is only slightly less than the estimated world population's energy consumption from primary energy sources, such as biomass and fossil fuels, in an entire year. The Earth's atmosphere, land surfaces, and oceans absorb a fraction of this incoming radiation, resulting in an increase in atmospheric, oceananic, and land mass heat, in addition to water evaporation. These heat and evaporation processes in turn drive the Earth's water cycle and produce atmospheric phenomena, including wind and weather patterns. Plants convert solar energy into chemical energy through the process of photosynthesis. Energy from the sun is critical to processes that sustain life on Earth. Directly capturing and utilizing even a small fraction of this vast energy source would offset fossil fuel consumption, which in turn would decrease the greenhouse gas (GHG) emissions that contribute to climate change.

Solar Energy Availability

Thermonuclear reactions in the sun result in the emission of radiation or light in all directions into space. The Earth intercepts a small fraction of this emitted radiation. At the top of the Earth's atmosphere, the amount of radiated power from the sun remains fairly constant, at 1,361 watts per square meter (W/m^2), varying only slightly throughout the year by plus or minus 3.5 percent because of the eccentricity of the Earth's orbit around the sun.

As this solar energy enters the Earth's atmosphere, some wavelengths of this light are absorbed, scattered, or reflected back into space by different molecules, such as the gases and water vapor in the atmosphere. For example, a process known as Rayleigh scattering results in the blue appearance of the sky. On clear days, with little cloud cover, the radiated solar energy reaching a flat surface on the Earth can exceed 1,000 W/m^2; however, on cloudy or overcast days, this power can be reduced to less than 100 W/m^2. Cloud cover, aerosols, dust, smoke, and suspended water droplets all reduce the transmittance of the atmosphere and decrease the amount of solar radiation reaching a given location.

For terrestrial solar energy applications, the amount of energy that falls on a given surface is of primary concern. The orientation of a surface relative to the incoming sun's rays must be considered when evaluating the solar energy available for a given application. The total solar radiation available to a surface at any orientation is a combination of two components: direct beam radiation and diffuse radiation. Because the incoming solar rays are nominally parallel, geometric relations can be used to estimate the amount of beam radiation available to a given surface. The surface's available direct beam radiation is related to its latitude, the tilt of the Earth on its axis (which depends on the day of the year), the time of day, the slope of the surface, and the orientation of the slope in the east–west direction. Maximum beam radiation on a clear day occurs when the incoming solar radiation is perpendicular to the surface of interest. Diffuse radiation refers to radiation that is scattered and reflected from the atmosphere, clouds, and the ground to a particular surface. The estimation of diffuse radiation availability depends upon the orientation of the surface, the scattering potential of the atmosphere, and the reflectance properties of the ground. On cloudy days, and when the surface is shaded from direct beam radiation, diffuse radiation is the primary source of solar energy availability.

Because the solar energy available at a specific location depends on many factors, historical data is often used to obtain more accurate estimates of energy availability during an average day. Solar energy availability can be estimated based on data collected from pyrheliometers and pyranometers. Pyrheliometers are instruments that track the sun and measure the available direct-beam radiation. Pyranometers obtain data on the total solar radiation or the diffuse solar radiation available to a fixed surface. Using historical data from such instruments located near a proposed installation site provides a more accurate measurement of the solar energy capture potential at that location. Such data can be modified using geometric relations and empirical observations to estimate the solar energy availability on a surface at any orientation.

Solar Energy Technologies

The wide variety of solar energy technologies are often broadly categorized as solar thermal or solar power technologies, depending upon the methods used to capture and convert solar energy into usable energy.

Photovoltaic solar power technologies use solar or photovoltaic (PV) cells that directly convert incident light into electrical energy. Currently, most PV cells and panels are produced in different forms from silicon, including monocrystalline silicon (c-Si) and polycrystalline silicon (poly-Si). Because of the physics of the photovoltaic effect and the construction of silicon PV panels, much of the incident solar energy is lost in the form of heat. Conversion efficiencies of incident solar radiation into usable electrical energy for such commercially available panels range from 12 to 20 percent. Thin films of cadmium telluride (CdTe), copper indium gallium selenide (CIGS), and gallium arsenide (GaAs) have also been used to produce PV cells with conversion efficiencies approaching 20 percent. Researchers have developed materials and solar panel configurations with optical light concentrations that have demonstrated up to 44 percent conversion efficiency.

Because of the intermittent nature of solar energy, a method of electrical storage is needed if electricity is required during times when sunlight is unavailable. Rechargeable battery systems can be configured to store the excess electrical energy; however, this decreases system efficiency, increases the complexity of the solar PV system, and greatly increases cost and maintenance requirements. Solar PV systems can be attached such that the system is connected to an electrical grid, with excess energy resupplied to the grid for consumption, a grid tied with battery storage, or a stand-alone unit with or without batteries for energy storage.

Hydrogen gas, generated from water and electrolysis with excess electrical energy, may also be used as an energy-storage medium. Hydrogen is of particular interest as a solar-generated fuel for transportation and mobile power applications that use fuel cells. Additional emerging solar power technologies related to PVs include dye-sensitized cells, organic polymer cells, thermoelectric modules, and photoelectrochemical cells.

Solar Thermal Technologies

Solar thermal technologies convert incoming solar radiation into thermal energy for heating or cooling applications. A solar water-heating system is an example of solar thermal technology. Solar energy collectors, in the form of flat plates or evacuated tube collectors, convert solar energy into heat. Heat from the collectors is used to heat the water. In low geographic latitudes, residential hot water temperatures can reach 140 degrees F (60 degrees C) with such solar heating systems. In a similar vein, solar concentrating technologies, using parabolic dish or parabolic trough reflectors, can greatly increase the temperature of a small solar collection area. Heated fluid at much higher temperatures can be used for industrial process heating.

Concentrated solar power (CSP) systems are large-scale, solar thermal energy technologies. CSP systems use large arrays of lenses or mirrors and advanced solar tracking systems to focus a large area of sunlight onto a small area. This concentrated heat is used as the heat source for a conventional power plant to generate electricity.

Solar energy can also be used in the water treatment process. Solar distillation technologies use the thermal energy from sunlight to evaporate saline water. By condensing and collecting the evaporated water, brackish water can be made potable. Similarly, solar water disinfection (SODIS) is a solar energy technology for water pasteurization. By placing polyethylene terephthalate (PET) bottles filled with water in sunlight, the water temperature increases to the pasteurization temperature, thereby killing organic pathogens and resulting in safe drinking water.

Passive solar thermal energy technologies can also be used to offset a portion of the energy required for building heating, ventilation, air conditioning, and lighting systems. Proper building design can allow for solar heating of a thermal mass to store heat. This stored thermal energy is released slowly to regulate, maintain, and offset the building's heating requirements. Similarly, proper shading techniques, such as the use of deciduous trees and building overhangs, can reduce a building's heat gain and cooling load. Passive solar ventilation systems or solar chimneys can be utilized to offset ventilation costs. As the chimney warms when exposed to solar radiation, the temperature of the air inside the chimney increases, creating an updraft that moves cooler air through the building.

A field of solar panels with tracking devices, designed to keep the panels pointed toward the sun, are arrayed at Nellis Air Force Base in Nevada, northeast of Las Vegas, December 2007. The 14-megawatt system will provide about a quarter of the base's electricity needs. The system is manufactured by SunPower, which manufactures the world-record 22.4 percent efficient Maxeon solar cell. At the end of 2011, the North American PV market was forecast to grow 33 percent quarter-on-quarter and 101 percent year-on-year.

Solar Energy Use

Currently, over 40 gigawatts (GW) of power are generated from solar PVs worldwide. Overall, the use of renewable solar energy technologies represents a small fraction of the total energy production worldwide. A primary obstacle to increased solar energy technology deployment is the high initial investment cost. Because of the high availability and relatively low cost of fossil fuel energy sources, the economic rewards for typical solar thermal energy technologies, such as domestic hot water heating, range from approximately 3 to 20 years depending upon the size of the installation and solar energy availability. Small scale, solar PV installations currently require from approximately 5 to more than 30 years for the economic gain to be realized. Such large initial capital expenditures and long repayment periods significantly reduce the incentives to implement this technology. As the cost of fossil fuels increases, government incentives for renewable energy technologies increase, and as the conversion efficiencies for solar energy technologies improve, such economic barriers to installation may decrease.

Solar's Impact on Climate Change

While the use of solar energy does not directly result in GHG emissions, the production and deployment of such technologies does. To understand the impact of such solar energy technologies on the environment, a life-cycle assessment must be performed. A common method of evaluating solar energy systems is to determine the energy payback time. The energy payback time of silicon PV panels is approximately three to five years, which indicates that a PV panel must operate for three to five years before it produces more energy than was required to make the panel. However, such an analysis considers only the energy directly used in the panel's fabrication. Additional energy is required for raw material extraction, transportation, and refining, as well as the costs incurred for transportation to and installation at the operation site. Further, the embedded energy in the fabrication and transport of supporting products, such as batteries, inverters, and interconnect equipment, is often not considered. Refinement and production also releases GHGs and environmentally damaging chemicals. Life-cycle analysis (LCA) studies of the entire PV production and installation process suggest that the energy and environmental payback time may exceed 5 to 20 years of installation operation and emit from 50 to 250 g. of CO_2 per kilowatt hour of energy produced. These estimates may vary widely depending upon the assumptions used in the LCA models. However, such results demonstrate that, although solar energy is a widely available resource, the embedded energy and life-cycle pollutant emissions must be considered when evaluating the overall effectiveness of solar energy technologies to influence climate change.

Stephen Keith Holland
James Madison University

See Also: Alternative Energy, Overview, Atmospheric Absorption of Solar Radiation; Economics, Cost of Affecting Climate Change; Life Cycle Analysis; Sunlight; Sunspots.

Further Readings

Bainbridge, D. and K. Haggard. *Passive Solar Architecture: Heating, Cooling, Ventilation, Daylight, and More Using Natural Flows.* White River Junction, VT: Chelsea Green Publishing, 2011.

Duffie, J. and W. Beckman. *Solar Engineering of Thermal Processes.* Hoboken, NJ: John Wiley & Sons, 2006.

Fthenakis, V., et al.. "Emissions From Photovoltaic Life Cycles." *Environmental Science and Technology*, v.4/6 (2008).

Messenger, R. and J. Ventre. *Photovoltaic Systems Engineering.* Boca Raton, FL: CRC Press, 2010.

Scheer, H. and A. Ketley. *The Solar Economy: Renewable Energy for a Sustainable Global Future.* London: Earthscan Publications, 2002.

Singer, Pete. "Photovoltaic Installations: Around the World" (December 14, 2011). Renewable Energy World. http://www.renewableenergyworld.com/rea/news/article/2011/12/photovoltaic-installations-around-the-world?page=2 (Accessed January 2012).

SunPower. http://us.sunpowercorp.com/about/the-worlds-standard-for-solar (Accessed January 2012).

Renewable Energy, Wind

As moving air, wind is a nondepleting source of renewable energy. Wind can be of devastating power. Storms and hurricanes are proof of strong winds having the power to uproot trees and destroy homes and other infrastructure. That wind power can be used as a source of energy has long been known to humanity. The utilization of wind energy is not a new technology; the generation of electricity from wind power took place many years ago. Electricity generation (energy conversion) from fossil fuels is largely contributing to greenhouse gas (GHG) emissions, primarily carbon dioxide (CO_2), which have been identified as the main driver of global warming and climate change. Wind is a clean energy source with the potential to replace other polluting, electricity-generation technologies.

How much energy the wind is carrying depends on several factors: the amount of wind energy flowing through a given area, or area swept by a turbine (A) during a fixed time; the time to pass the turbine (t); the wind's velocity before the turbine (v); and its density (ρ).

The wind's kinetic energy (E_{kin}) is calculated using the formula: $E_{kin} = \rho/2\ A\ v^3\ t$. The wind speed (v) is normally measured in meters per second (m/s) using an anemometer (wind speed meter). Practical wind power density (WPD) maps are established to provide the mean annual power available per sq. m of swept area of a turbine for different heights in watts per square meter (W/m^2) and serve as a reference as to the best location to install and operate wind turbines.

What Is Wind and How Is It Created?

Wind is produced by differences in air pressure, which are the result of the unequal heating of the Earth's surface by the sun (which in turn heats the air above it). This unequal heating of the Earth's surface is the result of the difference between outgoing and incoming radiation at high and low latitudes of the Earth, which rotates around a shifted axis.

Heated air expands and decreases in density. Following the second law of thermodynamics, the air then flows from areas of high pressure to areas of low pressure until the air pressure is balanced (and the entropy is maximized). The higher the pressure difference and gradient, the stronger the wind that seeks to balance the difference.

Because the Earth is rotating, the angular momentum is conserved with the wind, which is shifted along a longitudinal direction resembling a circular movement (Coriolis effect). On a local scale, the geographic and topographic properties of the Earth's surface largely influence how uniformly and consistently the wind is flowing.

Offshore and onshore winds are generated along the shores of large lakes and ocean beaches. Those winds blow very regularly as they are created by the different heat absorption and storage properties of the land and water surfaces. Likewise, mountain-valley breezes arise from the unequal heating properties of the mountain-valley topography. Surface wind speeds tend to be lower when the movement of air is obstructed by geographical features, vegetation, or buildings. Therefore, coastal and offshore sites are ideal locations for wind-turbine installations.

Wind Energy in the Past

Convincing accounts for the use of wind power technology exist from Al-Mas'udi (895–957 C.E.), who describes in his historic encyclopedia of 947 C.E. how Abu hu'lu'a, a Persian "builder of windmills," murdered the caliph in 644 C.E. because he did not agree with the high taxes that were levied upon him. This description of windmills in Seistan, which were vertical-axis grain mills at the Persian–Afghan border in present-day Iran, is therefore the first reliable account of windmill technology. Similar vertical-axis mills have survived in that region until modern times. At the same point in history, the Chinese already used vertical-axis wind wheels (structures made of bamboo with cloth sails) to pump water.

The traditional European-style windmills (horizontal axis) were probably invented independently of the vertical-axis wind wheels of Asia. Early accounts of post or trestle mills date back to 1180 C.E. in the Duchy of Normandy (present-day northern France). The technology quickly spread all over Europe and was applied to different purposes, the two most common of which were grinding work and pumping water.

The first systematic development aimed at utilizing wind power for the generation of electricity took place in Denmark. In 1891, Poul La Cour (1846–1908) built an experimental wind turbine in Askov, Denmark, which drove a dynamo to produce electricity (direct current). The generated electricity was used for the electrolysis of water to produce hydrogen gas and therefore stored the wind's energy in a readily accessible form. The hydrogen gas was then used in gas lamps. La Cour's wind turbines had, by today's measure, a modest power output ranging from 10 to 35 kilowatts (kW).

The first wind-powered turbine to provide electricity into an American electrical grid was set up in 1941 in Vermont. However, large-scale installation of wind power farms in the United States began only after the first oil crisis in 1973, primarily in California.

Modern Wind Turbines

Modern industrial wind turbines may be classified according to their rotor type, tip-speed ratio, work principles (vertical/horizontal), and power output, among other factors. The smallest type (in consumer products) is used for applications such as battery charging, camping, or auxiliary power.

The majority of large wind turbines produced to date have three fast-moving blades and are of the horizontal type—rotational shafts are parallel to the horizon, with blades perpendicular to it. The wind-turbine rotor converts the kinetic energy of the wind to rotational energy. The rotor blades are connected to the hub, which contains the blade pitch mechanism and bearings and connects the blades to the shaft. The shaft enters the nacelle, which contains the drive train consisting of the gearbox (speed increaser), generator (electricity convertor), and necessary control systems for speed, pitch, and yawing. The nacelle is mounted on a tower (tubular or lattice) and turned into the wind by one or many electric engines (the yaw system, or azimuth drive).

For energy-conversion efficiency, the rotor blades are of paramount importance. Very early windmills used a simple flat area, providing resistance to the wind and thereby being actuated by it. These simple-resistance blades or sails yielded a maximum conversion efficiency of about 12 percent. A more efficient actuation is achieved by using a blade design based on the aerodynamic principle, in which the wind creates a lifting force via uplift pressure. However, as Albert Betz (1885–1968) proved, the energy extractable from wind using a turbine is generally limited to 59.26 percent of the total energy contained in it. This is called the Betz limit. Output sizes of large turbines have steadily increased since the early designs of La Cour, and today feature standard offshore designs with up to 6 megawatts (MW) and standard land site designs of 2 to 3 MW, with up to 100 m of blade diameter. Single wind turbines are grouped in arrays, or "wind parks."

Challenges and Drawbacks

Compared to the power density of coal, oil, or uranium (per gram or volume), the wind energy that is available for instant conversion in a turbine is rather low. Additionally, wind is variable, both geographically and temporally. Yet, proper site location according to reliable WPD maps, together with the interconnection of multiple wind parks in different locations through the electricity grid, nevertheless may yield a stable base load electricity supply. In order to store the wind's energy in off-grid installations, other technologies must be used and/or adapted, as La Cour successfully demonstrated in 1891.

Early, smaller and fast-moving wind turbines were accused of causing up to 40,000 fatalities of birds and bats per year in the United States. Newer and larger turbines appear to cause fewer fatalities, even less when they are not set up in bird migratory paths. Tyler Miller and Scott Spoolman estimate that bird and bat fatalities because of those older wind turbines comprised only about 0.27 percent of all other recorded causes of their deaths (including glass windows, buildings, electrical transmission towers and lines, cats, cars, and trucks).

The Dabancheng onshore wind farm in the Xinjiang Uygur Autonomous region of China, September 2011. The wind farm features more than 300 turbines and a total installed power capacity of more than 500 megawatts (MW). The farm, which was more than 20 years in the making, uses a range of turbines from the older 20 kilowatt (kW) models to new 3 MW models. The farm was the brainchild of Wang Wenqui, former head of the Xinjiang Wind Energy Institute, who spent the later part of his life pioneering wind energy in China.

Rapid growth in the number of wind parks has led to a resistance in some countries that can be traced back to concerns about aesthetic values and the destruction of natural landscapes, limited noise, and shadow.

Additionally, the construction of a wind power installation may alter the local ecosystem structure through vegetation clearing and fragmentation, resulting in a loss of habitat and other problems. However, the largest environmental impact from wind turbines is determined earlier, during the production and assembly of its components. Carrying out a life cycle analysis (LCA) of specific wind-turbine designs enables the producer to identify and quantify the environmental effects of the different components and processes and offers the opportunity to further reduce the already comparably low environmental impacts.

Finally, connecting wind energy parks remains a challenge in many countries where those parks are located far from established grid connection points or the grid control is out of date, not connected to long-distance transfer technology, or unable to adequately react to variations in the electricity load supplied from wind turbines.

Clean Source of Sustainable Power

Sergio Pacca and Arpad Horvath showed that electricity from wind power produces the lowest amount of GHGs among the various renewable and fossil sources of energy (as gram of CO_2 per kWh) and posited that hydropower and wind power have a lower global warming effect (GWE), a combination of global warming potential and LCA, than other power plants. Wind power is sometimes called the most environmentally friendly way to generate electricity and can well satisfy present energy demand.

In 2005, C. L. Archer and M. Z. Jacobson showed that using modern wind turbines (at 80 m height, 77 m diameter, 1.5 MW output) capturing 20 percent of the wind energy at selected sites could satisfy all of the world's primary energy demand (6,995–10,177 million metric tons of oil equivalent, or Mtoe) and more than seven times the amount of electricity used in the world, which in 2004 was 1.8 terawatts (TW).

As of 2011, China has an installed wind power capacity of 44,733 MW and has thereby surpassed the United States (40,180 MW) and Germany (27,215 MW). Germany's goal is to generate 30 percent of its electricity demand from wind by 2030.

Sebastian Schulze
Brandenburg University of Technology

See Also: Alternative Energy, Overview; Atmospheric Composition; China; Energy, Climate; Green Economy; Pollution, Air; Renewable Energy, Overview.

Further Readings

Archer, C. L. and M. Z. Jacobson. "Evaluation of Global Wind Power." *Journal of Geophysical Research*, v.110 (2005).

European Wind Energy Association, eds. *Wind Energy—The Facts*. London: EarthScan, 2009.

Hau, Erich. *Wind Turbines: Fundamentals, Technologies, Application, Economics*. New York: Springer, 2005.

Hennicke, Peter and Manfred Fischedick. *Renewable Energies: The Energy Efficiency of Energy Systems*. Munich, Germany: Verlag C. H. Beck, 2007.

Manwell, James F., et al. "*Wind Energy Explained: Theory, Design, and Application*." Hoboken, NJ: Wiley, 2010.

Miller, G. Tyler and Scott E. Spoolman. "*Living in the Environment: Concepts, Connections, and Solutions*." Florence, KY: Cengage, 2009.

Needham, Joseph. *Science and Civilisation in China*. Vol. IV:2. Cambridge: Cambridge University Press, 2000.

Nelson, Vaughn C. *Wind Energy: Renewable Energy and the Environment*. Oxfordshire, UK: CRC/Taylor & Francis, 2009.

Pacca, Sergio and Arpad Horvath. "Greenhouse Gas Emissions From Building and Operating Electric Power Plants in the Upper Colorado River Basin." *Environmental Science and Technology*, v.36/14 (2002).

Shimkus, John. "The Top Ten Largest Wind Farms in the World" (March 10, 2011). Energy Digital. http://www.energydigital.com/top_ten/top-10-business/the-top-ten-largest-wind-farms-in-the-world (Accessed January 2012).

U.S. National Research Council of the National Academies. *Environmental Impacts of Wind-Energy Projects*. Washington, DC: National Academy of Sciences, 2007.

Renewable Energy Policy Project

The Renewable Energy Policy Project (REPP) is a nonprofit organization based in Washington, D.C. Its mission is to support the advancement of renewable energy technology and spur the growth of renewable energy. Emphases include technology development; domestic manufacturing of renewable technologies; policy analysis; the relationship between policy, markets, and public demand; the development of renewable energy growth strategies that are competitive and environmentally friendly; education and outreach; and promotion of renewable energy projects. Key projects have included the publication of various analytical tools and reports, including state reports outlining the potential local benefits of national renewable energy policies and online renewable energy discussion groups.

REPP was founded in 1995 with Alan Miller serving as its first executive director. It acquired the Center for Renewable Energy and Sustainable Technology (CREST) in 1999. The board of directors oversees REPP's activities through biannual meetings, while the executive director oversees day-to-day operations. The current executive director is George Sterzinger. Sterzinger's lengthy experience in the field of energy policy and regulation include a stint as commissioner of the Vermont Department of Public Service. REPP's board of directors represents a variety of participants in the field of renewable energy, including environmental organizations, environmental regulators, state and national government officials, multilateral development institutions, and companies involved in the financial sector.

The organization's internship program provides eligible candidates with three internships that are available on a rolling basis. Interns gain valuable training and experience in the field while working to develop analytical tools, models, and reports in a variety of formats. REPP receives annual or project-specific funding from a variety of organizations. Past donors have included the Energy Foundation, the Oak Foundation, the SURDNA Foundation, the Turner Foundation, the Bancker-Willimas Foundation, and the Joyce-Mertz-Gilmore Foundation. REPP receives government funding from the U.S. Department of Energy, the National Renewable Energy Lab, and the Environmental Protection Agency (EPA). Additionally, REPP often works with other related organizations, such as the Blue-Green Coalition.

The organization's research areas include hydropower, bioenergy, geothermal, wind power, photovoltaic, solar thermal, renewable hydrogen, and efficiency. Concerns include the relationships among policy, markets, and public demand in the increased use of renewable energy; the link between social and economic development and environmental issues; and the competitiveness and cost-effectiveness of renewable energy sources in energy markets. REPP's position is that renewable energy projects will involve resources that are under utilized or not utilized at all, provide local communities with economic benefits and job creation, and prove cost-effective by removing structural barriers. Another of REPP's core issues involves the use of renewable energies to stabilize U.S. carbon emissions because carbon is a greenhouse gas that contributes significantly to climate change and global warming. REPP advocates for renewable energy projects and production tax credits for such projects.

Strong Support for Manufacturing

Support for the manufacturing component of renewable energy projects has been one of REPP's core emphases. The group's research has determined that approximately 70 percent of the labor involved in the implementation of renewable energy projects, such as wind turbines and photovoltaic modules, occurs during the manufacturing phase as opposed to the installation and operation phase. Recommended goals for the renewable energy component manufacturing sector include transparency, access, development of the domestic manufacturing industry, federal support for local industry, and production and supply chain improvements. REPP facilitates these goals through research and technical assistance to states and regions.

REPP has also issued a variety of publications designed to inform government, industry, and the public and to promote renewable energy use. Its first report, "Environmental Imperative: A Driving Force in the Development and Deployment of Renewable Energy Technologies," was issued in 1996. Policy research includes issue and policy briefs, research reports, fact sheets, testimonials from public officials, and initiatives. REPP has also produced a buyer's guide and consumer directory of businesses providing renewable energy products and services.

The organization maintains a strong online presence through its Website, which includes a variety of informational and interactive formats to foster its goal of public education, promoting successful renewable energy programs, and providing easily accessible information and tools to renewable energy advocates. There are also a number of discussion lists and groups in a variety of renewable energy and policy fields. A community calendar provides information on events of interest. The online library provides access to both current and archived papers, as well as discussion articles, policy tools, and analytical reports. REPP is in the process of developing its Global Energy Marketplace (GEM) database.

REPP aids states in the preparation, implementation, and success of renewable portfolios and renewable portfolio standards mandating the generation of a specified percentage of electricity from renewable energy sources. REPP has created a series of state reports and associated Web-based products in association with United Steelworks and the Sierra Club. Each state report provides information on manufacturing potential calculations, assessments on the potential economic benefits of national renewable energy development, and a detailed listing down to the county level of related firms active in that state that could benefit. The goal is to help increase demand for renewable energy.

Marcella Bush Trevino
Barry University

See Also: Alternative Energy, Overview; Renewable Energy, Overview.

Further Readings

Evans, Robert L. *Fueling Our Future: An Introduction to Sustainable Energy*. New York: Cambridge University Press, 2007.

Renewable Energy Policy Project and Center for Renewable Energy and Sustainable Technology. http://www.repp.org (Accessed July 2011).

Resources

A resource is any item or substance that is in scarce supply and has some value. Resources are normally considered to be physical items, such as oil and natural gas. However, it is also possible to consider humans resources, since they are finite in number and are perishable under current technological conditions. Resources, when used in the context of computer or virtual environments, meanwhile, are inherently intangible in nature, although the hardware that produces them is not.

It is customary, when considering resources, to distinguish between those that are renewable and those that are nonrenewable. Resources such as oil are consumed in use and are, therefore, nonrenewable. However, in a number of other cases, it is possible to recreate or recycle some resources either in the original form or, at least, some components of the original. Glass and plastic bottles may, to some extent, be recycled into different forms, and so value is created from spent resources that appear to be valueless. Considerable effort has been expended in determining which resources may be recycled or recreated in this way and, in some countries, it has led to significant social change as people become accustomed to considering the issues involved and sorting out recyclable household waste. The process is also mirrored at the industrial level, especially when economic incentives are provided to encourage this behavior.

Improved technology has also provided two other means of increasing the stock of resources, or at least minimizing their depletion. The first is to employ more commonly occurring resources for more rare ones. This substitution may be seen in the prevalence of plastic bags, which are dispensed with alacrity at many retail outlets. More recently, the negative impact of those plastic bags has prompted the search for other materials that would be more environmentally friendly (more biodegradable). The process of beneficiation, on the other hand, is one in which technology enables the gathering or exploitation of resources that were previously considered to be too difficult or expensive to obtain. The search for coal deeper underground or in mines located underwater is an example of this, while the continuing demand for oil and the ability to extract it means that sources previously ignored have become of considerable strategic importance.

For example, water is considered a renewable resource, because once used, it can be returned to the circulatory system that returns it to use via evaporation and precipitation. However, the modern world has seen a growth in populations and demand for water, together with climate change, that has demonstrated the extent to which water resources are in fact insufficient for future use, given current trends for demand. It is possible to characterize the Middle Eastern wars between Israel and neighbors as the result of fighting for scarce water resources, while the conflict in Darfur in Sudan has been characterized as resulting from nomadic movements of people searching for water.

Moral Authority to Use and Deplete

Most human societies have developed with a religious basis that justifies mankind's prerogative to use the resources of the Earth for its benefit. Christianity, Judaism, and Islam, for example, have similar roots in a tradition that states a divine provenance for the world and the entire universe and the passing of responsibility for shepherding the world to humanity. Certain variations in scripture explain dietary rules for the different religions, and these have led to different uses of the land and the resources of the world. The same is true of those now rare religions that are believed to offer stewardship of the world to one specific group of people. The animism of the Mongols, for example, in common with that of certain other steppe peoples, was used to help justify the destruction of resources, including people, not immediately wanted or needed by khans and other leaders.

Buddhism stresses the endless cycle of birth, rebirth, and suffering in which souls are reincarnated in a variety of forms through the ages. Since souls could inhabit not only animate, but also inanimate objects, then it benefits people to take care of those items appropriately. They may be used in moderation, but not abused and used excessively. Other religious beliefs also confer upon humanity the right to use natural resources, but with certain limitations. The same is true of some moral creeds that have an environmental basis. Proponents of the Gaia hypothesis, for example, hold the resources of the Earth to be central to the successful existence of nature; consequently, husbanding of those resources is a central part of the successful functioning of society.

Belief systems based on nonreligious bases have not always been so favorable to the environment. Communism, for example, appropriates the resources of the Earth for the betterment of society, and has little to say about conservation of those resources. The impact on the environment by the Russian and Chinese communist parties has been among the most severe in the world. Similar levels of exploitation of resources, such as pollution and overlogging, for example, are also witnessed when private-sector, free-market interests have been able to gain access to resources. The *Tragedy of the Commons* by William Foster Lloyd framed the potential problem of a laissez-faire approach to the management of nature. The presence of democracy in a country, accompanied by fair and transparent policing of the laws, is one of the best means of ensuring that overexploitation of resources does not take place. The Indian economist Amartya Sen, who observed that no famine had ever occurred in a functioning democracy, originally noted this concept.

Diminution of Natural Resources

The history of cod fishing in the Atlantic Ocean is a graphic example of the abundance of resources available in past centuries and the way in which those resources have been enormously diminished within the last century. For hundreds and in some cases thousands of years, the ability of man to harvest resources, renewable resources in any case, was exceeded by the fertility of nature in replacing them. The development of industrialized harvesting techniques succeeded not only in depleting stocks of the fish, but also seriously damaged the environment, including the ecology in which the cod thrived. The diminution of this resource, in common with so many others, has been so severe that it is not possible to recreate a satisfactory understanding of the amount of the resource previously available. This makes it extremely difficult to identify means of returning to the status quo before overexploitation and, hence, it is not very likely that such a state could ever be attained. It would take an event as severe in its impact as World War II, which effectively prevented deep sea fishing in the Atlantic Ocean altogether for several years, for fish stocks to be replenished to any meaningful degree.

Overfishing has probably already destroyed ocean ecologies beyond repair, and the same is true of the logging of hardwood trees in the former rainforests of Thailand and Burma. Arguments persist over whether the production of oil and natural gas has yet peaked, or is at its maximum now, but the existing oil is not going to be replenished. Human society must prepare to live in a world in which many of the resources on which it had previously relied are no longer available.

Resource Allocation

Given that resources are, by definition, finite and scarce in nature, then there must be some mechanism to allocate different shares to different sets of people. Allowing everyone who has an interest in resource exploitation to do so as freely as desired will lead to disastrous depletion of the resource. Consequently, the basis of allocation must be determined.

In mature, democratic societies, coalitions of interests will help to set the agenda by which resources are allocated. This can be quite efficient in determining the share that each set of interests will gain, but has proved to be less successful in setting the amount of resource that may be allocated on a sustainable basis. Democratic debate must, consequently, be supplemented with a technical limitation determining overall exploitation in order to be viable. This is a superior approach to those that rely on market power (where resources go to those who can afford them and are denied to the poor), since these approcahes suffer from equity issues and, more relevantly, from the stress inflicted by inequality and the high probability that it will lead to social unrest and ultimately rebellion.

Systems that reward the rich at the expense of the poor rely, therefore, on the ability and willingness of the former to mobilize the threat of armed violence against the latter. Even so, social systems of this sort still rely upon the labor of the masses to produce goods and services to facilitate the lifestyle of the rich. Sequestering oxygen or water, therefore, which are essential for life, will provide short and possibly medium-term gains for the rich, but the system is not sustainable over the long term because it will lead to the deaths of so many of the poor. This might not threaten the survival of the system, but the reduction in production capacity will.

Irrespective of the means by which resource allocation is managed, it must be supplemented by attempts to determine the presence or creation of substitutes. Resource scarcity leads to inequality, and this reduces social stability. The promise of suitable substitutes at some stage in the future helps to alleviate the pressure that this builds.

John Walsh
Shinawatra University

See Also: Alternative Energy, Overview; Ethics; Oil, Consumption of; Oil, Production of; Religion.

Further Readings

Klare, M. T. *Resource Wars: The New Landscape of Global Conflict.* New York: Henry Holt, 2002.
Kurlansky, Mark. *Cod: A Biography of the Fish That Changed the World.* New York: Penguin, 1998.
Lloyd, W. F. *Lectures on Population, Value, Poor-Laws and Rent.* New York: Augustus M. Kelley, 1968.
Marsden, William. *Stupid to the Last Drop: How Alberta is Bringing Environmental Armageddon to Canada (and Doesn't Seem to Care).* Toronto, Canada: Alfred A. Knopf, 2007.
Sen, Amartya. *Development as Freedom.* New York: Anchor, 2000.
Shiva, Vandana. *Water Wars: Privatization, Pollution, and Profit.* Cambridge, MA: South End Press, 2002.

Resources for the Future

Resources for the Future (RFF) is a nonprofit, nonpartisan organization with headquarters in Washington, D.C. Founded in 1952 under the Truman administration, RFF initially focused on domestic issues, but later expanded to include international affairs. RFF was the first U.S. think tank devoted exclusively to environmental issues. The impetus for RFF came from the Columbia Broadcasting System's Board Chairman William Paley after he led a presidential commission charged with examining whether the United States was becoming overly dependent on foreign natural resources and commodities.

Today, the RFF board of directors consists of members of the business community, former state officials, academics, and leaders of environmental advocacy organizations. RFF has approximately 40 staff researchers composed of senior fellows, fellows, resident scholars, research assistants, and associates. In addition, RFF hosts visiting scholars from academia and the policy community. RFF scholars conduct research in dozens of countries and share their findings through seminars and conferences, congressional testimony, and global media. They publish in external peer-reviewed journals and several RFF publications, including discussion papers, reports, issue briefs, and *Resources* magazine. RFF Press offers hundreds of titles on environmental issues written by the organization's staff and outside experts. The online publication *Weathervane* is a guide to global climate policy. RFF *Connection* is an electronic newsletter that provides updates on events, research, and publications. The RFF's Seminar Series provides the Washington, D.C., community with a weekly forum in which scholars, journalists, advocates, and policymakers interact.

In 2010, RFF had an operating revenue of $11.4 million, of which nearly 76 percent came from individual contributions and private foundations. Approximately 18 percent was generated from government grants and the rest was withdrawn from a reserve fund created to support RFF's operations. This fund was valued at nearly $23 million in 2010.

RFF provides intellectual leadership in environmental economics. RFF scholars compile core knowledge on a range of environmental topics with the goal to contribute to scholarship, teaching, debate, and decision making. RFF internships and doctoral and postdoctoral fellowships train and support future leaders and scientists. Research methods are based in the social sciences and quantitative economic analysis, including cost–benefit tradeoffs, valuations, and risk assessments.

The organization's five major focus areas are energy and climate, health and environment, the natural world, regulating risk, and transportation and urban land. Nevertheless, the RFF staff conducts research on more than a dozen broad environmental areas, including waste management, resource management, and air quality. RFF provides ongoing support to many governmental and nongovernmental organizations, including the Intergovernmental Panel on Climate Change, which has included a number of current and former RFF staff. RFF is certified as a U.S. General Services Administration (GSA) Management, Organizational, and Business Improvement Services (MOBIS) contractor for consulting, survey, and facilitation services. MOBIS contractors assist the federal government in responding to new mandates and evolving practices. RFF researchers also analyze policies for state and local governments, as well as the business sector.

Critiquing U.S. Climate Policy

RFF has spent many years critiquing options for the design of U.S. federal climate policy. In June 2011, the findings of a two-year study suggested that federal agencies lacked the necessary policies and flexible institutional capacity to respond effectively to extreme climatic events and changing weather patterns. RFF's report, *Reforming Institutions and Managing Extremes*, outlines steps that the U.S. government can take to address climate impacts more effectively. The authors emphasize opportunities in identifying connections and synergies between adaptation and mitigation policies and other types of state policies across a number of different agencies and sectors. The report also stresses the need to increase equity and social justice in climate policies.

An important role of RFF is to provide stakeholders and policymakers with an understanding of policy options from which effective federal norms might be crafted. RFF researchers have estimated the costs of emissions abatement, calculated the benefits of mitigating climate change impacts, assessed the effect of the choice of discount rate for long-term policies, and characterized uncertainty in such analyses. In 2010, RFF and the National Energy Policy Institute released a comprehensive analysis of more than 30 available policy options for reducing U.S. fossil fuel consumption, as well as mitigating and adapting to climate change. With input from top academic experts, RFF analysts compared how different polices rank in terms of economic costs, political viability, and effectiveness in reducing greenhouse gas emissions and the number of barrels of oil consumed.

Mary Finley-Brook
Mary Brickle
University of Richmond

See Also: Climate Policy, U.S.; Policy, International; Resources; Sustainability.

Further Readings

Resources for the Future. *Annual Report.* Washington, DC: RFF, 2010.
Resources for the Future. http://www.rff.org (Accessed June 2011).

Revelle, Roger

An early predictor of global warming, Roger Revelle (1909–91) helped to start the scientific debate on the issue in the late 1950s. He challenged the accepted notion that global warming was countered by the absorption of carbon dioxide from the oceans. Revelle discovered that the particular chemistry of sea water hinders such absorption. Because of the respect that he earned among the scientific community, Revelle was regarded as a spokesperson for science whose advice on as diverse matters as world population, agricultural policies, education, and the preservation of the environment were held in high esteem.

Born in Seattle, Washington, on March 7, 1909, Revelle was raised in Pasadena, California, and soon stood out as a gifted student during his academic career. In 1925, Revelle enrolled at Pomona College with an interest in journalism, but later switched to geology as his major field of study. In 1928, Revelle met Ellen Virginia Clark, a student at the neighboring Scripps College and a grandniece of Scripps College founder Ellen Browning Scripps. The couple married in 1931.

Revelle obtained his bachelor's degree from Pomona in 1929, and then entered the University of California–Berkley to pursue his studies in geology. In 1931, his professor, George Davis Louderback, recommended him for a research assistantship in oceanography at the Scripps Institute of Oceanography in La Jolla, California. While at Scripps, Revelle took part in several expeditions on the *Scripps*, the institute's small research vessel. He was also a guest on ships of the U.S. Coast and Geodetic Survey and the U.S. Navy. In 1936, Revelle completed his dissertation, "Marine Bottom Samples Collected in the Pacific Ocean by the Carnegie on Its Seventh Cruise," and was awarded his Ph.D. He was immediately hired as an oceanography instructor at Scripps, under the directorship of Harald Sverdrup.

During World War II, Revelle served in the U.S. Navy as the commander of the oceanographic section of the Bureau of Ships and became head of their geophysics branch in 1946. His reputation and influence in the navy quickly grew during the war and enabled him to substantially influence the navy research program in oceanography. Revelle received an official commendation for this work from Secretary of the Navy James Forrestal after the war. After the conflict, he returned to Scripps in 1948 and directed it from 1951 to 1964. The scientist was involved in supervision of the first postwar atomic test on Bikini Atoll, Operation Crossroads. He led the oceanographic and geophysical components of the operation. His task was to study the diffusion of radioactive wastes and the environmental effects of the bomb at Bikini.

During his directorship at Scripps, Revelle was also appointed to other prestigious positions, such as chairman of the Panel on Oceanography of the U.S. National Committee on the International Geophysical Year (IGY). Scripps, which Revelle was constantly expanding thanks to his administrative skills, was initially designated as a participant in the IGY, but was later promoted to the main center in the Atmospheric Carbon Dioxide Program. As a result, Revelle's interest in the general carbon cycle and the solubility of calcium carbonate grew and he began a systematic research that engaged him for the rest of his life. The result of this interest was a famous 1957 article published in *Tellus*, a European meteorology and oceanography journal, which Revelle

Roger Revelle, who led U.S. Navy research in oceanography. His carbon dioxide (CO_2) research formed the basis of his predictions of global warming. He noted that the particular chemistry of sea water hinders the absorption of CO_2.

coauthored with Hans Suess, one of the founders of radiocarbon dating. The article demonstrated that carbon dioxide had increased in the air as a result of the use of fossil fuels.

Following this discovery, in 1963, Revelle took a leave of absence from Scripps and committed himself to public policy. Revelle was among the first in the scientific community to bring the subject of rising levels of carbon dioxide to the attention of the public as a member of the President's Science Advisory Committee Panel on Environmental Pollution in 1965. The committee, chaired by Revelle, published the first authoritative U.S. government report that officially stated that carbon dioxide from fossil fuels was a potential global threat. Revelle also founded the Center for Population Studies at Harvard University, and spent more than a decade as director. His primary interests were applications of science and technology to world hunger. In 1976, Revelle returned to the University of California–San Diego where he received the title of professor of science and public policy and joined the Department of Political Science.

As the chair of the National Academy of Sciences Energy and Climate Panel in 1977, Revelle concluded that about 40 percent of the anthropogenic carbon dioxide has remained in the atmosphere. It was produced two-thirds from fossil fuel, and one-third from the clearing of forests. In his role of spokesperson against the dangers of global warming, Revelle influenced public opinion on the carbon dioxide issue thanks to a widely read article published in *Scientific American* in August 1982.

His research emphasized the rise in global sea level and the melting of glaciers and ice sheets caused by the thermal increase of the warming surface waters. Through his international scientific contacts, Revelle circulated his research findings and fostered debates about his findings and the threatening environmental and social effects of increased atmospheric carbon dioxide. The scientist was an early advocate of governmental policy and action.

Revelle was a respected member of many academic, scientific, and government committees. He was science adviser to the secretary of the interior, president of the American Association for the Advancement of Science, and a member of the NASA Advisory Council. In November 1990, Roger Revelle received the National Medal of Science from President George Bush.

Luca Prono
Independent Scholar

See Also: Carbon Cycle; Carbon Dioxide; Global Warming; Greenhouse Gas Emissions; Navy, U.S.; Scripps Institution of Oceanography; Sea Level, Rising; Sverdrup, Harald Ulrik.

Further Readings

Morgan, Judith and Neil Morgan. "Roger Revelle: A Profile." http://scilib.ucsd.edu/sio/biogr/Morgan_roger_revelle.pdf (Accessed March 2012).

Weart, Spencer. *The Discovery of Global Warming*. Cambridge, MA: Harvard University Press, 2004.

Rhode Island

Rhode Island is a New England state sharing borders with Connecticut and Massachusetts and having 40 mi. of Atlantic seacoast. It is the smallest U.S. state by area, but is the second most densely populated, with over 1,000 inhabitants per sq. mi (386 per sq. km). Most of the state's population lives in urban and suburban areas. Narragansett Bay, an arm of the Atlantic Ocean, is an important geographical feature of the state as it includes many harbors and inlets, which facilitated trade in the colonial period and now are an important asset for tourist development. Over half the state is forested, and it has many small lakes and streams. The leading industries are electronics manufacturing and jewelry manufacturing, with agriculture playing a minor role. Major agricultural products include greenhouse and nursery goods (accounting for over half the state's agricultural income) and dairy products.

Rhode Island is already affected by global warming. Spring melt in New England currently occurs about two weeks earlier than it did 50 years ago, weather records show warming of average temperatures, and the number of safe days for outdoor ice skating and ice fishing have declined. In Providence, Rhode Island's capital, the average temperature has increased by 3.3 degrees F (1.83 degrees C) over the last 100 years, and this increase is expected to continue, along with an increase in the number of very hot days. Increased heat will also mean greater demands for Rhode Island's limited supply of freshwater for both agricultural and residential uses; currently, 35 percent of state residents depend on well water. The ocean level is expected to rise 1–12 in. (2.5–30 cm) in Rhode Island over the next 50 years, requiring expensive containment measures and causing some coastal properties to disappear entirely underwater or otherwise become unusable.

Air Pollution and Aquaculture

Rhode Island already has an air pollution problem (the number of smog days currently exceeds the legal limit), which is expected to continue in the future if the temperature continues to increase and nothing is done to reduce current sources of air pollution (including automobile emissions). Warmer temperatures provide an excellent environment for mosquitoes and ticks, and their populations are expected to increase, potentially resulting in more cases of insect-borne diseases such as West Nile virus, Lyme disease, and eastern equine encephalitis. Rhode Island's forests are also impacted by global warning; for instance, the black and red turpentine beetle, which damages and can kill pine trees, was previously known only in more southern states, but has now become quite common in Rhode Island.

A 1997 lobster die-off in Rhode Island was associated with the onset of a bacterial shell disease because of warmer ocean temperatures. Other fish and shellfish populations are also affected by global warming, and the New England fishing industry is expected to decline as ocean temperatures become too warm for current fish populations, particularly cod. Warmer water temperatures are also expected to curb the spawning of winter flounder and to favor diseases that affect oysters, scallops, and quahogs.

The Rhode Island Greenhouse Gas Stakeholder Project began in 2001 as a joint project among the Department of Environmental Management, State Energy Office, and governor's office. Stakeholders in this project published the Rhode Island Greenhouse Gas Action Plan in July 2002, outlining policies intended to reduce the state's greenhouse gas (GHG) emissions to 1990 levels, with further reduction by 2020 (to 10 percent below

In 1969, massive storm surges from Hurricane Carol raged through the Rhode Island Yacht Club. Decades later, the state is feeling the effects of global warming; spring melt in New England now occurs about two weeks earlier than it did 50 years ago.

1990 levels) and in the long term (to 75 percent or more below 1990 levels). The Greenhouse Gas Action Plan includes 49 consensus initiatives intended to lower the state's GHG emissions in the categories of buildings and facilities, transportation, land use, energy supply, and solid waste. Among the highest-priority items in the plan are a program to retrofit facilities heated with oil or natural gas to conserve those fuels, a program to encourage households to select the smallest appropriate appliances, an initiative to retrofit buildings heated by electricity to conserve fuel, creation of tax credits for energy efficiency, creation of a program to impose fees on consumers buying low-efficiency vehicles and offer rebates to those purchasing higher-efficiency vehicles, and requiring that a minimum percentage of retail electricity sold in Rhode Island be generated from renewable sources.

Sarah Boslaugh
Kennesaw State University

See Also: Forests; Massachusetts; Maximum Sustainable Yield; Pollution, Air; Sea Level, Rising.

Further Readings

Environment Council of Rhode Island. "Global Warming in Rhode Island: Warning Signs, Winning Solutions." http://www.dem.ri.gov/climate/pdf rigw .pdf (Accessed July 2011).

State of Rhode Island Department of Environmental Management. "Greenhouse Gas Project." http://www.dem.ri.gov/programs/bpoladm/stratpp/green hos.htm (Accessed July 2011).

Union of Concerned Scientists. "Backgrounder: Northeast." http://www.ucsusa.org/assets/documents/global_warming/us-global-climate -change-report-northeast.pdf (Accessed July 2011).

Richardson, Lewis Fry

Lewis Fry Richardson (1881–1953) was an innovative British mathematician, physicist, and psychologist who first tried to apply mathematical concepts to weather forecasting. Although his method for weather forecasting was not entirely successful during his lifetime, it was rediscovered with the advent of computers and formed the basis for computer-based weather forecasting. The recorded change over a given distance of temperature and wind (gradient) is named the Richardson number, after him.

Lewis Fry Richardson was born into a wealthy Quaker family in Newcastle-upon-Tyne on October 11, 1881. His mother, Catherine Fry, was the daughter of corn merchants, and his father, David Richardson, came from a family of tanners, a profession that he also took up. Lewis was the youngest of a large family of seven children. He attended Newcastle Preparatory School where he already showed his predilection for math, particularly the study of Euclid. In 1894, he went to Bootham School in York, an elite Quaker institution established in 1823. At this school, Richardson first combined his interest for math with science, and meteorology in particular. One of his teachers, Edmund Clark, was an expert in meteorology and greatly influenced Richardson. The institution also reinforced Lewis's pacifism, a value that had been taught to him by his parents and a fundamental tenet of Quakerism that led him to difficult career choices in his maturity. After leaving Bootham in 1898, Richardson spent two years in Newcastle at the Durham College of Science where he studied mathematics, physics, chemistry, botany, and zoology. Richardson completed his education at King's College, Cambridge, from which he graduated with a First Class degree in the Natural Science Tripos in 1903.

After graduation, Richardson was employed at many different posts. He worked in the National Physical Laboratory (1903–04, 1907–09) and the Meteorological Office (1913–16), and he was hired as a university lecturer at University College Aberystwyth (1905–06) and Manchester College of Technology (1912–13). In addition, he was a chemist with National Peat Industries (1906–07) and directed the physical and chemical laboratory of the Sunbeam Lamp Company (1909–12). He married Dorothy Garnett in 1909, and although they had no children of their own, they adopted two sons and a daughter.

Richardson was working for the Meteorological Office as superintendent of the Eskdalemuir Observatory at the outbreak of World War I in 1914. Because of his Quaker beliefs, he declared himself a conscientious objector and could not, therefore, be drafted into the military. This choice implied that he would never be able to qualify for university posts. While Richardson was not involved in military operations, from 1916 to 1919, he served in the Friends Ambulance Unit, attached to the 16th French Infantry Division, where his work earned him praise. After the war, Richardson returned to his position in the Meteorological Office, but had to resign from it in 1920 when the Meteorological Office became part of the Air Ministry. His pacifist beliefs could not allow him to continue to work for an institution that was part of the military. Richardson then went back to teaching. From 1920 to 1929, he headed the Physics Department at Westminster Training College, and from 1929 to 1940, he was principal of Paisley College of Technology and School of Art in Scotland. He retired in 1940 at the age of 59 to concentrate on research.

Richardson had a lifelong interest in the application of mathematics to meteorology. He was the first to apply the mathematical method of finite differences to the prediction of the weather in his study *Weather Prediction by Numerical Process* (1922). His method of finite differences was designed to solve differential equations, arising in his work on the flow of water in peat for National Peat Industries. As these methods allowed him to obtain highly accurate solutions, he decided to apply them to solve the problems of the dynamics of the atmosphere encountered while working for the Meteorological Office. The initial conditions were defined through observations from weather stations, and would then be used to solve the equations. Finally, a prediction of the weather could be made. Richardson's remarkable insight was ahead of its time, since the time taken for the necessary hand calculations in a pre-computer age took too long. Even with a large group of people working to solve the equations, the solution could not be found in time to be useful to predict the weather. Richardson admitted that it would take 60,000 people to have the prediction of tomorrow's weather before the weather actually arrived. In spite of this flaw, Richardson's work pioneered present-day weather forecasting.

Throughout his lifetime, Richardson published extensively on the application of mathematics to the weather and contributed to the theory of diffusion, specifically regarding eddy-diffusion in the atmosphere. For his scientific achievements, he was elected to the Royal Society in 1926. His deeply rooted interest in pacifism led him to apply mathematics to the study of wars and military conflicts. His results were published in three major books: *Generalized Foreign Politics* (1939), *Arms and Insecurity* (1949), and *Statistics of Deadly Quarrels* (1950). Richardson used mathematics to challenge the assumption that war was a rational national policy in the interests of a nation. He gave systems of differential equations governing the interactions between countries. Starting with the armament of two nations, Richardson constructed an idealized system of equations calculating the rate of a nation's military buildup as directly proportional to the amount of arms its rival has and also to the disputes toward the enemy. This rate is, instead, negatively proportional to the amount of arms it already has. Richardson died on September 30, 1953, in Kilmun, Argyll, Scotland.

Luca Prono
Independent Scholar

See Also: Climate Models; Weather.

Further Readings

Ashford, O. M. *Prophet—Or Professor? The Life and Work of Lewis Fry Richardson*. Bristol, UK: Adam Hilger, 1985.

Korner, T. W. *The Pleasures of Counting*. Cambridge: Cambridge University Press, 1996.

Lynch, P. *The Emergence of Numerical Weather Prediction*. Cambridge: Cambridge University Press, 2006.

Weart, Spencer. *The Discovery of Global Warming*. Cambridge, MA: Harvard University Press, 2004.

Rio+20

Rio+20 is the common name for the United Nations (UN) Conference on Sustainable Development, a global meeting to be held in Rio de Janeiro, Brazil, in June 2012. Its goal is to discuss and potentially create international agreements on policies and political institutions that can help generate economic development in ways that are beneficial, or at least not destructive, to the environment. It is the follow-up to the 1992 UN Conference on Environment and Development, also held in Rio de Janeiro.

Structured as a global conference, rather than an international treaty negotiation, Rio+20 is not designed to arrive at specific, legally binding agreements for participating countries. Rather, the conference is meant to be a public forum to discuss sustainable development problems and possible solutions and will be held at a level of prominence intended to attract the attention of world leaders and other powerful organizations around the world. Its organizers hope it will be the largest gathering of heads of state since the Copenhagen climate negotiations in December 2009. Because of this anticipated high level of engagement by world leaders, the conference could, however, produce political agreements by heads of governments, who could then convert those agreements into law by passing domestic legislation in their home countries.

Reviews and Themes

The conference is structured around two types of reviews and two areas of discussion. The reviews focus on covering problems in reaching sustainable development and what has been done in the past to solve those problems. The themes will focus on new solutions to sustainable development problems that have not previously been solved.

The first type of review concerns progress made so far to meet agreements reached at major international negotiations and summits on sustainable development issues. For example, this includes assessing how far countries have reduced greenhouse gas (GHG) emissions—something governments have agreed to do through negotiations at the UN Framework Convention on Climate Change (UNFCCC) meetings held over the last two decades. Governments will catalog gaps in the implementation of these agreements, areas of success, and overlaps in how commitments in one agreement may affect a country's ability to meet commitments in other agreements, such as how preserving forest plants and animals can also lead to carbon storage by conserving trees, thus helping to reduce climate change.

The second type of review considers new and emerging problems that were not in existence, not understood, or simply not addressed at previous international sustainable development summits. Topics to be considered include ocean acidification—caused by the oceans absorbing increasing GHGs—and its impact on ocean biodiversity and fishing economies. These types of emerging, interconnected issues, for example, where a problem in one area such as climate change makes another problem like species extinction worse, will receive special attention.

The first theme, as defined by the UN, concerns the green economy in relation to sustainable development and the eradication of poverty. While Rio+20 does not have a precise definition of the green economy, it is generally understood to mean economic development that is beneficial or restorative to the natural environment, or that at least does not harm the natural environment. Some countries involved in the Rio+20 negotiations have expressed concern that the green economy theme is meant to promote green protectionism, or the exclusion of some countries from global economic activity or trade on the basis of environmental or social protection rules. However, most countries in the Rio+20 process view the green economy theme as a way to facilitate environmentally sustainable economic development.

Some of the methods discussed for creating a green economy involve ways of generating energy without producing GHGs, including greater use of renewable energy sources such as wind and solar, as well as behaviors and technologies that will improve the energy efficiency of businesses and households. Other discussions concern the potential challenges, problems, and benefits of creating new indicators of progress toward sustainable development, such as possibly supplementing or replacing the economic indicator of gross domestic product with measures of human well-being and environmental health.

The second theme concerns institutional frameworks for sustainable development. This theme refers to the national and international organizations (institutions) that governments have created to realize sustainable development policies. For example, the World Bank is an international financial institution that sometimes makes grants and loans to countries to help them reduce their greenhouse gas pollution. The UN is another well-known institution. This theme is meant to foster debate on whether the current institutions are sufficient to achieve sustainable development, whether some should be eliminated, be combined, or have their responsibilities altered, or whether new institutions are needed.

For example, many nations in the Rio+20 process have discussed whether the UN Environment Programme, which reviews global environmental initiatives, should become the coordinating organization for all international environmental treaty implementation. Currently, many treaties operate as isolated institutions, without a central overarching body. According to the institutional framework discussion, a lack of coordination across institutions or a lack of international institutions with sufficient power to enforce sustainable development policies might prevent sustainable development from happening. Providing more power or more efficient organizational structures could then allow existing and newly developed sustainable development activities, such as those discussed under the green economy theme, to be more effective.

Preparatory Meetings

Although the term *Rio+20* officially refers only to the 2012 conference, a large number of preparatory activities, meetings, and negotiations are being undertaken before the Rio+20 conference to ensure that the reviews, thematic discussions, and any potential political agreements conclude by the end of the conference. Preparatory meetings began in May 2010. More than a dozen preparatory meetings, held both at the UN headquarters in New York City and in regional discussions throughout the world, will occur before governments meet in Rio. At these meetings, the themes and reviews are debated and discussed, and government representatives work to draft texts of potential agreements that may be finalized at the Rio conference.

Participation by Nongovernmental Groups

Like most UN meetings, people other than national government leaders also participate in the Rio+20 preparatory processes and will participate in the Rio+20 meeting. Collectively, these groups are called civil society. The United Nations recognizes nine types of civil society groups, termed *major groups*. These groups focus on nongovernmental advocacy and policy, children and youth, businesses and industry, indigenous peoples, farmers, subnational governments, science and technology, trade unions and workers, and concerns specifically related to women's interests. Various UN agreements and principles provide access and participatory rights to these major groups so that the conference's outcomes reflect the concerns and knowledge of all members of society, not just national governments.

History

Rio+20 is the second and follow-up to the 1992 UN Conference on Environment and Development, commonly known as the Rio Earth Summit. This meeting, considered to have been a significant achievement in international sustainable development cooperation because it coincided with the launch of three international environmental treaties, had nearly universal participation from governments worldwide. The Rio Conventions, as they are known, include the Framework Convention on Climate Change, the Convention on Biological Diversity, and the Convention to Combat Desertification. A statement on Forest Principles was also agreed to at the Earth Summit, although it failed to become an international treaty.

The Earth Summit also created the UN Commission on Sustainable Development (UNCSD), a yearly meeting to continue discussing agreements reached at the Earth Summit. This meeting focuses on reviewing detailed sustainable development topics such as mining, hazardous chemicals, and waste disposal. Rio+20 replaces the 20th meeting of the UNCSD. The Earth Summit also created a system to develop local and national programs to achieve the international sustainable development calls set at Rio. These programs are known as Agenda 21 activities.

In 1997, Rio+5 was held as a meeting of the UN General Assembly (UNGA) in New York City to review progress in meeting Earth Summit agreements. In 2002, Rio+10, held in Johannesburg, South Africa, and known as the World Summit on Sustainable Development, focused on partnerships between governments, businesses, educational institutions, and nonprofit organizations reaching Earth Summit goals. Future conferences or summits beyond Rio+20 are not automatically planned, as each meeting requires a decision by all of the UN countries through the UN General Assembly (UNGA). Approval for Rio+20 was agreed to by the UNGA in December 2009 during, but separate from, the climate negotiations in Copenhagen.

Kyle Brian Gracey
Global Footprint Network

See Also: Brazil; Green Economy; Policy, International; Sustainability; United Nations Framework Convention on Climate Change.

Further Readings

Stakeholder Forum. "Earth Summit History." http://earthsummit2012.org/beta/earth-summit-history (Accessed July 2011).

United Nations. "Rio+20—United Nations Conference on Sustainable Development." http://www.uncsd2012.org/rio20 (Accessed July 2011).

United Nations Environment Programme. "Green Economy." http://www.unep.org/greeneconomy (Accessed July 2011).

Risk

Risk is a concept that captures the probability and, in some instances, the potential severity of the occurrence of a negative outcome (i.e., being exposed to a hazard). There is much discussion surrounding the various risks associated with global warming and climate change, such as those related to the environment, ecosystem, human health, and the world economy. In this regard, various experts have used risk analysis to assess, manage, and communicate these associated risks.

Global warming occurs as a result of the accumulation of greenhouse gases in the atmosphere. These greenhouse gases occur both naturally (such as water vapor, carbon dioxide, methane, and ozone) and as a result of human activity (e.g., from the burning of fossil fuels, deforestation, and the use of chlorofluorocarbons and fertilizers). The latter has been the focus of a 2007 report from the Intergovernmental Panel on Climate Change (IPCC). This has led to a great deal of discussion surrounding the various policy implications that lie ahead.

There are several environmental risks associated with global warming and climate change—some of which have been noted by researchers worldwide as already occurring, and others that have been forecast. Climate change affects countries differently depending partly on their geographical location. The record high temperatures documented over the past two decades have resulted in early ice thaw on rivers and lakes. Furthermore, the rates of sea-level rise are expected to continue to increase as a result of both the thermal expansion of the oceans and the partial melting of mountain glaciers and the Antarctic and Greenland ice caps. It has been reported that increasing temperatures have also been the cause of many extreme weather events, such as heat waves, droughts, and wildfires.

Moreover, changes to ocean temperatures and wind patterns have resulted in more frequent and intense rain and ice storms, floods, and some natural disasters such as hurricanes and typhoons. An increase in the frequency and severity of disasters can lead to secondary effects such as massive mudslides—as was the case with Hurricane Mitch in 1998. They can also have far-reaching effects that can result in a loss of livelihood, displacement, as well as local and global migration, particularly for those living in communities in the most vulnerable areas (such as low-lying coastal areas and estuaries, alpine regions, and tropical and subtropical population centers). Additionally, these risks can be exacerbated for marginalized groups, as well as those living in more vulnerable communities (e.g., regions that are poverty laden or more crowded).

Human Health and Disease Risks

Extreme weather events can have disastrous effects on physical and mental health, as well as environmental health. The occurrence of droughts can lead to problems associated with water availability and quality (e.g., people sharing water with livestock). Similarly, heat-related effects such as exhaustion, cramps, heart attacks, stroke, and even death are possible outcomes as a result of heat waves. Furthermore, excessive rainfall and flooding are associated with the risk of injuries and death (as from drowning), as well as the spread of various water-borne diseases (via fecal-contaminated waterways and drinking supplies), and exposure to toxic pollutants (from nearby industrial sites and municipal sewage—as was the case with the Elbe flood that took place in 2002 in central Europe). The variation of risks associated with the transmission of infectious diseases as a result of extreme weather events have also been documented; however, their relation to global warming and climate change have not yet been conclusively reported in the literature.

The risks associated with the transmission of infectious diseases are dependent upon the kind(s) of weather event(s) that have occurred. As such, the reproduction and survival of disease-carrying vectors such as mosquitoes could be impaired by heavy rainfalls (such as flushing larvae from pooled water) or heightened by changes in climate and rainfall patterns. For instance, changes in climate have allowed vector-borne diseases such as malaria and dengue fever to survive in otherwise inhospitable areas (in higher elevations). Other vector-borne diseases (such as cholera, Ross River virus, and West Nile) and food-borne diseases (like the proliferation of bacteria in contaminated foods) are also at risk of occurring as a result of higher temperatures.

As noted by the Canadian Lung Association, climate change and the effects of it can lead to air quality problems such as those resulting from the increased burning of fossil fuels as a direct result of rising temperatures (e.g., increased use of air conditioners, refrigerators, and freezers); increases in forest fires; and increased mold growth as a result of elevated levels of precipitation. Associated health effects such as asthma and allergies, as well as other respiratory-related morbidity and mortality, are also of concern. The Canadian Lung Association provides a more comprehensive explanation of the connection between climate change, air quality, and respiratory health.

Global warming and climate change can impact both the balance and health of the ecosystem, which in turn puts human health at risk. A report by the UN Environment Programme provides an overview of the relationship between climate change and ultraviolet radiation, ozone depletion, terrestrial and aquatic ecosystems, and biogeochemical cycles. Changes and losses related to biodiversity (disruption in ecosystems and species extinction) are also of concern. In a 2007 article, Frederic Jiguet and colleagues note a number of studies that have shown that habitat degradation is taking place; particular reference is made to research on plants, butterflies, beetles, mammals, bumblebees, birds, coral reefs, and coral-dwelling fishes—however, there are a number of other habitats that may also be at risk. Moreover, based on findings from the first comprehensive assessment of extinction risk, the Natural Resources Defense Council notes, "more than one million species could be committed to extinction by 2050 if global warming pollution is not curtailed."

Diversity in species is important, because it aids in ecosystem services/functions (maintaining soil fertility and pollinating plants and crops); a change in ecosystem services/functions can have far-reaching implications (it can affect agro-ecosystems, marine systems, and freshwater, as well as the transmission of vector-borne diseases). Furthermore, there are also risks associated with nutrition, which is dependent on the state of agricultural output (such as changes in food productivity and associated pests that are involved in the transmission of diseases) as well as other food sources (like fisheries and mammals); this can have consequences on ecosystem and human health in the immediate area(s) and globally.

Economic Risks

As far as businesses are concerned, environmental risks are usually understood in terms of costs, managed by regulatory compliance, potential liability, and pollution release mitigation. Supply chain risks might occur, whereby suppliers may eventually pass the costs pertaining to carbon-related and more energy efficient alternatives (technology advancement) to customers. Also, businesses that generate significant carbon emissions could face similar litigation risks (lawsuits) as those experienced by tobacco, asbestos, and pharmaceutical industries that in turn may also put the company's reputation at risk.

Organizations that seek a competitive advantage in light of global warming and climate change may use a strengths, weaknesses, opportunities, and threats (S.W.O.T.) analysis to examine their organization's strengths and weaknesses in relation to the opportunities and threats posed by the environment in order to transform the various business-related threats into opportunities. For instance, organizations that are able to identify and implement new market opportunities for climate-friendly products and services may fare better than other organizations that are unable to achieve this. Additionally, organizations may want to take measures, such as quantifying their carbon footprint (the amount of CO_2 created by business operations), in order to show consumers that they are aware of their role in climate change—and then take measures to correct the identified shortcomings. This could lead to the development of new and profitable products that are also environmentally friendly, as well as an increase in consumer loyalty.

There are other far-reaching economic implications outside the immediate business arena. For instance, the costs associated with insurance (both for the customer and for insurance companies) may affect those living in more vulnerable areas. Moreover, it has been argued that the impacts of climate change are not evenly distributed; for instance, developing countries are not only at a geographical disadvantage (with high rainfall variability), but their economic livelihood is heavily dependent on their agricultural output (which is at risk with climate change). Furthermore, the implementation of more stringent regulations, as well as the adoption of better and safer technological alternatives and advancements, may not seem like a feasible option economically for developing countries.

Risk Analysis and Assessment

There are other various types of risks that need to be considered with respect to global warming and climate change. Risk analysis is a process that considers the scientific, social, cultural, economic, and political issues that shape the identification, evaluation, decision making, and policy implementation concerning risk. As such, risk analysis encompasses the assessment, management, and communication of risks, each in turn having a framework. In the context of environmental health, Annalee Yassi and colleagues have provided a comprehensive list of hazards that arise from both natural and anthropogenic (caused or induced by human activity) sources and explain the processes involved in the assessment and management of these risks—some of the hazards identified are applicable to global warming and climate change.

The risk assessment framework most frequently used in relation to environmental and human health follows the steps first identified by Lawrence E. McCrae in 1983; these steps include problem formulation, hazard identification, dose-response relationships, exposure assessment, and risk characterization. Once characterized, risks are sometimes ranked by organizations through the use of a risk matrix. Risk matrices are made up of rows and columns that denote the severity or impact of the hazard and the likelihood or probability of its occurrence, where the former is usually more subjective, and the latter is relatively objective.

Accurately predicting (through the use of modeling and other forecasting techniques) and assessing (via the five aforementioned steps) the environmental, ecosystem, and human health risks related to global warming and climate change is challenging because of the level of complexity and uncertainty involved. These factors can result for various reasons; among these are: incomplete, insufficient or inaccurate data; gaps and errors in observation; measurement error(s); lack of knowledge (inadequate or conflicting modeling, or unknowns pertaining to feedback effects); variation in assumptions; unforeseen circumstances; variability in natural conditions, exposures, and human activity; and contributions of natural and human-induced changes.

Risk Management Strategies

Both risk assessment and cost-benefit analysis (a tool used to determine whether or not costs outweigh benefits—the applicability of this tool has been a widely debated issue in the face of climate policy) usually drive risk management. As such, risk management is the process through which a regulatory agency decides which action(s) to take and which regulation(s) and policies to implement based on risk assessment estimate(s) and cost-benefit analyses. With respect to global warming and climate change, there are three risk management strategies that are frequently discussed in the literature—these are mitigation, adaptation, and geoengineering.

Mitigation strategies are concerned with taking actions that are aimed at reducing the extent of global warming and climate change. For instance, mitigation strategies might include adopting measures that reduce anthropogenic sources of climate change (e.g. limiting greenhouse gas emissions relating to fossil fuel combustion, or deforestation). On the other hand, adaptation strategies are aimed at decreasing vulnerability to global warming and climate change.

Adaptation solutions include insulating buildings (for heat-related illness); installing window screens (for vector-borne diseases); and constructing strong sea walls (for health and extreme weather events).

A third strategy utilizes geoengineering; this is essentially when large-scale manipulation of the environment takes place in an attempt to correct climate change. An example of geoengineering would include the (proposed) manipulation of the Earth's global energy balance by blocking a percentage of sunlight (via the use of superfine reflective mesh, or orbiting mirrors) in order to offset the doubling of carbon dioxide. However, approaches such as these carry great risks and are therefore largely debated.

Good risk communication includes the two-way exchange of information, concerns, and preferences between decision makers and the public in a manner where the mutual understanding of risks is achieved. The risks associated with global warming and climate change are numerous and have far-reaching implications; it is therefore imperative that all aspects of risk are carefully considered.

Ann Novogradec
York University

See Also: Climate Change, Effects of; Diseases; Ecosystems; Global Warming, Impacts of; Health.

Further Readings

Commission on Life Sciences National Research Council. *Risk Assessment in the Federal Government: Managing the Process*. Washington, DC: National Academy Press, 1983.

Klaassen, C. D. *Casarett & Doull's Toxicology: The Basic Science of Poisons*. New York: McGraw-Hill, 2001.

McCarthy, J. J., et al. *Climate Change 2001: Impacts, Adaptation, and Vulnerability*. Cambridge: Cambridge University Press, 2001.

Wright, Peter, et al. *Strategic Management Concepts*. Upper Saddle River, NJ: Prentice Hall, 1998.

Yassi, Annalee, et al. *Basic Environmental Health*. Oxford: Oxford University Press, 2001.

Romania

Romania is located in the southeastern part of central Europe, north of the Balkan Peninsula on the Lower Danube and bordering the Black Sea. The main topographical components of Romania are the Carpathian Mountains, Sub-Carpathians, Danube Delta, Black Sea hills, and plateaus, plains, and river meadows. The Danube is the second-largest river in Europe and the Danube Delta is one of the most magnificent deltas in the world. In 1991, it was declared both a Natural World Heritage and Ramsar site. It comprises 580,000 hectares, which represents 2.5 percent of Romania's surface area (the third-largest delta in Europe).

Romania has a transitional temperate–continental climate, with oceanic influences from the west, Mediterranean from the southwest, and excessive continental from the northeast. The 2005 convective season had a record of unusually severe weather events in Romania: record temperatures, flash floods, hail, intense cloud-to-ground lightning strikes, and many severe wind-related events like tornadoes, downbursts, waterspouts, and funnel clouds, mainly in the southeastern part of Romania. The hottest period in 107 years was in 2007, which had an average January temperature of 10.8 degrees F (6 degrees C) higher than the average between 1961 and 1990. In terms of precipitation, 2007 was a very dry year from April through July (the most important time for agriculture) and excessively wet during August through November (the time for cropping in agriculture).

In December 1989, Romania entered into a transition to a free-market economy. Previously, concepts like sustainable development and the human dimension of sustainability were not well known or understood, and were therefore

Bucura Lake in the Retezat Mountains in Romania. Global warming has been projected to increase the risk of flooding. The 2005 season had several unusually severe weather events in the country: record temperatures, flash floods, hail, and storms.

neglected. During the process of accession to the European Union (EU), Romania implemented concrete actions in 2005, such as the development of the National Strategy and Action Plan on Climate Change—adopted by governmental decisions—and has begun the implementation of the EU Emissions Trading Scheme as well as other climate-change-related actions promoted by the EU. Romania signed the United Nations Framework Convention on Climate Change (UNFCCC) in 1992 and was the first UNFCCC Annex I Party to ratify the Kyoto Protocol in 2001. Since then, Romania has been engaged in social, economical, and financial reform that has resulted in a real reduction of emissions.

Despite some differences in model results, according to the AR4 modeling of the Intergovernmental Panel on Climate Change (IPCC), an increase in the annual average temperature compared to the period from 1980 to 1990 is expected in Europe: 0.9–2.7 degrees F (0.5–1.5 degrees C) for the period from 2020 to 2029 and 3.6–9 degrees F (2.0–5 degrees C) for the period from 2090 to 2099.

Adaptation measures are proposed and undertaken for all affected sectors, of which agriculture is a priority because it pertains to the security of the food supply. The trend in greenhouse gas (GHG) emissions has been declining in all sectors. Accordingly, it is probable that Romania will meet its Kyoto Protocol commitments on the limitation of GHG emissions in the 2008 to 2012 commitment period, primarily because emissions reductions have occurred simply by indirect domestic policies and measures and the emergence of cleaner technologies. In 2009, Romania's GHG emissions, even without land use, land-use change, and forestry (LULUCF), have decreased 53.77 percent since the base year.

The Romanian coast has been subject to serious beach erosion problems for decades. The northern area, particularly the delta coast of the Danube, is most affected. In the last 35 years, the shoreline has retreated inland between 590 and 984 ft. (180 to 300 m) and 80 hectares per year of beach area has been lost. Erosion, together with storm events and rivers draining in low-lying coastal areas, is the main factor triggering coastal flood risk. Sea-level rise, although expected to be modest for the Black Sea, may lead to the inflow of salt water into the delta.

Ingrid Hartmann
University of Hohenheim

See Also: European Union; European Union Emissions Trading Scheme; Floods; Land Use, Land-Use Change, and Forestry.

Further Readings

Ministry of Environment and Forests, Romania. "Romania's Fifth National Communication on Climate Change Under the United Nations Framework Convention on Climate Change" (2010). http://unfccc.int/resource/docs/natc/rou_nc5_resbmit.pdf (Accessed July 2011).

Ministry of Environment and Sustainable Development, Romania. "Romania's Initial Report Under the Kyoto Protocol (2007)." http://unfccc.int/files/national_reports/initial_reports_under_the_kyoto_protocol/application/pdf/romanias_initial_report_under_the_kyoto_protocol.pdf (Accessed July 2011).

Ministry of Environment and Water Management, Romania. "National Action Plan on Climate Change of Romania (2005–2007)." http://unfccc.int/resource/docs/nap/romnap1.pdf (Accessed July 2011).

National Environmental Protection Agency, Romania. "Romania's Greenhouse Gas Inventory:

1989–2009 National Inventory Report" (2011). http://www.anpm.ro/upload/33020_Romanian%20 NIR-NGHGI%202011%20v.%201.3.pdf (Accessed July 2011).

Policy Research Corporation. "Romania" (2011). http://ec.europa.eu/maritimeaffairs/climate_change/ romania_en.pdf (Accessed July 2011).

Rossby, Carl-Gustav

Carl-Gustav Rossby (1898–1957) was a Swedish American meteorologist whose innovations in the study of large-scale air movement and introduction of the equations describing atmospheric motion were largely responsible for the rapid development of meteorology as a science. Rossby explained the large-scale motions of the atmosphere in terms of fluid mechanics and was one of the first scientists to notice the problem of global warming.

Rossby was born on December 28, 1898, in Stockholm, Sweden. When he was 20, he moved to Bergen, Norway, to study under pioneering atmospheric scientist Vilhelm Bjerknes at the Geophysical Institute. At that time, Bjerknes and his Bergen School were making great progress in laying the foundations of meteorology as a science with their breakthroughs in the polar front theory and air mass analysis. The center was the world's leading center of meteorological research. The young Rossby contributed his brilliant ideas to the development of the group's projects. Because of the impact of Bjerknes's guidance, Rossby, who had previously been interested in studying mathematics and astronomy, committed himself to meteorology.

Rossby moved to the United States in 1926, where he worked in Washington, D.C. He was employed as a fellow of the American-Scandinavian Foundation for Research at the U.S. Weather Bureau to explain the innovations of the polar front theory. While at the Weather Bureau, Rossby established the first weather service for civil aviation. The Weather Bureau was not a stimulating context for Rossby, who in 1928 became professor and head of the first department of meteorology in the United States at the Massachusetts Institute of Technology, Cambridge. At MIT, he made important contributions to the understanding of heat exchange in air masses and atmospheric turbulence. He also investigated oceanography to study the relationships between ocean currents and their effects on the atmosphere.

Rossby was given American citizenship in 1938. The following year, he became assistant chief of the Weather Bureau. In that capacity, Rossby was responsible for research and education and began his studies of the general circulation of the atmosphere. In 1941, he became chairman of the department of meteorology at the University of Chicago. Rossby carried out pioneering work on the upper atmosphere, proving how it affects the long-term weather conditions of the lower air masses. Measurements recorded with instrumented balloons had demonstrated that in high latitudes in the upper atmosphere, there is a circumpolar westerly wind that overlies the system of cyclones and anticyclones lower down. In 1940, Rossby developed the theory of wave movement in the polar jet stream. He demonstrated that long sinusoidal waves of large amplitude, now known as Rossby waves, are generated by perturbations caused in the westerlies by variations in velocity with latitude. Rossby also showed the importance of the strength of the circumpolar westerlies in determining global weather. When these are

Carl-Gustav Rossby with a rotating tank used to study atmospheric motion. Rossby explained atmospheric motion in terms of fluid mechanics and was one of the first scientists to notice the problem of global warming.

weak, cold polar air will sweep south, but when they are strong, they cause the normal sequence of cyclones and anticyclones. Rossby worked on mathematical models for weather prediction and introduced the Rossby equations, which with the introduction of digital computers in the 1950s, were of fundamental importance to forecastin the weather. During World War II, Rossby was in charge of training military meteorologists, and at the end of the war hired many of them to work in his department at the University of Chicago. Rossby served as president of the American Meteorological Society for 1944 and 1945, and laid the foundations for the Society's first scientific journal, the *Journal of Meteorology*.

Rossby and his Chicago Group were able to compile weather charts over periods of five to 30 days to extract the general features, and tried to analyze these using basic hydrodynamic principles. The group made radical simplifying suppositions, ignoring essential but transitory weather effects like the movements of water vapor and the dissipation of wind energy. Still, they began to conceptualize how large-scale features of the general circulation might arise from simple dynamical principles.

In 1950, Rossby returned to Sweden, but continued to visit the United States. In his home country, he worked with the Institute of Meteorology, which he founded in connection with the University of Stockholm. From 1954 to 1957, he was instrumental in arousing interest in atmospheric chemistry and the interaction of airborne chemicals with the land and the sea. On December 17, 1956, Rossby appeared on the cover of *Time* magazine and was praised for his key role in raising meteorology to the status of science. The piece also referred to a theory that Rossby was developing as a result of his interest in atmospheric chemistry. According to Rossby, the world's climate might be altered by solar heat trapped in the atmosphere because of a buildup of carbon dioxide. This was one of the first insights into the problem of global warming and paved the way for many researches in the field. Rossby was unable to fully develop this insight, as he died on August 19, 1957, just nine months after the *Time* article.

The Carl-Gustav Rossby Research Medal is the highest award for atmospheric science presented by the American Meteorological Society for outstanding contributions to the understanding of the structure or behavior of the atmosphere. Rossby was the second recipient of this prestigious award, when it was still called, Award for Extraordinary Scientific Achievement.

Luca Prono
Independent Scholar

See Also: American Meteorological Society; Global Warming; Waves, Rossby.

Further Readings

Laskin, David, "The Weatherman and the Millionaire: How Carl-Gustaf Rossby and Harry F. Guggenheim Revolutionized Aviation and Meteorology in America." *Weatherwise*, v.58/4 (2005).

Weart, Spencer. *The Discovery of Global Warming.* Cambridge, MA: Harvard University Press, 2004.

Royal Dutch/Shell Group

The Royal Dutch/Shell group is a major contributor to the release of greenhouse gases and has been among the leading oil companies to publicly embrace the need for sustainable development, including the need to address climate change. It is by most measures the world's second-largest oil company, with over 100,000 employees, operations in over 130 countries, 2006 production of nearly 3.5 million barrels of oil equivalent per day, and proven reserves of nearly 8.5 billion barrels of oil equivalent. Shell's 2006 income was $26.3 billion on revenue of $318 billion.

History

Shell Transport began in 1833 with a British shopkeeper importing oriental shells, leading to an export/import business importing oil. Royal Dutch Petroleum Company began producing petroleum in the Dutch East Indies. A partnership was formed in 1907, expanded rapidly, and was the main fuel supplier to the British in World War I, and the world's leading oil company by 1930. During this period, it also began developing its global network of service stations. Demand for petroleum exploded after World War II. During the 1960s, Shell strengthened its presence in the Middle East, and discovered reserves in the North Sea. The 1973 oil crisis led Shell to diversify into other energy sources, such as coal and nuclear power, with little economic success. Shell also acquired 50 percent of an Australian solar energy company, and began producing renewable softwoods that could be used for paper, construction, and fuel. Shell is the world's leading biofuels distributor.

After oil prices collapsed in 1986, Shell invested in research and development that led to huge improvements in drilling techniques, and began some of its most challenging offshore exploration. During the 1990s, high oil prices allowed Shell to further develop biomass technologies. Since 2000, Shell's greatest expansion has been in China. In 2005, the old partnership was dissolved, and one company was created, called Royal Dutch Shell.

Sustainable Development

The company experienced two major public image setbacks in 1995 related to sustainable development. After the British government approved Shell's plans to decommission the Brent Spar oil storage platform by sinking it in the North Sea, Greenpeace claimed that this would create large amounts of pollution. Shell argued that this would create the least environmental damage. Greenpeace activists protested, boarding the rig as it was being towed to the disposal area. Widespread media coverage followed, and resulted in huge negative publicity for Shell, spawning boycotts and even a firebombing of a Shell station. Shell eventually reversed its decision and towed the rig to port in Norway for dismantling. Dismantling revealed that claims of a negligible amount of oil in the rig were accurate, and Greenpeace later issued an apology.

That fall, Shell experienced another public relations fiasco. A wholly owned subsidiary, Shell Nigeria is the major international oil-producing company in Nigeria, with joint ventures with the Nigerian state-owned oil company and other multinational oil companies. Significant revenues from oil production go to the central government, with little benefit accruing to people in oil producing regions. Author Ken Saro-Wiwa, a member of the Ogoni tribe from the Niger Delta, led protests against the government for not using oil revenues to benefit the Ogoni, many Ogoni leaders for complicity with the central government, and Shell for substantial pollution from exploration and pipeline spills and gas flaring and for seeking security assistance from Nigerian military forces. Violence broke out, in which several Ogoni chiefs were killed, and Saro-Wiwa and eight of his associates were arrested, tried, and found guilty of murder.

International human rights activists regarded the charges as groundless, and the trial as unfair. Activists called on the Nigerian government to commute the sentences, and on Shell to use its influence to this end. Shell's CEO and the Shell Nigeria managing director appealed for clemency on humanitarian grounds, but Saro-Wiwa and his associates were executed 10 days after the verdict. Shell was heavily criticized for not doing more to attempt to influence the government to free Saro-Wiwa.

These events prompted Shell to expand its attention to sustainable development. The company has a Social Responsibility Committee that directs its sustainable development policies and performance, and it produces an extensive sustainability report annually stating that sustainable development is part of the duties of every manager at Shell. Every one of Shell's businesses is responsible for complying with corporate sustainable development policies and achieving unit-specific targets in this area.

Fuel Alternatives

Shell expresses a commitment to help meet the energy challenge by providing more secure and responsible energy. Shell is developing more environmentally friendly fossil fuel technologies like gas to liquids (GTL), which turns natural gas into cleaner-burning fuels. While they don't receive the same financial support, Shell also supports several renewable energies. Shell was the first energy company to build demonstration hydrogen filling stations in Asia, Europe, and the United States, and is one of the world's leading distributors of biofuels. However, Shell has not yet provided the financial investment needed for hydrogen expansion, and the hydrogen is derived from fossil fuels.

In 2006, Shell sold over 3.5 billion liters of biofuels, mainly in the United States and Brazil, enough to avoid over 3.5 million tons of CO_2 production. Shell believes that "first generation" biofuels are unreliable, requiring too much acreage to be planted to feedstocks, thus putting strain on the environment and food supply. Shell has therefore invested in "second generation" biofuels, such as the production of ethanol from straw, rather than corn. Shell claims that this second-generation biofuel could cut well-to-wheel CO_2 production by 90 percent, compared with conventional gasoline. In early 2007, Iogen, acquired by Shell in 2002, was one of six companies selected to receive funding under the U.S. Department of Energy's cellulosic ethanol program.

Shell invested in CHOREN Industries to create the first demonstration-scale biomass-to-liquids (BTL) plant, which came online in 2009. This process relies on the use of a woody feedstock, gasifies it, and then uses the Shell Middle Distillates Synthesis (SMDS) process to convert the gas into a high-quality fuel identical to GTL that can be blended with diesel fuel. If used at 100 percent concentration, it could also cut well-to-wheels CO_2 production by 90 percent compared with conventional diesel. Shell sold its share in CHOREN in November 2009. Shell also has small investments in solar and wind power. Currently, their financial impact on Shell is very small, and they are not seen as offering substantial room for growth.

Greenhouse Gas Reduction

In 1997, Shell started managing its CO_2 output with the goal of reducing total greenhouse gas emissions. In 2006, Shell facilities emitted 98 million tons of greenhouse gases, about 7 million lower than in 2005, and more than 20 percent below 1990 levels. Yet, Shell has target emissions limits of only 5 percent below the 1990 level for 2010. Shell claims that its standards are very aggressive, but agrees that they may need to be reconsidered if they have already been met. Most of Shell's reductions came from ending the venting of natural gas. Most of its anticipated reductions will continue to come from ending continuous flaring and increasing energy efficiency.

In 2006, Shell missed its annual Energy Intensity target, as it had underestimated how much extra energy would be required to produce more environmentally friendly low-sulfur fuels, and because of unplanned equipment shutdowns at several facilities that required extra energy to restart. In 2007, Shell launched a new energy efficiency effort that would make up for part of the increase. Its plan was to continue efforts to end continuous flaring by 2008, except in Nigeria where the target was 2009. Further greenhouse gas reductions are planned from energy efficiency improvements at refineries and chemicals plants.

Shell also states that it is trying to reduce its customers' CO_2 emissions. Shell's customers emit six to seven times (750 million tons of CO_2 in a typical year) more CO_2-consuming Shell products than Shell does producing them. Shell promotes the use of natural gas, which emits less CO_2 than coal, and is a more profitable part of Shell's business than coal. Shell has also patented a coal gasification technology that can reduce CO_2 emissions by up to 15 percent, compared to conventional coal-fired power plants.

Shell also actively supports governments in designing and implementing effective CO_2 trading schemes. They are part of the UN Partnership for Clean Fuels and Vehicles and the World Bank Clean Air Initiative in Asia to provide cleaner fuel and improve air quality in the developing world. Shell is also a member of the U.S. Climate Action Partnership created in 2007 by over 30 companies and environmental groups.

Shell's annual sustainability report is audited by an external review committee, which has praised Shell for its leadership and transparency in reporting, but has questioned whether the speed with which Shell is acting to tackle climate change is consistent with the urgent nature of the challenge. The committee specifically noted the expected rise in future emissions, the lack of published targets after 2010, how Shell will achieve future greenhouse gas reductions after it stops flaring, and the absence of adequate research and development fund allocation information to assess Shell's commitment to develop renewable energy sources and to greenhouse gas mitigation.

Gordon P. Rands
Tyler Sayers
Western Illinois University

See Also: BP; Fossil Fuels; Oil, Consumption of; Oil, Production of; Renewable Energy, Bioenergy; United Kingdom.

Further Readings

Remember Ken Saro-Wiwa. http://www.remembersarowiwa.com (Accessed March 2012).
Shell. http://www.shell.com (Accessed March 2012).
U.S. Climate Action Partnership. http://www.us-cap.org (Accessed March 2012).

Royal Meteorological Society

The Royal Meteorological Society is a British charity with the mission to "continue to be a world-leading professional and learned society in the field of meteorology." The society promotes collaboration with organizations that are active in Earth systems sciences. It supports the advancement of meteorological science, its applications, and its understanding not only for its professional members, but also for the wider community. The society also aims to provide scientific and reliable recommendations to policymakers on global warming and climate change. In February 2007, the society endorsed the *Fourth Assessment Report* of the Intergovernmental Panel on Climate Change (IPCC), although this endorsement has provoked internal debate. The society's statement on the IPCC's conclusions argued that government, scientific, and business communities and the general public should take global warming seriously and work to limit emissions of greenhouse gases (GHGs).

A council and its committees are responsible for running the society within the constraints of the royal charter. The council comprises a total of 21 officers and ordinary members of council elected at the annual general meeting. The president, elected for a two-year term, is supported by a vice president for Scotland and three other vice presidents, the treasurer, general secretary, four journal editors, four main committee chairmen, and ordinary members of council. The council convenes five times a year to consider applications for membership and supervise the running of the society through its honorary officers, committees, and permanent staff. The work of the council is largely organized by the recommendations made by its committees. The society staff are based at the society's headquarters in Reading, where committee meetings are normally held. The society's patron is the Prince of Wales. Its membership in 2008 consisted of more than 3,000 members worldwide.

Broadest Meaning of Meteorological Science

The Royal Meteorological Society checks the national qualifications of the profession and, under its royal charter, follows its mission to advance meteorological science. The society intends the phrase *meteorological science* in its broadest meaning.

Therefore, it seeks advancement in the day-to-day application of meteorological science in weather forecasting and in disciplines such as agriculture, aviation, hydrology, marine transport, and oceanography, as well as in the areas of climatology, climate change, and the interaction between the atmosphere and the oceans. The society disseminates the results of new research through its publications and supports the activities of researchers and professional meteorologists. It also provides a scientific forum for those whose work is connected to the weather or climate and to those who have a general interest in environment and the weather.

Because of this wide spectrum of services, the membership of the society includes a variety of figures, from professional scientists and practitioners to mere weather enthusiasts. This heterogeneous composition is reflected in the different membership levels. Associate fellows may be of any age and do not require any specific qualification in meteorology. Fellows normally require a formal qualification in a subject related to meteorology, plus five years experience, and must be nominated by two other fellows. The society has a number of regular publications: the monthly magazine *Weather*, the *Quarterly Journal of the Royal Meteorological Society*, *Meteorological Applications*, the *International Journal of Climatology*, and *Atmospheric Science Letters*.

The society was established on April 3, 1850, with the title of the British Meteorological Society and was later incorporated by royal charter in 1866, when its name was changed to the Meteorological Society. The privilege of adding "Royal" to the title was granted by Queen Victoria in

1883. In 1921, the society merged with the Scottish Meteorological Society.

In 1995, the Royal Meteorological Society developed a set of atmospheric dispersion modeling guidelines to encourage good practice in the use of mathematical atmospheric dispersion models, stressing the importance of selecting the most suitable modeling procedures and fully documenting and reporting the results of modeling assessments. The 1995 guidelines provided broad general principles of good practice for modeling studies applied across a wide range of modeling situations. The United Kingdom (UK) Atmospheric Dispersion Modeling Liaison Committee (ADMLC) commissioned an upgrade of the 1995 guidelines to account for the new developments in modeling techniques. The updated guidelines were completed and published in 2004.

The Royal Meteorological Society has specific resources on global warming for teachers and educational purposes. In 2009, the society launched the educational program Climate4Classrooms in partnership with the British Council and the Royal Geographical Society. The program developed curriculum resources in meteorology and climate change for educational institutions in the UK, China, Indonesia, and Mexico.

The society has acknowledged that the phenomenon is taking place and that it is the result of human activity. However, one of its former presidents, Chris Collier, and one of its leading researchers, Paul Hardaker, complained in 2007 about catastrophism and the "Hollywoodization" of weather and climate that only work to create confusion in the public mind. They argue for a more sober explanation of the uncertainties about possible future changes in the Earth's climate so as not to undermine scientists' credibility. According to both Collier and Hardaker, several organizations, including the American Association for the Advancement of Science (AAAS), have overplayed the evidence that the phenomenon of global warming is causing short-term devastating impacts. In spite of this internal debate, the society has increasingly campaigned to spread awareness of global warming and the strategies that all the different sectors of society can take to reduce its impact. Several of its national meetings and lectures have been devoted to topics such as extreme weather events, state-of-the-art research in black carbon emissions, the development of more ecological energy sources, and recommendations for sustainable development that are all linked to climate change.

Luca Prono
Independent Scholar

See Also: Climatic Data, Reanalysis; Climatic Research Unit; European Union; World Meteorological Organization.

Further Readings

Gosh, Pallab. "Caution Urged on Climate 'Risks.'" BBC News. http://news.bbc.co.uk/2/hi/6460635.stm (Accessed July 2011).

Royal Meteorological Society. http://www.rmets.org (Accessed March 2012).

Royal Meteorological Society. "The Royal Meteorological Society's Statement on the Inter-Governmental Panel on Climate Change's *Fourth Assessment Report*." http://www.rmets.org/news/detail.php?ID=332 (Accessed July 2011).

Russia

Recognizing the potential pros and cons of a changing climate, Russia's current climate policy focuses more on energy efficiency and implementation of new technologies, with a view to provide a sustainable economic growth and create new opportunities for technological innovation and employment. President Dmitry Medvedev maintained this as a priority for Russia.

Russia's climate policy framework can be summed up in these approaches: the Climate Doctrine signed by President Medvedev on December 17, 2009; various energy efficiency strategies; the November 2010 federal decree on implementing Joint Implementation projects and mandates under the Kyoto Protocol; and future initiatives.

Russia's Climate Doctrine

The Climate Doctrine is the official presidential direction that defines the goals, principles, tasks, and ways of implementing various national policies with regard to climate change for both the federal government and the private sector in Russia. It addresses the global nature of the problem and Russia's national interests and priorities, ensures openness of all climate policy, acknowledges the need for both domestic and international programs for research, assesses the winners and losers from climate change, and makes plans for adaptation. The doctrine aims to address these themes while taking into account the specific geographical and socioeconomic characteristics of the Russian Federation.

The strategic goals outlined in the doctrine are to strengthen and develop research on climate change, develop and implement operational and long-term activities with regard to climate change adaptation and the anthropogenic influence on climate, and participate in international initiatives in solving problems related to climate change.

The Climate Doctrine is being enhanced by the new governmental plan to increase energy efficiency by 40 percent by 2020.

Energy Efficiency Strategy in Russia

Russia is the third-largest energy consumer in the world, one of the largest fossil fuels suppliers, and fifth-largest greenhouse gas emitter. Russian officials are slowly becoming aware that in order to secure the national economy and be fully integrated into the global technological market, the country must diversify its economy toward more scientifically oriented products and processes, as well as move away from dependency on fossil fuel exports. Currently, the majority of the country's income comes from oil and gas exports, which is not viewed as a sustainable growth strategy, even in the near future.

Taking into account that the need for energy-generation independence, energy-efficient production processes, and development of an autonomous renewable energy generation are among the key topics discussed at the international arena, Russia aims to be a leader in technological modernization, energy efficiency improvements, and in the development of alternative energy. Russia is beginning to understand that switching to a more energy-efficient economy could have significant economic benefits, as current energy efficiency is still extremely low.

According to an analysis undertaken by McKinsey & Company in its study, Pathways to an Energy and Carbon Efficient Russia, between now and 2030 it is possible to achieve more than 60 billion Euros in savings (assuming €20 billion in investment, leading to a net gain of €40 billion over the same time period) through energy efficiency measures in the petroleum, gas, power, and heat sectors.

The last official message of the president to the Federal Assembly in November 2009 expressed broad support for Joint Implementation (JI) activities occurring in the country, as the president described burning of oil-associated gas as an example of inefficient use of energy. According to the World Wildlife Fund, Russia is the world leader in terms of the volume of oil-associated gas burning; the potential for reduction is enormous.

A good example of the potential for international cooperation is the strategic partnership between the two Russian state-owned companies and the German company Siemens in the area of renewable energy generation. Siemens is planning to install wind turbines with a total capacity of up to 1,250 megawatts (MW) in Russia by 2015, contributing to the national goal of a total capacity of roughly 5,000 MW by 2020. In order to manufacture most of the products in Russia, three joint ventures with local partners are planned, including the state industrial corporation JSC Rostechnologii and the state-owned company RusHydro, which oversees the operation of all hydropower plants in the country. This agreement will make Siemens a pioneer in the Russian market for renewable energies. In addition, a research center that will address energy efficiency issues in the industry is being created in Skolkovo Innograd.

Joint Implementation and the Kyoto Protocol

The Federal Statement of the Russian Federation On Measures to Implement Article 6 of the Kyoto Protocol to the UNFCCC stipulates the new national procedure for JI project approval.

In accordance with the United Nations Framework Convention on Climate Change (UNFCCC) data, Russia is hosting around 120 projects, which represent 61 percent of the total JI pipeline. Reuters reported in February 2010 that these projects are capable of reducing some 235 million tons of carbon dioxide (CO_2) by 2012.

In February 2010, Sberbank attracted bids from various companies, including the largest

Russian oil and gas companies. As a result, the green light was given to 15 clean energy projects, claiming emissions reductions in the volume of 40 million tons of CO_2 equivalent. This was above the initially settled limit for emissions reductions, but Sberbank decided to increase this amount for future proposals.

One of these 15 projects was recently submitted for registration at the UNFCCC and was intended for consideration at its upcoming meeting in October 2010. The project is located at the Shaturskaya Thermal Power Plant near Moscow and envisages the building of an additional electricity-generating unit using an energy-efficient combined cycle gas turbine (CCGT).

At the JI Conference held on September 6–7, 2010, in St. Petersburg, Presidential Advisor on Climate Alexander Bedritskiy stated that, "Although we started later than was expected, in general we can use a much larger potential. The results of the JI launch will be seen by the end of 2012; therefore, there is still some time left."

Unfortunately, none of the renewable energy projects such as biomass and wind were approved by Sberbank, whereas projects in the oil and gas industries made up almost 70 percent of the final list. A possible reason for this is that renewable energy projects perform relatively worse against the criterion of energy and environmental efficiency in terms of fewer carbon credits. The rejection of these projects is to some extent discordant with the government's outlined intention to increase the renewable share of the energy balance. However, the most important step forward in the process of adoption and implementation of the national JI approval procedure has been undertaken, which brings a certain optimism to project developers. The continuation of JI after 2012 will fully depend on the new climate change agreement, defining modalities and procedures for the project-based mechanisms.

During the November 2010 UNFCCC Conference of the Parties (COP 16) in Cancun, a decision was made to continue allowing emissions trading and project-based mechanisms under the Kyoto Protocol to be available to Annex I Parties as a means to meet their quantified emissions limitations. Several countries, including the Russian Federation, supported this statement in their official responses to the draft document.

Russia's Future Climate Change Position

The political position of Russia on the future climate change agreements has been dictated by its domestic strategy.

At the UNFCCC December 2009 Conference of the Parties (COP 15) in Copenhagen, Russia was determined to push forward strategies to reduce greenhouse gas emissions, claiming it could then meet its target of a 25 percent cut by 2020 against the baseline year of 1990. This target was largely criticized in the European press, which referred to an already-achieved reduction of 30 percent, and that there was in fact the potential for Russia to reduce emissions up to 40 percent by 2020. Despite of this conflict of opinion, Russia was perceived to be a reasonably constructive and cooperative contributor to the negotiations at that conference.

However, Russia's position in international negotiations has not always been this harmonious. In April 2009, President Medvedev said that the country would pull out of the Kyoto Protocol if a compromise could not be found concerning the reduction of carbon emissions.

India, China, and other emerging economies have not yet undertaken any quantitative obligations concerning the reduction of carbon emissions. The United States, which said it would not participate in the agreement until at least 2013, considers that climate change should be tackled more by modern technologies, not cuts in carbon emissions stipulated by the Kyoto Protocol.

In its official statement later at COP 16, the delegation of the Russian Federation reiterated this approach by stating that, "the adoption of commitments for the second commitment period under the Kyoto Protocol, as it stands now, would be neither scientifically, economically, nor politically effective ... given the aforementioned, Russia will not participate in the second commitment period of the Kyoto Protocol. However, we do believe that it would be judicious to continue to use Kyoto Protocol market mechanisms, including in a new agreement."

Russia will probably not participate in the new climate change agreement in case it is based on the current conditions of the Kyoto Protocol. However, it will support the continuation of the project-based mechanisms, and will not only participate in JI, but also very likely in the Clean Development Mechanism.

In Russia, the JI project selection is highly political and less project-participant supportive. There is a good chance, however, that the JI project approval process will be continuously revised in order to provide equal opportunities to all large- and small-scale projects to be selected for further submission to the UNFCCC.

With regard to Russia's future emissions reduction potential, as well as reduction achieved, Russia might obtain a favorable position in hosting JI projects and become a world leader in defining climate change policy after 2012.

Anastasia Northland
University of Miami

See Also: Afghanistan; Agriculture; Annex I/B Countries; Arctic and Arctic Ocean; Blizzards; Brazil Proposal; Climate Policy, U.S.; Coal; Cooperative Institute for Arctic Research; Deforestation; G8/G20; Georgia (Country); India; Intensity Targets; Tajikistan; Ukraine.

Further Readings

Bellona Foundation. "Medvedev Signs Russian Climate Doctrine as Negotiations in Copenhagen Get Back on Track" (December 17, 2009). http://www.bellona.org/articles/articles_2009/medvedev_signs_cliimate_doc (Accessed May 2011).

Climate Doctrine of the Russian Federation. "Main Contents, Information; RIA Novosti" (December 2009). http://eco.rian.ru/documents/20091217/199797341.html (Accessed May 2011).

Jha, Alok. "Climate Change Increased Likelihood of Russian 2010 Heatwave." *Guardian* (February 21, 2012) http://www.guardian.co.uk/environment/2012/feb/21/climate-change-russian-heatwave (Accessed April 2012).

RIA Novosti. "Medvedev Threatens Russian Withdrawal From Kyoto Agreement" (April 16, 2010). http://en.rian.ru/Environment/20100416/158607110.html(Accessed May 2011).

Russian Federation. "Notification of Russian Federation Federal Service for Hydrometeorology and Environmental Monitoring to the UNFCCC" (May 28, 2007). http://unfccc.int/files/meetings/ad_hoc_working_groups/kp/application/pdf/russia_lulucf1_may2010_eng.pdf (Accessed May 2011).

Russian Federation Ministry of Natural Resources. "Federal Statement of Russian Federation: National Registry." http://www.carbonunitsregistry.ru/eng-default.htm (Accessed May 2011).

Shuster, Simon. "Will Russia's Heat Wave End Its Global-Warming Doubts?" *Time* (August 2, 2010). http://www.time.com/time/world/article/0,8599,2008081,00.html#ixzz1rzN8p7m (Accessed April 2012).

Siemens. "Multi-Billion Euro Orders for Wind Systems in Russia" (July 19, 2010). http://www.siemens.com/press/en/pressrelease/2010/corporate_communication/axx20100783.htm (Accessed May 2011).

Solzhenitsyn, S., V. Klintsov, P.-A. Enkvist, T. Visek, Y. Solzhenitsyn, I. Shvakman, and K. Schneiker. *Pathways to an Energy and Carbon Efficient Russia.* New York: McKinsey & Company, 2009.

Rwanda

Rwanda is a landlocked country in central Africa sharing borders with Uganda, Tanzania, Burundi, and the Democratic Republic of the Congo. The country has a long border on Lake Kivu and includes or borders several other large water bodies such as the Mwongo, Nyabarongo, Ruzizi, and Akagera rivers; Lake Mugesera; and Lake Ihema. Rwanda became independent from Belgium in 1962, but has been the scene of recurring ethnic strife and civil war between the Hutu and Tutsi ethnic groups. Most of the country is savanna grassland, with mountains in the northwest.

The climate is temperate, with two rainy seasons (February through April and November through January), while snow and frost occurs in the mountain regions. Most of Rwanda's population lives in rural areas; the country's population density is the highest in Africa, with a high fertility rate (4.9 children per woman, the 23rd highest in the world) leading to a high population growth rate (2.8 percent, the 16th-highest in the world), further increasing pressure on the country's resources.

Rwanda's per-capita gross domestic product is $1,100, among the lowest in the world, and income is highly unevenly distributed. The Gini index is 46.8, the 35th highest in the world.

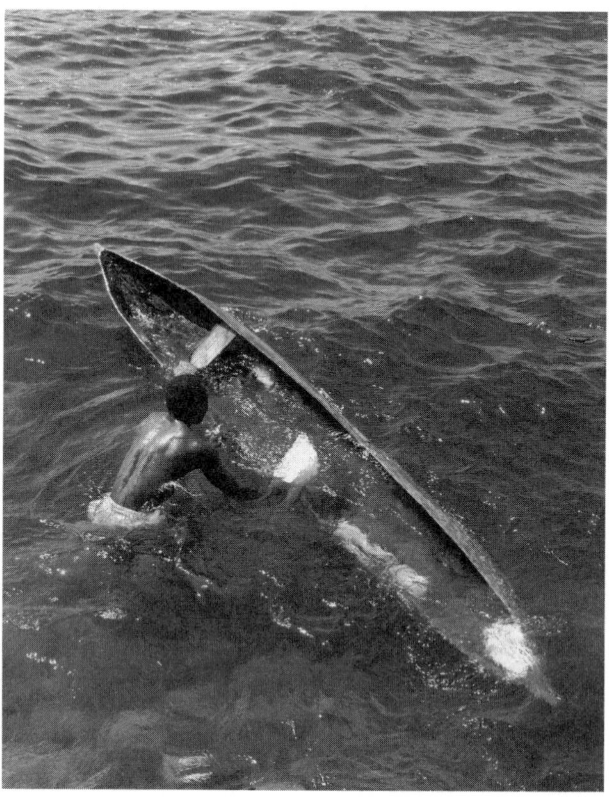

A fisherman in Lake Kivu readies his boat. By late 2012, Rwanda hopes to extract enough of the dissolved volcanic carbon dioxide and methane from the lake to obtain a third of its power needs, before the lake overturns from the explosive pent-up gas.

Rwanda rates poorly on measures of population health and well-being, such as infant mortality (64.04 deaths per 1,000 live births, the 25th highest in the world) and life expectancy (58.02 years, 193rd in the world), and has a high adult HIV prevalence rate (2.9 percent, the 25th highest in the world). As the median age is 18.7 years and 42.9 percent of the population is age 14 or younger, this young population structure complicates efforts to develop the country's economy and raise standards of living.

Meager Development, Low Emissions

Rwanda currently produces a low level of greenhouse gases because of the low level of development in the country, although this is likely to increase with increased economic development, despite strong government emphasis on developing renewable sources of energy. As of 2009, half of Rwanda's electricity was produced by hydropower, with the rest produced by thermal plants. Most fuel consumed in the country is biomass, including wood (57 percent), charcoal (23 percent), and peat and agricultural waste (6 percent), but the Rwandan government is engaged in a number of efforts to replace these sources with renewable forms of energy, including solar, wind, geothermal, and biofuel. Rwanda has no known reserves of oil, natural gas, or coal. Oil consumption has risen substantially over the past three decades, from 1,100 barrels per day in 1980 to 5,000 barrels per day in 2008. Electricity generation and installed capacity has been somewhat irregular over the same period, due in part to civil unrest within the country, while consumption was fairly stable between 1980 and 1996 and has risen somewhat irregularly since then, to 0.023 billion kilowatt hours in 2007. Carbon dioxide emissions from consumption of fossil fuels in 2009 was 0.72 million metric tons.

Rwanda is extremely vulnerable to climate change because most of the population is engaged in subsistence agriculture, which would be severely impacted by changes in the amount and pattern of rainfall and in temperature. Approximately 45 percent of the land is arable, and in 2009, Rwanda achieved self-sufficiency in food production. De-forestation is a serious problem—only about 7 percent of Rwanda's original forests remain today—as trees have been felled for fuel and the creation of agricultural land, resulting in increased vulnerability to flooding and erosion. The poverty of the country and its rapidly growing population complicate its ability to invest in the development of clean technologies, but sustainable policies are a high priority of the current government, which is engaged in programs of reforestation and development of renewable energy sources. Rwanda, like many sub-Saharan African countries, is not able to benefit substantially from the carbon trade programs included in the Kyoto Protocol because of their tiny share of the global market.

Sarah Boslaugh
Kennesaw State University

See Also: Agriculture; Biomass; Deforestation; Developing Countries; Methane Cycle; Volcanism; Vulnerability.

Further Readings

Ecosystem Restoration Association. "ERA Projects: Rwanda." http://www.eraecosystems.com/projects/africa/rwanda (Accessed July 2011).

Kanter, James. "Rwanda as an Example of the Dangers of Climate Change." *New York Times Green: A Blog on Energy and the Environment.* (June 11, 2008). http://green.blogs.nytimes.com/2008/06/11/rwanda-as-an-example-of-the-dangers-of-climate-change (Accessed January 2012).

Mininfra: Rwanda Ministry of Infrastructure. http://minifra.gov.rw (Accessed July 2011).

Rice, Xan. "Rwanda Harnesses Volcanic Gases From Depths of Lake Kivu." *Guardian* (August 16, 2010). http://www.guardian.co.uk/environment/2010/aug/16/rwanda-gas-lake-kivu (Accessed January 2012).

U.S. Energy Information Administration. "Country Information: Rwanda." http://www.eia.gov/countries/country-data.cfm?fips=RW (Accessed July 2011).

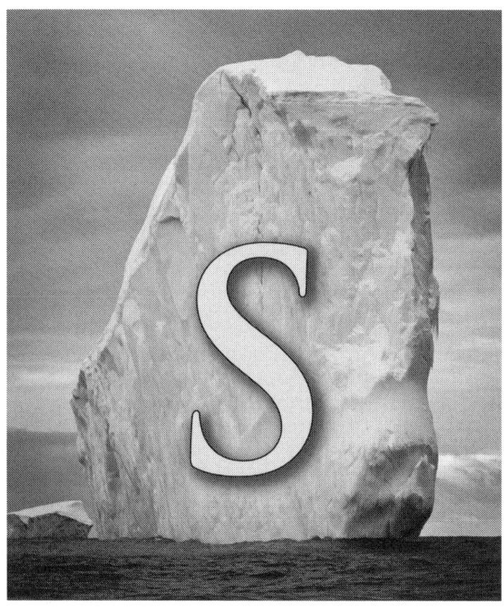

Saint Kitts and Nevis

The Federation of Saint Kitts and Nevis, a two-island nation in the West Indies, is the smallest sovereign state in the Americas, with a population of about 50,000 and a total area of 104 sq. mi. (269 sq. km). Saint Kitts has two official names, both Saint Kitts and Saint Christopher, but the first is more common. Roughly half of the population is urban. The islands were settled by Native Americans and occupied by the Spanish and the British before gaining independence in 1983. The relationship between the two islands has been tense since the 19th century, but Nevis's secession attempts have never garnered sufficient support.

The economy has traditionally been driven by the sugarcane that is well suited to the islands' volcanic soil, and sugar was the dominant export from the 1640s until the end of the 20th century. Low market prices led to a decline in the sugar industry, and many sugar plantations now lay dormant, especially in Saint Kitts. Elsewhere, cane fields have been burned to make room for new development. Tourism and offshore banking are now critical to the economy, as are apparel and electronics assembly for export. About 22 percent of the islands' land is arable, and another 3 percent is used for livestock. The increased importance of tourism and industry has resulted in per-capita carbon emissions more than doubling since 1990.

The islands are vulnerable to hurricanes and were hit very hard by Hurricane Georges in 1998; the gross domestic product took several years to recover from it. Rising water levels are not as problematic on the islands as elsewhere in the West Indies, but the local marine life, especially sea turtles, is put at jeopardy by rising water temperatures.

Bill Kte'pi
Independent Scholar

Se Also: Antigua and Barbuda; Hurricanes and Typhoons; Saint Lucia; Tourism.

Further Readings

Cato, James, ed. *Gulf of Mexico Origin, Waters, and Biota*. College Station: Texas A&M University Press, 2008.

Tunnell, John, et al., eds. *Coral Reefs of the Southern Gulf of Mexico*. College Station: Texas A&M University Press, 2007.

Saint Lucia

The island of St. Lucia, located in the Windward Islands and surrounded by the Caribbean Sea and the Atlantic Ocean, has a land area of 238 sq.

mi. (616 sq. km), a population of 173,765 (2009 census), and a population density of 672 people per sq. mi. (298 people per sq. km), making it the 39th most densely populated country in the world. About a third of the population lives in the capital, Castries, which is located on the sheltered western coast. About 8 percent of the land is arable and a further 5 percent is used for meadows and pasture. Some 41 percent of the exports of the country come from bananas, although historically, much of the income from the island was from its extensive sugar plantations.

St. Lucia has a number of offshore coral reefs, the most well known being those near Soufrière, in the southeast of the island. Just south of these are other coral reefs that also attract many tourists. There are worries about their preservation, with signs of coal bleaching resulting from the increase in water temperatures.

There is extensive public transport on the island that is maintained by private bus companies. However, there are 166 cars per 1,000 people in the country, the fifth highest in the West Indies after Puerto Rico, Antigua and Barbuda, Saint Kitts and Nevis, and Barbados, and slightly above that of Dominica. This is a contributing factor to carbon dioxide (CO_2) emissions, and there have been attempts to reduce traffic congestion in Castries.

As with many other developing economies, Saint Lucia has seen a rise in CO_2 emissions per capita, from 1.2 metric tons per person in 1990 to 2.2 metric tons in 1996, before falling to 1.4 metric tons in 1998, but rising again to between 1.9 and 2.1 metric tons per person over the next few years. The level rose rising steadily to 2.3 metric tons in 2007. All the CO_2 emissions in the country come from liquid fuel, and all electricity production in the country comes from fossil fuels.

The government of John Compton took part in the United Nations Framework Convention on Climate Change (UNFCCC), signed in Rio de Janeiro in May 1992, and ratified the Vienna Convention; two years later, Saint Lucia was represented at the Global Conference on the Sustainable Development of Small Island Developing States held in Barbados. The St. Lucia government of Kenny Anthony signed the Kyoto Protocol to the UNFCCC on March 16, 1998. The country ratified it on August 20, 2003, and it entered into force on February 16, 2005. The Saint Lucia government is one of the few in the world to establish its own climate change Website.

Justin Corfield
Geelong Grammar School
Robin S. Corfield
Independent Scholar

See Also: Climate Change, Effects of; Saint Kitts and Nevis; Transportation.

Further Readings

Environment News Service. "Vulnerable Caribbean Nations Prepare for Global Warming." http://ens-newswire.com/ens/jun2001/2001-06-04-02.asp (Accessed March 2012).

Gomes, P. I., ed. *Rural Development in the Caribbean*. London: C. Hurst, 1985.

Government of Saint Lucia. "Climate Change." http://www.climatechange.gov.lc (Accessed July 2011).

National Development Corporation. *St. Lucia: Advantages for Industrial Development*. St. Lucia, West Indies: NDC, 1973.

Saint Vincent and the Grenadines

Saint Vincent and the Grenadines, which achieved independence from Great Britain in 1979, is a chain of mostly small islands in the Caribbean Sea. Its area encompasses 150 sq. mi., (389 sq. km), 132 sq. mi. (344 sq. km) of which is the island of St. Vincent, which is about twice the size of Washington, D.C. Its population of 103,869 ranks 193rd in the world. The terrain is volcanic and mountainous, with nearly 18 percent arable land, and the climate is tropical, with a rainy season from May to November. There is one active volcano, La Soufriere, in the north of the island of St. Vincent, which poses threats of ash fall, as well as earthquake and eruption. The primary industries are tourism and agriculture (particularly banana production), both of which have been damaged in recent years by severe tropical storms; the country is located in the Atlantic hurricane belt. About half the population lives in urban areas. The gross domestic product (GDP) per capita in 2010 was $10,300 (107th in the world) and the government invests substantial resources in social programs; education expenditures in 2009 amounted to 6.6 percent of GDP, 20th highest in the world. However, the country has a high level of public debt, which was over 90 percent of GDP in 2010. Due to one of the highest out-migration rates in the world (minus 10.92 migrants per 1,000 population), Saint Vincent's population growth rate is negative, at minus 0.327 percent (among the lowest in the world).

Saint Vincent and the Grenadines is a relatively low consumer of petroleum and emitter of greenhouse gases due to its small area and population and limited industrial development. The country has no reserves of petroleum, natural gas, or coal. Petroleum consumption (entirely imported) has risen steadily over the past three decades, from 0.27 thousand barrels per day in 1980 to 1.70 thousand barrels per day in 2008. Electricity generation has risen in a similar fashion over the same period, and in 2007, Saint Vincent and the Grenadines generated 0.13 billion kilowatt hours of electricity and had an installed capacity of 0.4 gigawatt. Carbon dioxide emissions from the consumption of fossil fuels in 2008 was 0.23 million metric ton.

The population of St. Vincent and the Grenadines are particularly vulnerable to the effects of global warming due to the location and geography of the islands. Most of the islands are situated on steep slopes (often exceeding 45 degrees) and have little protection from wind and strong rainfall, leaving them prone to landslides and erosion. Since 1980, the country has been hit by eight hurricanes, the strongest (Hurricane Allen) in 1980. Landslides and coastal flooding are regular risks independent of major storms, and as residents depend on rainwater for agriculture and personal consumption, drought is also a regular risk; the period from 2009 to 2010 saw a major drought. There is little land-use planning on the smaller islands and construction codes are rarely enforced, resulting in a large number of buildings (both for local residents and the tourist trade) being built in low-lying coastal areas, which are at extreme risk due to storm surges.

In 2010, St. Vincent and the Grenadines were ranked 72nd in the Germanwatch Global Climate Change Risk Index in terms of risk for events due to climate change, with rising sea levels and an increase in the strength of hurricanes due to global warming as the most significant risks. More frequent heat waves and droughts are expected in future years, along with more intense rainfalls and flooding. Over 40 percent of the population is judged to be at risk of mortality from two or more natural hazards, ranking St. Vincent and the Grenadines in the top 60 globally for this type of risk.

Sarah Boslaugh
Kennesaw State University

See Also: Drought; Floods; Hurricanes and Typhoons; Sea Level, Rising.

Further Readings

U.S. Energy Information Administration. "Country Information: Saint Vincent/Grenadines." http://www.eia.gov/countries/country-data.cfm?fips=VC (Accessed July 2011).

World Bank Global Facility for Disaster Reduction and Recovery. "Disaster Risk Management in Latin America and the Caribbean: GFDRR Country Notes: St. Vincent and the Grenadines" (2010). http://www.gfdrr.org/gfdrr/sites/gfdrr.org/files/documents/St.VincentGrenadines-2010.pdf (Accessed July 2011).

Salinity

Two attributes of the oceans—temperature and salinity—determine the density of seawater, and the differences in density between the water masses in the world's oceans causes the water to flow in thermohaline circulation, thereby producing the greatest oceanic current on the planet.

Salinity is the distinct taste of seawater and is the result of the presence of dissolved salts (more than 85 dissolved constituents), of which chloride (Cl) and sodium (Na), the elements of common table salt, are the most abundant. The term *salinity* refers to the content of these dissolved salts, and has been defined as grams of dissolved salts per kg of seawater. Salinity has been expressed as parts per thousand (‰ or ppt), and more recently, by practical salinity units (psu). On average, 1 kg of seawater has 35 grams of dissolved salts, so its salinity content is 35‰, or 35 psu. The accuracy of most laboratory salinometers (see below) is about 0.001 psu. Thus, only those components with a concentration over 0.001‰ will contribute to such salinity estimates. Only 15 of the dissolved salts have concentrations above that limit.

A key observational result (known as the principle of Dittmar, after William Dittmar, a Scottish professor of chemistry; the principle of Maury, after Matthew Fontaine Maury, an American astronomer, oceanographer, and geologist; or the hypothesis of Forchhammer, after Johan Georg Forchhammer, a Danish mineralogist and geologist) is that the relative concentration between some of these most abundant salts is virtually constant over much of the world ocean. This finding indicates that the physical characterization of seawater is given by its temperature, pressure, and a single number reflecting the concentration of the most abundant components. Salinity is that number.

Measuring Salinity

In 1902, an international commission defined salinity as the total amount of solid material, in grams, contained in 1 kg of seawater when all the carbonate has been converted to oxide, the bromide and iodine have been replaced by chlorine, and all organic matter has been completely oxidized. With this definition in hand, the commission estimated the salinity of several seawater samples and the fixed relationship between salinity (S) and chlorinity:

$$S(‰) = 003 + 1.805\ Cl\ (‰)$$

This was known as Knudsen's equation, after Martin Hans Christian Knudsen, a Danish physicist (1871–1949), and was redefined in 1969 as

$$S(‰) = 1.80655\ Cl\ (‰).$$

Defining salinity in terms of chlorinity alleviates the practical difficulties of measuring salinity through evaporating water samples to dryness. For calibration purposes, artificial water with salinity almost equal to 35‰, known as Copenhagen water, is manufactured to serve as a reference. Copenhagen water has a chlorinity of 19.381‰. This approach requires the chemical titration of water samples usually obtained by Nansen bottles, named after Fridtjof Bedel-Jarlsberg Nansen, a Norwegian explorer and scientist (1861–1930), which are self-closing containers that collect water from different depths.

Pure water is a poor electrical conductor. However, the presence of dissolved salts greatly increases its conductivity, which in fact is a function of pressure, temperature, and the degree of ionization of the dissolved salts. In the second half of the 20th century, technical improvements in the measurement of the electrical conductivity of seawater led to the development of salinometers. Conductive salinometers measure the ratio between the conductivity of the sample against that of a reference sample of known salinity. Researchers using conductivity salinometers in the beginning were giving a salinity value of 35‰ to any sample having the same conductivity as the Copenhagen water, even though the mass of salt per kg of water was not guaranteed to be the same in both cases. This was because conductivity depends on the degree of ionization of the dissolved salts, not on the absolute mass of salt. In 1978, the salinity scale was redefined in terms of the conductivity ratio, K_{15}, between any given sample and the reference solution:

$$S(psu) = 0.0080 - 0.1692\ K_{15}^{1/2} + 25.3851\ K_{15} + 14.0941\ K_{15}^{3/2} - 7.0261\ K_{15}^{2} + 2.7081\ K_{15}^{5/2}$$

$$K_{15} = C(S,15,0)/C(KCl,15,0),$$

where $C(S,15,0)$ is the conductivity of the water sample at a temperature of 15 degrees C and atmospheric pressure, and $C(KCl,15,0)$ is the conductivity of a standard solution that contains 32.4356 g. potassium chloride (KCl) at the same temperature and pressure. The practical salinity unit is thus defined as a ratio of conductivities and has no physical units.

An alternative to conductivity salinometers are refractive salinometers, based on the fact that the speed of light through a medium depends on its density. In a refractive salinometer, a drop of sample water is placed on a prism. Because the water and the prism have different densities, light passing through the system is refracted at an angle that depends on the density (i.e., the temperature and salinity). On average, conductive salinometers have a precision of about 0.001 psu, whereas laboratory refractive salinometers have a precision of 0.06 psu, and handheld refractometers have a precision of 0.2 psu. Today's standard instrument for measuring both temperature and salinity is the Conductivity-Temperature-Depth profiler, which allows a quasi-continuous vertical sampling and a precision of 0.005 psu. Based on the same principle, conductivity/temperature instruments are mounted on autonomous profilers (e.g., Argo floats) and thermosalinographs that use near-surface water intakes of ships to continuously measure temperature and salinity.

A promising approach for remote sensing of salinity is microwave radiometry, measuring the emissivity or brightness temperature of the sea surface, because the dielectric constant of seawater depends on temperature and salinity. The largest sensitivity of the surface emissivity to salinity has been observed in the L-band (1.40–1.43 GHz). The Soil Moisture and Ocean Salinity mission of the European Space Agency, and the Aquarius-SAC/D mission of the NASA-Argentine Space Agency are the first two space missions designed to provide global, synoptic estimates of sea surface salinity with an accuracy of about 0.1 psu every 30 days, with a 100 to 200 km spatial resolution.

Processes Affecting Salinity

Since the *Challenger* expedition in 1877, when the chemical composition of seawater was first reported, no changes in the composition of seawater have been observed. Thus, it can be supposed that for the timescales pertinent to climate change, namely, decadal to centennial, salinity behaves as a conservative tracer. Thus, its time evolution is given by the three-dimensional transport of salinity by advection (as water parcels carry properties) or diffusion (tendency to smooth salinity gradients, even in still water). At the surface of the oceans (up to a depth where the turbulent action of the wind is balanced by the laminarity of the stable ocean stratification), salinity concentrations are also modified by the dilution/concentration resulting from mass fluxes through the air-sea interface, such as evaporation and precipitation, river runoffs, and the thawing/freezing of ice caps. In open oceans, the lowest values of salinity (below 30 psu) are found at high latitudes and at the mouth of the largest rivers. The highest salinities are found in the subtropics (over 35 psu), where evaporation dominates. In the tropics, which tend to be regions of strong precipitation, salinity is around 34 psu.

Salinity and Climate

The range of temperatures on Earth allows water to be present as a solid (ice) in ice caps and glaciers; liquid (water) in oceans, groundwater, lakes, and rivers; and gas (water vapor) in the atmosphere. The idealized path of a water molecule from one phase to the other is known as the hydrological cycle. The residence times, that is, the average time that the molecules spend in each phase, range from a few days (water vapor in the atmosphere), to several months (seasonal snow cover, rivers), to the thousands of years (oceans and groundwater). Changes in the hydrological cycle affecting precipitation, evaporation, ice cap thawing, and river runoff have the potential to change the salinity of the oceans. The reverse is also true, and salinity changes may have an imprint in the hydrological cycle after thousands of years.

The mechanisms by which salinity affects the hydrological cycle are numerous. Because of its role in density variations, salinity gradients contribute to ocean currents transporting heat, salt, microorganisms, and nutrients across the oceans. In regions of strong precipitation, a layer of low salinity may isolate the uppermost surface of the ocean from the cold ocean below, forming the barrier layer, which blocks the wind-stirring effects that cool the surface by mixing heat downward. This manifests as warmer sea surface temperatures, modifying the surface temperature gradients that drive surface winds. In the equatorial Pacific Ocean, such a phenomenon is of importance in the El Niño–La Niña cycles. Similar salinity effects also occur in the tropical Indian and Atlantic oceans and have potential feedbacks to the hydrological cycles in the region. Tropical surface anomalies may be advected to the deep convection regions, modulating the thermohaline circulation. One of the largest ocean climate events recorded in the Atlantic Ocean is the Great Salinity Anomaly, which lasted from 1968 to 1982. A salinity anomaly propagated over thousands of kilometers reached the Labrador Sea and perturbed the thermohaline circulation intensity. The origin and evolution of these anomalies is still not fully understood because of the historical lack of salinity observations, and studies of the mechanisms by which these salinity anomalies evolve are usually based on ocean and climate models.

Joaquim Ballabrera
UTM-CSIC
Raghu Murtugudde
University of Maryland
Jordi Font
CMIMA/CSIC

See Also: Climate Change, Effects of; El Niño and La Niña; Hydrological Cycle.

Further Readings

Emery, W. J. and R. E. Thomson. *Data Analysis Methods in Physical Oceanography.* Oxford: Elsevier, 2001.
Hill, M. N. *The Sea.* Vol. 2. Melbourne, FL: Krieger, 1982.
Levitus, S., et al. *World Ocean Atlas 1994.* Washington, DC: U.S. Department of Commerce, 1994.
Peixoto, J. P. and A. H. Oort. *Physics of Climate.* New York: American Institute of Physics Press, 1991.

A diver explores the coral reef at Fagatele Bay. Samoans, who for centuries have relied on coral reefs for protection, food, goods, and services, may face severe disruptions in lifestyles, livelihoods, and protection if climate change damages reefs.

Samoa

Samoa is an island nation in the South Pacific Ocean, consisting of several islands, including the two main islands of Upolu and Savai'i. Upolu is home to Apia, the political and cultural capital, while Savai'i is one of the largest Polynesian islands. Colonized by Europeans, the Samoan islands were divided into two jurisdictions at the turn of the 20th century: the eastern islands, which are today American Samoa; and the western islands, first controlled by the British, then the Germans (as German Samoa), and then New Zealand, before gaining independence in 1962 as the nation of western Samoa. In 1997, the constitution was amended to rename the nation Samoa. While Samoa and American Samoa share pre-20th century history, ethnicity, and cultural heritage, their modern ties are predicated upon their colonial heritage, with American Samoans traveling to Hawai'i, eating American food, and playing baseball, while Samoans have a close relationship with New Zealand and play rugby and cricket. Samoa is even changing its time zone at the end of 2011: the current time zone, 21 hours behind Sydney, was set in the 19th century for the convenience of American business associates on the west coast. After moving forward one day by shifting to the west of the International Dateline, Samoa's time zone will be 3 hours ahead of Sydney, making it easier to conduct business with Pacific countries such as Australia, New Zealand, and China.

The country has a population of under 200,000 in a land area of 1,093 sq. mi. (2,830 sq. km). About one-fifth of the country is arable and half is forested. Carbon dioxide emissions have been almost unchanged on a per-capita basis—since 1990, the level has stood at about 0.8 metric ton per person. Liquid fuels, which are used in household and factory generators, as well as power plants and transportation, account for nearly all emissions. Fossil fuels contribute 60 percent of Samoa's electricity, the remainder coming from hydroelectric sources.

Samoa has an equatorial climate, with temperatures staying close to the annual mean of 80 degrees F (26.6 degrees C) year round. The rainy season lasts from November to April. The major islands in Samoa have experienced land loss of about 1.5 ft. of shore per year since World War I. Several of the uninhabited islands are expected to become completely submerged if water levels continue to rise.

Samoa is a signatory to the United Nations Framework Convention on Climate Change and the Kyoto Protocol.

Bill Kte'pi
Independent Scholar

See Also: Alliance of Small Island States; Australia; China; Micronesia; New Zealand; Tonga.

Further Readings

Holmes, Lowell. *Quest for the Real Samoa*. New York: Bergin and Garvey, 1988.

Macpherson, Cluny and La'avasa Macpherson. *The Warm Winds of Change: Globalization in Contemporary Samoa*. Auckland, New Zealand: Auckland University Press, 2010.

San Marino

The landlocked republic, entirely surrounded by Italy, has maintained is independence from the Italian state. It has a land area of 24 sq. mi. (61.2 sq. km) a population of 31,716 (March 2010 est.), and a population density of 1,267 people per sq. mi. (501 people per sq. km), the 20th-highest density in the world. San Marino is a very prosperous country, with a gross domestic product per capita of $49,000 (2009 est.). As a result, it makes heavy use of electricity—air conditioning in the hot summers and heating for the winter, as well as regular domestic household and business uses. All electricity for San Marino is supplied by Italy, which produces some 80 percent of its electricity from fossil fuels.

Because of its geographical position, there are few data available for San Marino for greenhouse gas emissions; emissions are usually included under Italy, which has had carbon dioxide (CO_2) emissions per capita ranging from 6.9 metric tons per person in 1990, rising to 7.7 metric tons per person by 2003; and after rising briefly to 8 metric tons, it fell to 7.5 metric tons in 2008. Tourism is the major source of income, and most tourists come into the country in buses or by train from the nearby Italian town of Borgo Maggiore. The government took part in the United Nations Framework Convention on Climate Change (UNFCCC) signed in Rio de Janeiro in May 1992, and acceded to the Kyoto Protocol to the UNFCCC on April 28, 2010, the 190th national government to do so.

At a meeting of the UN General Assembly on October 17, 2003, Michela Bovi of San Marino spoke in support of the moves to encourage the use of renewable sources of energy. She also noted that, "San Marino had made significant progress in achieving new and environmentally safe energy sources." She went on to say, "In the future, it planned to introduce ambitious policies for solar, wind, hydro, biomass, ocean, and geothermal energy use. Wide and active cooperation among all countries to adopt new approaches to environmental development and preservation was essential for effective energy development."

Justin Corfield
Geelong Grammar School
Robin S. Corfield
Independent Scholar

See Also: Climate Change, Effects of; Tourism.

Further Readings

United Nations. "Climate Change, Disasters, Desertification Can Affect Security, Survival of

Developing Countries." Press Release. (October 17, 2003). http://www.un.org/News/Press/docs/2003/gaef3046.doc.htm (Accessed July 2011).
Valli, Andrea Suzzi. " General Information on the Phytosociological Study of Forest Vegetation in the Republic Of San Marino." *Studi Sammarinesi*, v.1 (1984).

São Tomé and Príncipe

São Tomé and Príncipe is the smallest country in Africa at 372 sq. mi. (964 sq. km), consisting of two major islands—São Tomé and Príncipe—and a number of smaller islands in the Gulf of Guinea off the west coast of Africa. It was a Portuguese colony used primarily for sugar, coffee, and cocoa plantations until achieving independence in 1975. The terrain is volcanic and mountainous, while the climate is hot and humid, with a rainy season from October to May. Only about 8 percent of the land is arable, yet much of the population is engaged in subsistence agriculture; agriculture contributes 14.7 percent of the gross domestic product (GDP). The population of 179,506 (July 2011 est.) has a young age structure, with 44.7 percent age 14 or younger, and the median age at 17.5 years. The total fertility rate is the 18th-highest in the world, at 5.08 children per woman, leading to a population growth rate of 2.05 percent, despite a high out-migration rate (9.33 migrants per 1,000 population in 2011) and fairly low life expectancy (63.11 years), which creates added strain on the country's limited resources. São Tomé and Príncipe's economy is heavily dependent on cocoa exports, but production has declined in recent years due to mismanagement and drought. The country must import most manufactured and consumer goods, all fuels, and a substantial amount of food, and is heavily indebted (although granted $200 million in debt relief, about two-thirds of its total debt, in 2000 under the Highly Indebted Poor Countries program. Per-capita GDP in 2010 was $1,800, ranking 194th in the world, and over 50 percent of the population lives in poverty.

São Tomé and Príncipe may benefit in the future from the development of petroleum resources in its territorial waters, which are currently being explored in a 60–40 split with Nigeria. Currently, however, it has no known reserves of oil, natural gas, or coal. Oil consumption has risen steadily from 1989 (0.24 thousand barrels per day) to 2008 (0.90 thousand barrels per day). Net generation of electricity has risen steadily over the past three decades, from 0.009 billion kilowatt hours (kWh) in 1980 to 0.04 billion kWh in 2007. Carbon dioxide emissions from fossil fuels in 2008 totaled 0.14 million metric tons.

Because of its undeveloped economy and low standard of living, São Tomé and Príncipe contributes relatively little to greenhouse gas emissions, even after taking the size of the country and its population into account. However, as an island nation it is extremely vulnerable to any rise in sea level, and as a country is largely dependent on agriculture and fishing for subsistence and trade, which are highly vulnerable to changes in temperature and precipitation. According to a report delivered to the United Nations Framework Convention on Climate Change, São Tomé and Príncipe's temperature is expected to experience an average increase of about 3.6 degrees F (2 degrees C) in temperature by 2100, along with a 15 percent decrease in precipitation. It was also noted that patterns of rainfall have already begun to change, with decreased precipitation in March through May and an increase in September through February, with an intensified fog season (December to February) and increased flooding.

About 20 percent of São Tomé and Príncipe's population are engaged in artisanal fishing (small-scale fishing with traditional techniques) and, along with others living in coastal communities, are particularly at risk from changes in ocean weather resulting in increased or less predictable storms. Most fishing is done with unmotorized canoes, and there is no reliable early warning system to convey information about approaching storms. Since 2006, the number of São Tomé and Príncipe fishermen lost at sea (240 per 100,000) is three times the global average for fishing as an occupation. Coastal communities have always been vulnerable to sea and river flooding, and these events are expected to increase with global warming.

Sarah Boslaugh
Kennesaw State University

See Also: Agriculture; Alliance of Small Island States; Floods; Sea Level, Rising.

Further Readings

U.S. Energy Information Administration. "Overview /Data: São Tomé and Príncipe." http://www.eia.gov/countries/country-data.cfm?fips=TP (Accessed July 2011).

World Bank. "Project Appraisal on a Proposed Grant From the Global Environment Facility Trust Fund in the Amount Of USD 4.1 Million to the Democratic Republic of São Tomé and Príncipe for an Adaptation to Climate Change Project" (May 2, 2011). http://www.thegef.org/gef/sites/thegef.org/files/repository/MMBC_PAD_November_6.pdf (Accessed July 2011).

Saudi Arabia

The Kingdom of Saudi Arabia, named for the ruling dynasty of the House of Saud, is the largest Middle Eastern country by area. It occupies most of the Arabian Peninsula, also called the Arabian subcontinent, the world's largest peninsula, forming the bridge between North Africa and Asia; it shares the peninsula with the much smaller countries of Kuwait, Qatar, Yemen, Oman, and the United Arab Emirates. The world's largest oil exporter in possession of the world's largest oil reserves, it also played a critical role in the development of Islam, as it includes both Mecca and Medina. The modern state dates to 1932, when Abdul-Aziz ibn Saud established his dynasty after 30 years of conquests, beginning with the capture of his ancestral home of Riyadh. Today, Riyadh has a population of nearly 5 million, with an additional 2 million in the greater metropolitan area.

Thanks to Saudi Arabia's oil revenues, the country enjoys a high standard of living. Fossil fuels are used for almost all of the country's electricity generation, contributing to its high carbon dioxide emissions. These levels rose quickly, from 12.1 metric tons per person in 1990 to 18.4 metric tons in 1993, but have since declined slightly to 16.31 metric tons.

About three-quarters of Saudi government revenue comes from oil exports and pays for a welfare state; the fortunes of the country rise and fall with oil. The oil industry accounts for nearly half of the country's gross domestic product. The country is also one of the largest consumers of energy in the world and the fastest-growing energy consumer in the Middle East, due in part to the prevalence of automobile ownership and heavy consumption of transportation fuels. Consumption increases also dovetailed with the electrification of rural areas, which happened relatively late: peak loads of electricity demand were 25 times as high in 2000 as in 1975 for primarily that reason. Increased consumption has resulted in the government beginning to stress energy conservation and sustainability, beginning in the 1990s, and becoming more

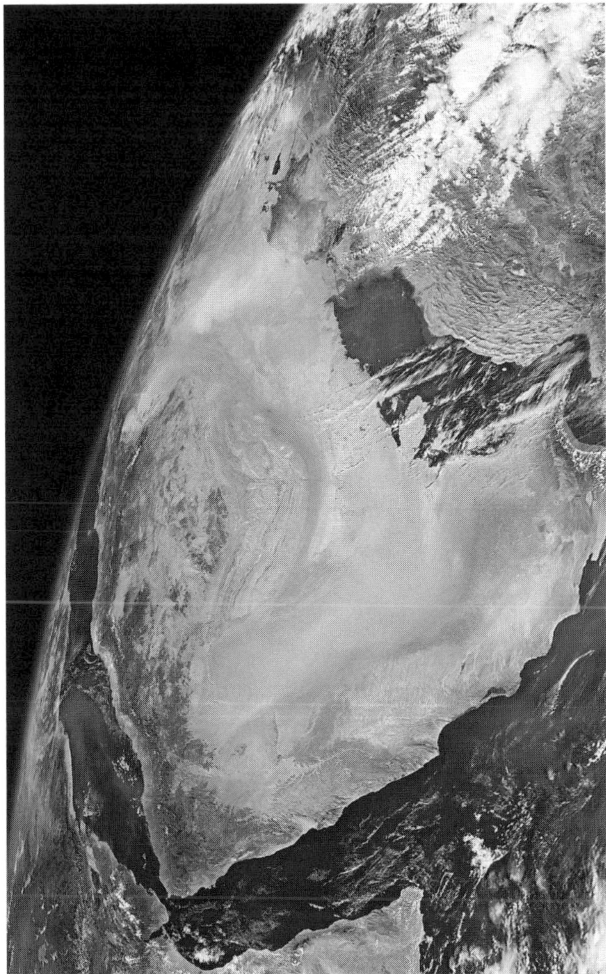

Massive sandstorms over Saudi Arabia, seen from space. A rise in global temperatures is likely to increase desertification; the country has only 2 percent arable land. A high standard of living and reliance on air conditioning drives heavy energy use.

serious recently. It has funded pilot projects in biofuels and other sustainable energy sources, but the emphasis is less on controlling emissions and more on the specter of running out of affordable fuel.

Inefficient usage is responsible for as much as 10 percent of Saudi Arabia's electricity consumption, and the electricity sector has not yet adopted time-of-use rate adjustments, nor encouraged simple habits like turning off the lights when exiting rooms. In this hot arid country, which faces further desertification in the near future as its 2 percent arable land dwindles, air conditioning is a major source of electricity consumption; as climate change drives temperatures up, the Saudi grid comes closer and closer to facing critical outages as peak demand exceeds supply. In 2009, the government announced energy-efficiency labeling requirements for high-consumption appliances like air conditioners, refrigerators, and washing machines.

The state-owned Saudi Aramco has announced that it wishes to become the world's largest solar energy provider. There is ample land available for solar farms, and other than dust storms, the peninsula faces little extreme weather to reduce solar energy collection. In 2010, the country took another step toward embracing renewable energy, as it announced the founding of the King Abdullah University of Science and Technology (KAUST, an energy research institute) and King Abdullah City for Atomic and Renewable Energy (KACARE). KACARE is a city within Riyadh that will demonstrate and develop nuclear and renewable energy technology and policy.

Bill Kte'pi
Independent Scholar

See Also: Deserts; G8/G20; Iraq; Jordan; Kuwait; Monsoons; Morocco; Oil, Production of; Organization of the Petroleum Exporting Countries; Qatar; United Arab Emirates; Yemen.

Further Readings

Cordesman, Anthony H. *Saudi Arabia Enters the Twenty-First Century: The Political, Foreign Policy, Economic, and Energy Dimensions.* Westport, CT: Praeger, 2003.

Jones, Toby Craig. *Desert Kingdom: How Oil and Water Forged Modern Saudi Arabia.* Cambridge, MA: Harvard University Press, 2010.

Lacey, Robert. *Inside the Kingdom: Kings, Clerics, Modernists, Terrorists, and the Struggle for Saudi Arabia.* New York: Viking Adult, 2009.

Woodward, Peter N. *Oil and Labor in the Middle East: Saudi Arabia and the Oil Boom.* Westport, CT: Praeger, 1988.

Schneider, Stephen H.

Stephen H. Schneider (1945–2010) was an American climatologist and professor in the Department of Biological Sciences, a senior fellow at the Center for Environment Science and Policy of the Institute for International Studies, and a professor by courtesy in the Department of Civil and Environmental Engineering at Stanford University from September 1992 until his death. Schneider was an outspoken advocate of the global warming theory since the 1980s and helped draw public attention to the issue of climate change. He argued for sharp reductions of greenhouse gas (GHG) emissions to combat the phenomenon. Schneider also served as an advisor to different U.S. administrations and federal agencies since the presidency of Richard Nixon. His research included modeling of the atmosphere, climate change, and the relationship of biological systems to global climate change. Schneider's latest efforts were devoted to communicating effectively to the general public the possible consequences and dangers of climate change. His book, *Science as Contact Sport: Inside the Battle to Save the Earth's Climate* (2009) exposes politicians' serious delay in taking seriously the warnings of the scientific community about the dangers of climate change.

Schneider was born on February 11, 1945, in New York. He received his Ph.D. in mechanical engineering and plasma physics from Columbia University in 1971. He investigated the role of GHGs and suspended particulate material on climate as a postdoctoral fellow at the National Aeronautics and Space Administration's Goddard Institute for Space Studies. He was appointed a postdoctoral fellow at the National Center for Atmospheric Research in 1972, and was a member of their scientific staff from 1973 to 1996, where he cofounded the Climate Project.

Although Schneider emerged as a public figure in the global warming debate in the 1980s, his interest in the subject dates back to the early 1970s, when he coauthored an article in *Science* titled, "Atmospheric Carbon Dioxide and Aerosols: Effects of Large Increases on Global Climate," which examined the competing effects of cooling from aerosols and warming from carbon dioxide (CO_2). The paper, however, predicted that CO_2 would only have a minor role and warned about a large possible decrease of the Earth's temperature. In the late 1970s, Schneider modified his position, stating that at that time, it was not possible to be certain whether the climate was cooling or warming. In *The Genesis Strategy: Climate and Global Survival* (1976), he wrote that "consensus among scientists today would hold that a global increase in atmospheric aerosols would probably result in a cooling of the climate; however, a smaller but growing fraction of the current evidence suggests that it may have a warming effect." In the mid-1970s, Schneider also founded the academic journal *Climate Change* and served as its editor-in-chief until his death on July 19, 2010.

Propelled Into the Limelight

With his 1989 book, *Global Warming: Are We Entering the Greenhouse Century?*, Schneider became a main figure in scientific debates about human effects on the environment. In clear language exempt from academic jargon, Schneider argued his case for global warming. The burning of fossil fuels, he claimed, causes a buildup of GHGs in the atmosphere. Those gases trap the solar energy reradiated by the Earth that would otherwise escape into space. This phenomenon could have devastating effects such as droughts, more frequent and more powerful tropical storms, and a rise in sea level. Schneider concluded that temperature change appears less noticeable than it would be otherwise, due to the capacity of the oceans to absorb heat. This conclusion made him aware of the necessity to improve climate models that would take into full account the interactions between atmosphere and oceans.

Schneider believed that it was crucial to involve the public in his research and frequently appeared in media events relating to environment. He tried to popularize complex scientific ideas to make them more accessible to the larger public. This earned him praise as well as criticism, as several of his colleagues charged him with trying too hard to get media attention for himself and his ideas. Still others described Schneider as an alarmist, as in their view, his statements were not always supported by evidence. Yet, Schneider constantly displayed his awareness that as a scientist he was ethically bound to tell the truth. At the same time, he was convinced that, to be effective, his ideas had to take hold of the public and its imagination. Schneider found that this double ethical issue could not be solved by any formula and that scientists had to find the right balance between being effective and being honest in their communications with the audience.

Schneider served in many key positions as an academic and a policymaker. He also received many honors for his research; contributed to 450 scientific papers, proceedings, legislative testimonies; edited books and book chapters, some 140 book reviews, editorials; wrote newspaper articles; and gave magazine interviews. He was a coordinating lead author in the Working Group II of the Intergovernmental Panel on Climate Change's *Third Assessment Report: Climate Change 2001* (TAR) and a coanchor of the *Key Vulnerabilities Cross-Cutting Theme for the Fourth Assessment Report* (AR4). Together with the other scientists of the IPCC and former U.S. Vice President Al Gore, Schneider was among the recipients of the 2007 Nobel Peace Prize for their efforts to disseminate knowledge about anthropogenic climate change and call for urgent actions. In addition to his publications on climate change, Schneider wrote *The Patient from Hell* (2005), a memoir on his personal battle against a particularly aggressive form of cancer, mantle cell lymphoma. For his ability to integrate and interpret the results of global climate research for both the academic community and general public, Schneider was awarded the MacArthur Fellowship in 1992.

For his furtherance of public understanding of environmental science and its implications for public policy, he also received, in 1991, the American Association for the Advancement of Science/Westinghouse Award for Public Understanding of Science and Technology. In 1998, he became a foreign member of the Academia Europaea, Earth, and Cosmic Sciences Section. He

Stephen H. Schneider at the Center for Environment Science and Policy of the Institute for International Studies. In his writings, Schneider argues his case for global warming using clear language and refrains from relying on academic jargon.

was elected chair of the American Association for the Advancement of Science's Section on Atmospheric and Hydrospheric Sciences (1999–2001). Schneider was elected to membership in the U.S. National Academy of Sciences in April 2002.

Luca Prono
Independent Scholar

See Also: Global Warming; Climate Change, Effects; Greenhouse Gas Emissions; Media, Books, and Journals; Stanford University.

Further Readings

Schneider, Stephen H. *Global Warming: Are We Entering the Greenhouse Century*? San Francisco, CA: Sierra Club Books, 1989.

Schneider, Stephen H. *Laboratory Earth: The Planetary Gamble We Can't Afford to Lose.* New York: HarperCollins, 1997.

Schneider, Stephen H. and Terry L. Root, eds. *Wildlife Responses to Climate Change: North American Case Studies.* Washington, DC: Island Press, 2001.

Schneider, Stephen H., Armin Rosencranz, and John O. Niles, eds. *Climate Change Policy: A Survey.* Washington, DC: Island Press, 2002.

Scripps Institution of Oceanography

The Scripps Institution of Oceanography is an interdisciplinary oceanographic institution housed at the University of California (UC), San Diego. Scripps is among the world's largest, oldest, and most renowned institutional centers for global ocean and Earth science research, education, and public service. It was founded in 1903, joining the University of California in 1912. The scientific areas encompassed by Scripps research and education include the atmospheric, biological, chemical, geological, and geophysical systems of the Earth and its oceans. Their mission statement is to seek, teach, and communicate scientific understanding of the oceans, atmosphere, Earth, and other planets for the benefit of society and the environment. They address their mission through research, academic and public education, and communication to government policymakers and private industry.

Scripps oversees over 300 international programs in the physical, chemical, biological, geological, and geophysical sciences in over 60 countries. The institution houses the Birch Aquarium and maintains an academic fleet of four oceanographic research vessels and the FLIP research platform, with a home port at Point Loma on San Diego Bay. A fifth research vessel will set sail in 2015. Scripps also encompasses a graduate Ph.D. degree-granting program in oceanography, marine biology, and the Earth sciences. Both Scripps and the larger UC San Diego campus promote sustainable campus operations.

Campus facilities include a hydraulics laboratory, visualization center, and analytical laboratory. Other research aids include ocean devices such as data-gathering floats, aircraft, laser-based and sound-imaging devices, satellites, and supercomputer networks. Scripps also enjoys access to the San Diego Supercomputer Center. The Scripps Library houses its extensive collections of marine life and geological specimens. Scripps and the University of California at San Diego have initiated or plan to initiate a variety of projects designed to create a more environmentally friendly campus. Scripps conservation and carbon-footprint reduction efforts include research and education on climate change and sustainability, energy efficiency and conservation practices, new construction practices, and recycling programs.

Research topics through its international programs include atmospheric-oceanic interactions, climate change and prediction, earthquakes and other natural disasters, drug-resistant diseases, energy alternatives, pollution, water shortages, the topography and composition of the seafloor, waves and currents, marine ecosystem biodiversity, marine chemistry, saving marine life, beach erosion, the marine food chain, marine genomics, and the geological evolution of ocean basins.

Climate Change Research

Climate change and sustainability has emerged as a key area of Scripps research. Scripps scientists seek to understand and explain climate change, as well as provide that information and knowledge to those applying it to social change. Scripps activities in this field include climate prediction and modeling, the use of algae for biofuels and carbon dioxide (CO_2) abatement, wildfire prediction, impacts of aerosols on the climate model, snowpack monitoring and its implications for water distribution and storage, sea-level rise and its impacts on coastlines, marine ecosystem resource management, and declining fish populations.

Scripps has participated in a number of key projects with implications for the study of climate change and global warming. Scripps researchers Tim Barnett and David Pierce conducted a study completed in 2005 alongside Krishna Achutarao, Peter Gleckler, and Benjamin Santer of the Lawrence Livermore National Laboratory's Program for Climate Model Diagnosis and Intercomparison that provided the first clear evidence of anthropogenic ocean warming. The study utilized both computer climate change models and observed data.

Scripps was a founding partner in the international World Ocean Circulating Experiment (WOCE) begun in 1990, one of the world's largest ocean-based scientific undertakings. The project studied world ocean circulation and its atmospheric interactions. Scripps has also joined leading international oceanographic institutions in the Partnership for Observation of the Global Oceans (POGO), promoting global oceanography and the implementation of an integrated international global ocean observing system.

The Scripps Center for Clouds, Chemistry, and Climate (C4) is coordinating the international Indian Ocean Experiment (INDOEX), studying the role of aerosols on the global climate. Scripps is also participating in the North Pacific Acoustic Laboratory (NPAL) Project, analyzing long-distance sound transmissions in the ocean as a possible method of determining global warming. Barnett is serving as coordinator of the Accelerated Climate Prediction Initiative (ACPI), a pilot project within the Climate Research Division sponsored by the U.S. Department of Energy to analyze the effects of climate change on western U.S. water resources, information that can then be globally extrapolated.

Scripps scientists search for past and present signs of climate change and global warming, as well as their known and potential impacts. Scripps chemist Dr. Charles David Keeling was the first scientist to show increases in atmospheric CO_2 levels due to human industrial activities such as the burning of fossil fuels. His data became known as the Keeling curve. Scripps scientists worked with colleagues from Washington State University to determine that the last ice age ended approximately 15,000 years ago due to abrupt, as opposed to gradual global warming, as had previously been believed. Scientists relied upon the analysis of samples of ancient air that had been encased in ice cores. Scripps scientists have also analyzed polar ice cores for other climate history information, such as the reconstruction of historical CO_2 levels and the impact of greenhouse gases and temperature on the atmosphere.

Scripps scientists have been leading participants in the prediction and study of El Niño and La Niña events, such as that of 1997 through 1998. These natural phenomena involve large warming or cooling of the tropical eastern Pacific Ocean's surface temperatures and often produce large global climate impacts. Scripps research has determined the early arrival of spring weather and a longer growing season in parts of the Northern Hemisphere beginning in the mid 1970s, a potential impact of global warming. Scripps research has discovered increased seasonal variations in CO_2 uptake by land-based plants in the Northern Hemisphere since the 1960s, a potential response to global warming. Scripps scientists have developed a number of new tools for measurement and data collection as part of their research, including the Spray autonomous underwater glider to study deep ocean global circulation. Scripps scientists pioneered the technological developments needed to monitor small atmospheric oxygen level change and to determine the chemical compositions of ocean currents to allow for tracking. The Argo program, begun in 2000, utilizes thousands of ocean floats that drift beneath the water's surface to gather subsurface data, which is transmitted to an Argo Data Center via satellite. Scripps is a participant in the Deep Space Climate Observatory (DSCOVR) satellite mission designed to collect information and monitor the Earth's climate from deep space.

Scripps emphasizes the importance of communication in the field of global climate change science, often participating in local community forums and events and lending their expertise to governments, businesses, and nongovernment organizations (NGOs) at local, national, and international levels. The collaborative UC Revelle Program supports this goal through the identification, accessibility, and clarity of scientific research related to policy issues to nonscientific audiences. Program partners include the UC Institute on Global Conflict and Cooperation and the UC San Diego Graduate School of International Relations and Pacific Studies.

Scripps seeks to educate the next generation of scientists and policymakers through its academic component. Scripps offers over 40 undergraduate courses in Earth and marine science and environmental systems, with an interdisciplinary approach and sustainability focus. At the graduate level, Scripps offers Ph.D. degrees in oceanography, marine biology, and the Earth sciences. Program areas are climate-ocean-atmosphere, geosciences of the Earth, oceans, and planets, or ocean biosciences. Curricular groups include biological oceanography, physical oceanography, marine biology, geological sciences, marine chemistry and geochemistry, geophysics, climate sciences, and applied ocean sciences. The National Research Council has ranked Scripps's oceanographic graduate program as first in faculty quality, distinction, and scholarly publications.

Marcella Bush Trevino
Barry University

See Also: Climatic Data, Oceanic Observations; National Academy of Sciences; National Science Foundation; Navy, U.S.; Oceanic Changes; Oceanography; University of California, Berkeley.

Further Readings

Marshall, John and R. Alan Plumb. *Atmosphere, Ocean and Climate Dynamics*. Burlington, MA: Academic Press, 2008.

Scripps Institution of Oceanography. http://www.sio.ucsd.edu (Accessed July 2011).

Vallis, Geoffrey K. *Climate and the Oceans*. Princeton, NJ: Princeton University Press, 2011.

Sea Ice

Sea ice is frozen ocean water. It forms primarily in and near the polar regions, although it can grow closer to the equator, as far as 40 degrees north and 55 degrees south latitude. Sea ice has a strong seasonal variability. In the Northern Hemisphere, the annual maximum extent occurs in late winter (March), covering about 5 million sq. mi. (15 million sq. km) on average. It then melts during spring and summer to an annual minimum extent of about 2 million sq. mi. (7 million sq. km) in September. In the Antarctic, the annual maximum is about 7 million sq. mi. (19 million sq. km) during September, and the annual minimum is about 1 million sq. mi. (3 million sq. km) in February or March. Overall, roughly 10 percent of the world's ocean area is covered with sea ice at some point during the year.

Sea ice typically grows to an average level thickness of 3 to 6.5 ft. (1 to 2 m) in the Antarctic and 10 to 13 ft. (3 to 4 m) in the Arctic. The ice is thinner in the Antarctic because most ice melts during the austral summer, whereas in the Arctic a significant fraction (~40 percent) remains through the summer and can grow over several years. A larger ocean heat flux at the bottom of the ice in the Antarctic also keeps the ice thinner.

However, thicker ice is not uncommon because of the effect of ice motion. Most sea ice is almost constantly in motion, mainly because of the force of winds and ocean currents; other factors include the Coriolis effect, the slope of the ocean surface, and the internal structure of the ice. The speed of sea ice motion varies considerably; it can move 30 mi. (50 km) or more in a day, although 1.2 mi. (2 km) per day is typical. The motion of the ice can result in convergence between different parts of the ice cover, causing the ice to pile up into features called ridges. Ridges may easily rise 16 to 33 ft. (5 to 10 m) above the surrounding level ice (and many tens of meters below the surface).

Role of Sea Ice in Climate

Sea ice plays an important role in climate. It has a much higher albedo than the unfrozen ocean, meaning that as much as 80 to 90 percent of the sun's energy is reflected by a snow-covered sea ice surface, whereas the unfrozen ocean reflects less than 10 percent of the sun's energy, resulting in much less energy absorption where ice is present. Sea ice is also a physical barrier between the ocean and atmosphere. This prevents the transfer of heat and moisture between the two and during winter. Thus, sea ice keeps the polar regions cooler and drier than they would be without ice. Sea ice also reduces fetch and dampens waves, limiting coastal erosion.

Because of its location near the poles, the thin nature of the ice cover, and its interaction with the ocean and the atmosphere, sea ice is a sensitive indicator of the climate state. Sea ice in the Arctic has been decreasing dramatically over the past several decades. Overall, the Arctic has lost approximately 30 percent of the average summer ice extent (areal coverage) since the late 1970s, and recent years have seen record or near-record low summer extents. September 2007 had the lowest extent on record, at 1 million sq. mi. (4.3 million sq. km), 23 percent below any previous September in the satellite record (since 1979) and likely the lowest in several hundred years and perhaps longer. Reductions of sea ice extent during winter are less, but are still significant. Septembers since 2007 have also been very low. In addition to declines in extent, the ice has thinned dramatically and has lost much of the older, thicker perennial ice cover than once used to dominate the Arctic Ocean. Now, most of the Arctic Ocean is covered by seasonal ice, or ice that forms each winter and melts during the subsequent summer. On the basis of current trends and projects by climate models, the Arctic is likely to be largely sea ice–free during at least part of the summer by 2050 or earlier. This reduction in sea ice has been linked to warming temperatures resulting from the anthropogenic emission of greenhouse gases, though other factors also play a role. Unlike in the Arctic, trends in Southern Hemisphere ice are showing a small increase, likely because of its remoteness relative to other continental land areas, the seasonal nature of the ice, changes in atmospheric circulation, and a greater ocean influence.

Changes in Arctic sea ice cover will have profound effects on climate, human activities, and wildlife, some of which are already being felt. Polar bears and other animals may be endangered, as well as the traditions of native communities. Less ice may also have benefits by opening up shipping routes and facilitating extraction of natural resources. Nonetheless, most effects are expected to be negative, and their implications for future climate will extend to regions far beyond the Arctic.

Walter N. Meier
National Snow and Ice Data Center
Julienne Stroeve
University of Colorado

See Also: Arctic and Arctic Ocean; Charismatic Megafauna; Climate Change, Effects of; Inuit; Marine Mammals; Oceanic Changes; Penguins; Polar Bears.

Further Readings

Johannessen, O. M., R. D. Muench, and J. E. Overland, eds. "The Polar Oceans and Their Role in Shaping the Global Environment." *American Geophysical Monograph*, v.85 (1994).

Lubin, Dan and Robert Massom. *Polar Remote Sensing*. Vol. 1. New York: Springer, 2006.

Untersteiner, Norbert, ed. *The Geophysics of Sea Ice*. New York: Plenum Press, 1986.

Wadhams, Peter. *Ice in the Ocean*. London: Gordon and Breach, 2000.

Sea Level, Rising

The rising sea level is a result of (1) the thermal expansion of the oceans, (2) the melting of glaciers and ice caps, (3) the melting of the Greenland and

Antarctic ice sheets, and (4) changes in terrestrial storage. The effects of sea-level changes include (1) the increased intensity and frequency of storm surges and coastal flooding; (2) the increased salinity of rivers, bays, and coastal aquifers as a result of saline intrusion; (3) increased coastal erosion; (4) the loss of important mangroves and other wetlands (the exact response will depend on the balance between sedimentation and sea-level change); and (5) the impact on marine ecosystems, that is, coral reefs.

The global sea level rose by about 393.7 ft. (120 m) during the several millennia that followed the end of the last ice age (approximately 21,000 years ago) and stabilized between 3,000 and 2,000 years ago. Sea-level indicators suggest that the global sea level did not change significantly from that time until the late 19th century. The instrumental record of modern sea-level change shows evidence for the onset of rising sea levels during the 19th century. Estimates for the 20th century show that the global average sea level rose at a rate of about 1.7 mm/year^{-1}.

Satellite data available since the early 1990s provides a more accurate sea-level measurement, with nearly global coverage. This decade-long satellite altimetry data set shows that since 1993, the sea level has been rising at a rate of around 3 mm/year^{-1}, significantly higher than the average during the previous half century. Coastal tide gauge measurements confirm this observation and indicate that similar rates occurred in some earlier decades.

Currently, sea-level rise is determined by two techniques: the use of tide gauges and satellite altimetry. Tide gauges provide sea-level variations with respect to the land on which they lie. To extract the signal of sea-level change due to ocean water volume and other oceanographic change, land motions need to be removed from the tide gauge measurement. Sea-level change based on satellite altimetry is measured with respect to the Earth's center of mass, and thus is not distorted by land motions, except for a small component due to large-scale deformation of ocean basins from GIA. The total 20th-century rise is estimated to be 0.17 m (0.12 to 0.22 m).

Rising sea levels are accelerating worldwide. Globally, 100 million people live within about one meter of present-day sea levels. The global average sea level rose at an average rate of 1.8 mm (1.3 to 2.3 mm) per year between 1961 and 2003 as a result of the global ocean temperature rise of 32.2 degrees F (0.10 degree C) from the surface to a depth of 2,297 ft. (700 m). This rate was faster between 1993 and 2003, about 3.1 mm (2.4 to 3.8 mm) per year. Whether the faster rate for 1993 to 2003 reflects decadal variability, or an increase in the long-term trend, is unclear. There is high confidence that the rate of observed sea-level rise increased from the 19th to the 20th century.

Projections for Sea-Level Rise

In 2001, the Intergovernmental Panel on Climate Change (IPCC) projected a sea-level rise from 3.5 in. to 34.6 in. (9 cm to 88 cm) between 1990 and 2100 and a global average surface temperature rise of between 35 and 42 degrees F (1.4 and 5.8 degrees C). In 2007, IPCC projections, based on different scenarios, predicted a sea-level rise from 0.007 in. up to 0.02 in. (0.18 mm up to 0.59 mm) by 2099. Toward the end of the 21st century, the projected sea-level rise will affect low-lying coastal areas with large populations. For instance, under the IPCC Special Report on Emission Scenarios (SRES) A1B scenario, by the mid-2090s, global sea level is rising at about 4 mm/year^{-1} and could reach 8.6 in. to 17.3 in. (0.22 m to 0.44 m) above 1990 levels. As in the past, sea-level change in the future will not be geographically uniform, with regional sea-level change varying within about plus or minus 5.9 in. (0.15 m) of the mean in a typical model projection. The cost of adaptation could amount to between 5 and 10 percent of gross domestic product (GDP). Projections are that mangroves and coral reefs will be further degraded, with additional consequences for fisheries and tourism.

For example, the Kangerdlugssuaq Glacier in Greenland is melting at a rate of 8.7 mi. (14 km) a year in comparison to just 3.1 mi. (5 km) a year in 1988. This loss will also have serious implications for sea-level rise. Some predict that within the next 100 years, ice cover in this region will completely disappear over summer, and that species living within it, such as polar bears, will be threatened. The complete melting of the Greenland Ice Sheet and the West Antarctic Ice Sheet would lead to a sea-level rise of up to 22.9 ft. and about 16.4 ft. (7 m and about 5 m), respectively. Factors that could accelerate polar ice sheet melting include more rapid heat absorption as the ice's white surface is replaced by darker wet ice, melting rates exceeding snowfalls, and direct contact of warmer seawater with the underside of ice sheets.

Economic, Natural, and Human Consequences

The potential socioeconomic consequences of rising sea levels are the direct loss of economic, ecological, cultural, and subsistence values through the loss of land, infrastructure, and coastal habitats. There will be an increased flood risk for people, land, infrastructure, and the aforementioned values. There will be other impacts related to changes in water management, salinity, and biological activities. A rise in sea level would inundate wetlands and lowlands, accelerate coastal erosion, exacerbate coastal flooding, threaten coastal structures, raise water tables, and increase the salinity of rivers, bays, and aquifers. Similarly, the areas vulnerable to erosion and flooding are also predominantly in the southeast, while potential salinity problems are spread more evenly throughout the coast. Such a loss would reduce available habitat for birds and juvenile fish, and would reduce the production of organic materials on which estuarine fish rely.

Some of the most important vulnerable areas are recreational barrier islands and spits such as those found within the Atlantic and Gulf of Mexico coasts. Coastal barriers are generally long, narrow islands and spits (peninsulas), with the ocean

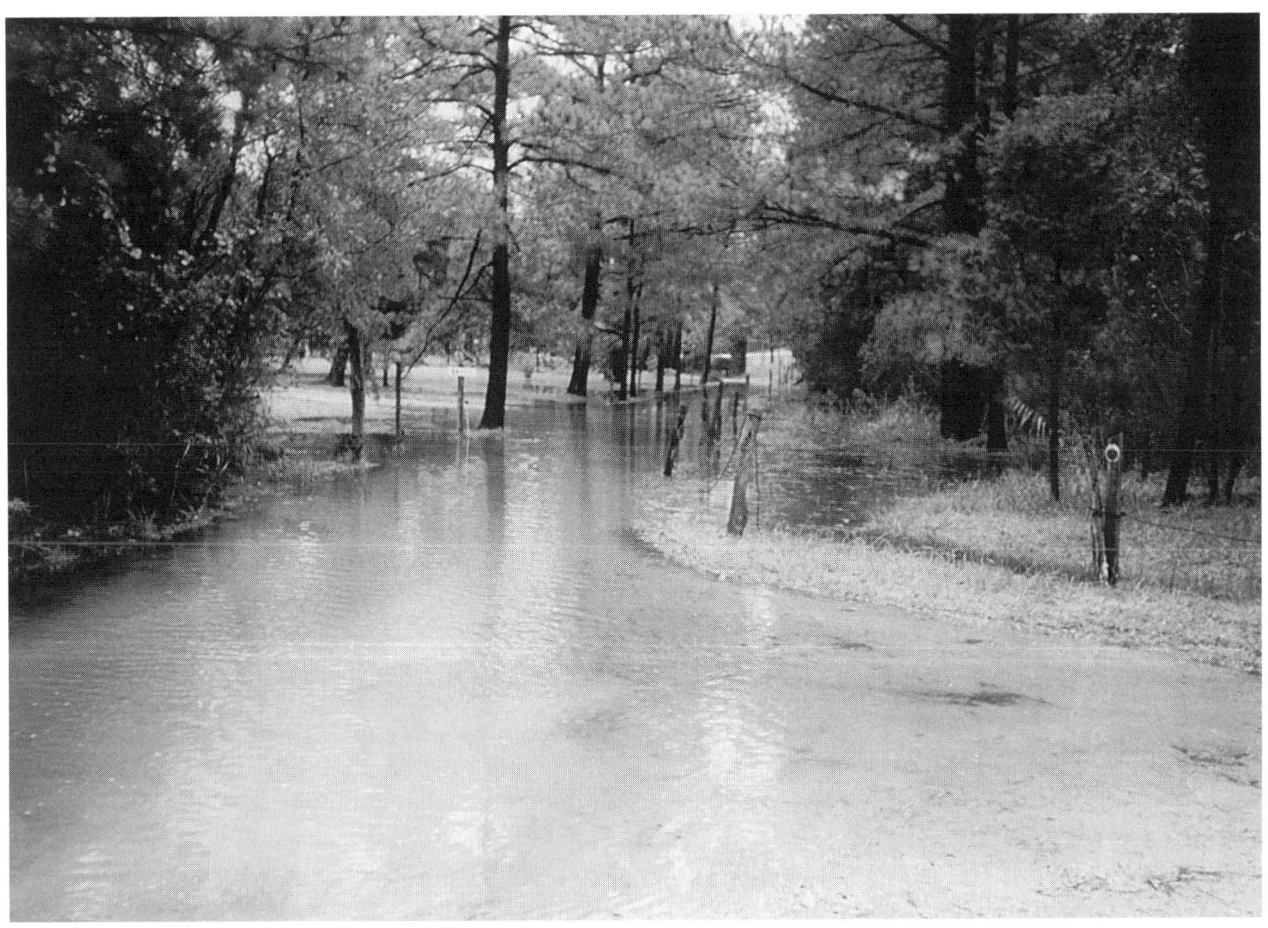

The Lower Patuxent River in Maryland, showing the flooding of low-lying areas by extreme high tides. If climate change causes sea level to continue to rise, this type of flooding will become increasingly common. Sea-level rise would also inundate wetlands, accelerate coastal erosion, threaten coastal structures, raise water tables, and increase the salinity of rivers, bays, and aquifers. Increased salinity on the coast would jeopardize habitats for birds and fish and damage the food supplies for estuarine fish.

on one side and a bay on the other. Typically, the oceanfront block of an island ranges from 6.6 ft. to 13.1 ft. (2 m to 4 m) above high tide, while the bay side is less than 3 ft. (1 m) above high water. Thus, even a 3 ft. rise in sea level would threaten much of this valuable land with inundation.

Erosion, moreover, threatens the high parts of these islands and is generally viewed as a more immediate problem than the inundation of their bay sides. While inundation alone is determined by the slope of the land just above the water, in 1962, Bruun showed that the total shoreline retreat from a rise in sea level depends on the average slope of the entire beach profile. For example, most U.S. recreational beaches are less than 100 ft. (30 m) wide at high tide, thus even a 1-ft. (30-cm) rise in sea level would require a response.

Finally, a rise in sea level would enable saltwater to penetrate farther inland and upstream in rivers, bays, wetlands, and aquifers, which would be harmful to some aquatic plants and animals, and would threaten human water usage. In Delaware, for example, salinity is a factor resulting in reduced oyster harvests.

Sea-level rise will cause coastal areas worldwide to become more vulnerable to flooding because higher sea levels provide higher bases for storm surges to build upon. In this context, a 3-ft. (1-m) rise in sea level would mean that a 100-year flood would occur every 15 years. Beach erosion will make land more vulnerable to a storm's waves, while higher water levels will increase the effects of flooding due by reducing coastal drainage. Sea-level rise will also raise water tables in various systems.

Japan is particularly vulnerable to the effects of sea-level rise; a 3-ft. rise (1-m) in sea level would increase the area situated below mean high water from 332 sq. mi. to 903 sq. mi. (861 sq. km to 2,340 sq. km). Future estimates show that this would affect up to 4.1 million people and cost $1.3 trillion. Egypt's Nile Delta is one of the world's most vulnerable areas to sea-level rise. Estimates are that about 30 percent of the city will be lost due to inundation, almost 2 million people will lose their homes, and approximately 195,000 jobs will be lost because of a predicted economic impact of over $3.5 billion over the next century. In Australia, a country where over 90 percent of the population lives near or within a 31-mile radius (50-km) of the coast, sea-level rise is projected to affect 21,748 mi. (35,000 km) of roads and rails. In the states of New South Wales and Queensland, 140,000 homes could be affected at a cost of $72 billion. Already large populations are exposed to coastal flooding in port cities, and across all cities approximately 40 million people (0.6 percent of the global population or roughly one in 10 of the total port city population in the cities within their project study) will be exposed to a coastal flood event every 100 years. The top 10 cities (in 2005) that have populations most vulnerable to sea-level rise include Mumbai, Guangzhou, Shanghai, Miami, Ho Chi Minh City, Kolkata, greater New York City, Osaka–Kobe, Alexandria, and New Orleans.

Overall, current science indicates that sea-level rise is tracking above the IPCC projections, which means that combined with accelerated ice loss from Antarctica and Greenland, the estimated sea-level rise by 2100 will be between 2.5 and 6.6 ft. (75 cm to 2 m). This will have significant ramifications worldwide on how humans work and live.

Mitigation Responses

The chapter on coasts in the 2001 United Nations Environment Programme *Handbook on Adaptation and Mitigation Methodologies* specifically outlines a suite of strategic responses to rising sea levels. It cautions, however, that before applying these strategies, policymakers must decide if their adaptation is to be autonomous or planned, reactive or proactive. There are three management responses to sea-level rise: retreat, accommodation, and protection. Retreat involves no effort to protect the land from the sea. The coastal zone is abandoned and ecosystems shift landward. This choice can be motivated by excessive economic or environmental impacts of protection. In an extreme case, an entire area may be abandoned.

Accommodation implies that people continue to use the land at risk, but do not attempt to prevent the land from being flooded. This option includes erecting emergency flood shelters, elevating buildings on piles, converting agricultural lands to aquatic uses such as fish farming, or growing flood- or salt-tolerant crops. Protection involves erecting hard structures such as sea walls and dikes, as well as soft solutions such as dunes and vegetation to protect the land from the sea so that existing land uses can continue.

Melissa Jane Nursey-Bray
University of Adelaide

See Also: Alliance of Small Island States; Antarctic Ice Sheets; Anthropogenic Forcing; Arctic and Arctic Ocean; Atlantic Ocean; Center for Ocean-Atmospheric Prediction Studies; Climate Change, Effects of; Climatic Data, Ice Observations; Climatic Data, Oceanic Observations; Cooperative Institute for Arctic Research; Current; Dangerous Anthropogenic Interference; Drift Ice; Glaciers, Retreating; Global Warming, Impacts of; Greenland Ice Sheet; Indian Ocean; Joint Institute for the Study of Atmosphere and Ocean; Marine Mammals; Modeling of Ocean Circulation; National Oceanic and Atmospheric Administration; Ocean Component of Models; Oceanic Changes; Pacific Ocean; Scripps Institution of Oceanography; Sea Ice.

Further Readings

Bindoff, N. L., et al. "Observations: Oceanic Climate Change and Sea Level." In *Climate Change 2007: The Physical Science Basis. Contribution of Working Group I to the Fourth Assessment Report of the Intergovernmental Panel on Climate Change*, edited by S. Solomon, D. Qin, M. Manning, Z. Chen, M. Marquis, K. B. Averyt, M. Tignor, and H. L. Miller. Cambridge: Cambridge University Press, 2007.

El-Raey, M., K. Dewidar, and E. El-Hattab. "Adaptation to the Impacts of Sea Level Rise in Egypt." *Mitigation and Adaptation Strategies for Global Change*, v.4 (1999).

Leafe, R., J. Pethick, and I. Townsend. "Realising the Benefits of Shoreline Management." *Geographical Journal*, v.164/3 (1998).

Neumann, James E. *Sea-Level Rise and Global Climate Change: A Review of Impacts to U.S. Coasts*. Arlington, VA: Pew Center on Global Climate Change, 2000.

Nicholls, R. J., S. Hanson, C. Herweijer, N. Patmore, S. Hallegatte, J. Corfee-Morlot, J. Château, and R. Muir–Wood. "Ranking Port Cities With High Exposure and Vulnerability to Climate Extremes: Exposure Estimates." In *OECD Environment Working Papers, No. 1*. Paris: OECD, 2007.

Seasonal Cycle

The seasonal cycles are generally thought of as spring, summer, autumn, and winter. The four seasons are based on the amount of sunlight reaching a hemisphere due to the Earth's angle relative to the sun. The elliptical orbit of the Earth prevents it from traveling at a constant speed. Furthermore, the main cause of seasonal effects as we know them is due to the angle of the Earth's rotation axis (23.5 degrees) with respect to its elliptical path around the sun. The Earth's orbit results in a hemisphere either being more or less titled toward the sun, which results in the seasonal cycles. For example, when a hemisphere is leaning toward the sun, the nights are shorter and the days are longer. In contrast, when a hemisphere is leaning away from the sun, the nights are longer and the days are shorter.

Some believe that the distance of the Earth relative to the sun results in warmer versus cooler temperatures. However, in the Northern Hemisphere, the Earth is closest to the sun in early January (91.6 million miles) and furthest away (94.8 million mi., or 152.5 million km) in early July. Even though it is 3.2 million mi. (5.14 million km) farther from the sun, the Earth averages about 4 degrees F (2.2 degrees C) warmer in July than January, because the Northern Hemisphere is 40 percent land and the Southern Hemisphere is 20 percent land. Both areas receive the same amount of sunlight, but land area heats up faster than water bodies. Thus, the amount of the land area is more of a predictor of temperatures than its closeness to the sun.

Seasons

There are two different views of the seasons: the astrological and the meteorological. Astrological seasons coincide with the Earth's transit over the equator (equinoxes) and the tropics of Cancer and Capricorn (solstices). For example, as the Earth rotates around the sun, the sun is positioned directly over the equator (0 degrees) around March 21 (vernal equinox), the tropic of Cancer (23.5 degrees north) around June 21 (summer solstice), the equator (0 degrees) around September 21 (autumnal equinox), and the tropic of Capricorn (23.5 degrees south) around December 21 (winter solstice). In Latin, the word *equinox* means "equal night." On the vernal equinox (around

March 21) and autumnal equinox (around September 21), the sun's energy is balanced between the Northern and Southern Hemispheres. On the equinox, day and night are approximately equal in length. The March equinox marks the start of spring in the Northern Hemisphere and the start of autumn in the Southern Hemisphere.

The term *solstice* means "stand still" in Latin. During the summer solstice in the Northern Hemisphere, the North Pole (location 90 degrees north) is tilted toward the sun (around June 21) and directly over the Tropic of Cancer. During the winter solstice, the North Pole is tilted away from the sun (around December 21) and directly over the Tropic of Capricorn. During the summer solstice, the Northern Hemisphere is positioned to receive extended periods of sunlight due to its position relative to the sun. The extended periods of sunlight coupled with shorter periods of darkness at night results in increased temperatures in the Earth's Northern Hemisphere. All locations in the Northern Hemisphere experience longer days, and locations north of the Arctic Circle (66.5 degrees north) experience a 24-hour period of sunlight on the summer solstice. In contrast, during the winter solstice in the Northern Hemisphere, the Earth is leaning away from the sun and the Earth receives a shorter period of sunlight due to its position relative to the sun. For example, all locations north of the Arctic Circle experience a 24-hour period of darkness on the winter solstice (December 21).

The seasons in the Southern Hemisphere are opposite to the seasons in the Northern Hemisphere. For example, spring in the Southern Hemisphere is equivalent to autumn in the Northern Hemisphere. The months of June, July, and August are generally the hottest months in the Northern Hemisphere, while December, January, and February are the hottest months in the Southern Hemisphere.

Each of the astrological seasons (spring, summer, autumn, and winter) are not equal in length. This is due to Earth's elliptical orbit around the sun. For example, the vernal equinox transit lasts 92.8 days; the summer solstice, 93.6 days; the autumnal equinox, 89.8 days; and the winter solstice period, 89 days. In the Southern Hemisphere, autumn and winter seasons are generally longer, and occur at the same time as the spring and summer seasons in the Northern Hemisphere. Differences in the length of seasons are due primarily to the Earth's elliptical orbit around the sun, where Earth tends to move faster when closer to the sun and slower when further away from the sun.

Meteorological Seasons

The meteorological seasons are different from the astrological seasons in the following ways: (1) seasons last 90 days each, (2) seasons begin about three weeks before the astrological seasons, and (3) seasons are based on temperature and heat lag and not the Earth's position. The four meteorological seasons in the Northern Hemisphere are: spring (begins March 1), summer (begins June 1), autumn (begins September 1), and winter (begins December 1). In the Southern Hemisphere, the four seasons are the opposite, with spring starting on September 1, summer on December 1, autumn on March 1, and winter on June 1.

Andrew Hund
Independent Scholar

See Also: Climate Zones; El Niño and La Niña; Hot Air; North Atlantic Oscillation; Simulation and Predictability of Seasonal and Interannual Variations; Southern Oscillation Index.

Further Readings

Geophysics Research Board. *Solar Variability, Weather and Climate.* Washington, DC: National Academy Press, 1982.

Green, Jen. *Weather and Seasons.* London: Wayland, 2008.

Kondratyev, K. and G. Hunt. *Weather and Climate on Planets.* New York: Pergamon Press, 1982.

Pierrehumbert, R. *Principles of Planetary Climate.* Cambridge: Cambridge University Press, 2011.

Seawater, Composition of

Seawater is a solution of salts of nearly constant composition, dissolved in variable amounts of water. It is denser than freshwater. It is risky to drink seawater because of its high salt content. More water is required to eliminate the salt through excretion than the amount of water that is gained from drinking seawater. Seawater can be turned into potable water by desalination processes or by diluting it with freshwater. The origin of sea salt is traced to Sir Edmond Halley, who in 1715 proposed that salt and other minerals were carried into the sea by rivers, having been leached out of the ground by rainfall runoff. On reaching the ocean, these salts would be retained and concentrated as the process of evaporation removed the water.

There are more than 70 elements dissolved in seawater as ions, but only six make up more than 99 percent of all the dissolved salts; namely, chloride (55.04 weight percent [wt%]), sodium (30.61 wt%), sulphate (7.68 wt%), magnesium (3.69 wt%), calcium (1.16 wt%), and potassium (1.10 wt%). Trace elements in seawater include manganese, lead, gold, and iodine. Biologically important elements such as oxygen, nitrogen, and iron occur in variable concentrations depending on utilization by organisms. Most of the elements occur in parts per million or parts per billion concentrations and are important to some positive and negative biochemical reactions. Properties such as salinity, density, and pH can be used to highlight the composition of seawater.

Salinity

Salinity is the amount of total dissolved salts present in 1 L. of water and is used to express the salt content of seawater. Normal seawater has a salinity of 35 g./L. of water; that is, 3.5 percent. The salinity of seawater is made up by the dissolved salts. Seawater is more enriched in dissolved ions than all types than freshwater. Salts dissolved in seawater come from three main sources: volcanic eruptions, chemical reactions between seawater and hot, newly formed volcanic rocks of spreading zones, and chemical weathering of rocks. Because of some chemical reactions between seawater and hot, newly formed volcanic rocks, the composition of seawater has been nearly constant over time. Salinity affects marine organisms because the process of osmosis transports water toward a higher concentration through cell walls. Marine plants and many lower organisms have no mechanism to control osmosis, which makes them very sensitive to the salinity of the water in which they live. The density of surface seawater ranges from 1,020 kg per cu. m to 1,029 kg per cu. m, depending on the temperature and salinity: the saltier the water, the higher its density. Seawater pH is limited to the range from 7.5 to 8.4 and increases with phytoplankton production. The speed of sound in seawater is about about 4,921 ft. (or 1,500 m) per second and varies with water temperature and pressure.

Carbon (IV) oxide in the sea exists in equilibrium with that of exposed rock containing limestone ($CaCO_3$). Seawater also contains small amounts of dissolved gases such as nitrogen, oxygen, carbon (IV) oxide, hydrogen, and trace gases. Water at a given temperature and salinity is saturated with gas when the amount of gas entering the water equals the amount leaving during the same time. Surface seawater is normally saturated with atmospheric gases such as oxygen and nitrogen. The concentrations of oxygen and carbon (IV) oxide vary with depth. The surface layers are rich in oxygen, which reduces quickly with depth to reach a minimum between 656 and 2,625 ft. (200 and 800 m) in depth. The amount of gas that can dissolve in seawater is determined by the water's temperature and salinity. Increasing the temperature or salinity reduces the amount of gas that can be dissolved. As water temperature increases, the increased mobility of gas molecules makes them escape from the water, thereby reducing the amount of gas dissolved.

The gases dissolved in seawater are in constant equilibrium with the atmosphere, but their relative concentrations depend on each gas' solubility. As salinity increases, the amount of gas dissolved decreases, because more water molecules are immobilized by the salt ion. Inert gases like nitrogen and argon do not take part in the processes of life and are thus not affected by plant and animal life, but gases like oxygen and carbon (IV) oxide are influenced by sea life. Plants reduce the concentration of carbon (IV) oxide in the presence of sunlight, whereas animals do the opposite in either light or darkness.

The world underwater is different from that above in the availability of important gases such as oxygen and carbon (IV) oxide. Whereas in air about one in five molecules is oxygen, in seawater this is only about four in every 1 trillion water molecules. Whereas air contains about one carbon (IV) oxide molecule in 3,000 air molecules, in seawater this ratio becomes four in every 100 million water molecules. Thus, carbon (IV) oxide is much more available in seawater than oxygen. All gases are less soluble as temperature increases, and particularly nitrogen, oxygen, and carbon (IV) oxide, which become about 40 to 50 percent less soluble with an increase of 45 degrees F (25 degrees C). When water is warmed, it becomes more saturated, resulting in bubbles leaving the liquid.

Akan Bassey Williams
Covenant University

See Also: Chemistry; Climate Change, Effects of.

Further Readings

Greene, Thomas F. *Marine Science: Marine Biology and Oceanography.* New York: Amsco, 2004.

The Royal Society. *Ocean Acidification Due to Increasing Atmospheric Carbon Dioxide.* London: Royal Society, 2005.

Senegal

Senegal is a West African country sharing borders with Mauritania, Mali, Guinea, Guinea-Bissau, and the Gambia, and has 330 mi. (531 km) of coastline on the Atlantic Ocean. The terrain is low, rolling plains rising to foothills, and the climate is tropic, with a hot rainy season from May to November and a dry season from December to April. Northern Senegal is in the Sahel region, a climatic zone of transition between the Sahara Desert and the savanna. The population of 12.6 million has a young age structure (43.3 percent age 14 or younger, median age 18 years) and a growth rate of 2.6 percent, due in large part to a high total fertility rate (4.78 children per woman). This population growth places an added strain on Senegal's resources. Senegal is heavily dependent on donor assistance and was granted debt relief under the International Monetary Fund Highly Indebted Poor Countries program in 2005. Its chief export industries are fishing, phosphate mining, and fertilizer production. Per-capita gross domestic product in 2010 was $1,900, ranking 192nd in the world, with unemployment and poverty rates of approximately 50 percent.

Senegal has no known oil reserves, but does have the capacity to refine petroleum (25.03 thousand barrels per day in 2009). Oil consumption has risen steadily over the past 30 years, from 17.4 thousand barrels per day in 1980 to 38 thousand barrels per day in 2008. Senegal began producing and consuming natural gas in 1993; in 2009, the country produced and consumed 2 billion cu. ft. annually. Senegal began consuming coal in 2004 (entirely imported) and in 2008 consumed 0.220 million short tons of coal. In 2007, Senegal generated 2.48 billion kilowatt hours (kWh) of electricity, consuming 1.97 billion kWh. In 2009, Senegal's carbon dioxide emissions from consumption of fossil fuels totaled 6.19 million metric tons.

Mean annual temperature in Senegal has increased by 1.6 degrees F (0.9 degree C) since 1960 for an average pace of 0.36 degree F (0.20

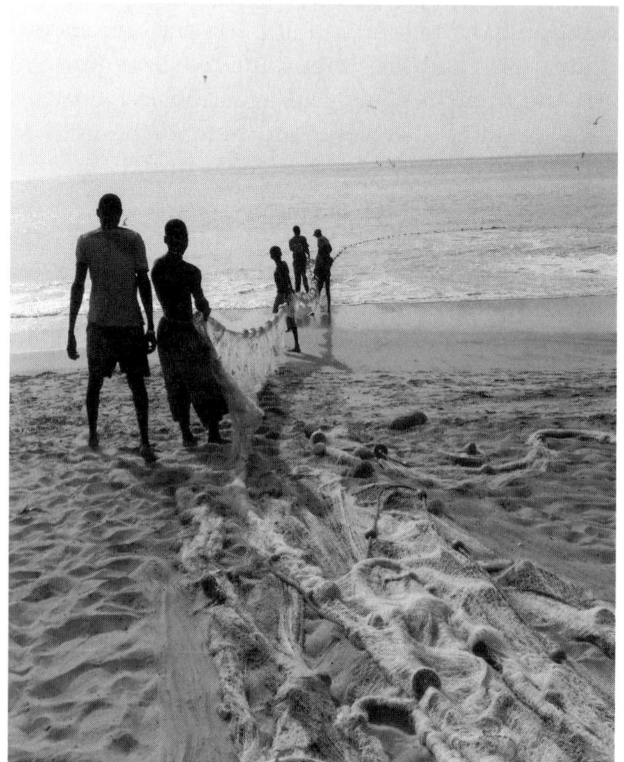

Net fishing in Nianing, near Dakar, Senegal. Senegal has experienced coastal erosion in some areas. Changes in rainfall, including irregular patterns coupled with heavy deluges and long dry periods, have negatively affected agriculture.

degree C) per decade. Rainfall has decreased by 10–15 mm per decade during the wet season in southern Senegal, and although unusually high rainfalls during the dry season have been observed in a few recent years, this is not part of a consistent trend. The mean annual temperature is expected to increase by 2–5.5 degrees F (1.1–3.1 degrees C) by the 2060s, and 3–8.8 degrees F (1.7–4.9 degrees C) by the 2090s, with faster warming in the interior regions than on the coast. The number of "hot" days (where the temperature exceeds that of the hottest 10 percent of days in the current climate of the region for the season in question) is expected to increase to 22–26 percent of all days by the 2060s and to 29–67 percent of all days by the 2090s. "Hot" nights are expected to occur on 27–51 percent of all nights by the 2060s and 37–70 percent of all nights by the 2090s, with the increase in both hot days and hot nights being greatest in the southern and eastern regions.

Global warming is already impacting the Sahel region, which includes northern Senegal. This semiarid region receives about 7.8 in. (20 cm) of rain annually, mostly from June through September, but changes in the pattern of rainfall are adversely affecting agriculture. Rainfall has become more irregular, with sudden heavy rain followed by several dry weeks. These heavy rains cause flooding and erosion, while the periods of no rain causes crops that have sprouted to fail. Water quality is also affected by lengthening dry periods, which creates health risks, particularly for cholera outbreaks.

Sarah Boslaugh
Kennesaw State University

See Also: Agriculture; Desertification; Floods.

Further Readings

Integrated Regional Information Networks, United Nations Office for the Coordination of Humanitarian Affairs. "Senegal: Climate Change Impacting Hard on Semi-Arid Sahel Nations." (December 7, 2005). http://www.irinnews.org/Report.aspx?ReportId=57490 (Accessed July 2011).

McSweeney, C., M. New, and G. Lizcano. "UNDP Climate Change Country Profiles: Senegal." ncsp.undp.org/sites/default/files/Senegal.oxford.report.pdf (Accessed July 2011).

U.S. Energy Information Administration. "Senegal" (June 30, 2010). http://www.eia.gov/countries/country-data.cfm?fips=SG (Accessed July 2011).

Serbia

Serbia is one of nine countries located in the Balkan Peninsula. Its longest border is with Romania at the northeast; with Bulgaria and Macedonia to the southeast; the still-disputed Kosovo to the south; Montenegro and Albania to the southwest; Bosnia, Herzegovina, and Croatia to the west; and Hungary to the north. With a total area of 29,912 sq. mi. (77,474 sq. km), Serbia has a population (not counting Kosovo) of 73,1055 (July 2011 est.) and a density of 278.42 people per sq. mi. (107.5 per sq. km). Some 31 percent of Serbia's area is covered by forests with an annual change rate of 1 percent, thus reforesting about 25,000 hectares per year. Located in a region of complex geography and politics, the country hosts a high level of endemism—rare, threatened and endangered species. Protected areas in Serbia cover 6 percent of its total surface. Earthquakes are the main source of natural hazard affecting the country.

With a gross domestic product per capita of $11,000 (2010 est.), Serbia's productive sectors are concentrated 13 percent in the agricultural sector, 22 percent in the industrial sector, and 65 percent in the services sector. Serbia had almost half the carbon dioxide (CO_2) emissions per capita in 2003 when compared to Europe: 4.9 thousand versus 8.5 thousand metric tons of CO_2. Most greenhouse gas (GHG) emissions are from the energy sector (79 percent) and from agriculture (14 percent), where 77 percent of the emitted GHGs are CO_2. Seeking to reduce its energy dependency, Serbia is in the process of harmonizing its energy legal framework with European Union standards. In 2010, 26,186 gigawatt hours (gWh) (70 percent) were generated by thermal power stations and 10,571 gWh (28 percent) from hydropower stations.

Karl-Heinz Gaudry
University of Freiburg
Giorgio Andrian
University of Novi Sad

See Also: Bosnia and Herzegovina; Bulgaria; Greenhouse Gas Emissions; European Union.

Further Readings

EarthTrends. "Climate and Atmosphere: Serbia." http://earthtrends.wri.org/pdf_library/country_profiles/cli_cou_891.pdf (Accessed May 2011).

Food and Agriculture Organization of the United Nations. "State of the World's Forests." Rome: FAO, 2011.

Mijuk, Goran. "Serbia Gets Energy Boost" (December 21, 2011). http://blogs.wsj.com/emergingeurope/2011/12/21/serbia-gets-energy-boost (Accessed December 2011).

United Nations Framework Convention on Climate Change. "GHG Emission Profiles for Non-Annex I Parties: Serbia." http://unfccc.int/ghg_data/ghg_data_unfccc/ghg_profiles/items/4626.php (Accessed May 2011).

Seychelles

The Republic of Seychelles is an Indian Ocean island country 932 mi. (1,500 km) east of Africa, northeast of the island of Madagascar. Originally uninhabited, the islands were used as a stopping point for trading ships and as a place for pirates to lay low until the French and British began to settle the area. The British assumed control in the early 19th century, governing Seychelles jointly with Mauritius until making it a separate crown colony in 1903. In 1976, Seychelles was granted independence and remained a member of the Commonwealth of Nations (Britain and its former holdings). The current constitution was adopted in 1993.

In 1990, Seychelles became the first African country to draw up a 10-year environmental management plan. It was the first serious commitment by the country to address environmental concerns, and was followed by its participation in the United Nations Framework Convention on Climate Change in 1992, the Vienna Convention in 1993, and the Kyoto Protocol in 1998.

The republic includes 115 islands, a total land area of 176 sq. mi. (455 sq. km) spread out over 251,000 sq. mi. (650,087 sq. km) of ocean and home to 87,000 people. Rising water levels and increasing water temperatures are expected to cause problems for the islands, and some of the smaller islands are expected to become completely submerged in the 21st century. One reason for the government's concern over environmentalism is the importance of ecotourism to the economy, and increased temperatures in the Indian Ocean have already led to coral bleaching on some of the Seychelles' coral reefs. The government has also made several efforts to protect the local sea turtles.

Temperatures stay close to 80 degrees F (26.6 degrees C) year round, with high humidity, especially from December to April (the rest of the year, the southeast trade winds help ease the humidity). Cyclones are rare, but the islands can occasionally experience heavy rainfall. The country has a somewhat high level of greenhouse gas emissions per capita because of its complete dependence on fossil fuels for electricity (including common gasoline-operated generators for commercial and residential use) and the increase in tourism since the 1990s. From 1990 to 2007 (the most recent year statistics are available), per-capita emissions rose from 1.6 to 7.47 metric tons.

Bill Kte'pi
Independent Scholar

See Also: Alliance of Small Island States; Comoros; Indian Ocean; Madagascar; Mauritius.

Further Readings

Carpin, Sarah. *Seychelles: Garden of Eden in the Indian Ocean.* New York: Odyssey Publications, 2005.

Scarr, Deryck. *Seychelles Since 1770.* New York: Africa World Press, 2000.

Sierra Leone

Sierra Leone is on the west African coast, covering 27,698 sq. mi. (71,740 sq. km), which is equivalent to an area the size of South Carolina. Sierra Leone has a population of 5.6 million (2011 est.) with nearly 1 million living in the capital,

Freetown. Sierra Leone ranked 158 (among 169 countries surveyed) in the United Nations Human Development Indicators (2010), in part because of an 11-year civil war from 1991 to 2002. The civil war also accelerated the depletion of natural resources, including diamonds, tropical timber, and wild game. In 2010, Sierra Leone had among the lowest per-capita gross domestic product in the world ($900).

The *Sierra Leone Millennium Development Goals Report 2010* reports that the incidence of poverty has fallen to 60 percent (against 70 percent in 2005); most of these poor are subsistence farmers. Products of economic significance to the country include cocoa, coffee, rice, palm oil, and fish. Alluvial diamond mining is the major source of hard currency earnings, whereas other extracted minerals include titanium ore, bauxite, iron ore, gold, and chromite. The geography is characterized by coastal mangrove swamps and wooded uplands inland, and the climate is tropical.

Environmental and Climate Challenges

A number of environmental issues are facing the country, including: deforestation of tropical timber, slash-and-burn agriculture, conversion of forest cover to pasture land for cattle grazing, increasing demand for fuelwood, mining, soil erosion, and overfishing. Sierra Leone is a party to the following international environmental agreements: biodiversity, climate change, climate change–Kyoto Protocol, desertification, endangered species, Law of the Sea, marine life conservation, ozone layer protection, ship pollution, and wetlands. It has signed, but not ratified, an international agreement on environmental modification. Climate change awareness in Sierra Leone is growing, but still low, due in part to environmental disasters, such as the deadly mudslides that occurred in Freetown in August 2009 and 2010.

The contributions that Sierra Leone makes to human-induced climate change are minimal compared with the rest of sub-Saharan Africa. In terms of carbon dioxide (CO_2) emissions, Sierra Leone has among the smallest carbon footprint in the world. For example, per-capita CO_2 emissions in 2007 were only 0.2 metric ton, compared to 19.3 metric tons for the United States and 55.4 for Qatar (the highest in the world). The burning of liquid fuels (petroleum products) represented 90 percent of the country's CO_2 emissions. Other non-CO_2 air pollution in Sierra Leone is low, compared to the rest of the continent and the world. Nitrogen oxide and carbon monoxide emissions (in 1995) were 64,000 and 1.38 million metric tons, respectively, making up just 0.007 and 0.008 percent of the totals for sub-Saharan Africa.

However, one area where Sierra Leone contributes to CO_2 emissions is through deforestation. Forest cover in Sierra Leone declined from 3.04 million hectares in 1990 to 2.75 million hectares in 2005, amounting to a 0.7 percent annual loss in forest cover. Of equal concern is the decrease in other woodland areas during that same time period, which dropped from 765,000 to 384,000 hectares as a result of conversation to grazing land for the rearing of livestock.

Climate change could have significant consequences on the people and the environment in Sierra Leone. With nearly 250 mi. (400 km) of coastline, significant areas, including the capital, could become more prone to flooding. A rising sea level would also destroy the extensive network of mangrove forests that cover much of the coastline. Climatic change leading to a shortened rainy season, especially inland, could accelerate the current conversion of tropical forests to grazing land for livestock. The intensification of rainfall could also accelerate soil erosion, instigate flooding and landslides, diminish water infiltration and lower groundwater tables, deplete soil quality, lower crop yields, and ultimately threaten people and their livelihoods.

Michael Joseph Simsik
U.S. Peace Corps

See Also: Climate Change, Effects of; Floods; Guinea; Liberia; Mauritania.

Further Readings

EarthTrends. "Country Profile: Sierra Leone." http://earthtrends.wri.org/text/climate-atmosphere/country-profile-159.html (Accessed June 2011).

Economist Intelligence Unit. "Country Report: Sierra Leone." *Economist* (2006).

International Monetary Fund. *Sierra Leone: Poverty Reduction Strategy Paper: Progress Report, 2008–10.* Washington, DC: IMF, 2011.

McCoy, John F. *Geo-Data: The World Geographical Encyclopedia*. Detroit, MI: Thomson-Gale, 2003.
Thomas, Abdul R. Sierra Leone Telegraph. "Sierra Leone Government Publishes Its Official Progress Report on the Achievement of its 'Agenda for Change.'" (July 22, 2011). http://www.thesierraleonetelegraph.com/articles/100559.htm (Accessed October 2011).
United Nations Information Service. "2011 Human Development Index Covers Record 187 Countries and Territories, Puts Norway at Top, DR Congo Last." http://www.unis.unvienna.org/unis/pressrels/2011/unisinf430.html (Accessed June 2011).
World Bank Development Indicators. "Sierra Leone." http://data.worldbank.org/country/sierra-leone (Accessed June 2011).

Simulation and Predictability of Seasonal and Interannual Variations

Measurements of changes in atmospheric molecular oxygen using a new technique shows that the oxygen content of air varies seasonally in both the Northern and Southern Hemispheres, and is decreasing from year to year. The seasonal variations provide a new basis for estimating global rates of biological organic carbon production in the ocean, and the interannual decrease constrains estimates of the rate of anthropogenic CO_2 uptake by the oceans. One example of research into variations are the interannual and interdecadal zooplankton population changes that have been observed in parallel with temperature (SST) changes at Helgoland, in the North Sea, over a period of 32 years.

Temperature determines the phenological timing of populations for each species in a unique way, as seen in multiannual regressions. Sign, inclination of the regressions of phenophases with temperature and determination coefficient, vary from species to species. Besides the limited predictability of annual temperature dynamics, the species specificity limits the predictability of future phenophase timing. However, the strong correlation of phenophase timing with preceding SST permits the prediction of the annual seasonality based on statistical models separately for each population.

The regressions determined in the correlation analyses are the first approach to the phenological prognoses, which the Senckenberg Research Institute calculates daily for 192 phenophases of zooplankton, including ichthyoplankton on a daily basis, and which it has published since April 2004 on the home page of the institute, at http://www.senckenberg.de/dzmb/plankton. The calculations are based on more than 30 years of weekly and more frequent sampling at Helgoland Roads (54°11'18" N 7°54' E), the only proper offshore island of the North Sea. Temperatures were provided by the Biologische Anstalt Helgoland and the German Weather Service.

Beyond the historic data used, current temperature measurements for the operative daily calculations are obtained from the Websites. They are corrected according to the historic deviations stemming from current temperature measurements for the position at Helgoland Roads, published on the Internet and weather report measurements, and are then used for the calculation of the minimum current error for phenophase prediction. The prognoses is exclusively restricted to temporal prognoses. Abundance predictions are not included.

Wulf Greve
German Centre for Marine Biodiversity Research

See Also: Oceanic Changes; Oceanography; Zooplankton.

Further Readings

Research Institute Senckenberg, German Centre for Marine Biodiversity Research. http://www.senckenberg.de/dzmb/plankton (Accessed March 2012).
"The Warming Trend at Helgoland Roads, North Sea: Phytoplankton Response." *Helgoland Marine Research*, v.58 (2004). http://www.awi.de/fileadmin/user_upload/Research/Research_Divisions/Biosciences/Shelf_Sea_Ecology/long-term_data/wiltshire_manly2004.pdf (Accessed March 2012).

Singapore

The southeast Asian city-state of Singapore will be impacted by global warming and climate change in many ways, thus it has begun preparing a variety of adaptation and mitigations strategies to address these effects. The small, low-lying tropical island nation is densely populated, with about 5 million residents living in an increasingly important global city. Despite its size, Singapore is an important player in the global financial community and has an export-oriented economy that revolves around the refining, petrochemical, and pharmaceutical industries. Much of the state's energy consumption and greenhouse gas emissions are directly attributable to these global exports, with the three oil refineries using about one-fifth of total energy consumed.

Currently, about 80 percent of the nation's energy production comes from natural gas. This is a large transition, since as recently as 2000, only 19 percent of Singapore's energy production came from natural gas, with the rest coming from fuel oil. While the quick transition from fuel oil to relatively clean-burning natural gas has lessened the city's carbon dioxide (CO_2) emissions, the city's renewable energy outlook is limited. Large-scale adoption of wind, hydropower, or geothermal energy is unrealistic due to physical geographic limitations. While waste-to-energy and solar energy production have some promise in Singapore, these energy sources are not yet price competitive with natural gas. Energy efficiency is Singapore's principle strategy for decreasing the city's carbon footprint.

Although Singapore contributes less than 0.2 percent of global CO_2 emissions, the island is particularly vulnerable to the negative effects of climate change. Like many island nations, Singapore's biggest climate change threats are related to

The Tanjong container port in Singapore, the busiest port in the world in terms of tonnage shipped. With a population of over 4.5 million people, it is noted for business, finance, manufacturing, exports, refining, and imports. As an island nation, Singapore is threatened by the potential for sea-level rise. Reclamation projects have built sea walls above internationally recommended levels. Rising heat in the city could also lead to heat-related illnesses among more vulnerable citizens.

rising sea levels. Currently, about three-quarters of Singapore's coastline is protected from coastal erosion and land loss by hard wall embankments. Since 1991, new reclamation projects have been building sea walls at least 49 in. (125 cm) above the highest recorded tide level—a level well above Intergovernmental Panel on Climate Change (IPCC) forecasts for the next 100 years. Higher sea levels will also make it more difficult for storm water to drain into the ocean, thus exacerbating inland flooding during tropical storms.

Another important climate change impact in Singapore is the effect of rising temperature in the city. Higher average annual temperatures may mean more severe hot weather extremes, leading to heat stress among the most vulnerable populations in the city. Additionally, the increase in hot weather may lead to an increase in the use of air conditioning and increased energy demand. In an effort to keep the city-state cool, Singapore has worked to encourage environmentally friendly building design and increase green cover within the city. Begun in 2005, Singapore's Building and Construction Authority introduced the Green Mark designation, which recognizes new buildings with environmentally friendly features. Since 2008, all new buildings must meet minimum Green Mark standards for energy efficiency. The National Parks Board in Singapore recognizes the cooling effect of urban green spaces and has worked to create community gardens, city parks, and green roofs on residential and commercial buildings.

There are many other ways that climate change might impact Singapore, including health, food security, and trade. With its tropical location, vector-borne diseases such as dengue fever are endemic to Singapore. Dengue cases rise during the warmer periods of the year, so Singapore has implemented a dengue control program of mosquito surveillance, control, research, and public education to reduce dengue transmission. Singapore's inability to produce food leaves the city vulnerable to rising food prices as agricultural markets shift in food-exporting countries. Some officials are concerned about the potential for a disruption in the city's trade primacy when new sea routes open up through the melting Arctic—thus potentially allowing ocean trade to bypass Singapore's ports.

While the Singapore government is aware that due to its small size, the city-state has a limited role in addressing climate change, it has remained active in the regional and international climate community. Singapore ratified the United Nations Framework Convention on Climate Change in 1997 and supported the Kyoto Protocol in 2006. Regionally, Singapore has supported efforts to preserve important carbon sinks in the tropical rainforests throughout southeast Asia, and in Indonesia in particular. Finally, Singapore has addressed climate change through efficient energy use and sustainable urban planning on its island nation.

Jeremy Bryson
Northwest Missouri State University

See Also: Alliance of Small Island States; Bangladesh; Climate Change, Effects of; Cyclones; G77; Global Warming, Impacts of; Green Buildings; Green Cities; Maldives; Oil, Consumption of; Philippines; Sea Level, Rising; Thailand; Tuvalu.

Further Reading

Elkington, John. "Singapore—You Drop It, You Break It, It's Finished." *Guardian* (January 26, 2011). http://www.guardian.co.uk/sustainable-business/sustainability-with-john-elkington/singapore-you-drop-sustainability-business (Accessed November 2011).

National Climate Change Committee. *Singapore's National Climate Change Strategy* (March 2008). http://www.app.mewr.gov.sg/data/ImgUpd/NCCS_Full_Version.pdf (Accessed June 2011).

Singer, S. Fred

Controversial atmospheric physicist, distinguished research professor at George Mason University, emeritus professor of environmental science at the University of Virginia, and founder of the Science and Environmental Policy Project (a policy institution on climate change and environmental issues), S. Fred Singer (1924–) is a leading skeptic of the scientific consensus on global warming. He points out that the scenarios pictured by most scientists are alarmist, that computer models reflect real gaps in climate knowledge, and he has stated that future warming will be inconsequential, or modest at most. He has also challenged the connection between ultraviolet-B radiation and melanoma and between secondhand smoking and lung cancer. Singer's critics have pointed out that the financial ties of his nonprofit organizations to tobacco and oil companies make Singer a clear case of conflict of interest.

Dr. Singer was born in Vienna on September 27, 1924. He did his undergraduate work in electrical engineering at Ohio State University and holds a Ph.D. in physics from Princeton University. He has served in numerous government and academic positions, such as acting as director of the Center for Atmospheric and Space Physics at the University of Maryland (1953–62); as special adviser to President Eisenhower on space developments (1960); as first director of the National Weather Satellite Service (1962–64); as founding dean of the School of Environmental and Planetary Sciences at the University of Miami (1964–67); as deputy assistant secretary for water quality and research, U.S. Department of the Interior (1967–70); as deputy assistant administrator for policy, U.S. Environmental Protection Agency (1970–71); as professor of environmental sciences, University of Virginia (1971–94); and as chief scientist, U.S. Department of Transportation (1987–89).

Nothing to Fear

To Singer, climate change is not something humans should fear. He argues that the climate has changed constantly throughout this and previous centuries and that people have always successfully adapted to it. In addition, he believes that humans can affect climate at a local level. Yet, whether they can cause global weather changes has still to be proved. Singer has repeatedly claimed that the atmosphere has not warmed up in recent decades. In fact, he has claimed that since 1979, it has slightly cooled down. Surface records that show increases in temperature are not, according to Singer, reliable sources of information, as thermometers tend to be placed in or very near to urban areas, which are traditionally warmer than other locations. Singer claims that models and observations about global warming do not agree. Although climatic models show that there should be an increase of about 1 degree F per decade in the middle troposphere, observations contradict these models. Singer is critical of arguments based on laboratory experiments, as the atmosphere is much more complicated and does not function under controlled circumstances. He recognizes that the increase in atmospheric CO_2 might lead to a slight warming, yet he says that this phenomenon is counterbalanced by increased evaporation of the oceans. The production of aerosols also causes cooling, which may counterbalance the effects of carbon dioxide. Yet, although Singer admits that aerosols last for a maximum of few weeks, but CO_2 stays for decades, he is critical of models emphasizing the role of aerosols in connection to carbon dioxide.

Singer argues that because aerosols are mostly emitted in the Northern Hemisphere, where industrial activities are rampant, we would expect the Northern Hemisphere to be warming less quickly than the Southern Hemisphere. Actually, according to such models, the Northern Hemisphere should be cooling. To him, however, the data show the opposite, as both the surface data and the satellite data agree that, in the last 20 years, the Northern Hemisphere has warmed more quickly than the Southern Hemisphere. This fact contradicts the whole idea that aerosols make an important difference and proves that aerosols cannot be invoked as an explanation for the discrepancies between models and observations.

High Insurance, Small Risk

Singer does not have much faith in computer models, which he describes as having been "tweaked" to produce the present climate and the present short-term variation. He also points out that the two dozen models presently used are not entirely consistent with each other. These models also fail to depict all types of clouds, which to Singer is a fundamental flaw. He compares the current concern over global warming and the urgent calls for action to buying insurance with a high premium against a risk that is small. The Kyoto Protocol, to Singer, is part of the high insurance premium. The reduction of energy use by about 35 percent within 10 years implies, according to Singer's estimate, giving up one-third of all energy use, using one-third less electricity, and demolishing one-third of all cars. In spite of accusations aimed at other scientists that they use an apocalyptic tone when describing global warming, Singer also uses apocalyptic overtones to describe the Kyoto scenario: "It would be a huge dislocation of our economy, and it would hit people very hard, particularly people who can least afford it."

To Singer, global warming is a big business, with governments pumping about $2 billion into climate research. Thus people have to justify this expenditure, which supports jobs and research. Yet, George Monbiot has emphasized that Singer has strong ties with oil and tobacco companies—a fact that constitutes a conflict of interest given his stance on CO_2 emissions and secondhand smoking: "In March 1993, APCO sent a memo to Ellen Merlo, the vice president of Philip Morris, who had just commissioned it to fight the Environmental Protection Agency: 'As you know, we have been working with Dr. Fred Singer and Dr. Dwight Lee, who have authored articles on junk science and indoor air quality (IAQ) respectively.'" Singer's Science and Environmental Policy Project also received multiple grants from ExxonMobil.

Luca Prono
Independent Scholar

See Also: Climate Change, Effects of; Climate Models.

Further Readings

Monbiot, George. "The Denial Industry." *Guardian* (September 19, 2006). http://www.guardian.co.uk/environment/2006/sep/19/ethicalliving.g2 (Accessed March 2012).

PBS. "What's Up With the Weather?" Interview with Dr. S. Fred Singer. http://www.pbs.org/wgbh/warming/debate/singer.html (Accessed March 2012).

Singer, S. Fred. *Global Climate Change: Human and Natural Influences*. St. Paul, MN: Paragon House, 1989.

Singer, S. Fred. *The Greenhouse Debate Continued.* San Francisco, CA: Institute for Contemporary Studies Press, 1992.

Singer, S. Fred. *Hot Talk, Cold Science: Global Warming's Unfinished Debate.* Oakland, CA: Independent Institute, 1997.

Singer, S. Fred. *The Scientific Case Against the Global Climate Treaty.* Oakland, CA: Science and Environmental Policy Project, 1997.

Slovakia

Slovakia is a landlocked country in central Europe lying between the Black and Baltic seas and sharing borders with the Czech Republic, Poland, the Ukraine, Hungary, and Austria. Slovakia was part of the Austro–Hungarian Empire until becoming part of Czechoslovakia after World War I. In 1993, by mutual agreement, the Czech Republic and Slovakia became separate countries. Slovakia joined the European Union in 2004 and the Euro area in 2009. Slovakia's climate is temperate, with cool summers and cold humid winters, and the terrain is lowlands in the south and rugged mountains in the central and northern regions. Much of the eastern two-thirds of Slovakia is forested. About 30 percent of the land is arable, and 3.5 percent of the population is engaged in agriculture. Most of Slovakia is drained by the Danube River and its tributaries. Annual precipitation averages 32 in. (824 mm) and annual freshwater withdrawal is 1.373 percent of the country's actual renewable water resources.

Slovakia's population of 5.5 million (July 2011 est.) makes it the 112th most populous nation in the world. The country has successfully transitioned from a centrally planned economy to one subject to market forces since 1993. The gross domestic product (GDP) in 2010 was $22,000, 58th highest in the world, and standards of living are generally high in terms such as literacy (99.6 percent), life expectancy (75.8 years), and access to improved drinking water sources and improved sanitation facilities (100 percent). The industrial sector amounts for 27 percent of employment and 35.6 percent of GDP, with primary exports including machinery and electrical equipment (35.9 percent), vehicles (21 percent) and base metals (11.3 percent).

Slovakia is a minor producer of oil, producing 10.6 thousand barrels per day in 2008 (88th highest in the world), far below its consumption of 85.9 thousand barrels per day (79th in the world). Slovakia produced 4 billion cu. ft. of natural gas in 2009 while consuming 217 billion cu. ft. The country has proven oil reserves of 0.01 billion barrels, and 1 trillion cu. ft. of natural gas. In the 1980s, Slovakia produced and consumed about 100 million short tons of coal annually, but this dropped sharply in 1992; as of 2008, the country produced 2.7 million short tons of coal and consumed 9 million. In 2007, Slovakia generated 26.5 billion kilowatt hours of electricity and consumed 26.8, and had an installed capacity of 7.35 gigawatts. Carbon dioxide emissions from burning fossil fuels in 2008 were 37.4 million metric tons.

In the 1970s and 1980s, when Slovakia was dominated by the Soviet Union, the country's air was polluted with high levels of sulfur and heavy metals—including lead and cadmium—and levels of atmospheric ozone and acidification increased. However, the air quality has significantly improved since 1991 due to restructuring of industry, increases in energy efficiency, and adoption of antipollution legislation. Over half the electricity generated in Slovakia comes from nuclear power plants, with thermal power plants representing about 30 percent of production.

The use of renewable energy sources increased by 50 percent from 1993 to 2008, and greenhouse gas emissions were reduced 33.9 percent over the same period. Total waste production decreased 43 percent from 2002 to 2009, although generation of municipal waste slightly increased during that period. About 80 percent of municipal waste is placed in landfills, and the level of material recovery is currently very low. The major change in land use from 2000 to 2009 has been the increase in urban and suburban areas and decline in agricultural area, with the resultant loss of carbon sinks, biodiversity, and soil as a food-generating resource.

Sarah Boslaugh
Kennesaw State University

See Also: Annex I/B Countries; Austria; Czech Republic; European Union; Forests; Hungary; Pollution, Air.

Further Readings

European Environment Agency. "Country Profile: Distinguishing Factors (Slovakia)." http://www.eea.europa.eu/soer/countries/sk (Accessed July 2011).

Food and Agriculture Organization of the United Nations. "FAO Country Profiles: Slovakia: Climate Change." http://www.fao.org/countryprofiles/index.asp?lang=en&iso3=SVK&paia=5 (Accessed July 2011).

U.S. Energy Information Administration. "Country Data: Slovakia" (June 30, 2011). http://www.eia.gov/countries/country-data.cfm?fips=LO (Accessed July 2011).

Slovenia

A Balkan nation in central Europe bordering both the Alps and the Mediterranean, Slovenia straddles the Slavic, Germanic, Uralic, and Romance worlds. Slovenia has been a part of the Roman Empire, the Holy Roman Empire, the Habsburg Empire, and Yugoslavia, before declaring its sovereignty in 1991. Today, it is the wealthiest Slavic nation-state, with the highest per-capita gross domestic product (GDP) of any of the new member states of the European Union (EU)—about $27,000. The service sector accounts for almost two thirds of GDP, followed by the industrial sector at about a quarter of GDP. The economy is not as stable as it could be, however; after the financial crisis began in 2007 and 2008, Slovenia's GDP shrunk more than any non-Baltic country in the EU other than Finland, about 7.33 percent. Unemployment exceeded 11 percent in 2011 and continued to rise. Further, Slovenia's wealth is unevenly distributed; while the west, which includes the capital city of Ljubljana, is more prosperous than the EU average, the southeast is considerably poorer.

About three-quarters of the country live in urban areas. About 12 percent of the country is arable, and twice that much is used for meadows and pasture. Forest covers 54 percent of the country's land area. Despite the large amount of forested area, Slovenia's prosperity and high standard of living has led to high per-capita greenhouse gas (GHG) emissions, the 50th highest in the world. Coal-burning plants mostly provide the 35 percent of the country's electricity that comes from fossil fuels. The remaining electricity is provided by the nuclear power plant at Krsko (35 percent) and hydropower (30 percent). Electricity and heat generation are responsible for 42 percent of the country's emissions.

Rising standards of living have resulted in greater use of electricity; a growing middle class means more consumer electronics use and greater industrial activity as consumers upgrade everything from stereo equipment to automobiles more frequently. The prevalence of car ownership has risen to about the same level as the United Kingdom: 365 cars per 1,000 people, and fuel is considerably cheaper than in western Europe. Transportation accounts for about 48 percent of GHG emissions. The train network is used predominantly by tourists, and the extensive public transportation isn't popular enough to significantly impact automobile usage, although multiple-automobile households and adolescents with their own cars are less common than in the United States.

Higher temperatures resulting from global warming have led to early thaws in the mountains, limiting the skiing season (important for tourism) and affecting agriculture. The government had early successes with cutting sulfur dioxide emissions in half from 1985 to 1995 and reducing nitrogen oxide emissions by 20 percent in the same period. Slovenia is a participant in the United Nations Framework Convention on Climate Change as of 1992, and signed the Kyoto Protocol in 1998 (it entered into force in 2005).

Bill Kte'pi
Independent Scholar

See Also: Annex I/B Countries; Austria; Croatia; European Union; Hungary; Organisation for Economic Co-operation and Development.

Further Readings

Luthar, Otto, ed. *The Land Between.* New York: Peter Lang, 2008.

Ramet, Sabrina and Danica Fink-Hafner, eds. *Democratic Transition in Slovenia.* College Station, TX: TAMU Press, 2006.

Smagorinsky, Joseph

An American meteorologist and the first director of the National Oceanic and Atmospheric Administration's Geophysical Fluid Dynamics Laboratory (GFDL), Joseph Smagorinsky (1924–2005) developed influential methods for predicting weather and climate conditions and lectured at Princeton for many years. With his decision to move the GFDL to Princeton, Smagorinsky made the university a leading center for the study of global warming.

Joseph Smagorinsky was born to Nathan Smagorinsky and Dina Azaroff. His parents were from Gomel, Belarus, but fled during the pogroms. Smagorinsky's father was the first to immigrate to the United States in 1913, settling in Manhattan's Lower East Side, where he opened a paint store. Three years later, he was joined by his wife and their children. Joseph was born on January 29, 1924, when the family was already living in the United States. Similar to his other three brothers, he worked in his father's paint store. He attended Stuyvesant High School for Math and Science in Manhattan. After high school, he expressed his wish not to stay in the family business and to go to college instead. As his intellectual skills had already become apparent, the whole family decided to support him in his decision. Smagorinsky earned his B.S. (1947), M.S. (1948), and Ph.D. (1953) at New York University. During his sophomore year there, he joined the Air Force and became a member of an elite group of recruits who had been selected for their talents in mathematics and physics. Because of his scientific interests, Smagorinsky was included in the Air Force meteorology program. As a part of the scheme, he was sent to Brown University to specialize in mathematics and physics for six months. Smagorinsky was then sent to the Massachusetts Institute of Technology to learn dynamical meteorology, under Ed Lorenz, the author of chaos theory. During World War II, Smagorinsky worked as a weather observer for the Air Force. In May 1948, Smagorinsky married Margaret Frances Elizabeth Knoepfel—one of the first female weather statisticians.

Choosing Meteorology

After the war, Smagorinsky concluded his studies. Although he had planned a career as a naval architect, the rejection of the Webb Institute led him to choose meteorology as a field. After a question-and-answer session with prominent Princeton meteorologist Jule Charney, Smagorinsky was invited to carry out the research for his doctoral dissertation at the Princeton Institute for Advanced Studies. In 1950, Smagorinsky was part of Charney's team of scientists who successfully solved Charney's equations on the Electronic Numerical Integrator and Computer, also known as ENIAC. This was a milestone event in modern meteorology, as it pioneered the use of computers for weather forecasting. At the Institute for Advanced Studies, Smagorinsky and Charney developed the technique of numerical weather prediction. This technique relied on data collected by weather balloon, which were then elaborated by computers according to the laws of physics. This enabled researchers to forecast the interaction of turbulence, water, heat, and other factors in the production of weather patterns.

After completing his doctorate, in 1953, Smagorinsky accepted a position at the U.S. Weather Bureau and was among the founders of the Joint Numerical Weather Prediction Unit. Two years later, at the suggestion of eminent meteorologist John von Neumann, the U.S. Weather Bureau created a General Circulation Research Section and appointed Smagorinsky to direct it. Smagorinsky conceived his task as the completion of the von Neumann/Charney computer modeling program. He wanted to obtain a three-dimensional, global, primitive-equation general circulation model of the atmosphere. The section was initially located in Suitland, Maryland, but was moved to Washington, D.C., where it was renamed the General Circulation Research Laboratory in 1959. In 1963, it became the GFDL before moving to Princeton University in 1968, where it is still located. Smagorinsky continued to serve as director of the lab until his retirement in January 1983.

Groundbreaking Models

Under Smagorinsky's directorship, the GFDL expanded, and Smagorinsky was able to attract respected international scientists such as Syukuro Manabe and Kirk Bryan to work there. The laboratory's work profoundly influenced the practice of numerical weather prediction around the world. Thanks to the GFDL's climate models, scientists have been able to assess more precisely humans' capabilities to affect climate change.

In his years at Princeton, Smagorinsky was also appointed visiting professor in geological and geophysical sciences at the university. As a member of the teaching staff, he helped to develop the Program in Atmospheric and Oceanic Sciences, a doctoral program in the Department of Geosciences that collaborates closely with the GFDL. After his retirement as director of the GFDL in 1983, Smagorinsky became a visiting senior fellow in atmospheric and oceanic sciences at Princeton until 1998.

Because of the connection established by Smagorinsky between Princeton and the GFDL, the university became a major center for the study of global warming. From the 1970s onward, scientists working under Smagorinsky created the first models illustrating how climate could change in the face of increasing levels of carbon dioxide in the atmosphere. These models provided the first modern estimates of climate sensitivity and stressed the importance of water vapor feedback and stratospheric cooling. Research at the laboratory also allowed the development of the first models coupling atmosphere–ocean climate for studies of global warming, establishing the important differences between "equilibrium" and "transient" responses to the growing levels of carbon dioxide.

Luca Prono
Independent Scholar

See Also: Climate Models; History of Meteorology; Weather.

Further Readings

Manabe, S., J. Smagorinsky, and R. F. Strickler. "Simulated Climatology of General Circulation With a Hydrologic Cycle." *Monthly Weather Review*, v.93 (December 1965).

Smagorinsky, J. "The Beginnings of Numerical Weather Prediction and General Circulation Modeling: Early Recollections." *Advances in Geophysics*, v.25 (1983).

Smagorinsky, J. "General Circulation Experiments With the Primitive Equations." *Monthly Weather Review*, v.91/3 (1963).

Smagorinsky, J. "On the Numerical Integration of the Primitive Equations of Motion for Baroclinic Flow in a Closed Region." *Monthly Weather Review*, v.86/12 (1958).

Smagorinsky, J. S. Manabe, and J. L. Holloway. "Numerical Results From a Nine-Level General Circulation Model of the Atmosphere." *Monthly Weather Review*, v.93 (1965).

Small Farmers

It is important to understand the implications of the climate change issue on small farmers, given the latter's critical role in food production in most developing countries. Smallholder farmers constitute a significant proportion of the world's population. According to an estimate by the International Fund for Agricultural Development (IFAD), approximately 50 percent of the population of developing countries was small farms. Some regions, such as sub-Saharan Africa, are estimated to have a proportion of its population to be as high as 70 percent. Moreover, smallholder farming remains essential in global food production. According to the Consultative Group on International Agricultural Research (CGIAR), hundreds of small farmers (with less than 2 hectares of land) and herders (with fewer than five large animals) feed the world's 1 billion poor people (those living on less than $1 per day). Small farmers are also known to be the main drivers in the cultivation of a large proportion of land and the production of food and cash crops in many developing countries.

While small farms have inherent characteristics that make them more resilient to environmental changes, relative to large-scale, intensive agriculture, these systems are also at risk of being severely impacted by climate variability. Often found in marginal environments, smallholder systems are generally characterized as both diverse and risk-prone. Small farmers practice complex production systems that use multiple combinations and integrations of plant and animal species and institutional arrangements for managing natural resources. This diversity in production allows them to cope with external disturbances, such as environmental and economic shocks. At the same time, most small farmers are described as poor rural producers who use mainly family labor, and whose principal source of income is the farm they manage. Their weak economic disposition and high dependence on land and natural resources make their existence highly susceptible to extreme environmental changes.

Small farming systems are based on traditional and local knowledge that have evolved by adaptive processes and passed down over long periods of time, usually through many generations. Some of the traditional farming techniques adapted by small farmers include, among others, use of drought-tolerant local varieties, water harvesting, multiple cropping, agroforestry, wild plant foraging, and opportunistic weeding.

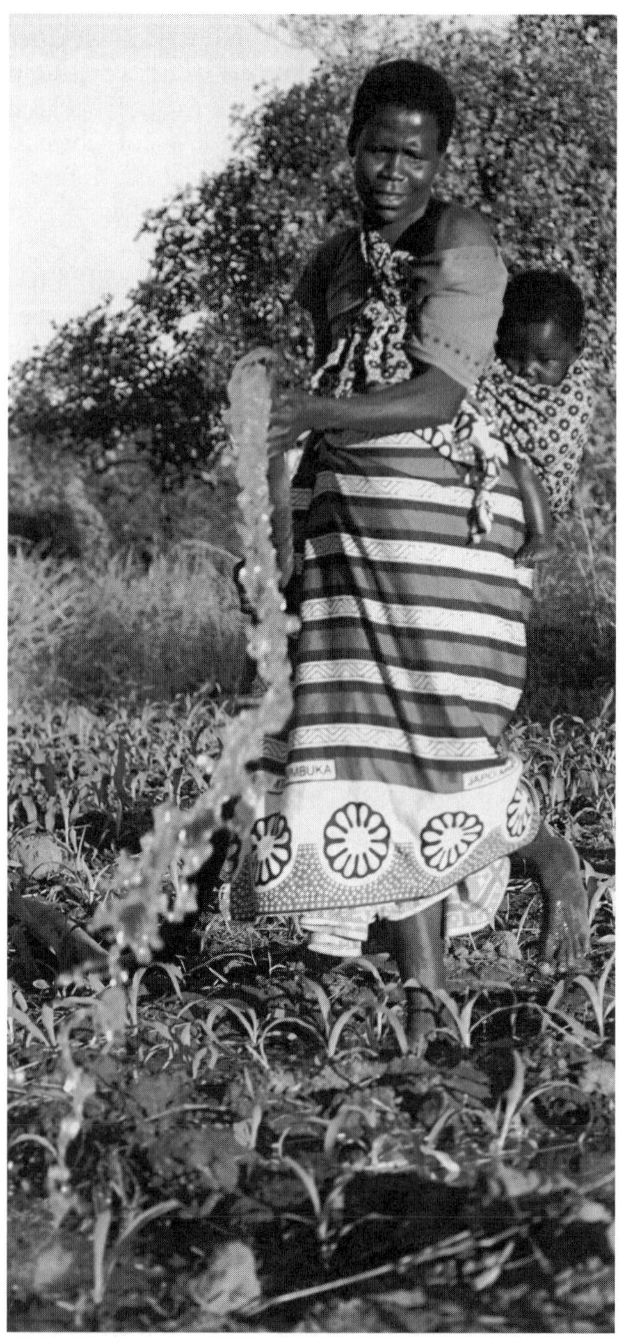

A Malawi woman waters her small field by hand. About 50 percent of the population of developing countries, which are the most threatened by climate change, are small farmers; that number reaches as high as 70 percent in some regions of Africa.

Advantages of Small Farming Systems

Small farming is believed to have the advantage over large, monocropping farming systems, especially when climate-change impacts render an area unsuitable for a single crop. Shifting from one crop to another is more difficult for large-scale farmers due to the greater risks associated with lack of knowledge and the need for large-scale changes in technology to accommodate the new crop being considered. On the other hand, small farmers are likely to see impacts in just some of their crops, but not in all, due to the diversity of crops planted.

For example, small farmers adopting cover crops, intercropping, and agroforestry in Central America suffered less damage from Hurricane Mitch in 1998 than their large-scale, monoculture counterparts. Overall, when accounting for climatic events in productivity, smallholder systems have the promise of achieving yields greater than that of large-scale, monoculture systems. Research reveals, for example, that systems intercropping sorghum, millet, and peanuts exhibit greater yield stability and lesser declines in productivity during extreme climatic events such as droughts than is the case of monocultures of sorghum or millet.

There are now a plethora of initiatives led by farmers' organizations and nongovernmental organizations in developing countries to reintroduce traditional small farming practices to the agriculture sector. These initiatives are designed to restore the ecological integrity of degraded lands, build resilience among households and communities, and improve rural livelihoods in marginal environments. Some of the exemplary cases are (1) the revival of old water harvesting systems (known as *zai*) in Burkina Faso and Mali; (2) diversification of plant production by the readoption of the traditional polyculture of corn, beans, and squash in the Mixteca region of Mexico; and (3) adoption of crop combinations of drought-hardy forage plants, leguminous trees, and short-term cash crops in the Sertao region of northeastern Brazil.

Vulnerability of Small Farmers

Notwithstanding that traditional small framing practices are intrinsically more resilient to climate-change impacts than large-scale, intensive monoculture systems, small farmers are still vulnerable to climate variability and extreme climatic events. Although these farmers have traditional and local knowledge of the weather and climate patterns affecting their lives, the rapid occurrence of novel climatic events and the persistent socioeconomic challenges faced by these populations can undermine the beneficial effects of traditional and local knowledge systems.

Many models predict that small farmers in developing and underdeveloped countries will disproportionately bear the impacts of climate change. A well-discussed model is that of P. G. Jones and P. K. Thornton, which in 2003 projected a 10 percent decrease in aggregate yields of maize in smallholder, rain-fed systems in Africa and Latin America by 2055. This model purports that the decrease in the yields of maize will affect approximately 40 million impoverished livestock keepers in Latin America and 130 million in sub-Saharan Africa. Changes in patterns of temperature, precipitation, and increased frequency of extreme climate variability such as droughts and floods are expected to result in loss of smallholder livelihoods as the likelihood of crop and livestock failures increases.

In Mozambique, for example, 90 percent of irrigation infrastructure was damaged, 20,000 heads of cattle were killed, and more than 113,000 small farm households lost their livelihoods due to massive floods in 2000. To a great extent, events such as this affect the livelihoods of small farmers, leading to the sale of agricultural and nonagricultural assets, indebtedness, out-migration, dependency on food relief, and the eventual inability to obtain good health and education.

The vulnerability of small farmers to climate-change impacts is largely determined by their low capacity to adapt to these environmental changes. Small farmers face the challenge of adapting to climate change due to several factors that influence collective action and memory, such as migration and the lack of extension services to disseminate and translate climate information to local needs. Based on a study conducted in Brazil, it was found that memories of extended droughts tend to decrease significantly after three years due to the above-mentioned factors. Over 50 percent of the small farmers interviewed in Brazil did not remember the drought of 1997 and 1998 and 40 percent of farmers have not changed their land-use practices to adequately respond to droughts. Moreover, even if small farmers do have access to information, most are still unable to take the necessary actions to curtail the impacts of climate change due to constraints related to lack of access to credit, agriculture inputs, drought-resistant crops, and farm technologies. Therefore, while small farming methods and strategies are encouraged to be invigorated and sustained, smallholder farmers should be given support to bolster their capacity to cope with climate-change impacts.

Marvin Joseph F. Montefrio
State University of New York
College of Environmental Science and Forestry

See Also: Adaptation; Agriculture; Global Warming, Impacts of; Livelihoods; Poverty and Climate Change; Vulnerability.

Further Readings

Altieri, M. A. and P. Kooharfkan. *Enduring Farms: Climate Change, Smallholders and Traditional Farming Communities*. Penang, Malaysia: Third World Network, 2008.

Brondizio, E. S. and E. F. Moran. "Human Dimensions of Climate Change: The Vulnerability of Small Farmers in the Amazon." *Philosophical Transactions of the Royal Society B: Biological Sciences*, v.363 (2008).

Herrero, M., et al. "Smart Investments in Sustainable Food Production: Revisiting Mixed Crop-Livestock Systems." *Science*, v.327 (2010).

Jones, P. G. and P. K. Thornton. "The Potential Impacts of Climate Change on Maize Production in Africa and Latin America in 2055." *Global Environmental Change*, v.13 (2003).

Leary, N., C. Conde, J. Kulkarni, A. Nyong, and J. Pulhin, eds. *Climate Change and Vulnerability*. London: Earthscan, 2008.

Leichenko, R. M. and K. L. O'Brien. "The Dynamics of Rural Vulnerability to Global Change: The Case of Southern Africa." *Mitigation and Adaptation Strategies for Global Change*, v.7 (2002).

Morton, J. F. "The Impact of Climate Change on Smallholder and Subsistence Agriculture." *Proceedings From the National Academy of Sciences*, v.104 (2007).

O'Brien, K. L. and R. M. Leichenko. "Double Exposure: Assessing the Impacts of Climate Change Within the Context of Economic Globalization." *Global Environmental Change*, v.10 (2000).

Snowball Earth

In the early 1960s, Brian Harland, a geologist at Cambridge University, observed that rocks on several continents, dating from the Neoproterozoic era (approximately 800–680 million years ago), contain glacial debris. Some of the glacial debris included carbonate rocks, which are known to form in the tropics (e.g., in the present-day Bahama Banks). This conclusion later gained additional support from paleomagnetic data. One potential explanation is that the entire Earth was covered by ice and snow during the Neoproterozoic. This is known as the "snowball Earth" hypothesis.

One early problem was understanding how a global ice age could have commenced. During the 1960s, the Russian climate scientist Mikhail Budyko used a computer simulation to establish that a runaway ice-albedo feedback effect could lead to global glaciation. The term *albedo* refers to the amount of the sun's energy that is reflected by the Earth's surface. As glaciers grow in extent, they reflect more of the sun's energy, which causes the atmosphere to cool. This in turn causes the glaciers to grow. Budyko showed that if the glaciers extended beyond a certain critical point, this ice-albedo feedback could lead to a global ice age.

A second obstacle was understanding how a global ice age could ever end once it began. In the early 1990s, Joseph Kirschvink of the California Institute of Technology observed that during a global ice age, the carbon cycle would shut down. Volcanoes sticking up through the ice cover would continue to add carbon dioxide to the atmosphere. Having nowhere else to go, the carbon dioxide would then accumulate over millions of years until a runaway greenhouse effect caused the ice to melt.

One important rival to the snowball Earth hypothesis is the high obliquity hypothesis. If the

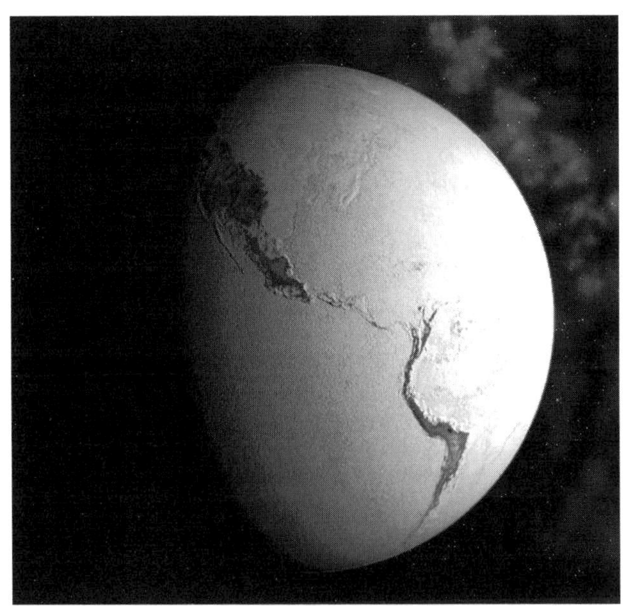

The snowball Earth hypothesis proposes that the Earth was entirely covered by ice in part of the Cryogenian period. This theory is challenged by paleontologists, who note that a true snowball Earth would have killed off almost all eukaryotic life.

tilt of the Earth's axis had been much different during the Neoproterozoic, the poles could have received more solar energy than the tropics. If so, it would be possible to explain the evidence for glaciers in the tropics, without supposing that the entire planet had frozen over.

In his widely cited 1992 paper, Kirschvink also proposed an explanation for banded iron deposits observed in Neoproterozoic glacial debris. Iron is not soluble in seawater in the presence of oxygen. During a true snowball Earth episode, the oceans would have become deoxygenated over time. Iron from thermal vents would build up in the seawater. Then, when the ice finally melted and oxygen was once again exchanged between the oceans and atmosphere, oxidized iron would have been left along with the debris from the retreating glaciers.

During the 1990s, two Harvard scientists, Paul Hoffman and Daniel Shrag, gathered additional, highly suggestive evidence that seemed to favor the snowball Earth theory. They found that in many places, the Neoproterozoic glacial debris occurs right below thick layers of carbonate rock (which are known as "cap carbonates"), and they showed how Kirschvink's proposal could account for this. During a snowball Earth episode, very large amounts of carbon dioxide would have built up in the atmosphere. As the ice receded and the carbon cycle resumed, large amounts of carbon would have been washed out of the atmosphere during storms and ended up in the form of carbonate rock on the ocean floor. More controversially, Hoffman and Shrag also studied the ratio of carbon-12 to carbon-13 isotopes in the cap carbonates. They argued that an unusual dip in the carbon isotope ratio signified a temporary shutdown of photosynthetic activity in the Earth's oceans.

Challenges to the Snowball Earth Theory

One potentially serious challenge to the snowball Earth theory comes from paleontology. Today, most geologists agree that there were at least two major ice ages during the Neoproterozoic: the Sturtian, around 750 million years ago, and the Varanger, around 590 million years ago. The second of these episodes occurred shortly before the Cambrian explosion of metazoan life. However, a true snowball Earth episode would have killed off nearly all eukaryotic life, and it is not clear that there was enough evolutionary time for life to recover from a global ice age. Some scientists have used computer models to show that softer versions of the snowball Earth episode might have been possible—for example, a mostly ice-covered planet with massive continental ice sheets in the tropics, but largely ice-free tropical oceans.

Although scientists generally agree that there was low-latitude glaciation during the Neoproterozoic, they continue to use a combination of fieldwork and numerical modeling techniques to work out the details. The snowball Earth scenario remains an intriguing live hypothesis.

Derek Turner
Connecticut College

See Also: Abrupt Climate Changes; Albedo; Carbon Cycle; Climate Feedback; Climate Models; Climate Thresholds; Computer Models; Earth's Climate History; Glaciology; Greenhouse Effect; Historical Development of Climate Models; Ice Ages; Ice Albedo Feedback; Ice Component of Models; Modeling of Ice Ages; Modeling of Paleoclimates; Ocean Component of Models; Paleoclimates.

Further Readings

Hoffman, P. F., A. J. Kaufman, G. P. Halverson, and D. P. Schrag. "A Neoproterozoic Snowball Earth." *Science*, v.281/5381 (1998).

Hoffman, P. F. and D. T. Schrag. "Snowball Earth." *Scientific American*, v.282/1 (2000).

Hyde, W. T., T. J. Crowley, S. K. Baum, and W. R. Peltier. "Neoproterozoic 'Snowball Earth' Simulations With a Coupled Climate/Ice Sheet Model." *Nature*, v.405/6785 (2000).

Kirschvink, J. L. "Late Proterozoic Low-Latitude Global Glaciation: The Snowball Earth." In *The Proterozoic Biosphere: A Multidisciplinary Approach*, edited by J. W. Schopf and C. Klein. Cambridge: Cambridge University Press, 1992.

Social Ecology

Social ecology is an ecological vision for the future developed by anarchist thinker Murray Bookchin. This theory is part of a left-wing tradition that rejects notions of hierarchy, domination, power, and place to advocate political reformism, or restructuring that will resolve basic issues of societal, gender, and environmental imbalance. Social ecology is based on the understanding that all our present ecological problems are a result of deep-seated social problems. As Bookchin states, "economic, ethnic, cultural, and gender conflicts, among many others, lie at the core of the most serious ecological dislocations we face today." Specifically, social ecologists argue that the chief source of ecological destruction is the capitalist system and its products, such as overconsumption, consumerism, and concomitant economic growth.

Trade for profit, industrial expansion, and the association of progress with corporate self-interest are among others. Bookchin argues, therefore, that to separate ecological from social problems underplays not only the sources of the environmental crisis, but also the interplay among all of these factors. Human beings must not downplay the importance of how they deal with each other as social beings. This, social ecologists argue, is the key to addressing the environmental crisis. The social ecological vision is to see a society that is based along social ecological lines. In this context, there are a number of principles that characterize social ecology.

First, a society based on social ecology would be one in which ecological regeneration would be inseparable from social regeneration. For example, social regenerative strategies might include the formation of ecocommunities and the adoption of ecotechnologies that establish a creative intersection between humanity and human nature. Spirituality, or what Bookchin calls regeneration of the spirit, is another principle, signifying the growth and development of a whole society. Such a society would be diverse and holistic in nature. Spirituality is defined as a natural phenomenon—one that focuses on the ability of humans to act as moral agents and actively promote the end to needless suffering, undertake ecological restoration, and foster aesthetic appreciation of all living things. Building on this spirituality will ensure the presence of liberty (in the sense of encouraging and nurturing individual and collective creativity, imagination, and personality) as a "continuum of natural evolution," resulting in a healthy society.

Social Ecology and the Environment

Social ecology goes further, however, addressing the deep structural failures within society, and seeks to redress the ecological effects that humans are having on the environment. Social ecology here seeks to change the definition of the very idea of a society based on hierarchy, class, and domination to one based on equity and the ethics of complementarity. In this case, social ecologists argue that humans must play a supportive role in maintaining the integrity of the planet. As such, they promote the establishment of community institutions that can embrace community-based ethical systems that in turn encourage the qualities of wholeness so integral to the social ecology vision. Societal structures that are supported by social ecology include confederal municipalism, in which municipalities conjointly gain rights to self-governance through the networks of confederal councils; empowerment of people; ecocommunities that are linked into the confederations of economy, fostering a healthy interdependence; and shared property.

Bookchin notes that "Social ecology calls upon us to see that nature and society are interlinked by evolution into one nature that consists of two differentiations: first or biotic nature, and second or human nature." By first nature, social ecologists refer to the way in which human beings are ultimately connected to their biological and evolutionary history. Second nature refers to the way in which humans produce or have a distinct social nature, as opposed to animals. As reflexive reflective beings, humans have the responsibility of being the voice of first nature. To understand and work within society, social ecologists argue, we must understand and embrace both natures. Consistent with anarchist theory, social ecologists see social hierarchy to be the enemy of natural order.

In his early work, *The Ecology of Freedom* (1971), Bookchin highlights a model of evolutionary human social development that suggests that social hierarchy first emerged in the Neolithic period ,with simple forms of governance within and between different social groups. Social ecologists perceive that humans are always rooted within their own biological evolutionary history. The separation of the current society of the human from the biological is a failure to think organically and to recognize wholeness. In this context, social ecologists argue that the human and the nonhuman must be seen as being part of an evolutionary continuum and, as such, we are in a state of continual becoming. Unlike deep ecology, which is underpinned by the belief that all organisms have intrinsic rights, proponents of social ecology argue that the environment has rights when and if they are conferred by humans.

Social ecologists also believe, however, that human beings have, by virtue of their innate creativity and powers of reason, intrinsic value. Bookchin maintains that the most ethical standpoint is for humans to understand different forms of hierarchy in nature, that is, different ecosystems, patterns, and orders, without implying that humans have the right to dominate those hierarchies. Social ecology attacks capitalism at a fundamental level, believing that modern capitalism is "structurally amoral and hence impervious to any moral appeals."

It believes that the driver for capitalism is to grow or die, and as such, that the system is inherently ecologically destructive. They advocate instead a society based on complementarity and mutual aid. In this way, social ecologists advance the need to redress the ecological effect humans have had on society by calling for social reconstruction based along ecological lines. The ethics of complementarity are based on a system in which human beings play a supporting role in upholding the integrity of the biosphere and the planet. Social ecologists argue that we have a moral responsibility to do this, as well as to enshrine the ethics of complementarity within social institutions in ways that will enable active participation of all in the process of reconstruction. In such a society, property would be shared, which would ultimately give rise to individuals for whom there is no separation between the individual and collective interest, the private and the personal, the political and the social.

Above All Else

Social ecologists stress the social causes and consequences of the degradation of the environment above all else. In this context, social ecologists seek answers to societal problems that are organic. They argue that as human forms of hierarchy have evolved, so has the human capacity to impose forms of domination on nature. Hence, it is the responsibility of the human race to redress the inequalities and problems attendant on the imposition of our own social order on nature. If social change in line with the organic elements ascribed to does not occur, social ecologists believe that the biosphere as we know it is heading toward complete destruction. Overall, the position of social ecologists has attracted critique from those who argue that their position is going too far in its interrogation of the link between society and the environment. Nonetheless, Bookchin's analysis of the relationship between society and the environment, and how this relationship constitutes and causes forms of domination—and hence ecological destruction—has played an important role in highlighting the nexus between social and natural dimensions of environmental decision making.

Melissa Jane Nursey-Bray
University of Adelaide

See Also: Climate Change, Effects of; Ecological Footprint; Public Awareness.

Further Readings

Bookchin, Murray. *The Ecology of Freedom*. Palo Alto, CA: Cheshire Books, 1982.

Bookchin, Murray. *Post-Scarcity Anarchism.* Berkeley, CA: Ramparts Press, 1972.
Bookchin, Murray. *The Rise of Urbanization and the Decline of Citizenship.* San Francisco, CA: Sierra Club Books, 1987.
Cudworth, Erica. *Environment and Society: A Reader.* London: Routledge, 2003.

Soil Organic Carbon

The Earth's terrestrial ecosystems store over 2,000 gross tons (Gt; 1 Gt = 1,015 g) of soil organic carbon (SOC), which is about four times more carbon than is stored in the atmosphere. Annually, soils release over 60 Gt carbon to the atmosphere, which is about 10 times that amount released by fossil fuel combustion. Warming can increase the rate at which fresh organic matter (e.g., recently senesced leaves, or fruits) decomposes, with the highest rates found where it is wet and warm. Less is known about how more thoroughly decomposed material (e.g., SOC) responds to changes in temperature, but it is assumed that global warming will increase SOC decomposition rates, with the sensitivity of SOC decomposition to warming determining the extent to which SOC storage will be altered by climate change. Because the sensitivity of SOC decomposition to temperature remains poorly quantified, it cannot be accurately predicted whether global soils will change from a net sink to a net source of CO_2 as the planet warms.

SOC serves many important ecosystem roles. Because the supply of organic carbon exerts a dominant control on the activity of soil heterotrophic organisms—from bacteria to insects—SOC is critical to regulating the structure and functioning of soil communities. Further, during SOC decomposition, large quantities of nutrients are released from organic to mineral forms, and so SOC provides a critical source of nutrients to growing vegetation. The amount of SOC can also affect the water-holding capacity of a soil, as well as water movement through soils. Despite these important roles, the tremendous complexity of SOC in natural and agricultural systems presents important challenges to quantifying SOC formation and decomposition, including the sensitivity of these processes to climate change. This complexity results from the fact that very large quantities of organic matter are cycled through soils annually, but only a very small fraction remains in soils, typically in a highly transformed state that can persist in soils for millennia as a result of chemical recalcitrance or protection by clay minerals.

Given the important effect that climate may have on SOC decomposition, it is critical that tools be developed to accurately predict how SOC formation, decomposition, and storage respond to climate change. Of particular importance is quantifying interactions among driving variables, as these interactions will influence responses in difficult-to-predict ways. For example, warming in cold and wet climates may result in the loss of SOC, as warming can dry out often anaerobic soils in which oxygen supply limits decomposition rates. In contrast, warming in temperate or tropical climates may have little effect on or even slow SOC decomposition rates, especially if moisture is limiting for the soil microbes responsible for decomposing SOC. Reducing uncertainty is important for accurately predicting how the terrestrial carbon cycle, and hence the climate, will respond to global warming.

Christian P. Giardina
Institute of Pacific Islands Forestry

See Also: Chemistry; Climate Change, Effects of; Soils.

Further Readings

Paul, E. P. and F. E. Clark. *Soil Microbiology and Biochemistry.* Maryland Heights, MO: Academic Press, 1996.
Zepp, Richard G. *The Role of Nonliving Organic Matter in the Earth's Carbon Cycle.* Hoboken, NJ: Wiley & Sons, 2001.

Soils

Soil contains pulverized rock, organic matter, and microorganisms, which convert the organic matter to humus. Soils range from loosely packed sandy soil, to finely packed, sticky clay soil. Between these two extremes is loam, with a high content of organic matter that makes it easy to cultivate. Soils are thin in the Mediterranean basin, and thick in northern Europe, Russia, and the American midwest and Mississippi delta. As the basis of agriculture and as a factor in the formation of climate, soils have shaped the destiny of humans.

Soils warm when absorbing sunlight, and cool when reflecting sunlight. Soil temperature coincides with an area's climate. In tundra, for example, the air warms in spring and summer much faster than the soil does. Only the outermost layer of the soil warms enough to thaw, and then for only a few months, before winter returns. Not all soils absorb and reflect sunlight at the same rate. Wet soil darkens, absorbing the most sunlight and gaining the most heat. Dark soil absorbs as much as 86 percent of sunlight, gray soil absorbs 80 percent, and light soil absorbs only 20 percent of sunlight, reflecting the rest into the atmosphere.

When water saturates soil, however, its color lightens. Being light, saturated soil reflects sunlight, yet the large amount of water that soil can hold increases the capacity of soil to absorb heat. Saturated soil absorbs and radiates heat slowly and in large quantities. Water evaporating from soil liberates heat, warming the surrounding air. Dry soil reflects more sunlight and heat than wet soil. Yet dry soil, because it contains little moisture, warms quickly on a cloudless day and becomes hot. At night, dry soil quickly cools, radiating heat back into the atmosphere.

The term *albedo* is the ratio of the amount of sunlight an object reflects to the amount it absorbs. A solid black object absorbs all sunlight and reflects none, and so has an albedo of zero. A white object absorbs no sunlight, reflects all of it, and has an albedo of one. The mean albedo of Earth is 0.36. Dark soil has an albedo between 0.1 and 0.2. Clay soil has an albedo between 0.15 and 0.35. Sandy soil has an albedo between 0.25 and 0.45, and light soil has an albedo between 0.4 and 0.5.

At most, sunlight penetrates only a few feet of soil. Infrared light, for example, penetrates to a depth of 3 ft (0.9 m). Because the amount of sunlight that soil receives varies during the day, soil temperatures vary. Soil temperatures also vary throughout the year. In Bridgewater, Massachusetts, for example, soil at a depth of one inch varied from a low of 35.2 degrees F (1.7 degrees C) in February to a high of 63.5 degrees F (17 degrees C) in August. Soil temperature therefore varies less than the temperature of air. Sunlight penetrates more deeply in rocky and wet sandy soil than in wet clay. Sunlight penetrates least in dry sandy soil.

Vegetation affects the capacity of soils to absorb sunlight. Plant cover decreases the albedo of light soil and increases the albedo of dark soil. Dense foliage blocks sunlight from reaching the soil, and so lessens the soil's absorption of sunlight and heat. Before the evolution of plants, soils must have been, along with the oceans, the major reservoirs of heat by absorbing sunlight. Like water, soil absorbs the most heat when the sun is overhead. Soil reflects increasing amounts of light as dusk approaches. Along with the oceans, the soil shapes the climate and influences the course of life.

Christopher Cumo
Independent Scholar

See Also: Agriculture; Albedo; Botanical Gardens; Climate; Plants; Sunlight.

Further Readings

Dobos, Endre. "Albedo." http://www.uni-miskolc.hu/~ecodobos/14334.pdf (Accessed March 2012).

Post, D. F., A. Fimbres, A. D. Matthias, E. E. Sano, L. Accioly, A. K. Batchily, and L. G. Ferreira. "Predicting Soil Albedo from Soil Color and Spectral Reflectance Data." http://soils.org/publications/sssaj/abstracts/64/3/1027 (Accessed March 2012).

Solar Energy Industries Association

The Solar Energy Industries Association (SEIA) was established in 1974 to "promote, develop and implement the use of solar energy in the United States." This trade association saw itself as the voice and champion of the solar industry to make solar a mainstream source of energy "by expanding markets, removing market barriers, strengthening the industry, and educating the public on the benefits of solar energy."

Today, the SEIA continues its mission, assisting members in building a profitable business and nurturing "a broad-based trade association supporting prompt, orderly, widespread, and open growth of solar energy resources now." The SEIA emphasizes the economic and business case for solar, rather than the climate benefits. However, its industry's environmental-neutral products present solutions to the problems of climate change. The SEIA's membership boasts 1,000 member businesses. Every dollar the association receives is applied to expand the U.S. solar energy market. Their stated objective is "to build a strong solar industry to power America with a specific goal of installing 10 gigawatts (GW) of solar capacity annually by 2015, enough to power 2 million households."

Climate Change and Solar Energy

According to the SEIA, each megawatt of photovoltaic solar power produced prevents 25,000 tons of air pollution over its useful life and reduces harmful particulate emissions from fossil fuel generation. Furthermore, as solar achieves higher market penetrations, it can make the electricity grid more reliable and secure by smoothing out the electricity demand curve and reducing the need for daily activation of costly, fossil fuel–burning peaker plants or the construction of polluting, base-load power plants to meet consistently higher peak loads.

Solar energy is pollution-free, produces no greenhouse gases (GHGs), and is fueled by an inexhaustible and renewable resource, the sun. Utility-scale solar power plants, which emerged as the fastest growing solar market segment in 2010, will generate extensive amounts of clean energy as part of a diverse energy portfolio that includes distributed generation, solar water heating, and other renewable sources, including wind. Such sources of energy are providing one of the quickest ways for states to meet their committed or mandated renewable energy portfolio standards.

While nuclear power has been touted by many of the large utility operators, such as the Southern Company, as more economical than solar installations, this is no longer the case. One study, produced by John O. Blackburn of NC WARN in 2010, illustrates that solar PV has joined the ranks of lower-cost alternatives to new nuclear plants. At least for the state of North Carolina, these alternatives include energy efficiency, wind power, solar hot water (displacing electric water heating), and cogeneration (combined heat and power).

Collaboration With Other Associations

For the past two years, the SEIA, in conjunction with the European Photovoltaic Industry Association (EPIA), sponsored a global solar initiative during the United Nations Framework Convention on Climate Change Conference of the Parties at COP 15 in Copenhagen and COP 16 in Cancun. At the most recent COP, more than 40 international solar organizations participated. They asked political leaders to support solar energy as a key solution to fighting climate change, reducing GHG emissions, and bringing clean electricity to both the developed and developing world. Reports were released that showed the per-country impact solar has in reducing emissions. The report from COP 16 highlighted that, "The solar industry is ready to provide the technological solutions and capacity to support climate mitigation measures in developing and developed countries alike. The obstacles continue to be political, not technical."

The combined solar targets provided by the participating countries in the report, *Seizing the Solar Solution: Combating Climate Change Through Accelerated Deployment* show that solar electric capacity by 2020 could reach 700 GW, with solar thermal capacity at 280 GW thermal by 2020. This is equivalent to a global reduction in carbon pollution by 570 megatons or 0.57 petagram, which is equivalent to eliminating the carbon emissions from the United Kingdom, Canada, or South Korea; taking 110 million cars off the road; or shutting down 100 coal plants.

The SEIA and its industry's environmentally neutral products present solutions to the problems of climate change, in concert with the overarching mission to help its members build a profitable business and nurture nationwide acceptance of solar energy.

Robert Karl Koslowsky
Independent Scholar

See Also: Intergovernmental Panel on Climate Change; International Solar Energy Society; Nuclear Power; Renewable Energy, Solar; Sunlight; Sunspots.

Further Readings

NC WARN. http://www.ncwarn.org (Accessed July 2011).

SolarCOP16. http://www.solarcop16.org (Accessed July 2011).

Solar Energy Industries Association. http://www.seia.org (Accessed July 2011).

U.S. Environmental Protection Agency. "Greenhouse Gas Equivalency Calculator." http://www.epa.gov/cleanenergy/energy-resources/calculator.html (Accessed July 2011).

Solar Wind

Two competing theories for global warming and its effect on Earth's changing climate persist today. The first theory suggests that the driver for global warming is the increasing amount of greenhouse gases dumped into the atmosphere as a result of humanity's burning of fossil fuels. The second theory posits that the solar wind and its associated magnetic field alters the Earth's cloud cover and adjusts the atmosphere's water vapor content, which leads to the steady temperature rise known as global warming.

The latter theory involves a stream of plasma, or high-energy charged particles, propelled from the sun's upper atmosphere. This stream of electrons and protons escapes the gravitational pull of the sun and creates the solar wind. It varies in speed from 190 to 500 mi. (306 to 805 km) per second and passes by Earth as the sun rotates in space. The solar wind affects Earth's magnetic field and, in turn, is believed to have a major effect on climate change.

Opponents to the greenhouse gas theory of global warming argue that increasing radiation activity from the sun over the past 300 years has been the primary culprit—not an increase of atmospheric carbon dioxide. Researchers believe that because the doubling of the sun's magnetic flux recorded in the 20th century had led to increased sunspot activity as it follows its periodic 11-year cycle, the ferocity of the solar wind and the overall brightness of the sun also increased.

Proponents of global warming who subscribe to increasing carbon dioxide emissions as the cause of the problem avoid citing work done by NASA's Goddard Institute for Space Studies (GISS) or other scientific evidence offering credence to the solar wind theory. Just like the greenhouse gas theory, the GISS climate model is used to show that changes in the solar wind throughout the ages have varied surface warming. Climate researchers determined that the sun has played a role in modulating the atmosphere's moisture content and its circulatory patterns, causing droughts in ancient times. Backing up these computer-generated data are a number of natural records that correlate with the model's projections. Lake sediment analysis, fire records, and tree-ring measurements from locations such as the Yucatan Peninsula, Mexico, and Peru, illustrate that periods of drought occurred during times of heightened solar output.

Increasing solar wind produces more ozone in Earth's upper atmosphere by breaking up oxygen molecules and heating the atmosphere. As a consequence, the circulation of the atmosphere is affected right down to the surface, which in general warms and reinforces existing rainfall patterns. Wet regions receive more rain, and dry

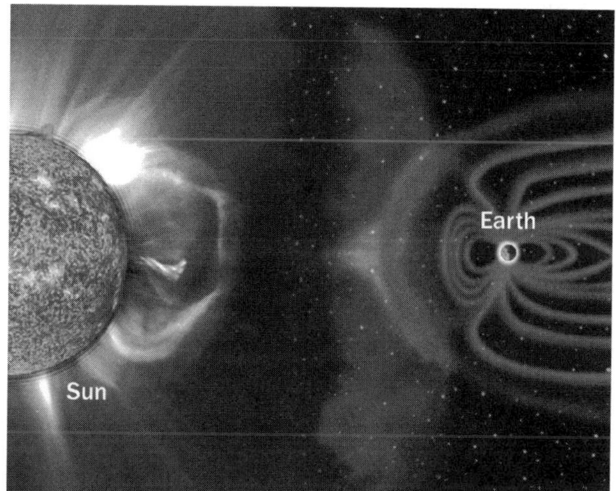

The Earth's magnetosphere, or magnetic field, protects it from most effects of solar wind and from solar storms. One global warming theory is that solar winds alter the Earth's cloud cover, adjusting the atmosphere's water vapor content.

regions become more susceptible to drought as the warmer air temperature pulls more moisture out of the soil. Droughts become more intense.

Although such facts are rarely disputed, the scope of the influence of solar wind is hotly debated. Some researchers state that the current period of global warming cannot be caused by the changes in solar output alone. Other researchers suggest that a double effect is in play, in which a more vigorous solar wind increases the global temperature, which in turn causes the oceans to warm, as made evident by melting sea ice. Because warm water absorbs less carbon dioxide, more of that greenhouse gas remains in the atmosphere. The debate boils down to whether all, some, or none of the burning of fossil fuels leads to global warming. Most scientists and the popular media believe that human activity adds so much greenhouse gas to the atmosphere that this round of global warming could be catastrophic to life on Earth.

Solar-focused satellites have been monitoring the sun since the 1970s. More recently, the Solar and Heliospheric Observatory and the Wind and Advanced Composition Explorer have kept their instruments trained on the sun to measure the sun's temperature, capture the ion content of the solar wind, and determine how the solar wind is accelerated. Recently, the twin STEREO spacecraft were launched to expand on and augment existing satellite measurements by tracking the sun together and reporting on its solar behavior. The Solar Terrestrial Relations Observatory became operational in 2007 and will provide researchers the first three-dimensional space forecasts associated with solar activity.

Whatever the conclusion as to the ultimate cause of global warming, the "third rock from the sun" will play a big role. The question for humanity is to determine which theory best describes global warming—greenhouse gas emissions or solar wind influences—and to develop policies to mitigate the effects on human civilizations.

Robert Karl Koslowsky
Independent Scholar

See Also: Climatic Data, Historical Records; Computer Models; Goddard Institute for Space Studies; Sunlight.

Further Readings

Shaw, Jane S. *Global Warming*. Farmington Hills, MI: Greenhaven, 2002.

Singer, Fred S. and Dennis T. Avery. *Unstoppable Global Warming: Every 1500 Years*. Lanham, MD: Rowman & Littlefield, 2007.

Solomon Islands

The Solomon Islands are a chain of islands east of Papua New Guinea and northeast of Australia in the South Pacific with a total area of 11,156 sq. mi. (28,896 sq. km) (slightly smaller than Maryland) and a coastline of 2,051 sq. mi. (5,313 km) They were a British protectorate from the 1890s, until achieving self-governance in 1976 and independence in 1978. The terrain is mostly rugged mountains with some coral atolls; less than 1 percent of the land is arable. The climate is tropical with recurring monsoons, and the islands include several active volcanoes, with the most recent eruption occurring in 2007. The Solomon Islands hold a strategic location in the South Pacific and were the site of heavy fighting during World War II. A high fertility rate (3.59 children per woman) and moderate life expectancy (74.2 years) drives a high population growth rate (2.22 percent), which increases pressure on the islands' resources.

The Solomon Islands have mineral resources (including lead, zinc, gold, and nickel), but these currently remain undeveloped and the economy is currently based largely on forestry, fishing, and agriculture, while most manufactured goods must be imported. Economic development has been hampered by government malfeasance and ethnic violence, as well as the physical isolation and diversity of cultures on the islands; the population of about 571,890 is estimated to speak about 60 dialects, and on many islands, traditional beliefs remain strong. Most of the population is engaged in subsistence agriculture, and in 2010, the per-capita gross domestic product was just $2,900. The Solomon Islands rank low in measures of development such as access to telephones, and much of the population does not have public health basics such as clean water or sanitation facilities. People in rural areas (about 80 percent of the population) have less access to improved water supplies and sanitary systems than those in urban areas.

The Solomon Islands have no petroleum, natural gas, or coal resources, and have no oil refinery capacity. Petroleum consumption has increased steadily over the past 30 years, from 0.7 thousand barrels per day in 1980 to 2.0 thousand barrels per day in 2009. Electricity generation has also increased steadily over that period: in 1980, 0.02 billion kilowatt hours (kWh) of electricity were generated; while in 2008, 0.08 billion kWh were generated. However, compared to other nations, these totals are quite low due to the undeveloped state of the Solomon Island's economy and its relatively small population (169th in the world in 2011). Total energy consumption in 2007 was 0.003 quadrillion Btu, ranking 201st in the world. Carbon dioxide emissions from consumption of fossil fuels in 2008 was low, at 0.23 million metric tons.

The Solomon Islands are part of the chain of the Melanesian Islands, including both volcanic and coral islands. Many are barely above sea level, particularly the coral atoll, and thus are extremely sensitive to any rise in ocean levels. If predictions of the International Panel on Climate Change are correct and the sea level rises by 6 in. to 3 ft. (15 to 91 cm) by 2100, many of the islands will simply disappear underwater and the population will migrate to the remaining islands, increasing resource pressures on them.

The Solomon Islands are home to a wide variety of plant, animal, and insect life, including over 230 species of orchids, hundreds of varieties of butterflies, and 20 species of mosquitoes, many of which would be endangered by climate change, either directly or through loss of habitat. Economic development is also threatening the natural state of the islands, as the tropical rainforests are harvested for timber and the resulting cleared land is used for agriculture. Soil erosion is a related problem, as the soil becomes drier due to loss of the forest cover and tree root systems and runoff pollutes the streams. Coral mining has destroyed some reefs, causing loss of habitat to marine species, and weakens coastal storm defenses. As many islanders depend on rainfall for their freshwater supply (many of the coral atolls have no groundwater), changes in amount and patterns of rainfall will also severely affect the human population.

Sarah Boslaugh
Kennesaw State University

See Also: Pacific Ocean; Sea Level, Rising.

Further Readings

Sevilla, Cherylee P. "The Solomon Islands: Headed for Self-Destruction?" Global Development Research Center. http://www.grdc.org/oceans/csevilla.html (Accessed July 2011).

South Pacific Applied Geoscience Commission. "Country Information: Solomon Islands." http://www.pacificwater.org/pages.cfm/country-information/solomon-islands.html (Accessed July 2011).

U.S. Energy Information Administration. "Country Information: Solomon Islands" (June 30, 2010). http://www.eia.gov/countries/country-data.cfm?fips=BP (Accessed July 2011).

Somali Current

The Somali Current can be found on the surface of the northern Indian Ocean, serving as a western boundary of this ocean. It is a movement of waters around the Indian Ocean, dispersing heat. Atmospheric circulation and ocean circulation together are the major mechanisms for global heat distribution. As atmospheric circulation defines large-scale air movements around the globe, ocean circulation refers to the patterned movement of particular waters.

In summer, a southwest monsoon blows upward from the east coast of the Horn of Africa. Carried along with the monsoon are the waters of the western Indian Ocean, moving in a northeast direction underneath, and powered by winds. These waters may reach speeds of 9 mi. per hour (14 km per hour). As the current reaches Somalia, the waters turn eastward. Some stay on near the Arabian Peninsula to form the East Arabian Current. Those that continue eastward eventually become the northeast monsoon during the autumn and winter, flowing southwest back to their origins. During the months of December and March, the Somali Current typically hovers between 5 degrees and 1 degree of latitude north of the equator, with this reach extending to span between 10 and 4 degrees latitude north during the central months of January and February.

The Somali Current is of interest because it creates an upwelling of cold water that is the only other region of such low surface temperatures within 10 degrees of the equator outside of Peru, and perhaps even colder. The cold surface temperatures around Peru are caused by the Peruvian or Humboldt Current, which is related to El Niño.

The waters of the Somali Current swirl into what is known as the Great Whirl, an eddy with a diameter that can reach 500 km (approximately 311 mi.), spinning in an anticyclonic direction. Anticyclonic direction is opposite to the Earth's rotation; in the Northern Hemisphere the eddy therefore spins clockwise. The upwelling occurs during the months of May through September, and can lower the surface temperature in the western Indian Ocean by up to 9 degrees F (5 degrees C). Ocean surface temperatures are an important data source for monitoring global warming; therefore, it is important to record the temperatures found during the northern (summer) swing of the Somali Current.

The Somali Current and other phenomena in the Indian Ocean were investigated at length in 1995 in a study that began in late 1994 and concluded in early 1996. It was an ambitious project that attempted to record all data related to the Indian Ocean during that year, and was undertaken by the World Ocean Circulation Experiment (WOCE) Indian Ocean Expedition.

Claudia Winograd
University of Illinois at Urbana-Champaign

See Also: El Niño and La Niña; Hadley Circulation; India; Indian Ocean; Modeling of Ocean Circulation; Monsoons; Oceanic Changes; Oceanography; Peru; Peruvian Current; Somalia; Thermocline.

Further Readings

Lutjeharms, J. R. E. *The Agulhas Current*. New York: Springer, 2006.

Nash, J. Madeleine. *El Niño: Unlocking the Secrets of the Master Weather-Maker*. New York: Warner, 2003.

Philander, S. George. *Our Affair With El Niño: How We Transformed an Enchanting Peruvian Current Into a Global Climate Hazard*. Princeton, NJ: Princeton University Press, 2006.

Somalia

Located in northeast Africa—the Horn of Africa—Somalia has a land area of 246,201 sq. mi. (637,657 sq. km), a population of 9.35 million (2010 est.), and a population density of 36 people per sq. mi. (14 people per sq. km). About 80 percent of the population is dependent on agriculture, although only 2 percent of the land is arable, with a further 69 percent used for meadows or pasture, mainly for low-intensity grazing of cattle, goats, and pigs. Some 14 percent of the land remains forested.

Because the country is underdeveloped, it has a very low level of electricity usage, and less than three cars per 1,000 of its population—one of the lowest levels of car ownership in the world, although three times that of its neighbor, Ethiopia. Carbon dioxide (CO_2) emissions per capita are among the lowest of any country in the world, even though accurate statistics have not been available for the last 10 years. Official statistics from 2001 record the entire electricity production for the country at 245 million kilowatt hours (kWh), with consumption levels at 228 kWh and 100 percent of all electricity coming from fossil fuels. It was estimated that CO_2 emissions were 0.1 metric ton per capita in 2007, the same level as Uganda and Tanzania, and with Afghanistan, Burundi, Chad, the Democratic Republic of the Congo, and Mali having lower per-capita rates of CO_2 emissions.

The effects of global warming and climate change on Somalia are expected to be extensive. The rising temperature is expected to make more of the arable land unusable for the growing of crops and to render pastureland even less productive than it is now. There is also the possibility of flooding in some low-lying parts of the country. The fishing industry has been devastated by overharvesting, and rising water temperatures are expected to make fishing even more difficult. This lack of legitimate opportunities for fishermen is often stated as one of the reasons for so many who are turning to piracy from the mid-2000s, endangering regional security. Speaking about the problem in Somalia, United Nations (UN) Secretary General Ban-Ki Moon stated, "Extreme weather events continue to grow more frequent and intense in rich and poor countries alike, not only devastating lives, but also infrastructure, institutions, and budgets."

The Somali government sent an observer for the UN Framework Convention on Climate Change (UNFCCC) that was signed in Rio de Janeiro in May 1992, and ratified the Vienna Convention in 2001. Because of the instability in the country, there have been few measures introduced to combat some of the effects of climate change. However, on July 26, 2010, the government signed the Kyoto Protocol to the UNFCCC, the 191st national government to do so.

Justin Corfield
Geelong Grammar School
Robin S. Corfield
Independent Scholar

See Also: Climate Change, Effects; Deserts; Floods.

Further Readings

Abbas, A. S. "The Health and Nutrition Aspect of the Drought in Somalia." Mogadishu: Somali Democratic Republic Ministry of Health, 1978.

Caputo, Robert. "Tragedy Stalks the Horn of Africa." *National Geographic*, v.184/2 (August 1993).

Climate Action. "Extreme Drought, Climate Change, and Security" (July 22, 2011). http://www.climateactionprogramme.org/news/extreme_drought_climate_change_and_security_in_somalia (Accessed July 2011).

EarthTrends. "Somalia: Climate and Atmosphere." http://earthtrends.wri.org/text/climate-atmosphere/country-profile-164.html (Accessed March 2012).

South Africa

South Africa's energy supply is dominated by coal, with 65.7 percent of its primary energy supply coming from this source. Crude oil follows, at 21.6 percent. Renewables and waste constitute 7.6 percent. Gas (2.8 percent), nuclear (1.9 percent), and hydropower (0.4 percent) make up the remaining primary energy supply. The dominance of coal in the energy supply portfolio is driven by South Africa's large coal reserves. Coal also assists in meeting liquid fuel requirements through the coal-to-liquid fuel processes conducted by Sasol, a South African energy company originally funded by the government that has been in the private sector since 1979. However, while large emissions generated from coal-powered plants is resulting in expansion into renewable energy sources such as nuclear, wind, and solar, coal is likely to remain the dominant fuel source for several generations.

Energy Mix: Coal and Renewables

South Africa has abundant coal reserves, estimated at 48 gigatons, which represents 5.7 percent of global reserves. South Africa is the fifth-largest producer of saleable hard coal in the world, after China, the United States, India, and Australia. 60 percent of South Africa's coal goes to Europe, while the rest goes to countries in the Pacific Rim. Coal is South Africa's third-largest export earner.

In the 1950s, for political and strategic reasons, the government began a program of reducing dependence of crude oil imports. In 1954, Sasol established the Sasol 1 coal-to-liquid fuel plant in Sasolburg. The process involves breaking down the coal and gas into a combination of hydrogen and carbon monoxide (called syngas) and then building these up through the Fischer–Tropsch process into hydrocarbons such as methane, petrol, diesel, paraffin, and polymers. Today, the synthetic fuel plants of Mossgas and Sasol supply about 35 percent of the final liquid fuel demand in the country, with the remaining supply coming from imported crude oil. South Africa now has the world's largest production of liquid fuels from coal and natural gas.

In 2004, South Africa began construction of the Medupi coal-fired power plant in the Limpopo province. The plant is the largest dry-cooled, supercritical, coal-fired power plant in the world and will add one-eighth to South Africa's current power generation, coming on line in 2012. Although dry-cooled plants are less efficient and more expensive to develop compared with water-cooled plants, constraints on water availability led

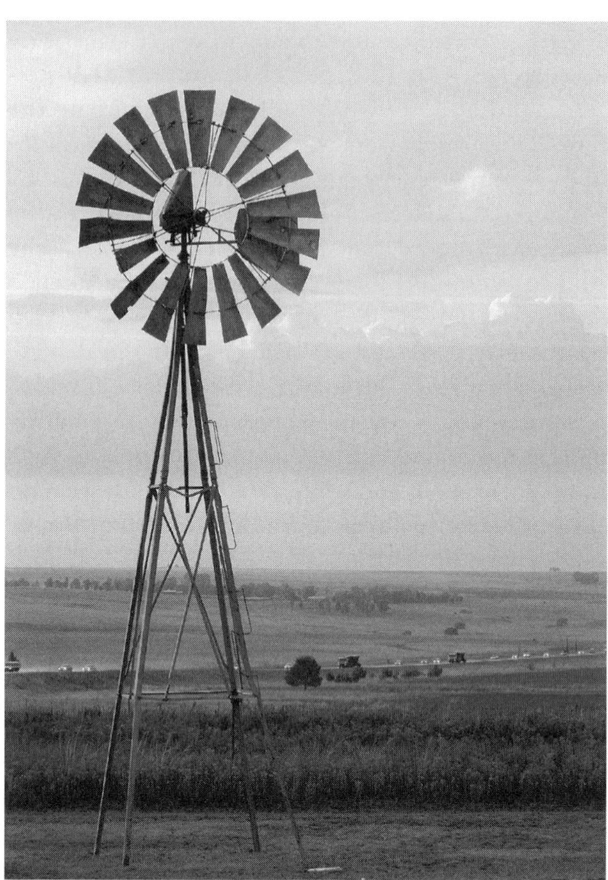

A windmill near Johannesburg: Wind resources, largely untapped, could be used to reduce carbon dioxide emissions from generating electricity. Onshore wind has an estimated potential to meet 1 percent of South Africa's electrical generation needs.

to the selection of supercritical, dry-cooled technology at Mendupi. A constraint on water resources is one of the factors leading to the promotion of more renewable energy technologies. Because of the environmental impacts of the Mendupi plant (including large CO_2 emissions and high water usage), the national government and the World Bank (which are assisting to fund the plant) have received some criticism. The agreement to go ahead with the Mendupi facility was viewed to be contrary to the South African government's policy commitments to reduce CO_2 emission levels.

South Africa is well placed to produce energy from biomass, wind, solar, small-scale hydropower, and waste. Currently, most of these sources remain untapped. Coastal regions have the greatest potential to produce wind energy. Onshore wind has an estimated potential to meet 1 percent of the required electricity in South Africa. Biomass wood fuel is a major energy resource for rural areas for domestic cooking. There is also potential for biofuel production from maize and sugarcane. Hydropower is viewed as having limited potential for expansion because of environmental impacts. South Africa has some of the best conditions for solar power, but currently, this source does not contribute any power to the national grid. Solar power can make significant contributions to rural electrification, and the Department of Energy is promoting solar PV panels for schools and clinics.

South African Emissions and Energy Plans

South Africa has made significant policy commitments to reduce CO_2 emissions, including the development of the first National Climate Change Response Strategy in 2004, which outlined principles and policy measures to mitigate and adapt to climate change. In 2005, a national white paper on renewable energy set a target of 4 percent of electricity supply from renewable energy by 2012. Targets for energy efficiency have also been set through the National Energy Efficiency Strategy in 2006 (updated in 2009), which aims for a 12 percent improvement by 2015. In January 2010, South Africa formally set a climate change mitigation targets of a 34 percent emissions reduction level below "business as usual" by 2020 and a 42 percent reduction by 2025. South Africa, along with China, India, Brazil, and the United States also signed an agreement in Copenhagen to reduce future emissions.

The Integrated Resource plan (IRP 2010) was approved by the government on March 17, 2011, and supersedes the 2001 Energy Security Master Plan. The IRP 2010 determines the demand profile for electricity over the next 20 years and outlines how this demand can be most effectively met from available sources. The planning scenario is based on growth in gross domestic product (GDP) averaging 4.5 percent over the next 20 years, which will require 41,346 megawatts (MW) of new capacity (excluding the capacity required to replace the decommissioned plant). The scenario assumes 3,420 MW of demand-side management programs.

While the IRP sees renewable energy making up 42 percent of all new electricity generated in

South Africa up to 2030, coal will continue to be a dominant source of electricity. Based on the IRP, in 2030, coal will make up 65 percent of electricity generation in South Africa, followed by nuclear (20 percent), hydropower (5 percent), gas (1 percent), and renewables (9 percent).

Aileen Anderson
Independent Scholar

See Also: Botswana; Coal; Congo, Democratic Republic of; Copenhagen Summit/Conference; Drought; Flexibility Mechanisms; G8/G20; G-77; Greenpeace International; Kyoto Protocol; Lesotho; Mauritius; Maximum Sustainable Yield; Namibia; Organisation for Economic Co-operation and Development; Renewable Energy, Overview; Swaziland; World Bank.

Further Readings

Davidson, O., et al. *Recommendations to the World Bank Group on Lending to South Africa for Eskom Investment Support Project That Includes a Large Coal Burning Power Station at Medupi.* Washington, DC: World Bank, 2010.

Electricity Governance Initiative of South Africa. *The Governance of Power. Shedding a Light on the Electricity Sector in South Africa.* Pretoria, South Africa: Idasa, 2010.

Maleka, Eyetwa Moses, Lindiwe Mashimbye, and Philip Goyns. *South African Energy Synopsis 2010.* Pretoria: South Africa Department of Energy, 2010.

Republic of South Africa. "Electricity Regulation Act (4/2006) Electricity Regulations on the Integrated Resource Plan 2010–2030." *Government Gazette* (May 6, 2011).

South Carolina

South Carolina is a state in the southeastern United States sharing borders with Georgia and North Carolina and with an Atlantic seacoast of 2,876 mi. (4,628 km). The state has considerable geographical variety, from the Blue Ridge Mountains to Atlantic beaches and subtropical coastal islands. It has a humid subtropical climate with hot summers, cold winters, and ample precipitation year-round. Tropical storms and hurricanes are a common occurrence in the summer months. South Carolina has a rich diversity of native wildlife, including 120 types of fish, 96 mammals, 313 birds, 72 reptiles, and 66 amphibians. Agriculture and forestry are important industries in the state, with an estimated economic impact of $33.9 billion per year: 92 percent of all land in South Carolina is used for either farming or forestry. Hunting, fishing, and wildlife viewing are also important contributors to the state's economy: in 2006, these activities produced nearly $2.1 billion in revenue and supported almost 40,000 jobs.

Annual temperature and rainfall have been monitored in South Carolina for almost 100 years, and over that time the state has become warmer and drier, particularly since 1957: from 1957 to 1991, the average annual temperature increased by nearly 1 degree F (0.56 degree C), while precipitation decreased by 3.2 in (8.12 cm) The Intergovernmental Panel on Climate Change estimates that by 2100, average temperatures in South Carolina could rise by 5.4 degrees F (9.7 degrees C) if global warming continues at its current pace. This rapid increase in temperature would have a severe effect on the state's pine forests, which are expected to be unable to adapt and would be largely replaced by grasslands and hardwood forests.

The coastal sea level in South Carolina has risen 9 in. (22.8 cm) in the past century and is predicted to rise an additional 19 in. (48 cm) in the next century if global warming continues unchecked. This would result in considerable beach erosion and the incursion of salt water into freshwater supplies. Rising coastal levels would also threaten coastal ecosystems such as that contained within the Cape Romain National Wildlife Refuge, home to species such as the red-cockaded woodpecker, brown pelican, and loggerhead sea turtle: a worst-case scenario projects the loss of over half of the area in this refuge.

Major storms, including hurricanes, have increased by about 50 percent since the 1970s, a change tied to increased average sea surface temperatures. These storms threaten coastal development, which will also be affected by rising sea levels. Insurance premiums for coastal properties have risen accordingly: in 2006, premiums increased 15 to 25 percent for properties on the South Carolina coast.

The human costs of global warming projected in South Carolina include increased heat-related illness and death (although a decrease in cold-related illness and death is also expected, it is expected to be less than the increase due to heat) and increased risks of shellfish-borne disease in coastal areas. Rising sea levels and increased erosion will threaten both homes and industries located on the coasts, while wetlands, which provide important habitats for many types of wildlife, will be increasingly lost to erosion and invasion of salt water. Changes in rainfall patterns will lead to increased runoff in some areas, including storm surge flooding, and increased drought in others. Changes in temperature and salinity may also favor invasive species over those native to the state, while weeds (rather than cash crops) are more favored by higher temperatures and increased carbon dioxide levels.

In 2007, South Carolina Mayors for Climate Change and Energy Leadership was formed by over 30 mayors from South Carolina towns with the purpose of increasing energy efficiency locally and pressing the state legislature to take meaningful action to combat climate change. Also in 2007, a bipartisan group of South Carolina state senators and over two-thirds of state representatives endorsed an open letter to the presidential candidates asking them to make climate change and energy policy a campaign priority. The South Carolina Department of Natural Resources (DNR) was created in 1994 through consolidation of several agencies, including the Wildlife and Marine Resources Department, Water Resources Commission, and Land Resources Commission. The DNR oversees a number of programs to safeguard the state's environmental resources, while also meeting the needs of the state's citizens. In partnership with the U.S. Fish and Wildlife Service, the DNR created a Comprehensive Wildlife Conservation Strategy in 2005. The South Carolina State Climatology Office, located within the DNR, is responsible for collecting and distributing information about climatology and meteorology within the state.

Sarah Boslaugh
Kennesaw State University

See Also: Atlantic Ocean; Forests; Hurricanes and Typhoons; Sea Level, Rising; Species Extinction.

Further Readings

National Wildlife Federation. "Global Warming and South Carolina" (January 30, 2009). http://www.nwf.org/global-warming/in-your-state.aspx (Accessed July 2011).

South Carolina Department of Natural Resources. www.dnr.sc.gov (Accessed July 2011).

South Carolina State Climatology Office. "The Impact of Climate Change on South Carolina." www.dnr.sc.gov/climate/sco/Publications/climate_change_impacts.php (Accessed July 2011).

Union of Concerned Scientists. "Backgrounder: Southeast" (2009). www.ucsusa.org/assets/documents/global_warming/us0global-climate-change-reportsoutheast.pdf (Accessed July 2011).

South Dakota

According to a National Assessment Synthesis Team report issued in 2001, South Dakota's geographical location near the center of the North American continent provides it with a high degree of sensitivity to climatic change, as most variations from mean weather conditions will be readily apparent. The state is located in a region that experiences all of the characteristics of the continental climate classification; according to John G. Lockwood in 1972, these include pronounced seasonality with long cold winters, hot summers, midlatitude cyclonic storms, and highly variable precipitation. A border running north to south, between the semihumid and semiarid precipitation regions of North America, divides the state (generally corresponding to the 100th meridian) and responds to changes, oscillating from east to west, in relation to overall global weather patterns. Thus, if there were significant changes in the global atmospheric system, such as warming resulting from anthropogenic or natural factors, climatic conditions would manifest themselves across the state.

A summary of historical climate research and recent climate modeling illuminates the possible consequences of climate change on South Dakota's people and landscape. The state's climate parameters are: Mean normal precipitation varies from 24.8 in. (620 mm) in the southeastern part of the state to less than 14 in. (350 mm) in the northwestern sector. Temperatures range from summer highs of 100 degrees F or more (37.8 degrees C) to minus 40 degrees F (minus 40 degrees C) in winter, with an average annual temperature of less than 39 degrees F (4 degrees C). Record high and low temperatures are 120 degrees F (49 degrees C) and minus 58 degrees F (minus 50 degrees C). Strong surface winds patterns, principally blowing from the north and northwest during the colder part of the year, also persist across the northern Great Plains, of which the state is a part. The region also experiences severe weather episodes such as tornadoes, hailstorms, and blizzards in their respective seasons. The state's Black Hills subregion is an anomaly to the general southeast–northwest precipitation and north–south temperature gradients, which vary temporally, with average wetter and warmer conditions prevailing at the higher elevations that support coniferous forest vegetation cover.

Definite Climate Shift

South Dakota's state climatologist Al Bender presented research in 2001 stating that the state had experienced a "definite climate shift" during the 1990s, recording an average of 2 to 3 in. (55 to 75 mm) of increased precipitation annually over the previous three decades. The agricultural effect of increased moisture facilitated the advancement of corn and soybean production westward toward the Missouri River, a feature that essentially divides the state into eastern and western sectors. Semihumid climatic characteristics associated with western Minnesota shifted about 200 mi. (320 km) to the west, according to Bender's 2001 report.

Conversely, during the current decade, the semiarid southwestern part of the state was in the grips of a multiyear drought resulting in agricultural losses and wildfires, in addition to increasing desertification conditions. The drought-affected area is the northeast extension of a multistate region in the western United States that has been experiencing prolonged, abnormally dry weather patterns. During 2010 and 2011, much more precipitation fell in the region and the Black Hills, and more normal conditions returned.

However, researched published by Carter Johnson in 2010 indicated that future global warming may adversely affect the eastern half of the state's wetlands and the surrounding area. Eastern South Dakota's geologic landscape was produced mainly by glacial deposits. The retreating ice left a terrain of ground moraines, outwash plains, end moraines, and a swath of tens of thousands of prairie potholes and larger kettles (shallow lake basins), which can be very productive waterfowl breeding areas when the moisture conditions are favorable. The area is a part of the Prairie Pothole Region (PPR) that covers parts of five other states and two Canadian provinces. With large-scale, Anglo-American settlement and development beginning in the second half of the 19th century, much of the area was transformed through drainage and intensive cultivation into a productive agricultural region, although a considerable acreage of natural wetlands remains. Western, or West River, South Dakota, was unglaciated during the Pleistocene epoch, and the landscape has been heavily influenced by the presence of the semiarid climatic regime. Contemporary agriculture focuses predominantly on cattle production and some row crops across vast areas of grasslands and rough terrain cut by a network of valleys and intermittent streams.

Johnson's 2010 research, based on a new climate modeling methodology (WETLANDSCAPE), concluded that future global warming might adversely affect the most productive remaining habitat for waterfowl in the PPR. A drier climate would shift from eastern South Dakota to areas to the east and much further north, locations currently less productive or where most wetlands have been drained. The 2010 research indicates that climatological warming will occur with negative impacts on both waterfowl habitat and commercial agriculture. The Intergovernmental Panel on Climate Change in 2007 concurred with other climate change research that predicted adverse consequences of global climate change in various areas of the world, especially in heavily populated or agriculturally productive regions of the Northern Hemisphere.

In sum, the climatological region in which South Dakota is located has always experienced a variable regime because of its geographical continentality. Temporally, the eastern half of the state was wetter in the latter 20th century. However,

Bruce Millett's 2009 report states that if the current global climate change forecasts become reality, the outlook for South Dakota's environment for the next several decades, especially the eastern portion, becomes problematic with the anticipation of less precipitation and warmer temperatures and their associated effects on land and life. According to experts, some degree of mitigation may be possible in the short run if initiated in time, but from a public policy perspective, the issue may stall because of political and philosophical differences.

Donald J. Berg
South Dakota State University

See Also: Blizzards; Montana; North Dakota; United States.

Further Readings

Bender, Al. "SD Monthly Climate Perspective" (February 15, 2001). http://climate.sdstate.edu/archives/feb/2001/ovr_fe01.htm (Accessed May 2011).

Johnson, W. Carter, et al. "Prairie Wetland Complexes as Landscape Functional Units in a Changing Climate." *BioScience*, v.60/2 (2005).

Lockwood, John G. "World Climatology: An Environmental Approach." New York: St. Martin's Press, 1974.

Millett, Bruce, et. al. "Climate Trends of the North American Prairie Pothole Region 1906–2000." *Climate Change*, v.93 (March 1, 2009).

National Assessment Synthesis Team. *Climate Change Impacts on the Unites States: The Potential Consequences of Climate Variability and Change.* Cambridge: Cambridge University Press, 2001.

National Defense University. *Research Directorate of the National Defense University Climate Change to the Year 2000: A Survey of Expert Opinion.* Minneapolis: University of Minnesota, 1978.

Southern Ocean

The global ocean influences the Earth's climate by storing and transporting vast amounts of heat, moisture, and carbon dioxide. Huge quantities of carbon are cycled annually among the biosphere (forests, grasslands, and marine plankton), the atmosphere, and the oceans. The oceans are the largest active reservoir of carbon, containing 50 times more carbon than the atmosphere. Of the 6 to 7 billion tons of carbon currently released into the atmosphere by human activities, approximately 3 billion tons remain in the atmosphere, 1 to 3 billion are absorbed by the oceans, and up to 2 billion appear to be absorbed by the terrestrial biosphere.

Oceanographers commonly refer to the oceanic region that surrounds the continent of Antarctica as the Southern Ocean. The northern boundary of the Southern Ocean is not well defined, but it coincides approximately with a broad zone of transition between the warm, saline surface waters of the subtropical regime and colder, fresher subantarctic waters, called the Subtropical Front, which occur between 40 degrees S and 45 degrees S. Using this definition, the surface area encompassed by the Southern Ocean represents approximately 29.7 million sq. mi. (77 million sq. km), or 22 percent of the global surface ocean. Its unique geography makes it a key player in global climate.

The Southern Ocean is the only ocean that encircles the globe unimpeded by a land mass. It is home to the largest of the world's ocean currents: the Antarctic Circumpolar Current (ACC). The ACC carries between 135 and 145 million cu. m of water per second from west to east along a 12,427-mi. (20,000-km) path around Antarctica, thus transporting 150 times more water around the globe than the total flow of all the world's rivers. By connecting the Atlantic, Pacific, and Indian Oceans, the ACC redistributes heat around the Earth and so exerts a powerful influence on global climate.

Near the Antarctic continent, the Southern Ocean is a source of cold, dense water that is an essential driving force in the large-scale circulation of the world's oceans. The cooling of the ocean and the formation of sea ice during winter increases the density of the water, which sinks from the sea surface into the deep sea. This cold, high-salinity water includes Antarctic Bottom Water and Antarctic Intermediate Water. Antarctic Bottom Water originates on the continental shelf close to Antarctica, spills off the continental shelf, and travels slowly northward, hugging the seafloor beneath other water masses, moving as far as the North Atlantic and North Pacific. Antarctic Intermediate Water is less saline and forms farther north, when cold surface waters sink beneath warmer sub-Antarctic waters at the Antarctic Convergence at about 55 degrees S. Together, these motions form a complex, three-dimensional pattern of ocean currents that extends around the globe, known as the thermohaline circulation, or "great ocean conveyor." The thermohaline circulation has a critical influence on climate by transporting heat efficiently around the globe and by controlling how much dissolved inorganic carbon is stored in the ocean.

At the sea surface, seawater exchanges gases such as oxygen and carbon dioxide with the atmosphere at the same time that it is being cooled. As a result, sinking water efficiently transfers changes in temperature, freshwater, and dissolved gases into the deep ocean 2.5 to 3 mi. (4 to 5 km) beneath the sea surface; in terms of carbon sequestration, this is called the solubility pump. Biological processes also play a role in the surface layer, where photosynthesis by single-celled marine phytoplankton can sequester carbon dioxide in the surface water and, through the process of sedimentation, transfer this organic carbon to deeper waters—the biological pump.

Below Maximum Capacity

The Southern Ocean is distinguished as a region of high levels of dissolved nutrients, but with modest rates of annual net primary production, so that the biological pump appears to be operating well below its maximum capacity. An interesting idea of recent years is that it may be possible to sequester much more atmospheric carbon if iron, an essential micronutrient, is added to the ocean to encourage the growth of marine phytoplankton, and thus stimulate the biological pump. The overall effect would be to lower the concentration of dissolved carbon dioxide in surface waters, allowing more atmospheric carbon dioxide to dissolve into the sea.

Understanding the global circulation and conditions under which surface waters penetrate into the deep ocean is critical for scientists estimating the timing and magnitude of climate change. At this time, the Southern Ocean is considered to be a net sink for atmospheric carbon dioxide; however, the magnitude of this sink has a high uncertainty, with mean annual estimates ranging between 0.5 and 2.5 billion tons. The degree of interannular variability in the Southern Ocean carbon sink, and its possible future response to climate change, is still poorly understood. However, climate model projections indicate that the Southern Ocean overturning circulation will slow down as the Earth warms. A decrease in the rate of overturning circulation will result in a decrease in the rate of carbon dioxide absorbed by the Southern Ocean, which represents a positive feedback and tends to increase the rate of climate change.

The presence of sea ice in the Southern Ocean is another factor that contributes to the Southern Ocean's important role in climate. Sea ice formation during the winter months is the largest single seasonal phenomenon on Earth, with approximately 7.7 million sq. mi. (20 million sq. km) of ice formed annually, effectively doubling the size of Antarctica. This has a profound effect on both regional and global climate processes. Because of its high albedo, sea ice reflects the sun's heat back into space, intensifying the cold. However, it can also act as a blanket, insulating against heat loss from the ocean to the atmosphere. Its yearly formation injects salt into the upper ocean, making the water denser and causing it to sink downward as part of the deep circulation. As ocean temperatures increase in response to the global warming, the amount of sea ice is expected to decrease; this has already been observed in the Arctic Ocean. The resulting decrease in the planetary albedo would act as a positive feedback, increasing the amount of energy from the sun absorbed by the Earth and tending to further increase the rate of climate change.

Al Gabric
Griffith University

See Also: Albedo; Antarctic Circumpolar Current; Arctic and Arctic Ocean; Carbon Cycle; Climate Models; Phytoplankton; Sea Ice; Thermohaline Circulation.

Further Readings

El-Sayed, Sayed Z. *Southern Ocean Ecology: The Biomass Perspective*. Cambridge: Cambridge University Press, 1994.

Furlong, Kate A. and Kate A. Conley. *Southern Ocean*. Edina, MN: ABDO, 2003.
Knox, George A. *Biology of the Southern Ocean*. 2nd ed. Boca Raton, FL: CRC, 2006.
Ramsey, A. T. and J. G. Baldauf, eds. *Reassessment of the Southern Ocean Biochronology*. London: Geological Society Publishing, 1999.

Southern Oscillation Index

Atmospheric and oceanic teleconnections identify linkages between climate variables occurring over large distances. On the order of weeks to years, these global associations are recognized by a disturbance in the transient wave field, resulting in distinctive patterns of climatic anomalies. They are also planetary-scale statistical relationships that follow the spatial interdependencies of atmosphere and ocean dynamics. The Southern Oscillation is one such atmospheric teleconnection that has been linked to global changes in general circulation features, affecting regional precipitation, temperature, and wind patterns.

First described extensively by British meteorologist Sir Gilbert T. Walker in the 1920s, the Southern Oscillation refers to the periodic exchange of mass across the equatorial Pacific that is recorded in sea-level pressure fluctuations between the eastern and western Pacific. Under normal conditions in the tropical Pacific, surface high (low) pressure prevails in the eastern (western) Pacific, with the easterly trade winds dominating surface wind and ocean flow. This pressure pattern, also known as the Walker circulation, tends to support rising air motions and convectional precipitation near eastern Australia, as well as sinking air motions and dry conditions near coastal northern Peru. Every two to seven years, this generalized atmospheric surface pressure pattern weakens as equatorial Pacific air pressure rises in the west and lowers in the east. This shift in the pressure field considerably weakens the trade winds and promotes the eastward movement of warm surface water across the tropical Pacific. The associated abnormal warming in the eastern Pacific is known as El Niño. Because the reversals in pressure and associated ocean temperature fluctuations are often simultaneous, this coupled climate variability between the tropical Pacific Ocean and atmosphere is often collectively referred to as the El Niño/Southern Oscillation (ENSO).

The mode and relative strength of the Southern Oscillation during a given time period is determined using one of several indices that signifies changes in the Walker circulation. A relatively simplistic and common method employed to gauge this change is the Southern Oscillation Index (SOI), which measures the monthly or seasonal sea-level pressure differences between two stations, one located in the central Pacific at Papeete, Tahiti (17.5° S, 149.6° W), and the other in the western Pacific at Darwin, Australia (12.4° S, 130.9° E). Negative (positive) SOI values result from abnormally low (high) pressure occurring in Tahiti and high (low) pressure occurring at Darwin, which tends to indicate an El Niño (La Niña) episode and warming (cooling) of the equatorial Pacific east of 180 degrees longitude.

However, low (high) SOI values have occurred independent of any equatorial sea surface temperature fluctuations, and may either precede or lag an El Niño or La Niña event. The sea-level pressures at these two tropical stations are negatively correlated and are associated with significant, yet contrasting shifts in regional temperature and precipitation patterns. Some of the most severe Australian summer droughts and heat waves (1982–83) have been associated with a strongly positive SOI.

The more common Tahiti-Darwin (T-D) SOI is calculated using monthly standardized sea-level pressure (SLP) values of the two stations. The T-D SOI is the standardized Tahiti–standardized Darwin SLP values divided by the monthly standard deviation. The standardized SLP of each station is found by subtracting the mean monthly SLP (using a 30 year average) from the actual SLP for a given month divided by the standard deviation. Although the T-D SOI can be calculated back to 1866, the Tahiti station poses some data quality issues and is generally considered more reliable since 1935. The standardized T-D SOI values since 1935 have a range from a minimum of minus 3.6 in February 1983 to a maximum of 2.9 in December 2010, with distinct cycles of predominantly negative or positive phases that differ markedly in strength and periodicity (Figure 1).

Values for the SOI (and similar climate indices that compare atmospheric components or conditions in geographically disparate locations) would likely change under modified climatic conditions such as global warming. Although the various global climate models (GCMs) project a warmer (and, in places, drier) world, these predictions do not show a spatially homogeneous pattern. The largest magnitudes of increased surface temperatures are forecast to occur over land surfaces, particularly in the Arctic and the Amazon basins. Surface temperatures over oceans, including the tropical Pacific, are predicted to be smaller; however, this is not thought to occur uniformly. Potential changes in the position of El Niño would, by extension, affect the Southern Oscillation. This prediction points to the possibility that Tahiti and Darwin might be not be the optimal locations to calculate the shift in atmospheric mass that is measured by the SOI in future climate scenarios.

Although the predictability of the Southern Oscillation has been subject to differences between observed and predicted time scales, researchers have shown that the current GCMs tend to place warm ENSO events within the observed two- to seven-year time scale. However, inconsistent results preclude a definitive answer to predicted changes in ENSO events in a warmer world. This lack of agreement may stem from the

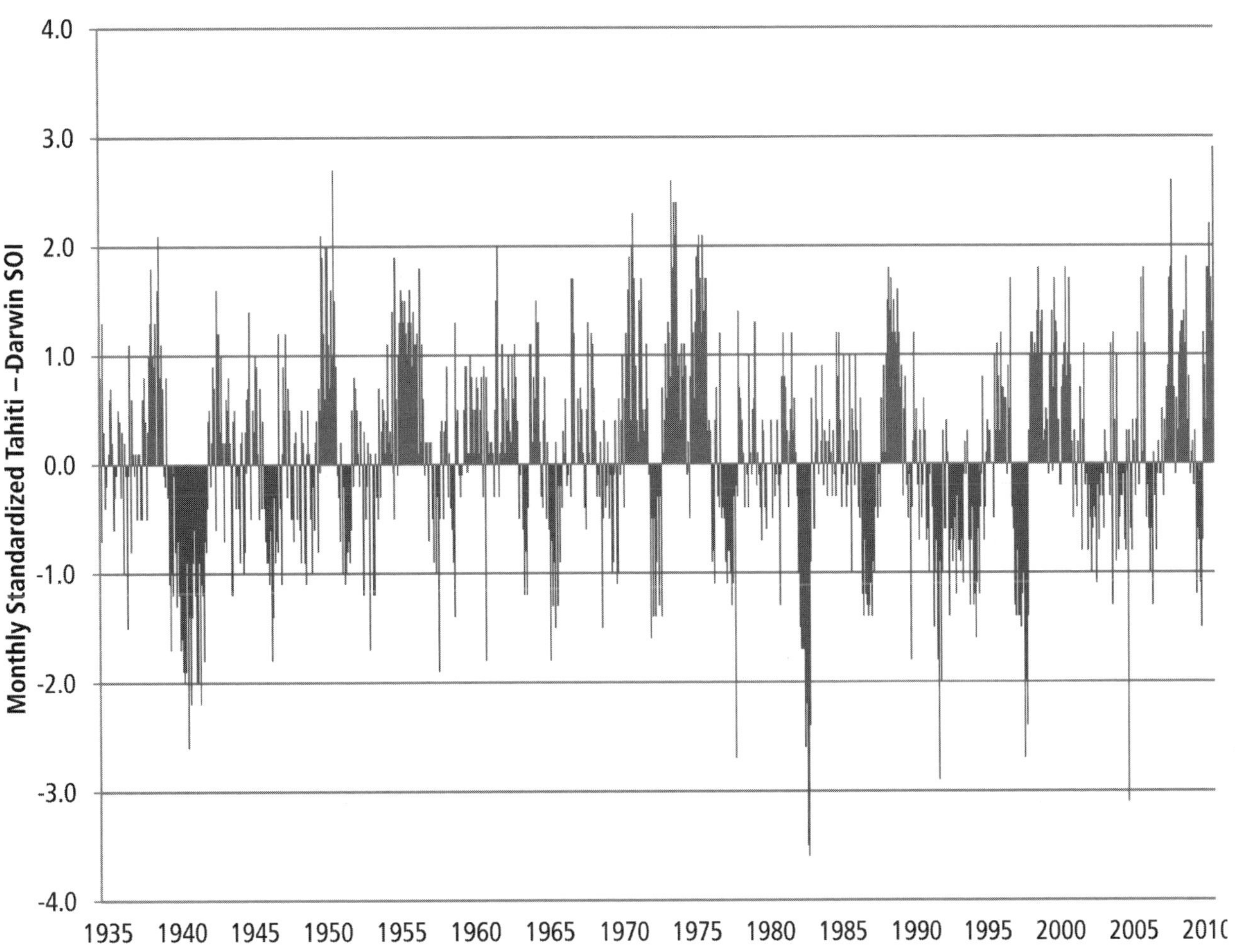

Figure 1 The Southern Oscillation Index (SOI) calculated as the monthly change in the standardized sea-level pressure (SLP) differences between Papeete, Tahiti, and Darwin, Australia, (T-D) from January 1935 to December 2010
Source: SOI data from the Climate Prediction Center (http://www.cpc.ncep.noaa.gov/data/indices), 2010.

possible errors involved in using GCMs to predict the Southern Oscillation.

Several sources of potential error with respect to prediction of Southern Oscillation events (and the veracity of forecasted SOI values) exist in the climate models. El Niño, which is based on the positive SST changes in the tropical Pacific, presents one of the potential obstacles in the assessment of the Southern Oscillation in the future. GCMs have shown mixed results when simulating the amplitudes and periodicities of sea surface temperature changes and may generate uncertainty in the Southern Oscillation component of the entire ENSO phenomenon. Thus, model-based simulations of the future necessitate a probabilistic approach. To better assess the spatial impacts that changes in the SOI would have may require the use of regional (smaller scale) climate models nested within global ones. This downscaling permits a more detailed view of a given location and its response to predicted changes.

Jill S. M. Coleman
Petra A. Zimmermann
Ball State University

See Also: El Niño and La Niña; North Atlantic Oscillation; Walker Circulation.

Further Readings

Philander, S. George, ed. *El Niño, La Niña, and the Southern Oscillation.* Maryland Heights, MO: Academic Press, 2006.

Wallace, John M. and Peter V. Hobbs. *Atmospheric Science: An Introductory Survey.* Maryland Heights, MO: Academic Press, 2006.

Washington, Warren M. and Claire L. Parkinson. *Introduction to Three-Dimensional Climate Modeling.* Sausalito, CA: University Science Books, 2005.

Spain

Spain is a southwestern European country, one of two (along with Portugal) occupying the Iberian Peninsula. Its territory also includes archipelagoes of islands in the Mediterranean (the Balearic Islands) and the Atlantic (the Canary Islands), and two autonomous cities in North Africa, outside of Morocco. The second-largest country in the European Union, Spain spans numerous geographic regions, including desert and grasslands, wetlands and forest, snow-topped mountains in the Pyrenees, and subtropical climate in the orange groves of Valencia.

Spain became a signatory to and ratified the Kyoto Protocol in 2002, which entered into force in 2005. That same year, it became the second highest generator of wind power when it added 1,764 megawatts (MW) of windpower to its total generation, exceeding by 27 MW and five years the 10,000 MW goal it had set for 2010. Wind power continues to be expanded in Spain.

In 2007, climate change was a continuing focus of new Spanish legislation. The minister of the environment introduced the Spanish Strategy for Climate Change and Clean Energy, Horizon 2007–2012–2020, listing five- and 13-year goals for moving forward in the country's commitment to climate change mitigation. Government buildings have been made more efficient, while blended biofuel is in use for official vehicles. Vehicle registration taxes in Spain are tied to emissions: low-emission vehicles are tax-exempt, and taxes on other vehicles increase according to their emission levels, as a way of preserving consumer freedom of choice while disincentivizing environmentally damaging behavior at both the consumer and (to a lesser degree) manufacturer level.

Spain has always been prone to drought, and experienced problematic droughts in the 1950s and 1960s, the 1990s, and again in the early 21st century. A National Water Plan, which would double the amount of water diverted from the Rio Ebro basin, met considerable controversy and protests, and was canceled four years later. Climate change is projected to decrease rainfall in most of Spain, especially in the spring and summer. Further, rainfall may be less evenly distributed, making both droughts and floods significant concerns. In 2005, for instance, the southwest coast of Spain was struck by Hurricane Vince, the first hurricane since 1842 to strike Spain. The impact was not severe: The ongoing drought had resulted in fires throughout the summer and early fall, and provincial reservoirs were replenished by the storm. Minor flooding struck many roads, while others were destroyed. By the time Hurricane Vince struck the peninsula, it had degenerated into a tropical depression, not a full-force storm; future hurricanes could possibly be more serious. Despite the unusual character of Vince—it formed in a part of the Atlantic not normally home to cyclonic formation—Tropical Storm Grace formed in the same area only four years later, an indication that what was once almost unheard of may have become more common due to changing conditions.

Climate models predict a 21st-century temperature increase of 6 to 12 degrees F (10.8 to 21.6 degrees C). Snow cover in the Pyrenees will be considerably reduced, impacting winter sports tourism, and summers could be much hotter and drier.

Bill Kte'pi
Independent Scholar

See Also: Afforestation; Andorra; Annex I/B Countries; Ecuador; European Union; Paraguay; Portugal; Renewable Energy, Overview; United Kingdom.

Further Readings

Williams, Mark R. *The Story of Spain: The Dramatic History of Europe's Most Fascinating Country.* San Mateo, CA: Golden Era Books, 2009.

Yates, Dorian. *Green Earth Guide: Traveling Naturally in Spain.* Berkeley, CA: North Atlantic Books, 2010.

Species Extinction

Changing climates increase the uncertainties of life for all organisms. A long-term warming trend would alter the distribution of life on the planet as colder habitats shrink and warmer ones expand. Some species would become more common, and others would become rarer. We cannot predict with any precision which species will become extinct—or when. Plants and animals that are highly adapted to already extreme (hot, cold, or dry) climates are most likely to be the first and most drastically affected.

A species is extinct once all known individuals of that type have died. Many interacting factors affect the survival of individual organisms, and therefore the persistence of their species. In general, extinction results when a species' requirements and abilities no longer match the resources and hazards in its environment.

For animals, these factors include food, water, and shelter from predators and weather extremes. For plants, they include water and nutrients and the action of herbivores and pollinators. Sometimes, a factor is critically important, like rainfall in a desert. It determines whether enough individuals will survive that a species can persist. If that "limiting factor" changes in some way, survival rates may rise or fall. If they fall far enough, extinction results. Climate is a major limiting factor for life on Earth. When it changes, life on Earth also changes. A continuing trend of global warming, cooling, or drying will lead to extinctions that might otherwise not occur as soon. We still know little about the precise climate limits or thresholds of most species. Because it is also difficult to predict precisely what the climatic conditions will be like in any given place at any given time in the future, it is even harder to predict which species will become extinct as a direct result of climate change, and when it will happen. In addition, because species interact and rely on each other in many ways, climate change produces many sometimes indirect or complex effects among them. This adds further layers of uncertainty to predictions about extinctions. For the most part, scientists can make only very general predictions about climate change and extinctions. This uncertainty leads many climate scientists and ecologists to conclude that humans would be wise to avoid or resist contributing to the uncertain risks of climate change, whenever it lies within our power to do so.

Climate, Biogeography, and Extinction

Climate change is complex. Tracking the local effects of regional or global change requires a great deal of data. Much of this information is now collected via remote sensing devices like radar and satellite-mounted cameras. So much data is collected that they can only be compiled into a usable form with very high-speed computers. However, those technologies are very recent. Scientists began collecting accurate and extensive climate measurements in the 18th century, recording data by hand.

Naturalists like Alexander von Humboldt and H. C. Watson first correlated climates and species distributions in the early 19th century. Thus began the study of biogeography.

Long before there were biogeographers, it was evident that different kinds of plants and animals occupied different kinds of places. Biogeography added mathematical precision to the folk knowledge that temperatures were lower at higher elevations and higher latitudes and that mountain ranges received more precipitation to windward than to leeward and were warmer on the sunnier slopes facing the equator. More climate and biogeographical data became available at the same time that cartography and species inventories were improving. All were necessary for accurately describing what lived where and for predicting what sorts of species would live in various places. Repeated inventories, measurements, and mapping were needed to show whether and how biogeography was changing.

Among the first patterns understood by biogeographers was that average temperatures on the Earth's surface changed with latitude. Temperatures tended to be low in polar regions and higher near the equator. At the same time, they saw that temperatures near sea level tended to be warmer than temperatures at higher elevations. They found that even near the equator, the tops of very tall mountains (such as the Andes) were as cold as the poles and discovered that the plants and animals of polar and alpine locations were very similar. As observations accumulated, biogeographers were able to begin mapping the ranges of different species. Climate measurements helped biologists determine the limits of heat, cold, and precipitation that various species could tolerate.

It was long debated whether species actually could become extinct. It was not until large, easily observed birds like the dodo and the great auk could no longer be found and the fossilized remains of large, otherwise unknown animals were discovered, that the fact of extinction was established. Not until the third quarter of the 19th century did it become clear that extinctions might regularly follow as the unintended consequences of intensive human activities. Climate changes traceable to human activities were hardly recognized for over another century, during which time the population doubled twice over. During roughly the same period, major technologies changed from being mostly animal, wind, and water powered to being combustion powered, using wood and fossil fuels.

Climate Dynamics and Life on Earth

Conditions on the Earth's surface and in its atmosphere have undergone many changes over time. Some of these changes were quite drastic and had proportionally drastic effects on living things. The Earth's climate has sometimes been much warmer than it is today, and at other times it has been much colder. This much can be inferred partially from recorded history, but more reliably from fossils and other geological evidence. Scientists have proposed many plausible explanations for these climate changes, but since the events cannot really be modeled in detail or replicated for study, they can only agree about general effects, rather than local specifics.

Paleontologists and others who study the evolutionary history of life on Earth have concluded that most of the species that ever inhabited the planet are now extinct. They have also estimated, as a sort of rule of thumb, that any given species, on average, persists for about a million years. Estimates of the total number of species that have existed on the planet range from tens of millions to hundreds of millions. Some of these species are known from the fossil record to have persisted much longer than a million years, and others for much shorter periods. Estimates and averages are only as good as the actual data and methods used to make them. Even if the data are reliable, the fossil record is far from complete, and different methods of analysis continue to yield different estimates.

Because the planet is changeable, or dynamic, extinction seems to be normal and inevitable for species, much as death is inevitable for individuals. At the same time, however, evolution also generates new species from some of the old ones, as the average characteristics of a population change and adapt to emerging conditions. Overall, there is still life on Earth because the rate at which species evolve has exceeded the rate at which they become extinct.

No one is credibly predicting that all life on Earth would end, and that all species would go extinct, as a result of human-caused global warming, but many scientists are concerned that any continuing trend in climate change would increase the rate of extinctions, changing life and perhaps making life more difficult or less interesting for humans as a result. Many people want to preserve life and to prevent extinctions of other species caused by human activities. Global warming is one of many environmental changes human activities may bring about. The combined effects of human activities, along with those of geological and even cosmic events, are complex. Among the extinctions that occur during the foreseeable future, very few will be blamed solely, or even mostly, on human-caused climate change. However, if apparent trends continue, climate change will probably contribute in some way—large or small—to almost any extinction that occurs.

Ecology and Climate Change

It is difficult to distinguish extinctions caused mostly or mainly by climate change from those caused by other factors, such as directly converting habitats to human uses. Conservation biologists have long considered habitat destruction to be the most likely cause of extinctions. Habitat has been described many ways, but it generally means an environment in which enough individuals of a particular species can survive and reproduce to keep their population from decreasing to zero. In other words, each species has a habitat, and each needs a persisting habitat to continue as a species.

Some habitats are more complicated than others, but all habitats can be thought of as having two general kinds of components. Biotic components are living things: all the other organisms that somehow affect the life of a plant or animal. Abiotic components are factors like terrain, minerals, water, sunlight, and temperature. Climate change can directly affect some of the abiotic components of a habitat. When particular places become warmer or cooler, or drier or wetter, the ability of any particular species to persist in that place also changes.

Some abiotic habitat components, such as temperature and humidity, will vary daily or seasonally. Organisms have to be able to tolerate the extremes of night and day, summer and winter, and wet and dry seasons. When the climate of a place changes to the point that one of an organism's tolerances is exceeded, a habitat literally ceases to exist.

Because of the shape of the Earth, less sunlight reaches the poles than the tropics. Habitats are limited by the climatic effects of latitude. If one could look down from space at the North Pole and see all the way to the equator, but still recognize all the land plants and animals, one would see that similar kinds of organisms are roughly arranged in a series of bands or zones centered on the pole, like a target.

Working out from the center, each zone is slightly warmer than the one immediately inside it. When the average global temperature falls, the polar center of the target expands and the hottest equatorial zones shrink or even disappear. This happened during the ice ages, when glaciers covered much of the Northern Hemisphere. Animals and plants had to change, migrate, or become extinct. When the average global temperature rises, the polar center shrinks, and each climate zone moves toward it. The icy center may even disappear, and the next zone takes its place. Meanwhile, entirely new, hotter zones may appear at the equatorial edge.

Because the atmosphere is less dense in the mountains than at sea level, all habitats are also limited by the climatic effects of elevation. Higher elevations are colder. Seen from directly above, a tall mountain has bands of similar plants and animals, just as the whole planet does. A general trend in climate change means that these bands move down and up the mountain, just as the latitude bands move toward and away from the poles.

Climate-Related Causes of Extinctions

Extinctions can occur gradually or suddenly. Large numbers of extinctions have sometimes occurred during relatively short periods of time. These "mass extinctions" resulted when a catastrophic event such as an asteroid impact suddenly made large areas of the Earth's surface, or its oceans, uninhabitable. The effects produced by such catastrophes probably included sudden, drastic climate changes, but not enough evidence has been found to say with certainty how great these changes were or exactly how long they lasted.

Changing climates affect the survival prospects of individual organisms. As a result, changing climates affect the survival and reproduction rates of whole populations and species. Populations may rise or fall as climates change. Some increases or decreases will be dramatic and obvious. Others will be almost unnoticeable. Almost all such population changes will result from combinations of many small changes, rather than a few catastrophic ones.

As average global temperatures increase (or decrease), populations will migrate to follow shifts in local conditions. Some organisms can do this quickly and easily. Many animals already migrate to follow seasonal changes in food and water supplies. There are rare exceptions, but many individuals, such as rooted plants, cannot move at all. Their populations can migrate only as seeds are dispersed and new individuals germinate and survive in newly suitable locations. Meanwhile, the old individuals, trapped in increasingly unsuitable locations, gradually die out. When populations cannot shift to new locations quickly enough, species may become extinct. Extinctions also follow when no new locations become available or when potentially suitable locations exist, but cannot be reached in time. One can easily imagine scenarios that include the extinction of plant species unable to disperse to new habitats. Because the phenomenon is so complex, scientists have been reluctant or unable to publish firm, reliable estimates of the numbers of species that could become extinct as a result of climate change, or to predict when such extinctions will occur.

Climate change is most likely to directly produce species extinctions in already extreme, barely survivable environments. These are the very cold,

Greater human density could affect survival rates of large Arctic mammals such as this caribou. With global warming, newly ice-free Arctic lands would potentially become available for mining, oil and gas development, and manufacturing activities. This would require an increase in human staffing and settlements, equipment, and infrastructure, which would eventually consume lands previously used for wildlife habitat. Migratory habits could become fragmented and the survival rates of birds and large mammals could be threatened.

hot, wet, dry, or chemically unusual places in which only relatively few types of highly specialized organisms can exist. Where such extreme conditions are climate induced, even small temperature changes can be highly significant. Organisms in extreme environments are likely to be living near the limits of physiological possibility. When extreme environments become more extreme, some organisms die. When extreme environments become too extreme, nothing can survive in them—but that is only part of the story.

When extreme environments become more moderate, more species can move into them, leading to increased competition for living space and other resources. They may become "too moderate" for specialists that have lost, or perhaps never evolved, ways to compete or escape in highly diverse and densely populated environments. In other words, given a trend of global warming, hot, dry environments may become hotter and drier, crossing some survival threshold of survival for desert-adapted species. Individuals of those species will have to emigrate or die. However, some hot, dry environments might become wetter, or cooler, or both, even as average global temperatures are rising. This will allow species adapted to the new, more moderate conditions to immigrate and to compete with, prey on, or infect the existing populations in unprecedented ways.

Climate Change and Species Extinctions

The direct effects of climate change are most likely to affect organisms of polar regions, mountaintops, and equatorial areas. Under a general trend of increasing temperatures, the very coldest climates—the Arctic and Alpine tundras—could disappear, and along with them would likely disappear at least some of the species adapted to tolerate them. As the warmer habitats move toward the poles and up the mountains, their species will follow. Those that cannot migrate or disperse as fast as their potential habitats are shifting will either have to evolve new climatic tolerances or die out.

When abiotic factors change, some habitats may contract, even to the point of disappearing altogether. Others may expand, and new ones may appear. Overall, they can be imagined as flowing slowly across the landscape, expanding in some directions, while retreating from others, sometimes forming and seemingly evaporating like puddles. The most obvious response for organisms that can move is to follow the changes in habitat or to find and occupy the new habitat. As long as enough individuals of a species can somehow keep up with these movements, their species may persist.

An expected direct effect of climate change with the potential for causing species extinction is a rise in sea level caused by the melting of polar and alpine glaciers. Large areas of low-lying coastal lands would be inundated by rising sea levels. In effect, some areas of terrestrial habitats would be converted to areas of aquatic habitats. Some low-lying oceanic islands would disappear, and along with them any land plants and animals that might be found nowhere else. Whether as a result of habitat inundation or other effects, species with very restricted ranges, called endemics, are likely to be more significantly affected by climate change than those with larger ranges.

Every species has different abiotic tolerances, so the edges of their potential habitats, based on moisture or temperature, rarely correspond exactly. Instead, these habitat edges usually overlap. Not only do they overlap, but climate change will affect each one differently, so different species habitats will move, grow, or contract at different rates. Not all species are equally mobile. This means that two species may experience different direct, abiotic affects in the same place. These differences create the possibility of many indirect effects of climate change as species interact in new ways and places.

Indirect Effects of Persistent Climate Change

Most effects on species resulting from any continuing climate change trend will be indirect. All animals rely on other species as sources of food. Many plants rely on insects and other animals to pollinate them or disperse their seeds. When different species come to depend on each other in predictable ways, their relationship is called a symbiosis. Symbioses range from pure exploitation, where only one species benefits, to cooperation or mutualism, where both species benefit. In many cases, such as those of internal parasites or intestinal bacteria, one organism actually becomes the entire habitat of another. Far more often, individuals of different species have no obvious interactions at all, but influence each other in much more subtle ways, such as by preying on another species' competitors, or its predators, or its pollinators, or by spreading its disease organisms.

The possibilities for changing species interactions are seemingly endless. Individuals of predatory species might find themselves able to range farther north, or higher into the mountains, where they will encounter potential prey species that have never seen them before. These prey animals may lack defensive or escape behaviors, and their populations may be significantly reduced. This does not mean that tropical cats like jaguars will decimate caribou herds. Most of the land animals in the world are insects, as are most of the predators. Humans are hardly aware of predation at the insect level, but it is cumulatively enormous and enormously influential.

Most insects are unable to regulate their body temperatures, except by seeking shelter. Flying insects have to meet minimum temperature requirements before their muscles work efficiently, allowing them to lift off. However, flying insects are highly mobile. Once aloft, they are often carried great distances by winds, sometimes to places where they normally cannot survive. However, if climates warm and their habitats move and expand, insects are likely to arrive in any newly suitable locations pretty quickly. If these pioneering insects are herbivores, they may find plants that have evolved no defenses against them. This could hasten the demise of individual plants and reduce populations that were physiologically capable of tolerating the direct effects of warmer temperatures.

Many plants rely on insects for pollination. Some plant populations could be reduced if predatory insects begin to survive in areas formerly unavailable to them because of climate factors and begin preying on local pollinators. If pollinators become too scarce, plant reproduction could be reduced to levels that cannot maintain a population. Both the plants and the pollinators could be affected.

Polar ice caps and alpine glaciers are composed of accumulating snow. If they melt, the resulting water is fresh, not salty. There is not enough freshwater in these sources to significantly dilute the world's oceans and change the fundamental chemistry of seawater. However, freshwater is less dense than salt water, and until the two mix, freshwater entering the oceans actually floats as a surface layer. The addition of massive amounts of cold, fresh water to the Arctic, North Atlantic, and north Pacific oceans, and to the south polar regions of the Atlantic, Pacific, and Indian oceans, would affect the way currents flow and nutrients circulate in these areas. This would affect the types and distribution of plankton, and thus all the many levels of oceanic food webs in those areas. Reduced plankton production would ultimately mean less prey for aquatic predators of polar seas such as polar bears, penguins, and some toothed whales, seals, and sea lions. Added to the direct effects of reduced pack ice, such changes could lead to the extinction of animals highly specialized for life under cold polar conditions that would no longer exist.

Hot deserts have fewer rivers, lakes, and ponds, but many of them have springs and small water courses that support endemic aquatic species including fishes, amphibians, reptiles, and many invertebrates and microorganisms. If these hot deserts become even hotter or drier because of climate change, these oases could literally dry up. In the process, numerous rare aquatic species that cannot move to other habitats (even if they existed) would become extinct in the process.

Aquatic species endemic to small tributary streams in any watershed face various new conditions when a region becomes drier or wetter. Neither trend is automatically beneficial. If it becomes drier, the smaller tributaries become ephemeral or intermittent, forcing fully aquatic species downstream into larger, more permanent waters, where they may encounter more (and larger) predators, at least for a time. If the region becomes wetter, the small tributaries will become larger, and the larger predators may move upstream. In high, steep terrain, the physical characteristics of the newest small tributaries may make them unsuitable for colonization.

In wet tropical areas, the effect of climate change will most easily be seen if it results in changes to the flow of atmospheric moisture to the region and, as a result, to the seasonality and overall amount of precipitation. At its simplest, a rainforest with less rain will gradually become another kind of forest, having fewer species requiring high moisture or seasonal inundation by floodwaters and more that tolerate drier conditions. As in all cases, if suitable habitat disappears or appears only at an unreachable distance, some species could become extinct. The complexity and diversity of tropical forests is such that not only some tree species, but also their dependent animals (and, in turn, their dependent animals), could become extinct in the process. Knowledge of the flora and fauna of these regions is insufficient to support any precise estimate of the number of species present, much less the number that could be affected by any particular degree of climate alteration.

Other Climate-Related Extinctions

In anticipation of a continued warming trend in polar regions, various countries are already positioning themselves to take advantage of ice-free Arctic seas and increasingly temperate high latitudes. Others are bracing for possible desertification in tropical grass and scrublands. Areas likely to experience intensified human use may have higher likelihood of species extinctions.

Increasing commercial ship traffic in Arctic waters would produce the same sorts of side effects that shipping has elsewhere. Leaks and spills of fuel and cargo oil would affect the biota of littoral zones. Ballast water exchange would further redistribute aquatic species, leading to new predation and competition among aquatic species without prior experience of each other.

Newly ice-free Arctic lands would potentially become available for mining, oil and gas development, and allied manufacturing activities. This will require an influx of workers and equipment, along with creation of the physical, economic, and cultural infrastructure needed to support them. Each activity entails a direct conversion of some existing habitat to human use. This could fragment the habitats of migratory birds such as snow geese and affect survival rates for large mammals such as caribou and musk oxen.

More ship traffic, mining, and oil exploration would encourage more permanent human settlements to service, and be serviced by, these industries, leading to a greater likelihood of chemical pollution and of sewage and solid waste management issues. Human population centers would encourage the establishment of human commensals and inquilines, such as dogs and cats, rats and mice, cockroaches and houseflies. Each potentially adds a new challenge to the persistence of Arctic species.

Under a continued warming trend, farmers in Europe, Asia, and North America would experience the same northward and upward shift in habitat bands affecting uncultivated plants and wild animals. For example, grain production will likely be possible farther north, in the Canadian "prairie provinces." This would require "sodbusting" of existing grasslands or logging of forests to convert them into farms, reducing or eliminating their habitat value to most wildlife. All the world's major crops—corn, soybeans, wheat, rice, and cotton, along with most every other valued plant—would become economically viable in new areas, while becoming impractical in others where they have been traditionally grown. The net effects on agricultural production are hard to estimate, as are the potential effects on other species.

Matthew K. Chew
Arizona State University

See Also: Agriculture; Antarctic Circumpolar Current; Arctic and Arctic Ocean; Atlantic Ocean; Botany; Charismatic Megafauna; Climate Zones; Conservation; Conservation Biology; Desertification; Deserts; Ecosystems; Geography; Glaciers, Retreating; History of Climatology; Ice Ages; Indian Ocean; Land Use; Marine Mammals; Modeling of Ocean Circulation; Modeling of Paleoclimates; Oceanic Changes; Pacific Ocean; Penguins; Phytoplankton; Plants; Polar Bears; Rainfall Patterns; Sea Level, Rising; Upwelling, Coastal; Upwelling, Equatorial.

Further Readings

Araújo, Miguel B., et al. "Reducing Uncertainty in Projections of Extinction Risk From Climate Change." *Global Ecology and Biogeography*, v.14/6 (November 2005).

Crowley, Thomas J. and Gerald R. North. "Abrupt Climate Change and Extinction Events in Earth History." *Science*, v.240/4855 (May 20, 1988).

Ebach, Malte C. and Raymond S. Tangney, eds. *Biogeography in a Changing World*. Boca Raton, FL: CRC Press, 2007.

Hassol, Susan Joy. *Impacts of a Warming Arctic: Arctic Climate Impact Assessment.* Cambridge: Cambridge University Press, 2004.
Peters, Robert L. and Thomas E. Lovejoy. *Global Warming and Biological Diversity.* New Haven, CT: Yale University Press, 1992.
Redman, Charles L., ed. *The Archaeology of Global Change: The Impact of Humans on Their Environment.* Washington, DC: Smithsonian Books, 2004.
Thomas, Chris D., et al. "Extinction Risk From Climate Change." *Nature*, v. 427 (January 2004).
Wyman, Richard L., ed. *Global Climate Change and Life on Earth.* London: Routledge, Chapman and Hall, 1991.

Sri Lanka

The Democratic Socialist Republic of Sri Lanka, formerly known as Ceylon, is a small, pear-shaped island of 25,332 sq. mi. (65,610 sq. km) lying off the southeastern coast of India. Sri Lanka has an extraordinary diversity of wildlife and vegetation because of its location near the equator and its remarkable range of terrain and climate. Although only a minute contributor to greenhouse gas emissions, the island is vulnerable to the effects of climate change.

Sri Lanka (pop. over 20 million in 2009) has a high population density (797 people per sq. mi., or 308 per sq. km) that creates increasing pressures and demands on natural resources. Around 25 percent of the population lives in urban or semi-urban areas, while 40 percent of the people are engaged in activities directly dependent on the environmental resource base. Land is Sri Lanka's most vital and heavily threatened natural resource.

Sri Lanka's carbon emissions per capita were 169, out of 214 countries measured worldwide in 2008, but more than doubled in the 10 years from 1995, largely in response to the country's increasing population. Like much of the rest of South Asia, Sri Lanka relies heavily on carbon-neutral biomass such as collected wood and animal waste for its domestic energy needs, particularly in rural areas. Biomass accounted for 80 percent of total residential energy consumption in 2005 and is expected to remain as high as 70 percent through 2020. Approximately 90 percent of Sri Lanka's industrial small- and medium-scale enterprises uses wood to produce thermal energy and to meet energy requirements.

Oil consumption more than doubled between 1990 and 2005 in response to a growing demand for transport fuels. Sri Lanka imports all of its daily crude oil consumption of 87,000 barrels, and in recent years, it has further increased oil imports to avoid overreliance on hydroelectricity for industrial power. Hydropower currently provides the majority of Sri Lanka's electricity, making the country vulnerable to changing rainfall patterns. In an effort to diversify, the Sri Lankan government is developing fossil fuel–fired power plants.

Sri Lanka's rich biodiversity includes an unusually large number of endemic species living in cloud forests, grasslands, and wetlands, as well as freshwater, coastal, and marine ecosystems. The island is both ecologically and economically vulnerable to climate change. Its famous tea plantations remain an important source of economic activity, and the island's rich cultural heritage, together with its tropical forests, beaches, and wildlife, make it a world-famous tourist destination. Climatic conditions and rich biodiversity are therefore key to maintaining Sri Lanka's economy. However, logging and population pressures continue to lead to deforestation and habitat loss. Large tracts of forest have been cut down for fuel wood or for timber export and have been replaced by rice, coconut, rubber, and coffee farms. Many species are in danger of extinction, including cheetahs, leopards, several species of monkeys, and wild elephants. Sri Lanka's coral reefs, already damaged by bleaching and the 2005 Asian tsunami, are being destroyed by human refuse and sewage and by dynamite fishing. Climate change–induced ecological stress will compound these environmental economic concerns.

Overall, the importance of environmentally sustainable behavior is underappreciated in the general population, as concerns such as poverty and recovery, after the long conflict between Tamil separatists and the Sinhalese government, are dominant. Even so, the government of Sri Lanka has taken action to conserve wildlife.

Over 13 percent of the land is protected, and the Sinharaja Forest Reserve, which protects the largest remaining stand of primary rainforest on the island, was declared a World Heritage Site in 1988. Sri Lanka has a long history of conservation, as it was the first country in the world to establish a wildlife reserve.

Harriet Ennis
Bootham School

See Also: Deforestation; Developing Countries; India; Rainfall Patterns; Tourism.

Further Readings

Energy Information Administration. "Country Analysis Briefs." http://www.eia.doe.gov/emeu/cabs/bhutan.html (Accessed May 2011).

United Nations. "Sri Lanka: Country Profile—Implementation of Agenda 21." http://www.un.org/esa/earthsummit/lanka-cp.htm (Accessed May 2011).

Stanford University

Stanford University is a private California university located in the city of the same name, close to Palo Alto. Stanford students, faculty, and employees are involved in a variety of initiatives designed to reduce the university's carbon footprint and combat global warming, among other environmental measures. Stanford offers degrees and courses in environmental fields and houses environmental projects and institutions, including the Center for Environmental Science and Policy (CESP), founded in 1998, the Woods Institute for the Environment, founded in 2004, and the Global Climate and Energy Project (GCEP).

Academic programs with an environmental focus are housed within the Department of Geological and Environmental Sciences. Graduate degrees awarded in the field are M.S., Engineer, and Ph.D. Foci include the application of Earth sciences to mineral, energy, and water resources. Stanford's Center for Environmental Science and Policy (CESP) directs the interdisciplinary Goldman Honors Program in environmental science, technology, and policy, and is affiliated with the Interdisciplinary Graduate Program on Environment and Resources. Student environmental groups include Students for a Sustainable Stanford, which played a major role in launching the Stanford Climate Change Campaign project.

Stanford employs a number of distinguished faculty, whose work have included environmental and climate change issues. Prominent among them was Stephen H. Schneider, Stanford professor and leading climate change researcher. Schneider served as a professor of biology, the Melvin and Joan Lane professor for Interdisciplinary Environmental Studies, and a senior fellow at the Woods Institute for the Environment, before his death in 2010. He was an active advocate for action on global warming, both within the political and scientific communities, as well as public forums. He was also the author of *Science as a Contact Sport: Inside the Battle to Save Earth's Climate* and was among the group of scientists honored with the 2007 Nobel Peace Prize alongside former Vice President Al Gore.

Stanford's CESP and Woods Institute for the Environment are affiliated. CESP fosters multidisciplinary research and networking within the academic and legislative communities. Their goal is to seek international solutions to environmental problems such as climate change and global warming through scientific and policy research. Educational outreach includes workshops, policy briefings, and publications. While CESP is not a degree-granting entity, it aids Stanford's undergraduate- and graduate-level academic programs. One key CESP initiative is the interdisciplinary Program on Energy and Sustainable Development, which studies the effect of energy production and consumption on sustainable development. The Woods Institute's main goal is the assessment of environmental science, technology, and policy at a variety of levels.

Climate Initiatives

Stanford University climate change and global warming initiatives include the 2002 establishment of the Global Climate and Energy Project (GCEP), a partnership between leading researchers, universities, and research institutions in the field of energy and private industry. GECP also seeks collaboration with those developing nations with fast-growth economies that most likely represent the highest future energy resource demands and subsequent emissions. GCEP's overall mission is to develop projects seeking to meet the world population's growing energy demands while maintaining environmental sustainability, efficiency, and cost effectiveness.

GCEP develops and oversees a variety of energy research programs involving a broad range of technologies and global energy resources and uses. Researchers carry out these programs both within the Stanford academic community and in collaboration with leading global institutions. These programs share the goal of successfully reducing greenhouse gas (GHG) emissions while existing as viable options in the energy marketplace. The project's international business sponsors are ExxonMobil, General Electric, Schlumberger, and Toyota. In addition to a project total investment of approximately $225 million, sponsor corporations also contribute technical expertise and energy-industry insights. An advisory board provides external oversight and guidance.

GECP's primary emphasis centers on research into commercial methods and technologies to reduce the GHG emissions of global energy systems. Specific goals include the identification of promising high-efficiency, low-emissions technologies for further research; the identification of barriers preventing the broad application of such technologies; research into technologies that will help overcome identified barriers and facilitate broad usage; and the publication of results to a wide audience in science, business, industry, government, media, and the public.

Some projects seek to develop more sustainable use of existing energy resources, such as carbon dioxide capture and sequestration technologies. Other projects focus on the potential impact of new methods of transforming, storing, and supplying energy and their potential environmental impacts. Past and planned project research areas include hydrogen and wind power; carbon capture energy distribution, storage, and infrastructure; biomass energy; solar energy; hydrogen production, storage, and use; nuclear energy; other renewable energy sources; advanced combustion; and advanced materials and catalysts.

GECP conducts activities in the two broad areas of analysis and research, which are designed to be complimentary. Analysis encompasses the assessment of processes and technologies under study for their potential to effectively provide energy while reducing GHG emissions. This includes both technology assessment and systems analysis, with work in collaboration with the GECP's integrated assessment project. Analysts determine the most promising ideas and projects in terms of viability and emissions reductions, as well as the best potential for integration of various technologies. Computational models provide a key tool for analysts.

Research encompasses the development of the science and technology necessary to meet the project's mission. GECP not only oversees current research but also continues to seek new research projects with an emphasis on those most likely to produce the greatest change, regardless of a high potential for failure. Educational outreach initiatives include Website maintenance, reports and other publications, and hosting events such as their annual energy symposia and technical area workshops.

Marcella Bush Trevino
Barry University

See Also: Antarctic Meteorology Research Center; Greenhouse Gas Emissions; Integrated Assessment; Media, Books, and Journals; Schneider, Stephen H.

Further Readings

Global Climate and Energy Project. http://gcep.stanford.edu (Accessed July 2011).

Schneider, Stephen H. *Science as a Contact Sport: Inside the Battle to Save Earth's Climate.* New York: Random House, 2009.

Stern Review

It has become commonplace to state that global warming emerged as one of the most serious risks for humanity. The scientific evidence has mounted over the last decade and is now overwhelming after the release of important global scientific assessments such as the 2007 *Fourth Assessment Report* by the Intergovernmental Panel on

Climate Change (IPCC). In terms of the human dimensions of these changes, the *Stern Review: The Economics of Climate Change*, released in 2006, is a major contribution in assessing the evidence and building understanding of the economics of climate change. It is based on an independent review commissioned by the Chancellor of the Exchequer in 2005 to inform both the chancellor and the prime minister about the nature of the economic challenges involved in such changes and how they can be met by the United Kingdom (UK) and the international community.

The review resulted in a comprehensive report with more than 700 pages coordinated by economist Nicholas Stern, chair of the Grantham Research Institute on Climate Change and the Environment at the London School of Economics. The report discusses the effect of global warming on the world's economy by examining the evidence of the economic impacts of climate change, the mitigation costs for stabilizing greenhouse gases (GHGs) in the atmosphere, and the complex policy challenges ahead to not only manage the transition to a low-carbon economy, but also ensure that societies can adapt to the consequences of a changing climate that can no longer be avoided due to GHG emissions.

International Collaboration is Required

Building upon an international perspective, the review assumes that climate change, as a phenomenon that is characterized by global causes and consequences, demands an international collective action to respond to it in an effective, efficient, and equitable way. To achieve this goal, the authors argue for a deep international collaboration in many areas such as carbon markets, research, technology, and the promotion of adaptation, particularly in developing countries. Besides being a global issue, climate change is addressed in the report through long time horizons and surrounded by uncertainty. It is also recognized as a unique challenge for the world's economy as its greatest and widest-ranging market failure ever seen.

The *Stern Review* implies that the effects of current societal responses and actions to mitigate climate change have only limited influence in the near future, because the global climate operates in long time scales. It is also difficult to predict with reasonable certainty the consequences of climate change in particular regions or places, although science has been advancing fast to understand the significance of the impacts predicted by these changes. The report asks for mitigation and the other plausible actions that are capable to reduce and stabilize GHGs emissions to be viewed as an investment, a cost incurred now and in the coming few decades to avoid the risks of very severe consequences of global warming in the future. One of the main assumptions carried out throughout the report is that if these investments are made wisely, the future costs for climate-related actions will be manageable and there will be a wide range of opportunities for growth and development along the range of these interventions. As a consequence, if policy is able to promote sound market signals, market failures can be overcome and mitigate many of the risks arising from unaddressed climatic changes at the global scale.

The conceptual framework applied in the report assumes the economic costs of the impacts of climate change as well as the costs and benefits of actions to reduce GHGs emissions, in three different ways. First, it not only considers the physical impacts of climate change on the economy, human life, and the environment, but also examines the resource costs of different technologies and strategies to reduce these emissions. Second, it uses economic modeling, including integrated assessment methodologies, to estimate the economic impacts of climate change by representing the costs and effects of the transition to low-carbon energy systems for the global economy. Third, it compares the current level and future pathways for the cost of impacts associated with an additional unit of GHG emissions coupled with the marginal abatement cost, to determine the costs associated with incremental reductions in units of emissions. Based on these different perspectives, the evidence gathered and analyzed by the *Stern Review* robustly supports the argument that the benefits of strong, early action considerably compensate for the costs of climate-related interventions in the future.

Postponing climate change responses and actions in the present has the potential to harm economic growth and the world's economy in the future, the review observes. The review also notes that the risks and damage arising from this inaction could be "on a scale similar to those associated with the great wars and the economic depression of the first half of the 20th century." It also highlights that many of the possible changes might be difficult or even impossible to reverse in the future. In this sense, the report argues for "pro-growth strategy for the longer term" that can be done "in a way that does not cap the aspirations for growth of rich or poor countries." The review's message highlights that if effective actions are taken early, the less costly they will be in the coming decades.

Positive and Negative Critiques

The *Stern Review* has received mixed reactions, not only from economists and scientists, but also from other sectors such as government representatives and the private sector. Regarding the positive responses, many see the report as a major contribution to raise awareness in relation to the issue of climate change through an economic perspective, opening up several opportunities in terms of new businesses, markets, and energy policy.

On the other hand, the report has also received various critical comments, particularly from other economists who are intrigued by the methods and models applied in the study (such as the discount rate applied and the treatment of uncertainty). One of the most common critiques is that the *Stern Review* overestimated the present value of climate change costs and underestimated the mitigation costs involved. In contrast, others have argued that the GHG emissions targets adopted by the *Stern Review* were very weak and that it underestimates the risks and the potential harms related to global climate change in the world's economy and in different economic sectors.

Since its release, several academic articles, as well as conferences, symposiums, and workshops related to the *Stern Review*, have been published and organized with the participation of its lead author and members of the writing team. Although some of the critiques remain unanswered, the principles underlying the analysis of the report are based on a solid, yet evolving literature. In this sense, although there is an overwhelming body of scientific evidence that indicates that climate change is a serious and urgent issue, progress has been made on a daily basis. Consequently, as one of the primary and comprehensive assessments of the economics of climate change, the *Stern Review* provides a fruitful avenue for new science and research activity in the climate change field, the findings of which should guide and support policymaking worldwide.

Rafael D'Almeida Martins
State University of Campinas

See Also: Economics, Cost of Affecting Climate Change; Economics, Impact From Climate Change; *Fourth Assessment Report*; Global Warming, Impacts of; Kyoto Protocol; Public Awareness.

Further Readings

Barker, T., et al. "Special Topic: The *Stern Review* Debate." *Climatic Change*, v.89/3–4 (2008).

Stern, N. *The Economics of Climate Change: The Stern Review.* Cambridge: Cambridge University Press, 2007.

Tol, R. S. J. and G. Yohe. "A Review of *Stern Review*." *World Economics*, v.7/4 (2006).

Weitzman, M. L. "A Review of the *Stern Review* on the Economics of Climate Change." *Journal of Economic Literature*, v.45/3 (2007).

Stockholm Environment Institute

In 1989, the Swedish government established the Stockholm Environment Institute (SEI) to develop an international environment/development research organization. The SEI now has over 20 years of research into local to global environmental and developmental issues. The institute is nonprofit and nonpartisan and can serve as a broker for complex developmental and social issues.

The SEI's mission developed from insights gained during the 1972 United Nations (UN) Conference on the Human Environment, which took place in Stockholm. Another element in the formation of the SEI and its mission was the work of the Brundtland World Commission for Environment and Development. The 1992 UN Conference on Environment and Development was a final factor in helping to develop the mission.

The SEI ranks as one of the top 10 environmental think tanks in the world. The SEI is noted for its rigor and objectivity in scientific analysis of its projects. It is especially well known for its work on the relationship of environmental issues to poverty. Sustainable development and the environment are its specialties, and it seeks to promote sustainable development through integrated science/policy analysis presented to decision makers. The SEI five-year strategy is to take a global view on challenges for the next decade to promote sustainable and responsible development in consultation with partners and world stakeholders.

It conforms to the principles of Agenda 21, the UN sustainable development action plan adopted in 1992, and as conventions on climate change, biological diversity, and ozone layer protection. It is following the pattern of the Beijer Institute in the effort to create a viable field of sustainability science, a field devoted to understanding the interactions between nature and society. Sustainability science seeks to help different societies transition to more sustainable futures.

Partnerships With Other Programs

Collaboration and intensive stakeholder involvement are integral to SEI projects. The SEI and the Greenhouse Gas Management Institute combine in the Carbon Offset Research and Education (CORE) Initiative, which promotes offset programs and policies that maximize their benefits and minimize their risks. The CORE team provides research findings to both the general public and policymakers. It also has a team for the aviation community. Overall, the SEI's programs include finding new approaches to development, mitigating the risk of climate change, managing environmental systems, and creating less vulnerable communities through the world.

The SEI is part of the Stockholm Resilience Center (SRC). Other members include Stockholm University's Centre for Transdisciplinary Environmental Research and the Beijer International Institute of Ecological Economics at the Royal Swedish Academy of Sciences. The SRC emphasizes cross-disciplinary research on socioecological systems, with a focus on the ability of those systems to handle change and continue to develop. It also provides research on biodiversity, environmental problems of the Baltic Sea, the history of humanity, and other areas. Funding is through the Foundation for Strategic Environmental Research, Mistra. Mistra investment in June 2006 totaled $30 million over 12 years. Between 2006 and 2009, the SEI received major funding from the Institutional Program Support element of the Swedish International Development Cooperation Agency (SIDA), which sought to make SEI a strong resource capable of providing state-of-the art analysis, synthesis, and governmental policy support. SIDA also provided startup resources and technical support. In 2011, SIDA agreed to continue funding the SEI at $25 million per year through 2014.

Internally, the SEI is organized into four teams that deal with overarching issues such as climate change, energy systems, governance, and vulnerability. They also deal with more specific matters such as air pollution and adequacy of water resources. Modeling for sustainability and assessment of vulnerability are particularly well received. SEI provides studentships, internships, international exchanges, and supervision of graduate degrees. It also maintains the Global Water Partnership Website.

The SEI has four administrative elements: SEI Stockholm, SEI Africa, SEI Asia, and SEI Oxford. SEI research centers are located in York and Oxford, the United Kingdom, as well as the United States, Tanzania, Thailand, Estonia, and Sweden. Each is largely autonomous, but all work within the five SEI programs and each seeks to have policy-relevant research. SEI–US, for instance, is a research affiliate of Tufts University, and SEI York is self-funded, but part of the University of York. The other centers are separate entities under SEI auspices. SEI Stockholm competencies are in biotechnology, energy, atmospheric environments, ozone-layer protection, and climate change. It provides vulnerability and risk assessments, as well as analyses of water resources, institutions and policies, and ecological sanitation.

John H. Barnhill
Independent Scholar

See Also: Greenhouse Gas Emissions; Nongovernmental Organizations; Sweden.

Further Readings

Carbon Offset Research and Education. "Welcome to CORE." http://www.co2offsetresearch.org (Accessed July 2011).
Stockholm Environment Institute International. http://sei-international.org/about-sei (Accessed July 2011).
Stockholm Resilience Centre. "Centre-Hosted Programmes and Networks." http://www.stockholmresilience.org/featurearchive/centrehostedprogrammesandnetworks.106.5686ae2012c08a47fb5800027440.html (Accessed March 2012).
Swedish South Asian Studies Network. "Stockholm Environment Institute, SEI." http://www.sasnet.lu.se/envirsth.html (Accessed July 2011).

Stommel, Henry

Henry Stommel (1920–92) was an American oceanographer and meteorologist whose theories on general circulation patterns in the Atlantic Ocean made him the creator of the modern field of dynamical oceanography. Stommel carried out a series of research studies and first suggested that the Earth's rotation is responsible for the Gulf Stream along the coast of North America. He theorized that its northward thrust must be balanced by a stream of cold water moving in the opposite direction beneath it. Carl Wunsch has described Stommel as "a transitional figure, being probably the last of the creative physical oceanographers with no advanced degree, uncomfortable with the way the science had changed, and deeply nostalgic for his early scientific days." Stommel has been praised for being both a creative theorist and an acute observer, who was willing to spend months at sea.

Stommel was born in Wilmington, Delaware, on September 27, 1920, into a family of extremely mixed background. His ancestors came from such different places as the Rhine Valley, Poland, Ireland, the Netherlands, England, and France, and they also had a trace of Micmac Indian. Henry's father, Walter, was a chemist born in northern Germany and trained in Darmstadt and Paris. During World War I, Walter Stommel emigrated to Wilmington, where he was employed by Dupont Chemical. While in the United States, he married Marian Melson. Their son Henry was born shortly after the marriage. Although the reason is not completely clear, perhaps because of anti-German sentiment following World War I, the family then moved to Sweden. Henry's mother, however, soon left Sweden with Henry and returned to Wilmington. Because of his mother's decision not to see her husband again, Henry and his sister Anne grew up in a single-parent family. When Henry was 5 years old, his mother moved with the two children to Brooklyn, New York, to live with her parents and other relatives. Marian supported the entire household thanks to her job as a fund-raiser and public relations officer at a hospital. Henry and his grandfather, Levin Franklin Melson, developed a meaningful relationship in a household dominated by women.

Stommel attended New York City's public schools. He spent one year at Townsend Harris High School, but finished high school at Freeport, Long Island, because his family had moved there. Thanks to his receiving a full scholarship, he was able to enroll at Yale University, from where he graduated in 1942. He remained at Yale for two years following graduation, teaching analytic geometry and celestial navigation in the U.S. Navy's V-12 program. He also spent six months at the Yale Divinity School, but his lifelong ambivalence toward religion made the ministry an unsuitable vocation for him. In 1944, renowned astrophysicist Lyman Spitzer suggested that Stommel apply for work at the Woods Hole Oceanographic Institution in Woods Hole, Massachusetts—an organization that was fast becoming a decisive part of the U.S. war effort. Stommel was recruited to work in acoustics and antisubmarine warfare, but disliked his assignment and tried to be employed in other areas.

In 1948, Stommel wrote "The Westward Intensification of Wind-Driven Ocean Currents," a paper that is unanimously regarded as constituting the starting point of dynamical oceanography. In it, he explained the Gulf Stream deductively by fluid dynamics. In particular, he discovered the mechanism (the latitudinal change of the Coriolis force on the rotating Earth) that produced the westward intensification of oceanic currents. Stommel proposed a global circulation

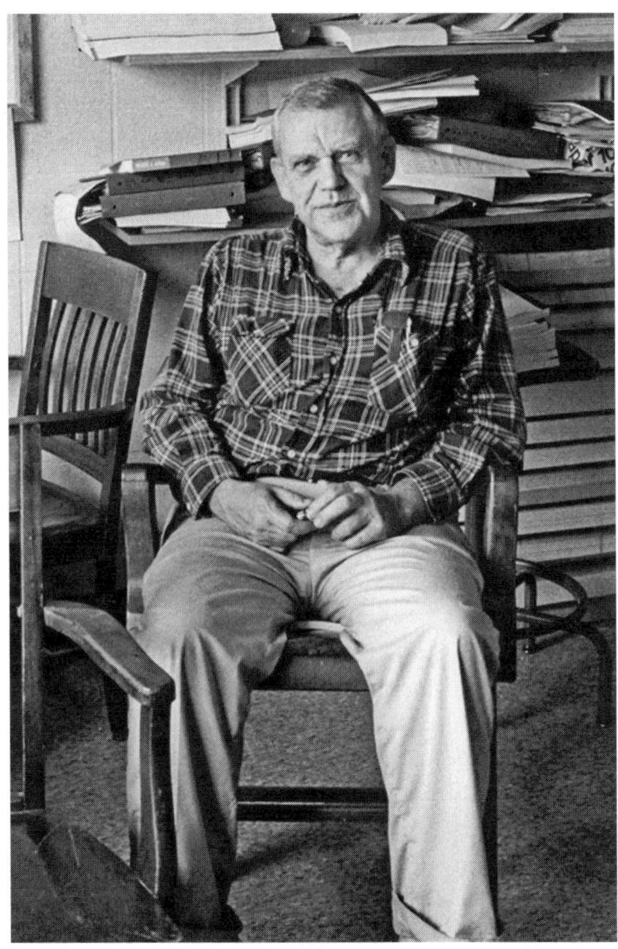

Henry Stommel theorized the circulation patterns of oceans and cold water moving in the opposite direction beneath them. His global circulation model was similar to a conveyor belt. He used the example of the Gulf Stream to explain general circulation.

model similar to a conveyor belt: surface water sinks in the far north to supply the deep, south-flowing current, and water rises in the Antarctic region to contribute a northward flow along the eastern coasts of North and South America. His important book *The Gulf Stream* was probably the first true dynamical discussion of the ocean circulation. He put the Gulf Stream in the wider context of the general circulation and paved the way for the development of thermocline theories. Stommel also concluded that changes in density caused by cooling and evaporation at the sea surface can be responsible for deep flows in the ocean. He was thus responsible for establishing the basic factors that helped to establish theories about global circulation. His thermocline theories stressed the role of oceans and sea currents in the definition of global climate and thus anticipated debates on global warming.

In December 1950, Stommel married Elizabeth Brown. The couple had three children. Although Stommel liked working at the Woods Hole Oceanographic Institution, he did not get on well with his director, Paul Fye. Therefore, he accepted an invitation to become a professor at Harvard University in 1959, lured by the prestige of the institution. He spent four unhappy years there, where his democratic ideals clashed with a rigid sense of hierarchies. After Harvard, Stommel went to work at the Department of Meteorology at the Massachusetts Institute of Technology (MIT). There, he worked with the most famous meteorologists of the day, such as Jule Charney, Norman Phillips, Edward Lorenz, and Victor Starr. Stommel worked enthusiastically with these scientists to improve theories of general circulation. He also worked on other important topics, such as the classification of estuaries and the effect of volcanoes on climate.

Stommel worked at MIT for 16 years as a professor of physical oceanography. He returned to the Woods Hole Oceanographic Institution when Fye retired, and continued to work there until his death on January 17, 1992. Stommel established several stations for the study of ocean currents, including the PANULIRUS station (begun in 1954) in Bermuda. He was elected to the National Academy of Sciences in 1962 and received the National Medal of Science in 1989.

Luca Prono
Independent Scholar

See Also: Current; Oceanography; Thermocline.

Further Readings

Hogg, N. G. and R. X. Huang, eds. *Collected Works of Henry Stommel.* Boston: American Meteorological Society, 1996.

Stommel, Henry. *The Gulf Stream: A Physical and Dynamical Description.* Berkely: University of California Press, 1958.

Wunsch, Carl. "Henry Stommel." Biographical Memoirs. http://www.nap.edu/readingroom/books/biomems/hstommel.html#FOOT1 (Accessed March 2012).

Stratopause

The stratopause is one of the layers into which the atmosphere is divided. It is the buffer region of the atmosphere that lies between the stratosphere and the mesosphere, from a height of about 31 to 34 mi. (50 to 55 km) above the Earth's soil. The atmospheric pressure is about 1/1000th of the pressure at sea level. In the stratopause, the temperature reaches a peak because of the heating generated by the absorption of ultraviolet radiation by ozone molecules in the stratospheric ozone layer. In this region, the catalytic cycles, which are less efficient at colder temperatures because of reduced O density, produce a significant ozone increase (~15 percent). Because of the considerable ozone presence in the stratopause, the understanding of this region is considered crucial to understanding the changes in climate and in the composition of the ozone layer. Above the stratopause, the temperature starts to decrease again with height as a result of the reduced solar heating of ozone.

The depletion of the ozone layer resulting from the emission of halogen atoms and the photodissociation of chlorofluorocarbon compounds is of particular concern to scientists, as the layer prevents the most harmful ultraviolet-B wavelengths from passing through the Earth's atmosphere. Near the stratopause, the ozone reduction is slightly smaller in the drier stratosphere because of the stronger temperature dependence of the drier atmosphere.

Studies of the temperature in the stratopause have also been important to assess the validity of global circulation models. For example, a study published in 2002 by the University of Illinois at Urbana-Champaign and the High Altitude Observatory of the National Center for Atmospheric Research in Boulder, Colorado, showed that wintertime warming caused by sinking air masses was not as strong as the models had assumed. The study employed lidar laser measurements and balloon observations made at the Amundsen-Scott South Pole Station from December 1999 to October 2001.

These measurements and observations were then used to calculate the monthly mean winter temperature profiles from the surface to about 63 mi. (110 km). The measured temperatures during midwinter in both the stratopause and mesopause regions were 20–30 degrees Kelvin colder than current model predictions. These differences were caused by weaker than expected compressional heating associated with subsidence over the polar cap. The study showed that the greatest difference occurred in the month of July, when the measured stratopause temperature was about 0 degrees F (minus 18 degrees C) compared with the about 40 degrees F (4.4 degrees C) predicted by the models.

Luca Prono
Independent Scholar

See Also: Atmospheric Composition; Atmospheric Vertical Structure.

Further Readings

Wallace, John M. and Peter V. Hobbs. *Atmospheric Science*. 2nd ed. Maryland Heights, MO: Academic Press, 2006.

Weilin, Pan, Chester S. Gardner, and Raymond G. Roble. "The Temperature Structure of the Winter Atmosphere at South Pole." *Geophysical Research Letters*, v.29/16 (2002).

Stratosphere

The stratosphere is a layer in the atmosphere that extends between about 9 to 31 mi. (15 to 50 km) in altitude. It is characterized by a vertical temperature structure that is nearly isothermal (no temperature change with altitude) in the lowermost stratosphere, and a pronounced inversion (increase of temperature with altitude) above. The stratosphere owes its name to the strong stratification, which is a consequence of this thermal structure.

The stratosphere plays an important role in the climate system. It contains the ozone layer, which shields the Earth's surface from harmful ultraviolet radiation and is responsible for the temperature of the stratosphere. Radiative processes in the infrared part of the electromagnetic spectrum also play an important role. Because in the stratosphere, chemistry, dynamics, and radiative processes operate under very different conditions than in the troposphere, the stratosphere is susceptible to climatic forcings in a different way than the troposphere. As a consequence, stratospheric processes play an important role for climate variability and change.

The stratosphere was discovered independently by Teisserence de Bort and Richard Assmann around 1900. It has been explored by balloon-borne observations since around the 1930s and by satellite observations since the 1970s.

The lower boundary of the stratosphere is the tropopause, the altitude of which varies with latitude (higher over the tropics than over the poles), season, and on a day-to-day scale related to weather systems. The upper boundary of the stratosphere is the stratopause. Below and above the stratosphere are the troposphere and mesosphere, respectively.

Because of its thermal structure (strong static stability), the circulation of the stratosphere is quasi-horizontal. The most important features of the zonal circulation are the vortices in the polar regions and the Quasi-Biennial Oscillation (QBO) in the tropics. The polar vortices form over both poles during the corresponding winter season and vertically extend through the entire stratosphere. The Arctic vortex is subject to strong variability on short timescales (during sudden stratospheric warmings, the vortex can break down completely within days) and on interannular timescales. The Antarctic vortex varies much less. The QBO is an oscillation of the zonal wind in the equatorial stratosphere, with changes from westerlies to easterlies and back to westerlies within approximately 28 months.

Compared with the zonal flow, the meridional circulation and associated vertical motion in the stratosphere are very weak, but are nevertheless important. The meridional circulation is caused by planetary waves originating from the troposphere, which break and dissipate in the stratosphere and thereby deposit momentum, decelerating the zonal flow and inducing a meridional flow component. The meridional flow is compensated for by vertical motion in the tropics and in the polar areas, forming a single meridional circulation cell, which in the context of trace gas transport is often referred to as Brewer-Dobson circulation. Air enters the stratosphere in tropical areas. On passing the tropopause, the air loses almost all of its moisture; hence, the stratosphere is very dry and mostly cloud free. In the stratosphere, the air slowly moves upward and poleward toward the winter hemisphere (the summer hemisphere is dynamically quiet). In the subpolar and polar region, the air descends and can eventually enter the troposphere in conjunction with midlatitude weather systems. The stratospheric meridional circulation has a turnover time of one to three years.

Chemically, the stratosphere is characterized by a layer of ozone (O_3) formed from atomic (O) and molecular (O_2) oxygen in the presence of ultraviolet radiation. Ozone can be destroyed by catalytic processes that involve radicals of chlorine, bromine, nitrogen oxides, or hydrogen oxides. The most important source of chlorine radicals are human-made chlorofluorocarbons (CFCs), which have caused a reduction of the ozone layer since the 1970s and the 1980s, and an enlargement of the Antarctic ozone hole (a substantial reduction of the total stratospheric ozone amount over Antarctica). The Montreal Protocol of 1987 and its amendments have led to a strong reduction in CFC emissions worldwide. However, because of the

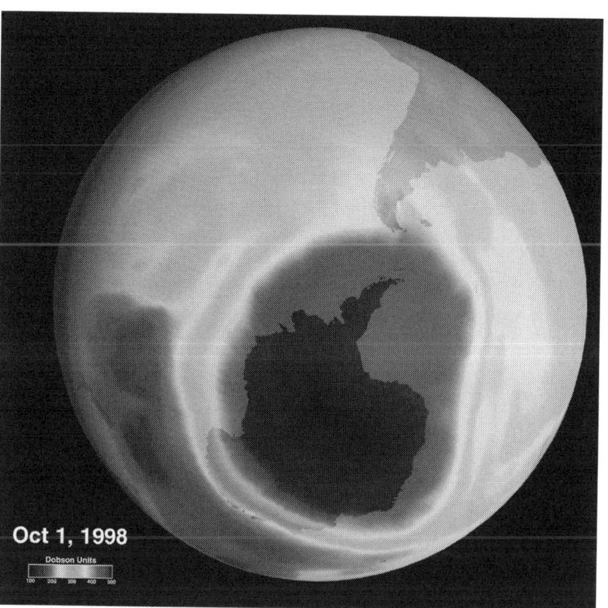

The large ozone opening over the poles (dark area) in October 1988. The discovery of this hole in 1985 led to the Montreal Protocol in 1987. Stratospheric ozone blocks harmful ultraviolet radiation produced by the sun.

long lifetime of CFCs, a full recovery of the ozone layer is only expected for the mid-21st century.

The anthropogenic greenhouse effect, as well as ozone depletion, cause a cooling of the stratosphere, whereas volcanic eruptions lead to warming. The stratosphere plays an important role for climate at the Earth's surface. Perturbations of the stratospheric circulation can propagate downward and affect weather at the ground. This provides a pathway through which some of the forcings can affect climate. For instance, it is now believed that part of the climate effect of strong volcanic eruptions operates via the change in stratospheric circulation induced by the heating effect of volcanic aerosols. Similarly, changes in solar irradiance could affect climate via stratospheric ozone chemistry and their subsequent effects on circulation.

Stefan Brönnimann
Institute for Atmospheric & Climate Science

See Also: Atmospheric Vertical Structure; Climatic Data, Atmospheric Observations; Mesosphere; Stratopause; Tropopause; Waves, Rossby.

Further Readings

Baldwin, Mark P. and Timothy J. Dunkerton. "Stratospheric Harbingers of Anomalous Weather Regimes." *Science*, v. 244 (2001).

Brasseur, Guy P. and Susan Solomon. *Aeronomy of the Middle Atmosphere*. New York: Springer, 2005.

Labitzke, Karin G. and Harry van Loon. *The Stratosphere. Phenomena, History, and Relevance*. New York: Springer, 1999.

World Meteorological Organization. *Scientific Assessment of Ozone Depletion: 2006*. Geneva: World Meteorological Organization, 2007.

Sudan

Located in northern Africa between Egypt and Eritrea, Sudan was, until July 2011, the largest country in Africa and the third-largest oil producer in sub-Saharan Africa. From the late 19th century until it became an independent nation in 1956, Sudan was jointly administered by Great Britain and Egypt. The linguistic, religious, racial, and economic divisions that have generated decades of ethnic tensions and civil war predate independence. The conflict over land and water deeply rooted in the history of Sudan has been exacerbated by climate change. Inadequate supplies of potable water, desertification, and drought are all problems in the nation that a 2009 World Bank study identified as the country at highest risk from the effects of climate change on agriculture. Neither the billions the nation earns from oil exports, nor its division into two nations on July 9, 2011, seem likely to end the devastations of climate change and armed conflict.

Geographically, Sudan ranges from arid and semiarid in the north to savannah in the central regions and equatorial in the south, but drought has been a dominant threat in all regions. The average rainfall has decreased significantly over the last 60 years. With 80 percent of its population dependent upon agriculture, such a decrease is disastrous. The country's first climate change report in 2003 predicted that the shorter growing seasons resulting from increased temperatures would lead to a significant long-term decline in the yields of staple millet and sorghum. Subsequent studies confirmed this prediction. Harvests for 2010 were improved, but vulnerability to climate change and political turmoil remain part of the scenario for 2011 and beyond.

Reduced rainfall not only leads to lower agricultural yields, it also contributes to desertification. According to the United Nations Environment Programme (UNEP), desert land has extended southward since the 1930s, and 25 percent of Sudan's agricultural land is at risk. The country has been battling desertification since the 1970s and was one of the first states to sign and ratify the UN Convention to Combat Desertification. Farming and livestock grazing have historically counted among the leading causes of desertification, and in recent years, refugee camps with their need for firewood have aggravated the problem. Land use and the overwhelming numbers of internally displaced persons and refugees are also factors in deforestation, making Sudan third among the countries of the world in deforestation. Sudan has lost 40 percent of its forests since the 1970s, and if current rates of land conversion to agricultural needs persist, the country will be completely deforested by 2060.

So severe are the climate-related problems of Sudan that a UNEP report released in 2007 claimed that environmental degradation and the symptoms of global warming are at root of the conflict in Darfur, one of the world's worst humanitarian crises. Many experts charge that such associations are simplistic and point out that Sudan has been engaged in civil wars for most of the period of its independence. Without denying the tragedy of Darfur, they note that the famine-related effects of the second civil war that broke out in 1983 led to more than 2 million deaths over two decades and over twice that number of displaced persons.

Country Splits in Two

The recognition of the Republic of South Sudan (RSS) as an independent country in July 2011 has not ended the conflict. The Abyei area—a 2,580,000-acre section between the Sudan and RSS—has a disputed border, and both countries still lay claim to the area. Conflicts are also heated over how the two countries will share oil reserves, 75 percent of which are held by RSS. Sudan has threatened to close down refineries and pipelines if they are unsatisfied with South Sudan's plans for sharing oil revenues. RSS president Salva Kiir has stated that his country can do without oil for the three years he estimates it will take to build their own infrastructure. Amid the conflicts, the two-nation area remains increasingly ill equipped to mitigate or adapt to climate change, and armed conflict and poor transportation infrastructure severely limit help from the international community.

Wylene Rholetter
Independent Scholar

See Also: Agriculture; Desertification; Drought.

Further Readings

One World. "Climate Change in Sudan." http://uk.oneworld.net/guides/sudan/climate-change (Accessed July 2011).

Polgreen, Lydia. "A Godsend for Darfur, or a Curse?" *New York Times* (July 22, 2007). http://www.nytimes.com/2007/07/22/weekinreview/22polgreen.html (Accessed July 2011).

United Nations Development Programme. "Sudan: National Adaptation Program of Action for Climate Change." http://www.sd.undp.org/projects/en1.htm (Accessed July 2011).

Sulfur Dioxide

Sulfur dioxide (SO_2) is an important component of the atmosphere, present as the result of both natural and human activity. Although it is a primary pollutant, causing respiratory irritation and damage to plants, it is the secondary pollutants produced from SO_2 that are particularly important in connection with global climate change. Sulfur dioxide is notorious as the cause of acid rain, but it is also a precursor to the formation of clouds. Hence, its release to the atmosphere is a major contributor to global dimming, a process that is thought to offset some of the effects of global warming.

There are, therefore, important implications of SO_2 release for the global climate change agenda. The reduction in SO_2 pollution in recent decades, stimulated by health concerns and by the effects of acid rain, is removing an unexpected and previously unidentified protection against increasing global temperatures. This illustrates the complexity of climate science that compounds the social and political responses to the threat of climate change.

Formation of Clouds

Once in the atmosphere, SO_2 is rapidly oxidized, ultimately producing sulfuric acid. Although this transformation is well known in the formation of acid rain, it also has broader climatological significance. The liquid sulfuric acid forms as an aerosol (tiny droplets suspended in the air), and this sulfuric acid aerosol attracts water vapor, which dissolves in the acid. In this way, the gas-to-liquid conversion of SO_2 to sulfuric acid brings about the nucleation of clouds: sulfuric acid aerosol is a cloud condensation nucleus (CCN).

Clouds play important roles in the atmosphere and in the climate, principally acting to transport water (and energy) between regions and affecting the Earth's radiation balance. Clouds have a very strong tendency to reflect sunlight (they have a high albedo) and absorb energy from the sun, so that the amount of cloud present in the atmosphere affects the amount of sunlight reaching the surface: more clouds result in a dimmer planet. This dimming effect of clouds is well documented.

Sulfur Dioxide and Climate Change

Furthermore, the influence of SO_2 and subsequent aerosol formation has been observed directly during volcanic eruptions. For example, the 1991 eruption of Mount Pinatubo in the Philippines released an estimated 20 megatons (20 million tons) of SO_2 into the atmosphere. The force of the explosion injected a large fraction of this material, along with dust particles, directly into the stratosphere, from which removal via rainout is very slow. The aerosol and clouds that formed as a consequence of this lasted for many years, with measurable effects on global temperatures. In the year following the eruption, the global average temperature reduced by 0.9 degree F (0.5 degree C), and even in 1993, the temperature was depressed by as much as 0.45 degree F (0.25 degree C).

In the lower atmosphere, the rate of SO_2 gas-to-liquid conversion is increased in the presence of other materials, notably particles such as soot, and this has an important effect on cloud condensation. The typically hydrophobic, water-repelling surfaces of soot particles catalyze the chemical reactions that convert SO_2 to sulfuric acid, so that the soot ends up coated with a water-loving, hydrophilic layer. The simultaneous emission of SO_2 and soot (e.g., from burning coal or diesel fuel) therefore increases the concentration of cloud condensation nuclei in the air, affecting both the amount and nature of cloud formation.

Because there are many more cloud condensation nuclei under these conditions than in the clean atmosphere, clouds form with smaller, more numerous droplets. More numerous particles means that the clouds reflect more light, and smaller droplets take longer to form raindrops. Hence, the clouds formed on sulphate/soot aerosol CCN are longer lived and have a higher albedo than ordinary clouds. In this way, SO_2 emissions in combination with soot increase the amount of incoming sunlight reflected away from the Earth, effectively dimming the planet's surface.

Sulfur dioxide pollution–related global dimming is thought to explain the slight global cooling trend in the period from 1950 to the late 1970s. With the recent legislation-driven decrease in emissions of SO_2 and soot particles from industry and transport in industrialized nations, there has been a steady rise in the amount of sunlight reaching the Earth. It is suspected that reducing this form of pollution is removing an effect that has been offsetting the full force of anthropogenic global climate change.

Christopher J. Ennis
University of Teesside

See Also: Aerosols; Albedo; Cloud Feedback; Pollution, Air; Volcanism.

Further Readings

Henson, Robert. *The Rough Guide to Climate Change: The Symptoms, The Science, The Solutions*. London: Rough Guides, 2006.
Wayne, Richard P. *Chemistry of Atmospheres*. Oxford: Oxford University Press, 2000.

Sulfur Hexafluoride

Sulfur hexafluoride (SF_6) is the most powerful of all the greenhouse gases recognized by the Kyoto Protocol and evaluated by the Intergovernmental Panel on Climate Change. Although its concentration in the atmosphere is low, the combination of a high global warming potential and a very long lifetime make emissions of SF_6 a considerable concern. It is primarily used as an electrical insulator in the high-voltage distribution network, and major industrial users are beginning to restrict the use and emission of SF_6.

The Kyoto Protocol requires developed nations to cut their emission of six greenhouse gases. The Intergovernmental Panel on Climate Change periodically assesses these gases and estimates a global warming potential (GWP) for each. The gasses, with their respective GWPs, are as follows: carbon dioxide (1), methane (21), nitrous oxide (310), hydrofluorocarbons (11,700), perfluorocarbons (9,200), and sulfur hexafluoride (23,900). In this list, HFC and PFC are groups of chemicals, and the value quoted is for the member of the group with the highest GWP. Despite the low atmospheric concentration of SF_6 (5.6 parts per trillion), its extremely high global warming potential and long lifetime (probably in excess of 1,000 years) mean that present emissions will have an effect on climate for a long time to come.

As SF_6 gas is denser (heavier) and more electrically insulating than either dry air or dry nitrogen, it is an ideal electrical insulating material. The gas is used extensively in electrical applications, and its principal use is in the electrical generation and high-voltage distribution industry. There are two specific advantages of SF_6 in these applications. First, its highly insulating character means that less space is needed between high-voltage components, so that equipment can be made significantly smaller than is possible when air or nitrogen are use as insulators. Second, gas-insulated switch gear using SF_6 rather than air demands a controlled environment, and the equipment is consequently more robust with regard to environmental pollutants and weathering than would be the case with simpler air-insulated equipment. In addition to these advantages, the gas is unreactive, nontoxic, and nonflammable. In the United States, the electric power distribution industry works on a voluntary basis with the SF_6 Emissions Reduction Partnership for Electric Power Systems to identify and implement technologies for reducing SF_6 emissions.

Another large-scale use of SF_6 is in magnesium metal manufacturing and casting. Magnesium metal is extremely reactive in air, particularly when hot or molten. The high density and low chemical reactivity of SF_6 make it a suitable choice as a protective gas layer preventing contact of the molten, highly reactive metal with oxygen and water in the air. A voluntary SF_6 Emission Reduction Partnership for the Magnesium Industry exists in the United States in association with the U.S. Environmental Protection Agency, which, together with the International Magnesium Association, is committed to eliminating SF_6 emissions from the industry by 2011.

SF_6 is also used in certain medical applications, including eye surgery and ultrasound scanning. Once again, it is the gas's high density and low toxicity that are used. In eye surgery, the gas is commonly used to form a plug to seal the retina during surgery. Its high density means that the gas stays in place and does not enter the blood at an appreciable rate. The density of the gas also makes it an excellent contrast agent in medical ultrasound scanning.

Similar to the perfluorocarbons, SF_6 is also used in the semiconductor industry, and there is concern about the growth of this industry leading to uncontrolled increases in the amount of SF_6 released to the atmosphere.

Tracer Gas Used by Climate Scientists

Paradoxically, because of its high chemical stability, low toxicity, and low natural abundance, SF_6 has been extensively used by atmospheric scientists as a tracer gas to understand the movements and mixing of air. The gas has, for instance, been injected into the exhaust plumes from power stations in an attempt to understand the origins of acid rain. In the United Kingdom, SF_6 tracer experiments have demonstrated that power stations are capable of delivering acid rain pollution to Scandinavia. For similar reasons, SF_6 is used to trace the movements of air within ventilation and air conditioning system tests. Recently, the gas was released on the London Underground in an attempt to understand the way toxic gases would spread throughout the system in the event of a terrorist attack.

Christopher J. Ennis
University of Teesside

See Also: Global Warming; Intergovernmental Panel on Climate Change; Kyoto Protocol; Perfluorocarbons.

Further Readings

Houghton, John. *Global Warming—The Complete Briefing*. Cambridge: Cambridge University Press, 2004.

Intergovernmental Panel on Climate Change. *Fourth Assessment Report, Working Group 1 Report, The Physical Science Basis*. http://www.ipcc.ch/publications_and_data/publications_and_data_reports.shtml (Accessed March 2012).

Wayne, Richard P. *Chemistry of Atmospheres*. Oxford: Oxford University Press, 2000.

Sunlight

Sunlight is the electromagnetic radiation given off by the sun. It is passed through the atmosphere to the Earth, where the solar radiation is reflected as daylight. Sunshine results when the solar radiation is not blocked. Sunlight is the primary source of energy to the Earth. It provides infrared, visible, and ultraviolet (UV) electromagnetic radiation with different wavelengths. Small sections of the wavelengths that are visible to the human eye are reflected as rainbow colors. Sunlight may be recorded using a sunshine recorder. Electromagnetic waves are waves that are capable of transporting energy through the vacuum of outer space and that exist with an enormous continuous range of frequencies known as the electromagnetic spectrum. The spectrum is divided into smaller spectra on the basis of interactions of electromagnetic waves with matter.

The longer-wavelength, lower-frequency regions are located on the far left of the spectrum, and the shorter-wavelength, higher-frequency regions are on the far right. Two very narrow regions within the spectrum are the visible light region and the X-ray region. The visible light region is a very narrow band of wavelengths located to the right of the infrared region and to the left of the UV region. Though electromagnetic waves exist in a vast range of wavelengths, human eyes are only sensitive to the visible light spectrum. The visible portion of the solar spectrum lies between 400 and

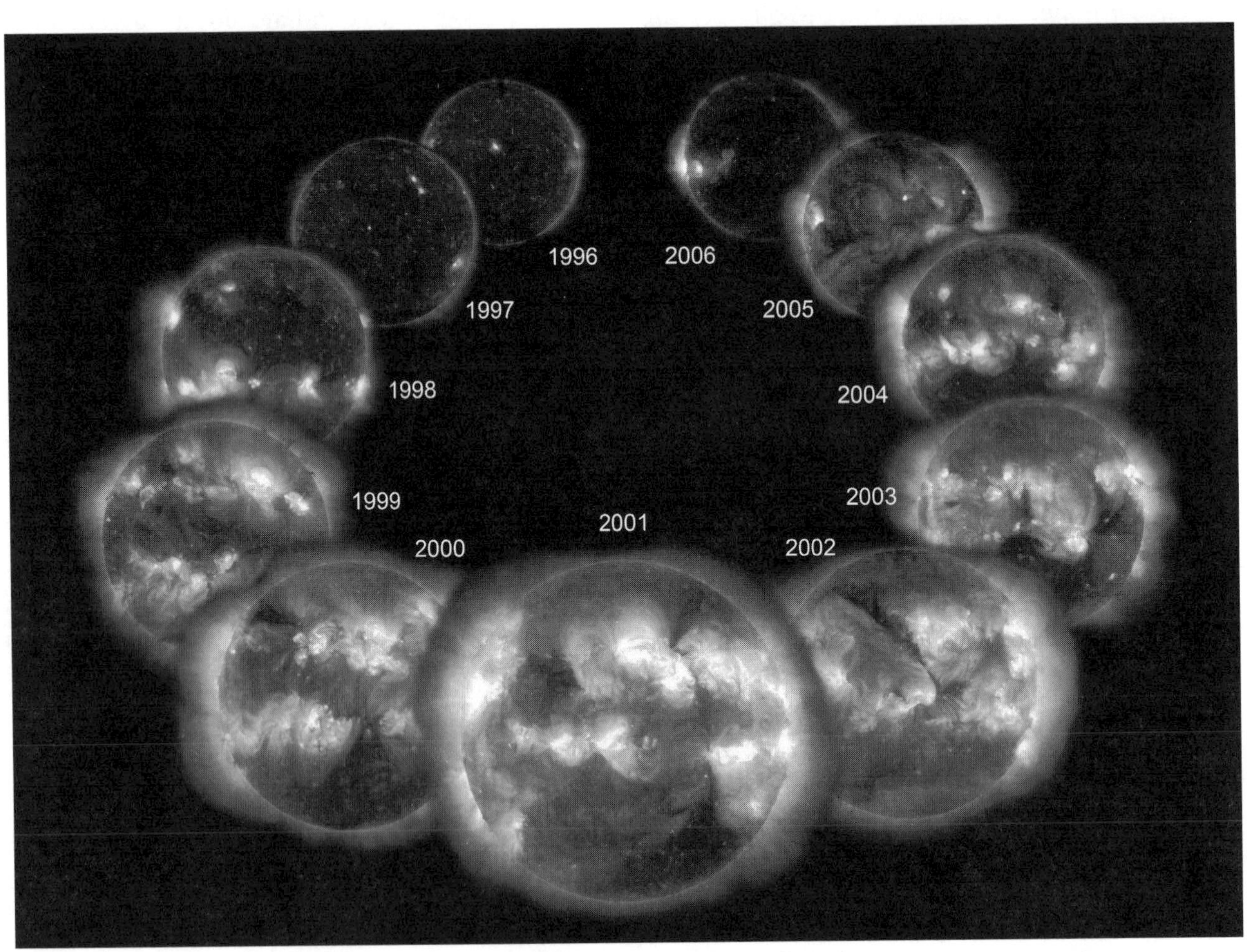

This composite by NASA shows that during periods of peak activity (front three images), sunspots, solar flares, and coronal mass ejections are more common and the sun emits slightly more energy than during periods of low activity (back images). The amount of sunlight energy that strikes Earth's atmosphere, called total solar irradiance (TSI), fluctuates by about 0.1 percent over the course of the sun's 11-year cycle, even though the soft X-ray wavelengths shown in this image vary by much greater amounts.

700 nm and separates the UV region of shorter wavelengths from the infrared region of longer wavelengths. A combination of waves results in white light. Red has the longest visible wavelength, whereas violet has the shortest. Waves longer than red are known as infrared, and waves shorter than violet are called UV.

The sun is the closest star to the Earth and the most closely studied. It is at the center of the solar system and accounts for about 99.8 percent of the mass of the solar system. The planets revolve around the sun. The sun is composed of hydrogen, helium, and other trace elements and goes around the center of the Milky Way galaxy at a distance of about 26,000 light years from the center of the galaxy. The amount of solar energy incident on the Earth's atmosphere is about 342 Watts per sq. m, based on the surface area of the Earth. Although the Earth's surface continuously radiates energy outward to space, only part of the surface area receives solar radiation at a time. Most of the solar energy incident on the Earth is in the UV region of shorter wavelengths. The sun is the source of heat that sustains life on Earth and controls the climate and weather. Only the sun's outer layers, which consist of the photosphere, chromosphere, and corona, can be observed directly. These three regions have different properties from one another, with regions of gradual transition between them. The sun has basically the same chemical elements as are present on the Earth. However, the sun is so hot that all of these elements exist in the gaseous state. Energy generated in the sun's core takes a million years to reach its surface. Solar energy is created deep within the core of the sun, where nuclear reactions take place.

Sunlight and the Danger of Skin Cancer

Every living thing exists because of the light from the sun. Sunlight is important in photosynthesis. For humans, UV light in small amounts is beneficial because it helps the body produce vitamin D from the UV region of sunlight. However, excessive exposure to sunlight is dangerous, as it can cause sunburns, skin cancer, and aging. UV light wavelengths are short enough to break the chemical bonds in skin tissue, and when the skin is exposed to sunlight, most skin will either burn or tan. The skin undergoes certain changes when exposed to UV light to protect itself against damage. The epidermis thickens, blocking UV light, and the melanocytes make increased amounts of melanin, which darkens the skin, resulting in a tan. Melanin absorbs the energy of UV light and prevents the light from penetrating deeper into the tissues. Sensitivity to sunlight varies according to the amount of melanin in the skin. Darker-skinned people have more melanin and therefore have greater protection against the sun's harmful effects. The amount of melanin present in a person's skin depends on heredity, as well as on the amount of recent sun exposure. Albinos have little or no melanin. The more sun exposure a person has, the higher the risk of skin cancers, including squamous cell carcinoma, basal cell carcinoma, and malignant melanoma. Actinic keratoses (solar keratoses) are precancerous growths, also caused by long-term sun exposure.

UV light, although invisible to the human eye, is the component of sunlight that has the greatest effect on human skin. Sunlight deficiency could increase blood cholesterol by allowing squalene metabolism to progress to cholesterol synthesis, rather than to vitamin D synthesis, as would occur with greater amounts of sunlight exposure. Larger amounts of UV light damage the body's DNA and alter the amounts and kinds of chemicals that the skin cells make.

UV light may also break down folic acid, sometimes resulting in a deficiency of that vitamin in fair-skinned people. UV light is classified into three types: UVA, UVB, and UVC, depending on its wavelength. Although UVA penetrates deeper into the skin, UVB is responsible for at least three-quarters of the damaging effects of UV light, including tanning, burning, premature skin aging, wrinkling, and skin cancer. The amount of UV light reaching the Earth's surface is increasing, especially in the northern latitudes. This increase is attributable to chemical reactions between ozone and chlorofluorocarbons that are depleting the protective ozone layer, creating a thinner atmosphere with some holes. The key to minimizing the damaging effects of the sun is avoiding further sun exposure. Damage that is already done is difficult to reverse.

Akan Bassey Williams
Covenant University

See Also: Atmospheric Absorption of Solar Radiation; Chemistry; Climate Change, Effects of; Health; Renewable Energy, Solar; Sunspots.

Further Readings

Hartmann, Thom. *The Last Hours of Ancient Sunlight: Waking up to Personal and Global Transformation*. New York: Three Rivers, 1999.

Ring, S. and Joseph Moran. *The Sun*. Mankato, MN: Coughlan Publishing, 2003.

Simon, Seymour. *The Sun*. New York: HarperCollins, 1989.

Sunspots

First viewed through a telescope by Galileo Galilei in 1610, sunspots are dark, active regions on the surface of the sun (the photosphere) that have a stronger magnetic field and cooler temperatures than the surrounding areas. The dynamo process (flow of solar plasma) within the exterior layers of the sun creates a magnetic field where convective motions transfer heat from the interior of the sun to its surface. As the magnetic field strengthens, the heat flow from the sun's interior is impeded, resulting in solar phenomena such as sunspots, solar flares, or coronal mass ejections. The regionally intense magnetic field inhibits convective heat flow, thereby reducing the temperature, energy emission, and brightness level, and giving sunspots their dark appearance. Fluctuations in solar irradiance and sunspot activity levels have been linked to global climate variability; however, much uncertainty exists regarding the mechanism and extent of their climate influence.

Sunspots are not uniformly dark (or cool), as most are characterized by two distinct regions, the umbra and penumbra. The umbra is the dark inner region of a sunspot with an average temperature around 6,700 degrees F (3,700 degrees C), about a third lower than the average photosphere temperature of 10,300 degrees F (5,700 degrees C). Based on Stefan-Boltzmann's law, the temperature difference between the umbra and photosphere means that the energy emitted from an average umbra region is only 20 percent of that emitted from the solar surface. Surrounding the umbra, the penumbra is a much warmer, lighter region with a fibrous edge that has temperatures somewhere between the umbra and solar surface. The average sunspot's size is comparable to the diameter of the Earth (about 8,000 mi. or 12,800 km) but some sunspots have been recorded with diameters 10 times larger.

The typical lifetime of a sunspot can be anywhere from a few days to a few months, but the average is about six days. The longest sunspot duration ever recorded was 134 days. The number of sunspots varies over a period of 11 years, going from a minimum to a maximum and back to a minimum in 22 years. This 11-year period is known as the solar cycle, or sunspot cycle, which is a periodic element of solar variation. Solar variation is the amount of change in radiation emitted by the sun and its spectral output over time. At the beginning of the solar cycle, sunspots begin to develop at midlatitude regions of the sun. However, they are transient (comprised of a leading spot and a trailing spot) and migrate toward the solar equator as the cycle continues. Sunspots typically emerge parallel to the solar equator (on either side) and appear in bipolar pairs with opposite directing magnetic fields.

Relationship to Solar Irradiance

The number of sunspots has a direct relationship to solar irradiance. Also known as the solar constant, solar irradiance is the average amount of incoming solar radiation per unit area received by the outer layer of Earth's atmosphere at a vertical angle. The solar constant varies on different time scales, but averages about 1,367 Watts per square meter (W/m^2). During a sunspot maximum, the solar disk contains the greatest amount of sunspots and an increased intensity of the sun's brightness due to additional magnetic energy emitted from the bright areas around the sunspots (faculae). Active sunspots produce more solar flares—great explosions on the solar surface—that produce geomagnetic storms on Earth from emitted X-rays and magnetic activity. Consequently, disturbances in power grids, radio communications, and satellite polarity increase during sunspot maximums. An increase in the northern and southern lights (Aurora borealis and australis) is also noticeable at these times.

Using a variety of techniques, sunspot cycles are actively examined for their potential influence on global climate change. Solar variability, including sunspot activity, has traditionally been analyzed using proxy data gathered from natural sources (e.g., tree rings, ice cores, and ocean sediments) or through surface-based telescopes. Satellites, outer-space-based telescopes (such as Hubble), and other technological advances enable solar activity to be routinely monitored by entities such as the National Oceanic and Atmospheric Administration Space Weather Prediction Center. These data are often put into climate models to identify the climate effects from solar variations. Solar radiation is found not to be significantly influenced by changes in the quantity of sunspots, but rather the change in their magnetic activity. The magnetic activity of sunspots, also varying during maximum and minimum periods, directly affects the amount of spectral irradiance received at the top of Earth's atmosphere.

Over a solar cycle, the amount of spectral irradiance, particularly involving ultraviolet (UV) radiation, may vary by more than 10 percent. These changes can influence the thinner, more sensitive layers of Earth's atmosphere, indicating potential climate change. However, Earth's climate is extremely complex due to the exchanges among its land, oceans, and atmosphere that makes it difficult to determine the climatic impacts from sunspot activity alone. Prolonged climate changes may result from sunspots but could also be influenced by concurring phenomena such as oceanic cycles (e.g., El Niño), volcanic eruptions, or anthropogenic causes.

Though it is certain that the sun influences the Earth's climate, there is still much debate as to the extent of sunspot effects on climate. Some evidence suggests that prolonged periods of below normal sunspot activity are associated with a cooler climate regime. During a period of low sunspot activity known as the Maunder Minimum (1645–1715), Europe and other world regions experienced a dramatic cooling period that significantly altered climate (such as a reduced growing season that caused widespread famine). The Maunder Minimum also occurred at the height of the Little Ice Age (c. 1550–1850), an extended episode of below-normal temperatures in the Northern Hemisphere. However, sunspot minimums may not be the only contributing factor in linking solar variability with climate. Solar irradiance changes (and ice ages) may be triggered by other astronomical causes such as the Milankovitch Cycles, a progression of changes in the Earth's eccentricity (orbit shape), obliquity (axial tilt), and procession (axis orientation).

Holly M. Widen
Jill S. M. Coleman
Ball State University

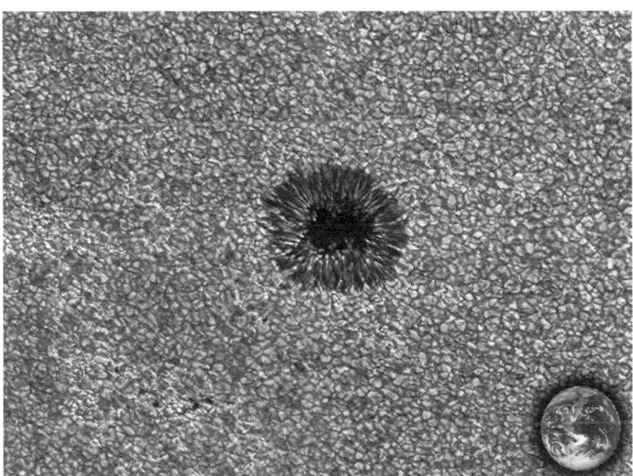

As seen in this solar telescope photo, the surrounding surface of a sunspot contains small, irregular granules that are the upwelled part of numerous local convection currents that carry hot hydrogen gas to the photosphere (Earth shown for scale).

See Also: Atmospheric Absorption of Solar Radiation; Bahrain; Climatic Data, Tree Ring Records; Little Ice Age; Milankovitch Cycles; Radiation, Absorption; Radiation, Ultraviolet; Radiative Feedbacks; Solar Wind; Sunlight.

Further Readings

Alexander, David. *The Sun*. Santa Barbara, CA: Greenwood/ABC-CLIO, 2009.

Balogh, A. A. and M. J. Thompson. "Introduction to Solar Magnetism: The Early Years." *Space Science Reviews*, v.144/1–4 (2008).

Hoyt, Douglas V. and Kenneth H. Schatten. *The Role of the Sun and Climate Change*. New York: Oxford University Press, 1997.

Jones, C. A., M. J. Thompson, and S. M. Tobias. "The Solar Dynamo." *Space Science Reviews*, v.152/1–4 (2010).

Suriname

The Republic of Suriname, a former Dutch colony, is the smallest sovereign state in South America. Most of the country's population of 490,000 people live along the coastal northern lowlands, in or near the capital of Paramaribo. Nearly the entire country—97 percent—is forested, although the timber industry is surprisingly small and most of the country is undeveloped. The country includes two main mountain ranges, the Bakhuys and the Van Asch Van Wijck. The undeveloped south includes numerous national parks and nature reserves. The Central Suriname Nature Reserve is a United Nations Educational, Scientific and Cultural Organization World Heritage Site, created in 1998 with Conservation International. It includes 6,178 sq. mi. (16,000 sq. km) of untouched tropical rainforest, both highland and lowland, and several granite domes.

Flooding and Disease

Suriname has a hot and humid tropical climate, with two long wet seasons from April to August and November to February. Rainfall, rising sea levels, and increased cyclonic activity are the major dangers it faces from global warming, as so much of the population is congregated along the coast. More severe or more frequent flooding could exacerbate the spread of insect-borne diseases like West Nile, dengue fever, and malaria. Greater amounts of rainfall could contribute to soil erosion, putting in jeopardy the country's small amount of arable land and the food security of the rural areas. In Paramaribo, higher temperatures and hotter summers would increase the amount of ground-level ozone, and severe summer highs would increase the prevalence of heat stroke, tax freshwater resources, and increase the load on the power grid as air conditioning attempts to compensate.

Suriname's economy is dependent on bauxite mining and exports, agriculture (employing 25 percent of workers; rice and bananas are the principal exports), tourism, and the country's gold and oil reserves. About two-thirds of the country's electricity comes from hydropower, with the remainder coming from fossil fuels. Liquid fuels account for 95 percent of the country's carbon emissions, which rose from 4.5 metric tons per person in 1990 to 4.78 metric tons in 2007, the most recent date for which information is available.

Bill Kte'pi
Independent Scholar

See Also: Alliance of Small Island States; Brazil; Guyana; Jamaica.

Further Readings

McCarthy, Terrence J. *A Journey Into Another World: Sojourn in Suriname*. Tucson: Wheatmark, 2010.

Westoll, Andrew. *The Riverbones: Stumbling After Eden in the Jungles of Suriname*. Toronto: Emblem Editions, 2008.

Sustainability

The language of sustainability emerged during the 1970s, though the concept was introduced as sustainable development in 1980 in the World Conservation Strategy and was popularized in 1987 by the World Commission on Environment and Development (also known as the Brundtland Commission, after its Norwegian chairperson Gro Harlem Brundtland). Today, there are numerous definitions of sustainability, but what is to be sustained? Is it the planet, particular environments, individual species, current lifestyles, certain rates of economic growth, a specific level of profit?

Sustainability, especially as constructed in mainstream definitions of sustainable development, is very similar to the concept of conservation espoused by American forester Gifford Pinchot in the late 19th century. Conservation emphasized using natural resources wisely, not depleting nonrenewable resources, ensuring that all American men received a fair share of the distribution of benefits, and that consideration be given to the needs of their descendents.

The Concept of Sustainable Development

Sustainable development globalizes the discourse. The World Commission on Environment and Development report in 1987 defined sustainable development as "development that meets the needs of the present without compromising the ability of future generations to meet their own needs." This definition is used today in many parts of the world by governments, businesses, environmental groups, and educators. The history of the concept, and the specific term, mean that it may be interpreted as a repackaging of environmental management. The managerial focus and faith in technological progress evident in this definition of sustainability mean that it is critiqued by more radical sustainability advocates.

In Australia, the term *ecologically sustainable development* emerged as a unique approach as a result of the power of major environmental groups in Australia in the early 1990s. In 1992, ecologically sustainable development was defined as "using, conserving and enhancing the community's resources so that ecological processes, on which life depends, are maintained and the total quality of life, now and in the future, can be increased." This terminology and definition, which arose as a result of the political power of environmental groups in the early 1990s in Australia, highlights the dependence of all life on ecological processes (e.g., thermodynamics, hydrological cycles, nutrient cycles).

The Australian definition leans toward what has been termed strong sustainability, meaning that humans should not be substituting human-made capital for natural capital. In contrast, weak sustainability advocates substitution, provided the total store of capital is not diminished. Critics of the weak sustainability approach point out that this is what has been happening for thousands of years, leading to the destruction of the environment. Other critics reject the notion of turning nature into "natural capital," and therefore do not engage in the strong versus weak sustainability debates.

Implementation: Working Out the Idea

The concept of sustainable development was the basis for a massive conference in Rio de Janeiro in 1992 that was chaired by Maurice Strong and attended by 178 governments, including 118 heads of state. The United Nations Conference on Environment and Development (UNCED, otherwise known as the Earth Summit) was the five-year follow-up to the release of the Brundtland Report. The conference attempted to move from debates about the notion of sustainability and sustainable development to working out how to implement this idea. The idea of expanding the global economy, although controversial, was accepted within sustainable development discourses because development was seen as necessary to overcome poverty. Sustainable development was intended to allow economic growth to continue, but to make this growth greener. Growth was seen as essential for developing countries and developed countries, so as to facilitate trade and help the poorer countries of the world. This concept of sustainability was compatible with that of the newly founded business organization, the World Business Council for Sustainable Development, which was influential in shaping the idea of sustainable development and how it would be implemented.

Implementation has been the focus of subsequent conferences in New York (1997) and Johannesburg (2002), and in the ongoing work of the United Nations Commission on Sustainable Development. Many countries, states/provinces, and local governments, as well as some businesses, have also introduced departments focused on implementing sustainability within their organization. Implementation is challenging because there are many barriers to implementing sustainable development. These include corporate cultures, countervailing market signals, and jurisdictional issues. Another issue is that although the temporal emphasis within the concept of sustainability is apparent, the spatial or geographical scale is unclear. This has led to various scales of analysis and implementation, including concepts such as sustainable lifestyles, sustainable cities, and sustainable regions.

The UNCED conference in 1992 produced five important documents, including Agenda 21, which was a 40-chapter document outlining the actions needed to implement sustainable development. Chapter 28 highlighted the important role of local government in implementing the concepts introduced at the global level. This led to the development of Local Agenda 21 (LA21). At the Johannesburg Conference in 2002, LA21 was relaunched as Local Action 21, which is the second decade of this program, containing a focus on action and implementation.

Oyster aquaculture in Tomales, California. In June 2011, the U.S. Department of Commerce and NOAA released national sustainable aquaculture policies in order to meet demands for healthy seafood, create jobs, and restore vital ecosystems.

Future Planning

There are also different ways of conceptualizing sustainable development vis-à-vis sustainability. Some authors present sustainability and sustainable futures as the goals to be reached by a process called sustainable development. Other authors maintain a distinction between sustainable development and ecological sustainability on the basis of their approach to existing structures and institutions. Sustainable development is seen as more of a reformist approach by advocates who primarily support the existing institutions but want them to be greener, whereas those activists and authors who emphasize sustainability or ecological sustainability often question the structures that perpetuate unsustainable practices.

Given the adoption of legislation related to sustainability, most governments are planning for sustainability. However, many of the differences in various concepts of sustainability can be attributed to the relative weight given to the economic, social, cultural, and environmental components of sustainability. The differences are also caused by the perception of how these components fit together.

There are two main approaches to conceptualizing sustainability, with numerous variations on these approaches. The dominant, mainstream representation of sustainable development that emerged from the Brundtland Commission, and that has been adopted by many governments and business groups throughout the world, is the balanced approach. Although the notion of balance has been largely discredited in scientific ecology, it is still a powerful metaphor within the environmental literature. In many models of sustainable development, balance is achieved by the construction of three circles of equal size to represent the economy, society, and environment. At the intersection of these three equal-sized circles is sustainable development.

In contrast, more radical advocates of sustainability may posit a hierarchical approach, in which the hierarchy may vary between models. It often includes ecological considerations at its base, followed by society (because there would be no society without an environment), and then a smaller economy (because there would be no economy without society). Variations may include the use of thermodynamic processes to support biochemical cycles that allow ecosystems to flourish, eventually reaching human social and individual scales.

Some environmental groups avoid using the term *sustainable development*, partly because of its perceived co-option. Other groups use the term *sustainability*, whereas some groups attempt to avoid this language altogether. The challenge for sustainability advocates is to be able to implement something that moves humankind and the rest of the planet away from a state of being unsustainable at a rate that is needed to avoid catastrophe.

Phil McManus
University of Sydney

See Also: Australia; Conservation; Culture; Norway; Resources; World Business Council for Sustainable Development.

Further Readings

Ecologically Sustainable Development Steering Committee. *National Strategy for Ecologically Sustainable Development.* Canberra: Australian Government Publishing Service, 1992.

Purvis, Martin and Alan Grainger, eds. *Exploring Sustainable Development: Geographical Perspectives.* London: Earthscan, 2004.

World Commission on Environment and Development. *Our Common Future.* Oxford: Oxford University Press, 1987.

Sverdrup, Harald Ulrik

Harald Ulrik Sverdrup (1888–1957) was a Norwegian meteorologist and oceanographer known for his studies of the physics, chemistry, and biology of the oceans and is considered the founding father of modern physical oceanography. Sverdrup explained the equatorial countercurrents and helped develop the method of predicting surf and breakers. A unit of water flow in the oceans was named after him by oceanographic researchers: 1 sverdrup (Sv) is equal to the transport of 1 million cu. m of water per second. The American Meteorological Society honored him with the Sverdrup Gold Medal, which recognizes researchers for outstanding contributions to the scientific knowledge of interactions between the oceans and the atmosphere.

Sverdrup was born on November 15, 1888, in Sogndal, Sogn, Norway, into an ancient and respected family of university lecturers, lawyers, politicians, and Lutheran ministers. His father Johan was a teacher and, following the family tradition, became a Lutheran minister of the State Church of Norway. In 1894, his father became minister in the island district of Solund, about 40 mi. (64 km) north of Bergen, and then moved to Rennsö near Stavanger. In 1908, he became professor of church history in Oslo. Because of his father's different jobs, Sverdrup spent much of his boyhood in various sites in western Norway and was taught by governesses until he was 14 years old. At that age, he went to school in Stavanger. During his adolescence, Sverdrup experienced conflicts between his interest in natural science and the religious background of his family. It was particularly difficult for him to reconcile the concept of evolution with his religious upbringing.

Because he was not aware of the possibility to study science at university, he first opted for the classical curriculum in 1903. Within this field, his major interest became astronomy. Sverdrup left the gymnasium with honors and spent a year in Oslo preparing for university preliminary examinations. Military service was compulsory at the time, so he decided to combine it with his scientific education, enrolling at the Norwegian Academy of War. This training was combined with the study of physics and mathematics. The physical training that he received while at the academy was extremely useful for his survival during his later long arctic expeditions.

When Sverdrup entered university, he decided to major in astronomy. In 1911, he was offered an assistantship with Professor Vilhelm Bjerknes, the preeminent Norwegian meteorologist and founder of the Bergen School, which allowed him to enter one of the brightest scientific circles in the country. The Bergen School was supported by an annual grant that Bjerknes received from the Carnegie Institution of Washington, almost from its founding. Sverdrup initially planned to continue his research in astronomy, but he became increasingly interested in meteorology and oceanography, and thus changed his major. When, in 1912, Bjerknes went to Germany to work at the University of Leipzig as professor and director of the new Geophysical Institute, Sverdrup followed him, remaining in Germany from January 1913 to August 1917. He also continued his thesis for the University of Oslo and received his doctorate in June 1917 on a published paper on the North Atlantic trade winds.

Career Highlights

In July 1918, Sverdrup joined Roal Amundsen's expedition in the Arctic, on the *Maud*, as a chief scientist. Although the planned duration of the expedition was from three to four years, it lasted for seven and a half years. Sverdrup did not return to Norway until December 22, 1925.

He was enthusiastic about the experience and defined the most interesting period as the eight months between 1919 and 1920 spent in Siberia, living with nomadic reindeer herders, the Chukchi. Sverdrup's Arctic expedition was interrupted for six months between 1921 and 1922, when the meteorologist had the chance of spending a profitable period of time at the Carnegie Institution. From the Arctic expedition, Sverdrup gained better understanding of the basic physical oceanography of currents. He argued that the effect of the Earth's rotation, a fundamental aspect of the dynamics of the oceans, is best observed in the polar regions, because it reaches its greatest level there. On his return to Norway, Sverdrup married Gudrun Bronn Vaumund.

By 1926, Sverdrup was a well-established scientific researcher and was offered the chair of meteorology at Bergen, which had been previously held by Bjerknes. The Carnegie Institution had also offered Sverdrup a permanent position twice, but the Norwegian scientist refused the American offer and took up the position at Bergen. In this capacity, Sverdrup edited the scientific report of the *Maud*. He also continued to collaborate with the Carnegie Institution. In 1931, he led the scientific group in the Wilkins-Ellsworth North Polar Submarine Expedition, during which valuable information was gathered, despite its failure to achieve the chief goal of the expedition, the submarine exploration of the Arctic in the *Nautilus*.

In 1936, Sverdrup accepted the position of director of the Scripps Institute of Oceanography in La Jolla, at the University of California, remaining there for almost 12 years. During his tenure as director, Sverdrup expanded the Scripps Institute, making it an institute with a research program, and developing closer ties between Scripps and the University of California, Los Angeles. During World War II, Sverdrup was involved in the U.S. war effort, although he did not directly work for the University of California Division of War Research. He worked on problems related to forecasting surf conditions for military beachhead assaults. His current and wave forecasting methods were applied by military weathermen to predict landing conditions for Allied invasions.

Sverdrup was a central figure in the postwar development of oceanography and allied sciences. He served on many scientific committees after the war, and his contributions to science were increasingly recognized. He was elected to the National Academy of Sciences in 1945. He joined the Executive Committee of the American Geophysical Union in 1945 and presided over the American Geophysical Union Oceanography Section. In 1946, he became president of the International Association of Physical Oceanography. He chaired the Division of Oceanography and Meteorology at the 1946 Pacific Science Conference. Sverdrup returned to Norway in 1948, where he worked as a professor of geophysics at the University of Oslo until his death on August 21, 1957.

Luca Prono
Independent Scholar

See Also: History of Climatology; Oceanography.

Further Readings

Nierenberg, William A. "Harald Ulrik Sverdrup." In *Biographical Memoirs of the National Academy of Sciences*, v.69 (1996).

Sverdrup, Harald Ulrik. *Among the Tundra People*. San Diego, CA: University of California, San Diego, 1994.

Swaziland

The Kingdom of Swaziland is landlocked between its neighbors, South Africa and Mozambique. It has a land area of 6,704 sq. mi. (17,363 sq. km), a population of 1.18 million (2009 est.), and a population density of 176.6 people per sq. mi. (68.2 people per sq. km). About 24.1 percent of the population live in urban areas. Some 11 percent of the land is arable, with much of it used for subsistence farming, and also for growing maize, cotton, rice, sugarcane, and citrus fruits. In addition, 62 percent of the country is used as meadows or pasture for low-intensity grazing of cattle, sheep, and goats. About 6 percent of the country is forested, and there is a significant logging industry. Sugar and wood pulp are the country's major exports.

Because the country is largely undeveloped, there is a relatively low use of electricity, with a significant component used for heating in winter. Electricity production in Swaziland comes from fossil fuels (55.8 percent) and hydropower (44.2 percent), with most of it imported from South Africa. In terms of its carbon dioxide (CO_2) emissions, Swaziland's per-capita emissions were 0.5 metric ton in 1990, dipping to 0.1 metric ton in 1993, but rising steadily to 0.92 metric ton per person by 2003, which the country maintained up to 2007. All the CO_2 emissions in Swaziland are attributed to the use of solid fuels, with most of the electricity generation, as well as residential and business heating, fueled by coal or wood. This has resulted in significant per-capita emissions of carbon monoxide. In an attempt to reduce this, the Maguga Hydro Power Station began construction in 2004 and was officially opened in May 2011 by King Mswati III. When it reaches its peak, it is expected to produce all the electricity required for Swaziland, although the country already has hydropower stations at Ezulwini, Dwaleni, and Mbabane. At that point, Swaziland should be able to export electricity to South Africa, earning crucial income for the Swzai government.

As a result of global warming and climate change, Swaziland has seen the effects of water shortages for some of its crops, such as rice, and desertification of some areas previously used for farming. There have been a number of droughts, and some farmers have had major difficulties in growing enough crops, leading to hardships in rural areas. The spring and summer of 2007 was particularly bad, with the country experiencing a major drought. This led the World Food Programme to calculate that 400,000 of the population in that year needed some form of food aid, which was twice the level of the previous year. Life expectancy in the country is very low, at 41.4 for females and 40.4 for males.

Other effects on the climate have seen cooler winters and some violent storms, with Manzini in central Swaziland experiencing its first tornado in 2005. Even during the rainy season in September and October, the amount of precipitation has lessened, leading to further problems. This has been reflected in more problems for the agricultural sector, resulting in a significant fall in gross domestic product per capita at a time when the country has been reeling from its increased exposure to the human immunodeficiency virus and acquired immune deficiency syndrome (HIV/AIDS) pandemic. The Swaziland government took part in the United Nations Framework Convention on Climate Change (UNFCCC), signed in Rio de Janeiro in May 1992, and signed the Vienna Convention in the same year. On January 13, 2006, the country accepted the Kyoto Protocol to the UNFCCC, the 155th country in the world to do so, with it coming into force on April 13, 2006.

Justin Corfield
Geelong Grammar School
Robin S. Corfield
Independent Scholar

See Also: Climate Change, Effects; Desertification; Disease; Drought; Lesotho.

Further Readings

EarthTrends. "Swaziland: Climate and Atmosphere." http://earthtrends.wri.org/text/climate-atmosphere/country-profile-172.html (Accessed October 2011).

Gamedze, Mduduzi. "Climate Change Vulnerability and Adaption Assessments in Swaziland." http://unfccc.int/files/adaptation/adverse_effects_and_response_measures_art_48/application/pdf/200609_swaziland_iva_abstract.pdf (Accessed July 2011).

IRIN. "Swaziland: Facing Climate Change." http://www.irinnews.org/report.aspx?reportid=73337 (Accessed July 2011).

Meadows, Michael E. and Timm M. Hoffman. "Land Degradation and Climate Change in South Africa." *Geographical Journal*, v.169/2 (2003).

Sweden

A northern European country bordering the Baltic Sea, Sweden has made climate change mitigation a top political priority. The Swedish government has made strong political commitments, both short term and long term, to limit the impact and in order to adapt to climate change through heavy investments of nearly 5 billion Swedish crowns (€549 million) between 2009 and 2011. At the European Union (EU) level, Sweden is committed to the joint emissions reduction of 8 percent by 2010 and a minimum reduction of 20 percent by

2020, 30 percent within a global climate change agreement framework. It is expected that Sweden will meet or exceed its Kyoto Protocol targets, as well as European reduction levels.

Although a relatively sparsely populated country of only 9 million people, Sweden's gross domestic product ranks 33rd in the world, and the country's economic growth has been significantly higher than the European average from 1999 to 2009. The economy is predominantly based in the service sector, and Sweden has a large natural resource base of timber, hydropower, and iron ore. Sweden currently relies on hydropower and nuclear power as primary sources of energy; most of the national greenhouse gas (GHG) emissions come from the transportation sector, followed by manufacturing and construction, and electricity and heating. Although the government had previously planned to phase out nuclear power plants, in 2009, it announced that existing reactors would be replaced. Energy demand has been relatively stable since the 1990s, and Sweden has been heavily investing in renewable energy in order to increase its share of energy production.

Sweden and Scandinavia are expected to experience a temperature rise higher than the global average as a result of climate change, and precipitation patterns will also change. Recent models predict that plant and animal life will be especially sensitive to the expected changes in Sweden's climate. As a result of climate change, the winters are expected to be milder and a great risk of flooding in lakes and watercourses is expected as precipitation increases in all seasons except summer, when the climate will be warmer and drier.

In addition to its international commitments, the Swedish government is implementing a package of national initiatives for climate change mitigation and adaptation. The Swedish government has set the goal of becoming the world's first oil-free country by 2020, introduced instruments in the sectors of energy and transportation, and tightened environmental and tax policies since the early 1990s. Beyond its participation in the EU Emissions Trading Scheme, Sweden has set a target of reducing GHG emissions by 40 percent in 2020, or around 20 million tons of carbon dioxide (CO_2) equivalents. National GHG emissions have been declining, from 74.6 million tons of CO_2 equivalent in 1980 to 54 million tons in 1990 to a current level of 47.7 million tons in 2007. This declining trend in GHG emissions has primarily been driven by the use of renewable energy technology and development of sustainable transportation systems. Currently, Sweden is expected to achieve an emissions reduction of 11.5 million tons per year by 2020, in addition to meeting the EU targets for increasing energy efficiency by 20 percent and providing 20 percent of the energy supply from renewable energy sources.

The national climate strategy relies heavily on the use of economic instruments, regulatory legislation, and the setting of targeted initiatives such as the procurement of technology and differentiated investment grants. Since 2002, a series of more targeted measures have been enacted by the Swedish government, the largest of which is an increased CO_2 tax and a series of climate awareness initiatives and subsidies for special investments. Other instruments, such as the establishment of frameworks for development, are set to ensure the maximum development of initiatives such as district heating, carbon-free electricity generation, and the expansion of sustainable mobility networks. The main focus for domestic reductions are targeted at housing, waste management, agriculture and forestry, and other areas of industry, and additional measures developed so far also include green investments in developing and other foreign countries. In Sweden, municipalities are primarily held responsible for enforcing environmental regula-

On September 18, 2009, the world's first BioDME production plant was built at the Smurfit Kappa paper mill in Piteå, Sweden. It has a capacity of about 1,600 gal. per day using forest residue feedstocks. BioDME/methanol is produced from biomass.

tions at the local level. The three action plans in place include the following targets: a 50 percent supply of energy from renewable energy sources by 2020; a vehicle fleet independent of fossil fuel energy by 2030; zero net GHG emissions by 2050; and a 10 percent of renewable energy use in the transportation sector.

Internationally, Sweden is investing in emissions reductions in developing countries in Africa, Asia, eastern Europe, and Latin America. One-third of the target of 40 percent emissions reductions have taken place abroad through the Clean Development Mechanism (CDM) and Joint Implementation (JI) programs as part of the United Nations Kyoto Protocol agreement. In addition, Sweden is a party to the international agreements on climate change and is leading the International Commission on Climate Change and Development.

Cary Yungmee Hendrickson
University of Rome

See Also: Carbon Tax; Clean Development Mechanism; Joint Implementation; United Nations Framework Convention on Climate Change.

Further Readings

European Biofuels. "BioDME/Methanol." http://www.biofuelstp.eu/methanol.html (Accessed January 2012).

European Environmental Agency. "Distinguishing Factors (Sweden)." http://www.eea.europa.eu/soer/countries/se (Accessed May 2011).

Swedish Commission on Climate and Vulnerability. "Sweden Facing Climate Change" (2007). http://www.sweden.gov.se/sb/d/574/a/96002 (Accessed May 2011).

Swedish Environmental Protection Agency. "Sweden's Climate Policy." http://www.naturvardsverket.se (Accessed May 2011).

Switzerland

Switzerland has a long-standing tradition of environmental awareness and protection and has been active in bringing the debate on climatic change to the forefront of international environmental affairs. Switzerland has been a strong supporter of the Intergovernmental Panel on Climate Change since its inception in 1988 and an important negotiator for the United Nations (UN) Framework Convention on Climate Change and its series of conferences of the parties. Climate research is high on the agenda of Swiss academia, with recognized expertise in paleoclimate reconstructions, climate modeling, and impacts studies.

The keen awareness of Switzerland to climatic change is the result of many climate-driven changes in the Alpine environment that are already perceptible, such as the retreat of mountain glaciers, permafrost degradation at high elevations, subtle changes in vegetation, and increased costs resulting from extreme events. Swiss climate is a patchwork of local regimes resulting from the competing influences of contrasting climates that converge upon the region (the Mediterranean, continental, Atlantic, and polar systems) and the influence of topography on local flows, temperature, and precipitation. Temperatures have risen by up to 3.6 degrees F (2 degrees C) in many parts of Switzerland since 1900—well above the global average 20th-century warming of about 1.2 degrees F (0.7 degree C).

Future climatic change in the Alps will be a complex aggregate of decadal- to century-scale forcing factors related to the North Atlantic Oscillation, the Atlantic Multidecadal Oscillation, and the anthropogenic greenhouse effect. According to the future pathways of greenhouse gas (GHG) emissions by 2100, regional climate models suggest that by then, Swiss winters could warm by up to 9 degrees F (5 degrees C) and summers by over 12.6 degrees F (7 degrees C); in parallel, precipitation is projected to increase in winter and sharply decrease in summer. Strong heat waves similar to the 2003 European event are likely to become the norm by 2100, and both drought and intense precipitation are projected to increasingly affect the country.

Impacts From Climate Change

Climatic change will have significant impacts on both the natural environment and socioeconomic activities in all parts of the country. Alpine glaciers may lose between 50 and 90 percent of their current volume, and the average snow line will rise by 492 ft. (150 m) for each degree of warming. Hydrological systems will respond in quantity and seasonality to changing precipitation patterns and to the timing of snow melt in the Alps, with a greater flood potential in spring and drought potential in summer and fall. More extreme events will trigger frequent slope instabilities, and at high elevations, melting permafrost will compound these problems. The distribution of natural vegetation will change as plants seek new habitats with similar climatic conditions to those of today. A rapidly warming climate will result in a loss of mountain biodiversity, as not all species will be capable of adapting to the speed and amplitude of projected change.

The direct and indirect effects of a warming climate will affect important economic sectors such as winter tourism, hydropower, agriculture, and the insurance industry, which will be confronted with more frequent natural disasters. With changing water resources will come increasing rivalries between economic sectors competing for a dwindling resource, or a resource that may no longer be available at critical times of the year; for example, between the energy and agricultural sectors. Climate-related health risks (allergies, pollution) are expected to increase, with consequent economic effects resulting from prolonged morbidity and absenteeism.

Swiss climate policy is currently aligned upon the European Union policy that aims for an upper limit of global warming of 3.6 degrees F (2 degrees C) over pre-industrial values. In early 2011, the Swiss parliament approved the principle of cutting back domestic carbon emissions by 20 percent as a first step toward a post-Kyoto framework for GHG abatement measures. While a 3.6 degrees F (2 degrees C) cap on global warming may be considered a laudable policy option, such a global increase is likely to translate into a greater level of change in the Alps, as has been observed over the past 150 years. As a consequence, the severity of impacts on sectors such as health, water, agriculture, ecosystem services, and energy may not necessarily be alleviated in view of the expected regional changes in mean and extreme climates.

Martin Beniston
University of Geneva

See Also: Austria; Climate Change, Effects of; Liechtenstein; Public Awareness.

Further Readings

Beniston, Martin. *Climatic Change and Impacts: A Focus on Switzerland*. New York: Kluwer Academic/Springer, 2004.

Syria

Syria is a Middle Eastern country sharing borders with Turkey, Iraq, Jordan, Lebanon, and Israel. Syria has 111 mi. (193 km) of coastline on the Mediterranean Sea and is crossed by the Euphrates River. The terrain is mostly semi-arid and desert plateau, with a narrow coastal plain and mountains in the west. The climate is mostly desert with hot, dry summers and mild, rainy winters along the coast. About one-fourth of the land is arable, and agriculture uses about 90 percent of the total water in Syria, of which about 47 percent is surface water shared with neighboring countries from the Tigris, Euphrates, and Orontes rivers. Desertification and rising salt levels already threaten the productivity of some of this land. Per-capita gross domestic product in 2010 was $4,800, ranking 153rd in the world, with 11.9 percent of the populating living below the poverty line (2006 est.) and an unemployment rate in 2010 of 8.3 percent.

Syria is the only significant oil producer in the eastern Mediterranean and occupies a key geographic position with regard to energy transit. In 2010, Syria had 2.5 billion barrels of oil reserves, mainly near the Iraq border and along the Euphrates River; and 8.5 trillion cu. ft. of natural gas reserves, mainly in the central and eastern areas of the country. Oil production in Syria has declined for years since the 1990s (peak production was 583,000 barrels per day in 1996), recently stabiling to 400,000 barrels per day (2009), but the country is currently a net importer of oil and petroleum products. Syria's refining capacity as of July 2010 was 240,000 barrels per day, with studies underway to examine the feasibility of adding a new refinery with a capacity of 100,000 barrels per day. Natural gas production in 2009 was 213 billion cu. ft., which is expected to double by 2010. The Arab Gas Pipeline links Egypt with Jordan, Syria, and Lebanon, and is expected to expand in the future to include Turkey, Iraq, and Iran. Electricity production in 2007 was 36.5 billion kilowatt hours (kWh), ranking 57th in the world, and electricity consumption was 27.35 kWh.

Syria has achieved considerable economic growth in recent years, averaging more than 5 percent annually from 2004 to 2008. However, climate change threatens future growth. Although food security has improved steadily from the 1970s, it is lower than the international average, and rising food prices due to diminished productivity from increased temperatures and decreased rainfall caused by global climate change, as well as rising oil prices (which would increase the cost of fertilizer as well as fuel for mechanized agriculture) would threaten these gains. Urban households would suffer more than rural one from higher commodity prices, and the rural nonfarm population (including Syria's Bedouin population) would be hardest hit of all.

Syria already suffers from recurring droughts, a problem expected to increase under a global warming scenario. Droughts lower or eliminate agricultural yield and affect the national economy by causing a rise in food prices, which has the effect of increasing poverty, even among the nonfarm population. A 2009 report from the International Institute for Sustainable Development found that some 160 Syrian villages were abandoned in 2007 and 2008 due to climate change, primarily linked to water scarcity, and predicted that struggles for access to water resources could lead to armed conflict in the Middle East. The report predicts hotter, drier, and less predictable climate in the Middle East, leading to food insecurity, water shortages, and large-scale population movements.

Sarah Boslaugh
Kennesaw State University

See Also: Agriculture; Desertification; Food Production.

Further Readings

Breisinger, Clemens, et al. "Global and Local Economic Impacts of Climate Change in Syria and Options for Adaptation." International Food Policy Research Institute (June 2011). http://www.ifpri.org/sites/default/files/publications/ifpridp01091.pdf (Accessed July 2011).

El-Atrache, Talal. "160 Syrian Villages Deserted Due to Climate Change.'" Agence France–Presse (June 2, 2009). http://www.google.com/hostednews/afp/article/ALeqM5jXbS8a3ggiMm4ekludBbmWQMb-HQ (Accessed July 2011).

U.S. Energy Information Administration. "Country Analysis Brief: Syria" (June 2010). http://www.eia.gov/countries/cab.cfm?fips=SY (Accessed July 2011).

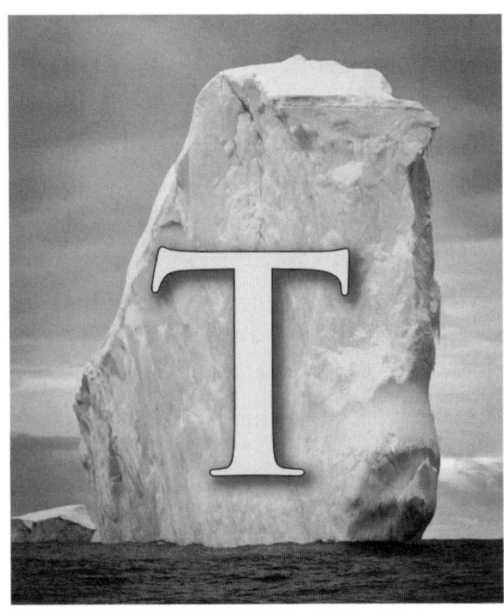

Tajikistan

Tajikistan is a landlocked country in Asia that was part of the Soviet Union from 1925, achieving independence in 1991. The climate in most of the country is continental, with mild winters and hot summers, and semiarid to polar climate in the Pamir Mountains in the southeast. About 6.5 percent of the land is arable, and agriculture occupies about half the labor force and contributes 19.2 percent of the gross domestic product (GDP). The primary crop is cotton. The per-capita GDP in 2010 was $2,000, among the lowest among the former Soviet republics; in 2009, 53 percent of the population lived below the poverty line. A high number of Tajik citizens (some estimate 1 million, out of a total population of 7.6 million) work abroad, primarily in Russia, and remit some of their earnings home.

Almost three-quarters of Tajik citizens live in rural areas; many of them have little to no access to an adequate energy supply. Most (95 percent) of Tajikistan power generation comes from hydroelectricity. The Sangtuda I hydropower dam was completed in 2009 and the Sangtuda II dam is scheduled for completion in 2012, while another dam currently in the planning stages, the Roghun dam, will be the tallest in the world if completed. However, the distribution infrastructure is old, and decreased water flow in the winter months means that the plants operate at as little as 30 percent of capacity during these times, resulting in regular energy crises in the winter. In rural areas, many citizens resort to burning biomass and fossil fuels in the absence of a reliable energy supply, leading to loss of forest cover, soil degradation, and deterioration of interior air quality.

Sharp Decline in Energy Production

Oil production in Tajikistan has declined sharply over the past three decades, from 1.31 thousand barrels per day in 1992 to 0.22 thousand barrels per day in 2009. Natural gas production has also declined, from 3.53 billion cu. ft. in 1992 to 1.342 billion cu. ft. in 2009. Tajikistan is a net importer of both petroleum and natural gas, importing 38.78 thousand barrels per day of petroleum and 7 billion cu. ft. of natural gas in 2009. Tajikistan has substantial coal deposits, but production and consumption of coal has declined sharply since 1990. Electricity production in 2007 was 0.17 quadrillion Btu (ranking 85th in the world) and consumption was 0.28 quadrillion Btu (also 85th worldwide). Carbon dioxide (CO_2) emissions from consumption of fossil fuels was 6.89 million metric tons in 2008, ranking 113th in the world.

A recent report indicated that even where the energy supply in Tajikistan is adequate, energy is often used inefficiently, so that Tajikistan uses twice as much energy to produce a unit of GDP compared to the average for the world, and three times as much as developed countries. To address this, in 2011, Tajikistan crafted an Energy Efficiency Master Plan that regards energy efficiency as a source of energy supply and as a new basis of energy strategies and policies. Tajikistan also has high potential for developing other renewable sources of energy, including solar, wind, and biogas.

Tajikistan enjoys 280–330 sunny days annually, with intensity of solar radiation ranging from 280 to 925 megajoules per sq. m in Piedmont regions and from 360 to 1120 megajoules per sq. m in the highlands. Expert estimates are that Tajikistan could provide 10–20 percent of its energy from solar sources, with each solar station avoiding 0.30–0.35 ton of CO_2 emissions annually. Wind energy could prove a useful supplement to hydropower in some regions of Tajikistan, with the strongest winds in the highland regions, where mean annual wind speed is in the range of 5–6 m per second. A few biogas generators exist in Tajikistan, and the most promising sources of biogas are anaerobic fermentation of manure and biomass conversion of cotton residues. Hydropower is already the primary source of electricity generation in the country, but distribution could be improved by development of small hydropower plants in mountain regions.

Sarah Boslaugh
Kennesaw State University

See Also: Energy Efficiency; Renewable Energy, Solar; Renewable Energy, Wind.

Further Readings

Novikov, Victor, George Petrov, and Umed Karimov. "Use of Renewable Energy in Tajikistan" (June 2003). http://www. Iiasa.ac.at/Research/ECS/IEW 20003/Abstracts/2003A_novikov.pdf (Accessed July 2011).

United Nations Development Programme. "Tajikistan." http://www.undp.fj/index.php?option =com_content&task=view&id=429&itemid=129 (Accessed July 2011).

U.S. Energy Information Administration. "Country Information: Tajikistan" (June 30, 2010). http:// www.eia.gov/countries/country-data.cfm?fips=TI (Accessed July 2011).

Tanzania

Tanzania is one of the largest and poorest countries in east Africa. It is located between 1 degree and 11 degrees south of the equator, and largely experiences a tropical climate. It is projected that Tanzania will face enormous socioeconomic challenges resulting from global climate change. This is because many sectors of its economy, including agriculture and hydroelectric power generation, are water dependent. Additionally, an increase in global climate temperature is expected to lead to sea-level rise, thus inundating low-lying coastal towns and areas. Another possible effect of global climate change on Tanzania will be an increase in drug-resistant strains of human and animal diseases. How Tanzania will be able to adapt and cope with these inevitable climate changes is vital to the future of the country.

Rain: The Vital Ingredient

Tanzania's economy is heavily dependent on rainfed agriculture, making it susceptible to climate variability. Furthermore, climate fluctuations are anticipated to affect the country's main food crops, which include maize, beans, and millet. Additionally, rising global temperatures would equally affect its cash-crop production of coffee and cotton. A short-term approach for mitigating the effects of climate change on food supply would be through agricultural extensification, but increasing human populations makes it unsustainable. Also, agricultural extensification, particularly in the rangelands and forestlands, would exacerbate droughts, torrential rains, and floods. Kenya, Tanzania's neighbor to the north, has already recorded falling crop yields, partially attributed to global climate change.

Water resources would also be affected by climate variability, despite the important role it plays on the country's domestic and economic interests. For example, an estimated 90 percent of Tanzania's electricity comes from hydroelectric power, which is dependent on the country's major rivers. Furthermore, domestic water supplies would be greatly affected by climate change. Rural communities without piped water would have to travel longer distances to get to water sources. Of major concern are women and children, who would be directly impacted because of

Lions in Tanzania stay close to their water source. Lion attacks have increased in rural Tanzania, the increase mirroring the dramatic rise in population, which grew by nearly 50 percent between 1988 and 2002 and encroached on the lions' habitat. Water resources would also be affected by climate variability, creating competition for water amid the growing population and threatening agriculture and hydropower. Rural communities without piped water would have to travel longer distances to get to water sources.

their role as the main water providers in rural areas. Although water availability varies regionally, the Mount Kilimanjaro water catchment area will continue to lose its ice caps and glacier, consequently affecting water availability in the region. Substantial reduction in water for domestic and agricultural use would directly result in increased human–wildlife conflicts, mainly in areas within and surrounding the country's major national parks, such as the Serengeti. An increase in global climate change would also lead to sea-level rise, inundating and destroying historical monuments, cultural artifacts, towns, and cities, particularly the city of Dar es Salaam and the islands of Zanzibar and Pemba. Similarly, mangroves and coral reefs, the primary breeding grounds for fish, would be extensively destroyed, therefore affecting the country's fishing industry. Disease incidences, including malaria and sleeping sickness, are expected to rise.

Effects of global climate change in Tanzania are likely to be exacerbated because the country is poor compared to other nations. In the meantime, Tanzania faces a dilemma on appropriate approaches for mitigating anticipated climate changes. Crop diversification, adoption of crop varieties that are not rainfall dependent, and adoption of new livestock varieties could help to ensure that Tanzania is self-reliant and self-sufficient. In addition, building more dams, particularly in arid and semiarid lands, would also ensure that water is available for domestic and agricultural use during dry seasons. Tanzania also needs to embrace and adopt clean energy sources, such as solar power and wind energy, in order to reduce its overdependency on hydroelectric power. It is also important to undertake nationwide reforestation projects, mainly in deforested areas, to reduce flooding and retain water in the soil.

Peter Kimosop
Bowling Green State University

See Also: Burundi; Kenya; Reducing Emissions From Deforestation and Forest Degradation (UN-REDD); Rwanda; Somalia; Stockholm Environment Institute.

Further Readings

Agrawala, S., et al. *Development and Climate Change in Tanzania: Focus on Mount Kilimanjaro*. Paris: Environment Directorate and Development Co-operation Directorate, 2003.

Hulme, L., et al. "African Climate Change: 1900–2100." *Climate Research*, v.17 (2001).

Maganga, F., et al. "Domestic Water Supply, Competition for Water Resources and IWRM in Tanzania: A Review and Discussion Paper." *Physics and Chemistry of the Earth*, v.27 (2002).

Mortimore, M. and W. Adams. "Farmer Adaptation and 'Crisis' in the Sahel." *Global Environmental Change*, v.11 (2001).

Paavola, J. "Livelihoods, Vulnerability, and Adaptation to Climate Change in Morogoro, Tanzania." *Environmental Science & Policy*, v.11 (2008).

Reardon, T. and S. Vosti. "Links Between Rural Poverty and the Environment in Developing Countries: Asset Categories and Investment Poverty." *World Development*, v.23/9 (1995).

Tata Energy Research Institute

The Energy and Resources Institute (TERI) began in 1974 as the Tata Energy Research Institute. Motivated by concerns about finite, nonrenewable energy resources and pollution, Darbari Seth, a chemical engineer working for Tata Chemicals, proposed a research institute dedicated to the collection and dissemination of information regarding energy production and utilization. Chairman R. J. D. Tata actively supported the institute, and TERI was formally established in Delhi in 1974. By 1982, TERI expanded to include research activities in the fields of energy, environment, and sustainable development. As the scope of activities continued to widen, TERI maintained its acronym, but was renamed the Energy and Resources Institute in 2003.

TERI relies on entrepreneurial skills to create benefits within and beyond India. TERI's staff of over 700 conducts research and provides professional support to governments, institutions, and corporations worldwide. They currently operate through divisions: Earth Sciences and Climate Change, Resources and Global Security, Sustainable Development Outreach, Social Transformation and Youth Services, Water Resources, Energy–Environment Policy, Energy–Environment Technology Development, Environmental and Industrial Biotechnology, Biotechnology and Management of Bioresources, Industrial Energy Efficiency, Information Technology and Services, Regulatory Studies and Governance, Human Resources, Technology Dissemination and Enterprise Development, and Decentralized Energy Solutions. Each division has a mandate that links to other divisions, and each division also has subsidiary components. For example, the Earth Sciences and Climate Change Division focuses on cutting-edge climate change research and sustainable development to inform policymaking for India and internationally. To this end, this division includes the Center for Global and Environmental Research (CGER), Center for Environmental Studies (CES), and the Forestry and Biodiversity (F&B) areas.

Examples of TERI-led research projects vary widely and include basic and applied research to study water and climate change issues; village-scale solar energy systems; microbial bioremediation of oil spills and oil sludge deposits; design and dissemination of biomass gasifiers; wasteland reclamation and biodiesel production through *jatropha curcas*, an inedible, oil-bearing crop; e-waste recycling; and design and implementation of green buildings and ecovillages.

The Asian Development Bank (ADB) declared TERI a clean energy knowledge hub in 2006. As evidence of its dedication to clean energy, TERI established the Green Rating for Integrated Habitat Assessment (TERI–GRIHA), the first of its kind in India. Other work in this area includes research and training initiatives to advance large-scale use of renewable and clean energy, energy efficiency, and responding to climate change in Asia and the Pacific region. While large-scale projects with international cooperation remain key to many of TERI's goals, TERI also promotes the empowerment of disadvantaged populations and the generation of employment through small-scale entrepreneurial endeavors.

TERI's global leadership in efforts to mitigate the threat of climate change has been further endorsed by the election of Dr. Rajendra K. Pachauri, TERI's director general since 1981, as chairman of the Intergovernmental Panel on Climate Change (IPCC) in April 2002. In October 2007, the IPCC and Al Gore jointly received the Nobel Peace Prize for their contributions to understanding and educating about global climate change.

Although spawned by the Tata Group—the largest conglomerate in India, accounting for 96 companies operating in over 40 countries and exporting to 140 countries—TERI operates as a nonprofit, nongovernmental organization. TERI's work is sponsored by over 900 organizations (Tata Group included), and more than 200 organizations from 43 countries serve as partners in TERI projects.

The institute established TERI University in 1998, which became a Deemed University in India in 1999. The university offers degree programs only at the masters and Ph.D. levels, and faculty and students participate in research conducted by the institute.

TERI continues to collect, generate, and make available a wide range of publications on issues related to its areas of research and training. Given its long history of publication, the institute established TERI Press, which has entered the higher-education textbook market. TERI also has a Library Information Center that serves as a repository for publications (hard copy and electronic) of information linked to its mission and divisions.

During its relatively brief existence, TERI has expanded to include research, training, and support efforts throughout India and claims to be the only developing-country institution to have established a significant presence in North America, Europe, the larger Asian continent, Japan, Malaysia, and the Middle East.

Jennifer E. Coffman
James Madison University

See Also: Climate Change, Effects of; India; Intergovernmental Panel On Climate Change; Japan; Malaysia; Nongovernmental Organizations; Oil, Production of; Renewable Energy, Biomass.

Further Readings

Bhattacharya, P. K. and S. Ganguly. "The TERI ENVIS Centre: An Indian Information Centre on Renewable Energy and the Environment." *Information Development*, v.22/3 (2006).

Energy and Resources Institute. http://www.teriin.org (Accessed July 2011).

TERI University. "TERI 2009/2010 Annual Report." http://www.scribd.com/doc/52737003/3/Teri-University (Accessed December 2011).

Technology

Technology is defined as applying science to manipulate or change the human environment. Although it is usually thought to involve some form of machinery or physical equipment, technology can just as effectively be intangible in form, such as with management technology, which provides different ways of understanding how resources, including people, may be organized for more efficient production or operation. Historically, technology changed and developed very slowly around the world. However, in more recent years, the improvements in infrastructure—and particularly in communications—have meant that technological advances have increased at an ever-quicker rate. Meanwhile, the dissemination of that technology has spread around the world, although there are still many hundreds of millions of people too poor to benefit from it. Nevertheless, for most people, especially in the Western world, life and society have been completely transformed by the technologies that have emerged over the last two decades.

Concurrent with the rise of technology is the growth of cities, such as Hong Kong, China, above. One of the principal causes of increased atmospheric heat is the presence of cities, especially large cities, across most of the inhabited world. It has been suggested that cities contribute to global warming, as they are constructed to maximize population density, tend to lack greenery, and absorb heat and energy. Some technological remedies would be to paint buildings white, create more water areas, and plant more grass.

Because the rate of change of new technological innovation continues to accelerate, life and society in the future are as difficult to predict now as they would have been a few decades in the past. Because of its prevalence in society, its ability to reduce the time needed for generally undesirable domestic tasks, and its ability to improve leisure opportunities, among other attributes, most people welcome technology and believe it to be beneficial to their lives.

However, there are still individuals and groups of people who may, perhaps for ideological or religious reasons, reject the use of technology. Because global climate change has come to be associated with the use of technology and the energy required to power so much of it, technology as a whole has come to be regarded by some people as an enemy that must be resisted and eradicated. In the extreme case, there are people who believe that only by returning to a form of society in which all forms of technology are rejected can humanity survive the forthcoming environmental crisis.

Philosophers such as Michael Foucault, meanwhile, consider technology a tool, most commonly used by the powered elites of society to suppress the masses. They would point out that the introduction of technology is customarily followed by the imposition of restrictions that prevent the majority of people from accessing the benefits of that technology. For example, Internet technology in China is regularly used to spy on the activities of ordinary people and keep their discussions heavily monitored. This is an instance of technology being used to suppress people and maintain the existing architecture of power. In contrast, it is possible to argue that the very same technology actually represents a liberation of people because of the many new ways it enables people to communicate with each other and share information.

Most people tend toward a more moderate position, recognizing the better lifestyle that some aspects of technology provide and they are unwilling to abandon these forms, while accepting the need for greater efficiency in the use of resources. They would be unwilling to voluntarily choose to forego the use of that technology to cause some future effects to abate. In other words, people will not vote for significant reductions in the use of personal technologies to reduce future damage caused by climate change. To change their minds, some activists believe that it is necessary to startle or scare people into realizing what sort of changes are likely in the future. Those more skeptical of future changes, meanwhile, accuse such activists of regularly committing this act and concluding, as a result, that all calls for changes in behavior are overstated, and possibly politically motivated. This argument has been successfully deployed, in that it has muddied the waters of debate, and has reduced the likelihood of future changes in behavior.

Role of Technology

Technology can be employed to abate current and future climate change in a number of ways. At a large or macrolevel, there are plans to place enormous mirrors into orbit around the Earth so that they reflect light energy from the sun back out into space, reducing atmospheric temperature. At a medium or mesolevel is the attempt to develop new and cleaner technologies in terms of energy production, which would reduce carbon emissions. At a microlevel, there are the efforts to reduce resource use inefficiencies by such means as recycling waste products, turning off unattended electronic devices, and generally developing technologies to mitigate future climate change. The extent to which it is possible to abate future climate change by action at the microlevel is not clear, and estimates vary widely.

Turning off unused computers or televisions that are idle will save some energy and reduce carbon emissions, but many people believe that this amount of saved energy will be dwarfed by the increased amount of emissions resulting from the rapid and rather dirty industrialization occurring, particularly in India and China. There is an argument that because the effects elsewhere are so great, there is no point trying to reduce emissions on a personal level. This argument does not bear rational examination. First, the reduction of the rate of acceleration of global warming is a necessary and important thing; second, the people and governments of India and China (and most of the rest of the world) are also aware of the problems of global climate change and are willing to do what they can to bring about changes in their lives.

Other arguments suggest that changes at the microlevel can have significant changes in the extent of future climate change. Noted skeptical environmentalist Björn Lomborg, for example, has argued that one of the principal causes of increased atmospheric heat is the presence of cities, especially large cities, across most of the inhabited world.

Cities are built or have organically grown to maximize population density, and as such are dry areas without much greenery or standing water. Further, many of the buildings or infrastructure within cities are dark in nature, and as a result absorb a great deal of energy in a needless fashion. The result is that cities are several degrees F/C or more hotter than surrounding areas. This problem fuels increased use of air conditioning systems and other energy use (e.g., for refrigerators and other cooling devices), which leads to a vicious cycle.

Lomborg argues that low-technology solutions can reduce the urban effect, such as painting buildings white, introducing more water areas, and replacing some tarmac with grass. Taking these steps may reduce the temperature in city areas by several degrees. This would have a knock-on effect, too, as in many cases, political priorities are determined by urban electorates. As a consequence, demonstrating that technology—even comparatively low-technology solutions—can lead to a measurable improvement in quality of life that might then lead to more positive sentiment toward the use of technology to improve future lifestyles.

Search for Alternative Energy Sources

Because the use of hydrocarbon fuels leads to carbon dioxide emissions and is the largest contributor to global climate change, it follows that finding alternative forms of energy that do not emit carbon to the same extent would represent the best means of abating future climate change. Further, the world's reserves of hydrocarbon fuels are finite, and there is a need to develop alternatives if current and projected lifestyles are to be maintained in the future. It is not clear exactly when the reserves of oil will be depleted under current trends of usage, as it is possible (although increasingly less likely) that significant areas of previously unexploited reserves will be found. Improvements in technology have made it possible to profitably extract existing reserves that have to date been too difficult or expensive to obtain. In addition, as the price of oil increases in general—and continues to increase as demand increases—with respect to supply, more and more known, but problematic reserves will become commercially viable. Already, using oil-soaked earth in parts of Canada that were previously prohibitively expensive to process has become viable because of the effect of supply and demand. Other difficult-to-access reserves will, likewise, become more viable.

Although the figures are controversial and contested, it seems likely that all oil reserves will be exhausted within about 120 years, based on current rates of consumption. It is possible, although far from certain, that peak oil production has already been reached. Although declining supply relative to demand will stimulate some increased efficiency of use of oil, new forms of energy will need to be developed within the next few decades.

Clean technologies, such as wind and wave power, are expected to make up an increasing part of a portfolio of alternative sources of energy. They already contribute significantly to the energy production of some European countries, although there are problems with NIMBY-ism—or Not in My Back Yard-ism—as people complain that wind turbines are unsightly and noisy. Solar panels have been used with moderate success in many parts of the world, particularly those areas with high levels of sunshine. However, improvements in this technology are still needed. Photovoltaic cells have been used to collect power from sunlight, and water pipes have been heated by placing them in sunlight, but there is a need for more integrated solutions to ensure that a higher level of efficiency is achieved and to reduce initial start-up costs, which may be high. This is likely to come about through regulation, rather than market influences.

Other technologies that might be developed include the tapping of geothermal energy, which has been used for thousands of years in a nonsystematic manner, in the form of hot springs. There remains considerable scope for the further development of the use of geothermal energy on a more systematic basis. Hydroelectricity is a further form of alternative energy that has been used effectively on many rivers. Countries such as China plan a massive increase in the number of hydroelectric stations, with attendant dams, on the rivers passing through its territory. This includes plans for as many as 12 dams on the Mekong River, for example. However, this form of energy is problematic because of the effect on people living downstream of reduced flow of water and because of the effect of building large dams on the populations living in the vicinity, many of whom must be resettled—some of them forcibly. The ownership of a river passing through the territory of more than one country is also a problematic issue.

The one means of producing energy that is available without the consideration of geography is that of nuclear power. The technology involved is to employ certain heavy elements such as plutonium and uranium, which undergo atomic decay on a largely predictable manner, releasing considerable amounts of energy at the same time. A more advanced approach is to employ nuclear fission, which involves causing atoms to collide with each other so as to release more subatomic particles such as neutrons, and hence more power. The amount of power that is converted to electricity and may be released through these processes is limited only by the availability of the appropriate heavy elements. These elements are scarce, and the existing amounts are controlled, although not always effectively. There is a need for control because of the radioactive nature of the substances involved, which makes them very dangerous to life, as well as the possibility that they may be used to create highly destructive nuclear bombs.

Many governments are planning to increase, perhaps quite significantly, their reliance on nuclear power. This causes problems because of the threat of an accident releasing nuclear material, as happened at Chernobyl in the Ukraine, causing thousands of deaths. The use of power plants in known earthquake zones is of particular concern. In addition, depleted uranium or other material, which is no longer productive, remains dangerously radioactive for many thousands of years, and it is not clear where that waste may be safely stored over the long term. There is also the potential problem that nations developing nuclear power plants might also employ this knowledge and technology to develop nuclear weapons. One consequence of this is that there is widespread public concern about the use of nuclear energy and opposition to it in democratic countries. Even so, improvements in the technology of safety suggest that more governments will wish to augment their nuclear power production capacity and begin building new plants, knowing that it takes several years or more between deciding to construct such a plant and when power from it is ready to enter the grid.

Macrolevel Technology Approaches

Various macrolevel technologies have been suggested as a means of reducing climate change. These are generally very expensive and time-consuming to create and maintain; hence, they tend to be regarded as something of a last-ditch attempt. These technologies include placing large mirrors or reflective dust in space to absorb or reflect away the sun's energy, or to use huge series of tubes leading to the ocean floor, through which excess heat energy may be circulated. Many of these ideas derive from the United States, which has a long tradition of optimism in terms of technology and also large firms and organizations possessing the kind of capital necessary to develop and implement such solutions. Because none of these solutions has, to date, been operationalized, it is not yet clear whether all or any of them will in fact be viable. Nevertheless, companies are starting to realize the market opportunities emerging for green or clean technologies. At the individual level, technology will be consumed by households to meet their mandated requirements; for example, in terms of recycling or energy use reduction. At the state level, in contrast, public-sector support of megaprojects can provide sustained funding for a number of years sufficient to underwrite very large research and development operations. Because these technologies are not yet proven feasible and may be extremely expensive in practice, it would be much more cost-effective to make extensive use of micro- and mesolevel technologies immediately, rather than waiting for a last-ditch attempt to maintain the planet as a place for human life. This will require some sacrifices in the short and medium terms.

Technology and Regulation

The Montreal Protocol of 1987, which helped to resolve the problem of atmospheric ozone depletion, demonstrated the ability of states to work together to solve transboundary environmental issues effectively. Market-based attempts to achieve similar goals, for example through creating markets in tradable carbon emission permits, have foundered without a strong institutional shaping of the rules of the market and supervision of its activities. Technologists around the world have created numerous efficiencies that would help abate future climate change, but that will only be implemented when state regulation requires it to be introduced. Just as the numbers of deaths from road traffic accidents was reduced (proportionate to the amount of traffic) after governments introduced legislation requiring safety belts to be worn, so too have buildings become more resistant to earthquakes after stricter building codes were introduced.

In countries such as Germany, new regulations concerning energy production enable households under certain conditions to produce and sell their power, such as that generated from solar power. Good regulations provide appropriate incentives to encourage people to behave in the desired way, and disincentives to dissuade people from behaving in undesired ways. The extent to which this can affect behavior and make measurable changes to energy use may be seen in California, where state-level legislation, transparently introduced over a sustained period, has ensured that energy use per capita has not increased in a number of years, despite the significant amount of electronic consumer goods owned per household.

John Walsh
Shinawatra University

See Also: Alternative Energy, Overview; Renewable Energy, Solar; Technology Transfer.

Further Readings

Foucault, Michel. *Politics, Philosophy, Culture: Interviews and Other Writings, 1977–84*. London: Routledge, 1990.

Hausladen, Gerhard, Michael de Saldana, Petra Liedl, and Christina Sager. *Climate Design: Solutions for Buildings That Can Do More With Less Technology*. Boston, MA: Birkhauser, 2005.

Justus, Debra and Cédric Philibert. *International Energy Technology Collaboration and Climate Change Mitigation: Synthesis Report*. Paris: OECD, 2005.

Leary, Neil, James Adejuwon, Vicente Barros, Ian Burton, Jvoti Kulkarni, and Rodel Lasco, eds. *Climate Change and Adaptation*. London: Earthscan, 2007.

Lomborg, Björn. "Paint It White." *Guardian* (November 18, 2007). http://commentisfree.guardian.co.uk/bjrn_lomborg/2007/11/paint_it_white.html (Accessed June 2008).

Technology Transfer

Technology lies at the heart of the development process of any country. Given that the lion's share of technologies, including climate-related technologies, still originates from developed countries, technology transfer (TT) from developed to developing countries assumes enormous significance for developing countries. Endeavors on the part of developing countries to follow a low-carbon development trajectory are also contingent, in large measure, upon technology transfer from developed countries. The United Nations Framework Convention on Climate Change (UNFCCC) recognizes the need for technology transfer in various provisions (e.g., Article 4.5) and over the years has undertaken several initiatives toward implementing them, albeit with very limited headway. With the Bali Action Plan (2007), the issue moved to center stage.

According to the 1985 Draft International Code of Conduct on the Transfer of Technology negotiated under the aegis of the UN Conference on Trade and Development (UNCTAD), technology transfer refers to the transfer of systematic knowledge for the manufacture of a product, or the application of a process, or for the rendering of a service and does not extend to the transactions involving the mere sale or mere lease of goods. While the UNFCCC does not define technology transfer, the definition of technology transfer that is most frequently referred to in the climate change context is the one developed by the Intergovernmental Panel on Climate Change (IPCC). The IPCC defines technology transfer as a broad set of processes covering the flows of knowledge, experience, and equipment for mitigating and adapting to climate change among different stakeholders such as governments, private sector entities, financial institutions, nongovernmental organizations (NGOs), and research and education institutions.

The treatment of technology transfer in this definition is much broader than that in any article of the Framework Convention. While the convention focuses only on international technology transfers, particularly from developed countries to developing countries, the IPCC definition not only covers all kinds of international technology transfers (including developed to developing country technology transfer), but also encompasses within-country technology transfer among different stakeholders. However, it is international TT, and more specifically developed to developing country TT, that has turned out to be the most contentious dimension of the concept.

Notably, the UNCTAD definition, developed specifically in the context of international technology transfer, emphasizes the knowledge element of technology transfer. It indicates that the mere import of goods (e.g., equipment) should not be regarded as technology transfer. In a similar vein, the IPCC definition does not confine its scope to equipment or the hardware element of a technology; it also includes software elements such as knowledge and experience (i.e., the knowledge dimension of a technology). According to IPCC, technology transfer comprises the process of learning to understand, utilize, and replicate the technology, including the capacity to choose and adapt it to local conditions and integrate it with indigenous technologies.

Internalized and Externalized Transfers

Modes of TT can be broadly classified into two categories: (1) those that are internalized, namely from a transnational corporation (TNC) to foreign affiliates under its control; and (2) those that are externalized (i.e., those between independent firms). Internalized modes can take place either through the establishment of a new affiliate in a host country or through the acquisition of a local firm that can be turned into a suitable recipient. Externalized modes of transfer take a variety of forms such as minority joint ventures, franchising, capital goods sales, licenses, technical assistance, and subcontracting or original equipment manufacturing arrangements.

While internalized modes necessarily involve TNCs, externalized ones may also involve TNCs selling technologies on contract. Not all modes of technology import are equally conducive to indigenous learning. Internalized transfers come highly packaged with complementary factors to ensure their efficient deployment. The retention of technology and skills within the network of a TNC may restrain deeper learning processes and spillovers into the local economy, especially where the local affiliate is not developing research and development capabilities.

Externalized transfers tend to call for a greater learning effort by the recipient. Given the potential for greater learning, deepening, and externalities, externalized modes are often regarded as superior to the internalized mode because developing countries increase their technological capabilities. Thus, where a choice exists between internal transfers to foreign affiliates or external transfers to local technology recipients, a developing country's government may wish to intervene to affect the terms of a transfer associated with each modality, for example, where incentives are offered to TNCs for the transfer of advanced technical functions. However, it has been argued that in practice, the externalized modes also often suffer from serious flaws because of various imperfections in the technology market and the dominance exercised by technologically advanced firms, particularly in light of the strengthened intellectual property rights (IPRs) protection under the World Trade Organization (WTO) Agreement on Trade-Related Aspects of Intellectual Property Rights (TRIPS).

In the climate change context, the potential of the Clean Development Mechanism (CDM) as a vehicle for technology transfer has been underscored. However, empirical findings differ on the role that the CDM has been playing as a vehicle for TT.

Intellectual Property Rights

Commercial technology is usually exploited through the application of IPRs, which helps to increase the technology's value to its owner by creating relative scarcity through legally restricted access to it. According to proponents of strong intellectual property regimes, IPRs can serve as an important support for technology markets characterized by the problems of appropriability and asymmetric information. Without adequate protection from leakage of new technical information, firms would be less willing to provide it on open technology markets. In an environment of weak IPRs, they may choose not to transact at all or may opt for other strategies that may have adverse implications for TT, such as offering older-generation technologies or keeping the information within the firm by dealing only with subsidiaries.

According to a counterview, however, the availability (and enforceability) of IPRs will, by no means, create a sufficient incentive for TT to occur, and instead can sometimes make TT more problematic. A weak intellectual property regime may offer local firms in the recipient countries some options for imitating foreign technologies and reverse engineering products—options that would be rather limited if IP protection is strengthened. Moreover, inventors, by virtue of their lead times and IP protection, may be expected to sell technologies at a price higher than marginal cost, which is less than optimal socially for the recipient country, at least in a static sense. High levels of IP protection can also deepen negotiating imbalances in IP licensing in favor of the licensor, which often results in the imposition of abusive practices that restrain competition. The one-size-fits-all approach (as adopted by the WTO TRIPS regime) has been criticized, as far as the role of IP in technology development and transfer is concerned. Although there exists a large body of empirical literature on the relationship between IPRs and TT, the findings turn out to be diverse and often ambiguous.

Barriers to TT may vary from one technology to another and may manifest themselves differently in different countries. According to the IPCC, there is no preset answer to enhancing TT. The identification, analysis, and prioritization of barriers should be country specific. It is important to tailor action to the specific barriers, interests, and influences of different stakeholders in order to develop effective policy tools.

Kasturi Das
Research and Information System for Developing Countries

See Also: Asia–Pacific Partnership on Clean Development and Climate; Bali Roadmap; Technology.

Further Readings

Correa, C. M. "Can TRIPS Agreement Foster Technology Transfer to Developing Countries?" In *International Public Goods and Transfer of Technology*, edited by K. Maskus and J. H. Reichman. Cambridge: Cambridge University Press, 2005.

Das, K. "Technology Transfer Under the Clean Development Mechanism: An Empirical Study of 1,000 CDM Projects, Working Paper 014." In *Governance of Clean Development Working Paper Series*. East Anglia, UK: University of East Anglia, 2011.

Metz, Bert. *Methodological and Technological Issues in Technology Transfer: A Special Report of IPCC Working Group III*. Cambridge: Cambridge University Press, 2000.

United Nations. "Transfer of Technology." In *UNCTAD Series on Issues in International Investment Agreements*. Geneva: United Nations, 2001.

Tennessee

An inland southeastern state, Tennessee was the first constitutional government formed west of the Appalachians, which dominate the eastern half of the state. Admitted to the union in 1796, it had a reputation for wildness and iconoclasm throughout the 19th century as the birthplace of era-defining, military and political figures like Andrew Jackson, Sam Houston, and Davy Crockett.

The state encompasses 42,143 sq. mi. (109,149 sq. km), 926 sq. mi. (2,398 sq. km) of which are inland water. Major rivers include the Mississippi, Tennessee, and Cumberland, and the topography of the state includes plateau, valleys, mountains, and forested ridges. Winters are mild and summers are hot and muggy, except in the Blue Ridge Mountains, which are cooler and drier most of the year. Snowfall is common in the winter and blizzards are rare, but snow tends not to accumulate, except in the mountains. Heavy March and April rains lead to annual floods as the water joins snow melting up north.

The west is dependent on agriculture, especially cotton, corn, livestock, and tobacco, which is also grown in the east. In the Nashville basin, livestock and dairy farming dominate the agricultural sector. Coal mining is a major part of the economy, and much of Tennessee's electricity is generated by coal-fired plants, as well as hydropower and nuclear power.

In the 21st century, Tennessee temperatures are expected to rise by 4 to 5 degrees F (2.2 to 2.7 degrees C). Summer precipitation could increase by as much as 50 percent (depending on the climate model), which could result in regular severe flooding in eastern Tennessee and the major urban centers of Nashville, Memphis, and Chattanooga. Ground-level ozone could severely increase in those urban centers, and heatstroke and other heat-related health problems could also increase. Higher stream flows and lake levels would benefit hydropower and dilute pollutants, but increased runoff of pesticides and fertilizers could dangerously contaminate the river basins of western Tennessee. Flooding would also increase the spread of insect-borne diseases like West Nile virus and malaria. The altered climate of Tennessee would more greatly favor pine and oaks over eastern hardwoods in the forested areas, and wildfires would become more severe and common.

Tennessee is a participant in Rebuild America, an energy-saving organization that assists school systems and governments.

Bill Kte'pi
Independent Scholar

See Also: Alabama; Coal; Drought; Kentucky; Mississippi; Ohio.

Further Readings

Bales, Stephen Lyn. *Natural Histories: Stories From the Tennessee Valley.* Knoxville: University of Tennessee Press, 2007.

Bergeron, Paul, Jeannette Keith, and Stephen Ash. *Tennesseans and Their History.* Knoxville: University of Tennessee Press, 1999.

Moore, Harry L. *Geologic Trip Across Tennessee.* Knoxville: University of Tennessee Press, 1994.

Tertiary Period

The Tertiary period (ca. 66.4 to 1.8 million years ago [Ma]) was an interval of enormous geologic, climatic, oceanographic, and biologic change. It spans the transition from a globally warm world of relatively high sea levels to a world of lower sea levels, polar glaciation, and sharply differentiated climate zones. Over the past decade, however, it has become increasingly clear that Tertiary climatic history was not a simple unidirectional cooling driven by a single cause, but a much more complicated pattern of change controlled by a complex and dynamic linkage between changes in atmospheric carbon dioxide (CO_2) levels and ocean circulation, both probably ultimately driven by tectonic evolution of ocean–continent geometry. Although satisfactory explanations for many aspects of Tertiary climate history are available, many areas remain incompletely understood.

The early Tertiary (Paleocene and most of the Eocene epochs, ca. 66–50 Ma) was characterized by a continuation of Cretaceous warm equable climates extending from pole to pole. Global temperatures may have been as much as 18–22 degrees F (10–12 degrees C) higher than present, and pole-to-equator temperature gradients were about 9 degrees F (5 degrees C) during the Paleocene, as compared with about 45 degrees F (25 degrees C) today.

The Paleocene–Eocene boundary (about 54 Ma) was marked by a geologically brief episode of global warming known as the Paleocene–Eocene thermal maximum (PETM), characterized by an increase in sea surface temperatures of 9–11 degrees F (5–6 degrees C), in conjunction with ocean acidification, a decline in productivity, and a large and abrupt decrease in the proportion of isotopically heavy terrestrial sedimentary carbon in the oceans. The PETM is thought to have lasted only about 170,000 to 220,000 years, with most of the temperature and isotopic change occurring in the first 10,000 to 20,000 years. Its causes remain unclear, but it was probably associated with dissolution of methane hydrates on the ocean floor, which would then have caused greenhouse warming. Possible triggers for this hydrate release include an increase in volcanism, leading to an increase in atmospheric CO_2 and consequent sudden initiation of greenhouse warming; a change in ocean circulation; or massive regional submarine slope collapse.

Global temperatures warmed still further during the early Eocene, reaching their highest levels of the past 65 million years during an interval sometimes called the early Eocene climatic optimum (52–50 Ma). Global cooling began during the early middle Eocene (ca. 50 Ma) and accelerated rapidly across the Eocene–Oligocene boundary (ca. 34 Ma), at which time Antarctic continental glaciation began. This shift is frequently referred to as a change from a greenhouse to an icehouse climate regime, and it is one of the most fundamental reorganizations of global climate known in the geological record.

Initiation of Antarctic glaciation has long been attributed to the tectonic opening of Southern Ocean gateways, especially the Drake Passage between South America and the Antarctic Peninsula, which allowed establishment of the Antarctic Circumpolar Current and the consequent isolation of the southern continent from warmer low-latitude waters. This has been questioned recently, however, as a result of the redating of the formation of these gateways, as well as modeling results that point to a greater role for reduced atmospheric CO_2.

Most estimates of early Cenozoic atmospheric CO_2 range between two and five times the present values in the middle to late Eocene and then decline rapidly during the Oligocene to reach approximately present levels in the latest Oligocene. This decline in CO_2 may in turn have been at least partly a result of the tectonic uplift of the Tibetan plateau, beginning around 40 Ma, leading to increased rates of chemical weathering. Levels of CO_2 remained relatively constant throughout the Miocene, suggesting that the substantial climate changes during this time were driven by other factors, including changes in weathering or ocean circulation.

Global temperatures warmed again in the late Oligocene, followed by a brief (ca. 200,000 years) but deep glacial interval at the Oligocene-Miocene boundary (ca. 24 Ma). Temperatures then stabilized or slightly increased (punctuated by several more brief glacials), leading to what is sometimes referred to as the mid-Miocene climatic optimum around 17 to 15 Ma, during which time deep water and high-latitude sea surface temperatures were 11–18 degrees F (6–10 degrees C) warmer than at present. The causes of this warming are not clear, but they may have been related to increased northward oceanic heat transport in the North Pacific brought via intensified currents primarily triggered by narrowing of the Indonesian seaway in the western equatorial Pacific.

Another major cooling occurred between 14.2 and 13.7 Ma and is associated with increased production of cold Antarctic deep waters and a growth spurt of the East Antarctic Ice Sheet, leading to an increased latitudinal temperature gradient and drying in midlatitudes. A further episode of aridity occurred between 8 and 4 Ma. This cooling trend continued into the Quaternary period, with a short warming interval in the early to mid-Pliocene (ca. 5–3.2 Ma), characterized by warmer sea and air temperatures across much of the North Atlantic region.

Northern Hemisphere ice sheets first expanded about 3.5 Ma, with a major pulse of growth occurring 2.5 Ma, at which time the Earth is usually said to have passed over a thermal threshold, initiating the latest ice age, in which mode the planet is still today. The initiation of Northern Hemisphere glaciation has been attributed to completion of the formation of the Central American isthmus at around 3.5 Ma, which deflected warm low-latitude currents flowing westward from Africa northward into the Gulf of Mexico and through the Florida straits to join the Gulf Stream. This strengthened Gulf Stream then transported more moisture to high latitudes, where it supplied an increase in snowfall, leading to increased albedo and temperature decline.

Warren D. Allmon
Paleontological Research Institution

See Also: Earth's Climate History; Ice Ages; Paleoclimates.

Further Readings

Francis, J. E., et al., eds. *Cretaceous-Tertiary High-Latitude Palaeoenvironments.* London: Geological Society, 2006.

Keller, G. and N. MacLeod. *Cretaceous-Tertiary Mass Extinctions: Biotic and Environmental Changes.* New York: W. W. Norton & Co, 1996.

Texas

In a post–World War II climate of mass consumption, urban disinvestment, and the emerging dominance of automobiles as the preferred mode of transportation, Texas and its economy grew dramatically. Fleeing postindustrial urban decay and the loss of manufacturing economies, millions of Americans and immigrants flocked to the wide-open and nonunionized spaces of the southwest United States. Home to almost 25 million residents, the state of Texas ranks second only to California in total population, contains six of the top 20 most populated cities, and experienced the largest net population growth of any state in the nation.

Driven by this continuing growth in population, diversifying economic development, persistent low-density suburban development, almost exclusive reliance on the automobile for transportation, and increasingly hot, dry summers, the demand for cheap and plentiful energy—and the emissions of greenhouse gases (GHGs)—is growing rapidly. Texas leads the United States in GHG emissions not only because of its reliance on existing coal-burning power plants for electricity generation but also because of methane emissions from booming natural gas production in the Barnett shale and elsewhere.

In a recent study analyzing urban sprawl, Fort Worth/Arlington and Dallas metro areas rated 10th and 13th on the list of the nation's 83 most sprawling urban areas. Having increased by 30 percent over the past 10 years, the 2000 Environmental Protection Agency Ozone National Ambient Air Quality Standard report estimated that Texas vehicle miles traveled would increase between 2007 and 2030 by over 44 percent, and the Dallas/Fort Worth, Houston, and San Antonio metropolitan regions would remain in nonattainment for ground-level ozone. These urban areas are experiencing significant population growth, continuing sprawl development, and increasingly severe highway congestion that contribute significantly to their climate change effect. The consequences for Texas of impending climate change are serious, especially given the already extreme nature of much of its regional weather and the continuing reluctance of the state government to take aggressive climate change mitigation action.

Consequences of Climate Change

Texas exhibits a wide variety of climates within its boundaries, from subtropical in the southeast to high desert in the north and west. It is famous for its scorching summers and sunshine that, in local parlance, will "peel the chrome off a trailer hitch." Although all regions are likely to experience an increase in mean annual temperature (both daily maximums and minimums) and increasing shortages of freshwater, other challenges faced by the state from climate change differ as a function of geography. Texas' 370 mi. (595 km) of coastline will experience higher sea levels and resulting

The Dallas metro area is rated 13th on the list of the nation's 83 most sprawling urban areas. Texas is home to almost 25 million residents and is experiencing the largest net population growth of any state in the United States. Economic development, low-density suburban development, heavy reliance on automobiles, the demand for cheap and plentiful energy, methane emissions from natural gas production, and increasingly hot, dry summers are all driving up the state's rate of greenhouse gas emissions.

beach erosion, saltwater infiltration, and subsidence. Increased water temperatures in the Gulf of Mexico may result in more frequent and widespread algal blooms toxic to indigenous fish and plant species. Warmer ocean and Gulf waters will also contribute to the intensity, if not the frequency, of coastal storms and hurricanes.

Climate change will also result in a redistribution of rainfall across the state, significantly affecting agricultural economies and freshwater supplies available to increasingly urban populations. As of May 2011, almost 50 percent of the state is experiencing "exceptional" drought, up from less than 20 percent in 2009. As it continues to drain its aquifers to meet the needs of growing populations and economies, Texas increasingly relies on freshwater captured in surface reservoirs. A future that is markedly warmer and drier in many regions of the state will jeopardize these supplies. Increasing temperatures will contribute to already pronounced urban heat islands, resulting in increased frequency of heat-related illnesses and deaths, the introduction of new disease vectors, and the increased severity of isolated weather events, especially thunderstorms bearing isolated, flooding rains.

Agricultural and forestation patterns and productivity, staples of regional and state economies, will be disrupted not only by the changes in rainfall, increased uncertainty in available irrigation water, and higher mean temperatures, but also by a changing variety of natural weeds and pests that will migrate northward as the climate warms. Infestations of insects new to Texas crops will result in reduced crop productivity and, as farmers try to respond, a likely increase in the number and environmental toxicity of herbicides and pesticides.

Renewable Energy Sources

Although it confronts serious climate challenges across the state, Texas is also blessed with sources of renewable energy that have only begun to be exploited. The largest portion of its electricity needs is generated by coal-fired and nuclear power plants. Recently, the state backed away from approving the construction of as many as 11 new coal-fired power plants and is in the process of redefining its energy policies. As technological advances reduce the price and increase the efficiency of solar energy generation devices, Texas, especially in its western reaches, will be able to capitalize on the abundant radiant energy provided by the sun. In addition, wind-generated electricity is being produced in increasingly economical quantities by western Texas wind farms. Although a transmission infrastructure is evolving to supply the state's urban demand, Texas ranks first in wind power generation among U.S. states. It is also a leading producer of biodiesel transportation fuel. In 2010, the Electric Reliability Council of Texas reported an increase of 30 percent from 2009 in the amount of electricity generated from renewable sources.

State and Local Action

While the Texas Commission on Environmental Quality provides a voluntary greenhouse gas (GHG) mitigation reporting program for political jurisdictions and businesses, it continues to resist state and national climate change regulation. Texas is one of several states to legally challenge Clean Air Act regulation of carbon dioxide (CO_2) emissions by the U.S. Environmental Protection Agency as directed by the U.S. Supreme Court in *Massachusetts v. Environmental Protection Agency*.

In the absence of mandatory, coordinating climate protection policy at the federal or state levels, many Texas cities have joined municipalities in other states in a variety of nongovernmental organization–led initiatives to reduce their carbon footprints and to lobby for action on climate change and related environmental issues at both the state and federal levels. Among other actions taken by Texas municipalities, 31 Texas cities—including Austin, Dallas, and San Antonio—have signed the U.S. Conference of Mayors Climate Protection Agreement, committing their respective cities to CO_2 reductions similar to those contained in the Kyoto Protocol. Austin, Dallas, and San Antonio are also among the 15 Texas cities that are members of the Cities for Climate Protection, a global campaign of local and regional entities led by ICLEI Local Governments for Sustainability. Through process implementation and performance monitoring, these cities are committed to a rigorous accounting and reduction of their GHG emissions.

Kent L. Hurst
University of Texas at Arlington

See Also: BP; Drought; Hurricanes and Typhoons; Land Use; Monsoons; Oil, Production of; Renewable Energy, Wind.

Further Readings

Electric Reliability Council of Texas. "2010 Annual Report." http://www.ercot.com/news/press_releases/2011/nr-05-13-11 (Accessed June 2011).

Ewing, Reid, Rolf Pendall, and Don Chen. *Measuring Sprawl and Its Impact*. Washington, DC: Smart Growth America, 2002.

ICLEI Local Governments for Sustainability. "Cities for Climate Protection Campaign." http://www.icleiusa.org (Accessed June 2011).

Texas Commission on Environmental Quality. "Inventory of Voluntary Actions to Reduce Greenhouse Gases." http://www.tceq.state.tx.us (Accessed June 2011).

Texas Department of Transportation. "TxDOT has a Plan: Strategic Plan for 2007–2011." http://www.dot.state.tx.us (Accessed March 2012.

U.S. Census 2010. http://2010.census.gov/2010census (Accessed June 2011).

U.S. Conference of Mayors. "Climate Protection Agreement." http://usmayors.org/climateprotection/agreement.htm (Accessed June 2011).

U.S. Court of Appeals for the District of Columbia Circuit. "*Coalition for Responsible Regulation, Inc., et al. v. Environmental Protection Agency*: Petition for Review of Environmental Protection Agency Final Orders." https://www.oag.state.tx.us/newspubs/releases/2011/052311endangerment_brief.pdf (Accessed June 2011).

U.S. Environmental Protection Agency. "How Will Climate Change Impact the EPA Region 6 Area?" (1997). http://www.epa.gov/region6/climatechange/impact-in-r6.htm (Accessed September 2007).

Thailand

Thailand, a country located in the southeastern region of mainland Asia, is bordered to the north by Myanmar and Lao, to the east by Lao and Cambodia, to the south by Malaysia and the Gulf of Thailand, and to the west by the Andaman Sea. The total land area of the country is approximately 198,070 sq. mi. (513,000 sq. km). The country is divided into five regions: north, northeast, central, east, and south. The northern region is generally mountainous, while the northeast is on a high plateau. The central region is generally flat and fertile, while the southern region, a peninsula, constitutes most of the country's coastline.

Thailand's latest population estimate stands at 65 million, with the majority dwelling in rural, agricultural areas. Agriculture is the primary sector, employing 49 percent of the population and contributing to 10 percent of the gross domestic product (GDP). Thailand's 1,988-mi.- (3,200-km-) long coastline brings its tourism and fishery sector alive, providing 6 percent of GDP and a livelihood to 10 percent of its population.

Thailand has recognized the significance of climate change by becoming a member of the United Nations Framework Convention on Climate Change (UNFCCC) on December 28, 1994, and later ratifying the Kyoto Protocol on August 28, 2002. In comparison to other industrialized countries, Thailand was able to cut down the level of greenhouse gas (GHG) by 0.6 percent, while contributing less per-capita emission than the world's average. Thailand's contribution to GHG emissions is only 0.8 percent of the world's total. However, emissions doubled between 1991 and 2002.

The effects of climate change on Thailand include floods, droughts, higher surface temperature, severe storms, and sea-level rise. Weather patterns during the past decade have fluctuated from severe droughts to severe floods. Crops are therefore threatened, along with coastal tourism. Precipitation has also been in decline, causing water shortages. Intense rainfalls ands more out-of-season storms have led to worse flooding over the past 10 years.

Thailand's primary export and major grain is rice, which is an essential component of the country's culture and economic system. With climate change, the impact on the rice industry would be high, especially since decreasing grain yields are directly proportional to increasing temperatures.

Bangkok, the country's capital, is home to 15 percent of the total population and serves as the social, political, and economic center of Thailand. The city has been sinking at the rate of about 4 in. (10 cm) per year. Coupled with rising sea levels, this situation puts the city at risk. Construction

Driving during a monsoon in Thailand. The Boxing Day tsunami in Thailand in 2004, which resulted in the deaths of an estimated 8,200 people, including many hundreds of foreign tourists. The high death toll has been partially blamed by some experts on climate change, as hydrological disasters have reportedly increased in the last decades, both in magnitude and frequency.

of a flood prevention wall around the capital has therefore been a paramount adaptation effort.

Bangkok's per-capita carbon emissions (7.1 tons) is much higher than London, which is 5.9 tons per capita. The majority of these emissions—around 84 percent—are from energy use and transportation. To mitigate its emissions, the city adopted a target to source 8 percent of its energy requirement from renewable sources by 2011 and increase it to 35 percent by 2020. In addition, public and private organizations have signed the Bangkok Declaration on Mitigation of Climate Change aimed at reducing the city's contribution to climate change through energy consumption reduction, GHG emissions reduction, lifestyle changes, and raising public awareness. Through this declaration, Bangkok seeks to achieve a 15 percent reduction in its GHG emissions below currently projected 2012 levels. In order to meet this target, Bangkok has been encouraging commuters to use public transportation. It also began to improve mass transit systems, construct bike lanes, promote the use of efficient lighting and appliances, encourage recycling and reusing, and implement surcharges on gasoline.

On the national front, Thailand's National Committee on Climate Change has been playing an important role in drawing up a national strategy to address climate change issues. Thailand has adopted a basic policy of "no regrets" in the choice of options to mitigate GHGs. It has hosted several activities and projects through the United Nations' Joint Implementation and Clean Development Mechanism and has participated actively in research and development of climate change issues with other countries.

Laurence Laurencio Delina
Independent Scholar

See Also: Abrupt Climate Changes; Adaptation; Carbon Sequestration; Climate Change, Effects of; Drought; Forests; Global Warming, Impacts of; Kyoto Protocol; Land Use; Preparedness; Rainfall Patterns; Renewable Energy, Overview; Vulnerability.

Further Readings

Bachelet, D., et. al. "Climate Change in Thailand and Its Potential Impact on Rice Yield." *Climatic Change*, v.21 (1992).

Royal Government of Thailand. *Thailand's Initial National Communication Under the United Nations Framework Convention on Climate Change.* Bangkok: Office of Environmental Policy and Planning, Ministry of Science and Technology, 2000.

Yusuf, A. A and H. Francisco. *Climate Change Vulnerability Mapping for Southeast Asia.* Singapore: Economy and Environment Program for Southeast Asia, 2009.

Thermocline

The thermocline is the region of the ocean where temperature decreases most rapidly with increasing depth. It separates the warm, well-mixed upper layer from the colder, deep water below. A thermocline is present throughout the year in the tropics and middle latitudes. It is more difficult to discern in high latitudes, where temperature is more uniform with depth. The presence of a very shallow thermocline in the eastern equatorial Pacific Ocean has important implications for global climate.

The thermocline exists because the ocean absorbs most of the sun's heat in a shallow layer near the surface. The heat absorbed from the sun increases the temperature of the surface relative to that of the deep ocean, maintaining the thermocline. This is in contrast to the atmosphere, where a much larger portion of incident solar radiation passes through to the Earth's surface.

Two important properties of the thermocline are its depth and its strength, or how rapidly temperature decreases with increasing depth. The thermocline's depth is influenced by the winds at the surface of the ocean. In the Atlantic and Pacific oceans, surface winds push warm surface water away from the equator toward the poles, bringing the thermocline close to the surface at the equator.

Water that diverges at the equator accumulates in the subtropics, increasing the depth of the thermocline there. The thermocline is generally 82 to 656 ft. (25 to 200 m) deep in the equatorial regions and up to 3,281 ft. (1,000 m) deep in the subtropics.

The thermocline is strongest in the tropics and weakest in high latitudes. This reflects the fact that the surface temperature of the ocean generally decreases from the tropics to the poles, whereas the temperature of the deep ocean is nearly the same at all latitudes. As a result, the temperature contrast between the upper ocean and the deep ocean is greatest in the tropics. The temperature can drop by as much as 18 degrees F (10 degrees C) in less than 164 ft. (50 m) in the tropical thermocline.

In the extratropical oceans, the strength and depth of the thermocline vary from season to season. There is a main thermocline throughout the year, between 656 and 3,281 ft. (200 and 1,000 m). During summer, the sun heats the ocean's surface more strongly than in winter. Most of the additional heat is absorbed in a very shallow surface layer, generating a sharper "seasonal" thermocline above the main thermocline. The seasonal thermocline is similar to the tropical thermocline in terms of its strength and depth. It erodes in the winter as the surface cools relative to the temperature in the main thermocline.

Tropical Oceans

The existence of a strong and shallow thermocline in the tropical oceans has important implications for climate. In the equatorial Pacific Ocean, westward surface winds lead to an accumulation of warm surface water in the west, depressing the thermocline there and raising it to near the surface in the east. The shallow thermocline in the east enables cold, nutrient-rich water to be mixed upward into the surface layer. Every few years, the thermocline in the eastern equatorial Pacific deepens in association with an El Niño event. The mixing of cold, nutrient-rich thermocline water into the surface layer is reduced, the surface temperature of the eastern equatorial Pacific Ocean increases, and biological productivity decreases. The warmer surface temperatures associated with El Niño affect atmospheric circulation in the tropics and alter weather patterns throughout the world.

The depth of the eastern equatorial Pacific thermocline has varied significantly in association with changes in global climate. For example, during the early Pliocene period (between 4.5 and 3 million years ago, the most recent period with global temperatures significantly higher than today), the eastern Pacific thermocline was much deeper than it is today, much like it is during a modern El Niño event.

Gregory R. Foltz
University of Washington/Joint Institute for the Study of the Atmosphere and Ocean

See Also: El Niño and La Niña; Mixed Layer; Wind-Driven Circulation.

Further Readings

Philander, S. George, James R. Holton, and Renata Dmowska. *El Niño, La Niña, and the Southern Oscillation*. San Diego, CA: Academic Press, 1989.

Pickard, George and William Emery. *Descriptive Physical Oceanography*. Oxford: Butterworth–Heinemann, 1990.

Thermodynamics

The science of thermodynamics, a branch of physics, aims to describe transformations in energy. Thermodynamics is comprised of three laws. The first holds that energy can neither be created nor destroyed. Energy in various forms may be transformed into heat (thermal energy) and heat may be transformed into another form of energy, so long as the total energy in the system remains constant. The second law states that entropy, a measure of the amount of energy dissipated as heat, increases over time in a closed system. The conversion of energy into heat increases the entropy of a system and the dissipation of heat likewise increases the entropy of a system. The third law states that as temperature approaches absolute zero, the theoretical minimum temperature in the universe, entropy approaches a maximum.

Three Laws of Thermodynamics

The first law of thermodynamics accounts for the relative constancy of the climate, averaged over long durations. Were Earth simply a reservoir for energy in the form of sunlight, it would heat up to a very high but finite temperature. Earth does not heat up to this magnitude because it radiates heat back into space. The dissipation of energy as heat, according to the second law of thermodynamics, describes the Earth's shedding of radiant energy received from the sun as heat. This law, functioning as a heat accountant, is at the heart of understanding the role of heat in determining the climate. The third law of thermodynamics does not operate as long as the sun generates energy. Rather, the third law anticipates the end of the universe. The sun will one day burn out. Bereft of its heat, Earth's climate will be eternally cold, as its temperature approaches absolute zero. Not only will the sun be extinguished, but all stars in the universe will one day burn out. The heat from these stars will dissipate in all directions in the universe, bringing the temperature, uniform throughout the universe, near absolute zero.

The science of thermodynamics traces the origin of energy in the solar system to the sun. Energy from the sun is the basis of Earth's climate, but not all sunlight reaches Earth. The thermosphere lies 190 mi. (306 km) above Earth's surface, and is the outermost layer of the atmosphere. It absorbs ultraviolet light so efficiently that its temperature rises as high as 570 degrees F (299 degrees C). This conversion of the sun's radiant energy into thermal energy obeys the second law of thermodynamics. The next layer of the atmosphere, the mesosphere, is 50 mi. (80 km) above Earth. Its temperature, cooler than the thermosphere, is 200 degrees F (93 degrees C). Carbon dioxide (CO_2) in the mesosphere absorbs infrared light as heat, and that light radiates from Earth back into space. CO_2 molecules absorb a portion of this light before it reaches space. The larger the number of CO_2 molecules, the more heat they will absorb. The heating of the atmosphere by the absorption of infrared light causes the greenhouse effect, the warming of Earth's climate. Beneath the mesosphere is the ozone rich stratosphere, roughly 15 mi. (24 km) above Earth. The ozone in the stratosphere blocks some 90 percent of sunlight from reaching Earth. Ozone, like the thermosphere, absorbs ultraviolet light. Beneath the ozone layer is the troposphere, a variable layer 5 mi. (8 km) thick at the poles and 20 mi. (32 km) thick at the equator. The troposphere holds water vapor, which absorbs both infrared and ultraviolet light, heating the atmosphere. These layers of the atmosphere both absorb and radiate heat. The heat that they radiate either scatters into space or reaches Earth.

Earth absorbs sunlight, chiefly at the equator. This sunlight, in the form of heat, moves to the poles through the currents of the oceans and air. This distribution of heat from an area of greater concentration (the equator) to a region of lesser concentration (the poles) obeys the second law of thermodynamics. Heat supplies the energy for the movement of the oceanic and air currents, which in turn transform the potential energy of stasis into the kinetic energy of motion. On an idealized Earth on which the oceans and air distributed heat evenly throughout the planet, heat would reach thermodynamic equilibrium, the point at which entropy would be at a maximum. Earth is much less efficient than this idealized model. For all the motion of the oceanic and air currents, heat nevertheless concentrates at the equator, which is always warmer than the poles. The waters at the equator hold enormous amounts of heat. Because the oceans liberate their heat slowly, heat accumulates at the equator and is slowly transferred toward the poles.

In accord with the second law of thermodynamics, entropy would increase as heat moves from equator to poles, but the sun continuously adds heat to Earth, keeping the equator warmer than the poles. Entropy does not increase because the equator remains warmer than the poles. Without the oceanic and air currents, heat would accumulate at the equator and would not circulate to cooler regions of Earth. The currents therefore perform an important function in carrying heat from the equator to temperate and cold latitudes.

Earth and the atmosphere reflect roughly one-third of the sunlight they receive and radiate the other two-thirds into space. Earth sheds the same amount of heat as it receives, keeping Earth on average at 60 degrees F (16 degrees C). By contrast, outer space, which has no atmosphere to absorb heat, is much colder, at minus 454 degrees F (minus 270 degrees C). Earth absorbs sunlight as ultraviolet and visible light and continually radiates it back into space as infrared light.

Earth also reflects light back into space. The oceans reflect half the sunlight they receive, whereas ice and fresh snow reflect 90 percent. In accord with the second law of thermodynamics, entropy decreases when Earth absorbs heat, and increases when the oceanic and air currents diffuse heat to other regions of the planet. Similarly, entropy increases when Earth reflects light back into space, thereby dissipating heat.

Entropy is least in equatorial waters because they retain heat and slowly liberate it to other regions of Earth. Heat is not evenly distributed in equatorial waters, as thermodynamic equilibrium would suggest. In holding heat, the oceans at the equator moderate the climate, keeping lands near them warmer than inland stretches of territory.

The land warms four times faster than the oceans; the air warms faster still. Land and air also radiate heat faster than the oceans. The climate of a desert underscores the rapidity of heating and cooling on land. Temperatures in a desert rise rapidly during the day, often surpassing 100 degrees F (38 degrees C). At night, a desert cools with equal speed, dipping as low as freezing. In accord with the second law of thermodynamics, entropy decreases as a desert absorbs heat and increases as it dissipates heat.

Warm climates hold heat not only in water and land, but also in air. Warm air holds more moisture than cool air in the form of water vapor, a greenhouse gas. Water vapor holds more heat than CO_2, methane, or other greenhouse gases. Water in all three phases absorbs and emits heat. Ice absorbs the least heat and reflects the most sunlight back into space. Liquid water and water vapor are efficient reservoirs of heat.

The laws of thermodynamics work because Earth and its atmosphere absorb and radiate heat. The absorption and radiation of heat give Earth its distinctive characteristics and its ability to sustain life.

Christopher Cumo
Independent Scholar

See Also: Climate; Greenhouse Effect.

Further Readings

Li, Tim, Timothy F. Hogan, and C. P. Chang. "Dynamic and Thermodynamic Regulation of Ocean Warming." *Journal of Atmospheric Sciences*, v.57 (2000).

Ozawa, Hisashi, Atsumu Ohmura, Ralph D. Lorenz, and Toni Pujol. "The Second Law of Thermodynamics and the Global Climate System: A Review of the Maximum Entropy Production Principle." *Review of Geophysics*, v.41 (2003).

Thermohaline Circulation

The warm surface waters of the tropical oceans occupy a very shallow layer, approximately 328 ft. (100 m) deep that floats on the far colder water of the deep ocean that reaches to depths of 3.1 mi. (5 km). The temperature and salinity of the deep ocean is so uniform that its water must originate in cold, high latitudes where surface waters sink into the deep ocean and then spread across the globe. Ultimately, this water must rise back to the surface and return to the regions of sinking, thus constituting a conveyor belt. Because the motion has strong north–south and up–down components, it is referred to as a meridional overturning cell. For the water to sink, it must be dense, which means that it must be cold and saline. Hence, the circuit away from and then back to the regions of sinking, depending on density and salinity gradients, is known as the thermohaline circulation. Its depiction in documentary films such as Al Gore's *An Inconvenient Truth* has brought the oceanic conveyor belt to the public's attention.

At first, oceanographers speculated that the thermohaline circulation is symmetrical about the equator, with water sinking in polar regions and rising to the surface in lower latitudes. In reality, the circulation is very asymmetrical. Sinking in the northern hemisphere is limited to the northern Atlantic Ocean and is absent from the Pacific and Indian oceans because those two oceans are insufficiently saline at the surface. Confirmation of this asymmetry is available in carbon-14 measurements of the age of a parcel of seawater, the time since it was last at the ocean surface. The results show that the deep water is youngest in the northern Atlantic and oldest in the northern Pacific, where its age approaches 1,000 years. The surface flow toward the northern Atlantic affects a northward transport of heat that contributes to the temperate climate of western Europe.

Oceanic density's dependence on both temperature and salinity means that the increase in density associated with low temperatures in high latitudes could be countered by a flux of freshwater onto the ocean's surface, for example, when glaciers melt. If the water becomes too buoyant, it no longer sinks, the conveyor belt stops, and oceanic circulation experiences radical changes. Such changes, which may have altered climate in Earth's very distant past, are shown vividly in the movie *The Day After Tomorrow*.

Geochemical measurements provide a wealth of information about the thermohaline circulation. Unfortunately, there is also much misinformation concerning the role of that circulation in Earth's climate. The reason is a lack of information about key aspects of the thermohaline circulation. Although it is known where the surface waters sink into the deep ocean, where that water returns to the surface is at present a puzzle. The water has to be heated as it rises into warmer layers close to the surface. The required rate of heating is far greater than that which measurements show to be available in much of the oceans. Hence, the cold water probably rises in relatively small regions of strong turbulent mixing, for example over ridges on the ocean floor. It is possible that the waters that sink in high latitudes rise back to the surface in high latitudes where they can do so without requiring significant heat. Once in the surface layers, the motion of the water is strongly under the influence of the winds that drive intense currents such as the Gulf Stream and Kuroshio. Those shallow, swift currents, which would be present even in the absence of a thermohaline circulation, are also involved in the transport of heat to high latitudes, and therefore in the maintenance of temperate climates in high latitudes. Do fluctuations in the intensity of the Gulf Stream indicate imminent changes in the climate of western Europe? What is the relative importance of the wind-driven and thermohaline circulations in the poleward transport of heat today? Did changes in that transport contribute to climate changes in Earth's distant past, and contribute to the recurrent ice ages, for example? The disagreements among scientists regarding answers to these questions reflect the uncertainties in what is known about the thermohaline circulation.

This circulation is of central importance to the global climate because it maintains remarkably uniform conditions in the deep ocean, which has a high concentration of the greenhouse gas carbon dioxide. It is so slow that water particles take approximately 1,000 years to travel from the North Atlantic to the deep northern Pacific; measurements of current conditions provide only limited information about possible changes in that circulation. Observations of past climates similarly provide limited information. As a result, predictions of future changes in this circulation that will accompany global warming are also uncertain.

S. George Philander
Princeton University

See Also: Biogeochemical Feedbacks; Chemistry; Climatic Data, Atmospheric Observations; Meridional Overturning Circulation; Salinity.

Further Readings

Barreiro. M., A. Fedorov, R. Pacanowski, and S. G. Philander. "Abrupt Climate Changes: How Freshening of the Northern Atlantic Affects the Thermohaline and Wind-Driven Circulations." *Annual Review of Earth and Planetary Science*, v.36 (2008).

Woodman, Matthew Raymond Henry. *The Thermohaline Circulation: Its Importance in Climate Changes*. Reading, UK: University of Reading, 2000.

Thermosphere

The Earth is surrounded by a blanket of air called the atmosphere. The atmosphere is a thin layer of gases that envelope the Earth. The gases are held close to the Earth by gravity and the thermal movement of air molecules. Life on Earth is supported by the atmosphere, solar energy, and the magnetic fields. Five layers have been identified in the atmosphere, using thermal characteristics, chemical composition, movement, and density. The atmosphere is divided into the troposphere, stratosphere, mesosphere, thermosphere, and exosphere. The thermosphere, from the Greek word for heat (*thermos*), is the fourth atmospheric layer from Earth, separated from the mesosphere by the mesopause. It begins about 50 mi. (80 km)

The northern lights occur in the thermosphere, which is the fourth atmospheric layer from Earth. It is the first layer of the atmosphere that is exposed to the sun's radiation, so it is the hottest layer of the atmosphere.

above the Earth and is the layer of the atmosphere directly above the mesosphere and below the exosphere. The lower part of the thermosphere, from 50 to 342 mi. (80 to 550 km) above the Earth's surface, contains the ionosphere, which is the region of the atmosphere that is filled with charged particles. Beyond the ionosphere, extending out to perhaps 6,214 mi. (10,000 km), is the exosphere.

Hottest Layer in the Atmosphere

The Earth's thermosphere is the layer of the atmosphere that is first exposed to the sun's radiation, and so is first heated by the sun; it is the hottest layer of the atmosphere. Within the thermosphere, temperatures rise continually to well beyond 1,832 degrees F (1,000 degrees C). In the thermosphere, ultraviolet radiation causes ionization. At these high altitudes, the residual atmospheric gases sort into strata according to their molecular mass. Thermospheric temperatures increase with altitude as a result of the absorption of highly energetic solar radiation by the small amount of residual oxygen present. Temperatures in the thermosphere are highly dependent on solar activity. Radiation causes the air particles in this layer to become electrically charged, enabling radio waves to bounce off and be received beyond the horizon.

The few molecules that are present in the thermosphere receive extraordinary amounts of energy from the sun, causing the layer to warm to high temperatures. Air temperature, however, is a measure of the kinetic energy of air molecules—not of the total energy stored by the air. The air is so thin that a small increase in energy can cause a large increase in temperature. Because the air is so thin within the thermosphere, such temperature values are not comparable to those of the troposphere or stratosphere. Again, because of the thin air in the thermosphere, scientists cannot measure the temperature directly. Instead, they measure the density of the air by how much drag it puts on satellites, and then use the density to determine the temperature.

Although the measured temperature is very hot, the thermosphere would actually feel very cold to humans because the total energy of the few air molecules residing there would not be enough to transfer any appreciable heat to our skin. In addition, it is so near vacuum that there is not enough contact with the few atoms of gas to transfer much heat. A normal thermometer would read significantly below 32 degrees F (0 degree C). The dynamics of the lower thermosphere are dominated by the atmospheric tide, which is driven in part by the very significant diurnal heating.

The atmospheric tide dissipates above this level because molecular concentrations do not support the coherent motion needed for fluid flow. The International Space Station has a stable orbit within the upper part of the thermosphere, between 199 and 236 mi. (320 and 380 km). The northern lights also occur in the thermosphere.

Akan Bassey Williams
Covenant University

See Also: Atmospheric Composition; Atmospheric Vertical Structure; Aurora; Climate Change, Effects of; Mesosphere; Stratosphere; Troposphere.

Further Readings

Johnson, R. M. and T. L. Killeen. *The Upper Mesosphere and Lower Thermosphere.* Washington, DC: American Geophysical Union, 1995.

Windows to the Universe. http://www.windows.ucar.edu (Accessed March 2012).

University Corporation for Atmospheric Research. http://www2.ucar.edu (Accessed December 2011).

Thunderstorms

A thunderstorm is a localized storm that is produced by a cumulonimbus cloud and always contains thunder and lightning. Thunderstorms form in conditionally unstable environments, meaning there is cold, dry air aloft over warm, moist surface air. This causes the air to become buoyant and allows for rising air motion. A lifting mechanism is also needed to start the air moving. Such lifting mechanisms include surface heating, surface convergence, lifting due to mountains, or lifting along frontal boundaries.

The heat and the humidity of the summertime can often produce what are called ordinary thunderstorms or air mass thunderstorms. These are the type of the thunderstorms that seem to suddenly "pop up," last less than an hour, and are rarely severe. A severe thunderstorm is defined by the National Weather Service as having 3/4-inch-diameter hail and/or surface winds exceeding 58 mph and/or producing a tornado. Ordinary thunderstorms also do not usually have excessive vertical wind shear, meaning that the wind speed or direction does not change greatly with height.

Stages of a Thunderstorm

Thunderstorms usually go through a series of stages, from birth to decay. The first stage is known as the cumulus stage, which is dominated by updrafts. The updrafts bring in warm, moist air, which then cools and condenses as it rises. When the clouds further develop and precipitation starts to fall, a downdraft is produced. This marks the beginning of the mature stage, which is the most intense. During this stage, the strong updraft is still present, supplying the warm, moist air, but the strong downdraft is also evident. The gust front is located at the boundary of the updraft and the downdraft. This is an area where the wind velocity rapidly changes. Eventually, the downdraft will cut off the supply of warm, moist air in the updraft. When this occurs, typically 15–30 minutes after the mature stage, the thunderstorm will start to weaken and enter the dissipating stage due to the deprivation of energy from the updraft.

If the vertical wind shear increases, this allows for the thunderstorm to tilt. Therefore, the downdraft is less likely to cut off the updraft, which allows the thunderstorm to persist for a longer period of time. Sometimes, the downdraft can slide underneath an updraft, which can produce multiple-cell thunderstorms, or simply multicell storms. If the vertical wind shear becomes extremely strong, the shear can produce a large rotational thunderstorm known as a supercell, thunderstorms that last longer than an hour, are often severe, and can produce tornadoes. The strong wind shear creates horizontal spin, which can then rotate vertically when the updraft encounters the vortex.

Thunderstorms can occur as a line of multiple-cell thunderstorms known as squall line thunderstorms. These usually form along or slightly ahead of a cold front. The line of thunderstorms can extend over 500 mi. (800 km) and often exhibit severe characteristics. When thunderstorms occur in a large circular pattern, they are known as a Mesoscale Convective Complex (MCC), a large, convectively driven system that usually lasts more than 12 hours and covers more than 386,000 sq. mi. (100,000 sq. km). Many thunderstorms are embedded within the MCC and often form during the summer in the Great Plains. As warm, moist air is brought in from the Gulf of Mexico, the tops of the very high clouds cool rapidly by emitting radiation into space. This makes the atmosphere very unstable and allows for the MCCs to generate and persist. Since MCCs are usually located underneath weak, upper-level winds, they tend to travel very slowly, which can cause locally heavy rains and flooding events.

All thunderstorms have lightning, the electrical discharge, thunder, and resulting shockwave produced by extreme heating. Lightning has a temperature of approximately 54,000 degrees F (30,000 degrees C), which is five times hotter than the surface of the sun. Lightning occurs during the mature stages of thunderstorms and can appear within a cloud, connect from one cloud to another, or travel from cloud to ground. Most lightning strikes are within a cloud.

Worldwide, it is estimated that 50,000 thunderstorms occur every day and over 18 million occur per year. Thunderstorm frequency is the most common in the tropics, especially near the Intertropical Convergence Zone (ITCZ), an area of low pressure near the equator. Lower frequencies occur in drier regions near 30 degrees north/south, which is dominated by the subtropical high pressure, and also in the polar regions. In the United States, thunderstorm activity is predominantly in the southeast, with a maximum located over Florida. Florida has over 90 days of thunderstorms per year due to the convergence of wind from the Gulf of Mexico and the Atlantic Ocean.

Greenhouse Gases and Thunderstorms

Thunderstorms release a massive amount of latent heat energy through condensation. This is a major mechanism for the Earth to transfer heat from areas of energy surplus near the equator to areas of energy deficit toward the poles. Increased greenhouse gas (GHG) emissions are causing global temperatures to rise. With increased global

Thunderstorm over Toronto, Canada. Worldwide, an estimated 50,000 thunderstorms occur daily, with over 18 million occuring annually. Storm frequency is the most common in the tropics. Some models and climate change researchers predict that with increasing global temperatures, increased evaporation rates will create more clouds. The relationship between global warming and increased thunderstorm activity, however, is not well established or understood; scientists especially lack consensus on the role of aerosols and radiation.

surface temperatures, it is expected that more clouds will be produced because of increased evaporation rates. However, the potential impacts of increased cloud coverage and what it means for potential rainfall and thunderstorm activity is not fully understood. The degree to which the increased aerosols and clouds will reflect solar energy back into space is the main discrepancy. Some scientists think that global warming will increase evaporation rates, thereby producing more precipitation. Others have speculated that the increased clouds and aerosols in the atmosphere will greatly reduce the amount of radiation reaching the Earth. As a result, the lower and midlevels of the atmosphere will become warmer, therefore reducing evaporation rates. Since the thermal gradient (the difference in temperature) will be reduced from the surface to the atmosphere, the reduced evaporation rates could then potentially make for drier conditions.

Kevin T. Law
Marshall University

See Also: Agriculture; Climate Change, Effects of; Clouds, Cumulus; Cyclones; Doldrums; Floods; Hot Air; Hurricanes and Typhoons; Intertropical Convergence Zone; Jet Streams; Midwestern Regional Climate Center; Monsoons; Precipitation; Rain; Sunlight.

Further Readings

Ahrens, C. Donald. *Meteorology Today*. Belmont, CA: Brooks/Cole, 2007.
Houghton, John. *Global Warming*. Cambridge: Cambridge University Press, 2004.
Mackenzie, Fred T. *Our Changing Planet*. Upper Saddle River, NJ: Prentice Hall, 2003.

Togo

The Togolese Republic is a small West African country that is predominantly dependent on agriculture. About 38 percent of the country is arable, and another 4 percent is used for livestock. Approximately 28 percent of the country is forested. Controlled by various indigenous tribes in the Middle Ages, the region was part of the Slave Coast from the 16th to 18th centuries, a major European trading center. It was governed by both Germany and France, before gaining independence in 1960.

Nearly all electricity in Togo—98 percent—is generated by fossil fuels. The remainder comes from hydropower. Some of Togo's electricity is imported from Ghana. The country is fairly poor, and the low standard of living keeps carbon dioxide emissions low. Per-capita emissions rose from 0.2 metric ton in 1990 to 0.38 metric ton in 2003, but dropped back down to 0.21 in 2007. About two-thirds of emissions come from the use of liquid fuels, with the remainder generated by the manufacture of cement. There is little public transport in Togo, so automobile usage is higher than normal; a single train line connects the cities of Lome and Blitta. About 35 percent of emissions come from transportation, 37 percent from manufacturing and construction, and only 14 percent from the generation of electricity and heat.

Lome, the capital city, lies on the coast of the Gulf of Guinea and is vulnerable to rising sea levels resulting from climate change. In the north, increased desertification is expected in the 21st century as temperatures rise. Togo is a signatory to the United Nations Framework Convention on Climate Change and the Kyoto Protocol.

Bill Kte'pi
Independent Scholar

See Also: Automobies; Benin; Burkina Faso; Desertification; Ghana.

Further Readings

Mendonsa, Eugene. *West Africa: An Introduction to Its History, Civilization, and Contemporary Situation*. Durham, NC: Carolina Academic Press, 2002.
Piot, Charles. *Nostalgia for the Future: West Africa After The Cold War.* Chicago: University of Chicago Press, 2010.

Tonga

Located in the South Pacific, the Kingdom of Tonga has a land area of 289 sq. mi. (748 sq. km), with a population of 100,000 (2006 est.) and a population density of 396 people per sq. mi. (153 people per sq. km). The country consists of 169 islands, but only 36 of these are permanently inhabited. With 24 percent of the country listed as arable, some 6 percent is used for meadows and pasture. About 12 percent of the country is forested; Tonga has a very restricted timber industry program.

Tonga also has a low per-capita rate of carbon dioxide (CO_2) emissions, with 0.8 metric ton per person in 1990, rising to 1.12 metric tons by 2003, and then rising again to 1.7 metric tons in 2007, ranking Tonga 133th in the world in per-capita emissions. Although electricity production in the country is low, it is all generated from fossil fuels. All the country's CO_2 emissions come from liquid fuels, accounting for electricity production as well as the use of automobiles and small household or business generators. There are 174 automobiles per 1,000 people in Tonga, the 59th-highest in the world, the highest level of car ownership of any of the Pacific islands.

Global warming and climate change are already having a major effect on Tonga, although the government—especially its Department of Environment—was previously slow to recognize the problems that it faced and to acknowledge the cause of these issues. This has resulted in no comprehensive action plan. Overall, Tonga has seen flooding in parts of the country, including a number of the uninhabited islands, and the very real risk of large parts of the country being lost as the water level rises. In addition, there is a problem concerning the alienation of arable land, deforestation leading to soil erosion, and offshore coral reef bleaching and loss of marine life. It is also expected that rising temperatures will lead to more insect-borne diseases such as dengue fever, as well as food-borne diseases. The former accounted for six deaths in 2003 and four the following year.

The Tonga government took part in the United Nations Framework Convention on Climate Change (UNFCCC), signed in Rio de Janeiro in May 1992 and ratified in 1998, in the same year as the ratification of the Vienna Convention. The Tonga government expressed a position on the Kyoto Protocol to the UNFCCC on January 14, 2008, the 177th national government to do so.

Justin Corfield
Geelong Grammar School
Robin S. Corfield
Independent Scholar

See Also: Climate Change, Effects of; Floods.

Further Readings

EarthTrends. "Tonga: Climate and Atmosphere." http://earthtrends.wri.org/text/climate-atmosphere/country-profile-181.html (Accessed October 2011).

Nunn, Patrick and Eric Waddell. "Implications of Climate Change and Sea Level Rise for the Kingdom of Tonga." Apia, Western Samoa: South Pacific Regional Environment Programme, 1992.

Smitz, Paul. *Samoan Islands and Tonga*. London: Lonely Planet, 2006.

Toronto Conference

Scientists from various international organizations, such as the World Meteorological Organization in Geneva, met with their peers in groups at various locations for three years. Following the signing of the United Nations Vienna Convention on the Protection of the Ozone Layer (1985) and the Villach Conference (1985), these meetings helped to develop the basis for further action. From the discussions at these meetings, a scientific accord on the main aspects of how much climate warming can be expected emerged. The confluence of this emerging consensus and other events led to the Toronto Conference in 1988.

The scientists' efforts gained the support of the United Nations, the World Meteorological Organization, the Canadian government, and other international organizations. The scientists then came together in Toronto, Canada, from June 27 to 30, 1988. In attracting national policymakers as well as 300 scientists from 46 countries and organizations, this conference became the first such international conference to combine science and policy.

Titled the International Conference of the Changing Atmosphere: Implications for Global Security, the meeting highlighted atmospheric issues in a comprehensive way. The concern for the potential damage to the planet was compared with the consequences of nuclear war, and the scientific consensus at the conference astonished its chair, Stephen Lewis, who was then Canada's ambassador to the United Nations. Lewis also brokered the strongly worded final declaration. Identifying the existing situation as "an unintended, uncontrolled, globally pervasive experiment," the conference statement claimed that the consequences of this experiment would be second only those of a global nuclear war.

Recognizing that attempts to address issues affecting the atmosphere as a whole had been fragmentary to date, the Toronto Conference took a more global approach. The initiative was to integrate the existing Vienna Convention (1985) and the 1979 Geneva Convention on Long-Range Transboundary Air Pollution and to provide a basis for including issues that had not yet been addressed or recognized. Such an integrated approach to considering the atmosphere as a whole would conceivably permit a more complex approach to interrelated issues and solutions. As such, this initiative raised the possibility of a comprehensive law of the air.

Recommendations

The comprehensive approach and wide representation enabled attention to be paid to scientific, economic, and social concerns. The attendees generated specific calls for action to governments, industry, and nongovernmental organizations. Working groups within the conference made specific recommendations to address a wide range of issues that were relevant to the health of the global atmosphere.

Issues that were recognized were those that arose both directly out of usage of the atmosphere and indirectly through human effects on land and water. The atmospheric effects of the manner and form of human settlement—including the increasing urbanization of populations and acid rain—were directly relevant. Indirect atmospheric effects resulted from the full range of human activities, including food production, industry, energy usage, trade, and investment.

Changing climate and human effects on coastal and marine resources were also pertinent. Human decision making involving forecasting, uncertainty, futures, and geopolitical issues—higher-order considerations resulting from the integration of programs and legal issues—were also addressed.

The precise form of a global pact was debated, with the Canadian government favoring the concept of an international law of the air. Canadian Prime Minister Brian Mulroney pointed out that the groundwork for such an approach exists in the Montreal Protocol to protect the ozone layer and in the impending international protocol on nitrogen oxide control. Norwegian Prime Minister Gro Brundtland recommended a global convention on protection of the climate.

The meeting recommended a global pact to protect the atmosphere and a world atmosphere fund to facilitate global solutions, which recognized differential issues in usage and effects. For instance, the different historical consumption of and contribution to the atmosphere of already industrialized nations, and those in the process of industrializing, would be balanced by having the fund financed in part by taxes on fossil fuels consumed in industrialized nations. The proposed atmosphere fund would then be used partly to provide economic assistance to developing countries pursuing environmentally friendly strategies, such as reducing deforestation.

The delegates concluded that immediate action was imperative to address ozone depletion, global warming and sea-level rise, and acidification by atmospheric pollutants. The potential role of nuclear power as a clean energy source was debated, but no official recommendations emerged. Reduction of other greenhouse gases, substances that deplete the ozone layer, and acidifying emissions were recommended.

Specifically on the issue of global warming from greenhouse gases and climate change, the conference reached a consensus on the likelihood of a rise in the global mean temperature of between 2.7 and 8 degrees F (1.5 and 4.5 degrees C) by about 2050, but not on whether such warming has begun. The conference statement called for a 20 percent cut in present (1988) levels of global carbon dioxide emissions by 2005, about half of which could be achieved through conservation, leading to an eventual cut of 50 percent. This statement was possible as a result of the participation of governments that voluntarily committed to cutting carbon dioxide emissions by 20 percent by 2005. This became the "Toronto target" for greenhouse gas emissions and went beyond the emissions targets recommended by most later international conferences, as well as the 1992 UN Framework Convention on Climate Change and the core goal of Kyoto.

The Toronto Conference was also influential in other developments. The Intergovernmental Panel on Climate Change (IPCC), an international grouping of over 300 of the world's best climate scientists charged with peer reviewing and reporting on the latest international science, effects, and responses to climate change, had been formed just before the conference. The conference was instrumental in promoting the IPCC and in the eventual appointment of Swedish scientist Bert Bolin to head it.

Public Awareness

Discussions at the conference also led to the allocation of resources to the World Climate Programme and other global research institutions, to the support for technology transfer solutions, to the advocacy of reduction of deforestation, and to raising public awareness of issues related to the atmosphere.

As a follow-up to June's Toronto Conference, a smaller meeting of legal and policy experts was held in Ottawa, Canada, from February 20 to 22, 1989, to begin developing an international accord for the protection of the atmosphere. The 80 participants, acting in a personal capacity, constituted a broad spectrum of experts and officials from developed and developing countries, nongovernmental organizations, and academic institutions. They discussed the legal and institutional framework for dealing with emerging atmospheric problems, agreed where possible on the basis of an umbrella convention framework, and identified areas of possible disagreement.

With the 1982 Law of the Sea as a precedent, the meeting recommended that one or more international conventions, such as a law of the atmosphere and a narrower climate change convention with appropriate protocols were urgently needed, especially to limit greenhouse warming. The statement from this meeting presented early drafts of the proposed documents.

The law of the atmosphereapproach was criticized as being more unrealistic than the narrower climate change convention and did not receive much attention from subsequent negotiators. In carrying the ideas from the Toronto Conference forward, the Ottawa meeting proposed broad terminology for atmosphere and atmospheric interference; discussed the obligation of nations to protect the atmosphere, recognizing the relationship between the atmosphere and other aspects of the environment; recognized the need to balance of development internationally; and proposed an international notification process for harmful activities, liability, compensation, and dispute resolution mechanisms, as well as details of the Atmospheric Trust Fund.

Features that were eventually included in the 1992 Climate Convention included the approach of a framework treaty that deals with the central issues, with protocols for particular details. This would be similar to the Vienna Convention with the Montreal Protocol. Two years after the Toronto Conference, the IPCC issued assessments that provided the basis for the United Nations Framework Convention on Climate Change in 1992, followed by the Kyoto Protocol in 1997.

Lester de Souza
Independent Scholar

See Also: Climate Change, Effects of; Villach Conference; World Meteorological Organization.

Further Readings

Environment Canada Library. http://www.ec.gc.ca (Accessed March 2012).

Zaelke, D. and J. Cameron. "Global Warming and Climate Change: An Overview of the International Legal Process." *American University Journal of International Law and Policy*, v.5/2 (1990).

Tourism

The relationship between tourism and global warming is paradoxical: global warming has become a threat to tourism, yet tourism remains a major cause of global warming. This vicious circle is well known to all stakeholders of the tourism industry, but implementing meaningful change has proven difficult because of three types of resistance: politico-economic resistance (from policymakers in regions and countries that rely heavily on tourism as a source of income), commercial resistance (from the tourism industry), and sociocultural resistance (from tourists who are not ready to change their behavior).

Several factors account for the considerable development of tourism since World War II: growing affluence, longer holidays, cheaper transportation, the availability of preorganized packaged tours, and the development of an industry catering both to mass tourists and independent travellers. The subsequent increase in demand has resulted in an exponential rise in visitor numbers, both domestically (within countries) and internationally (especially from developed countries to developing countries).

Although domestic tourism is statistically much more important (e.g., it accounts for 99 percent of all U.S. tourism and 85 percent of all Australian tourism), international tourism is easier to measure (through a simple head count at borders); in addition, international tourism corresponds much more to the mainstream imagery of tourism, such as an island-hopping cruise in the Caribbean, a romantic holiday in Paris, or a big game safari in Kenya. According to the World Tourism Organization, the number of international tourists increased from a mere 25 million in 1950 to 800 million in 2005. This number is predicted to double and reach 1.8 billion by 2020, as more and more people want to travel. They may well know that they contribute to global warming and climate change, and some may feel a pang of guilt and remorse, but their desire to travel is stronger.

Climate is a key resource for tourism: Favorable climatic conditions are key attractions for tourists, be it to ski in the mountains, to relax on a beach, or to experience nature. As soon as climatic conditions fluctuate and become less predictable, tourism demand is affected and tourist flows move elsewhere: Tourism, as a geographic phenomenon, is fickle and versatile. The mass media occasionally run stories about tourism hot spots that are victims of climate change and see their tourism appeal decrease; examples abound from all across the world, from less snowfall and shorter skiing seasons in Aspen, Colorado, or in Chamonix in the French Alps to damage to coral reefs and rising ocean water in Australasia, not to mention hurricanes that affect island resorts and the cruise industry.

These media stories are not just anecdotes or isolated incidents: They are part of a wider concern already well documented in the tourism literature, both in the academic literature (with seriously researched case studies, a nascent modelization of the relationship between climate change and tourism, and an increasing number of specialists, such as the Canadian Daniel Scott, the Dutch Bas Amelung, and the French Jean-Paul Ceron) and in the professional literature (industry publications such as professional bodies' reports and newsletters, as well as travel guides for tourists).

Warmer temperatures at ski resorts: As soon as climatic conditions fluctuate and become less predictable, the tourism demand is affected and tourist flows move elsewhere; tourism, as a geographic phenomenon, is fickle and versatile. Negative effects include increased storm surge or beach erosion that threatens tourist destinations on low-lying islands. However, in places like Banff National Park, Alberta, Canada, any longer warm-weather tourism season is expected to have a positive impact on visitation.

Environmental Awareness and Tourism

International tourism organizations endeavor to raise awareness, harness energies, and articulate realistic plans of action. In 2003, the World Tourism Organization held its first Summit on Climate Change and Tourism. This resulted in the Djerba Declaration on Tourism and Climate Change; signatories encouraged all governments to act to control climate change. The Djerba Declaration also asked the travel industry to adjust its activities to minimize climate change, and it invited consumer associations and the media to further raise public awareness, both at destinations and in generating markets.

Taking place in 2007 in Davos, Switzerland, the Second International Conference reviewed and emphasized the key aims and intentions of the Djerba Declaration, strengthening its urgency and exploring concrete ways for tourism to respond to climate change, while still ensuring tourism development as a tool of economic growth and sociocultural well-being. At another level, in 2007, the World Travel and Tourism Council launched an international campaign on the same topic, calling for an open and mature dialogue on issues of tourism, climate change, and the environment; the campaign ran full pages in major publications such as the *Daily Telegraph*, *Newsweek*, and the *Wall Street Journal*, as well as travel trade media around the world.

Tourism professionals are aware of their responsibilities with regard to the environment, but they also know how much tourism contributes to the world economy. In 2007, the tourism industry globally represented 231.2 million jobs, which corresponded to 8.3 percent of total employment—one in every 12 jobs worldwide. By 2017, this figure is expected to rise to 262.6 million jobs. Through their statements and declarations, tourism policymakers at all levels (local, regional, national, and supranational) and from all sectors (public, private, and voluntary) show that they understand the seriousness of the situation with regard to global warming and climate change, but tourism is still one of the world's largest economic sectors.

Tourism is not just about holidays and recreation, it is a powerful economic force that cannot be obliterated; rather, it has to be managed by implementing ways to maximize its benefits and minimize its costs. Through the concept of sustainable tourism, sustainable development has now become a key agenda in tourism, covering a range of social, cultural, and environmental issues, including references to climate change and global warming; yet, solutions and ways forward are difficult to find. Simply controlling the number of international flights and limiting the amount of international tourists may not be a viable alternative at all; such a short-term measure could have devastating effects on many regions and countries where the economies are dependent, if not overreliant, on tourism income; for instance, small island states such as the Maldives, the Netherlands Antilles, and the Seychelles.

In 2007, global tourism generated over $7 trillion, and this number is likely to double within the next decade. The ongoing democratization of tourism means that more and more people can afford to travel and readily do so, even when they are aware of the effect on climate change and global warming; their arguments are usually twofold: first, that their own individual contribution is minimal; and second, that tourism is only one cause of climate change and global warming.

Many factors account for the ongoing increase in tourists' numbers: technological developments (epitomized by the superjumbo A380, with its capacity of 850 passengers), market deregulation (leading to more competition, which keeps prices low, especially in the airline industry), and the multiplication of specialized niche markets (such as sports tourism, senior tourism, gay tourism, or industrial tourism, making the demand more fragmented, but also easier to target and satisfy). Neither financial penalties (tourism ecotaxes imposed on airlines or at the destination) nor ethical appeals (campaigns asking would-be tourists to reconsider because of their carbon footprint) are proving effective deterrents for what is increasingly regarded as a right and not a privilege. Even the most vocal critics of tourism like to travel to conferences (business tourism) and go away on holidays (leisure tourism), which weakens the arguments of the antitourism lobby.

Threats to Tourism

Climate change poses several risks to tourism; not only direct risks (climate variability affecting immediate demand as well as tourists' comfort and safety in the short term), but also indirect risks (such as causing damage to ecosystems or reducing water supplies, which may jeopardize tourism in the long term). This is ironic, inasmuch as tourism as a whole is partly responsible: by definition, tourism relies on methods of travel that generate air pollution (greenhouse gas emissions from vehicles that transport tourists, especially aircraft), so by its very existence, tourism heavily contributes to climate change. Rather than attempting a difficult—if not impossible—balancing act, specialists have identified two strategies. The first strategy involves innovation (e.g., with regard to sources and production of energy) and disseminating best practice in terms of sustainable development (e.g., through benchmarks and rewards). This first approach takes tourism in its wider industrial context, applying to tourism some policies and measures from other sectors, such as building and manufacturing.

The second strategy involves analyzing how climate change affects tourism to proactively restructure the tourism industry, both in terms of tourism demand and tourism supply; for instance, extremely hot temperatures in summer in seaside destinations may lead to a decrease of the demand in summer, but to higher rates in other warm times of the year, such as warmer winter periods. As seasonality has always been a plague of the tourism industry, this climate change may eventually prove beneficial, though it will require adapting and revisiting established patterns. This second approach considers tourism as a specific and idiosyncratic system, although some related sectors, such as agriculture, local transport, and the entertainment industry, are also likely to be affected (a phenomenon conceptualized as backward linkages). The two strategies are not mutually exclusive: they can be combined, as they are both underpinned by a mix of idealism and pragmatism (adaptation and mitigation are two concepts used by scholars and policymakers alike).

Because of the intrinsic diversity of the tourism industry (exemplified by differing tourists' needs and expectations, from a backpacking teenager touring Europe to jetsetters staying in exclusive resorts), there may not be a one-size-fits-all solution. Case studies of destinations, ranging from the Fijian archipelago to Banff National Park in Canada, show that each region needs to develop methodologies and planning scenarios to anticipate changes in tourism demand and distribution, while remembering that it is not only the complex tourism system that is affected by global warming and climate change, but also local communities. Collaboration between partners and agencies is heralded as a necessary mechanism; a good illustration is the official cooperation between the World Tourism Organization and the World Meteorological Organization.

In 2006, the Expert Team on Climate and Tourism was jointly set up by both agencies. Such meetings of experts can result in sharing intelligence to help with research projects, as there is a wide recognition that decisions need be evidence based. The tourism industry is aware of the problems posed by global warming and climate changes, and it wants to be part of the solution.

Loykie L. Lomine
University of Winchester

See Also: Economics, Cost of Affecting Climate Change; Transportation.

Further Readings

Becken, S. and M. Patterson. "Measuring National Greenhouse Gas Emissions From Tourism as an Important Component Towards Sustainable Tourism Development." *Special Issue: Journal of Sustainable Tourism: Climate Change and Tourism*, v.14/4 (2006).

Gossling, Stefan and Michael C. Hall, eds. *Tourism and Global Environmental Change*. London: Routledge, 2005.

Hall, Michael C. and James Higham, eds. *Tourism, Recreation and Climate Change*. Clevedon, UK: Channel View Publications, 2005.

Matzarakis, Andreas and Chris de Freitas, eds. *Proceedings of the First International Workshop on Climate, Tourism and Recreation: Report of a Workshop Held at Porto Carras, Neos Marmaras, Halkidiki, Greece, 5–10 October 2001*. Milwaukee, WI: International Society of Biometeorology, 2001.

Scott, Daniel, et al., eds. *Climate, Tourism and Recreation: A Bibliography*. Waterloo, Ontario, Canada: University of Waterloo, 2006.

Trade Winds

The trade winds are a large-scale component of Earth circulation, occupying most of the tropics, straddling the equator between approximately latitude 30 degrees N and latitude 30 degrees S, with a seasonal shift of the entire trade wind belt system about 5 degrees of latitude northward during summer (July) and southward during winter (December).

In the Northern Hemisphere, warm equatorial air rises and flows north toward the pole, the Coriolis effect (caused by the Earth's rotation) deflects the current, and as the air cools, it descends, blowing southwestward from the northeast. In the Southern Hemisphere, warm equatorial air rises and flows south toward the pole, the Coriolis effect deflects the current, and as the air cools, it descends, blowing northwestward from the southeast. The rising air is associated with deep atmospheric convection, heavy precipitation, and weak wind speeds, with an influence on global weather patterns. Air heated by the sun rises and releases moisture through rain and thunderstorms. Once the air cools, it descends as drier air. In the equatorial low, the air rises and travels aloft to the subtropical highs, where it then sinks.

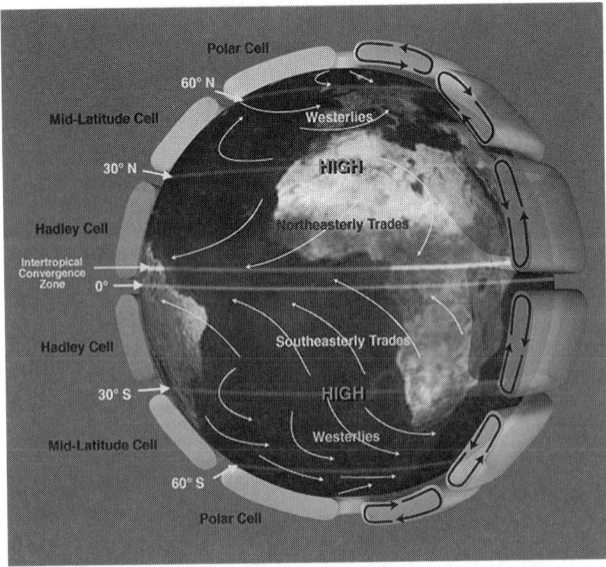

A NASA diagram of climatic zones and trade winds. In the tropics, the easterly, consistent trade winds dominate. These winds are under the influence of El Niño, La Niña, and the development of hurricanes and cyclones.

Mariners called these reliable wind currents for sailing the trade winds, or westerlies. The name *trade winds* comes from an old sailing term meaning that the winds could be counted on to blow steadily from the same direction at a constant speed. The trade winds, or easterlies, carried air from east to west at low latitudes and are less regular over land areas than they are over the oceans. The trade winds meet at the Intertropical Convergence Zone. The doldrums (downward branch of the Hadley Cell, named for George Hadley, whose 1735 paper linked rising air and the Earth's rotation in causing the trade winds) are the calm winds at the Intertropical Convergence Zone in the area between latitude 5 degrees N and latitude 5 degrees S, where a sailing ship might not move because of the calm winds. In satellite imaging, the Intertropical Convergence Zone appears as a band of clouds. The strength and position of the Intertropical Convergence Zone influences tropical and global weather patterns.

Air temperature differences across the Earth's surface (both land and water) create winds, with warm air being lighter than cold air. Near the equator, the sun heats the sea surface, causing the warm air at the surface to rise and be replaced by the trade winds blowing from subtropical high pressure systems into equatorial low-pressure troughs. The trade winds blow steadily for days and are among the most consistent on Earth. When trade winds move over warm tropical waters, they pick up moisture and bring heavy rainfall to the windward-facing slopes of mountainous areas, contrasting with the downward motion of dry air that creates desert areas on land. Because the area of Earth between the Tropic of Cancer and Tropic of Capricorn, lying at approximately 23 degrees latitude on either side of the equator, receives more solar heat than the rest of the Earth, the warm air creates clouds and rain, with thundershowers there almost every day.

The influence of the trade winds on weather and climate is seen with El Niño, La Niña, and the development of hurricanes and cyclones. The differences in pressure and temperature between the two sides of the Pacific are caused by the trade winds; air blowing from east to west pushes water, making the sea level higher in the western Pacific, and makes cold water rise toward the surface, making the eastern Pacific approximately 14 degrees F (7.7 degrees C) cooler than the western Pacific.

During El Niño years, the eastern Pacific sea surface is warmer, and the Intertropical Convergence Zone is closer to the equator, causing rainfall over the Pacific. The warm surface temperature is associated with reversed air pressure patterns and decreasing strength of trade winds, so more water stays in the eastern Pacific off the coast of South America. With the rain pattern shift eastward, the western Pacific can become drier over India and much of southeast Asia. A similar pattern sets up in the Atlantic, resulting in extreme drought in the eastern United States and reduced tropical storm development in the Atlantic Ocean.

During La Niña years, the trade winds are stronger than normal, causing more cold water to rise to the ocean surface. The cooler surface temperature is associated with a rain pattern shift westward. The eastern areas thus become drier, with an increased probability of flooding from monsoons in both India and much of southeast Asia.

Hurricanes (Atlantic) and cyclones (Indian Ocean) are tropical storms of low-pressure cells. Formation of hurricanes in the Atlantic comes from solar heating of water off the West African coast along the Intertropical Convergence Zone, with high cumulus cloud formation in the low-pressure area along the edge. These systems are pushed westward by the trade winds, and the rotation is set in motion by the Coriolis effect. A similar pattern sets up in the Pacific, causing cyclones.

Lyn Michaud
Independent Scholar

See Also: Coriolis Force; Doldrums; El Niño and La Niña; Winds, Westerlies.

Further Readings

Farrar, Tom and Robert Weller. "Where the Trade Winds Meet: Air-Sea Coupling in the Inter-Tropical Convergence Zone." http://www.oar.noaa.gov/spotlite/archive/spot_pacs.html (Accessed March 2012).

Fishman, Jack and Robert Kalish. *The Weather Revolution: Innovations and Imminent Breakthroughs in Accurate Forecasting*. New York: Plenum Press, 1994.

National Weather Service Southern Region Headquarters. "Effects of ENSO in the Pacific." http://www.srh.noaa.gov/jetstream/tropics/enso_patterns.htm (Accessed June 2008).

Transportation

Transportation can be defined as the movement of people, goods, and services from one place to another. Its system consists of fixed facilities, flow entities, and control systems that permit the free flow and efficient movement of people and goods from place to place across geographical boundaries. Generally, there three forms of transportation exist: road, water, and air transportation.

Three Types of Transportation

The major road transport systems are the vehicular transport system and the rail transport system. The vehicular transport system comprises the different grades, sizes, and types of automobiles, and the rail transport system comprises the train system. Other forms of road transport systems include motorcycles and bicycles. These are single-track, two-wheeled vehicles. While motorcycles are powered by small internal combustion engines, bicycles are human driven. In developing countries, where motor vehicles as a means of transportation are both expensive and unavailable for rural transportation, motorbikes and bicycles are the major means of transportation. In addition, in the major cities of developing countries and in China and Japan, the population of cars and other automotive means of road transportation has created heavy traffic, resulting in traffic jams.

People in these areas often resort to motorbikes or bicycles for quick navigation. Unlike industrialized nations, where cities have good transportation systems and networks, people in rural areas and cities of developing countries employ motorbikes and bicycles as mass transit systems. They also use these bicycles and motorbikes to transport commuters, thereby creating their own avenues of employment. In addition to two-wheeled motorcycles and bicycles, there are also three-wheeled bikes.

Electric cars, also known as e-cars, are in development as well. This type of automobile runs on electric batteries or other forms of electrically charged cells to power its engine. Although electric cars were driven in the 19th century, they are presently being redeveloped for efficiency and effectiveness and are currently not as common as other automotive vehicles that run on liquid fuels.

Road traffic is a major contributor to environmental degradation and global warming. It provides the largest net contribution to warming, through large emissions of CO_2 and significant emissions of ozone and soot. Soot particles emitted by diesel engines have the ability to absorb sunlight, thus heating up the climate. Total warming from road traffic is about 0.19 watt per sq. m (W/m^2), forming about 7 percent of the total climate forcing due to an increased concentration of ozone, soot, and greenhouse gases (GHGs).

Water transportation comprises various aquatic vehicles such as ships, boats, ferries, canoes. Air transportation, in turn, comprises the different grades, shapes, and sizes of airplanes and helicopters. Ocean liners, boats, tankers, and cargo ships are other means of transportation that contribute to GHGs. Ships (whether large or small) and boats run on internal combustion engines that use diesel fuel. Diesel is a hydrocarbon containing between 10 to 15 carbon atoms covalently bonded to hydrogen. However, diesel fuel is a type of fossil fuel that contains sulfur. Its combustion leads to the emission of GHGs that include sulfur dioxide and carbon dioxide (CO_2). The sulfur dioxide can react with water vapor to form acid rain.

Air traffic, as a sector in the transportation industry, also shows a trend toward increased environmentally damaging emissions. Airplanes fly in the upper edges of the atmosphere where the air is rarefied, and the planes release large quantities of GHGs. CO_2, the main constituent in the exhaust gases, slowly descends into the lower altitude. However, the large number of planes flying across the sky has caused the average amount of these GHGs to increase. CO_2 and other GHGs get caught up in the stratosphere where they become much more potent than at the Earth's surface; these gases block radiant energy from reaching the planet. Thus, the global warming effect of air traffic pollution in the stratosphere is very high. In addition, the NO_x gases emitted have an especially large effect on ozone formation. More recent research reports on air traffic suggest that the occurrence of ice clouds, called *cirrus clouds*, at higher altitudes is increasing in areas with heavy air traffic because the vapor trails left by aircraft at high altitudes, under certain meteorological conditions, can expand. These clouds, found at altitudes of between 5 and 7 mi. (8 and 12 km), have a warming effect on the climate because their greenhouse effect is stronger than their cooling effect through the reflection of light. This is a result of the low temperature at this height.

Transportation Emissions

Each one of these motorized forms of transportation runs on fossil fuels of crude oil distillates and coal, except for e-cars, bicycles, and human-powered tricycles. When used to power an internal combustion engine, fossil fuels emit gaseous byproducts of CO_2, carbon monoxide, and sulfur dioxide. In addition, cars emit methane and nitrous oxide. Worldwide, the amount of GHG emitted by these vehicles is very high because of factors that include an increase in the number of vehicles, volume of passengers, and freight traffic. The level of contribution of CO_2 from transportation alone varies from state to state and country to country. In 1990, Japan's contribution was approximately 19 percent. CO_2 levels of the United States doubled between 1960 and 2001—specifically, from 2 billion metric tons in 1960 to almost 5.7 billion metric tons in 2001, accounting for more than a 100 percent increase. Over 20 percent of this increase in emissions is linked to transportation. Transportation accounts for 40 percent of volatile organic compounds, 77 percent of carbon monoxide, and 49 percent of nitrogen oxide emissions in the United States.

In Canada, transportation is the largest single anthropogenic source of outdoor air pollution. On average, each of the several million vehicles registered across the country emits approximately 5 metric tons of air pollutants and gases annually. This trend is the same for all industrialized nations and several developing nations, such as Nigeria in West Africa, because of increasing population growth and a rapid rate of economic growth that results in increased automobile and other transportation uses. This increase has been accompanied by greater emissions of air pollutants and GHGs because of the type of engines in use and the nature of the fuels in place.

Combustion engines emit nitrogen oxides (NOx), carbon monoxide, and unburned hydrocarbons capable of chemical transformation in the atmosphere, creating other gaseous matter such as ozone. Ozone is a triatomic molecule consisting of three atoms of oxygen; it is an allotrope of oxygen, but is much less stable. Its instability makes it a strong oxidizing agent, having the ability to decompose to oxygen in the atmosphere within 30 minutes. In its physical undiluted state at standard temperature and pressure, it is a pale blue, odorless gas. In the troposphere, ozone acts as a GHG and has a radiative forcing of about 25 percent that of CO_2. Around the Earth's surface, it poses a regional air pollution problem by damaging human health and agricultural crops.

Residual fuel oils, particularly the heavy oil used aboard ships, contain sulfur, which reacts with atmospheric water and oxygen to produce sulphate particles and sulfuric acids, also known as acid rain. Acid rain lowers soil and freshwater pH, resulting in damage to the natural environment and causing chemical weathering. Its ability to increase the reflection of part of the sunlight that should come into the Earth's surface creates a cooling effect.

Global Warming Implications

The various gases emitted by these engines pose serious challenges to the environment. CO_2, the most prevalent of the GHGs, is a colorless, odorless gas with a covalent bond between its atomic constituents. It is the most potent GHG and is highly atmospherically stable; it has a life of over 100 years and is able to absorb radiations below the visible light spectrum, which traps heat attempting to escape from the Earth's surface, thereby causing an increase in the surface temperature of the planet. It has a radiative forcing of 1.5 watts per square meter (W/m^2) and is regarded as the most powerful GHG because of its long atmospheric stability period.

The presence of a ton of CO_2 emitted into the atmosphere thus has a deleterious environmental warming effect for more than 100 years. As a result, the potential consequences of the increasing anthropogenic emission of this gas from transportation and other sources has been, and may continue to be, the subject of important climate change debates around the world.

Methane (NH_4), in contrast, is a covalent compound, colorless, and odorless. It is not as stable as CO_2, but has a stronger effect as a GHG than CO_2. Its stability period in the Earth's atmosphere is 10 years. It absorbs infrared radiation and affects tropospheric ozone. Its global warming potential, when compared with CO_2 over a 20-year period, is 72. Methane may not be as popular a GHG as CO_2, but its effects on the climate are stricter than CO_2. It is, however, rated as the second potent GHG after CO_2, with the exception of water vapor, because of its short life and the quantity of it found in the Earth's atmosphere.

Another gas emitted from the burning of fossil fuels is nitrous oxide. It is a colorless, nonflammable, sweet-smelling gas having two nitrogen atoms and one oxygen atom covalently bonded together. When released into the atmosphere, nitrous oxide is the third-largest GHG contributor to global warming, and has more of an effect than an equal amount of CO_2 per equal mass of CO_2. Nitrous oxide is 296 times stronger a GHG than CO_2. It attacks ozone in the stratosphere, increasing the amount of ultraviolet light entering the Earth's surface. This ultraviolet light has deleterious effects on the human immune system, as well as the eye and the skin. It causes sunburn, inflammation, immunosuppression, tanning, and the accelerated aging of the skin.

The other aforementioned gaseous pollutants from transportation are carbon monoxide and sulfur dioxide (SO_2). Carbon monoxide is a colorless, odorless, and tasteless gas, formed by the thermal composition of excess carbon with oxygen. It consists of a carbon atom covalently bonded to an oxygen atom. It is emitted in motor vehicle exhaust, having gone through an incomplete internal combustion process of burning excess fossil fuels in the presence of oxygen. It is a toxic air pollutant. Its reaction in the atmosphere with some atmospheric constituents like the hydroxyl radical (OH-) can increase the amounts of atmospheric methane and tropospheric ozone, thus causing an indirect forcing effect. It has the ability to combine with oxygen in the atmosphere to make CO_2, thereby contributing to greenhouse effects and global warming. Apart from this environmental degradation, CO has deleterious effects on human health. Exposure to excess carbon monoxide can lead to heart and respiratory problems and has an effect on pregnancy, the central nervous system, and the heart, to mention a few adverse effects.

Sulfur dioxide, in contrast, is a covalently bonded chemical compound made up of an atom of sulfur and two atoms of oxygen. It is anthropogenically produced from the combustion of coal and petroleum. It is a colorless gas, smells similar to burning sulfur, and is able to undergo serial combination to form sulfuric acid, which is an acid rain with corrosive tendencies that is found in the atmosphere. Sulfur dioxide is also toxic and has caused damage to humans in the past. However, its atmospheric cooling effect as a regulatory measure on the global warming effects of GHGs on the Earth's surface is limited by its short life span of not more than a week.

With the present systems of transportation in use the world over, the rate of emissions of these harmful gases will continue to increase and their deleterious effects will become severe, as anthropogenic emissions of CO_2, methane, and nitrous oxide are among the major causes of global warming. Gaseous pollutants of carbon monoxide and sulfur dioxide also contribute to dangerous effects on humans and the environment. However, various reports of the Intergovernmental Panel on Climate Change exist, each pointing to the damaging effects of such GHGs as CO_2, methane, and nitrous oxide. Transportation's contribution to the emission of these gases is high and determining what must be done to reduce these emissions must be a global exercise that involves the scientific community. Transportation's effects on global warming and environmental pollution cannot be reduced without a redesign of the present system of engines vis-à-vis the available fuel systems. The emissions from modes of transportation can only increase as the number of transport methods increases.

Mitigating the emission of GHGs from transportation involves proper planning and encouragement of the use of e-cars, mass transit systems, bicycles, and human-powered tricycles; proper urban planning will also reduce sprawling, which in turn will reduce travel time and thereby minimize GHG emissions. Furthermore, encouraging the use of public transport will reduce the quantity of GHGs emitted per traveling passenger. Depending on the type of vehicles and the engine system, the average mass (in grams) of CO_2 emitted per passenger km varies and can be over 200 g. per passenger km. Thus, whatever limits travel time and distance can also limit the amount of CO_2 emitted per km.

Oluseyi Olanrewaju Ajayi
Covenant University

See Also: Automobiles; Carbon Footprints; Climate Change, Effects of; Food Miles; Fossil Fuels; Oil, Consumption of; Tourism.

Further Readings

Myhre, M., E. J. Highwood, K. P. Shine, and F. Stordal. "New Estimates of Radiative Forcing Due to Well Mixed Greenhouse Gases." *Geophysical Research Letters*, v.25/14 (1998).

Partington, J. R. *A Short History of Chemistry*. 3rd ed. Mineola, NY: Dover, 1989.

Sawyer, C. N., P. L. McCarty, and G. F. Parkin. *Chemistry for Environmental Engineering and Science*. 5th ed. New York: McGraw-Hill, 2003.

Wang, M. *The Greenhouse Gases, Regulated Emissions and Energy Use in Transportation (GREET) Version 1.5*. Argonne, IL: Argonne National Laboratory, 1999.

Triassic Period

The Triassic period is the geologic time period that extends from about 251 to 199 million years ago. This is the first period of the Mesozoic era, following the Permian and preceding the Jurassic period. Both the start and end of the Triassic are marked by major extinction events. During the Triassic period, both marine and continental life showed an adaptive radiation, beginning from the starkly impoverished biosphere that followed the Permian–Triassic extinction. The first flowering plants may have evolved during the Triassic, as did the first flying vertebrates, the pterosaurs. The Triassic period is further separated into Early, Middle, and Late Triassic epochs.

During the Triassic period, almost all the Earth's land mass was concentrated into a single supercontinent centered more or less on the equator, known as Pangaea. This supercontinent began to rift during the Triassic period but had not yet separated.

The Triassic climate was generally hot and dry, forming typical red bed sandstones and evaporites. There is no evidence of glaciation at or near either pole. The polar regions were moist and temperate—a climate suitable for reptile-like creatures. Pangaea's continental climate was highly seasonal, with very hot summers and cold winters. It probably had strong, cross-equatorial monsoons. The interior of Pangaea was hot and dry during the Triassic period. This may have been one of the hottest times in Earth history. Rapid global warming at the very end of the Permian may have created a super hothouse world that caused the great Permo-Triassic extinction.

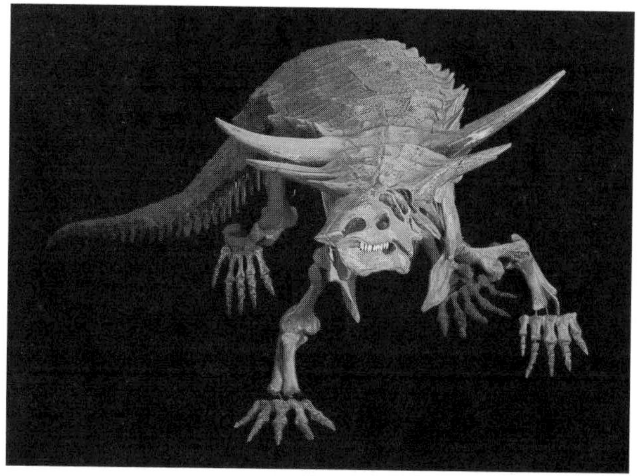

An Aetosaur of the Triassic period, constructed from bones found in the Petrified Forest in Arizona. The Triassic climate was generally hot and dry. There is no evidence of glaciation around either of the poles; rather, they were moist and temperate.

The Permian–Triassic extinction event, also known as the Great Dying, was an extinction event that occurred 251.4 mya (million years ago). This was the Earth's most severe extinction event, with up to 96 percent of all marine species and 70 percent of all terrestrial vertebrate species becoming extinct. There are several proposed mechanisms for the extinction event, including both catastrophic and gradualistic processes, similar to those theorized for the Cretaceous extinction event. The former include large or multiple impact events, increased volcanism, or sudden release of methane hydrates from the seafloor. The latter include sea-level change, anoxia, and increasing aridity. Evidence that an impact event caused the Cretaceous–Tertiary extinction event has led naturally to speculation that an impact may have been the cause of other extinction events, including the Permian–Triassic extinction. Several possible impact craters have been proposed as possible causes of this extinction event, including the Bedout structure off the northwest coast of Australia and the Wilkes Land crater of east Antarctica. In each of these cases, the idea that an impact was responsible has not been proven and has been widely criticized. If impact was a major cause of this extinction event, it is possible or even likely that the crater no longer exists. Seventy percent of the Earth's surface is sea, so an asteroid or comet fragment is over twice as likely to hit sea as to hit land. There is evidence that the oceans became anoxic toward the end of the Permian. There was a noticeable and rapid onset of anoxic deposition in marine sediments around east Greenland near the end of the Permian. The most likely causes of the global warming that drove the anoxic event was a severe anoxic event at the end of the Permian, causing sulphate-reducing bacteria to dominate the oceanic ecosystems and causing massive emissions of hydrogen sulfide, which poisoned plant and animal life on both land and sea. These massive emissions of hydrogen would have severely weakened the ozone layer, exposing much of the life that remained to fatal levels of ultraviolet radiation.

Pangaea's formation would also have altered both oceanic circulation and atmospheric weather patterns, creating seasonal monsoons near the coasts and an arid climate in the vast continental interior. Marine life suffered very high, but not catastrophic rates of extinction after the formation of Pangaea—rates almost as high as in some of the Big Five mass extinctions. The formation of Pangaea seems not to have caused a significant rise in extinction levels on land, and in fact, most of the advance of Therapsids and the increase in their diversity seems to have occurred in the late Permian, after Pangaea was almost complete. Thus, it seems likely that Pangaea initiated a long period of severe marine extinctions, but was not directly responsible for the Great Dying and the end of the Permian.

The possible causes, which are supported by strong evidence, appear to describe a sequence of catastrophes, each one worse than the previous. The resultant global warming may have caused perhaps the most severe anoxic event in the oceans' history. The oceans became so anoxic that anaerobic sulfur-reducing organisms dominated their chemistry.

Fernando Herrera
University of California, San Diego

See Also: Global Warming; Paleoclimates; Tertiary Period.

Further Readings

Beerling, D. "CO_2 and the End-Triassic Mass Extinction." *Nature*, v.24 (2002).
Kerr, R. A. "Paleontology: Biggest Extinction Hit Land and Sea." *Science*, v.289 (2000).
University of California Museum of Paleontology. "Triassic Period." http://www.ucmp.berkeley.edu/mesozoic/triassic/triassic.html (Accessed March 2012).

Trinidad and Tobago

The southern Caribbean island nation of Trinidad and Tobago has a 2011 population estimated at over 1.2 million. Most of the population resides on the island of Trinidad. Trinidad and Tobago's status as a small island developing state heightens its vulnerability to the impacts of global warming and climate change. The impacts of climate change will exacerbate the environmental degradation already caused by past inadequate government environmental policies and poor land and water resource management. Trinidad and Tobago is a member of numerous regional and international climate change mitigation and adaptation programs and is a signatory to the United Nations Framework Convention on Climate Change (UNFCCC) and the Kyoto Protocol.

Environmental concerns related to climate change include diminished reservoirs and water resources, saltwater intrusion, rising sea levels and loss of coastal lands and settlements, rising air and water temperatures, changing rainfall patterns, and more frequent and severe extreme weather events such as floods, droughts, and storms. Rising sea temperatures have already reduced the population of endangered leatherback turtles. The impacts of climate change also threaten the country's efforts at poverty reduction, economic growth, and sustainable development.

Although Trinidad and Tobago contributes far less greenhouse gas (GHG) emissions than developed nations, its carbon dioxide emissions levels per capita are among the world's highest. The nation's economy relies heavily on the production and export of petroleum and related products. Other contributions include gas flaring, the use of national coal deposits for energy production, cement manufacturing, and other industry. Domestic contributions include widespread air conditioning, automobile use, and the lack of sufficient public transportation. Global warming also poses a potential threat to the country's vibrant tourism industry.

Trinidad and Tobago's national government has actively joined in regional efforts at climate change mitigation and adaptation, as well as related issues such as energy security. The government is a member of the Caribbean Community (CARICOM) and supports CARICOM's commitment to the 2009 Liliendaal Declaration on Climate Change and Development and the position of the Alliance of Small Island States (AOSIS). These commitments include regional unification in international negotiations and stabilization of GHG emissions to prevent a global temperature rise above 2.7 degrees F (1.5 degrees C) from pre-industrial levels. Trinidad and Tobago participated in the 1994 Global Conference on the Sustainable Development of Small Island Developing States and is committed to its Barbados Programme of Action.

The country also participated in the resulting Caribbean Planning for Adaptation to Climate Change (CPACC) project, the subsequent Adapting to Climate Change in the Caribbean (ACCC) project of 2001–04, and the Mainstreaming Adaptation to Climate Change (MACC) project of 2004–07. Regional project achievements include the creation of regional and national climate change adaptation policies; the implementation of public-awareness campaigns; the establishment of networks for monitoring sea level, coral reefs, and climate changes; and the development of coastal and marine resources inventories. Trinidad and Tobago also supported the development of the Caribbean Community Climate Change Centre, which opened in 2004.

Trinidad and Tobago's Ministry of Housing and Environment oversees national government climate change mitigation and adaptation efforts. A draft National Climate Change Policy and strategies for GHG emissions reductions are in place. Trinidad and Tobago participated in the UNFCCC and has signed and ratified the Kyoto Protocol. The government supports UNFCCC objectives but believes that equitable, international climate change agreements must recognize the needs of developing nations for sustainable economic growth and outside financial and technological assistance. Other government initiatives include green building construction, energy efficiency, and the development of renewable energy resources such as solar and wind power.

Trinidad and Tobago's national government has partnered with the United Nations Development Programme (UNDP) and local nongovernmental organizations since the mid-20th century on a variety of environmental and economic initiatives. The Global Environment Facility/UNDP Small Grants Programme has provided financial assistance. Efforts have included reforestation, sustainable and organic agriculture, energy security, pollution reduction, endangered species protection, land management, solid-waste management, national environmental policy development, and assistance in meeting international agreements such as the Kyoto Protocol.

Marcella Bush Trevino
Barry University

See Also: Alliance of Small Island States; Developing Countries; United Nations Development Programme.

Further Readings

Lizcano, G., C. McSweeney, and M. New. "UNDP Climate Change Country Profiles: Trinidad and Tobago" (October 30, 2008). http://ncsp.undp.org/sites/default/files/Trinidad_and_Tobago.oxford.report.pdf (Accessed October 2011).

Nath, Shyam and John L. Roberts. *Saving Small Island Developing States: Environmental and Natural Resource Challenges*. London: Commonwealth Secretariat, 2010.

Tropopause

The tropopause is the boundary region dividing the troposphere, the lowest layer of the atmosphere, from the overlying stratosphere. Since the tropospheric and stratospheric air masses have rather distinct features, in correspondence to each surface location, the tropopause height is the level in the vertical where abrupt changes in the physical and chemical properties of the atmosphere are observed.

Three different definitions are typically adopted. The thermal tropopause is related to the change of the sign of the vertical derivative of the temperature (lapse rate), which is negative in the troposphere and positive in the stratosphere.

The World Meteorological Organization defines the tropopause as the lowest level where the absolute value of the temperature lapse rate decreases to 2K/km or less, with the average lapse rate between this level and all higher levels within 1.2 mi. (2 km) not exceeding 2K/km. The dynamical tropopause is defined in terms of sharp changes in the potential vorticity (much higher in the stratosphere), which measures stratification and rotation of the air masses. An abrupt increase (decrease) with height of the ozone (water vapor) mixing ratio indicates the presence of the chemical tropopause. In spite of the necessity of choosing phenomenological thresholds, the three definitions of the tropopause are quite consistent.

Tropical Tropopause

Typically, the tropopause height decreases with latitude, at around 3.7 mi. (6 km) near the poles and 11 mi. (18 km) near the equator. Whereas radiative and convective processes with time scale of the order of one week to one month basically determine the properties of the tropical tropopause; in the midlatitudes, a relevant role is also played by baroclinic-fuelled extra-tropical cyclones, having a typical time scale of a few days, where the tropopause readjusts its height to effectively act as a stabilizing mechanism limiting the growth of the weather perturbations. The tropopause is not a hard boundary: Exchanges of tropospheric and stratospheric air occur through various mechanisms, including vigorous thunderstorms and midlatitude perturbations.

The globally averaged tropopause height tends to increase if the troposphere warms up and/ or the stratosphere cools down, and the height change is approximately proportional to the difference between the tropospheric and stratospheric temperature changes. Therefore, the mean tropopause height can act as a robust indicator of climate change. Recent climate simulations have shown that the estimated increase after 1979 of about 492 ft. (150 m) may be primarily explained by anthropogenic causes, namely the stratospheric cooling driven by ozone depletion and the tropospheric warming driven by increases in greenhouse gas concentration. Considering natural processes, episodic short-lived and strong reductions of the globally averaged mean tropopause height are caused by large explosive volcanic eruptions, which warm the troposphere and cool the stratosphere.

Valerio Lucarini
University of Bologna

See Also: Atmospheric General Circulation Models; Atmospheric Vertical Structure; Cyclones.

Further Readings

Santer, B. D. "Contributions of Anthropogenic and Natural Forcing to Recent Tropopause Height Changes." *Science*, v.301/5632 (2003).

World Meteorological Organization. "Meteorology: A Three Dimensional Science." *WMO Bulletin*, v.6 (1957).

Troposphere

On the basis of thermal characteristics, the atmosphere is normally subdivided into four major vertical layers: the troposphere, stratosphere, mesosphere, and thermosphere. The troposphere makes up the lowest of these layers, extending from the surface to a global average height of 7.5 mi. (12 km). Coined in 1908 by French scientist Leon Philippe Teisserenc de Bort, the name troposphere is derived from the Greek word *tropos*, meaning to turn, mix, or change. The term aptly describes the extensive vertical mixing and stability changes of this layer, which generates clouds, precipitation, and other meteorological events. For this reason, the troposphere is commonly referred to as the weather sphere.

Depth, Temperature, and Height

The depth of the troposphere is relatively thin, yet it contains approximately 80 percent of the atmosphere's mass. Because the atmosphere is compressible, air molecules are more compact closer to the surface, thereby increasing the density and pressure of the air at lower altitudes. The relationship between density and pressure with altitude is nonlinear, falling at a decreasing rate with increasing altitude. In the lower troposphere, the rate of pressure decrease is about 10 millibars for every 330-ft. (100-m) increase in elevation.

Temperature in the troposphere generally decreases with height, contrasting considerably between its lower and upper boundaries. Temperature in this layer is largely affected by the radiant energy exchanges from the underlying surface and insolation intensity. The global average temperature at the surface is 59 degrees F (15 degrees C), but decreases to around minus 82 degrees F (minus 63 degrees C) at the top of the troposphere. On the basis of mean tropospheric depth, the average rate of temperature decrease is 3.6 degrees F per 1,000 ft. (6.5 degrees C per km), a measurement known as the normal lapse rate. This rate represents average global conditions, deviating substantially depending on latitude, time, and local modifications. The actual temperature change with height is the environmental lapse rate, which is measured remotely using satellites, or directly using Radiosondes (a balloon-borne instrument package). Eventually, temperature ceases to decline with height, transitioning into a zero lapse rate region (or isothermal layer), where temperature is neither increasing nor decreasing. This shift demarcates the boundary between the troposphere and the stratosphere, known as the tropopause.

The mean height of the tropopause can have considerable spatial and temporal variability. In the tropics, the depth of the troposphere is around 16 km (10 mi.), but near the poles, the depth dwindles to about 8 km (5 mi.) or less. The tropopause also varies seasonally, with higher heights occurring during the summer than the winter. Warm surface temperatures occurring at low latitudes and high sun periods encourage vertical thermal mixing, thereby extending the depth of the troposphere. Accordingly, the environmental lapse rate in these regions continues to remain positive (i.e., temperature decreases with height), and tropopause temperatures are typically lower in the tropics than for high latitudes. Occasionally, the tropopause is difficult to discern because of extensive mixing between the upper troposphere and the lower stratosphere.

This situation is common in portions of the midlatitudes, usually defining the location of jet streams (a narrow belt of high-velocity winds often in excess of 185 km per hour, or 115 mi. per hour) that steer midlatitude cyclones. Because the height of the tropopause is dependent on the average temperature of the troposphere, temperature changes in this layer can influence the location of extratropical storm tracks and cloud depth.

Embedded frequently within the troposphere are thin sublayers in which the temperature actually increases with height, known as temperature inversions. Radiation inversions result from nocturnal surface cooling. Under certain ambient conditions (e.g., cloudless nights), terrestrial radiation loss to space is enhanced and the ground (and air above) cool rapidly, thereby establishing a shallow inversion layer. Conversely, subsidence inversions occur from mid-upper tropospheric processes that produce areas of sinking air warmed by compression; hence, lower tropospheric temperatures are actually colder than those aloft. This setting tends to stabilize the air, inhibiting vertical mixing and cloud growth.

A semipermanent sublayer of the troposphere is the planetary boundary layer (PBL), a section directly influenced by surface daily conditions. Comprising typically the lowest 1 km (3,300 ft.) of the troposphere, the PBL is characterized by turbulence generated by frictional drag from the surface beneath and rising thermals (heated air parcels). The depth of the PBL amplifies and diminishes with the daily solar cycle, such that the greatest thickness is during the day when the atmosphere is most turbulent.

Evidence suggests that the troposphere has undergone a significant rate of warming during the past century. The tropospheric temperature trend in the latter half of the 20th century is estimated at a 0.18 degree F (0.10 degree C) increase per decade, similar to the surface temperature rate change. Higher temperatures mean increased surface evaporation and tropospheric water vapor content. As a consequence, cloud cover has also shown an increase, and extratropical precipitation in the Northern Hemisphere has increased 5 to 10 percent since 1900.

Other climate-forcing agents (e.g., anthropogenic-induced greenhouse gas emissions) can alter the Earth's radiation balance and may also explain the upward temperature trend. For instance, tropospheric ozone (O_3), a greenhouse gas and surface pollutant, has increased by nearly 35 percent since the preindustrial era.

Jill S. M. Coleman
Ball State University

See Also: Atmospheric Boundary Layer; Atmospheric Vertical Structure; Tropopause.

Further Readings

Intergovernmental Panel on Climate Change. *Climate Change 2001: The Scientific Basis*. Oxford: Cambridge University Press, 2001.

Lutgens, Frederick K., Edward J. Tarbuck, and Dennis Tasa. *The Atmosphere: An Introduction to Meteorology*. 10th ed. Upper Saddle River, NJ: Prentice Hall, 2006.

Oke, Timothy R. *Boundary Layer Climates*. 2nd ed. London: Routledge, 1987.

Tsunamis

Tsunamis (sometimes called seismic sea waves) are large sea waves that are created by underwater earthquakes, volcanic eruptions, or even nonseismic events such as landslides and meteorite impacts. They are not tidal waves, which are

Seven years after the deadliest tsunami in history ruptured through the Indian Ocean and claimed 230,000 lives, a sailboat lies among debris in Ofunato, Japan, following a 9.0 magnitude, record-breaking earthquake and subsequent tsunami in March 2011. The massive, 23-ft. tsunami destroyed much of the Pacific coast of northern Japan; the official human toll was 15,839 dead and 3,647 missing. A nuclear crisis soon ensued after the Fukushima Daiichi Nuclear Power Station exploded and failed. Three other plants reported problems.

waves created by the gravitational influence of the sun and moon. The word *tsunami* is Japanese for "harbor wave."

Tsunamis are not easily seen on the open water, since they have extremely long wavelengths on the order of tens of kilometers. The speed of the wave is directly related to the depth of the water; therefore, as water depth decreases, the tsunami moves slower. As the waves propagate toward the coast, the speed will decrease, but the amplitude or the height of the waves can achieve extraordinary levels. Tsunamis lose energy as they approach the coast, but they still have incredible amounts of energy as they often cause beach erosion and undermine trees and other types of coastal vegetation. The fast-moving water is capable of flooding several hundreds of meters inland, well above normal flood levels, and destroying buildings and other structures. Tsunamis can extend to heights well above sea level, sometimes in extreme cases as high as 100 ft. (30 m).

Causes of Tsunamis

Volcanic activity and earthquakes are the prime causes of tsunamis. When the sea floor starts to buckle, the overlying water will begin to displace. As the sea floor rises and sinks, the displaced water will form waves due to the effects of gravity. Most of the major earthquakes occur at plate boundaries. There are three different types of plate boundaries. A divergent boundary takes place where two plates move away from each other. At this type of boundary, volcanoes will form, out of which molten material will flow. Weaker, shallow-focus earthquakes can also occur along these boundaries. Divergent boundaries are very common in the mid-ocean, such as the Mid-Atlantic Ridge, the East Pacific Ridge, the Mid-Indian Ridge, and the Southeast Indian Ridge. Convergent boundaries occur where two plates moving in opposite directions collide. One plate is denser and will subduct underneath the other. These subduction zones are a very common location for earthquake activity. There are three different types of convergent boundaries: oceanic–continental, oceanic–oceanic, and continental–continental convergence.

At an oceanic–continental convergent boundary, the oceanic crust is denser and will subduct underneath the continental crust. Volcanoes will form along the continental boundary, while deep trenches will form off the coast. Shallow-focus earthquakes often form along these subduction zones, such as along the west coast of South America. At an oceanic–oceanic convergent boundary, two ocean plates collide, forming a volcanic island arc on the ocean floor. Examples of this type of boundary include the Aleutian Islands, the Mariana Islands, and Japan. At a continental–continental convergent boundary, two continental plates collide, typically forming huge mountain ranges, such as the Himalayas or the Alps. Under this type of convergence, volcanic activity is rare, but earthquake activity is very common. The final type of boundary is called a transform boundary, where two plates slide past each other. Transform boundaries occur along vertical fractures called faults, which are noted for great magnitudes of earthquake activity. Most faults are found near mid-oceanic ridges, but can also extend through continents, as evidenced by the San Andreas Fault in California.

Tsunamis can be formed by anything that displaces a large volume of water from its equilibrium state. When earthquakes or volcanoes generate tsunamis, water is displaced due to the uplift, or subsidence, of the sea floor and water column. Sometimes, submarine landslides, which are common with large earthquakes and volcanic collapses, can displace great volumes of water. However, these types of water disturbances occur from above, rather than from below. Tsunamis derived from these types of mechanisms usually do not last long and have minimal impacts on coastlines.

The most recent deadly tsunami struck Tohoku, Japan, on March 11, 2011 (now known as the Great East Japan Earthquake), after a 9.0 magnitude earthquake (on the Richter Scale, devised to estimate the amount of energy released in an earthquake) struck off the east coast of Japan. Over 15,000 deaths were reported and several thousand were injured or missing. Waves were reported to be as high as 124 ft. (38 m) and went approximately 6 mi. (10 km) inland. Nearly 250 mi. (400 km) from the epicenter, near Tokyo, waves were approximately 4 ft. (1.3 m) high. The waves from this tsunami propagated across the Pacific and could be felt in countries as far away as Chile (over 10,500 mi., or 17,000 km) with waves about 6.5 ft. (2 m) high. The Early Earthquake Warning system from the Japan Meteorological Agency issued the warning shortly after the major earthquake and saved many lives. Thousands were left homeless, and other infrastructural damage included the cooling failure of two major nuclear power plants.

Another deadly tsunami was the Asian tsunami that occurred after the 2004 Indian Ocean earthquake on December 26. The epicenter (the location at the Earth's surface directly above the focus of the earthquake) took place off the coast of Sumatra, Indonesia. The magnitude of this earthquake was estimated at between 9.1 and 9.3 on the Richter scale. This made it the fourth most powerful recorded earthquake since 1900. The earthquake lasted almost 10 min. and was the second most powerful earthquake ever recorded on a seismograph (an instrument that measures seismic waves from an earthquake). Waves were reported as high as 100 ft. (30 m) and it was the deadliest earthquake in recorded history, with approximately 230,000 lives lost, mostly in the countries of Indonesia, India, Sri Lanka, and Thailand.

Prior to this event, the 1782 Pacific Ocean tsunami was the deadliest in recorded history, with an estimated 40,000 casualties in the South China Sea. Other powerful tsunamis include the 1883 tsunami after the eruption of Krakatoa, a volcanic island in Indonesia, and the 1908 tsunami that occurred in the Mediterranean Sea near Messina, Italy.

How Does Global Warming Contribute?

There has been speculation about the possible effects of global warming after the 2004 tsunami, with proponents suggesting that the increase in average temperature allows the atmosphere and the oceans to gather energy, which may cause more earthquake activity. Critics, however, claim that if this were the case, there would be more of a correlation between El Niño and tsunamis, since El Niño warms the ocean over an active region with many plate boundaries. Therefore, links to global warming and tsunamis are unsubstantiated because it is difficult to associate what is happening at the surface of the ocean with the depth at which the focus of the earthquakes takes place. However, increased sea levels due to global warming may make coastal areas more susceptible to tsunamis. Over the last 100 years, sea levels have risen about 7 in. (18 cm) per year. However, considering the magnitude of such devastating tsunamis as the 2011 Japanese earthquake and tsunami, which had waves over 124 ft. (38 m), this would only constitute an approximate 0.5 percent increase.

Kevin T. Law
Marshall University

See Also: Cyclones; Floods; Glaciers, Retreating; Hurricanes and Typhoons; Indonesia; Japan; Volcanism; Vulnerability.

Further Readings

Abbott, Patrick L. *Natural Disasters*. 6th ed. New York: McGraw-Hill, 2007.

Easton, Thomas A. *Taking Sides: Clashing Views in Science, Technology, and Society*. 7th ed. New York: McGraw-Hill/Dushkin, 2005.

McKnight, Tom and Darrel Hess. *Physical Geography*. Upper Saddle River, NJ: Prentice Hall, 2005.

Tunisia

Tunisia is located in North Africa between longitudes 7 and 12 degrees east and latitudes 32 and 38 degrees north. Due to its geographical position and general orientation of the main relief, Tunisia is influenced in the north by the Mediterranean, with the south under the influence of the Sahara. The center is under the joint effect of these two elements, which grants its particular high climatic variability, making the country also particularly vulnerable to desertification.

The north of the Tunisian Dorsal is characterized by hot, dry summers and mild, relatively rainy winters. The center of the country and the Gulf of Gabes have a semiarid climate, with relatively high temperatures and modest rainfall, between 7.8 and 15.7 in. (200 and 400 mm) per year. The rest of the country witnesses a desert arid climate characterized by high temperatures of high amplitudes and disparate rainfall rarely exceeding 3.9 in. (100 mm).

Hydrous Stress Situation

According to international standards, Tunisia is in a hydrous stress situation close to a shortage, sharpened by a high anthropic pressure. Climate changes can result in harmful effects on water resources, ecosystems dependent on water, and the various economic activities that require large quantities of water, such as agriculture and tourism. A projected increased frequency of rains resulting from torrential storms and downpours may result water streaming over land, rather than being absorbed by the soil, triggering erosion and desertification. Besides its agricultural systems management policy that Tunisia has launched for many years, an adequate policy of struggle against desertification has been realized by the ratification of the desertification convention.

Coastal water resources will realize direct effects due to climate warming and indirect effects following sea-level rise, which would damage the aquifer coastal formations and other underground freshwater reserves by the intrusion of sea water, especially since the anthropic pressure on these underground water resources is very high.

The humid littoral areas are especially vulnerable to accelerated ASLR sea-level rise. In general, for the case of Tunisia, the most vulnerable humid places will be the lagoons, sebkhas, and lowest coastal marshes, which will mostly be annexed to the sea domain. The littoral forests seem relatively less vulnerable to ASRL, except the coastal oasis, where ASLR could result in a retreat of the coastal line and an increased salinization of the littoral ground water, which would be detrimental to the growth of palms.

Islands with higher relief will be less affected by rising sea elevations, whereas the flatter islands will be highly affected by ASLR—for instance, the Jerba Islands as well as the Kerkenna Islands, where erosion risks are likely to increase and salinization will continue. This also negatively impacts the stability of historical buildings and economic gains from the tourism industry. The most important impact for this archipelago will be recorded in the maritime tides and the sebkhas. Sea-level rise, even of some decimeters, could result in their permanent annexing to the sea.

The inventory of greenhouse gases (GHGs) for 1994 shows a relatively limited contribution by Tunisia to the greenhouse effect in comparison with other nations. The net anthropogenic GHG emissions of Tunisia are 23.4 million tons of carbon dioxide equivalent (TE-CO_2e), which represent 2.66 TE-CO_2e per capita or 1.8 TE-CO_2e per $1,000 of gross domestic product (GDP). The energy sector contributes the most and will increase until 2020, according to estimates, but will be offset by land use, land-use change, and forestry (LULUCF) and a reduction of agricultural emissions (see Table 1).

Tunisia signed the United Nations Framework Convention on Climate Change (UNFCCC) in 1992 and ratified it in July 1993. It ratified the Kyoto Protocol in 2002. Its only National Communication under the UNFCCC as a Non-Annex I country dates back to 2001, and most data are even older, mostly from 1994 and 1997.

Table 1 Anticipated GHG emissions in Tunisia: reference scenario (1,000 TE-CO_2e)

Total/Source	1997	2010	2020
Energy	17,010	31,636	48,993
Industrial Processes	3,265	7,409	12,068
Agriculture	6,440	7,522	8,746
Land-use change and forestry (LUCF)	-2,744	-7,209	-12,785
Emissions due to LUCF	3,952	3,917	3,596
Absorptions due LUCF	-6,696	-11,126	-16,381
Wastes	1,182	4,678	5,338
Total gross emissions	31,849	55,162	78,741
Total net emissions	25,153	44,036	62,360

In the beginning of 2011, a revolution started by civil society turned over the previous government of former President Zine El Abidine Ben Ali, which will also influence ongoing climate negotiations.

Ingrid Hartmann
University of Hohenheim

See Also: Algeria; Forests; Land Use, Land-Use Change, and Forestry; Mauritius.

Further Readings

Ministere de l'Environnement et de l'Amenagement du Territoire. "Programme d'Action National de Lutte Contre la Désertification" (June 1998). http://www.ddc-as.org/images/stories/Countries/NAPs/tunisia.pdf (Accessed July 2011).

Republic Of Tunisia Ministry Of Environment And Land Planning. "Initial Communication of Tunisia Under the United Nations Framework Convention on Climate Change" (October 2001). http://unfccc.int/resource/docs/natc/tunnc1esum.pdf (Accessed July 2011).

United Nations Development Programme. "Portfolio of Projects for Reducing Greenhouse Gas Emissions in Tunisia: An Overview" (January 2002). http://unfccc.int /resource/docs/natc/tunnc1add.pdf (Accessed July 2011).

Turkey

Turkey, situated between Europe and Asia, is a fast-developing country with a population of approximately 74 million. Undergoing rapid population growth and industrialization, Turkey has required the use of its natural resources and energy sources to keep pace with development.

For the last decade, the government of Turkey has been responsive to calls for multilateral climate change negotiations. Turkey ratified the United Nations Framework Convention on Climate Change (UNFCCC) in May 2004 and the Kyoto Protocol in August 2009. By 2012, the government of Turkey hopes to create substantial climate change–related policies and clean development projects in partnership with external nongovernmental organizations, with the goal of fulfilling its responsibilities under the UNFCCC.

Turkey is located in the eastern Mediterranean basin, a region belonging to the highest climate change risk group, according to the Intergovernmental Panel on Climate Change *Fourth Assessment Report*. Statistical records have demonstrated that climate change has been occurring in Turkey. A general upward trend in average temperatures has been observed in southern parts of the country, as well as a general increasing trend in maximum temperatures during summer months. Precipitation levels indicate a decreasing trend, especially during the winter months, with significant drops in precipitation levels correlated with the North Atlantic Oscillation in all seasons, except summer. Overall, climate change has worsened Turkey's already existing environmental problems of droughts, floods, and wildfires.

Governance, Adaptation, and Mitigation

In 2006, Turkey's greenhouse gas emissions (GHGs) and carbon dioxide (CO_2) emissions were 331.8 million tons equivalent carbon (MTec) and 273.7 Mt, respectively. In 2009, Turkey's CO_2 emissions decreased to 299.1 (MTec). The government of Turkey, while considering the importance of climate change and its consequences globally, has acknowledged the need to recognize the individual circumstances of developing countries as parties to UNFCCC. During the 2001 Seventh Conference of the Parties (COP 7) meeting, Turkey was given special recognition as an Annex I Party with conditions different than that of other Annex I countries based on its historical GHG emissions, per-capita GHG emissions, and basic economic and social indicators. Turkey subsequently ratified the Kyoto Protocol on August 2009.

Currently, Turkey's main commitments to the UNFCCC are the submission of regular reports; National Communication and GHG inventories, and implementation of policies relevant to climate change mitigation, adaptation, research, education, and public awareness.

During the 2010 UNFCCC COP 16 meeting in Cancun, Turkey recognized its geographic location as one of the most vulnerable regions and maintained the need for continued support in the areas of technology and capacity building for climate change mitigation and adaptation. Presently, some of the climate change mitigation and adaptation projects initiated in the country include: Enhance Turkey's Capacity to Adapt to Climate Change, funded by the UN Millennium Development Goals Achievement Fund; Sectoral Mitigation Potential of GHG Emissions Reduction and Related Costs, funded by and in partnership with the State Planning Organization (SPO); and Developing the Capacity of Turkey to Participate Efficiently in the International Climate Change Negotiations and Voluntary Carbon Markets, funded by the SPO and the UN Development Programme. The World Bank's Global Environment Facility (GEF) is involved as the major funding source for climate change–combating projects in Turkey under the guidelines of the UNFCCC.

Overall, Turkey's short- to medium-term development goals concerning climate change mitigation and adaptation focus on energy, transportation, industry, waste management, land-use change, and forestry. The Turkish government has created laws and policies to reduce fossil fuels as a major source of energy, improve power plants, expand and improve transportation systems, and increase energy efficiency in cement, iron, and steel plants. The Turkish government is also committed to energy recovery from its landfills in the form of methane. As of 2009, 51.73 megawatts of energy were generated from landfills. In addition, Turkey has set an ambitious afforestation campaign, with a goal of 2.3 million hectares of land in a five-year period (2008–13). As a result of this campaign, the government hopes to sequester 181.4 million tons of CO_2 by 2020.

Ezgi Akpinar-Ferrand
University of Cincinnati

See Also: Climate Change, Effects of; Global Environment Facility; Intergovernmental Panel on Climate Change; United Nations Development Programme.

Further Readings

Republic of Turkey, Ministry of Energy and Natural Energy–Environment. http://www.enerji.gov.tr/index.php?dil=en&sf=webpages&b=enerji_cevre_iklim_EN&bn=218&hn=&id=40720 (Accessed July 2011).

Republic of Turkey, Ministry of Environment and Forestry Office of Climate Change. "National Activities of Turkey on Climate Change." http://www.iklim.cob.gov.tr/iklim/Files/Raporlar/National%20Activities%20of%20Turkey%20on%20Climate%20Change.pdf (Accessed July 2011).

United Nations Development Programme. "Helping Turkey to Attain Environmental Sustainability Working for Poverty Reduction and Improving Livelihoods of People." http://www.undp.org.tr/publicationsDocuments/Helping_Turkey_to_attain_environmental_sustainability.pdf (Accessed July 2011).

United Nations Framework Convention on Climate Change. "UNFCCC COP 16 Submission From Turkey." http://unfccc.int/resource/docs/2010/awglca13/eng/misc08.pdf (Accessed July 2011).

Turkmenistan

Located in central Asia, Turkmenistan is a largely desert country that lies north of the Kopet-Dag Mountains, between the Caspian Sea in the west and Amu–Darya River in the east. Turkmenistan covers 188,456 sq. mi. (488,100 sq. km) and had a population of just over 5 million in 2007. Despite its extensive oil and gas resources, Turkmenistan remains a poor, predominantly rural country, with the majority of population relying on intensive agriculture in irrigated oases, and it is extremely vulnerable to climate change.

Up to 80 percent of Turkmenistan is desert. The country has a distinctive continental climate with average annual air temperature ranging from 53 to 62 degrees F (12 to 17 degrees C) in the north to 59 to 64 degrees F (15 to 18 degrees C) in the southeast. The absolute maximum temperature is 118 to 122 degrees F (48–50 degrees C) in the central and southeast Kara-Kum. The highest amount of rainfall is observed in the mountains, up to 15.6 in. (398 mm) in Koyne-Kesir. The least amount of rainfall is less than 3.7 in. (95 mm) above the Kara-Bogaz-Gol Bay.

Meteorological data reveal an increase of annual and winter temperatures in Turkmenistan since the beginning of the past century. The mean annual temperature has increased by 0.11 degree F (0.6 degree C) in the north and by 0.07 degree F (0.4 degree C) in the south since 1931. The number of days with temperature higher than 104 degrees F (40 degrees C) has increased since 1983. Climate models predict temperature increases in Turkmenistan of 5.4 to 7.2 degrees F (3 to 4 degrees C) by the middle of the 21st century. Precipitation projections are highly uncertain, but given the existing aridity and high interannual and interseasonal variability of the climate of Turkmenistan, even a slight temperature increase is likely to deepen the existing water stress in the region.

During the last few decades, Turkmenistan has experienced widespread changes in land cover and land use following the socioeconomic and institutional transformations of the region, catalyzed by the collapse of the Soviet Union in 1991. The decade-long drought events and steadily increasing temperatures in the region came on top of these institutional transformations, affecting the long-term and landscape-scale vegetation responses.

Turkmenistan signed and ratified the United Nations Framework Convention on Climate Change in 1995. In January 1999, Turkmenistan ratified the Kyoto Protocol and published the *Initial Communication on Climate Change*. Covered under the Kyoto Protocol's Clean Development Mechanism, Turkmenistan can trade carbon credits with the countries that fall under Joint Implementation.

The major sources of carbon emissions in Turkmenistan include oil and gas extraction, petroleum refining, the chemical industry, and motor transportation, concentrated mainly in Ashgabat, Turkmenbashi, Balkanabat, Mary, Turkmenabat, and Dashoguz. Turkmenistan has taken some steps to reduce these carbon emissions, such as the Green Belt Project, a massive tree-planting drive throughout the country; modernization of Turkmenbashy and Seyidi refineries to conform to modern ecological standards; and the removal of cement factories from inhabited areas.

However, the widespread poverty, recent decline in the educational system, misuse of hydrocarbon revenues, and high economic dependence on cotton production and export leave Turkmenistan particularly vulnerable to high climatic variability, desertification, and droughts.

Elena Lioubimtseva
Grand Valley State University

See Also: Climate Models; Deserts; Drought; Kyoto Protocol; Natural Gas; Oil, Production of.

Further Readings

Glanz, M. H. "Water, Climate, and Development Issues in the Amu Darya Basin." *Mitigation and Adaptation Strategies for Climate Change*, v.10/1 (January 2005).

Lioubimtseva, E., R. Cole, J. M. Adams, and G. Kapustin. "Impacts of Climate and Land-Cover Changes in Arid Lands of Central Asia." *Journal of Arid Environments*, v.62/2 (July 2005).

Lioubimtseva, E. and G. M. Henebry. "Climate and Environmental Change in Arid Central Asia: Impacts, Vulnerability, and Adaptations." *Journal of Arid Environments*, v.73/11 (November 2009).

United Nations Framework Convention on Climate Change. *Turkmenistan: Initial National Communication on Climate Change.* Geneva: UNFCCC, 1999.

Tuvalu

Tuvalu has been independent since 1978, joining the United Nations in 2000. The resident population was estimated at 10,544 people in 2011, living on nine small, low-lying atoll islands. The total land area is only 9.4 sq. mi. (24.4 sq. km); including the maritime area of the Exclusive Economic Zone (EEZ), the total national area of Tuvalu is 289,576 sq. mi. (750,000 sq. km). Tuvalu does not have substantial natural resources.

Its low-lying coral atoll island soil is of poor quality, alkaline, shallow, and with low water-holding quality. Added to these environmental challenges are a lack of rainfall, lack of natural resources, the occurrence of cyclones, and the threat of sea-level rise. About 44 percent of the population lives on urban Funafuti, and the internal migration rate is expanding. The overcrowded urban situation has meant that having a sufficient freshwater supply is difficult.

Vulnerable Nation

A small and fragmented Pacific Island nation, Tuvalu is increasingly vulnerable to cyclones, flooding, and sea-level rise, accompanied by environmental threats such as land loss, loss of vegetation, coastal erosion, soil salinization, and intrusion of saltwater into the atolls' freshwater lenses and groundwater. Loss of vegetation and decreased access to freshwater impacts on the food and water security of the population. Furthermore, two incidences of flooding have been particularly severe for the urban island of Funafuti—one in 1977 and another in 1993. Other climatic effects include coral bleaching in reef systems. Tuvalu's vulnerability, however, is also because of its impoverished economic situation, underemployment, aid dependency, and a lack of natural resources for export; fragmentation and isolation of the islands; and a lack of capacity to support its population. This combination of challenges makes sea-level rise a problematic development issue. Current estimations of average sea-level rise in the South Pacific are 0.7 mm per year. However, in relation to sea-level rise, the land of Funafuti Island is sinking, a problem that is accelerated by tectonic movement. As a consequence, and including indications of tide-gauge data, the sea-level rise in the Funafuti area has been estimated by J. A. Church and colleagues in 2006 at 2 ± 1 mm per year.

Tuvalu entered international negotiations on climate change in 2002, when Prime Minister Koloa Talake suggested pursuing legal actions on the international level against greenhouse gas emissions leading to global warming and the consequences of sea-level rise for Tuvalu. The National Summit on Sustainable Development in Tuvalu in 2004 emphasized the need to promote awareness and strategies for adaptations on the national level. This has been followed up with Tuvalu's National Strategy for Sustainable Development Plan for 2005 to 2015. An appeal to the United Nations Climate Change Conference in Cancun, Mexico, resulted in financial support for climate change adaptation, and in May 2011, negotiations began toward a Tuvalu National Climate Change Policy and Joint Climate Change Adaptation and Disaster Risk Management National Action Plan (JNAP).

Migration as a final solution is not desired by Tuvaluans. Existing migration schemes for Tuvalu are small. Differing from many other Pacific nations, Tuvalu, similarly to Kiribati, does not have the privilege of free access to one of the Pacific Rim countries. Current schemes allow 75 Tuvaluans to migrate each year to Australia and New Zealand. The strong relationship of Tuvaluans to their land, however, makes migration a last resort. Some publications have suggested that Tuvalu and Tuvaluans are being publicly victimized when they are represented as potential places of disaster and as environmental refugees, respectively. Voyeuristic "disaster tourism" has become popular and provides a great challenge for Tuvaluans, because it fuels such victimization and increases air pollution.

Maria Borovnik
Massey University

See Also: Australia; Climate Change, Effects of; Floods; Kiribati; New Zealand; Pacific Ocean; Pollution, Air; Population; Preparedness; Rainfall Patterns; Resources; Sea Level, Rising; Tourism.

Further Readings

Central Intelligence Agency. "The World Factbook: Tuvalu." https://www.cia.gov/library/publications/the-world-factbook/geos/tv.html (Accessed July 2011).

Church, J. A., N. J. White, and J. R. Hunter. "Sea-level Rise at Tropical Pacific and Indian Ocean Islands." *Global and Planetary Change*, v.53 (2006).

Connell, J. "Losing Ground? Tuvalu, the Greenhouse Effect and the Garbage Can." *Asia Pacific Viewpoint*, v.44/2 (2003).

Farbotko, C. "The Global Warming Clock is Ticking to See These Places While You Can: Voyeuristic Tourism and Model Environmental Citizens on Tuvalu's Disappearing Islands." *Singapore Journal of Tropical Geography*, v.31 (2010).

Montreux, C. and J. Barnett. "Climate Change, Migration, and Adaptation in Funafuti, Tuvalu." *Global Environmental Change*, v.19 (2009).

NcNamarra, K. E. and C. Gibson. "'We Do Not Want to Leave Our Land': Pacific Ambassadors at the United Nations Resist the Category of 'Climate Refugees.'" *Geoforum*, v.40 (2009).

Shen, S. and F. Gemenne. "Contrasted Views on Environmental Change and Migration: The Case of Tuvaluan Migration to New Zealand." *Intenrational Migration*, v.49/1 (2011).

Tyndall, John

John Tyndall (1820–93) was born in Leighlinbridge, County Carlow, Ireland, on August 2, 1820. After working as a surveyor and a mathematics teacher, he attended the University of Marburg in Germany, where he received his Ph.D. In 1854, he became a professor of natural philosophy at the Royal Institution in London (a scientific research center founded in 1799). In 1867, he was made superintendent of the institution, taking over from Michael Faraday.

Tyndall's most well-known scientific studies included the nature of sound, light, and radiant heat and observations on the structure and movement of glaciers. Glaciers had become a scientific area of interest during that time because in the 1830s, Louis Agassiz (considered the father of glaciology) had discovered that a large portion of Europe and North America had once been covered with ice.

Tyndall developed an interest in meteorology as a result of his love for mountain climbing. He studied alpine glaciers and took meteorological measurements on Mont Blanc in the Alps. Using a spectrophotometer he designed, Tyndall studied the absorption of infrared light (at the time called radiant heat) by atmospheric gases. Infrared light is felt as heat and has wavelengths of approximately 0.7 to 1.0 μm. Visible light, by comparison, has wavelengths of approximately 0.4 to 0.7 μm.

Some of the invisible gases (oxygen, nitrogen, and hydrogen) were transparent to radiant heat, whereas water vapor and carbonic acid (now known as carbon dioxide) absorbed and reemitted infrared light, thereby warming the atmosphere close to the Earth. From these experiments, Tyndall realized that water vapor, carbon dioxide, and ozone—even in small quantities—were the best absorbers of heat radiation, and he later speculated on how changes in these gases could correlate to climate change.

Greenhouse Gases

Among the various atmospheric gases in the troposphere, Tyndall discovered the importance of water vapor and carbon dioxide as greenhouse gases that trap heat on the surface of the Earth. Without these atmospheric gases to trap heat, the heat would rapidly radiate back into space, and the Earth would be much colder. During cold nights, there is an enhanced chance of fog or dew in the mornings because the moisture in the air (water vapor) condenses into droplets as the air cools. In deserts—hot and dry climates—there is a lack of water vapor in the air, and the sand radiates heat easily into space. Changes in the atmospheric levels of gases also produce changes in the climate.

In the 1860s, Tyndall began to suggest that slight changes in the atmospheric composition could bring about climatic variations. Tyndall was the first scientist to explain the ice ages as caused by greenhouse effect. In particular, he noted that variations in water vapor resulted in a change in the climate and realized the importance of the greenhouse effect in maintaining ecosystems necessary for life. He thought that changes in the composition of the atmosphere may have produced all changes in the climate.

Using the laws of thermodynamics, Tyndall proposed that changes in carbon dioxide cause an initial change in temperature, and when the humidity changes accordingly, the change in humidity leads to a second change in the temperature.

During the Victorian era, he was the contemporary and friend of other important scientists. In addition to his scientific research, he promoted scientific education of the public with scientific demonstrations and advocated using hands-on laboratory experimentation to teach science. In 1874, Tyndall used his address before the British Association for the Advancement of Science to proclaim rational thought and skepticism as the aim and superiority of science.

In the 1870s, he made a lecture tour of the United States that included his lab demonstrations. Over the course of his professional life, he was the recipient of numerous scientific awards and recognition. In addition to his dynamic public speeches using laboratory experimentation, he published a wide range of papers, treatises, and books. Digital versions of John Tyndall's books include *Lectures on Light Delivered in the United States in 1872–1873*, first published in 1873; *Hours of Exercise in the Alps*, first published in 1871; and *The Glaciers of the Alps*, first published in 1861.

John Tyndall was a 19th-century physicist renowned for his work on the absorption of heat by atmospheric gases. In the 1860s, he suggested that slight changes in the atmospheric composition could bring about climatic variations and even cause ice ages.

John Tyndall died on December 4, 1893, from an overdose of chloral hydrate. At the inquest, his wife testified that she mistakenly gave him chloral hydrate instead of his normal medication, leading to his death.

Tyndall Effect

The Tyndall effect, a scientific principle on the dispersion of light beams through colloidal suspensions or emulsions, was named for him. The importance of his research continues, as his work continues to shape scientific research on climate change.

The Tyndall Centre, with headquarters at the School of Environmental Sciences at the University of East Anglia, founded in 2000, is named after John Tyndall, because he was one of the first scientists to recognize the Earth's natural greenhouse effect and observed, identified, and proposed the climate effects of the radiative properties of atmospheric gases. The mission of the Tyndall Centre is to research and educate policymakers on climate change, to develop and apply research methods for climate change, and to promote international dialogue on managing future climate change.

In support of this goal, research at the Tyndall Centre includes action to provide information through data collection, interpretation, and modeling to assess possible scenarios of the effect of human and natural causes on climate change and to disseminate this information by encouraging international discussion and policymaking. By using empirical research, the Tyndall Centre proposes protection of coastline ecosystems, city-scale emissions testing, and accountability for contributions to climate change. The Tyndall Centre focuses on this objective with a strategy to investigate and identify behavior modification and education opportunities that promote sustainable approaches to limiting human-induced climate change.

Lyn Michaud
Independent Scholar

See Also: Atmospheric Composition; Carbon Dioxide; Glaciology; Greenhouse Effect; Greenhouse Gas Emissions.

Further Readings

Bowen, Mark. *Thin Ice: Unlocking the Secrets of Climate in the World's Highest Mountains.* New York: Henry Holt, 2005.

Earth Observatory. "John Tyndall." http://www.earthobservatory.nasa.gov/Features/Tyndall (Accessed March 2012).

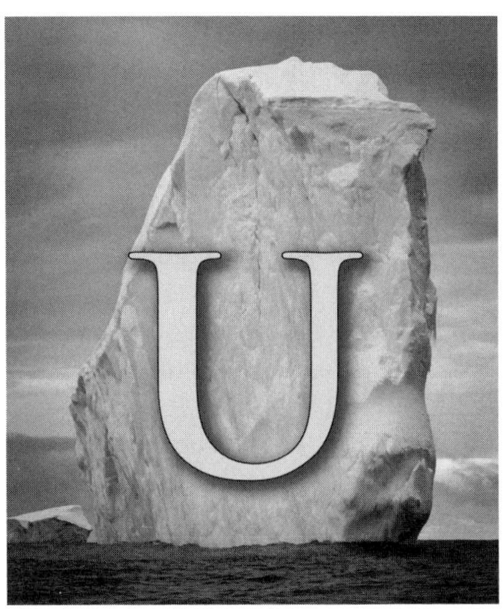

Uganda

The mid-eastern African nation of Uganda has a population of over 34 million (2011 est.). While most of Uganda's population is rural, its urban population is growing at a faster rate. Rain-fed agriculture is the dominant economic sector, with approximately 40 percent of the population living below the poverty line. Uganda's greenhouse gas (GHG) emissions sources include deforestation and land degradation, poor waste disposal, and energy usage. The industrial sector and small mining industry are pollution sources. Developing countries like Uganda, with its largely rural, agricultural economy, contribute less to GHG emissions that fuel global warming, but are more vulnerable to the effects of climate change. Annual rainfall patterns and levels have become increasingly unpredictable, resulting in more frequent and intense extreme weather events such as droughts, floods, and lower crop yields.

The effects of climate change are increasing the strain on Uganda's natural resources. Deforestation, land degradation, and poor agricultural practices contribute to Uganda's GHG emissions. Uganda enjoys greater per-capita renewable water resources than many African nations, but these are increasingly threatened by climate change and related factors such as population growth, urbanization, economic development, and environmental degradation. Women and children often travel great distances daily to obtain potable water. Rivers, lakes, and wetlands represent approximately 18 percent of the country's surface area. Nearby Lake Victoria provides a key source of domestic and industrial water supply, as well as hydropower and fishing. Scientists theorize that receding ice caps within western Uganda's Rwenzori Mountains could completely melt by 2025, drying up a source of freshwater.

Rain-fed agriculture in regions such as Katine dominates the Ugandan economy, employing over 80 percent of the workforce. Staple crops include cassava, soya, and rice. Animal husbandry is a key economy in the country's "cattle corridor," but is threatened by conflict over reduced water supplies. Subsistence farmers, livestock herders, and fishermen are among the most vulnerable to the agricultural disruptions associated with climate change impacts, such as less predictable rainfall patterns, soil erosion, natural disasters, and extreme weather events. Inability to time planting and crop damage have resulted in lowered crop yields and fish stocks. The broader population will suffer from increased food shortages, lowered food security, malnutrition, and endemic disease outbreaks such as malaria and diarrhea, already large-scale problems throughout much of the country.

Energy generation and usage is another key source of Uganda's GHG emissions. More than

90 percent of the country's energy needs are supplied by burning biomass such as firewood and charcoal. Other sources include petroleum products such as kerosene. Less than 10 percent of the population has access to electricity, most of those in urban areas. Renewable energy utilization is a potential future source of GHG emissions reductions for a nation almost devoid of internal fossil fuel sources. The government implemented its Renewable Energy Policy for Uganda in 2007. Hydropower is an important, although underutilized, source of Uganda's energy needs. The Ugandan government is planning future hydropower expansion through its Hydropower Development Master Plan. The government features solar power in its efforts to increase rural electricity access through the Uganda Photovoltaic Pilot Project for Rural Electrification. Energy demands are rising due to economic development programs and population growth.

Uganda is a signatory nation to the United Nations Framework Convention on Climate Change and the Kyoto Protocol. The Ugandan government developed a National Adaptation Plan of Action to respond to the rising threat of climate change impacts. Other government programs related to climate change include the Water Action Plan and Disaster Preparedness and Management Strategy. Many parliamentary members also belong to the Parliamentary Forum on Climate Change, created in 2008. Uganda has developed a pioneering solid waste–composting program for several cities and towns under the Kyoto Protocol's Clean Development Mechanism (CDM). The program is designed to reduce methane emissions from solid waste disposal.

Obstacles to climate change mitigation and adaptation in Uganda include low public-awareness levels, low government priority and institutional capacity, lack of strong comprehensive policies and a legal framework, and limited skills, technology, and financial resources. Programs aimed at rural transformation, poverty reduction, economic development, and industrialization can drive rising energy demands and increase GHG emissions. These include the government's Poverty Eradication Action Plan (PEAP).

Marcella Bush Trevino
Barry University

See Also: Developing Countries; Poverty and Climate Change; Small Farmers.

Further Readings

Mangheni, Margaret Najjingo, Mulondo Ssenkaali, and Fred Onyai. "Rural Development and Environment in Uganda." In *Environment, Development, and Sustainability: Perspectives and Cases From Around the World*, edited by G. Wilsa, et al. New York: Oxford University Press, 2010.

Purkitt, Helen E. *African Environmental and Human Security in the 21st Century*. Amherst, NY: Cambria, 2010.

Ukraine

Europe's second-largest country in land area, Ukraine borders the Black Sea and is located between Poland, Romania, and Moldova in the west and Russia in the east. The country gained its independence in 1991 after the collapse of the Soviet Union. While energy consumption and greenhouse gas (GHG) emissions have dropped since the country became independent, the country has one of the most energy-intensive economies in the world, using energy about three times less efficiently than European Union countries on average. Ukraine is the 20th-largest emitter of carbon dioxide (CO_2) from fossil fuel burning and consumption. Other environmental concerns include inadequate supplies of potable water, air and water pollution, deforestation, and radiation contamination in the northeast from the 1986 accident at the Chernobyl nuclear power plant.

Forward From the Soviet Legacy

Many of Ukraine's environmental problems can be attributed to Soviet industrialization and years of inadequate investment in green technology and energy efficiency. The Ukrainian economy was the second-largest in the Soviet Union. Independence and the transition to a market economy caused heavy poverty and an inflation rate that set a record in 1993 for the highest in the world in a single year. Industries that had been pouring GHGs into the atmosphere ceased production. In the year before independence, Ukraine emitted

719,464 gigagrams of CO_2 equivalent (CO_2e). By 2009, emissions had dropped nearly 62 percent. The dramatic decrease has worked to Ukraine's advantage. As a signatory to the Kyoto Protocol, Ukraine is committed to reduce emissions by 20 percent, but the starting point for the calculations of targeted reductions is data from 1990. The dramatic fall in Ukraine's emissions in the years following 1990 allows the country to increase its emissions, even as it meets its target. At the United Nations climate change summit in 2009, the nation's pledge to reduce emissions by 20 percent from the 1990 level was in real terms a grant for a 75 percent increase over current levels. Ukraine can also profit from the sale of carbon credits to countries or companies that cannot meet commitments to reduce emissions, as the country did in 2009 when it sold $300 million worth of carbon credits to Japan. However, the benefits are likely to be in the short term, since a developing economy and increased industrial production will increase emissions of GHGs. In the first three months of 2011, Ukraine's CO_2 emissions increased by nearly 10 percent over the same period in 2010.

The Black Sea, bordering the south of the Ukraine, is the largest natural reservoir of sulfate dioxide. In the anoxic conditions, some microorganisms are able to use the sulfate to oxidate organic material, producing hydrogen sulfide and carbon dioxide.

Ukraine plans to increase the percentage of coal in the country's energy mix. President Viktor Yanukovych announced in 2011 that the nation's increased coal production would increase to 90 million tons per year by 2015, 20 million more tons than the country produced in 2009. The increase in coal power will be accompanied by an increase in GHG emissions, doubling CO_2 emissions by 2030. Highly dependent on imported energy, particularly Russian gas and oil, Ukraine has the eighth-largest coal reserves in the world. Coal provides Ukraine with a secure energy supply. The country also plans to increase its use of nuclear energy. Almost half of Ukraine's energy consumption in 2009 came from nuclear plants, a pattern the government expects to maintain. Since demand for electricity is expected to increase from 173 kilowatt hours (kWh) in 2009 to 307 billion kWh by 2020, and 420 billion kWh by 2030, nuclear capacity must also increase. Current plans call for 22 new nuclear reactors by 2030, along with extending the life of existing reactors. Critics point out that Ukraine lacks a unified national system for dealing with radioactive waste and spent nuclear fuel, as required by nuclear legislation. They also note that the occurrence of failures, such as minor emissions and leaks and the appearance of cracks and short circuits, is increasing among aging nuclear plants. Water use for nuclear plants is also a concern, both in terms of availability in areas where water for drinking and agriculture are in short supply and in terms of contaminating water sources.

Renewable energy plays a small part in Ukraine's energy balance. Approximately 7 percent of the country's energy came from hydropower plants in 2009. There has also been some investment in wind farms, and the use of biomass, especially for heat, is common practice in rural areas. The country has adopted targets to increase the use of renewable energy, but progress has been slow. In February 2011, Ukraine became a member of the

European Energy Community, aligning the country with the energy policies of the European Union. The move imposed a legal obligation to increase usage of renewable energy and energy efficiency and to address climate change concerns.

Wylene Rholetter
Independent Scholar

See Also: Coal; Kyoto Protocol; Nuclear Power; Russia.

Further Readings

Central Intelligence Agency. "The World Fact Book: Ukraine." https://www.cia.gov/library/publications/the-world-factbook/geos/up.html (Accessed July 2011).

International Energy Agency. "Ukraine Energy Policy Review." http://www.iea.org/Textbase/npsum/Ukraine2006SUM.pdf (Accessed July 2011).

United Nations Development Programme. "Ukraine." http://www.undp.org.ua/en/energy-and-environment (Accessed July 2011).

United Arab Emirates

The United Arab Emirates (UAE) is a federation of seven emirates that merged in 1971 and 1972, forming the third-largest economy in the Middle East, with the second-highest gross domestic product per capita. The UAE economy is diversified relative to other Middle Eastern countries, with only about 25 percent dependent on oil and gas output. The UAE shares borders with Oman and Saudi Arabia and has 818 mi. (1,318 km) of coastline on the Persian Gulf and Gulf of Oman. Most of the land is desert, with mountains in the east; less than 1 percent of the land is arable.

The UAE is a major producer of oil and natural gas and has embraced enhanced oil recovery (EOR) technologies to increase the extraction rate of mature oil projects. The UAE has 97.8 billion barrels of oil reserves (seventh-largest in the world) and in 2010 produced about 2.81 million barrels per day and exported about 2.32 million barrels per day, primarily to Japan (40 percent of the export total) and other Asian nations. The UAE has 214.4 trillion cu. ft. of proven natural gas reserves, seventh-largest in the world, with most of these reserves located in Abu Dhabi. In 2009, the UAR produced 5.1 billion cu. ft. of natural gas per day, but domestic consumption (primarily for electricity generation) also outstripped production for the first time (with a net deficit of 361 billion cu. ft.).

The UAE's electric production capacity of 18.747 gigawatts is strained at peak seasonal times, although service interruptions have primarily been due to lack of natural gas feedstock, rather than insufficient production capacity. Plans are underway for a region-wide electric grid among the UAE, Kuwait, Qatar, Bahrain, Saudi Arabia, and Oman, with the purposes of reducing power outages and increasing power exchange. Total carbon dioxide (CO_2) emissions from consumption of fossil fuels in 2008 was 198.80 million metric tons, ranking 26th in the world.

Global Warming Steps

In 2009, the Environment Agency–Abu Dhabi (EAD) filed a Climate Change Policy for Abu Dhabi Emirate and is currently working with the Ministry of Environment and Water to develop a similar policy for the UAR. A 2010 report by the EAD reported that numerous changes are expected if global warming continues. In the coastal zone, degradation of fishery habitats, inland migration in sabkhat areas, reduction of the mangrove forest, reduced productivity of sea grass, and greater bleaching frequency of coral reefs are predicted. For dryland ecosystems, expected changes include reduced biodiversity, increasing levels of aridity, displacement or disappearance of some ecosystems, and increasing drought.

In 2005, the UAE became the first major oil-producing nation to ratify the Kyoto Protocol, and is taking a number of steps to reduce greenhouse gas emissions. The Abu Dhabi National Oil Company has adopted a strategic objective of zero flaring (burning off waste gas or oil during production and testing) and reduced flaring from 7.5 million cu. m per day in 1995 to 2.5 million cu. m per day in 2004. The Masdar Initiative, launched in 2006, includes a $15 billion investment in renewable energy projects (including solar, wind, and hydrogen power), sustainable development, carbon reduction and management, research and development, education, and manufacturing. Since January 2008, all buildings in the UAE are required to meet environmental standards, and Dubai is developing a master plan to improve energy efficiency that will include increasing public transportation and introduce demand-side management of electricity production. In 2009, the UAE contracted with a consortium led by Korea Electric Power Corporation to design, build, and operate nuclear power plants in the UAE; in the same year, the 123 Agreement between the United States and the UAE called for peaceful nuclear cooperation and enhanced international standards of safety, security, and nuclear nonproliferation.

Sarah Boslaugh
Kennesaw State University

See Also: Oil, Production of; Organization of the Petroleum Exporting Countries; Renewable Energy, Overview.

Further Readings

Embassy of the United Arab Emirates. "Energy in the UAE." http://www.uae-embassy.org/uae/energy (Accessed July 2011).

Environment Agency Abu Dhabi. "Study Reveals the Effects of Climate Change on the UAE" (January 14, 2010). http://www.ead.ae.en/news/climate .change.in.the.uae.aspx (Accessed July 2011).

U.S. Energy Information Administration. "Analysis: United Arab Emirates" (January 2011). http:// www.eia.gov/countries.cab.cfm?fips=TC (Accessed July 2011).

United Kingdom

The United Kingdom (UK) of Great Britain and Northern Ireland has a land area of 94,526 sq. mi. (244,821 sq. km), with a population of 62.2 million (mid-2010 est.) and a population density of 661.9 people per sq. mi. (255.6 people per sq. km). London, the capital and 16th-largest city in the world, has a population density of 12,892 per sq. mi. (4,978 per sq. km). About 25 percent of the land is devoted to agriculture, with a further 46 percent used for meadow or pasture; approximately 10 percent of the land is forested.

Traditionally, most of the electricity generation in the UK has come from coal, which has been mined in parts of Scotland, Yorkshire, Nottinghamshire, and South Wales. The continued use of coal and oil—Britain has made use of the North Sea oil fields since the 1970s—has meant that 73.2 percent of Britain's electricity generation was, in 2001, still coming from fossil fuels (including natural gas) with 23 percent coming from nuclear fuel and only 1.5 percent from hydropower. Until the late 19th century, coal was used for heating in houses throughout the country, but because of the famous London smog, the use of coal was banned for heating households and was replaced by coke. This has also been phased out in many cities and towns. Although recent governments have tried to use nuclear power more extensively, this move has been widely opposed by many people who are concerned about the safety of nuclear power, with political pressure over the location of the various nuclear power stations.

The UK ranks 43rd in terms of its carbon dioxide (CO_2) emissions per capita, with 10 metric tons in 1990, falling steadily to 9.2 metric tons by 1998, then rising to 9.79 metric tons by 2004, before gradually falling to 8.5 metric tons per person in 2008. A third of all CO_2 emissions in the country are from the generation of electricity, due to the relatively wealthy population and cold winters, as household heating is a significant factor. Household air conditioning is still relatively rare, but is common in many office complexes. Of the country's CO_2 emissions, about 27 percent comes from transportation through the heavy use of private automobiles, with large traffic jams and tailbacks in London and many other major cities and motorways; approximately 17 percent is generated for residential use; and 15 percent comes from manufacturing and construction. In terms of the source of these emissions, 27 percent comes from solid fuels, with 36 percent from liquid fuels, and 35 percent from gaseous fuels, and 1 percent from gas flaring.

Impacts of Global Warming

There have been many effects of global warming and climate change on Britain. Because statistics have been collected there since the 18th century, it has been easier to study the changes. The number of cold days has steadily decreased, with an average of four days per year above 68 degrees F (20 degrees C) for most of the period since 1772, but 26 days above that temperature in 1995. October 2001 was the warmest October in central England, with four of the five warmest years in the previous three and a half centuries recorded in the 1990s and early 2000s. One study by dendrologists has shown that oak trees have experienced earlier leafing as the climate gets warmer.

In addition to rises in temperature, there have also been widespread floods, with the events of October and November 2000 resulting in the flooding of some 10,000 houses at a cost of about $1.5 billion. This was the worst flooding in Britain since the events of 1947, with the melting of a six-week snowpack, although some World War II damage to canals leading into the River Thames worsened the situation. Floods were subsequently recorded in 1968, 1993, and 1998, with those in 2000 following the wettest autumn since records were first collected in the late 1660s. Although floods are not unknown in Britain—the River Thames flooded again in 2003 and 2006—in June and July 2007, there were much more serious floods. These caused damage estimated at $3 billion, with Northern Ireland experiencing floods on June 12 and East Yorkshire and the Midlands hit three days later. Over the next five weeks, large parts of Berkshire, Gloucestershire, Oxfordshire, and South Wales were also inundated, with rainfall amounts in June 2007 reaching twice the June average. Some areas of the country received the average monthly precipitation in just one day. June 2007 was one of the wettest months ever in recorded British history. There were also floods in the fall of 2008, and floods in 2009 took place in November and December.

Floods causing an estimated $3 billion in damages hit the United Kingdom in 2007, including this flooding in Sheffield. Some areas received the average monthly precipitation in just one day. The event was followed by floods in 2008 and 2009.

The worry has been that floods, which took place twice per century on average, are now taking place every three to five years. That they are not worse is only because of the construction of the Thames Barrier in the 1970s, which has prevented any serious floods from occurring in London since the events of January 1928, March 1947, and 1968. This increase in floods—and especially their severity—is believed to be linked to global warming and climate change.

During the 1990s, a detailed survey of plant species in the country showed that the date of the first flowering of 385 British plant species had advanced by an average of 4.5 days when compared with the previous four decades. The activity of plant flowering is particularly sensitive to temperatures in the month before the plant flowers, indicating that plants have become sensitive to the changes in temperature, with those that flower in spring the most responsive. In terms of fauna, British birds have steadily expanded their ranges northward, with more birds that had previously only been found in the south of the country being spotted in northern England and Scotland. Over the last 25 years, some birds have expanded the northern margins of their ranges by about 12 mi. (19 km). Another study of birds has shown that between 1971 and 1995, some 32 percent of the 65 species in the study have started laying eggs an average of 8.8 days earlier each year. In addition, frogs, toads, and newts have begun spawning between 9 and 10 days earlier than had been the case 20 years earlier.

Government and Mitigation Efforts

The British government of John Major took part in the United Nations Framework Convention on Climate Change (UNFCCC) signed in Rio de Janeiro in May 1992. The following government of Tony Blair signed the Kyoto Protocol to the UNFCCC on April 29, 1998, ratifying it on May 31, 2002; it entered into force on February 16, 2005.

The UK Climate Strategy introduced in 1994 had the objective of keeping greenhouse gas (GHG) emissions of CO_2, methane, and nitrous oxide at 1990 rates: CO_2 efforts focused on incentives to business and home users to conserve energy; methane efforts focused on reducing landfill through a landfill levy and a greater regulatory environment, as well as limiting methane emissions from coal production; and nitrous oxide efforts worked through technological innovations in the manufacture of nylon. The introduction of three-way catalytic converters was planned to reduce carbon monoxide, especially from car exhausts, by up to 50 percent. The reorganization of large power stations was expected to reduce nitrous oxides by 35 percent. This is particularly important, because the UK has a high per-capita ownership of passenger cars, with 458 per 1,000 people, the 24th highest in the world, but below many other European countries such as Austria, Germany, Ireland, France, Belgium, Spain, and Sweden.

In November 2000, the UK's Climate Change Policy, which was formulated following the UN Conference on Environment and Development, was formally launched. In 2004, the UK was the eighth-largest producer of carbon emissions, with the country responsible for about 2.3 percent of the world's total coming from fossil fuels. The plan drawn up by the Blair government was not just to cut emissions back to 12.5 percent less than the 1990 rate from 2008 until 2012, as agreed by the Kyoto Protocol, but also to reduce these levels to 20 percent lower than the 1990 rate by 2010. The methods used by the British government to reduce carbon emissions largely hinged on encouraging business to improve its use of energy, cut back on emissions from cars by providing better public transportation, promote energy efficiency in homes, and get agriculture to reduce emissions. In 1990, CO_2 emissions per capita were 10 metric tons, rising slightly for the next two years, falling to less than 10 metric tons, reaching 8.9 metric tons in 2007, and finally 8.5 metric tons in 2008.

As worry about the effects of global warming and climate change received much publicity in the British press, the Campaign against Climate Change was founded in 2001 to oppose the rejection of the Kyoto Protocol by U.S. President George W. Bush. Although it had small beginnings, on December 3, 2005, the campaign organized a large rally in London, and another took place on November 4, 2005. By that time, there were also a number of other pressure groups, including Stop Climate Chaos. Formed as a coalition of a number of other groups, including the Campaign against Climate Change, Stop Climate Chaos was also organizing protests by September 2005. This was to lead to the I Count campaign's attempt to get governments around the world to introduce measures to prevent world temperatures from rising more than 3.6 degrees F (2 degrees C).

On June 21, 2006, royal assent was given to a parliamentary bill that became the Climate Change and Sustainable Energy Act 2006, introduced to the British Parliament by Mark Lazarowicz, a Scottish Labor member of parliament. The bill encourages microgeneration installations to reduce the use of large power stations, reduce CO_2 emissions, and provide fuel for poor people who are unable to afford to heat their residences. The impetus from this led to the drafting of the Climate Change Bill, which was published on March 13, 2007, based heavily on the measures suggested in the I Count campaign. It aimed to reduce the UK's carbon emissions for 2050 to 60 percent of the level for 1990, with an intermediate target range of 26 to 32 percent by 2020. The bill was initially criticized for failing to include international aviation and shipping, but quickly gained cross-party support, although it has not been passed into law. The British government has also been very keen on establishing a GHG allowance trading regime, although plans for this are still being drawn up.

In December 2009, the British government played a major role in the Copenhagen climate negotiations summit. A week before the start of the conference, about 20,000 people took part in a march in London, calling on the British to force developed countries to reduce their carbon emissions by 2020 by 40 percent and provide loans and grants to less developed countries to help them to adapt to climate change. Addressing the conference, British Prime Minister Gordon Brown stated "we have made a start," but wanted a legally binding agreement, which was not achieved at the conference. David Cameron, British prime minister as of May 2010, has not commented much on climate change, but stated his support for the Climate Change Act of 2008.

Justin Corfield
Geelong Grammar School

See Also: Automobiles; Climate-Gate; European Union; European Union Emissions Trading Scheme; Floods; Ireland; Kyoto Protocol; Norway; Royal Dutch/Shell Group; Royal Meteorological Society.

Further Readings

Carrington, Damian. "David Cameron Must Speak Out on Climate Change, Says Top Scientist." *Guardian* (June 29, 2011).

Furniss, Charles. "Dossier: Flooding in the UK." *Geographical*, v.78/6 (June 2006).

EarthTrends. "United Kingdom: Climate and Atmosphere." http://earthtrends.wri.org/text/climate-atmosphere/country-profile-189.html (Accessed October 2011).

Great Britain Department of the Environment. *Climate Change: The UK Programme: The United Kingdom's Report Under the Framework Convention on Climate Change.* London: H.M. Stationary Office, 1994.

Henson, Robert. *The Rough Guide to Climate Change.* London: Rough Guide, 2006.

Hulme, Mike and John Turnpenny. "Understanding and Managing Climate Change: The UK Perspective." *Geographical Journal*, v.170/2 (June 2004).

McCormick, John. *Environmental Policy in the European Union.* New York: Palgrave, 2001.

Monbiot, George. "Then ... And Now." *Guardian* (June 30, 2005). http://www.guardian.co.uk/environment/2005/jun/30/climatechange.climatechangeenvironment1 (Accessed July 2011).

O'Riordan, Tim and Jill Jäger, eds. *Politics of Climate Change: A European Perspective.* London: Routledge, 1996.

Parry, M. L. *The Potential Effects of Climate Change in the United Kingdom* London: H. M. Stationary Office, 1991.

Parry, M. L., et al. *Review of the Potential Effects of Climate Change in the United Kingdom: Second Report.* London: H.M. Stationary Office, 1996.

van der Wurff, Richard. *International Climate Change Politics: Interests and Perceptions—Comparative Study of Climate Change Politics in Germany, the United Kingdom, and the United States.* Amsterdam: Universiteit van Amsterdam, 1997.

Wissenburg, Marcel. *European Discourses on Environmental Policy.* Aldershot, UK: Ashgate, 1999.

Wordsworth, A. and M. Grubb. "Quantifying the UK's Incentives for Low Carbon Investment." *Climate Policy*, v.3 (2001).

Yamin, Farhana, ed. *Climate Change and Carbon Markets: A Handbook of Emissions Reduction Mechanisms.* London: Earthscan, 2005.

United Nations Conference on Trade and Development

The United Nations Conference on Trade and Development (UNCTAD) played an important role in the implementation of the Kyoto Protocol to the United Nations Framework Convention on Climate Change (UNFCCC, or Rio Treaty) in 1992. It acted primarily through the Earth Council Alliance. In UNCTAD's words, its "main role in addressing this global challenge is to help developing countries master the trade and development implications and take advantage of emerging trade and investment opportunities."

The Earth Council Alliance plans and facilitates research, conferences, consulting, and publications for developing countries to assist in improving their greenhouse gas (GHG) emissions and in using the Clean Development Mechanism (CDM) under the Kyoto Protocol. CDM projects are important vehicles for directing investment from many sources into new technology and GHG-reducing facilities.

UCTAD offers an extensive catalog of publications and guides to the CDM system, as well as informative publications about evolving trade related to the cap-and-trade markets. Two guides have been published that a number of countries have used to write their laws and regulations and organize their programs related to global warming: *The Clean Development Mechanism Guide 2009* and the *Trade and Development Opportunities and Challenges Under the Clean Development Mechanism.*

Missions and Initiatives

UNCTAD's mission is organized into activities for the following:

- Economic analysis on trade and climate change interface, for example by assessing trade and development impacts of specific emissions-reduction proposals
- Development of training material on the rules of the CDM so that a considerable number of developing nations can attract investment via CDM toward energy development projects
- Organization of international policy forums to discuss the interface and mutual supportiveness of trade and climate change policy at the international, regional, and national levels
- Participation in intergovernmental and technical meetings to showcase UNCTAD's activities related to biofuels production, domestic use, and trade, including assessment of nontechnical barriers related to trade in biofuels.

The close relationship between controlling global warming and increasing the use of biofuels has been highlighted by UNCTAD's Clean Fuels Initiative (CFI), begun in 2005 at an international conference. It is important that developing countries mate CDM projects with expanded investments in biofuels, such as methane and ethanol from agricultural sources, in their national sustainability programs.

The work of the CFI helps assess the potential of specific developing countries to engage in the growing worldwide production, use, and trade of biofuels. It looks at the possible opportunities and impacts on domestic energy policies, food security, environmental management, job creation, and rural development. It deals with trade flows, tariff regimes, and market access and entry issues affecting international trade in biofuels.

In 2007, the activities of CDM and CFI were paired with a Carbon Neutral Initiative announced by UNCTAD's secretary general. The organization is assessing its GHG emissions footprint and introducing practices to become carbon neutral. This will involve not only offsetting credits for the emissions from its Geneva offices, but also ways to neutralize the significant carbon-fuels emissions produced during travel by staff and UNCTAD conference participants.

UNCTAD's place in the UN economic and financial policy process was highlighted in 2010, when Secretary-General Ban Ki Moon appointed the secretary-general of UNCTAD, Dr. Supachai Panitchpakdi of Thailand, to join a very select group of prime ministers and finance ministers on a commission to make recommendations for financing the climate change activities of the UN from 2012 to 2020. The commission made recommendations for the post-2012 period to the Kyoto Protocol Conference of Parties in Puebla, Mexico, in December 2010, for financing $100 billion in development funds.

The Earth Council's research and training program has brought a high level of awareness and current capabilities to developing countries. The council's staff have concentrated on disseminating updates on the carbon markets, in particular the European Union Emissions Trading Scheme, including describing the role of prices in moving CDM projects into production.

UNCTAD also shares responsibility for expanding the global cap-and-trade system to Africa and South America, regions that have lagged far behind Asia as locations for CDM investments and carbon trading. With other UN and global agencies, such as the UN Environment Programme and the UN Development Programme, the World

Bank, the International Emissions Trading Association, and many nongovernmental organizations (NGOs), UNCTAD has brought together officials of African countries in a series of All-Africa Carbon Forums, such as the one in March 2010.

In the lead-up to the highly anticipated Fifteenth Conference of the Parties (COP 15) in Copenhagen in 2009, a new interest in forestry for reducing GHG emissions and climate change was building. In a UNCTAD-sponsored study published in 2009, Aaron Cosbey of the International Institute for Sustainable Development in Canada focused on the emerging system titled Reducing Emissions from Deforestation and Forest Degradation (UN-REDD) in developing countries. The program, created in June 2008, was designed as a fund for voluntary contributions by Annex II Parties and others to disburse funds in support of country-level actions and capacity building. The emissions reductions generated from the projects were intended to be available as compliance credits by Annex I Parties, since the Clean Development Mechanism does not cover avoided deforestation. Overall, the idea of the program is as a pilot to test methodologies and find ways in which to link it to the UNFCCC process in fulfillment of the Bali mandate.

UNCTAD's influence through such work is evident in this new stage of UN activities to protect forests as key contributors to reducing GHG emissions. Such cooperative planning has produced the expansive system of finance, sponsorship, expertise, and project implementation designated REDD+, which combines systematic involvement of local populations in the REDD projects and protection of social, cultural, and traditional conditions as forests become integrated into the global economy.

Since 1997, UNCTAD has become well established as a meeting place for motivated global planners, educators, technical specialists, economists, and others to support developing people in the complex transition from traditional to modern economies. These important responsibilities have been enhanced by a central concern for assisting all other actors in attacking the sources and effects of climate change and global warming.

Alan B. Reed
University of New Mexico

See Also: Conference of the Parties; G77; Reducing Emissions From Deforestation and Forest Degradation (UN-REDD); United Nations Development Programme; United Nations Environment Programme; World Bank; World Trade Organization.

Further Readings

Cosbey, Aaron. "Developing Country Interests in Climate Change Action." unctad.org/en/docs/ditcbcc20092_en.pdf (Accessed December 2011).

United Nations Conference on Trade and Development. "Carbon Markets and Beyond: The Limited Role of Prices and Taxes in Climate and Development Policy." unctad.org/en/docs/gdsmd pg2420084_en.pdf (Accessed December 2011).

United Nations Development Programme

The United Nations Development Programme (UNDP), a division of the United Nations, was established in 1965 and operates in 166 countries. The organization is headquartered in New York City and funded by voluntary contributions of its member nations. UNDP focuses on solutions to global and national development issues and is largely known for its work on Millennium Development Goals (MDGs), which consist of eight international development goals that all 192 UN member states and numerous international organizations agreed to accomplish by 2015. MDGs include eliminating extreme poverty, reducing child mortality rates, fighting disease epidemics, and establishing a global partnership for development. UNDP maintains that eliminating poverty and dealing with climate change go "hand-in-hand." As a result, UNDP is taking an active role in projects concerning climate change.

UNDP and Climate Change

Corresponding to Agenda 21, accepted at the UN Conference on Environment and Development in 1992, UNDP sought to integrate an environmental perspective into its projects at all levels of its organization. Subsequently, through its collaboration with the Global Environment Facility (GEF), UNDP assisted over 138 countries to prepare their national strategies for the UN Framework Convention on Climate Change (UNFCCC) and the Convention on Biological Diversity. Following the Kyoto Protocol in 1998, UNDP also played an important advocacy role by supporting sustainable energy use, clean development, and the lowering of greenhouse gas emissions in its development projects. The World Energy Assessment, a partnership among UNDP, the Department of Economic and Social Affairs, and the World Energy Council, was further initiated in 1998 to provide a technical and scientific foundation for energy assessment for various stakeholders. UNDP continues to fund and support climate change adaptation and mitigation projects in 140 nations with the goal of green, low-emission, and climate-resilient development.

Millennium Development Goals

MDGs present a framework to the UN system to collaborate comprehensibly toward the common goal of eradicating extreme poverty, diminishing child mortality rates, fighting disease epidemics, and creating a global partnership for development. Climate change, according to UNDP, affects the MDGs, with particular impacts on agricultural production and food security, access to water, infrastructure, women, human health, and environmental sustainability. UNDP observes that receding forests and changing precipitation patterns exacerbate poverty and undermine the future of nations. For example, studies show that in Ethiopia, children exposed to drought during early childhood are 36 percent more likely to be malnourished, undermining the country's human capital. Due to climate change, UNDP estimates that up to 600 million people in Africa could be subjected to malnutrition as agricultural systems collapse, and an additional 1.8 billion people could face water scarcity in Asia. The organization indicates that a person from a developing country is 79 times more likely to experience climate disaster than someone from a developed country. Statistics of this nature are a warning to UNDP and to the international community that decades of development work could be undermined as a result of climate change.

Comprehensive Response to Climate Change

UNDP aims to help least developed and developing countries in response to climate change by strengthening their capacity to adapt, and in particular, by trying to reduce poverty so that people can better cope with climate-related problems.

For climate change projects, UNDP partners largely with GEF, the operating entity of the financial system of UNFCCC and a division of the World Bank. The UNDP–GEF partnership helps nations prepare their National Adaptation Programs of Action (NAPAs), which identify and propose projects for urgent climate change adaptation needs within their national poverty programs; supports various nations in creating disaster-management legislations (such as Bangladesh and Madagascar); and assists developing countries with the creation and funding of climate change capacity-building projects. At the community level, UNDP and the GEF Small Grants Programme help local communities to implement projects of various sizes. UNDP projects use the Least Developed Countries Fund (LDCF), Special Climate Change Fund (SCCF), and Strategic Priority on Adaptation (SPA). SPA was set up under the GEF Trust Fund, and the LDCF and the SCCF were established under the UNFCCC. To date, LDCF made contributions of $332.6 million and SCCF $180 million to climate change–related projects. However, as discussed during the 2009 UNFCCC Conference of the Parties in Copenhagen and in the subsequent negotiation meetings, more financing is needed for climate change adaptation projects.

As of November 2010, the UNDP–GEF portfolio, funded by LDCF, SCCF, and SPA, included projects totaling $773 million. Of this amount, $607 million was obtained through cofinancing. At present, the UNDP–GEF portfolio includes climate change adaptation projects in 52 countries. Nearly half of these projects target Africa, and out of these, close to one-third of the available funds focus on west and central Africa, where the impact of climate change is the most severe.

Ezgi Akpinar-Ferrand
University of Cincinnati

See Also: Adaptation; Climate Change, Effects of; Global Environment Facility; United Nations Framework Convention on Climate Change.

Further Readings

United Nations Development Programme. *Adapting to Climate Change: UNDP-GEF Initiatives Financed by the Least Developed Countries Fund, Special Climate Change Fund and Strategic Priority Adaptation.* Framingham, MA: Wing Press, 2011.

United Nations Development Programme. *Annual Report of the Administrator for 1998: DP/1999/15.* New York: UNDP, 1999.

United Nations Development Programme. "Fast Facts: Climate Change and UNDP." http://www.undp.org/publications/fast-facts/FF-climate.pdf (Accessed July 2011)

United Nations Environment Programme

The United Nations Environment Programme (UNEP) is the United Nations' designated entity to address environmental issues at the global and regional level. Its mandate is to "coordinate the development of environmental policy consensus by keeping the global environment under review and bringing emerging issues to the attention of governments and the international community for action."

The only UN body to have its headquarters in Africa, it is based in Nairobi, Kenya. It also has six regional offices and various country offices, with the Nairobi headquarters responsible for overall coordination and governance. UNEP supports a growing network of centers, such as the UNEP Collaborating Centre on Energy and Environment (UCCEE), Global Resource Information Database Centre (GRID), and UNEP World Conservation Monitoring Centre (UNEP–WCMC). In addition, it cooperates closely with an increasing number of global and regional multilateral environmental agreements (MEAs), including the UN Framework Convention on Climate Change (UNFCCC) and its designated bodies.

Still, UNEP does not administer all of these MEAs; the United Nations Framework Convention on Climate Change (UNFCCC), for example, also has its secretariat. There has been a long debate on exactly how much of a coordinating role UNEP could and should have. The currently UNEP-administered environmental conventions include the Ozone and the Stockholm Persistent Organic Pollutants Secretariats. In addition, together with the World Meteorological Organization, UNEP established the Intergovernmental Panel on Climate Change (IPCC) in 1988. UNEP is also one of several implementing agencies for the Global Environment Facility (GEF) and is also a member of the UN Development Group.

Environmental Policy Initiatives

UNEP is well known for its environmental policy work. It collects environmental data, conducts scientific research, and publishes many reports, atlases, and newsletters (including the Global Environment Outlook, the GEO data portal), providing analysis and information for policymakers and the concerned public. Climate change is one of the six thematic priorities of UNEP, but for 20 years, it has been working with countries on climate initiatives and, at the global level, on climate policy and science. Reducing Emissions from Deforestation and Forest Degradation (UN-REDD), land-use change, and science and outreach were identified as three areas on which it focuses its climate activities. These activities include the following:

- *Ecosystems-based adaptation*: the use of biodiversity and ecosystem services as part of an overall adaptation strategy to help people and communities adapt to the negative effects of climate change at local, national, regional, and global levels.
- *Clean Technology Readiness*: assisting developing countries to make low-carbon and efficient energy economic growth.
- *REDD+*: reducing emissions from deforestation and forest degradation and associated efforts to conserve, sustainably manage, and enhance forest carbon stocks.

In addition, in its recently published *Green Economy Report*, which received great political attention both from developing and developed countries and is seen as a major input to the upcoming UN Conference on Sustainable Development (the Rio+20 Summit, to take place in Rio de Janeiro in 2012), UNEP places great emphasis on the need for decarbonization and low-carbon growth as part of a new, more sustainable economic growth pathway around the world. Next to the *Green Economy Report*, the whole UNEP Green Economy Initiative is seen as an important contribution to the upcoming 2012 Rio+20 Summit, which will focus on green economy (including decarbonization) in the context of sustainable development and poverty eradication.

As a link to its Green Economy Initiative work, in 2010, UNEP produced the *Emissions Gap* report, which generated a great deal of support at the 16th Conference of the Parties (COP 16) in Cancun, Mexico. It informs governments and the wider community on how far a response to climate change has moved based on the pledges associated with the Copenhagen Accord of 2009, and what might be achieved in terms of limiting a global temperature rise to 3.6 degrees F (2 degrees C) or less in the 21st century, and in terms of setting the stage for a green economy.

Groundwork

UNEP also works on the ground, together with the National Cleaner Production Centres (with the UN Industrial Development Organization) and UNEP's 10 regional networks of ozone officers, which have helped ensure the success of the Montreal Protocol. It promotes renewable energy technology through the development of nationally appropriate mitigation actions linked with its Green Economy Advisory Services and specific programs such as the Solar Loan Program in India. Furthermore, it assists the exchange of best practices linked with and feeding into the Climate Technology Centre and Network under the UNFCCC.

In addition, it supports countries in accessing the Adaptation Fund and contributes to the UN-REDD work program, helping to provide REDD readiness to countries that have requested support, but no program exists due to lack of funds.

UNEP's climate-related science efforts primarily focus on supporting work on non-carbon dioxide gases (black carbon, hydroflourocarbons, and near-term forcers) and promoting climate science, including through an initiative called the Programme of Research on Climate Change Vulnerability, Impacts, and Adaptation.

With major challenges to address in the field of environment (and related to climate change), UNEP is not a principal organ of the UN, nor a specialized agency of it. As a program, it has a relatively weak position in the UN system, as it does not have universal membership, nor guaranteed funding. Its coordination role is also weak over the MEAs, even though its mandate includes facilitating the coordination of UN activities on matters concerned with the environment. Competitive personal interests, members, and funding issues make it difficult.

The need to strengthen international governance and the related role of UNEP became high on the political agenda after the Intergovernmental Panel on Climate Change *Fourth Assessment Report* in February 2007, when French President Jacques Chirac read the Paris Call for Action, supported by 46 countries, asking for the UNEP to be replaced by a new and more powerful UN Environment Organization (UNEO). The motion for this organization today is not as strong, due to lack of international political will (the 46 countries included all European Union nations, but not the United States, Russia, or China). Still, the need to reform the existing sustainable development structure and revisit the role of the UNEP and its overview over the MEAs are already well-established ideas internationally, and are also reflected in the second theme of Rio+20 (the International Framework for Sustainable Development).

Gyorgyi Gurban
PPKE-JAK Law School

See Also: Emissions, Baseline; Integrated Assessment; Intergovernmental Panel on Climate Change; International Energy Agency; Joint Implementation; Montreal Protocol; Rio+20; Sea Level, Rising; World Resources Institute; World Trade Organization.

Further Readings

Biermann, Frank and Steffen Bauer, eds. *A World Environment Organization*. Aldershot, UK: Ashgate, 2006.

Jabbour, Jason, et al. *UNEP Year Book 2009: New Science and Developments in Our Changing Environment*. Geneva, United Nations, 2009.

United Nations Conference on Sustainable Development. http://www.uncsd2012.org/rio20 (Accessed July 2011).

United Nations Environment Programme. "Climate Change." http://www.unep.org/climatechange (Accessed July 2011).

United Nations Environment Programme. "Emissions Gap Report" (2010). http://www.unep.org/publications/ebooks/emissionsgapreport/pdfs/EMISSIONS_GAP_TECHNICAL_SUMMARY.pdf (Accessed July 2011).

United Nations Environment Programme. "Green Economy Report" (2011). http://www.unep.org/greeneconomy/GreenEconomyReport/tabid/29846/Default.aspx (Accessed July 2011).

United Nations Environment Programme. "30 Ways in 30 Days: Inspiring Action on Climate Change and Sustainable Development" (2011). http://www.unep.org/pdf/30ways.pdf (Accessed July 2011).

United Nations Environment Programme. "UNEP Six Priority Areas Factsheets: Climate Change" (2011). http://www.unep.org/pdf/UNEP_Profile/Climate_change.pdf (Accessed July 2011).

UN-REDD. http://www.un-redd.org (Accessed July 2011).

United Nations Framework Convention on Climate Change

The United Nations Framework Convention on Climate Change (UNFCCC) is a multilateral (international) environmental treaty that came alive at the UN Conference on Environment and Development (UNCED, or the Rio Summit), together with the other two Rio Conventions: the Convention to Combat Desertification and the Convention on Biological Diversity.

The UNFCCC was opened for signature in May 1992 and entered into force in March 1994. Currently, it has 195 parties (194 states and the European Union as a regional, international organization). The United States is also a party to the UNFCCC, and as such is obligated to meet its goals.

The convention, however, sets no mandatory limits on greenhouse gas (GHG) emissions for individual countries and contains no enforcement mechanisms; as such, under international law, it can be considered "soft" legislation, which is legally nonbinding. Instead, the treaty provides for updates, or protocols, that set mandatory emission limits. Principal among them is the Kyoto Protocol, which has obligatory targets, as well as a unique enforcement mechanism. The Kyoto Protocol, however, is only understandable in relation to the UNFCCC, which sets its framework, along with the key objectives, principles, roles, and tasks of key players and countries (the diversification between Annex I and Annex II countries and Non-Annex countries).

The UNFCCC recognized the seriousness of the problem of a changing climate and set as its key objective to stabilize GHG concentrations in the atmosphere at a level that would prevent dangerous anthropogenic interference with the climate system, in a timely manner, before serious food security and economic problems arose. Its guiding principles are the right of future generations and the precautionary principle, the principle of common but differentiated responsibilities and proportionality.

For key obligations, above many others, it stipulates for all parties (but taking into account the common but differentiated responsibilities with leadership of the developed countries) the need to adopt national policies on the mitigation of climate change. It also obliges Annex I countries to establish national inventories of GHG emissions and removals. These were later used to create the 1990 benchmark levels for accession of Annex I countries to the Kyoto Protocol and for the commitment of those countries to GHG reductions. These updated inventories are regularly submitted by Annex I countries.

Actions, however, were only obligations under "soft law" commitments and were aimed primarily at industrialized countries (developed countries and countries in transition), with the intention of stabilizing their emissions of GHGs at 1990 levels by 2000. The parties agreed in general that they would recognize common but differentiated responsibilities in the meaning of greater responsibility for reducing GHG emissions in the near term on the part of developed/industrialized countries, which were listed and identified in Annex I of the UNFCCC.

Controversial Sticking Points

The most controversial and debated issue of the UNFCCC is exactly this principle, which many claim may have been misused. Annex I countries (industrialized countries and economies in transition) and Annex II countries (developed countries that pay for the costs of developing countries) bear all the responsibilities under the UNFCCC (and under the mirroring Kyoto Protocol), while countries that are not included in any of the Annexes are deemed to be developing countries, without any responsibility to mitigate climate change.

Under the convention, developing countries are not required to reduce emission levels unless developed countries supply enough funding and technology. Setting no immediate restrictions for them under UNFCCC is under the assumption that GHG reductions would be a restriction to their development.

However, developing countries may volunteer to become Annex I countries when they are sufficiently developed, but this option has not been widely used, even though many countries went through great economic development during the recent decade. The countries that are most criticized for showing no leadership in climate change mitigation and no ambition to follow obligatory targets under an international regime are the "emerging economies": Brazil, India, and China, with the latter under the most pressure, as it is one of the greatest GHG emitters next to the United States and the European Union (EU). While negotiating the next (post-Kyoto) regime, following the current system, one of the politically most sensitive and debated questions is exactly how these emerging economies will take their fair share in the mitigation efforts. This was one of the key topics planned for the November–December 2011 Durban Conference of the Parties (COP 17).

Conferences of the Parties

Parties under the convention have been meeting annually in Conferences of the Parties (COP) to

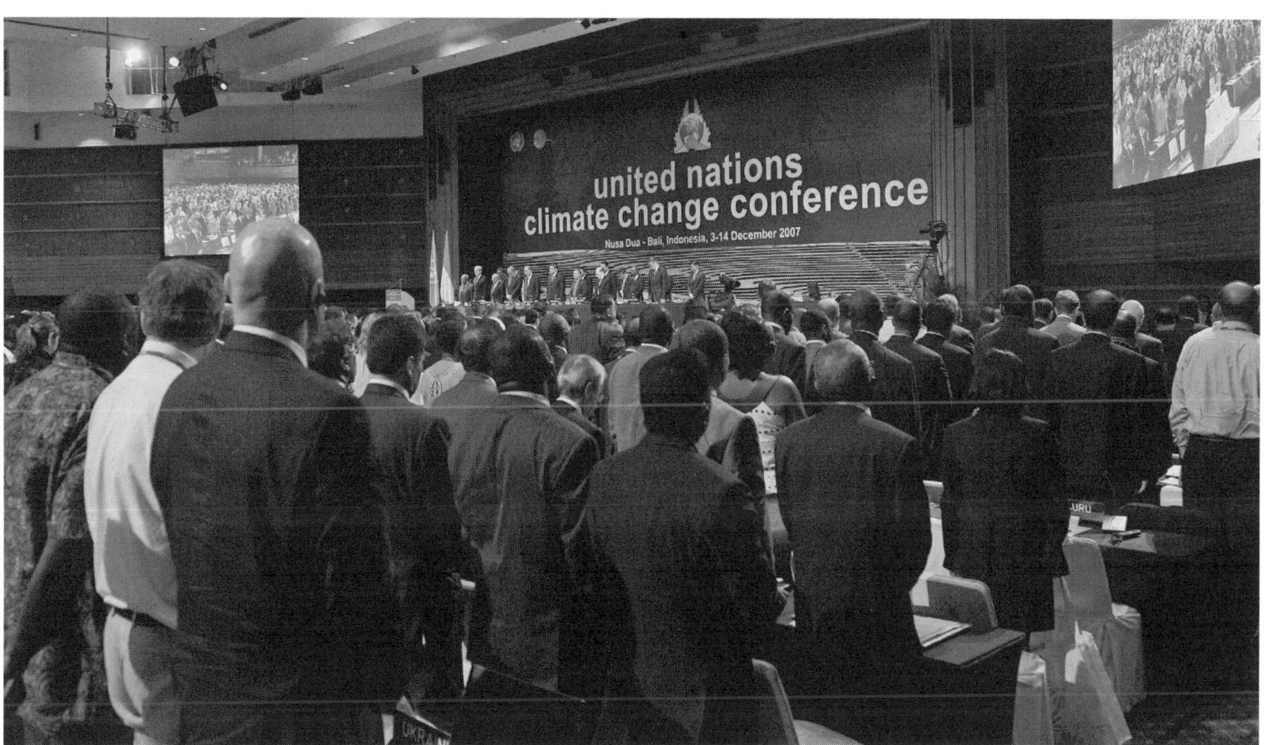

A moment of silence was observed for the victims of the Algiers bombing during the United Nations Climate Change Conference in Bali, Indonesia, December 3–14, 2007. The gathering, also called COP 3/MOP 3, brought together representatives of over 180 countries. It culminated in the agreement on the timeline and negotiation structure for the post-2012 regime, the adoption of the Bali Action Plan, and the establishment of the Ad Hoc Working Group on Long-term Cooperative Action Under the Convention.

assess progress in dealing with climate change, and from the mid-1990s, to negotiate the Kyoto Protocol to establish legally binding obligations for developed countries to reduce their GHG emissions. From 2005, the conferences have met in conjunction with Meetings of Parties of the Kyoto Protocol (MOP), and parties to the convention that are not parties to the protocol can participate in protocol-related meetings as observers.

The first COP took place in 1995 in Berlin, Germany, which was followed in 1996 in Geneva, Switzerland, by COP 2, which accepted the Intergovernmental Panel on Climate Change (IPPC) findings and called for legally binding targets. COP 3, in 1997, was a breakthrough in Kyoto, Japan, where the Kyoto Protocol was adopted, with specific GHG obligations for Annex I countries: the establishment of the Kyoto mechanisms (Joint Implementation, the Clean Development Mechanism, and emissions trading). Under the Protocol, the United States has no specific obligations (seen as major constraint of the protocol), because even though President Clinton signed it, the later Bush administration rejected it. During COP 4 in Buenos Aires, Argentina, in 1998, a two-year Plan of Action to advance efforts and to devise mechanisms for implementing the Kyoto Protocol was adopted, to be completed by 2000. In 1999, in Bonn, Germany, COP 5 stayed rather on a technical level, without conclusions, while COP 6 in 2000, in The Hague, Netherlands, focused on the issue of forests and agricultural lands. Because of other sensitive political issues, including funding and meeting of commitments, COP 6 did not bring a breakthrough, either.

In 2001, COP 7 in Marrakesh, Morocco, delivered the needed technical agreements and was considered the major document that laid out the specific, technical rules for the flexible mechanisms of the protocol. These agreements became known as the Marrakesh Accords. Several conferences took place in New Delhi, India; Milan, Italy, and Buenos Aires, Argentina, in the intervening years before the next major conference (COP 11) took place in Montreal, Canada, in 2005. As the Kyoto Protocol came into force, it was also the first Meeting of the Parties (MOP 1) to the Kyoto Protocol since the initial meeting in Kyoto in 1997. The Montreal Action Plan is an agreement hammered out at the end of the conference to "extend the life of the Kyoto Protocol beyond its 2012 expiration date and negotiate deeper cuts in greenhouse gas emissions."

COP 12/MOP 2 in Nairobi, Kenya, in 2006, resulted in the adoption of a five-year plan of work to support climate change adaptation by developing countries and agreed on the procedures and modalities for the Adaptation Fund.

COP 13/MOP 3, in Bali, Indonesia, in 2007 culminated in the agreement on the timeline and negotiation structure for the post-2012 regime and saw the adoption of the Bali Action Plan. It also established the Ad Hoc Working Group on Long-term Cooperative Action under the Convention (AWG–LCA) and amended the New Delhi Work Programme. The following COP 14/MOP 4, in Poznan, Poland, in 2008, agreed on the principles of financing for the least developed countries and on the importance of forest protection in climate change–related efforts.

The most awaited COP 15/MOP 5 in 2009, in Copenhagen, Denmark, did not result in its aim of a legally binding, post-2012 global agreement, but it did achieve long-term action and resulted in ambitious voluntary political commitments both from developed and developing countries.

The breakthrough, building on the achievement of Copenhagen, came in Cancun, Mexico, where at COP 16/MOP 6 the parties made the Copenhagen "deal" legally binding, followed it up with further voluntary commitments both for mitigation and financing climate-related action in the developing world. Developed countries have committed, in the context of meaningful mitigation actions and transparency on implementation, to a goal of jointly mobilizing $100 billion per year by 2020 to address the needs of developing countries.

Lingering Issues After Cancun

There are several main issues to be followed up after Cancun. The first includes the implementation of the Cancun agreement. The second involves the question of the post-2012 period and whether it should include a second commitment period under the Kyoto Protocol. Many developing countries are only able accept a post-2012 period with this condition, while most developed nations cannot agree with this and can only foresee a single legally binding instrument to all parties, with the

EU in the middle. The next is to clarify the assumptions and the conditions related to the attainment of developed countries' mitigation commitments, as well as to understand the diversity of developing countries' mitigation actions, their underlying assumptions, and the support needed for the implementation of these actions. There is also the need for an agreement on a detailed work program for these processes, the need for an agreement on further rules of accounting in the Cancun Agreements, the need to make them operational by completing the work program for the development, and the necessity of work on the first enhanced national communications from Annex I Parties and biennial reports from Non-Annex I Parties (to be submitted no later than January 1, 2014, to ensure provision of information for the 2015 review). The process of designing the Green Climate Fund also needs to be undertaken.

Administrative Structure

The UNFCCC is also the name of the UN Secretariat charged with supporting the operation of the convention, with offices in Haus Carstanjen, Bonn, Germany. From 2006 to 2010, the head of the secretariat was Yvo de Boer; on May 17, 2010, his successor, Christiana Figueres from Costa Rica, was named. The secretariat, augmented through the parallel efforts of the IPCC, aims to gain consensus through meetings and the discussion of various strategies.

The UNFCCC and the Kyoto Protocol are serviced by the secretariat, also known as the Climate Change Secretariat, with a mandate laid out in general terms in Article 8 of the convention.

The main functions of the secretariat are to make practical arrangements for sessions of the convention and protocol bodies; to monitor implementation of the commitments under the convention and the protocol through collection, analysis, and review of information and data provided by parties; to assist parties in implementing their commitments; to support negotiations, including through the provision of substantive analysis; to maintain registries for the issuance of emission credits and for the assigned amounts of emissions of parties that are traded under emission trading schemes; to provide support to the compliance regime of the Kyoto Protocol and to coordinate with the secretariats of other relevant international bodies, notably the Global Environment Facility (GEF) and its implementing agencies (the UN Development Programme, UN Environment Programme, and World Bank), the IPCC, and other relevant conventions.

Specific tasks include the preparation of official documents for the COP and subsidiary bodies, the coordination of in-depth reviews of Annex I Party national communications, and the compilation of GHG inventory data.

The growth in technical work needed since the adoption of the Kyoto Protocol, including reporting guidelines and the land use, land-use change and forestry sector, is leading to a trend of increased technical expertise within the secretariat.

The secretariat is institutionally linked to the UN without being integrated in any program and is administered under UN rules and regulations.

Its head, the executive secretary, is appointed by the UN secretary-general in consultation with the COP through its bureau, and currently holds the rank of assistant secretary-general.

The executive secretary reports to the secretary-general through the under-secretary-general heading the Department of Management on administrative and financial matters, and through the under-secretary-general heading the Department for Economic and Social Affairs on other matters.

Gyorgyi Gurban
PPKE-JAK Law School

See Also: Annex I/B Countries; Bali Roadmap; Berlin Mandate; Conference of the Parties; Copenhagen Summit/Conference; Declaration of The Hague; Intergovernmental Panel on Climate Change; Kyoto Protocol; Marrakesh Accords; Montreal Protocol; Rio+20; Toronto Conference; Vienna Convention; Villach Conference.

Further Readings

International Institute for Sustainable Development. "Reporting on the COPs." http://climate-l.iisd.org/events/seventeenth-session-of-the-conference-of-the-parties-to-the-unfccc-and-seventh-meeting-of-the-parties-to-the-kyoto-protocol (Accessed July 2011).

United Nations Framework Convention on Climate Change. http://unfccc.int/resource/docs/convkp/conveng.pdf (Accessed July 2011).

United Nations Framework Convention on Climate Change. "The Copenhagen Accord." http://unfccc.int/files/meetings/cop_15/application/pdf/cop15_cph_auv.pdf (Accessed July 2011).

United States

As the world's largest economy, the United States is also one of the largest consumers of fossil energy sources and has above average per-capita emitters of greenhouse gases (GHGs). Over the past decade, emissions have continued to grow, largely as a result of an expanding transportation sector and reliance on coal in the energy sector. While the recent economic downturn has seen a reverse in emissions growth, this overall trend has caused the United States to be widely portrayed as a laggard in the regulation of climate change. Such criticism has not only been leveled against the federal strategy to address domestic GHG emissions, which has been based on a fragmented array of voluntary measures and incentives, executive regulation, and issue-specific legislation; it also extends to the international level, where U.S. positions have also faced intense scrutiny ever since the United States withdrew from the Kyoto Protocol to the United Nations Framework Convention on Climate Change (UNFCCC) in 2001.

Background

According to data released in December 2010, domestic U.S. GHG emissions amounted to 6,633.2 megaton carbon dioxide (CO_2) equivalent in 2009, a figure that was 7.3 percent higher than emissions levels in 1990. In absolute terms, only China emits a larger volume of GHGs each year. Primary energy consumption by source is dominated by oil, natural gas, and coal, jointly providing more than 75 percent of the primary energy need. An important trend in recent years has been the discovery of substantial natural gas reserves in domestic shale formations, allowing for a shift away from coal in electricity generation, therefore reducing the carbon intensity of U.S. energy.

Public opinion about climate change and relevant policy responses has proven volatile. While nationwide surveys suggest that a majority of Americans consider global warming a serious or very serious problem, fewer believe that global warming should be a priority for government action. Global warming remains a distinctly partisan issue, as regional differences between the midwest and the coastal states reverberate at the federal level and powerful interest groups continue to portray emission reductions as costly and economically harmful. Finally, the U.S. climate debate has again turned to the question of whether anthropogenic climate change exists. Although concern about energy security and high oil prices has been a more successful driver of popular support, it provides a limited basis only for policies to mitigate climate change or adapt to its impacts.

As was widely expected, the U.S. midterm elections of November 2, 2010, resulted in a solid victory of the Republican Party, which successfully captured a majority in the House of Representatives and eroded the majority of the Democratic Party in the Senate. Likewise, gubernatorial elections in six states saw Democratic incumbents replaced with a Republican contender, and a majority of states now have a Republican governor. These election results have dampened the prospects for future action on climate change, both at the federal level and in a majority of states. Notwithstanding this election outcome, President Barack H. Obama has affirmed the U.S. commitment to reduce GHG emissions by 17 percent below 2005 levels until 2020, as pledged internationally under the Copenhagen Accord. With public concern focused on the ongoing economic crisis and more immediate challenges, however, support for climate change policies will continue to be cautious and unpredictable.

Federal Legislation

Given the scope and complexity of climate change, a formal act of Congress as the federal legislature is generally considered the most suitable vehicle for a comprehensive policy response. While efforts to pass relevant legislation date back more than a decade, the United States—unlike a majority of industrialized nations—currently has no dedicated climate legislation in force. Prospects for congressional legislation improved with the 2006 elections, when the Democratic Party—which has traditionally been more likely to favor measures against climate change—secured a majority in both houses of the 110th Congress. Ensuing personnel changes had a profound impact on the political dynamic underlying climate legislation. Chairing all relevant committees with jurisdiction over GHG mitigation policies, the Democratic leadership also created a Select Committee for Energy Independence and Global Warming in the House of Representatives in March 2007, illustrating the new political weight afforded to climate change in Congress.

Despite a substantial increase in climate policy activity, the 110th Congress did not see passage of comprehensive climate legislation. With expanded Democratic majorities in both houses after the 2008 elections, the 111th Congress set the first milestone when the House of Representatives adopted the Clean Energy and Security Act of 2009 (ACES) by a narrow margin of 219 to 212 votes. Also known as the Waxman–Markey Bill after its authors, once in force, this legislation would have placed limits on GHG emissions from a large section of the U.S. economy and would have introduced a combined energy-efficiency and renewable-electricity standard. Specifically, it called on electric utilities to meet 20 percent of their electricity demand through renewable energy sources and energy efficiency by 2020, established energy-saving standards for new buildings and appliances, and mandated CO_2 emission reductions from major domestic sources by 17 percent by 2020, 42 percent by 2030, and 83 percent by 2050 over 2005 levels. To help achieve these objectives at reduced cost, an economy-wide emissions trading system was to be phased in starting in 2012.

In keeping with the U.S. legislative system, however, the foregoing bill could not enter into force until a counterpart bill was adopted by the Senate and both versions were reconciled in a Conference Committee. Unlike the House of Representatives, where the Committee on Energy and Commerce holds sole responsibility for initiating climate legislation, jurisdiction in the Senate is divided across a number of committees. More importantly, where passage of a bill only requires a simple majority in the House of Representatives, voting rules in the Senate call for 60 votes to close the debate on a bill and proceed to a substantive vote. Even within the Democratic caucus, several moderate senators from heavily industrialized and agricultural states expressed reluctance to pass legislation they perceived as costly and damaging to their states' economies. Despite numerous attempts to reach bipartisan compromise, it therefore proved impossible to secure the required level of support among Democrats and moderate Republicans in the Senate.

On July 22, 2010, the Democratic leadership of the U.S. Senate conceded failure to pass comprehensive climate legislation during the 111th Congress. As a result, any hopes to see comprehensive federal climate legislation in the United States in the foreseeable future suffered a decisive blow. When the 112th Congress convened on January 3, 2011, the legislative docket was cleared, erasing all previous progress and requiring both houses to commence efforts anew in the current legislative period. Given the outcome of the recent midterm elections, prospects for successful climate legislation on the federal level remain greatly diminished going forward, and recent legislative initiatives launched by the Republican majority in the House of Representatives have instead been geared at preventing climate action by the federal administration.

Federal Administration

In the absence of congressional action against climate change, the dynamic has lately shifted somewhat to the federal administration. Although the executive branch had been fairly restrained on action against climate change in the past, Obama's win in the presidential election of November 4, 2008, heralded a major shift in the climate and energy policies of the U.S. administration.

With the adoption of the American Recovery and Reinvestment Act on February 17, 2009, the newly elected president sponsored an unprecedented economic stimulus plan mandating new investments and tax credits exceeding $787 billion. Of the appropriated funds, around $50 billion were earmarked for the energy sector alone, with $11 billion allocated to a Smart Grid Investment Program, $6.3 billion to State Energy Efficiency Programs, $6 billion to renewable energy loans, and $2.4 billion to an Advanced Battery Grants Program and other electric vehicle projects.

At the same time, the president directed the Environmental Protection Agency (EPA) to regulate CO_2 and other GHGs under the existing Clean Air Act (CAA). Established on December 2, 1970, as an independent agency of the U.S. government, the EPA implements federal legislation on a broad range of environmental subject matters, frequently adopting substatutory regulations and supervising their implementation by federal and state authorities. On the issue of climate change, however, the EPA had not become significantly involved until a landmark decision of the Supreme Court in the case of *Massachusetts et al. v. Environmental Protection Agency et al.* declared that GHGs are pollutants and hence fall within the jurisdiction of the EPA. By December 7, 2009, acting on the request of the president, the EPA formally adopted an Endangerment Finding under Section 202 of the CAA stating that anthropogenic climate change threatens the environment and public health, a prerequisite for the adoption of rules to limit GHG emissions from mobile and stationary sources.

Responding to this mandate, the administration announced a plan to integrate federal fuel economy standards under the Energy Policy and Conservation Act, called CAFE standards, with federal vehicle emissions standards under the Clean Air Act, increasing these to an average of 35.5 mpg by 2016. Overall, this would translate into an emissions limit of 250 g of CO_2 per mile by 2016. A joint final rule was adopted by the EPA and the National Highway Safety Administration (NHTSA) on April 1, 2010, and its requirements will begin applying on October 1, 2012.

On June 26, 2009, Rep. Edward J. Markey announced the narrow passage of the American Clean Energy and Security Act of 2009 by the U.S. House of Representatives. However, the bill later failed in the U.S. Senate amid concerns over enormous costs, questionable provisions such as energy-efficient loan standards for Fannie Mae and Freddie Mac, and unclear definitions of "renewable biomass" and "green" banking centers. The bill claimed provisions for growing jobs, cutting emissions, and investing in clean technologies.

For stationary sources, a comprehensive rule on Mandatory Reporting of Greenhouse Gases issued by the EPA, adopted October 30, 2009, requires more than 10,000 facilities throughout the United States—accounting for nearly 85 percent of U.S. GHG emissions—to report their emissions of eight categories of GHGs on an annual basis. In force since January 1, 2010, this rule requires certification of the emissions inventory and imposes penalties for failure to report. While it does not impose reduction targets, this rule yields vital information for the design and implementation of further mitigation measures.

Since January 1, 2011, new or substantially modified emitters have also been subject to a permitting requirement under the CAA, with a tailoring rule adopted on May 13, 2010, ensuring that these requirements apply only to the largest stationary sources of GHGs. While this system of operating permits allows the EPA to elaborate guidance for implementing state authorities on how to define best available control technologies (BACTs) for each covered source, it does not provide the means to specify GHG emission standards. Rather, the CAA provides additional pathways through which to regulate GHG emissions from stationary sources, including, most importantly, national ambient air quality standards (NAAQS) and performance standards for new and existing stationary sources (NSPS), with the latter considered more suitable for nontoxic GHGs.

Exercising this prerogative, the EPA signed decrees on December 23, 2010, requiring the agency to propose new source-performance standards and emission guidelines for GHG emissions from refineries and electric generating units fired with natural gas, oil, and coal; two sectors that, when combined, account for approximately 40 percent of U.S. emissions.

Standards would apply directly to modified and new facilities, while emission guidelines for existing facilities would need to be implemented by state authorities. As the EPA has stated, actual emission reductions under these rules will not be required from existing facilities before 2016. Judicial proceedings filed against these requirements by several states and the private sector have also introduced a measure of uncertainty here, however.

States

A narrow focus on the federal level and difficulties in passing legislative and regulatory action risks overlooking a vibrant landscape of mitigation activities already underway at the subfederal level. A number of initiatives have developed in states and cities geared toward assessing and tracking emissions, coordinating research, establishing baselines for carbon sequestration, and promoting renewable-electricity generation.

Across the country, a majority of states have adopted or are currently developing strategies to reduce their GHG emissions. Among these, California has traditionally been a frontrunner, with pioneering measures on the promotion of renewable energy sources, energy efficiency, and mitigation of transportation emissions. On September 27, 2006, California adopted groundbreaking legislation with the intention of cutting state GHG emissions to 1990 levels by 2020. Known as the Global Warming Solutions Act of 2006, this legislation directs the California Air Resources Board (CARB) to establish a system for GHG emissions reporting and to monitor and enforce compliance. Although it does not mandate specific measures to reduce GHG emissions, the legislation authorizes the state board to adopt market-based compliance mechanisms, such as emissions trading.

Under the act, regulations to implement such a compliance program needed to be developed by January 1, 2011, in order for the program to begin in 2012. On June 30, 2007, a Market Advisory Committee (MAC) issued a final report recommending design options for an emissions trading system, including broad coverage of all major emitting sectors, a "first-seller approach" to capping emissions associated with imported electricity, and a mixed approach of free allocation and auctioning of allowances. Further, a Scoping Plan approved by CARB on December 11, 2008, took up these recommendations and specified the relationship of an emissions-trading system to other GHG reduction actions, such as direct regulations, alternative compliance mechanisms, monetary and nonmonetary incentives, and voluntary actions. In December 2010, CARB approved the design of a cap-and-trade system, paving the way for its implementation from 2012. Electricity production, as well as imports and large industrial facilities emitting more than 25,000 metric tons of CO_2 per year will be covered by this system. Although judicial action has delayed implementation of the program, the election of a strong advocate of climate policy as governor and simultaneous rejection of a referendum that would have prevented implementation altogether suggest that California will remain a frontrunner of climate policy in the United States.

Regional

Given the benefits of coordinated action, climate change has also become a subject of subfederal cooperation between states, resulting in three programs covering multiple jurisdictions and partly involving participants and observers from neighboring countries: the Regional Greenhouse Gas Initiative (RGGI) in the northeast and mid-Atlantic, the Western Climate Initiative (WCI) on the west coast, and the Midwest Regional Greenhouse Gas Reduction Accord (MGGA).

Although these agreements differ in coverage and scope, they all share the objective of harnessing carbon markets for climate policy objectives. Over time, such initiatives may converge and become mutually linked in a nationwide policy framework evolving from the "bottom up." Should efforts to adopt federal climate legislation fail or be deferred due to resistance in a minority of states, regional action may hence provide an important fallback option to still achieve meaningful action in the United States. Conversely, action at the regional level may improve the case for federal legislation as a way of pre-empting a regulatory patchwork and ensuring uniform conditions throughout the U.S. economy.

Regional Greenhouse Gas Initiative

Operational since January 1, 2009, the RGGI is a cooperative effort by 10 U.S. states in the northeast and mid-Atlantic to limit GHG emissions from the electricity sector. Connecticut, Delaware, Maine, Maryland, Massachusetts, New Hampshire, New Jersey, New York, Rhode Island, and Vermont are all signatories to a Memorandum of Understanding (MoU) released on December 20, 2005, that sets out common objectives and design elements of its regional CO_2 Budget Trading Program. Under the MoU, these 10 states have committed to initially stabilizing and later reducing CO_2 emissions from large electricity generators. Overall emissions are broken down into state allocations specified in the MoU.

Each participating state was required to adopt the necessary rules for implementation of the Budget Trading Program. In order to ensure consistency across states, a Staff Working Group (SWG) consisting of state officials issued a Draft Model Rule on August 15, 2006, providing a template for state legislation. As a result, the individual trading programs in each of the 10 participating states are linked through CO_2 allowance reciprocity. Accordingly, regulated entities will be able to use a CO_2 allowance issued by any of the 10 participating states to demonstrate compliance with the state program governing their facility. Overall, the 10 individual state programs function as a single regional compliance market for carbon emissions, creating the first mandatory, market-based GHG emissions reduction program in the United States.

While the Model Rule creates a uniform framework for the Budget Trading Program, it also leaves states with flexibility regarding various design features, including allowance allocation. Under the MoU, participating states agreed to allocate a minimum of 25 percent of allowances to support consumer benefit programs, with the remaining 75 percent left for states to decide. Auctions are conducted in regular intervals on an electronic platform, pursuant to a uniform auctioning format. In the first auction, held on September 25, 2008, 12.5 million allowances were sold to 59 bidders at a clearing price of $3.07 per allowance.

Early compliance trading started out at significantly higher prices, yet emissions data subsequently released by the EPA revealed the system to be overallocated, translating into limited emissions constraints and a high supply of allowances in the market. With emissions trending lower due to the economic recession, the main fundamental drivers—allowance allocation and emissions trends—have exerted continued downward pressure on prices, but administrators in the RGGI system have already announced a future review of the reduction target to address the excess supply.

Western Climate Initiative

On February 26, 2007, the Western Climate Initiative (WCI) was launched to develop regional strategies to address climate change. It currently brings together Arizona, California, Montana, New Mexico, Oregon, Utah, and Washington, and the Canadian provinces of British Columbia, Manitoba, Ontario, and Quebec. These states and provinces have adopted a regional goal of lowering GHG emissions by 15 percent below 2005 levels by 2020. In July 2010, the WCI released a detailed program design for the emissions trading system, serving as the central compliance instrument in the program. Starting on January 1, 2012, the system only covered emissions from large downstream emitters, notably electricity, industrial processes, and industrial and commercial sources. From January 1, 2015, however, coverage is set to extend to upstream emissions from fuel combustion for transportation purposes and at residential, commercial, and industrial facilities, to the extent that these are not already covered.

Once implemented, coverage could extend to nearly 90 percent of the emissions in the region, representing over 20 percent of the U.S. economy. Initially, at least 10 percent of the allowances will be auctioned, rising to a minimum of 25 percent by 2020. Participating states and provinces aspire to a higher auctioning percentage over time, possibly rising to 100 percent. With a view to ensuring a consistent and strong price signal, the first 5 percent of allowances auctioned by each partner will have a minimum price. If part of the allowance is not purchased at or above the minimum price, a fraction will be retired. Additionally, no more than 49 percent of emissions reductions may be achieved through offsets.

Midwestern Greenhouse Gas Accord

On November 15, 2007, the governors of nine midwestern states and one Canadian premier signed the Midwestern Greenhouse Gas Accord (MGGA) at the Midwestern Governors Association Energy Security and Climate Change Summit. Currently, Iowa, Illinois, Kansas, Michigan, Minnesota, and Wisconsin and the province of Manitoba are members, and Indiana, Ohio, South Dakota, and the province of Ontario are observers. Under the accord, members pledge to establish reduction targets for emissions of six GHGs and implement a common mitigation framework in state laws.

Additionally, the accord states that a carbon market should be operational within 30 months of its signing and calls for the future emissions trading system to link to other regional or global carbon markets to reduce leakage and increase market efficiency. On June 8, 2009, an advisory group charged with developing appropriate targets and designing a regional trading system recommended that participating jurisdictions reduce their emissions by 20 percent below 2005 levels by 2020 and by 80 percent below 2005 levels by 2050. Given the legislative activities at the federal level, however, the advisory group indicated its preference for a federal emissions trading system and merely formulated provisional recommendations for a regional cap-and-trade system if efforts toward implementing a federal system stall.

Municipal Action

On February 16, 2005, the date when the Kyoto Protocol entered into force, the mayor of Seattle, Gregory J. Nickels, launched the U.S. Mayors Climate Protection Agreement. Its objective was to encourage at least 141 U.S. cities to adopt the reduction objective agreed upon for the United States under the Kyoto Protocol prior to its withdrawal: a GHG emissions reduction of 7 percent below 1990 emissions levels by the 2008 to 2012 period. Specifically, participating cities committed to strive to meet or beat the Kyoto Protocol targets in their communities, urge their state governments and the federal government to enact necessary policies, and urge the U.S. Congress to pass bipartisan legislation to establish a federal emissions trading system.

By July 1, 2011, 1,050 mayors, representing in excess of 88 million citizens, had signed the Climate Protection Agreement. In 2007, the U.S. Conference of Mayors launched the Mayors Climate Protection Center to administer and track the agreement. While it appears that few signatories to the agreement will achieve the Kyoto Protocol reduction target by 2012, the agreement has prompted several cities to launch policy initiatives aimed at reducing municipal GHG emissions, including energy-efficiency improvements to city buildings and transportation fleets, expansion of public transportation networks, renewable energy mandates, new building codes with efficiency requirements for residential and commercial structures, urban development plans that discourage vehicle use, and tax incentives and grants for community groups that take additional steps to reduce their GHG footprints.

Outlook

Despite a new administration that elevated climate change and energy sustainability to one of its central areas of concern, preoccupation with the cost of climate policy, intensified by the recession of 2008 and 2009, ultimately prevented passage of comprehensive climate legislation in the U.S. Congress. Instead, the U.S. response to climate change remains highly fragmented, drawing on federal, regional, and local efforts to achieve the official U.S. target of a 17 percent GHG reduction below 2005 levels by 2020.

Climate change will remain a highly politicized issue in the United States, preventing the emergence of a shared vision on suitable policy responses. The pace and scope of the domestic debate has direct implications for any international engagement by the United States, with uncertainties at the national level directly translating into the international negotiations on a future climate regime. Without strong commitments entered by the United States, other regions will find it more difficult to justify support for strong domestic and international action. Future developments in the United States may thus be the single most important condition for meaningful progress in the global struggle against climate change.

Michael Mehling
Ecologic Institute

See Also: Bush Administration, George H. W.; Bush Administration, George W; Carter Administration; Clean Air Act, U.S.; Climate Policy, U.S.; Environmental Protection Agency, U.S.; Midwestern Regional Climate Center; National Academy of Sciences; National Aeronautics and Space Administration; National Association of Energy Service Companies; National Center for Atmospheric Research; National Oceanic and Atmospheric Administration; National Science Foundation; Obama Administration; United Nations Framework Convention on Climate Change; United States Global Change Research Program; Western Regional Climate Center.

Further Readings

Goubet, C. *United States: Regulating Greenhouse Gases Under the Direction of the EPA*. Paris: CDC Climate, 2010.

Harrison, K. "The Road Not Taken: Climate Change Policy in Canada and the United States." *Global Environmental Politics*, v.7 (2007).

Hitt, Greg and Stephen Power. "House Passes Climate Bill." *Wall Street Journal* (June 27, 2009). http://online.wsj.com/article/SB124602039232560485.html (Accessed January 2012).

Light, A., et al. *Prospects for U.S. Climate Policy: National Action and International Cooperation in a Changed Political Landscape*. Washington, DC: Friedrich Ebert Foundation, 2010.

Monast, J., et al. *Avoiding the Glorious Mess: A Sensible Approach to Climate Change and the Clean Air Act*. Durham, NC: Nicholas Institute for Environmental Policy Solutions, 2010.

Profeta, T., et al. *The U.S. Climate Policy Debate: How Climate Politics Are Moving Forward on Capitol Hill and in the White House*. Washington, DC: German Marshall Fund of the United States, 2008.

United States Global Change Research Program

The United States Global Change Research Program (USGCRP) began as a presidential initiative in 1989 and was passed into law by Congress through the Global Change Research Act of 1990, which called for "a comprehensive and integrated United States research program which will assist the Nation and the world to understand, assess, predict, and respond to human-induced and natural processes of global change." The USGCRP mission is to build a knowledge base that informs human responses to climate and global change through coordinated and integrated federal programs of research, education, communication, and decision support.

During the past two decades, the United States, through the USGCRP, has made the world's largest scientific investment in the areas of climate warming and global change research and has played a key role in the development of reports for the Intergovernmental Panel on Climate Change.

The USGCRP was known as the U.S. Climate Change Science Program from 2002 through 2008; the George W. Bush A]administration changed its name to the Climate Change Science Program (CCSP) as part of its U.S. Climate Change Research Initiative. The rationale behind this was the vision of a "nation and the global community empowered with the science-based knowledge to manage the risks and opportunities of change in the climate and related environmental systems." President Bush established priorities for climate change research to focus on scientific information that could assist the evaluation of strategies to address global change risks. During the Bush administration, one of the CCSP's cornerstones was the creation of 21 synthesis and assessment products to provide information to help policymakers make better decisions. Since its inception, the USGCRP has supported research and observational activities in collaboration with several other national and international science programs.

Governance, Activities, and Publications

The USGCRP is made up of 13 departments and agencies in the United States. It is overseen by the Subcommittee on Global Change Research under the Committee on Environment and Natural Resources, overseen by the Executive Office of the President and facilitated by an Integration and Coordination Office. The diversity of agencies include the Agency for International Development, U.S. Department of Agriculture, U.S. Department of Defense, and U.S. Department of Transportation.

The USGCRP's 13 participating agencies coordinate their work through interagency working groups that span a wide range of interconnected issues of climate and global change. These groups include a group on education, which serves as a forum in the USGCRP for the development of a national climate and climate change education strategy for USGCRP member agencies and related agency education activities. Another working group studies land-use and land-cover change, which is a key component of the carbon cycle. The last of the seven working groups is Observations and Data, which attempts to answer the question, "How can we provide seamless, platform-independent, timely, and open access to integrated data, products, information, and tools with sufficient accuracy and precision to address climate and associated global changes?"

The working group activities have led to major advances in several key areas, including: observing and understanding short- and long-term changes in climate, the ozone layer, and land cover, and identifying the impacts of these changes on ecosystems and society. The USGCRP also develops decision-support systems, including assessments and other tools and information to support adaptation and mitigation decision making, and coordinates them in a distributed fashion across the program.

The USGCRP runs the U.S. Global Change Research Information Office, which provides access to data and information on climate change research and technologies, and global change-related educational resources. The program and its member departments and agencies have also commissioned a number of reports to help guide its current activities and future planning. Some recently published reports include: *Restructuring Federal Climate Research to Meet the Challenges of Climate Change* (2009), *Global Climate Change and Extreme Weather Events: Understanding the Contributions to Infectious Disease Emergence* (2008), and *Earth Science and Applications from Space: National Imperatives for the Next Decade and Beyond* (2007.)

Political Controversies

Because of the politically polarizing nature of the organization, the USGCRP has been at the source of several controversies between the two major political parties in the United States. During the Bush administration, the USGCRP operated in a political atmosphere of the belief that scientific investigation was necessary before policies should be implemented. During that time, the USGCRP faced the challenge of navigating between administration officials who were skeptical of the general scientific consensus about greenhouse gases, and officials who were skeptical about almost every aspect of the science that the administration related to climate change. As a result, during this time, the USGCRP was under more scrutiny than other federal scientific coordination programs.

The National Research Council (NRC) audited the USGCRP several times. Their 2004 review concluded that "the Strategic Plan for the U.S. Climate Change Science Program articulates a guiding vision, is appropriately ambitious, and is broad in scope" and "the [USGCRP] should implement the activities described in the strategic plan with urgency." During this time, it was also recommended that USGCP should expand its traditional focus on atmospheric sciences to better understand the impacts, adaptation, and the human dimension of climate change.

However, a 2007 review was more critical, stating, "Discovery science and understanding of the climate system are proceeding well, but use of that knowledge to support decision making and to manage risks and opportunities of climate change is proceeding slowly." The NRC was particularly harsh about the program's failure to engage stakeholders or advance scientific understanding of the impacts of climate change on human well-being. Looking to the future of the program, a 2008 NRC report put forward a set of research recommendations similar to those adopted by the USGCRP in the 2008 strategic plan.

The Global Change Research Program has occasionally been eviscerated for the alleged suppression of scientific information. The most visible incident occurred in March 2005, when Rick S. Piltz resigned from USGCP after he claimed that there was severe political interference with scientific reports. In his exiting statement, he said, "I believe ... that the administration ... has acted to impede forthright communication of the state of climate science and its implications for society." Piltz further charged that the Bush administration had suppressed the previous National Assessment on Climate Change by systematically deleting references to the report from government scientific documents and suppressed sea-level rise mapping studies.

Statement on Climate Change

As an organization dedicated to the study of climate changes on the human environment, the USGCRP, through the Global Change Research Act of 1990, defines global change as: "Changes in the global environment (including alterations in climate, land productivity, oceans, or other water resources, atmospheric chemistry, and ecological systems) that may alter the capacity of the Earth to sustain life."

Caitlin M. Augustin
University of Miami

See Also: Global Warming, Impacts of; United States; World Health Organization.

Further Readings

U.S. Climate Change Science Program. "Revised Research Plan for the U.S. Climate Change Science Program" (2008). http://www.climatescience.gov/Library/stratplan2008/CCSP-RRP-FINAL.pdf (Accessed July 2011).

U.S. Global Change Research Information Office. "First National Assessment of the Potential Consequences of Climate Variability and Change." http://www.gcrio.org/NationalAssessment/index.htm (Accessed July 2011).

U.S. Global Change Research Program. "U.S. Government Agencies Participating in the USGCRP" (October 20, 2008). http://www.usgcrp.gov/usgcrp/agencies (Accessed July 2011).

University Corporation for Atmospheric Research

The University Corporation for Atmospheric Research (UCAR) is a nonprofit consortium of 76 member universities (plus 25 academic affiliates and 52 international affiliates) engaged in atmospheric research. Sponsored by the National Science Foundation, UCAR is also funded by federal agencies such as the National Oceanic and Atmospheric Administration, National Aeronatucis and Space Administration, the Federal Aviation Administration, the U.S. Departments of Energy and Defense, and the Environmental Protection Agency. It was established in 1959 to support research in the atmospheric sciences, and manages the National Center for Atmospheric Research in Boulder, Colorado. Of UCAR's over 1,560 staff members, over 1,000 work at NCAR;

the rest are engaged in administration or community programs.

Member universities are North American universities with doctoral degrees in atmospheric sciences or a related field. International affiliates offer similar degrees outside of North America, while academic affiliates are North American universities that award bachelor's or master's-level degrees in appropriate fields. Private sector members collaborate with UCAR on research, fund specific projects, and act on UCAR governance boards. In addition to fostering and funding research, UCAR also manages an Office of Education and Outreach to communicate with the public, especially students, about Earth systems science.

UCAR operates numerous programs. The Cooperative Program for Operational Meteorology, Education, and Training (COMET) program is operated jointly with the National Weather Service to promote the understanding of new weather technology and mesoscale meteorology. The Constellation Observing System for Meteorology Ionosphere and Climate (COSMIC) program supports satellite and ground-based observations and research and is engaged in the FORMOSAT-3/COSMIC mission with Taiwan, in which six micro satellites were launched in 2006, providing data to the COSMIC Data Analysis and Archival Center in Boulder. The Global Learning and Observations to Benefit the Environment (GLOBE) program is an environmental education program aimed at primary and secondary schools, with participating schools in 100 countries.

The Joint Office for Science Support (JOSS) supports community research. It often acts as a consultant for federal agency officials or academics planning field experiments and monitoring projects, helping to project costs and assemble resources. JOSS also assists with facilitating networking connections among researchers and government agencies and helps plan and conduct conferences, workshops, and meetings. It frequently offers logistical and administrative support to research institutions engaged in climate change studies and produces a series of general-interest monographs about climate change called Reports to the Nation on Our Changing Planet.

Bill Kte'pi
Independent Scholar

See Also: Climatic Data, Atmospheric Observations; Integrated Science Program; National Center for Atmospheric Research; University Corporation for Atmospheric Research Joint Office for Science Support.

Further Readings

Henson, Bob. "The First 14." *UCAR Quarterly* (Winter 2008–09).

University Corporation for Atmospheric Research. http://www2.ucar.edu (Accessed June 2011).

University Corporation for Atmospheric Research Joint Office for Science Support

The Joint Office for Science Support (JOSS) of the University Corporation for Atmospheric Research is formed by a group of professional and skilled technical and administrative specialists whose mission is to serve and support the scientific community, with particular attention to the areas of atmospheric and related sciences. It is a joint effort of the National Center for Atmospheric Research (NCAR), National Science Foundation (NSF), and National Oceanic and Atmospheric Administration (NOAA).

Many of the research projects sponsored by JOSS have to do with climate change and its effect on human existence. JOSS currently offers administrative support to the Carbon Cycle Science Program Office, Climate Program Office, U.S. Global Change Research Program, and U.S. Climate Variability and Predictability Research Program. JOSS also collaborated closely with the Global Atmospheric Research Program (GARP) for its Alpine (1979) and monsoon (1982) experiments in Switzerland and India, respectively. Together with the NOAA Office of Global Programs, JOSS publishes the monograph series *Reports to the Nation on Our Changing Planet.* The books are targeted to a general, nonspecialized audience and aim to supply them with authoritative information on climate change.

JOSS headquarters are located in Boulder, Colorado. The NSF and the NOAA, as well as other U.S. agencies, private sources, and international organizations, are the main financial supporters of JOSS. Before 2005, JOSS was divided into two groups: the Field Operations and Data Management (FODM) group and the Program Support Operations (PSO) group. In 2005, the FODM staff moved to NCAR's Earth Observing Laboratory, and PSO/JOSS remained in the University Corporation for Atmospheric Research and retained the name JOSS.

JOSS collaborates with the scientific community to organize and conduct scientific programs and events such as meetings, workshops, and conferences in productive and cost-effective ways. The office is responsible for almost 500 scientific meetings every year. JOSS supports planning efforts, research programs, field experiments, and data-management activities worldwide. Through JOSS, academics can develop projects that would be too large to manage with their own university departments. The organization offers a wide range of services. It can act as a consultant for both individual investigators and research managers and funding-agency officials planning extensive geophysical field experiments and monitoring projects.

JOSS also supports the creation of professional networks between scientific researchers and key people in other institutions and agencies, the U.S. Department of State, and other national governments. JOSS does not work only for established scholars; it also puts particular emphasis on the education of young scientists in their early careers. The activities of the office for U.S. researchers build on an effective and extensive knowledge of several regions of the world. Equally, JOSS can advise scientists from all over the world about research activities in the United States and elsewhere.

The office team organizes field trips through detailed budgeting, site surveys, and logistical and operational support. It establishes staff and operations centers for field projects and directs daily operations such as supplying ground support for aircraft, ships, radar, and other observing platforms and systems, and hiring and training local workers. Scientists and governmental agencies rely on these international and domestic meetings to plan future research or share information and opinions. Several important JOSS-supported meetings have focused on a wide range of topics related to climate change and how the phenomenon impacts on the daily lives of different groups of people. For example, JOSS supported a large meeting of Native Americans from across the United States. Representatives from various tribes came together to share their opinions about how climate affects their lives and how climate issues are forcing them to adapt their old ways to the modern world. JOSS has also supported several large, politically significant meetings in Washington, D.C., designed to help atmospheric scientists share their knowledge, research findings, and predictions with agencies that have the power to influence government decisions.

Indian Ocean Experiment

From 1995 to 2001, JOSS took an active part in the Indian Ocean Experiment (INDOEX). The experiment began from the assumption that regional consequences of global warming depend critically on the potentially large cooling effect of another pollutant known as aerosols. These aerosols scatter sunlight back to space and cause a regional cooling effect, thus causing uncertainty in predicting a future climate. The Indian Ocean Experiment addressed the complex influence of aerosol cooling on global warming by collecting in situ data on the local cooling effect of sulfate and other aerosols. The project's goal was to study natural and anthropogenic climate forcing by aerosols and feedbacks on regional and global climate. The International Global Change Research Program considered this issue to be of critical importance. INDOEX measured long-range movements of air pollution from south and southeast Asia toward the Indian Ocean during the dry monsoon season from January to March 1999. Surprisingly high pollution levels were observed over the entire northern Indian Ocean toward the Intertropical Convergence Zone at about 6 degrees south. Agricultural burning and especially biofuel use were shown to enhance carbon monoxide concentrations. Fossil fuel combustion and biomass burning caused high aerosol loading. The experiment pointed out that the growing pollution in the region gave rise to extensive air-quality degradation, which had local, regional, and global implications, including a reduction of the oxidizing power of the atmosphere. JOSS was also involved in GARP experiments, as well as in the establishment of the international research institute on El Niño and its consequences.

JOSS's strategic plan for the future confirms its close collaboration with the research community in the atmospheric and related sciences. The agency intends to implement a more cost- and time-effective policy by exploiting the potential of new technologies, and to apply these technological advances to meetings, conferences, and even field trips. Virtual scientific meetings and remote participation in campaign operations are two of the ways in which JOSS aims to expand and advance research within the scientific community.

Luca Prono
Independent Scholar

See Also: Climatic Data, Atmospheric Observations; Integrated Science Program; National Center for Atmospheric Research; University Corporation for Atmospheric Research.

Further Readings

Indian Ocean Experiment. http://www-indoex.ucsd.edu (Accessed July 2011).
Joint Office for Science Support. http://www.joss.ucar.edu/index.html (Accessed July 2011).

University of Alaska

The University of Alaska system has three main campuses: the University of Alaska Fairbanks (UAF), University of Alaska Anchorage, and University of Southeast, and 15 satellite campuses. Beyond individual researchers at various campuses, all four global warming or climate change research centers, institutes, and groups are found at the UAF campus. The four centers are the Alaska Climate Research Center (ACRC), the International Arctic Research Center (IARC), Center for Global Change and Arctic System Research (CGCASR), and the Alaska Center for Climate Assessment and Policy (ACCAP).

Under the direction of the Geophysical Institute, the ACRC conducts secondary data gathering, storage, and report analysis. The ACRC is funded by the State of Alaska under Title 14, Chapter 40, Section 085. The IARC, a joint venture between Japan and the United States, is an internationally focused research center that includes 20 research groups and over 60 international scientists. The IARC mission is to determine whether climate change is human-made or natural, what data points are needed to make this determination, and the possible effects of climate change.

In 2007, a national and international controversy sprang out of the IARC when Director Syun-Ichi Akasofu claimed that the 2,500 peer-reviewed Intergovernmental Panel on Climate Change (IPCC) report was methodologically flawed. Akasofu claimed that the IPCC did not have a natural greenhouse gases control to account for their assertion of a ~1 degree F (0.6 degree C) human-made climate change over the past 100 years. Professor Akasofu also claimed that carbon dioxide (CO_2) was not responsible for climate change if the researchers would have included natural greenhouse gases. The alternative theory proposed by Dr. Akasofu was that the Earth was still in recovery from the Little Ice Age. In a public response to criticism from the academic community, Akasofu noted that "Since I am not a climatologist, all the data presented in my notes are found in papers and books published in the past; that is why I do not want to publish my notes as a paper in a professional journal." Although not a climatologist, Akasofu's speculation about climatology has been widely circulated among groups that do not agree with the mainstream scientific assessment, as presented by the IPCC, about anthropogenic global warming.

The Center for Global Change and Arctic System Research was founded in 1990 with the goal of fostering interdisciplinary Arctic and sub-Arctic research in arctic biology, atmospheric chemistry, climatology, engineering, geophysics, hydrology, natural resources management, social sciences, and marine sciences to better understand global change in the Arctic. There are two subgroups and a 1,042-page Arctic research study by the Center for Global Change and Arctic System Research affiliates. One subgroup is Globe Learning and Observations to Benefit the Environment (GLOBE), founded in 1997.

GLOBE is an international group that develops hands-on environmental science curriculum for K–12 students, teachers, and scientists. Another subgroup, founded in 1994, is the University of Alaska–National Oceanic and Atmospheric Administration (UA–NOAA) Cooperative Institute for Arctic Research (CIFAR). CIFAR is one of 13 national university-based NOAA institutes. It focuses on atmospheric and climate research and modeling.

The UA–NOAA studies focus on arctic haze, marine science, fisheries, and sea ice research. The last subgroup is a research study called the Arctic Climate Impact Assessment (ACIA). The ACIA report was prepared for the Fourth Arctic Council Meeting in Reykjavik, Iceland, in November 2004. Some of the findings from the report note that over the past 50 years, the mean surface air temperature has increased between 3.6 and 5.4 degrees F (2 and 3 degrees C) and late summer ice has decreased by 15 to 20 percent over the past 30 years. In addition, between 1961 and 1998, North American glaciers lost 108 cu. mi. (279 cu. km) of ice.

The ACCAP was founded in 2006 with a mission to determine the biophysical and socioeconomic effect of climate change within Alaska and to improve regional, local, and Alaskan ability to create policies that address the changing climate. The ACCAP works in affiliation with the NOAA–Regional Integrated Sciences and Assessments, Institute of Northern Engineering, International Arctic Research Center, Institute for Socioeconomic Research at the University of Alaska Anchorage, and Alaska Climate Research Center.

Andrew Hund
Independent Scholar

See Also: Cooperative Institute for Arctic Research; Education; Inuit.

Further Readings

Akasofu, S. "The Recovery from the Little Ice Age (A Possible Cause of Global Warming) and the Recent Halting of the Warming (The Multi-Decadal Oscillation)." http://people.iarc.uaf.edu/%7Esakasofu/little_ice_age.php (Accessed April 2012).

Alaska Center for Climate Assessment and Policy. http://www.uaf.edu/accap (Accessed July 2011).

Alaska Climate Research Center. http://climate.gi.alaska.edu/index.html (Accessed July 2011).

Arctic Monitoring and Assessment Programme. *Arctic Pollution Issues: A State of the Arctic Environment Report*. Oslo: AMAP, 1997.

Center for Global Change and Arctic System Research. http://www.cgc.uaf.edu (Accessed July 2011).

Hassol, Susan. *Impacts of a Warming Arctic: Arctic Climate Impact Assessment*. Cambridge: Cambridge University Press, 2004.

Intergovernmental Panel on Climate Change. http://www.ipcc.ch (Accessed July 2011).

International Arctic Research Center. http://www.iarc.uaf.edu (Accessed July 2011).

University of California, Berkeley

The University of California, Berkeley, is the premier public research university in the United States, with 48 of its 52 academic programs (92 percent) ranked among the top 10 in the country by the U.S. National Research Council. Berkeley counts 66 Nobel Laureates among its faculty, researchers, and alumni—the sixth most of any university in the world; 19 have served on its faculty. Commonly referred to as UC Berkeley, Berkeley, and Cal, the university's academic excellence is sustained by a $2.3 billion endowment. Berkeley was founded in 1868 and is the oldest of the 10 University of California campuses. The current chancellor is Robert J. Birgeneau, who has filled this role since 2004.

Berkeley is a comprehensive university offering more than 7,000 courses in 130 academic departments and 80 interdisciplinary units organized into 14 colleges and schools, offering nearly 350 degree programs. The university awards over 5,500 bachelor's degrees, 2,000 master's degrees, 900 doctorates, and 200 law degrees each year. With 35,838 students and 2,082 faculty in the fall of 2010, the student–faculty ratio is 17 to 1—among the lowest of any major university. Berkeley is one of the most selective universities in the country. For the 2010–11 academic year, 50,312 people applied to undergraduate programs at Berkeley and only 4,109 freshmen were enrolled.

In 2008, the Association of Research Libraries ranked Berkeley as the top public university library in North America. Collectively, Berkeley's 32 libraries are the fifth-largest academic library system in the United States, surpassed only by the U.S. Library of Congress and the libraries of Harvard, Yale, and the University of Illinois. As of 2011, Berkeley's library system contained 11,087,687 volumes and maintained over 90,000 serial publications.

Berkeley faculty, alumni, and researchers have a distinguished record in both the physical and social sciences. Awards include 221 American Academy of Arts and Sciences Fellows, 7 Field Medals, 83 Fulbright Scholars, 384 Guggenheim Fellows, 45 MacArthur Foundation Genius Grant recipients; 87 members of the National Academy of Engineering, 132 members of the National Academy of Sciences, 11 Pulitzer Prize winners, 92 Sloan Fellows, 15 Turing Awards, 9 Wolf Prizes, and 20 Academy Award winners.

Environmental Science Research

As part of its academic excellence, Berkeley is also a leader in environmental research. In recent years, Berkeley's faculty, alumni, and researchers have increasingly focused their attention and efforts on issues related to climate change. One example of this is the work of Dr. Steven Chu, who graduated from Berkeley with a doctorate in physics in 1976 and later taught in the same department. Dr. Chu directed the Lawrence Berkeley National Lab. While doing research at Berkeley, Dr. Chu conceived of a global glucose economy, a form of a low-carbon economy in which glucose from plants could be used for biofuel. For this and other innovative thinking, Dr. Chu was awarded the Nobel Prize in 1997. In January 2009, Dr. Chu was selected by President Obama to become the 12th energy secretary. As energy secretary, he has become a vocal advocate for combating climate change and is encouraging a move away from fossil fuels and more research on alternative energy.

There are over 100 individual undergraduate and graduate programs at Berkeley that focus on the environment, in addition to dozens of top research centers. The university is also active in research concerning global warming and climate change, with several centers involved in research and advocacy of these and related issues.

One such center is the Energy Biosciences Institute (EBI), which was created in 2007, in part because of the research efforts and ideas of Dr. Chu. EBI is a 10 year, $500 million collaborative project between the University of California, Berkeley; the University of Illinois; and Lawrence Berkeley National Laboratory. Additional funding comes from British Petroleum (BP). As the world's first research institution solely dedicated to the new field of energy bioscience, EBI is focusing on the research and development of next-generation biofuels, as well as various applications of biology to the energy sector. Currently, more than 300 researchers at EBI are working on, among other things, how cellulose from plants can be broken down into sugars that can then be processed into biofuel. Also being studied is fossil fuel bioprocessing, which is the use of microbes to transform hydrocarbons and reduce environmental contamination. There has been some criticism of this initiative on a number of fronts, including

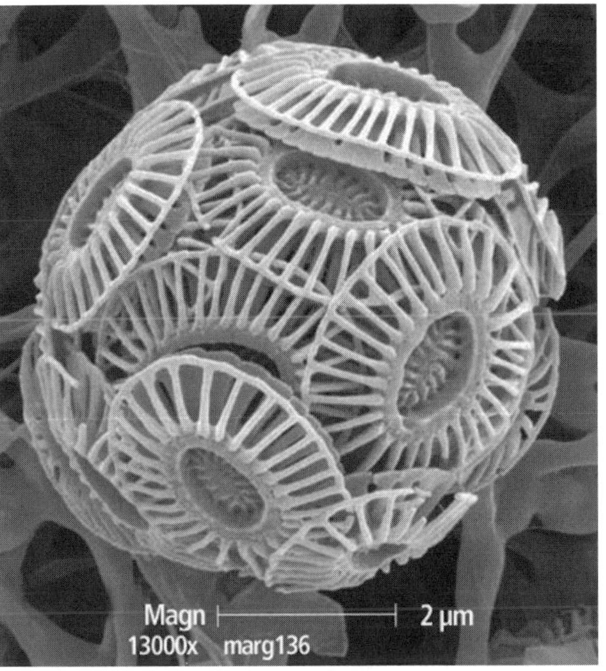

The calcium carbonate shell of Emiliania huxleyi, an ocean microbe that cycles carbon and reduces environmental contamination. The University of California, Berkeley, is studying microbes like these through its Energy Biosciences Institute.

concerns about the influence of large, private corporations on public universities, and concern about the influence and motives of oil companies in particular on public research.

Another example of the university's research facilities involved in climate change research is the University of California Climate Change Center, which was established in 2003 by the California Energy Commission to undertake a broad program of scientific and economic research on climate change in California. The center has sites at both the Berkeley and San Diego campuses. The Berkeley Center, based at the Goldman School of Public Policy, focuses on economic and policy analysis, whereas the site in San Diego (at the Scripps Institute of Oceanography) focuses on physical climate modeling. Several other departments on the Berkeley campus are involved in the work of the center, including the Department of Agricultural and Resource Economics, Department of City and Regional Planning, Department of Civil and Environmental Engineering, Graduate Group in Energy and Resources (ERG), and Environmental Energy Technologies Division of the Lawrence Berkeley National Laboratory.

The ERG is an interdisciplinary academic unit of the Berkeley campus that was created in 1973 to develop, transmit, and apply critical knowledge to enable a future in which human material needs and a healthy environment are mutually and sustainably satisfied. ERG conducts programs that include graduate teaching and research on energy issues, resources, development, human and biological diversity, environmental justice, governance, global climate change, and new approaches to thinking about economics and consumption.

Begun in 1980, the University of California Energy Institute (UCEI) is a multicampus research unit of the University of California system with a mission to foster research and educate students and policymakers on energy issues. The Center for Global Metropolitan Studies (GMS) is a campus initiative to foster interdisciplinary collaboration to investigate and address problems and opportunities posed by global metropolitan growth and change. The Berkeley Institute of the Environment (BIE) was established in 2005. It brings together and helps enhance diverse campus programs and research units by making research tools and knowledge accessible across disciplinary lines to address complex environmental problems, while fostering collaboration and thinking about critical environmental problems. The Institute of Transportation Studies (ITS) is one of the world's leading centers for transportation research, education, and scholarship. Research areas include transportation sustainability, future urban transit systems, and environmental effects. The University of California Berkeley Transportation Sustainability Research Center (TSRC) was formed in 2006 to combine the research forces of the (ERG, UCEI, GMS, BIE, and ITS). The TSRC is a multicampus unit that supports research, education, and outreach.

Other campus activities related to global warming include the research and advocacy work of the College of Engineering. For example, on August 2, 2007, the college released a blueprint for fighting global warming by reducing the amount of carbon emitted when transportation fuels are used in California. This low-carbon fuel standard is designed to stimulate improvements in transportation fuel technologies and is expected to become the foundation for similar initiatives in other states, as well as nationally and internationally.

There is also a Chancellor's Advisory Committee on Sustainability that promotes environmental management and sustainable development on the Berkeley campus. The mission of the committee is to engage the campus in an ongoing dialogue about reaching environmental sustainability and to integrate environmental sustainability with existing campus programs in education, research, operations, and public service. The committee is charged with advising the chancellor on matters pertaining to the environment and sustainability as they directly relate to the university.

Social Action

Berkeley's tradition of student political action has also merged with global climate change issues. For example, in March 2007, students organized the California Campus Climate Challenge Summit to learn about global warming, climate change, and methods for influencing policy change via student activism. Such activities also highlight the social justice aspects of issues concerning global warming and the desire by some to create an environmental and social movement to help raise awareness about this issue.

Michael Joseph Simsik
U.S. Peace Corps

See Also: Carbon Sequestration; Education; Geoengineering; Phytoplankton; Transportation.

Further Readings

"The Alternative Choice: Steven Chu Wants to Save the World by Transforming Its Largest Industry." *Economist* (July 2, 2009).

American Library Association. http://www.ala.org (Accessed June 2011).

Berkeley Endowment Management Company. http://www.berkeleyendowment.org (June 2009).

Dalton, Rex. "Berkeley's Energy Deal With BP Sparks Unease." *Nature*, v.445 (February 15, 2007).

Energy Resources Institute. http://www.energybiosciencesinstitute.org (Accessed June 2011).

Fisk, Edward B. *Fiske Guide to Colleges 2005*. 21st ed. Naperville, IL: Sourcebooks, 2005.

Owens, Eric. *America's Best Value Colleges*. New York: Random House, 2004.

University of California, Berkeley. "Brief History of the University of California, Berkeley." http://www.berkeley.edu/about/history (Accessed June 2011).

University of Colorado

In 2007, the National Science Foundation ranked the University of Colorado (CU) system first of 150 public universities in federally funded environmental science research. The university also obtains funds from the U.S. Departments of Commerce, Energy, and the Interior, the Environmental Protection Agency (EPA), and the National Aeronautics and Space Administration (NASA). In 2008, the three campuses of the system combined for a university record $661 million in grants.

In 2008, CU-Boulder received $32 million from NASA for a five-year program to develop and operate a database for tracking world frozen weather patterns, such as changes in sea ice, glaciers, snow, and ice sheets. Another NASA grant was established that year to build a $34 million solar instrument package for a global climate change monitoring satellite. A $12 million grant from NASA to the CU atmospheric sciences department funds a study of tropical storms, clouds, and ozone over Panama and Costa Rica to chart climate change.

Other departments researching global change and environmental science include chemistry and biochemistry, atmospheric and oceanic sciences, civil environmental and architectural engineering, geography, geological sciences, and environmental studies.

The Institute of Arctic and Alpine Research (INSTAAR) in 2007 reported that melting glaciers and ice caps were causing 60 percent of the rise in sea levels, while melting ice sheets in Greenland and Antarctica contributed 40 percent. Nearer to home, at the INSTAAR's Mountain Research Station at Niwot Ridge, Colorado, researchers documented long-term environmental changes in glacial lakes, wetlands, tundra, and forest.

For two decades, CU's Cooperative Institute for Research in Environmental Sciences has used hundreds of CU students and researchers to record environmental change from the Arctic to the Antarctic and from Boulder to Costa Rica. A 2008 study indicated that Baffin Island ice caps were half the size they were half a century earlier and would probably be completely gone by 2050. Director Konrad Steffen records melting of the Greenland Ice Sheet through a battery of ice-based climate stations. The university's National Snow and Ice Data Center documented in 2007 that the Arctic sea ice during the September minimum was 460,000 sq. mi. (1,191,394 sq. km) less than the previous minimum.

In January 2008, Jim Maslanik of the Colorado Center for Astrodynamics Research reported that the oldest and thickest Arctic sea ice was almost totally gone, and that the remaining perennial ice was dangerously thin.

In October 2011, the university population center and the National Center for Atmospheric Research (NCAR) Resilient and Sustainable Cities Project co-hosted a workshop for 10–15 experts and interested scientists to examine the science of environmental factors in human decisions to migrate. Of particular interest were urbanization and climate change.

A 2011 report documented declining world and U.S. news coverage in the aftermath of the spike for the December 2009 Copenhagen meetings. Sociologists at CU defined news organizations, political players, and the general public as similar to ecosystems with limited carrying capacity. Competing issues, such as elections and the economy, were cited for deflecting attention from climate change. Another factor was that the public and media felt that climate change lacked the dramatic interest it once had.

Sea Level Controversy

The University of Colorado is considered a leader in climate change research, particularly in changes in the world's ice fields and the impact of ice melt on society. However, when the university's Sea Level Research Group in May 2011 decided to add 0.3 mm per year to its actual measurements of sea levels, the decision led critics to charge the group with manufacturing data. James M. Taylor of the Heartland Institute claimed that the result was an exaggeration, a sea-level rise that was fictitious. The director of the center, which was widely depended on for data, justified his team's decision to add the thickness of a fingernail to each year's data by noting that land masses were still rising in the aftermath of the ice age and increasing the amount of water capacity in ocean basins. He noted that the 0.3 mm accounted for the growing size of ocean basins and consequent expansion of ocean volume. The technical term is glacial isostatic adjustment.

Taylor and climate scientist John Christy of the University of Alabama-Huntsville rejected the adjustment, with Christy noting that the volume of water and the sea level were not the same concept. Christy and Taylor argued that measurement should be along the coasts, where climate change scientists claimed the water was rising.

The University of Colorado at Boulder was founded in 1876 at the base of the Rocky Mountains. The natural beauty of the area draws many students interested in pursuing degrees in environmental studies and policies. The University of Colorado received nearly $661 million in research funding in 2008, which shattered previous records and reaffirmed the institution's position as a leading research university. The funds were earmarked for work in renewable energy, global climate studies, and other fields.

The director of the Colorado center countered that removing the effect of rising land masses would actually produce a larger rise in the rate of sea-level change. As a compromise, the center director mentioned the possibility of putting both sets on the Website. Either way, the result was an inch, while consensus estimates are a sea-level rise of 2 to 4 ft.

John H. Barnhill
Independent Scholar

See Also: Antarctic Ice Sheet; Arctic and Arctic Ocean; Glaciers, Retreating; Greenland Ice Sheet.

Further Readings

Daily Climate. "2010 in Review: The Year Climate Coverage 'Fell Off the Map'" (January 3, 2011). http://wwwp.dailyclimate.org/tdc-newsroom/2011/01/climate-coverage (Accessed July 2011).

Institute of Behavioral Sciences, University of Colorado. "CUPC–NCAR Workshop on Migration, Urbanization, and Climate Change." http://www.colorado.edu/ibs/CUPC/short_courses/NCAR_migration_climate (Accessed July 2011).

Lott, Maxim. "Changing Tides: Research Center Under Fire for 'Adjusted' Sea-Level Data." Fox News (June 17, 2011). http://www.foxnews.com/scitech/2011/06/17/research-center-under-fire-for-adjusted-sea-level-data (Accessed July 2011).

Scott, Jim. "CU Researchers Fan out Across Globe to Probe Climate Change." http://www.colorado.edu/news/reports/climatechange (Accessed July 2011).

University of Colorado Office of the President. "CU Reaches Record Level of Research Funding" (August 7, 2008). https://www.cu.edu/content/cu-reaches-record-level-research-funding (Accessed July 2011).

University of East Anglia

The University of East Anglia (UEA) is home to one of the world's leading climate institutions for research into natural and anthropogenic climate change: the Climatic Research Unit (CRU). The CRU was the first to begin establishing a historical record in sufficient detail to explore processes, interactions, and change in the Earth over time; previously, interest tended to be only theoretical.

The CRU was established at UEA in Norwich, United Kingdom, in 1972. Its founding director was Hubert Lamb, one of the first to go against the prevailing view of the 1960s that aside from seasonal variability, the climate was generally constant over the years. Through the 1970s, the CRU worked on establishing the history. Then in 1979, the CRU hosted Climate and History, an interdisciplinary conference that proved pivotal to the future of historical climatology and the impact of climate on societies. A second conference in 1998 continued the study of the past, but emphasized the future. The decision to establish climate research at a university was crucial, and the CRU was critical in making the field something more than an academic backwater.

Setting the Pace

The CRU set the pattern for serious climate research and made it a political issue. Since the 1980s, the CRU has been a leader among those contending that global warming is real and worrisome. The current view is that climate changes on scales of time that are relevant to humanity and human economic and social systems.

The 30 scientists and students in the Climatic Research Unit have created several datasets in wide use by the world's climate researchers. One of the most important is the global temperature record that monitors the climate; others are models and statistical software. The CRU's missions are to explore past climate history and its influence on humanity, examine climate change in the current century, and evaluate the impacts of climate change on the future. Research is both pure and applied, and funding comes almost completely from outside sources, such as academic funding councils. Other sources include government, charitable foundations, nongovernmental organizations, and business and industry.

The CRU participates in teaching in the School of Environmental Sciences, particularly the MS in climate change. It has close ties within the school with other groups such as the Centre for Social and Economic Research on the Global

Environment. It is a turn-to source for the media on matters of policy and science. Staff also publish in peer-reviewed and popular journals and edit bulletins and newsletters.

The CRU is a major source of primary data to the world's climate scientists. CRU's temperature dataset, which began in 1978 and continues today, shows a global warming of 1.44 degrees F (0.8 degree C) over 157 years. CRU is also developing a precipitation database, which gives researchers around the world basic data for their studies. The CRU also has a tree-ring database developed with several world institutions that gives the CRU a world reputation in dendroclimatology.

Climate-Gate

In 2009, East Anglia became the center of a controversy after leaked e-mails appeared to show that researchers were knowingly manipulating climate change data, a scandal that quickly became known as Climate-Gate. The e-mails were intercepted by hackers, who then attempted to upload them to the Website RealClimate, which notified CRU of their possible security breach later that day. Two days later, the e-mails began circulating on public Websites, including Wikileaks, and were eventually reported by the mainstream media. The e-mails were internal to the CRU, particularly to research department head Phil Jones.

Emerging in the e-mail dialogues was David Holland, an electrical engineer with 40 years of experience, who had demanded prosecution of the scientists for violating disclosure laws. Holland's requests for documentation under the Freedom of Information Act, and the scientists' private responses to his e-mails, were among the leaked correspondence. Holland's complaint led the Information Commisioner's Office to investigate the CRU and other institutions tied to it.

The leaked e-mails contained rude language and threats of violence, including one against a Cato Institute senior fellow in environmental studies, Patrick J. Michaels, who had earlier been trying to warn about the private activities going on at CRU. In the more than 1,000 e-mails from scientists at the CRU, Michaels claimed that they show a pattern of exclusion of dissenting views and evidence, and obstructing critical climate data from the public.

Even some environmentalists wanted resignations at the CRU, but Jones refused, noting that the CRU findings were consistent with those based on other datasets such as the National Oceanic and Atmospheric Administration and National Aeronautics and Space Administration Goddard Institute for Space Studies datasets.

To defuse the controversy, the scientists agreed to publish the entire datasets. The CRU had to first obtain releases from nonpublication agreements. CRU acknowledged that paper and magnetic tape raw data had been discarded in the 1980s during a relocation of the CRU, but noted that at that time, climate change was not the large issue it became a generation later. Adjusted data based on the paper and tapes was still available, however. Skeptics criticized the data as evidence that CRU was saying "trust us," rather than allowing full debate.

The eventual report of the House of Commons Science and Technology Select Committee verified that no fraud had been committed by Jones and his staff and that allegations of manipulating data and of subverting the peer-review process were without merit. Still, the committee suggested to CRU and UEA the adoption of a transparent policy of disclosure of data under Freedom of Information requests. At the conclusion of these inquiries, the UEA scientists were cleared of all charges, and Jones was reinstated at CRU as director of research. East Anglia remains a leader in collecting and disseminating documentation of global warming.

John H. Barnhill
Independent Scholar

See Also: Climate-Gate; Climatic Data, Nature of the Data; Global Warming Debate; United Kingdom.

Further Readings

Cato Institute. "Cato's Pat Michaels at Center of 'Climategate' Controversy That Rocks Climate Change Establishment" (January/February 2010). http://www.cato.org/pubs/policy_report/v32n1/cpr32n1-4.html (Accessed December 2011).

Climate Research Unit. "History of the Climatic Research Unit" http://www.cru.uea.ac.uk/cru/about/history (Accessed July 2011).

Hickman, Leo and James Randerson. "Climate Sceptics Claim Leaked Emails are Evidence of Collusion Among Scientists." *The Guardian* (November 20, 2009). http://www.guardian.co.uk/environment/2009/nov/20/climate-sceptics-hackers-leaked-emails (Accessed July 2011).

Kelemen, Peter. "What East Anglia's E-Mails Really Tell Us About Climate Change." *Popular Mechanics* (December 18, 2009). http://www.popularmechanics.com/science/environment/climate-change/4338343

Leake, Jonathan. "Climate Change Data Dumped." *The Times* (London) (November 29, 2009). http://www.timesonline.co.uk/tol/news/environment/article6936328.ece (Accessed July 2011).

Mendic, Robert. "Climategate: University of East Anglia U-Turn in Climate Change Row." *Telegraph* (November 28, 2009). http://www.telegraph.co.uk/earth/copenhagen-climate-change-confe/6678469/Climategate-University-of-East-Anglia-U-turn-in-climate-change-row.html

University of Hawai'i

Hawai'i is the only state in the tropics, the only state completely surrounded by water, and is now a claimant to being the first area of the world adversely affected by climate change. Hawai'i anticipates that its future will include a rising sea level, changes in precipitation, and greater vulnerability to storms. At a 2009 climate change conference hosted by the University of Hawai'i (UH), local researchers and U.S. Geological Survey (USGS) experts painted a portrait of Hawai'i's future: hotter, drier, and more susceptible to coastal erosion and severe rain events. Illustrating the differing effects of climate change, Hawai'i is in a band of latitude that has experienced a century of diminished rainfall as the world on average has seen an increase. The university takes the stance that to sustain and recover requires science and engineering to adjust to real problems, such as changing recharging of its aquifers, more frequent and intense floods, and public-health and food-security problems. Models and solid data are critical for predicting, anticipating, and adapting.

Major Climate Change Research

The flagship campus of the UH system and the major site for climate change research is UH–Manoa. Its Center for Island Climate Adaptation and Policy (ICAP) is an interdisciplinary research facility providing real-world solutions for Hawai'i's private- and public-sector decision makers. It is the link between the university and the broader community for adaptation and accommodation to climate change.

ICAP researchers include ocean science, Hawai'ian studies, law, and planning specialists. Also solicited on an as-needed basis are topic experts from other departments in the university. The center draws on science experts within the university for climate data and models. The center coordinates projects requested by outside agencies. Law school input to the center comes in the area of policy white papers and science-based implementation of current law. Among the agencies benefiting from the university's expertise is the Hawai'i Greenhouse Gas Emissions Reduction Task Force, which has a mandate to create equitable and workable regulations to meet targets set by Act 234. The center also reviews the state statutes and county codes for laws where climate issues might have an influence.

Planning experts assist the center in incorporating their climate-sensitive views into infrastructure and planning projects to create model projects for

Divers clean the outside of SeaStation 3000, an open-ocean aquaculture project operated by the University of Hawai'i. Climate change research is carried out by the university's Center for Island Climate Adaptation and Policy, in Manoa.

adaptive strategies and to partner with green-technology groups and other foundations for developing hazard mitigation approaches. The Hawai'i Coastal Zone Management Program produced in 2006 an update to the Hawai'i Ocean Resources Management Plan, which deals statewide with protection and enhancement of the condition of ocean resources. The OMRP is not specifically concerned with climate change, but the OMRP Working Group advocates adoption of approaches that mitigate climate change's impacts. The working group in 2009, with aid from UH ICAP, produced A Framework for Climate Change Adaptation in Hawai'i, a set of guidelines for state planners. Issues of concern include shoreline erosion, coastal hazards, ocean acidification, scarcity of freshwater, sea-level rise, and more frequent and stronger storms. Another UH element involved in the three-year plan is the Sea Grant.

The university center also uses Hawai'ian knowledge to incorporate the indigenous perspective, an often-overlooked knowledge compiled over thousands of years. Language and other indigenous sources help to integrate ancestral wisdom into modern science.

Climate change research at UH does not consist solely of work on Hawai'i. A team went to Antarctica in 2009 with North Carolina State and Raytheon Polar partners in Food for Benthos on the Antarctic Continental Shelf 2 (FOODBANCS2) to collect samples for lab examination of the impact of climate change on the Antarctic peninsula.

Charles Fletcher, a professor and former chair of the geology and geophysics department at UH, published an essay in a 14-week series in the *Honolulu Civil Beat* in 2010 based on excerpts from a book by Craig Howes, *The Value of Hawai'i: Knowing the Past, Shaping the Future*, which examines Hawai'i's weather history and goes into detail about the problems that Hawai'ians will face from climate change. Fletcher outlines observations that the air temperature is rising, rain and stream flows are diminishing, sea level and sea temperatures are rising, and the ocean is becoming more acidic. His essay asks: As water becomes scarcer and flash flooding intensifies, who decides where and what to build, and which ecosystems die in favor of flood-control channels? Hawai'ians were able to comment on the series and discuss issues such as the 1980s flash floods that caused millions of dollars of damage, swept houses off their foundations, and displaced tens of thousands of people.

For UH professors, the remaining on climate change remains wondering what will manage the future—the public good or private interests?

John H. Barnhill
Independent Scholar

See Also: Global Warming, Impacts of; Hawai'i; Nauru; Precipitation; Rain; Sea Level, Rising; Volcanism.

Further Readings

Fletcher, Chip. "The Value of Hawai'i: Climate Change by Chip Fletcher." *Honolulu Civil Beat* (October 11, 2010). http://www.civilbeat.com/articles/2010/10/11/4988-the-value-of-Hawai'i-climate-change-by-chip-fletcher (Accessed July 2011).

Hawai'i Coastal Zone Management Program. "A Framework for Climate Change Adaptation in Hawai'i." (November 2009). http://Hawai'i.gov/dbedt/czm/ormp/reports/climate_change_adaptation_framework_final.pdf (Accessed December 2010).

Howes, Craig. *The Value of Hawai'i: Knowing the Past, Shaping the Future*. Honolulu: Biographical Research Center and University of Hawai'i, 2010.

Kershner, Jessi. "A Framework for Climate Change Adaptation in Hawai'i." (December 13, 2010). Climate Adaptation Knowledge Exchange. http://www.cakex.org/case-studies/2683 (Accessed July 2011).

Letman, Jon. "Climate Change in Hawai'i: Caught Between a Rock and a Big Wave." (December 17, 2009). Green Chip Stocks. http://www.greenchipstocks.com/articles/climate-change-Hawai'i/602 (Accessed July 2011).

University of Hawai'i School of Law, Center for Island Climate Adaptation and Policy. http://www.law.Hawai'i.edu/icap (Accessed July 2011).

University of Maryland

The University of Maryland (UMD) refers to the University of Maryland, College Park. UMD is the flagship institution of the University System of Maryland. A large public research university chartered in 1856, today, UMD has a myriad of institutions and research centers that unite diverse disciplinary climate change studies with strong partnerships with the U.S. government in climate policy, economics, physical science, and technology development.

Research Institutions and Academics

The School of Engineering houses several institutes focusing on low-carbon energy. Evolving out of a 1980s energy laboratory at UMD, the Center for Environmental Energy Engineering focuses on high-efficiency and alternative-energy conversion systems for buildings and transportation. The UMD Energy Research Center (UMERC) pulls expertise from almost every college on campus in the areas of energy storage, energy efficiency, renewable and nuclear energy, carbon capture and storage, and climate and energy policy. UMERC is the umbrella organization for many other climate-related institutions on campus.

The Joint Global Change Research Institute (JGCRI) was established in 2001 between UMD and the Pacific Northwest National Laboratories (PNNL, U.S. Department of Energy). JGCRI develops models to simulate economic and physical impacts of global-change policy options, employing partial and general equilibrium models and an agricultural systems model. JCGRI's work is regularly used by the Intergovernmental Panel on Climate Change (IPCC), with several JGCRI experts repeatedly selected as major contributors to the IPCC's Assessment Reports.

UMD's Earth System Science Interdisciplinary Center (ESSIC) operates jointly with NASA to study interactions of the atmosphere, oceans, and land. ESSIC also hosts the Cooperative Institute for Climate Studies (CICS), which partners with the National Oceanic and Atmospheric Administration (NOAA) on a range of projects to study and communicate climate variability and change. NOAA's new National Center for Weather and Climate Prediction is being built at a UMD public/private Research and Technology Park called M-Square. The park is intended to attract government and private partnerships, drawing on campus expertise. Climate change is one strategic focus area.

The Center for the Use of Sustainable Practices is one of several affiliated centers on campus that researches and disseminates sustainable practices in building design and urban planning. Other institutes with components of climate research include the Center for Technology and Systems Management, Institute for Philosophy and Public Policy, and Center for Society and the Environment Many individual departments and academic programs have strong climate focuses in teaching and research. A Nobel Laureate in economics, professor Thomas Schelling is a member of UMD's School of Public Policy and has written on climate and energy policy choices for several decades. Professor Ruth Defries, formerly of UMD's Geography Department, received a MacArthur Genius Grant in 2007 for her work at the university on satellite imaging to understand deforestation impacts on the global carbon cycle.

Campus Climate Initiatives

In recent years, UMD has been acclaimed for campus initiatives to increase climate awareness among the student population, stimulate academic initiatives and research partnerships, and reduce the campus environmental footprint. The university is a charter signatory to the American College and University Presidents' Climate Commitment. The university's Climate Action Plan, commissioned in 2007, committed the university to a 60 percent reduction in emissions by 2025 (below a 2005 baseline) and climate neutrality by 2050.

The university Sustainability Council began meeting in September 2009 to advise the administration, Office of Sustainability, and campus community on integration of sustainability and carbon reduction into UMD policies and activities. The university employs a natural gas cogeneration power plant to produce heating for nearly the entire campus and about half the campus's annual electricity consumption. Solar arrays are being added to several campus buildings, including a 631 kW public/private university partnership built next to the campus. A campus geothermal project is also under consideration. A series of new residential and nonresidential campus buildings meet U.S. Green Building Council standards for LEED Gold or Silver certification. Comprehensive energy surveys were conducted in many campus buildings to determine the most effective carbon-mitigation strategies.

Student activism has contributed to UMD's climate initiatives. A successful 2007 student referendum recommended that the university offset fossil energy use through the purchase of Renewable Energy Credits. In 2010, UMD ranked sixth among U.S. universities in volume of RECs purchased, amounting to 26 percent of the university's total electricity usage. The first national Powershift Conference was held on the UMD campus in 2007. In that year, UMD's entry to the U.S. Department of Energy Solar Decathlon won second place globally.

Phillip Matthew Hannam
UNEP-Tongji Institute of Environment

See Also: Education; Maryland; National Oceanic and Atmospheric Administration; Renewable Energy, Geothermal; Sustainability.

Further Readings

University of Maryland Newsdesk. *UM Climate Change Experts*. http://www.newsdesk.umd.edu/experts/hottopic.cfm?hotlist_id=105 (Accessed July 2011).

University of Maryland Office of Sustainability. *Campus Sustainability Report 2010*. http://www.sustainability.umd.edu/documents/2010_Campus_Sustainability_Report.pdf (Accessed July 2011).

University of New Hampshire

Located in Durham, the University of New Hampshire was founded as the College of Agriculture and Mechanic Arts in 1866 and was reorganized as the University of New Hampshire in 1923. The student population is roughly 12,000 undergraduates and 2,500 graduate students. It offers the standard liberal arts programs, as well as marine research facilities, including the Anadromous Fish and Aquatic Invertebrate Research Laboratory, Jackson Estuarine Laboratory, Coastal Marine Laboratory, Jere A. Chase Ocean Engineering Laboratory, and Shoals Marine Laboratory.

The university's Climate Change Research Center (CCRC) is an atmospheric research center with faculty teaching a number of undergraduate- and graduate-level classes on atmospheric science, Earth system science, atmospheric chemistry, and atmospheric aerosol and precipitation chemistry. The CCRC also provides teaching materials and training for K–12 teachers and offers undergraduate internships. Areas of faculty research include the impact of human activity on atmospheric properties, airborne sciences, biosphere–atmosphere exchange, ice course and air–snow exchange, and climate assessment of the local region.

The Earth Sciences Department is a multidisciplinary department blending geology, hydrology, ocean mapping, oceanography, and geochemical systems specialization. Field work is offered in the western United States, the Himalayas, Indonesia, Pakistan, China, Mexico, Antarctica, Greenland, and at sea in the Pacific and Indian oceans. The Natural Resources and Earth Systems Science Program is a graduate program offering doctoral degrees in Earth and environmental science or natural resources and environmental studies, founded on interdisciplinary work in atmospheric science and policy issues.

The university has received numerous awards for its commitment to sustainability; for example, it sources nearly 25 percent of its food for the dining hall from local producers. The president of the university joined other college presidents in the American College and University Presidents Climate Commitment to minimize emissions on campus and integrate sustainability throughout the university experience and community. The Office of Sustainability is the oldest endowed sustainability program in the country, founded in 1997, and focuses on sustainability in research, operations, and the curriculum. A dual major in sustainability will be offered in the fall of 2012.

Bill Kte'pi
Independent Scholar

See Also: New Hampshire; Sustainability.

Further Readings

Aber, John, Tom Kelly, and Bruce Mallory, eds. *The Sustainable Learning Community*. Durham: University of New Hampshire Press, 2009.

Dube, Kate and Jeff Lewis. *University of New Hampshire Off the Record.* Pittsburgh: College Prowler, 2012.
Sperduto, Dan and Ben Kimball. *The Nature of New Hampshire.* Concord: New Hampshire Press, 2011.

University of Oklahoma

The University of Oklahoma (OU) in Norman was founded in 1890 by the Oklahoma Territorial Legislature and today is an internationally recognized research institution with an annual enrollment of over 30,000 students. David L. Boren is the current president, having held that position since 1994. OU has 21 colleges and offers 163 majors at the baccalaureate level, 166 at the master's level, 81 at the doctoral level, and 27 at the doctoral professional level, in addition to its 26 graduate certificates. The Health Sciences Center is one of only four in the country to include seven professional schools, with colleges of Allied Health, Dentistry, Medicine, Nursing, Pharmacy, Public Health, and Graduate Studies. OU has over 2,400 full-time faculty members and an annual operating budget of $1.5 billion. Since 1994, research and sponsored programs have more than doubled and in FY 2010 the budget for research and sponsored programs was over $261 million, including $118 million at the OU Health Sciences Center. The Carnegie Foundation classifies OU in its highest tier of research activity.

The Department of Geography and Environmental Sustainability at OU offers undergraduate (B.A., B.S.) and graduate (M.A., Ph.D.) programs that integrate knowledge of human society and culture with an understanding of the biosphere and the Earth's physical system to provide a holistic perspective on the relationships between humans and their environment. The department has about 75 undergraduate majors and 55 graduate programs. Specific research projects germane to environmental warming include the Oklahoma Wind Power Initiative, the Pacific Rainfall Program, the Schools of the Pacific Rainfall Climate Experiment, and the Land Cover Land-Use Change Group.

The School of Environmental Engineering and Environmental Science is the most highly funded department ($4.35 million in 2010) within the College of Engineering; the department offers degrees in environmental engineering and environmental science at the undergraduate (B.S.) and graduate (M.S.) levels. Affiliated research centers include the Water Technologies for Emerging Regions Center (WaTER Center) and the Center for the Restoration of Ecosystems and Watersheds. The Sarkeys Energy Center includes six interdisciplinary institutes drawing faculty from the colleges of Geosciences, Arts and Sciences, Business, Law, and Engineering. Projects at the Sarkeys center focus on developing technology and programs relevant to the energy industry.

Environmental Commitments

In 2007, President Boren signed the American College and University Presidents' Climate Commitment, signifying OU's commitment to reducing greenhouse gas emissions. Numerous organizations on campus are dedicated to improving sustainability on campus, including the Environmental Concerns Committee, OUr Earth, and the OU Sustainability Committee. OU celebrates Green Week annually to raise campus awareness of environmental and sustainability issues. In 2010 and 2011, the *Princeton Review* named the University of Oklahoma a green college, signifying that it was in the top 20 percent of universities in environmental and sustainability initiatives and commitments.

OU has introduced a number of initiatives to its dining services program to reduce its environmental impact: These include increasing the use of locally grown and produced products, organic beef, cage-free eggs, and organic, fair-trade coffee; replacing Styrofoam cups with paper; and upgrading to Energy-Star-rated appliances. The university's custodial services has adopted many environmentally friendly products, including phosphate-free glass cleaner; toilet paper made from 100 percent post-consumer-recycled materials; paper towels that include at least 30 percent post-consumer-recycled materials; and a biorenewable, multipurpose cleaner. The university is in the process of upgrading energy efficiency across the campus. Steps in this process include the installation of low-flow toilets, photoluminescent exit signs, motion sensors for vending machines, and occupancy sensors for office lighting.

All of the university's printing has used soy-based inks for the last decade, and all stationery is printed on recycled paper. OU has a campus recycling program for aluminum, plastics, paper, and batteries. In fiscal year 2009, the university recycled 932.55 tons (846 metric tons) of material, a 20 percent increase from fiscal year 2008. University Fleet Services includes 21 vehicles that run on compressed natural gas, 61 flex-fuel vehicles, three gasoline/electric hybrids, 55 electric vehicles, and 24 biodiesel vehicles. OU has major investments in wind power, and in 2008 signed an agreement that by 2013 all electricity purchased from Oklahoma Gas and Electric Company would come from renewable sources.

Sarah Boslaugh
Kennesaw State University

See Also: Energy Efficiency; Green Design; Greenhouse Gas Emissions; Oklahoma; Renewable Energy, Wind.

Further Readings

American College and University Presidents' Climate Commitment. http://www.presidentsclimatecommitment.org (Accessed June 2011).

Ellisor, Larry. "Princeton Review Names OU 'Green' College." *Oklahoma Daily* (April 22, 2011). http://oudaily.com/news/2011/apr/22/princeton-review-names-ou-green-college (Accessed June 2011).

University of Oklahoma. http://www.ou.edu (Accessed June 2011).

University of Washington

The University of Washington (UW) was founded in 1861 in Seattle, making it one of the oldest universities on the U.S. west coast. Today, it is an internationally recognized research university with campuses in Tacoma and Bothell, both founded in 1990, as well as Seattle. In the 2009–10 academic year, 48,022 students were enrolled at UW (34,523 undergraduates, 11,592 graduates, and 1,907 professionals) with an additional 63,178 registrations for extension courses. Over 29,000 faculty and staff members were employed at UW. The university has 16 colleges and schools and confers over 12,000 degrees annually; in the 2009–10 academic year, UW awarded 8,458 bachelors degrees, 2,988 masters degrees, 686 doctoral degrees, and 495 professional degrees. In fiscal year 2010, UW received $1.3 billion in research funding, including more National Institute of Health American Recovery and Reinvestment Act awards than any other U.S. university.

Focus on Climate Change

Several schools and departments within UW focus on topics related to global warming and climate change. The College of Built Environments (CBE) offers undergraduate and graduate degrees in architecture, construction management, landscape architecture, and urban design and planning as well as a college-wide Ph.D. on the built environment, which incorporates the study of sustainable systems and prototypes, and an interdisciplinary Ph.D. program in urban design and planning, which includes consideration of the interactions of the physical, social, economic, and natural environment. Current research centers within CBE include the Urban Ecology Research Laboratory, Design Machine Group, Institute for Hazards Mitigation Planning and Research, Northwest Center for Livable Communities, and Green Futures Research and Design Lab.

The Department of Geography offers four tracks of study at the undergraduate level, including environment, economy, and stability (focusing on how economic activity, environmental policy, and sustainability interact) and globalization, health, and development (focusing on the influence of geography in affecting global health and development). The department also offers M.A. and Ph.D. programs. The Participatory Geographic Information Systems for Transportation project works with regional agencies and stakeholders to study how geographic information systems and Internet technology can improve transportation decision making.

The College of Engineering offers undergraduate- and masters-level study in a variety of fields, including aeronautical and astronautical engineering, bioengineering, chemical engineering, civil and environmental engineering, computer engineering, electrical engineering, industrial systems, materials science and engineering, mechanical engineering, and human-centered design and engineering. Research centers associated with the department include the Advanced Materials for Energy Institute, Bioenergy Program, Northwest National Marine Renewable Energy Center, Institute of Advanced Materials and Technology, Transportation Northwest (TransNow), Intelligent Transportation Systems, and Washington State Transportation Center.

The College of the Environment, founded in 2009, is a hub for education (at the bachelors, masters, and doctoral level) and research related to the interaction of humans and their environment. The college is composed of 11 core units: the School of Aquatic and Fishery Sciences, Department of Atmospheric Sciences, Department of Earth and Space Sciences, Program on the Environment, School of Forest Resources, School of Marine and Environmental Affairs, School of Oceanography, Friday Harbor Laboratories, Joint Institute for the Study of Atmosphere and Ocean, Washington National Aeronautics and Space Administration (NASA) Space Grant Program, and Washington Sea Grant Program. Other research institutes, facilities, and field stations associated with the college include the Environmental Institute, the Program on Climate Change, and the Center for Sustainable Forestry.

UW has received many honors for environmental leadership from organizations and publications such as the Sustainable Endowment Institute, *Sierra Magazine*, the *Princeton Review*, Salmon Safe, and *Forbes Magazine*, and was the 2010 winner of the Green Washington Award in the Academic/Government category. In 2009, UW created an Institutional Climate Action Plan with the goal of achieving carbon neutrality, and has instituted a number of programs to reduce waste and increase sustainability. Ten buildings on the UW campus are Leadership in Energy and Environmental Design (LEED) certified, and another 22 LEED projects are underway. Over half of UW's waste is diverted to recycling and composting facilities, rather than landfills. Improvements in energy efficiency at the campus data center have reduced energy use by 26 percent. UW has also increased recycling of paper products (including 100 percent of copy and print paper), has reduced paper consumption by 30 percent, and is transitioning to using only 100 percent recycled paper for copiers and printers.

Sarah Boslaugh
Kennesaw State University

See Also: Carbon Footprint; Education; Oceanography; Washington.

Further Readings

College Sustainability Report Card. "University of Washington." http://www.greenreportcard.org/report-card-2011/schools/university-of-washington/surveys/campus-survey#climate (Accessed June 2011).

Reno, Steve. "Government/Academic: University of Washington, Seattle University, King County GreenTools." *Seattle Business* (October 2010). http://www.seattlebusinessmag.com/article/governmentacademic-university-washington-seattle-unversity-king-county-greentools (Accessed June 2011).

University of Washington. http://www.washington.edu (Accessed June 2011).

Upwelling, Coastal

Coastal upwelling occurs when water along a coastline flows offshore and deeper water—usually relatively cool, rich in nutrients, and high in partial pressure of carbon dioxide—flows upward to fill its place. Upwelling areas are notable for their effect on carbon cycling, as upwelling not only brings dissolved inorganic carbon to the surface, where it is released into the atmosphere, but also stimulates phytoplankton blooms that further remove some of that carbon through photosynthesis; a small percentage of this bloom also sinks in the form of organic matter (organic carbon) to deep water and becomes buried in sediment, creating a long-term carbon sink. There is considerable interest among carbon-cycle scientists regarding the reciprocal interactions between upwelling systems and climate change. Although such upwelling can in principle occur along any coastline, marine or freshwater, some marine coastlines (e.g., Peru, the western United States, northwest Africa, and southwest Africa) are renowned for their annual upwelling events that are the source of major blooms of diatoms and dinoflagellates, which become the base for extensive marine food webs and coastal fishing industries.

Influence on Carbon Cycling

In the past several decades, major research programs have developed around the influence of coastal upwelling ecosystems on ocean carbon cycling and atmospheric carbon dioxide, how natural climate change (such as glacial–interglacial cycles) has affected coastal upwelling and associated biological productivity over a range of time scales, and how human-induced climate change is affecting coastal upwelling rates and timing and the associated fisheries.

Carbon dioxide exchange between coastal surface water and the atmosphere varies considerably in time and space. Because the pattern is complicated and dynamic relative to the number of direct measurements, considerable uncertainty lingers regarding the net carbon flux through the system over the course of a year. In general, outgassing occurs near the coastline, where upwelled water outcrops at the surface. This water is often rich in carbon dioxide arising from the respiration of organisms ingesting organic matter that sank from the surface to deeper water (which may be the sea bottom along the continental shelf). As upwelled water moves from shore, phytoplankton bloom in response to dissolved nitrogen, phosphorus, and other nutrients and begin to use up some of the dissolved inorganic carbon, reducing the partial pressure of carbon dioxide.

Because this process occurs over a period of several weeks, the rate of uptake of dissolved inorganic carbon also changes through time, so that net outgassing will occur early in an upwelling event, gradually changing to net ingassing. Much of the phytoplankton is recycled in the surface layer, prolonging the bloom, but some of the nutrients and carbon escape the system through the fecal material of heterotrophs feeding on the phytoplankton. The nutrients of remineralized organic matter that sink may come to the surface in future years through upwelling, or the organic matter may sink below the depth of upwelled water into the deep sea, or get buried in sediment. The latter two processes can take carbon out of the atmosphere for thousands or millions of years, respectively. Although these processes occur in other aquatic areas, enough of the global ocean carbon flux in a given year occurs through coastal upwelling zones to affect atmospheric carbon dioxide.

The strength and direction of surface winds that drive coastal upwelling vary over a broad spectrum of time scales. Changes in global heat retention through time affect the potential for temperature gradients that influence wind speed, and the distribution of land masses and topographic features such as mountains affect coastal shape, coastal currents, sea level and coastal profile, and atmospheric circulation patterns. Temperature and precipitation patterns and sea level, among other variables, affect nutrient distribution in the oceans. All of these affect upwelling strength, biological productivity, carbon burial, and net effect on the global carbon cycle. Much research has been dedicated to understanding upwelling changes during glacial–interglacial cycles, tracking responses to changed wind speeds and to lowered sea level, and therefore steeper coastal profiles. Other research has examined how to predict occurrences of upwelling in, for example, the Mesozoic, under the assumption that upwelling is responsible for the accumulation of some petroleum deposits.

Both models and empirical observations of several coastal upwelling areas, such as off the coast of California and northwest Africa, suggest that atmospheric warming is leading to greater rates of upwelling. This increase is driven by a greater land–ocean temperature gradient and therefore greater wind speeds. This can lead both to greater outgassing of carbon dioxide (if not balanced by increased productivity) and loss of certain fish that cannot maintain their population position because of higher offshore current velocities.

Robert M. Ross
Paleontological Research Institution

See Also: Benguala Current; Carbon Cycle; Mesozoic Era; Oceanic Changes; Peruvian Current; Upwelling, Equatorial.

Further Readings

Barth, John A., et al. "Delayed Upwelling Alters Nearshore Coastal Ocean Ecosystems in the Northern California Current." *Proceedings of the National Academy of Sciences*, v.104/10 (2006).

Chavez, F. P. and T. Takahashi. "Coastal Oceans." In *The First State of the Carbon Cycle Report*. SOCCR. http://cdiac.ornl.gov/SOCCR (Accessed March 2012).

McGregor, H. V., et al. "Rapid 20th-Century Increase in Coastal Upwelling off Northwest Africa." *Science*, v.315/5812 (2007).

Upwelling, Equatorial

Equatorial Upwelling (UE) is upward water's motion in the upper layer of the equatorial ocean. It occurs when a persistent easterly wind is blowing over the equatorial zone. Maximum upward velocity in the UE occurs just at the equator.

The EU is a result of a permanent divergence of a westward surface South Equatorial Current in the narrow equator vicinity forced by the southeast trade wind. Divergence of the westward current at the equator is caused by the change of the sign of the Coriolis force between the Northern and Southern Hemispheres. As a consequence of divergence, the upper thermocline becomes shallower at the equator. Strong permanent equatorial divergence also causes an intense entrainment of more cold water of thermocline into the upper mixed layer because associated vertical velocity in the equatorial thermocline is typically $\sim 10^{-5}$ m per sec. This leads to cooling of the upper mixed layer. As a result, the sea surface temperature is about 1.8 degrees F (1 degree C) lower in the vicinity of the equator than in the interior equatorial ocean outside of it.

Location of Pure Equatorial Upwelling

Pure UE occurs in the narrow vicinity of the equator, just within the divergent zone. Because of the slope of equatorial thermocline in a zonal direction (the thermocline is deeper in the western equatorial Atlantic and Pacific oceans than in the eastern) and the generation of coastal upwelling in the eastern equatorial oceans, UE manifestation, as relatively cold surface water, is more pronounced just in the upper layer of the eastern equatorial oceans. Therefore, such cooler sea-surface water looks like a long, thin tongue along the equator, spreading from the eastern equatorial oceans. There is also high biological activity in the vicinity of this relatively cold tongue.

The thickness of the UE is restricted by the upper boundary of equatorial undercurrent because the eastward current is accompanied by equatorial convergence and, hence, downward water motion. That is why this thickness varies from about 330–660 ft. (100–200 m) (in the western equatorial Atlantic or Pacific oceans, respectively) to 33–66 ft. (10–20 m) (in the eastern equatorial Atlantic and Pacific oceans, respectively).

The UE is quite a persistent phenomenon in the Atlantic and Pacific oceans because the westward surface South Equatorial Current occurs there in the equator's vicinity almost throughout the entire year. However, the UE intensity varies from season to season and from year to year. Seasonally, it is at a maximum in the equatorial Atlantic and Pacific when the South Equatorial Current intensifies, following a seasonal cycle of the southeast trade wind (with some delay, which does not typically exceed a month); that is, in boreal late summer to early fall. Interannual variations of UE are mostly due to the El Niño/La Niño phenomena, especially in the Pacific Ocean. Just before the development of an El Niño event (the anomalous warming of the upper layer in the equatorial Pacific), the southeast trade wind dramatically weakens and UE is over.

In contrast, during a La Niño event (a cold episode in the equatorial Pacific Ocean), UE is strongly developed as a result of anomalous intensification of the southeast trade wind, and hence the South Equatorial Current. Interannual variability of UE in the equatorial Atlantic follows to Pacific variability with some delay, which is typically not more than a few months. However, the magnitude of interannual UE variations in the Atlantic Ocean is not as large as in the Pacific Ocean. A seasonal cycle prevails in the equatorial Atlantic, where the magnitude of seasonal UE variations is two to three times larger than interannual variations.

In the Indian Ocean, UE (as a persistent phenomenon) occurs only in boreal winter, when the

Cold-water upwelling (UE) in the Gulf of Tehuantepec: This image of the Isthmus of Tehuantepec in Mexico shows sea surface temperatures observed by the Moderate Resolution Imaging Spectroradiometer (MODIS) on NASA's Aqua satellite. Pure UE occurs in the narrow vicinity of the equator, just within the divergent zone. The UE's cooler sea-surface water appears as a long, thin tongue along the equator, spreading from the eastern equatorial oceans. There is also high biological activity in this vicinity.

northeast monsoon has been developing. The UE is most pronounced in the western part of this basin. Seasonal UE variability is at maximum just in the Indian Ocean. Interannual UE variability in the Indian Ocean is controlled by Indo-Ocean Dipole, which is the inherent Indo-Ocean mode interrelated with the Pacific interannual variability (the El Niño/La Niño phenomena), as can be seen in the 2007 results from researchers Swadhin Behera, Toshio Yamagata, Alexander Polonsky, and colleagues.

Low-frequency (decade-to-decade) variability of the southeast trade wind and/or northeast monsoon would generate quasi-equilibrium Upwelling Equatorial variations. A more (less) intense southeast trade wind and northeast monsoon would lead to more (less) intense Upwelling Equatorial.

Alexander Boris Polonsky
Marine Hydrophysical Institute, Sevastopol

See Also: Atlantic Ocean; Climatic Data, Oceanic Observations; Equatorial Undercurrent; Indian Ocean; Mixed Layer; Pacific Ocean; Thermocline; Trade Winds; Upwelling, Coastal.

Further Readings

Behera, Swadhin and Toshio Yamagata. "Influence of the Indian Ocean Dipole on the Southern Oscillation." *Journal of the Meteorological Society of Japan*, v.81/1 (2003).

Kraus, Eric, ed. *Modelling and Prediction of the Upper Layers of the Ocean*. New York: Pergamon Press, 1977.

Polonsky, Alexander, Gary Meyers, and Anton Torbinsky. "Interannual Variability of Heat Content of Upper Equatorial Layer in the Indian Ocean and Indo-Ocean Dipole." *Physical Oceanography*, v.21/1 (2007).

Uruguay

The Oriental Republic of Uruguay is one of the most economically developed countries in South America. Most of Uruguay is hilly grassland, river systems, and a short Atlantic coast. The lack of mountain ranges makes the country susceptible to rapid weather changes, including droughts and flooding. The grasslands are used to graze cattle and sheep, a large portion of the country's agricultural sector.

Global warming may actually benefit Uruguay, particularly the livestock industry, as the grasslands flourish under warmer temperatures and higher concentrations of atmospheric carbon dioxide. Current models predict a likely increase of 2 degrees F (1.1 degrees C) by 2050. Rising sea level is of greater concern. Nearly half of Uruguay's otherwise sparsely distributed population lives in the coastal city of Montevideo, home to the government and most of the country's industry. The 1.5- to 3-ft. (0.45- to 0.91-m) increase in sea level in the 21st century, which has been predicted, would damage or destroy the city's sewage and water system, along with much of its commercial and residential real estate.

Uruguay's per-capita carbon emissions have remained stable, and the nation is generally considered a carbon-neutral country. Its emissions increased from 1.8 metric tons per person in 1997 to 1.86 metric tons in 2007, a fairly minor increase that is probably due to a slight economic expansion and greater use of automobiles. Most of Uruguay's emissions come from liquid fuel use, with 8 percent from the local manufacture of cement. Uruguay is a signatory to the United Nations Framework Convention on Climate Change and was one of the first countries to submit a greenhouse gas inventory. Its sustainable practices efforts date to the early 1980s, and its laws regulating soil management and the sequestration of carbon are responsible for its carbon neutrality.

Bill Kte'pi
Independent Scholar

See Also: Agriculture; Argentina; Brazil; Paraguay; Reducing Emissions From Deforestation and Forest Degradation (UN-REDD); World Trade Organization.

Further Readings

Climate Hot Map. "The Impact of Global Warming in South America." http://www.climatehotmap.org/samerica.html (Accessed March 2012).

U.S. Department of State. "Uruguay." http://www.state.gov/r/pa/ei/bgn/2091.htm (Accessed March 2012).

Utah

The sixth most urbanized state in the United States, Utah, located in the southwest, is a geographically diverse state of canyons, desert, salt flats, and mountains. The population of the state increased 23.8 percent between 2000 and 2010, making Utah the third fastest-growing state in the United States. The growing population is putting pressure on the state's water resources. According to a U.S. Geological Survey, only Nevada has higher per-capita water usage. Scientists warn that global warming could intensify water concerns in the decades ahead. The Intergovernmental Panel on Climate Change estimates that average temperatures in Utah could rise about 6.75 degrees F (3.75 degrees C) by 2100 if global warming continues unabated, bringing drought, floods, and wildfires. In the first decade of the 21st century, Utah took positive actions to address global warming concerns, earning the American Council for an Energy-Efficient Economy's (ACEEE) recognition as a "most improved state," but 2010 saw state legislators pass a resolution that challenged the mainstream scientific consensus on global warming.

Global Warming Impacts

A 2011 report released by the U.S. Department of Interior predicted that the Colorado River basin—of which Utah is a part—will suffer an 8.5 percent reduction in its water supply by midcentury. Despite more than a 2 percent increase in precipitation, also part of the report, temperatures warmer by 5 to 7 degrees F (2.7 to 3.8 degrees C) will bring more rain than snow and an evaporation increase that will more than account for additional precipitation. Utah uses almost 5 billion gallons of water a day; just over 80 percent goes to farms, which increasingly sell a portion of their water rights to cities, where two-thirds of that water is used for landscaping. Utah has set a goal of a 25 percent reduction in individual use of water by 2050.

In 2005, Utah's gross greenhouse gas (GHG) emissions were rising at a faster rate than those of the nation as a whole, increasing about 40 percent from 1990 to 2005. The state accounted for around 68.8 million metric tons of carbon dioxide (CO_2) emissions in 2005. The combustion of fossil fuels for electricity generation used in-state and for transportation accounted for 61 percent of these emissions. Utah produced 2 percent of the nation's coal in 2007 (the most recent year for which detailed statistics are available). The industry also accounted for approximately 4,700 jobs—1,900 directly, with another 2,800 through ripple effects—and $196 million for 2007. Coal-based electrical generation accounts for more than 80 percent of Utah's electricity. The state's coal-powered electric plants put out nearly 35 million metric tons of CO_2.

Utah joined the Western Climate Initiative (WCI), a regional effort to reduce GHG emissions and address climate change, in May 2007. In August, the WCI announced its regional, economy-wide GHG emissions target of 15 percent below 2005 levels by 2020, or approximately 33 percent below business-as-usual levels. In October 2007, the Governor's Blue Ribbon Advisory Council on Climate Change (BRAC), established in August 2006 to assess the policy options available to Utah for addressing climate change, released its final report. BRAC recommended nearly 60 high-priority actions in five areas: agriculture/forestry, cross-cutting issues, energy supply, residential/commercial/industrial, and transportation/land use. In March 2008, Utah adopted a voluntary renewable portfolio goal that encourages utilities to produce 20 percent of their energy from renewable sources by 2025. The goal requires that utilities pursue cost-effective renewable energy. Over the following months, Utah increased its budget for energy-efficiency programs and set goals for energy efficiency and renewable energy. In 2009, the state moved from 23rd to 12th in the American Council for an Energy-Efficient Economy's ranking of most energy-efficient states.

In 2010, *Forbes* ranked states in costs, current economic climate, growth prospects, labor supply, quality of life, and regulatory environment, and named Utah the best state for business. The same year, the Utah state legislature introduced approximately 20 states' rights bills that aimed at exempting the citizens of Utah from various federal infringements, from healthcare to firearms. In line with this turn was the joint resolution that called on the U.S. Environmental Protection Agency to cease its regulation of GHG emissions until the claims of climate scientists could be substantiated. Although the resolution is nonbinding and it provoked opposition within the legislature and the citizenry, its passage suggests that a majority in Utah are skeptical of the mainstream consensus on climate change and are concerned about the impact that emissions regulations will have on the state's economy. The House also passed a resolution demanding that the governor leave the WCI. Utah has rejected participation in WCI's cap-and-trade policy, and Governor Gary Herbert did not attend the 2010 meeting of the group.

Wylene Rholetter
Independent Scholar

See Also: Carbon Dioxide; Coal; Environmental Protection Agency, U.S.; Utah Climate Center.

Further Readings

American Council for an Energy Efficiency Economy. "Utah State Energy Efficiency Policy Database." http://www.aceee.org/sector/state-policy/utah (Accessed July 2011).

Loomis, Brandon. "Climate Change to Sap Utah Water Supply." *Salt Lake Tribune* (April 26, 2011). http://www.sltrib.com/csp/cms/sites/sltrib/pages/printerfriendly.csp?id=51690000 (Accessed July 2011).

U.S. Energy Information Administration. "Utah." http://www.eia.gov/state/state-energy-profiles-analysis.cfm?sid=UT (Accessed July 2011).

Utah Department of Environmental Quality. "Governor's Blue Ribbon Advisory Council on Climate Change: Final Report." http://www.deq.utah.gov/BRAC_Climate/final_report.htm (Accessed July 2011).

Utah Climate Center

The Utah Climate Center (UCC) aims to disseminate climate data and information and to use its scientific expertise in effective and engaging communications with the general public. As part of its mission, the center designs new products to meet the present and future needs of key sectors in Utah's economy, such as agriculture, industry, and tourism, and provides resources for the region's natural resources, government, and educational organizations. In doing so, the center responds to the challenges brought about by climate change. As part of Utah State University, the center has been recognized as a state center by the American Association of State Climatologists.

Much of the climate information requested from the center comes from published records and computerized databases. Published records in print form extend through 2006 and are available at the National Climatic Data Center (NCDC). Many original historical data records for Utah for the 19th and early 20th centuries have been transferred to and archived in the Utah State University Merrill–Cazier Library. The Utah Climate Center serves as an official repository for both published climate data records and official publications from the NCDC, encompassing several decades, as part of an official agreement with the NCDC.

The weather and climate data provided by the Utah Climate Center are elaborated with the cooperation of the National Oceanic and Atmospheric Administration, Federal Aviation Administration, and other federal, state, and local authorities. The Utah Climate Center strives to provide quality climate data.

Data Gathering and Communications

The Utah Climate Center gathers and archives climatic data from 22 networks throughout the state. It also monitors and compiles information from networks used by the Forest Service, University of Utah, Department of Agriculture, the Bureau of Land Management, Natural Resource Conservation Service, and other government departments. Data such as maximum and minimum temperature, precipitation, evaporation, evapotranspiration (a measure of water lost from the soil as a result of transfer to plants, rather than straight evaporation), and solar radiation are collected from various sites. The center collects 57,000 pieces of information from hundreds of locations each day. The earliest data at the UCS date back to the 1870s and 1880s, although most of the stations were put in place in the 1900s.

In the past, the Utah Climate Center has provided information through paper via the postal service, fax, or electronic mail, but it now uses a global information system search facility. The center has updated its Website, which now provides real-time weather data to users. It also constantly updates its Climate DataSet map page and adds new weather-observing networks. The center's visibility and communication with the public has also been enhanced by the partnership with Utah Public Radio, to which it contributes daily weather forecasts, and by regular articles featured in the *Herald Journal*.

The center was initially reluctant to assess global warming. In 1998, Director Donald T. Jensen stated that any signals of global warming in Utah "have been lost in the noise of other temperature fluctuations ... It's hard to find any real evidence here because temperatures here have always bounced up and down." However, with the passing years, the center has expressed views that are more aligned with the general scientific opinion on global warming, and climate change has become one of its main foci of research. Robert Gillies, the present director of the center, has pointed out that "the massive and growing scientific evidence has convinced the atmospheric science community that climate change is occurring, and is the result of human activities, specifically the release of greenhouse gases." In addition, the center recognizes that, with the exception of the Arctic, areas that intersect with the state's borders, such as the American west and the Colorado basin, are among those warming most rapidly in the world. The UCC took part in the 2007 Governor's Blue Ribbon Advisory Committee on Climate Change, supplying the chapter "Climate Change and Utah: The Scientific Consensus" in the state report, *Climate Change and Utah: The Scientific Consensus*.

Luca Prono
Independent Scholar

See Also: Climatic Data, Nature of the Data; Global Warming Debate; Utah.

Further Readings

Gillies, Robert. "Letter to the Editor: Statement on Climate Change From the Utah Climate Center." *Hard News Café* (February 7, 2007). http://hardnews1.ansci.usu.edu/archive/feb2007/020707_climateletter.html (Accessed July 2007).

Utah Climate Center. "Climate Change and Utah: The Scientific Consensus." http://climate.usurf.usu.edu/news/111708Sec-A-1_SCIENCE_REPORT.pdf (Accessed July 2011).

Uzbekistan

Uzbekistan is a landlocked country in central Asia that became part of the Soviet Union in 1924 and achieved independence in 1991. Uzbekistan has an arid climate, with limited precipitation in most of the country and hot summers and cold winters. Mean precipitation ranges from 3.1 to 7.8 in. (80 to 200 mm) annually in the desert plains to 23.6 to 31.4 in. (600 to 800 mm) in the highland mountain regions, with most rain falling in the winter and spring months. Natural pasture occupies 40 percent of Uzbekistan, and cropland covers an additional 12 percent. Agriculture is an important part of the economy; livestock (primarily beef and cow's milk) provides about 40 percent of the cash value of agricultural production, while crops provide the other 60 percent. The major cops are cotton and wheat, and because of Uzbekistan's dry climate, over 85 percent of cropland is irrigated. Forty percent of Uzbekistan's exports come from agriculture, primarily cotton; Uzbekistan is the second-largest exporter of cotton in the world, after the United States.

Over half of Uzbekistan's soil is salinized, which is leached from the soil and deposited in local groundwater systems through irrigation, an effect that is increased when water is reused downstream. Erosion is already a significant problem; about half of the irrigated land is subject to wind erosion and overgrazing, and poor crop management has increased the problem. Agrochemicals were overused during the Soviet period, and the land still suffers from this legacy.

Uzbekistan's gross domestic product (GDP) declined from 1991 to 2000, but has since rebounded, with an annual growth rate of 6.2 percent from 2000 to 2007. However, the per-capita GDP in 2010 was still relatively low at $3,100, and inflation of 15 percent in 2010 was among the highest in the world. Official unemployment is just over 1 percent, but many sources estimate it is 20 percent higher, and 26 percent of the population lived below the poverty line in 2008.

Oil and Natural Gas

Uzbekistan has 594 million barrels of proven oil reserves, ranking 47th in the world, and 1.841 trillion cu. m of proven natural gas reserves, ranking 19th in the world. The country is a significant producer and exporter of both oil and natural gas, producing 140,000 barrels per day of oil and exporting 6,104 barrels per day in 2009. It also produced 67.6 billion cu. m of natural gas in 2008 (the 13th-highest in the world) and exported 15 billion cu. m of natural gas in the same year Uzbekistan produced 44.8 billion kilowatt hours (kWh) of electricity in 2009, consuming 40.1 billion kWh, while exporting 11.52 billion kWh and importing 11.44 billion kWh. Total carbon dioxide emissions from fossil fuel consumption in 2008 amounted to 127.12 million metric tons, ranking 35th in the world.

The World Bank ranks Uzbekistan sixth highest in terms of climate change vulnerability out of the 28 countries in the Europe and central Asia region. An increase in mean annual temperature of 3.4–4.3 degrees F (1.9–2.4 degrees C) is predicted by 2050, with the greatest warming in winter and spring. An increase in mean annual precipitation of 15–18 percent is also predicted, with the greatest increase coming in the summer months. Glaciation has decreased by about one-third in the central Asia region over the past century, reducing the amount of freshwater available for irrigation. The water deficit at the Aral Sea basin is expected to increase from 0.47 cu. mi. (2 cu. km) in 2005 to 2.63–3.11 cu. mi. (11–13 cu. km) in 2050. The agricultural sector will be particularly hard hit by expected changes due to global warming as the environment is predicted to become more arid. Increased evapotranspiration is expected to offset projected increases in precipitation, thus increasing demands on already stressed water resources. Changes in temperature will also expose livestock and crops to new diseases and pests. On the positive side, a longer growing season should lead to increased productivity in some areas of the country, particularly in the northern regions.

Sarah Boslaugh
Kennesaw State University

See Also: Agriculture; Glaciers, Retreating; Natural Gas; Oil, Production of.

Further Readings

U.S. Energy Information Administration. "Country Profile: Uzbekistan." (June 30, 2010). http://www.eia.gov/countries/country-data.cfm?fips-UZ (Accessed July 2011).

World Bank. "Uzbekistan: Climate Change and Agriculture Country Note" (September 2010). http://siteresources.worldbank.org/ECAEXT/Resources/258598-1277305872360/7190152-1303416376314/uzbekistan-countrynote.pdf (Accessed July 2011).

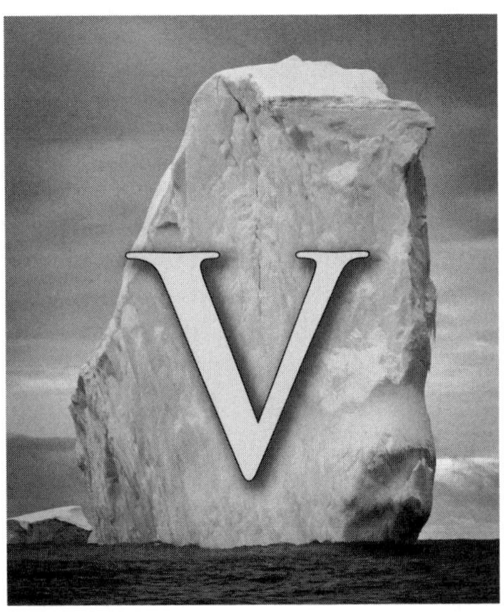

Validation of Climate Models

The climatic system is constituted by four intimately interconnected subsystems—atmosphere, hydrosphere, cryosphere, and biosphere—which evolve under the action of macroscopic driving and modulating agents, such as solar heating, Earth's rotation, and gravitation. The climate system features many degrees of freedom, which make it complicated, and nonlinear interactions taking place on a vast range of time-space scales accompanying sensitive dependence on initial conditions, which makes it complex. The climate is defined as the set of statistical properties of the observable physical quantities of the climatic system.

The evaluation of the accuracy of numerical climate models and the definition of strategies for their improvement are crucial issues in the Earth system scientific community. On one hand, climate models of various degrees of complexity constitute tools of fundamental importance to reconstruct and project in the future the state of the planet and to test theories related to basic geophysical fluid dynamical properties of the atmosphere and the ocean, as well as of the physical and chemical feedbacks within the various subdomains between them. On the other hand, the outputs of climate models, and especially future climate projections, are gaining an ever-increasing relevance in several fields, such as ecology, economics, engineering, energy, and architecture, as well as for the process of policymaking at a national and international level. Regarding influences at the societal level of climate-related findings, the effects of the *Fourth Assessment Report* of the Intergovernmental Panel on Climate Change (IPCC4AR) are unprecedented.

The validation or auditing—overall evaluation of accuracy—of a set of climate models is a delicate operation that can be decomposed in two related, albeit distinct, procedures. The first procedure is intercomparison, which aims at assessing the consistency of the models in the simulation of certain physical phenomena over a certain time frame. The second procedure is verification, the goal of which is to compare the models' outputs with corresponding observed or reconstructed quantities. Difficulties emerge because there are three different kinds of attractors: the attractor of the real climate system, its reconstruction from observations, and the attractors of the climate models. Depending on the timescale of interest and on the problem under investigation, the relevant active degrees of freedom (mathematically corresponding to the separation between the slow and fast manifolds) needing the most careful representation change dramatically. For relatively short timescales (below 10 years), the atmospheric degrees of freedom are active, whereas the other subsystems can be considered essentially frozen. For longer timescales (100–1,000 years), the ocean dominates the dynamics of climate, whereas for even longer timescales (over 5,000 years), continental ice sheet changes are the most relevant factors of variability. Therefore, the scientific community has produced different families of climate models, spanning a hierarchical ladder of complexity, each formulated and structured for specifically tackling a class of problems.

Coupled Global and Regional Climate Models

Whereas most considerations are quite general, the coupled global climate models (GCMs) and regional climate models (RCMs) are currently used for the simulation of the present climate and for the analysis of climate variability up to centennial scales. In these models, whereas the dynamical processes of the atmosphere and of the hydrosphere are represented within a wide framework of numerical discretization techniques applied to simplified versions of thermodynamics and Navier-Stokes equations in a rotating environment, the continental ice sheets are typically taken as fixed parameters of the system. In contrast, the subscale processes, which cannot be explicitly represented within the resolution of the model, are taken care of through simplified parameterizations.

Several crucial processes, such as radiative transfer, atmospheric convection, microphysics of clouds, land-atmosphere fluxes, ice dynamics, and eddies and mixing in the ocean, as well as most of those controlling the biosphere evolution, undergo severe simplifications. With time, the formulation of the GCMs has developed through

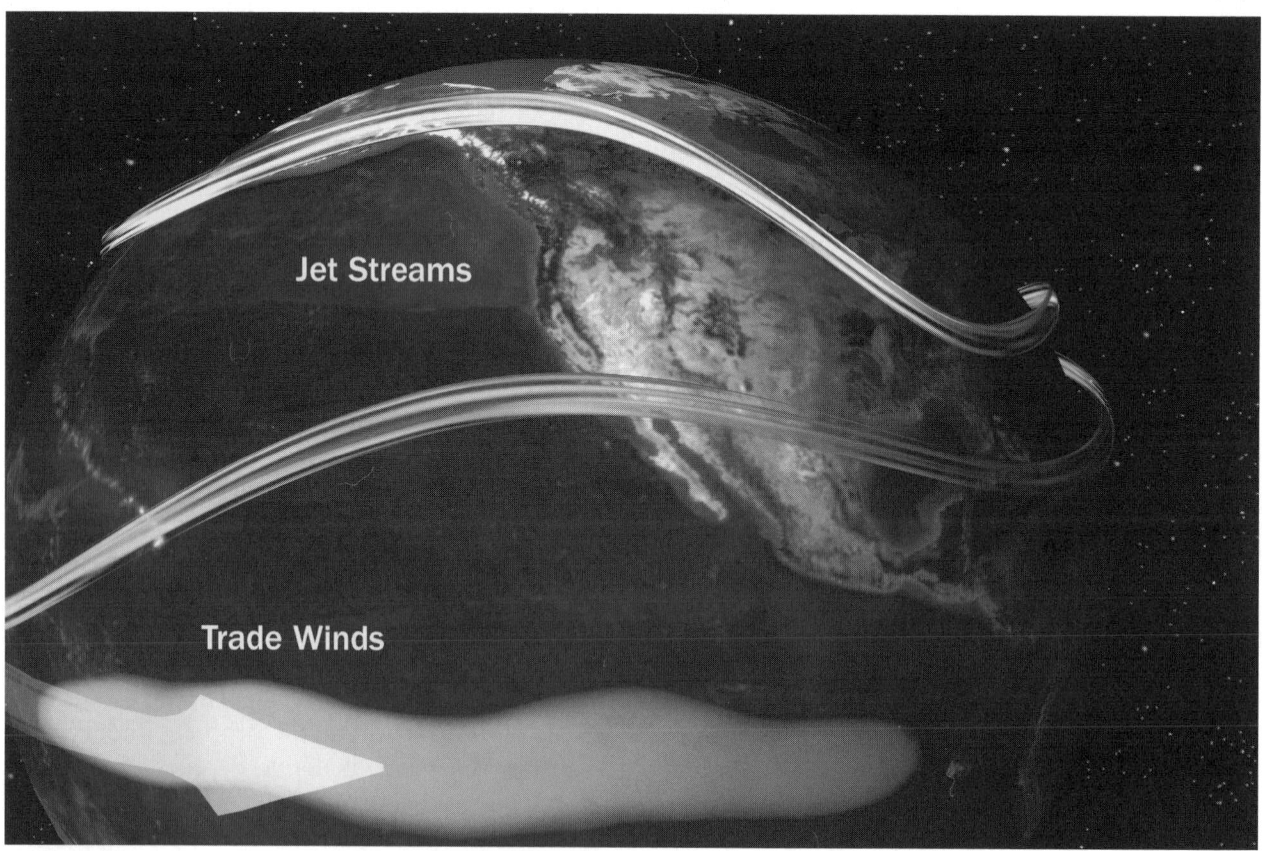

A NASA diagram of a strong El Niño striking surface waters in the Pacific Ocean. Warm water anomalies develop (indicated by shape at the bottom) and westerly winds weaken, allowing the easterly winds to push the warm water against the South American coast. To capture the differences in the representation of specific physical processes, it is necessary to use process-oriented metrics as indexes for model reliability. Examples of these metrics include major features of atmospheric variability, such as El Niño–Southern Oscillation.

refinements to the spatial resolution, ameliorations of numerical schemes, and improved parameterizations, as well as through the inclusion of a larger and larger set of processes, such as aerosol chemistry and interactive vegetation, which are relevant for the representation of system feedbacks and forcings.

In addition, limited-area climate modeling faces the mathematical complication of being a time-varying boundary conditions problem, as RCMs are nested into driving GCMs. Therefore, RCMs tend to be enslaved on timescales, depending on the size (and position) of their domain and in principle, the balances evaluated over the limited domain are constrained at all times. Therefore, climate reconstructions and projections performed with an RCM can critically depend on the driving GCM. Other, more technical issues arise from the delicate process of matching the boundary conditions at the models' interface, where rather different spatial and time grids have to be joined.

Model results and approximate theories can be tested only against past observational data of nonuniform quality and quantity, essentially because of the space and the timescales involved. The available historical observations sometimes feature a relatively low degree of reciprocal synchronic coherence and individually present problems of diachronic coherence, as a result of changes in the strategies of data gathering with time, whereas proxy data, by definition, provide only semiquantitative information on the past state of the climate system. Extensive scientific effort is aimed at improving the quality and quantity of climatic databases. In particular, the best guess of the atmospheric state of roughly the last 50 years has been reconstructed by two independent research initiatives, through the variational adaptation of model trajectories to all available meteorological observations, including the satellite-retrieved data, producing reanalyses.

Given all the above-mentioned difficulties, as well as the impossibility—because of the entropic time arrow—of repeating world experiments, the validation of GCMs is not epistemologically trivial, as the Galilean paradigmatic approach cannot be followed. Validation has to be framed in probabilistic terms, and the choice of the observables of interest is crucial for determining robust metrics able to effectively audit the models. Recently, the detailed investigation of the behavior of GCMs has been greatly fostered and facilitated, as some research initiatives have been providing open access to standardized outputs of simulations performed within a well-defined set of scenarios. Relevant examples are the project PRUDENCE (RCMs) and the PCMDI/CMIP3 initiative (GCMs included in the IPCC4AR).

One aim—from the end-user's viewpoint—is checking how realistic the modeled fields of practical interest are, such as surface temperature, pressure, and precipitation. In these terms, current GCMs feature a good degree of consistency and realism when considering present climate simulation, and they basically agree on short-term climate projections, down to seasonal averages on continental scales. When decreasing the spatial or the temporal scale of interest, the signal-to-noise ratio of climatic signals—both observative and model generated—typically decreases, so that the validation of GCMs in control runs and climate change simulations becomes more difficult, even if improvements are observed over time by state-of-the-art models. Statistical and dynamical—provided by nested RCMs—downscaling of climatological variables enlarges the scope of model validation. In particular, RCMs provide a better outlook on small-scale and nonlinear processes, such as surface–atmosphere coupling, precipitation, and effects of climate change on the biosphere.

However, the above-mentioned quantities can hardly be considered climate state variables, whereas strategies for model improvement can benefit from understanding the differences in the representation of the climatic machine among GCMs. The comparison of the statistical properties of bulk quantities defining the climatic state, such as top-of-the-atmosphere energy fluxes, tropospheric average temperature, tropopause height, geopotential height at various pressure levels, tropospheric average specific humidity, and ocean water structure, allows the definition of global metrics that constitute robust diagnostic tools. Moreover, to capture the differences in the representation of specific physical processes, it is necessary to use specialized diagnostic tools—process-oriented metrics—as indexes for model reliability.

Examples of these metrics are major features of atmospheric variability, such as tropical and extratropical cyclones; detailed balances, such as water vapor convergence over continents or river basins; teleconnection patterns, such as El Niño–Southern Oscillation or Madden–Julian Oscillation; or oceanic features, such as the overturning circulation and Antarctic current intensity. The latter approach may be especially helpful in clarifying the distinction between the performance of the models in reproducing diagnostic and prognostic variables. Even if improvement is ongoing and promising, in these more fundamental metrics describing the climatic machine, current GCMs do not generally feature a comparable degree of consistency and realism at a quantitative level, and further investigations on basic physical and dynamical processes are needed.

Limit on Realistic Simulations

Because the goal of a climate model is to reproduce the most relevant statistical properties of the climate system, the structural deficiencies, together with an unavoidably limited knowledge of the external forcings (uncertainties of the second kind) intrinsically limit the possibility of performing realistic simulations, especially affecting the possibility of representing abrupt climate change processes. The uncertainties of the initial conditions (uncertainties of the first kind), constituting, because of the chaotic nature of the system, probably the most critical issue in weather forecasting, are not in principle so troublesome—assuming that the system is ergodic—when considering the long-term behavior, where "long" is evaluated with respect to the longest timescale of the system. Nevertheless, to avoid transient behaviors, which may induce spurious trends in large-scale climate variables on the multidecadal and centennial scales, it is crucial to efficiently initialize the slowest dynamical component of the GCMs, namely, the ocean. The validation of GCMs requires considering such uncertainties and devising strategies for limiting their influence when control run and, especially, climate change experiments are performed.

As for taking care of possible issues related to initial conditions, often an ensemble of simulations, where the same climate model is run under identical conditions from a slightly different initial state, allows a more detailed exploration of the phase space of the system, with a better sampling—on a finite time—of the attractor of the model. Some climate models have recently shown a rather encouraging ability to act as weather forecasting models, thus featuring encouraging local, in-phase space properties. Although such an ability gives evidence that short timescales' physical processes are well-represented, it says little on the overall performances when statistical properties are considered.

The structural deficiencies of a single GCM and the stability of its statistical properties can be addressed, at least empirically, by applying Monte Carlo techniques to generate an ensemble of simulations, each characterized by different values of some key uncertain parameters characterizing global climatic properties, such as climate sensitivity. Therefore, in this case, sampling is performed by considering attractors that are parametrically deformed.

To describe synthetically and comprehensively the outputs of a growing number of GCMs, recently it has become common to consider multimodel ensembles and to focus the attention of the ensemble mean and the ensemble spread of the models, taken respectively as the (possibly weighted) first two moments of the models outputs for the considered metric. Then, information from rather different attractors is merged. Although this procedure has advantages, especially for GCMs intercomparison, such statistical estimators should not be interpreted in the standard way—with the mean approximating the truth and the standard deviation describing the uncertainty—because such a straightforward perspective relies on the (false) assumptions that the set is a probabilistic ensemble, formed by equivalent realizations of given process, and that the underlying probability distribution is unimodal.

Valerio Lucarini
University of Bologna

See Also: Abrupt Climate Changes; Atmospheric Component of Models; Atmospheric General Circulation Models; Biogeochemical Feedbacks; Chaos Theory; Climate; Climate Models; Climate Sensitivity and Feedbacks; Climatic Data, Historical Records; Climatic Data, Proxy Records; Climatic

Data, Reanalysis; Intergovernmental Panel on Climate Change; Modeling of Paleoclimates; Ocean Component of Models.

Further Readings

Held, Isaac M. "The Gap Between Simulation and Understanding in Climate Modeling." *Bulletin of the American Meteorological Society*, v.86/11 (2005).

Lucarini, Valerio. "Towards a Definition of Climate Science." *International Journal of Environment and Pollution*, v.18/5, 2002).

Lucarini, Valerio, Sandro Calmanti, Alessandro Dell'Aquila, Paolo M. Ruti, and Antonio Speranza. "Intercomparison of the Northern Hemisphere Winter Mid-Latitude Atmospheric Variability of the IPCC Models." *Climate Dynamics*, v.28/7–8 (2007).

Peixoto, Josè P. and Abraham H. Oort. *Physics of Climate*. New York: American Institute of Physics, 1992.

Saltzmann, Barry. *Dynamic Paleoclimatology.* Maryland Heights, MO: Academic Press, 2002.

Solomon, S., et al., eds. *Climate Change 2007: The Physical Science Basis. Contribution of Working Group I to the Fourth Assessment Report of the Intergovernmental Panel on Climate Change.* Cambridge: Cambridge University Press, 2007.

Vanuatu

An archipelago of 83 islands lying between New Caledonia and Fiji in the South Pacific, Vanuatu was previously known as New Hebrides before its independence in 1980. One of the world's smallest countries, Vanuatu is also among those countries most vulnerable to climate change. The country is already experiencing signs of climate change. Subsistence crops have been affected by severe drought, and rising sea levels continue to erode the shoreline. Water supplies are threatened, a particular problem for a country where potable water is already a pressing concern. Increasing global warming will intensify the country's high vulnerability to natural disasters such as cyclones, floods, earthquakes, landslides, tsunamis, and volcanic eruptions. Vanuatu has been classified by the United Nations as a least developed country since 1995. The country is facing additional stresses on land use and water sources from a growing population because its economic growth sectors are highly sensitive to climate change, and its ability to recover from those stressors is low.

Prime Concerns: Water and Agriculture

Most of Vanuatu's population lacks easy access to potable water. Its urban centers and outer islands are dependent on groundwater. Warmer temperatures are likely to increase the demand for water, even as decreased rainfall and increased evaporation reduce the rate of groundwater recharge. Contamination of drinking water from sedimentation and saltwater intrusion in shallow ground aquifers could pose an additional threat to the availability of potable water.

Water is also of prime concern to the agricultural sector. Approximately 80 percent of Vanuatu's population is engaged in subsistence or small-scale farming of coconuts or other cash crops, and agriculture accounts for about 20 percent of the country's gross domestic product. Any changes in rainfall, in terms of amounts and distribution, could have severe effects on agricultural production. Drought would mean less water for agricultural purposes, but more frequent and intense rainfalls during planting seasons—another effect of global warming—could damage seedlings, hinder growth, and foster plant pests and diseases. Crop production has decreased as a result of higher temperatures, more frequent dry seasons, and greater variation in rainfall. Pest infestation has become a greater problem. However, it is the likelihood of the increased frequency and intensity of cyclones that farmers find most troubling, since high winds and heavy rainfall can lead to widespread and acute crop damage. Subsistence farming is particularly vulnerable to these changes, and its failure would compromise the nation's food security.

Coastal erosion is another primary concern for Vanuatu. A substantial portion of the country's 80 percent rural population is concentrated on the coast. Port Vila, the economic and commercial center of Vanuatu; and Luganville, the nation's second-largest city, are located on the perimeter of the major islands, only a few meters above sea level. Much of the country's road network is also located on the perimeter. Even small increases in sea level and storm surges can pose serious threats to Vanuatu's infrastructure. Coastal erosion and inundation have proven severe in some areas, posing danger to human life and coastal settlements and foreshadowing larger threats if climate changes remain unmitigated. One Vanuatu village has already experienced the costs of global warming. In 2005, the 100 inhabitants of Lateu became one of the first communities forced to move as a result of global warming. The high coral reef that had been the villagers' defense against high tides and waves no longer protected them, and the coastline was eroding from 7 to 10 ft. (2 to 3 m) per year. When their homes were repeatedly swamped by surges and large waves linked to climate change–driven storms, the villagers, with the help of the international community, were relocated to higher ground in the interior of Tegua, one of Vanuatu's northern provinces.

The costs, present and potential, exacted from Vanuatu are disproportionate to their contribution to global warming. The country's carbon dioxide emissions per capita in 2007 (the most recent year for which numbers are available) were less than 0.5 ton (0.5 metric ton.) per capita. U.S. emissions for the same period were 19.7 tons per capita. Despite its scant contribution to global emissions and its vulnerabilities to the impacts of climate change, Vanuatu has set a goal of generating 33 percent of its electricity from renewable sources by 2013. Renewable energy is not a new idea in the country. As early as 1985, a study by the World Bank and others considered the use of hydroelectric and geothermal energy. Current plans include a geothermal plant near the hot pools of Takara Village scheduled to open in 2014.

Wylene Rholetter
Auburn University

See Also: Agriculture; Alliance of Small Island States; Pollution, Water; Sea Level, Rising.

Further Readings

National Advisory Committee on Climate Change. "Republic of Vanuatu National Adaptation Programme for Action." http://unfccc.int/resource/docs/napa/vut01.pdf (Accessed July 2011).

U.S. Department of State. "Background Note: Vanuatu." http://www.state.gov/r/pa/ei/bgn/2815.htm (Accessed July 2011).

Venezuela

The Bolivarian Republic of Venezuela is a major oil-producing state and a founding member of the Organization of Petroleum Exporting Countries (OPEC). Venezuela is the fifth-largest member of OPEC in terms of annual oil production. According to the International Energy Agency, the country had a production level of 2.2 million barrels per day in June 2010. Oil generates about 80 percent of the nation's export revenues. The state-owned Petróleos de Venezuela (PDV) is one of the world's largest oil companies. PDV's subsidiary, Citgo, refines crude oil in Texas for U.S. markets. In addition to sales to the United States, Chinese oil purchases and other trade with Venezuela is growing. Venezuela's large reserves of natural gas remain largely untapped.

Although contributing to global warming, the country's greenhouse gas (GHG) emissions have been estimated at less than 1 percent of annual global levels. The president of Venezuela, Hugo Chavez, is a strong critic of the limited action in wealthy countries to address climate change, which he identifies as primarily linked to capitalist economies. Chavez has repeatedly stated at international forums that industrialized countries must acknowledge their historic responsibilities as major emitters of GHGs. Simultaneously, Venezuela provides low-cost oil to many Latin American and Caribbean countries under an agreement called PetroCaribe. Venezuela guarantees favorably priced petroleum for PetroCaribe partners utilizing subsidies, exchanges for goods and services, and interest-deferred financing for oil purchases. PetroCaribe may stall a transition to alternative energy sources among member countries.

The Venezuelan population has regular access to subsidized gasoline at prices considerably lower than the global market value, which may encourage inefficiency in fuel use and is an expensive policy to maintain. At the same time, the Venezuelan government has signed numerous international environmental accords and uses oil revenue to finance domestic social and environmental programs. Venezuela is promoting energy efficiency in urban areas and renewable energy initiatives in rural zones. Over 100 million incandescent light bulbs have been replaced with fluorescent alternatives. More than 800 photovoltaic systems

Angel Falls in Venezuela is the world's highest free-falling, freshwater waterfall, with a drop of 2,648 ft. The water eventually feeds into the mouth of the Caroni River, 77 km downstream, where the Guri hydroelectric dam is located.

provide renewable energy to schools, clinics, cafeterias, and security stations in isolated, frontier, and indigenous areas. The Tree Mission program, started in 2006, promotes agroforestry and productive sustainable alternatives for rural populations. Over 18,000 hectares have been reforested with trees intercropped with agricultural products such as coffee.

Venezuela has nearly 48 million hectares of forests, representing 5.55 percent of the forests in Latin America and the Caribbean and 1.25 percent of the world's forests. A system of protected areas covers 70 percent of the national territory. Venezuela is among the top 20 countries in terms of endemism and hosts more than 20,000 different plant species. Its diverse climatic and biogeographical regions cover a range of elevations. There are 1,740 mi. (2,800 km) of coastline, including vast mangrove swamps and numerous islands. There has been a documented retreat of glaciers in the Sierra Nevada range. The glacier on Pico Bolivar may completely disappear during the next decade.

Because of insufficient research, there are a number of uncertainties about potential impacts of climate change in Venezuela. Some pressing concerns include rising sea level on the islands and mainland coast, and the risk of increased mortality from diseases with mosquito vectors such as malaria, dengue, and yellow fever. The consequences of climate change on reservoirs, especially Camatagua and Guri, and declining rates of agricultural production as a result of both droughts and floods, are alarming. In December 1999, Venezuela experienced its highest monthly rainfall in a century and subsequent landslides and flooding led to the deaths of 30,000 people. The majority of the country's electricity comes from the Guri hydroelectric dam, one of the largest dams in the world, but a 2010 drought brought blackouts and power rationing.

Plans for several wind projects are currently being developed. The wind projects have been cited as demonstrating Venezuela's recognition of the need to begin to transition to a post-oil era, while also acknowledging the challenges that climate change may create for the production of hydroelectric power. Venezuela does not host any Clean Development Mechanism projects. The country ratified the United Nations Framework Convention on Climate Change in 1994 and the Kyoto Protocol in 2005.

Mary Finley-Brook
Mary Brickle
University of Richmond

See Also: Floods; Forests; Oil, Consumption of; Oil, Production of.

Further Readings

Chavez, Hugo. "Venezuelan President's Speech on Climate Change in Copenhagen." http://www.venezuelanalysis.com/analysis/5013 (Accessed June 2011).

Duran, Alexis and Lelys Guenni. "Probabilistic Estimation of Climate Change in Venezuela Using

a Bayesian Approach." *Revista Colombiana de Estadística*, v. 33 (July/December 2010).
Petroleum of Venezuela. http://www.pdvsa.com (Accessed June 2011).

Vermont

Known as the Green Mountain State, Vermont underwent widespread reforestation following farm abandonment in the mid-1800s. In addition to the existence of several land trusts, the Green Mountain Club has protected more than 55 mi. (88 km) along the hikers' Long Trail. There is a strong environmental movement in the state with numerous local groups and chapters of national organizations. Vermont hosts one of the nation's leading environmental law and policy programs at the Vermont Law School. The University of Vermont and Middlebury College are leaders among institutions of higher education with campus sustainability initiatives.

Average temperatures across the northeast of the United States have risen more than 1.5 degrees F (0.83 degree C) since 1970. The average length of the annual growing season has been extended by eight days and bloom dates have changed. Winters are warming most rapidly, with a 4 degree F (2.22 degrees C) increase between 1970 and 2000. Scientists have documented quicker thawing of lakes, known as ice-out dates, as well as earlier runoff from mountains.

Warming trends concern Vermonters because of the postential loss of economic revenue from nature tourism during autumn foliage and the winter skiing season. Data have suggested that maple production from the sugar maple (*Acer saccarum*), the state tree, may be vulnerable to climate shifts. Under several climate change models, this species would entirely shift out of the United States. Studies suggest that seven of 80 eastern tree species may decline by as much as 90 percent in the next century if temperature changes continue at current rates.

In May 2011, the governor of Vermont, Peter Shumlin, created a climate change cabinet to coordinate efforts across state agencies, provide information, reduce state dependency on fossil fuels, develop alternative energy sources, and increase energy efficiency. In terms of greenhouse gas (GHG) emissions, Vermont's energy portfolio is one of the greenest in the nation. Nearly one-third of Vermont's energy is created from a mixture of in-state renewables, ranging from methane capture from manure—which provides supplemental income to dairy farmers—to wind, solar, and biomass. Net metering and tax incentives make it cost-effective for Vermonters to generate some of their electricity using small-scale renewable energy systems.

Energy from Hydro-Quebec, an extensive series of dams across Quebec Province in Canada, provides a third of Vermont's energy needs. While hydropower is a renewable energy source, the creation of the dams has been controversial due to the extensive flooding of indigenous lands. In 2010, former Vermont Governor Jim Douglas and Quebec Premier Jean Charest signed a memorandum of understanding to extend Vermont's purchase of electricity from Hydro-Quebec from 2012 to 2038.

The Vermont Yankee nuclear power plant, online since 1972, provides approximately 35 percent of Vermont's energy requirements. Its license is scheduled to expire in 2012. In 2010, the Vermont Senate voted 26–4 not to renew the plant's permit, citing radioactive leaks, misstatements in testimony by plant officials, and other problems. In March 2011, the Nuclear Regulatory Commission granted Vermont Yankee permission to stay open until 2032, a 20-year extension from its initial closure date. The Louisiana-based company Entergy, which bought the plant in 2002, filed a federal lawsuit in April 2011 to try to change the state law that gives the Vermont legislature power to decide the end date of operations. Meanwhile, less than half of the estimated $1 billion needed to decommission the plant has been assigned.

State energy efficiency programs, including weatherization grants and low-interest loans, assist homeowners and small businesses to reduce energy use in buildings. Efficiency Vermont, the nation's first statewide provider of energy-efficiency services, is operated by an independent, nonprofit organization. The program, funded by a charge on users' electric bills, provides technical advice, financial assistance, and design guidance to make homes, schools, and businesses more energy efficient.

Vermont has joined regional climate change mitigation initiatives. In 2001, Vermont joined a Climate Change Action Plan reaching across the international border. The New England governors and the eastern Canadian premiers have committed to a long-term GHG emissions reduction target of 75 to 85 percent below 2001 levels by 2050.

This goal represents a more significant decrease in emissions than the Kyoto Protocol. Vermont is also part of the Regional Greenhouse Gas Initiative (RGGI), the first market-based regulatory program in the United States to reduce GHG emissions. Ten northeastern and mid-Atlantic states have capped and will reduce carbon dioxide emissions from the power sector 10 percent by 2018. RGGI members' emissions credits are regularly auctioned.

Mary Finley-Brook
Mary Brickle
University of Richmond

See Also: Energy Efficiency; Greenhouse Gas Emissions; Renewable Energy, Overview.

Further Readings

Union of Concerned Scientists. "Vermont." In *Confronting Climate Change in the U.S. Northeast: Science, Impacts, and Solutions.* Cambridge, MA: NECIA/UCS, 2007.

Vermont Agency of Natural Resources. "Climate Change Team." http://www.anr.state.vt.us/anr/climatechange (Accessed June 2011).

Wald, Matthew L. "Vermont Senate Votes to Close Nuclear Plant." *New York Times* (February 24, 2010).

Vienna Convention

The Vienna Convention for the Protection of the Ozone Layer was drawn up in March 1985. The convention was initially developed out of international concern over the loss of ozone in the stratosphere. Subsequently, the Montreal Protocol in 1987 strengthened the convention, which was further adjusted and amended on June 29, 1990.

The concern over ozone was linked to the phenomena of climate change. Eventually, the issue of the ozone stabilized, and the issue of climate change evolved into a larger international concern.

From Ozone to Broader Concerns

The attention to the ozone issue relied on scientific understanding of the effects of chlorofluorocarbons in the context of the atmosphere. Scientific research invested in developing increasingly complex models of the atmosphere to more accurately address climate issues. Over time, concern over the issue of the loss of ozone in the stratosphere declined, as international efforts appeared to have improved the condition of the stratospheric ozone layer. Ozone levels in the troposphere continue to be a contributor to local pollution.

As the concern over stratospheric ozone declined, the broader concern over climate change that was noted in the convention continued and assumed a proportionately larger role. The improved science used to address the ozone issues then also became valuable in addressing the more complex climate change matter.

The development of the convention and climate change regulations through the United Nations Environment Programme have been closely associated with the science and models of the atmosphere. The science was addressed in related conferences and workshops, including the Villach Conference in 1985. Eventually, the convention led to the formulation of the UN Framework Convention on Climate Change.

The preamble to the convention indicates the generally accepted views of the international community. This preamble mentions the relationship between the environment, its modification, and the potentially harmful effects on human health.

The awareness noted in the preamble is an acknowledgement that human activity can and does modify the environment. It also permits the recognition of the association between modifying the environment and its effect on human health. This awareness then establishes the possibility of regulating human activity.

Human activity is also associated with relevant scientific and technical considerations. There is a recognized need for research and systematic observations of the phenomena. The use of science and technology is an important part of the convention and its implementation.

The intended result of the convention is twofold. First, the twin concerns are to protect human health and the environment. The particular focus is the concern about adverse effects to both humans and the environment, which result specifically from modifications of the stratospheric ozone layer.

The convention is not concerned with all modifications of the ozone layer, only those that have adverse effects on human health and the environment. In fact, any modification of the ozone layer that has a positive effect on human health and the environment would be favorable. Under the definitions in Article 1, "adverse effects" refer to changes in the physical environment or biota, including changes in climate. The relevant changes are those that have significant deleterious effects on human health, on ecosystems whether natural or managed, and on "materials useful to mankind."

To achieve the intended result, the preamble raises some significant considerations. The reference to principle 21 of the Declaration of the UN Conference on the Human Environment highlights the context for the convention, which includes the sovereignty of each country, also known as a state, over its own domain and resources. It also underlines the responsibility of each state not to cause damage beyond its national jurisdiction or to other states.

In such circumstances, a state's performance can then be evaluated in the international context against its sovereign jurisdiction and its effect beyond its boundaries. The sovereignty of each state also necessarily involves the recognition of cooperation between states. It is through cooperation and action that states can be brought within the scope of international law and the charter of the United Nations.

The preamble discloses its relatively recent formulation when it includes nonsovereign entities in its discussions. It generally acknowledges international and national organizations; their work and studies are pertinent to the scope of the convention. An example of such an organization noted in the convention is the World Plan of Action on the Ozone Layer of the UN Environment Programme.

Another significant feature of the convention is the attention to developing countries in the global effort. The preamble generally recognizes the circumstances and particular requirements of developing countries. This recognition eventually leads to the more complex form of differentiated contributions to total human activities and their consequences.

The agency responsible for the convention is the Ozone Secretariat of the UN Environment Programme. In 1988, the assessment panel process was initiated under the Montreal Protocol, Article 6, and four panels were established. These are the panels for scientific, environmental, technology, and economic assessments. In 1990, the panels for technical assessment and for economic assessment were merged into the Technology and Economic Assessment Panel.

Conference of the Parties

Under Article 6 of the convention, a Conference of the Parties was established. The Conference of the Parties to the Convention, in its decision VCV/2 of its fifth meeting, acknowledges and encourages the collaboration of the three assessment panels with other entities. These entities link the convention with other international entities involved in addressing climate change issues. The other entities named in this decision are the Intergovernmental Panel on Climate Change, the Subsidiary Body on Scientific, Technical and Technological Advice under the UN Framework Convention on Climate Change, the International Civil Aviation Organization, and the World Meteorological Organization.

These entities were in addition to those recognized in Article 6, which included the World Health Organization and International Atomic Energy Agency. Within limits, it also allowed for any state not a party to this convention and for any body or agency, whether national or international, governmental, or nongovernmental, to be represented at the Conference of the Parties by observers. The convention is specifically referred to in the preamble of the UN Framework Convention on Climate Change, signed in 1992, which came into force in 1994.

Lester de Souza
Independent Scholar

See Also: Atmospheric Composition; Climate Change, Effects of.

Further Readings

Bryk, Dale S. "The Montreal Protocol and Recent Developments to Protect the Ozone Layer." *Harvard Environmental Law Review*, v.15/1 (1991).

Lang, W., H. Neuhold, and K. Zemanek, eds. *Environmental Protection and International Law.* London: Graham & Trotman/Nijhoff, 1991.

Paterson, M. *Global Warming and Global Politics.* London: Routledge, 1996.

Social Learning Group, Massachusetts Institute of Technology. *Learning to Manage Global Environmental Risks, Volume 1: A Comparative History of Social Responses to Climate Change, Ozone Depletion, and Acid Rain.* Cambridge, MA: MIT Press, 2001.

Weiss, Edith Brown. "International Law." In *Encyclopedia of World Environmental History*, edited by Carolyn Merchant and Shepard Krech. New York: Routledge, 2003.

Vietnam

Vietnam is a coastal country on the South China Sea in southeast Asia, on the Indochina Peninsula. The 13th most populous country in the world, it has a population of about 90 million people. Ho Chi Minh City, the capital, with 3.5 million people, is the 49th-largest city in the world. Once controlled by China and later colonized by the French, the current state of Vietnam dates to the North Vietnamese victory over South Vietnam in 1975. Diplomatic relations with most of the world were reestablished by 2000, after a series of political and economic reforms, and Vietnam is part of the World Trade Organization. It has been enjoying considerable economic growth since 2000.

Hydropower

Vietnam has invested considerable resources into hydropower, taking advantage of the three major rivers in the country: the Red, Mekong, and Pearl rivers. Hydropower accounts for about 60 percent of the country's electricity generation. The remainder comes from fossil fuels. Electricity generation is responsible for about 25 percent of carbon dioxide emissions; 38 percent comes from transportation, and 30 percent from manufacturing and construction. Automobile usage is common, though there is public transportation in the cities. Emissions rose from 0.3 metric ton per capita in 1990 to 1.29 metric tons in 2007. Most of that rise was due to the boom in tourism beginning in the 1990s, leading to increases in construction, air conditioning of hotels and businesses, electronics usage, aviation fuel consumption, and taxi usage. About 80 percent of the country's emissions come from fuel; manufacturing is a minor concern.

The northern part of the country is predominantly highlands and the Red River delta. The south consists of coastal lowlands and forest. Forty percent of the country is mountainous. The Red River delta is more highly developed and densely populated than the Mekong delta. Monsoon winds blow from the Chinese coast from November to April, keeping the winter moist. In the south, the climate is primarily tropical all year long, with lows in the 60s F in the winter and prevailing humidity throughout the year. The northern mountains and plateaus get much colder, close to freezing in the winter, but also experience hotter summers. Vietnam is the 16th most biodiverse country in the world, and 16 percent of the world's species are represented in the country.

Vietnam is a signatory of the United Nations Framework Convention on Climate Change, and ratified the Kyoto Protocol in 2002; the Kyoto Protocol entered into force in 2005.

Bill Kte'pi
Independent scholar

See Also: Cambodia; Korea, North; Korea, South; Laos; Renewable Energy, Overview; Species Extinction; Thailand.

Further Readings

Sterling, Eleanor Jane, Martha Maud Hurley, and Le Duc Minh. *Vietnam: A Natural History.* New Haven: Yale University Press, 2007.

Templer, Robert. *Shadows and Wind: A View of Modern Vietnam.* New York: Penguin, 1999.

Villach Conference

The conference held in Villach, Austria, from October 9 to 15, 1985, was the result of the continuing work of several international entities. The background to the conference and the starting point for internationally cohesive attempts to understand the issues related to the stratospheric ozone layer depletion and climate change have been traced to the United Nations (UN) Conference on Human Development in Stockholm in 1972. The technical, scientific understanding of the possibility of human-induced effects on the ozone layer and climate change developed in the diplomatic context as complementary efforts linked through a reliance on common research initiatives.

A World Climate Conference held in Geneva in 1979 continued the efforts of the 1972 UN Conference on Human Development and led to the World Climate Programme. The World Meteorological Organization, UN Environment Programme, and International Council of Scientific Unions collaborated to hold a series of workshops that have come to be known as the Villach Conference.

The UN Environment Programme Ad Hoc Working Group of Legal and Technical Experts for the Elaboration of a Global Framework Convention for the Protection of the Ozone Layer was established by the UN Environment Programme Governing Council decision 12/14, Part I. The first part of the fourth session of the working group was held at the Palais des Nations, Geneva, from October 22 to 26, 1984.

The Working Group was also informed of the collaborative effort of the UN Environment Programme, World Meteorological Organization, and International Council of Scientific Unions to hold a major scientific conference to assess the carbon dioxide/ozone and climate question in October 1985 at Villach, Austria, with the support of the government of Austria.

Intersection of Ozone and Climate Change

Located at the intersection of the ozone and climate change issues, the Villach Conference became immediately significant to the Working Groups of both UN endeavors and to the overall development of international initiatives to address atmospheric environmental issues. A workshop on chlorofluorocarbons at Villach initiated the processes that led to a protocol to the Vienna Convention, which had been signed earlier that year, in March 1985. The same conference augmented the UN Environment Programme's role in determining the assessment of the greenhouse gas/climate issue. In this regard, the October 1985 Villach Conference was significant.

The participants at the Villach Conference were a small group of environmental scientists and research managers in nongovernmental organizations. The dominant contribution of these participants was their expertise in climate modeling. With accreditation from recognized scientific institutions such as the International Institute for Applied Systems Analysis and Harvard University, the results of this series of workshops carried considerable weight internationally. Even before the conference, a majority of the scientists who attended had publicly advocated what they considered an imperative to respond to a perceived threat to planetary climate stability within a strategy that was consistent with sustainable development, which was eventually incorporated in the Brundtland Report. Through affiliations with nongovernmental institutions, the scientists had improved modeling techniques that they relied on, and that led them to generally agree with the conclusions and recommendations of the conference.

Results of the Conference

At the Villach Conference, the scientific community in attendance arrived at an initial consensus as to the technical features of greenhouse gases, the depletion of the stratospheric ozone layer, and the chemical reactions that were relevant.

The general conclusion of the scientists and participants at the conference was that they could anticipate an unprecedented rise of global mean temperature in the first half of the 21st century. The scale and actual increase in global mean temperature was expected to be higher than any rise in the record of the planet's history. To mitigate the perceived events, the participants recommended a strategy that relied on technical and science-based research to establish target emission or concentration limits. In doing so, they sought to regulate the rate of change of global mean temperature within specific parameters.

Another result of the work by the participants at Villach was the establishment of the Advisory Group on Greenhouse Gases in 1986. This group was established to ensure continued academic and public interest in the effect of rising levels of greenhouse gases on the ozone layer and on climate change. The Advisory Group on Greenhouse Gases was jointly sponsored by the World Meteorological Organization, UN Environment Programme, and International Council of Science Unions.

The Brundtland Report, published in 1987, popularized sustainable development and advocated the development of a low-energy economy. This publication included a section on energy authored by Professor Gordon Goodman, who was by then a prominent member of the Advisory Group on Greenhouse Gases.

Work by the Advisory Group on Greenhouse Gases subsequently led to the 1988 Conference on the Changing Atmosphere: Implications for Global Security in Toronto, Canada, which called for 20 percent reductions in carbon dioxide (CO_2) emissions. The Advisory Group followed that up by preparing the Meeting of Legal and Policy Experts in February 1989 in Ottawa, which recommended an umbrella consortium to protect the atmosphere. The 1988 Toronto Conference then led to the establishment of the Intergovernmental Panel on Climate Change, with the mandate for continuing international research on climate change phenomena.

Lester de Souza
Independent Scholar

See Also: Atmospheric Composition; Climate Change, Effects of.

Further Readings

Abrahamson, Dean Edwin, ed. *The Challenge of Global Warming*. Washington, DC: Island Press, 1989.

Paterson, Matthew. *Global Warming and Global Politics*. London: Routledge, 1996.

Skodvin, T. *Structure and Agent in the Scientific Diplomacy of Climate Change: An Empirical Case Study of Science-Policy Interaction in the Intergovernmental Panel on Climate Change*. Boston, MA: Kluwer, 2000.

Virginia

The Commonwealth of Virginia is a southern Atlantic state on the eastern seaboard of the United States. Virginia is the 35th-largest state in the country, with an area of over 42,000 sq. mi. and an estimated population of about 7.9 million. Cities in the commonwealth are considered independent and function in the same manner as counties. Virginia was founded on May 13, 1607, at Jamestown, the first permanent English settlement in North America, and is one of the original 13 colonies of the American Revolution. Virginia joined the Union on June 25, 1788, and was the 10th state to ratify the U.S. Constitution. More U.S. presidents (eight) have come from Virginia than from any other state. Republican Governor Bob McDonnell assumed office in January 2010.

Virginia's geography varies considerably across the state. Elevations in Virginia range from sea level to nearly 6,000 ft. (1,828 m) at the summit of Mt. Rogers. The state is divided into six distinct geographical regions: the ridge and valley, between the Blue Ridge Mountains to the east and the Appalachian and Allegheny plateaus to the west; the Shenandoah Valley, located within the ridge and valley; the Blue Ridge Mountains between the ridge and valley to the west and the

Virginia has five distinct climate regions due to significant topographic differences and the influence of the Atlantic Ocean. The regions east of the Blue Ridge and the southern area of the Shenandoah Valley are considered a humid subtropical climate.

Piedmont region to the east; the foothills, between the Piedmont and the Blue Ridge mountains; the Piedmont between the Blue Ridge Mountains to the west and the Tidewater region to the east; and the Tidewater region between the fall line to the west and the Atlantic coast, including the eastern Shore.

Virginia's climate is considered mild for the United States, but can be quite variable due to significant topographic differences and the influence of the Atlantic Ocean. The regions east of the Blue Ridge, as well as the southern part of the Shenandoah Valley, are considered a humid subtropical climate (Koppen climate classification *Cfa*). In the mountainous areas between the Allegheny and Blue Ridge mountains in the west, the climate becomes humid continental (Koppen climate classification *Dfa*).

Variability in Climate

Virginia has five distinct climate regions: the Tidewater, Piedmont, northern Virginia, western mountains, and southwestern mountain regions. Climate varies significantly across the state due to differences in topography, the influence of the Atlantic Ocean and the Gulf Stream in the east, and the complex pattern of rivers and streams. Much precipitation that Virginia receives results from storms associated with low-pressure (cyclone) systems—warm and cold fronts that typically move from west to east across the state, curving northwestward as they reach the Atlantic Ocean. Virginia also experiences occasional tropical storm activity, most often in early August and September, and typically through the venue of the mouth of the Chesapeake Bay. Tropical storms can provide significant amounts of precipitation. In recent years, development trends in northern Virginia extending out from Washington, D.C., have created an urban heat island effect characteristic of other large urban areas.

In Virginia, emissions of carbon dioxide, one of the primary greenhouse gases that alter climate, have risen by over 30 percent since 1990, driven by economic growth, development patterns, and increased transportation trends. Human-induced climate change is expected to have many effects on Virginia's weather, wildlife, food production, and water supplies. Sea level in the Mid-Atlantic region is estimated to rise several inches in the coming decades, threatening low-lying areas and coastal developments. Changes in precipitation and temperature regimes have the potential to disrupt agriculture and forestry. Hurricanes make the coastal area of Virginia vulnerable. Potential increased tropical storm activity that might result from climate change would carry the threat of fiscal and ecological damage to Virginia's communities.

Virginia has begun to address the issue of climate change in several ways. It is one of several states to have completed (in 2007) a comprehensive Climate Action Plan, manifested as the Virginia Energy Plan. The process of developing a climate action plan identifies cost-effective opportunities by which a state may reduce emissions of the greenhouse gases (GHGs) that alter climate. In May 2007, Governor Timothy Kaine announced that Virginia had joined the Climate Registry, a state-sponsored initiative to standardize methods to record and measure GHG emissions. Virginia adopted a voluntary renewable portfolio standard and set a goal of reducing GHG emissions by 30 percent by 2025 through energy conservation and renewable energy actions. In December 2007, Governor Kaine established the Governor's Commission on Climate Change to prepare a plan for Virginia that identifies ways to reduce GHG emissions. The commission examined actions in four arenas: adaptation and sequestration, built environment, electric generation and other stationary sources, and transportation and land use. The commission released a report on December 15, 2008, listing a set of recommendations for the state; this report is publicly available on the Virginia Department of Environmental Quality's Website.

Karin Warren
Randolph College

See Also: Atlantic Ocean; Climate; Gulf Stream; United States; Washington.

Further Readings

Commonwealth of Virginia. http://www.va.us (Accessed June 2011).

Commonwealth of Virginia Department of Mines, Minerals, and Energy. *The Virginia Energy Plan*. Richmond: Commonwealth of Virginia, 2007.

Hayden, Bruce P. and Patrick J. Michaels. *Virginia's Climate*. Charlottesville: University of Virginia Climatology Office, 2007.
Virginia Department of Environmental Quality. http://www.deq.virginia.gov (Accessed June 2011).

Volcanism

Members of the scientific community largely concur that the Earth is undergoing a change in climate and that global warming is occurring at an increasing rate. This acceleration in the late 20th century is mainly a result of carbon dioxide (CO_2) emissions generated by human activity. Carbon dioxide acts like a glass barrier over the Earth, preventing heat from leaking into the environment and thus creating a greenhouse effect. In its latest report, the United Nations Intergovernmental Panel on Climate Change (IPCC) shows that greenhouse gases are an integral factor in global warming, more so than natural causes such as solar activity and volcanoes.

Influence on Weather

Evidence suggests that volcanism, which refers to phenomena such as the outward flow of pyroclastic materials and the upsurge of gas and steam connected to the movement of molten rock, can influence short-term weather and may have an effect on long-term climate change. Similar to human activity, volcanism leads to both global warming and cooling. The effect of volcanism on climate depends on the interaction between the sun's heat and the volcanic debris. Scientists believe that ongoing volcanic eruptions have maintained the Earth's temperate climate for millions of years and are responsible for the gases in today's atmosphere. Volcanoes that erupt explosively can send particles many miles away from the volcano and high into the stratosphere where the Earth's ozone is concentrated. In addition, the debris can be dispersed for months or even years.

Volcanic dust blown into the atmosphere reduces the sunlight that reaches the Earth's surface and cause temporary cooling; the degree of cooling is dependent on the volume of dust and the duration is dependent on the size of the dust particles. Additionally, the strength of gases can vary greatly among volcanoes. Water vapor is typically the most abundant volcanic gas, followed by CO_2 and sulfur dioxide. Other principal volcanic gasses include hydrogen sulfide, hydrogen chloride, and fluorine.

Volcanoes that discharge great quantities of sulfur compounds affect the climate more significantly than those that release only dust. In fact, the greatest volcanic effect on the Earth's short-term weather patterns is caused by sulfur dioxide gas. When sulfur dioxide and other volcanic gases mix with oxygen and water vapor in the presence of sunlight, the result is vog, or volcanic smog. Vog poses a health hazard by exacerbating respiratory conditions. Sulfur dioxide in the atmosphere is often transformed into sulfur trioxide which, when combined with water, forms sulfuric acid that reflect the sun's heat and triggers cooling of the Earth's surface. Acid rain contains greater than normal amounts of sulfuric and nitric acids, and when deposited on the Earth's surface, becomes a critical environmental problem that can affect lakes, streams, forests, and the inhabitants of these ecosystems.

Volcanoes also discharge water and CO_2 in large quantities in the form of atmospheric gases; in the atmosphere, these gases can absorb and retain heat radiation emanating from the ground. Estimates suggest that water makes up to 99 percent of gas in volcanic expulsions. This short-term warming of the air causes water to become rain within a matter of hours or days, and it causes the CO_2 to dissolve in the ocean or to be absorbed by plants. The majority of the heat energy connected to global warming exists in the ocean. If the oceanic depth at which heat is stored is decreased, then global temperature increases are expected to be greater than predicted.

Volcanic eruptions combined with human-made chlorofluorocarbons (CFCs) also can contribute to ozone depletion. CFCs were developed in the early 1930s; because they were nontoxic, nonflammable, and met a number of safety criteria, CFCs were used in industrial, commercial, and household applications such as refrigeration units and aerosol propellants. In February 1992, however, following evidence that CFCs contributed to depletion of the ozone layer, the U.S. government announced plans to phase out the production of CFCs by December 1995. Members of the Montreal Protocol in 1992 followed suit and agreed to an accelerated phaseout by the end of 1995.

The ozone layer, which rests in the stratosphere and begins at 7.5 mi. (12 km) above the Earth, is a shield that protects living beings from ultraviolet-B (UV-B), the sun's most harmful UV radiation. In high doses, UV-B can lead to cellular damage in plants and animals. Scientists believe that global warming will lead to a weakened ozone layer. As the Earth's surface temperature rises, the stratosphere will become colder, slowing the natural repair process of the ozone layer. Decreased ozone in the stratosphere results in lower temperatures. Unlike ozone depletion created by human-made CFCs, which will take decades to repair, scientific theories indicate that as volcanic activity diminishes, the damaged caused by the volcanoes ias gradually repaired as volcanic activity diminishes.

Yasur volcano on Vanuatu is one of the world's most active volcanoes, erupting dozens of times a day for about 800 years. Similar to human activity, volcanism can influence short-term weather and may have an effect on long-term climate change.

Finally, hydrogen fluoride gas can concentrate in rain or on ash particles, contaminating grass, streams, and lakes with excess fluorine. Excess fluorine in grass and water supplies can poison the animals that eat and drink at contaminated sites and eventually causes fluorosis, which destroys bones. In fact, excessive fluorine can lead to a major cause of injury and death in livestock during ash eruptions.

Today, millions of people live near active or potentially active volcanoes. The area around the Pacific Ocean in the Ring of Fire, known as the Cascade Volcanic Arc, accounts for about 75 percent of the world's volcanoes. In 1991, Mount Pinatubo in the Philippines emitted about 22 million tons of sulfur dioxide, which combined with water to form sulfuric acid, decreasing global temperatures for approximately a year. Additionally, in May 1980, Mount St. Helens, located in the state of Washington in the Cascade Volcanic Arc, released approximately 520 tons (472 million metric tons) of ash into the atmosphere. This volume of ash can have a short-term cooling effect hundreds or even thousands of miles away. Very cold temperatures leading to crop failures and famine in North America and Europe, followed the eruption of Mt. Tambora, Indonesia, in 1815.

One of Iceland's largest volcanoes, the Eyjafjallajökull, showed signs of activity in March 2010 after almost 200 years of dormancy. The eruption on April 16 disrupted air traffic in northern Europe and beyond as volcanic ash spread across northern and central Europe. Although it was marked as the worst travel disruption during peacetime, ash from the Eyjafjallajökull volcano reached 55,000 ft. (16,764 m), which was less than the almost 78,000 ft. (23,774 m) reached by ash from Mt. Pinatubo. Because volcanic ash generally leaves the environment within a few years or even months, Eyjafjallajkull was expected to have a short-term impact on climate. Scientists believe that the temporary cooling of the planet by a volcanic eruption does not offset CO_2 from the burning of fossil fuels.

Robin K. Dillow
Oakton Community College

See Also: Chlorofluorocarbons; Climatic Research Unit; Colombia; Global Warming Debate; Hawai'i; Iceland; Japan; Pollution, Air; Renewable Energy, Geothermal; Saint Vincent and the Grenadines; Solomon Islands; Tsunamis.

Further Readings

National Oceanic and Atmospheric Administration and National Geophysical Data Center. "Significant Volcanic Eruptions Database." http://www.ngdc.noaa.gov/nndc/servlet/ShowDatasets?dataset=102557&search_look=50&display_look=50 (Accessed July 2011).

"Nowhere to Turn for Climate Change Deniers." *New Scientist*, v.14/5 (April 2007).

Parfitt, Liz and Lionel Wilson. *Fundamentals of Physical Volcanology.* Malden, MA: Blackwell Science, 2008.

Schmincke, Hans-Ulrich. *Volcanism.* New York: Springer, 2005.

Union of Concerned Scientists. http://www.ucsusa.org (Accessed July 2011).

von Neumann, John

John von Neumann (1903–57) was a Hungarian-born American mathematician who contributed important theories to all the different branches of the discipline. His discoveries influenced quantum theory, automata theory, economics, and defense planning. Von Neumann was one of the founders of game theory and, along with Alan Turing and Claude Shannon, was one of the conceptual inventors of the stored-program digital computer. He is mostly remembered in the field of global warming for his elaboration of general circulation models and for his 1955 article in *Fortune*, in which he stated that "microscopic layers of colored matter spread on an icy surface, or in the atmosphere above one, could inhibit the reflection-radiation process, melt the ice, and change the local climate." This is one of the earliest conceptualizations of the problem of global warming.

John von Neumann was born János Neumann in Budapest, Hungary, on December 28, 1903, into a wealthy and completely assimilated Jewish family. His father was a banker, and his mother originally came from a family who made a fortune selling farm equipment. John was a child prodigy and was initially educated by private tutors in mathematics and foreign languages. In 1911, he entered Budapest's most prestigious school, the Lutheran Gymnasium. When Bela Kun established his revolutionary government, von Neumann's family fled Hungary and briefly emigrated to Austria. Kun's government failed after only five months. Because it was mainly composed of Jews, von Neumann's family, no matter how hostile to the regime, was blamed, together with many other Jews, for the brutality of the revolutionary government. In 1921, von Neumann completed his education at the gymnasium and his father strongly advised him to take up a career in business and not in mathematics, where von Neumann's talent was already obvious. His father was afraid that mathematics would not allow von Neumann to lead a wealthy and comfortable existence. As a compromise, it was decided that von Neumann would study both chemistry and mathematics. In spite of the strict quotas for Jewish entry to university, von Neumann succeeded in attending the University of Budapest for mathematics and the University of Berlin for chemistry. He later switched to the Swiss Federal Institute in Zürich, from where he graduated with a degree in chemical engineering in 1925. The following year, he obtained a doctorate in mathematics from the University of Budapest.

Von Neumann quickly gained a reputation in set theory, algebra, and quantum mechanics. In particular, his paper "An Axiomatization of Set Theory" (1925) was read and appreciated by the famous mathematician David Hilbert. Because of Hilbert's interest, von Neumann worked at the University of Göttingen in 1926 and 1927. Von Neumann was then employed as a *privatdozent* ("private lecturer") at the universities of Berlin (1927–29) and Hamburg (1929–30). At a time when Europe was characterized by political unrest and totalitarianism, he was invited to visit Princeton University in 1929 to lecture on quantum theory. He was then hired at Princeton as visiting professor the following year, although teaching was not one of his strongest assets, and when the Institute for Advanced Studies was founded there in 1933, he was appointed to be one of the original six professors of mathematics, a position he retained for the rest of his life. In 1930, von Neumann married Mariette Koevesi. They had one child, Marina, who later became an economist. The couple separated amicably in 1937, and the following year von Neumann married his childhood sweetheart Klara Dan, with whom he remained until his death. After his appointment at the Institute for Advanced Studies, von Neumann resigned from all his German academic positions. As a Jew, he could not work for a Nazi state. He prophetically stated that if Adolf Hitler remained in power, it would ruin German science for a long time.

During World War II, von Neumann worked for the Manhattan Project at the invitation of its director, Robert Oppenheimer. His expertise in the nonlinear physics of hydrodynamics and shock waves was instrumental in the design of the Fat Man atomic bomb dropped on the Japanese port of Nagasaki. Von Neumann argued against dropping the bomb in Tokyo on the Imperial Palace. In the postwar years, von Neumann continued to work as a consultant to government and industry. Starting in 1944, he contributed important insights for the development of the U.S. Army's ENIAC computer, initially designed by J. Presper Eckert, Jr. and John W. Mauchly. In a crucial contribution, von Neumann modified the ENIAC to run as a stored-program machine. von Neumann did not invent the computer, but he invented the software that made it run. The ENIAC machine was a combination of electronic hardware and punch-card software that allowed it to be employed for a variety of uses, including weather forecast. Von Neumann then campaigned to build an improved and faster computer at the Institute for Advanced Study. The institute's machine, which began operating in 1951, used binary arithmetic, where the ENIAC had used decimal numbers. Von Neumann's publications on computer design (1945–51) caused a clash with Eckert and Mauchly, who wanted to patent their contributions, and paved the way for the independent construction of similar machines around the world. Von Neumann's intuition for a single-processor, stored-program computer became the accepted standard.

As a consultant for RAND Corporation, von Neumann was given the task of planning a nuclear strategy for the U.S. Air Force. In this capacity, he was a strong advocate of nuclear weapons, and he became one of President Dwight Eisenhower's top advisers on his nuclear deterrence policy. Von Neumann was diagnosed with bone cancer in 1955. In spite of his deteriorating health, he continued to work and in 1956, he was honored with the Enrico Fermi Award. He converted to Roman Catholicism just before his death on February 8, 1957, in Washington, D.C.

Luca Prono
Independent Scholar

See Also: Climate Models; Education; Technology.

Further Readings

Aspray, William. "The Mathematical Reception of the Modern Computer: John von Neumann and the Institute for Advanced Study Computer." In *Studies in the History of Mathematics*. Vol. 26, edited by Esther R. Phillips. Washington, DC: Mathematical Association of America, 1987.

Bochner, Salomon. "John von Neumann." *National Academy of Sciences Biographical Memoirs*, v.32 (1958).

Dieudonné, J. "von Neumann, Johann (or John)." In *Dictionary of Scientific Biography*, edited by Charles C. Gillespie. New York: Charles Scribner's Sons, 1981.

Tropp, H. S. "John von Neumann." In *Encyclopedia of Computer Science and Engineering*, edited by Anthony Ralston and Edwin D. Reilly, Jr. New York: Van Nostrand Reinhold, 1983.

Weart, Spencer. *The Discovery of Global Warming*. Cambridge, MA: Harvard University Press, 2004.

Vostok Ice Core

Russia's Vostok station is located in east Antarctica. The Vostok station holds the record for the lowest temperature ever recorded at minus 129 degrees F (minus 89 degrees C). Soviet researchers began deep drilling at the Vostok Station in 1980. The ice cores brought to the surface in segments provide information (chemistry, structure, and inclusions) about climate conditions, similar to tree ring samples. The information from air bubbles allows for measurement of the atmospheric concentration of greenhouse gases (e.g., carbon dioxide, methane, nitrogen, helium, sodium, and organic carbon). Besides presenting an extraordinary human effort, spanning two decades in one of the most inhospitable places on Earth, the drilling at Vostok has produced one of the richest scientific treasure troves of all time. Previously, analysis revealed tracking between carbon dioxide and temperature, and that the magnitude of carbon dioxide swings could account for the magnitude of temperature swings.

Scientific Treasure Trove

The first hole drilling stopped in 1985 because of problems. A second hole drilled with French-Russian cooperation produced an ice core 2,083 m long, or 1.33 mi. With a climate record of 160,000 years, drilling on this hole ended in 1990. A third hole was drilled, with collaboration among Russia, France, and the United States. The drilling reached a depth of 2.25 mi. (3.6 km) and in January 1998 produced the deepest ice core recovered at the time (now exceeded by the European Project for Ice Coring in Antarctica)—11,886 ft. (3,623 m) deep, containing a climate record of 420,000 years, for a total of four climate cycles. Drilling stopped at this depth because the researchers were recovering accretion ice refrozen to the bottom of the glacier, indicating the presence of an underlying lake, and did not want to put themselves in danger from the release of pressurized lake water or risking contamination of the lake. Researchers have found microbes in the glacial ice from the Vostok core and four times more in the glacial-accretion ice transition, suggesting that the underground lake contains microbes and organic carbon.

Polar snowfall can be preserved in annual layers within an ice sheet to provide a climate record. These layers can be studied to develop an accurate picture of the climate history, extending over long time periods (the deepest Vostok core extends over a 400,000-year time frame). Impurities (volcanic debris, sea salt, organic material, and interstellar particles) are also deposited with snow, making those layers distinctive.

Air bubbles trap gases in the ice and allow for testing to determine the air's composition at distinctive periods in the climate record. Water pockets may also become trapped the deeper the ice core is, and closer to the underlying rock or water. Researchers can determine the composition of water in comparison with heavy water isotopes to indicate environmental temperature; cold periods are those with moisture removed from the atmosphere.

Studies on the second Vostok core showed a correlation between carbon dioxide and temperature over the past 160,000 years, and provided evidence linking climate change with the greenhouse effect. The trapped air bubbles provided gas isotopes, and when compared with the temperature variations, matched up to show that greenhouse gases were the primary driver of climate change over time. The climate variation of ice ages also matched solar records.

Initial Vostok studies, when combined with later research, provide an inclusive representation of the multiple factors involved in climate change by using a multidisciplinary approach to climate change research, using astronomical tables, chemistry, and physics. The Vostok ice cores indicate periods of ice ages, contain gases for comparison with temperature changes, and highlight the last ice age of 8 degrees F (4.4 degrees C) cooler than the present, taking place about 18,000 years ago.

Vostok's cores have provided significant evidence of greenhouse gas variations driving climate change and have provided information for the modeling of future climate changes in relation to greenhouse gas concentrations.The third Vostok core, recovered in 1998, provides additional confirmation and extends the historical record through the four most recent glacial cycles, showing that increased concentrations of greenhouse gases have forced the temperature higher and can be compared with the geological record of the same time frames.

The Vostok ice core has become the standard for creating timescales from cores recovered from other parts of the world. Researchers are able to plot the isotopes of their samples with similar isotope ratios of the known sample to provide an accurate time period reference.

Lyn Michaud
Independent Scholar

See Also: Greenhouse Gas Emissions; Milankovitch, Milutin; Paleoclimates.

Further Readings

Bowen, Mark. *Thin Ice: Unlocking the Secrets of Climate in the World's Highest Mountains.* New York: Henry Holt, 2005.

Inman, Mason. "The Dark and Mushy Side of a Frozen Continent." *Science* (July 6, 2007).

World Data Center for Paleoclimatology. "Vostok Ice Core Data." http://www.ncdc.noaa.gov/paleo/icecore/antarctica/vostok/vostok.html (Accessed March 2012).

Vulnerability

Vulnerability research embraces an array of different definitions for vulnerability. Vulnerability to climate change as defined by the Intergovernmental Panel on Climate Change (IPCC) is "the degree to which a system is susceptible to, or unable to cope with, adverse effects of climate change, including climate variability and extremes," and "the vulnerability of a given system or society is a function of its physical exposure to climate change effects and its ability to adapt to these conditions."

Vulnerability is the degree of fragility of a person, group, community, or area toward defined hazards. It also encompasses the idea of response and coping, since it is determined by the potential of a community to react and withstand a disaster. The vulnerability of an area varies both at the geographical and social level. Population belonging to different socioeconomic strata have differentiated sensitivities and hence vulnerabilities for a natural disaster.

Evolution of Vulnerability Concept

Vulnerability science has evolved considerably in recent years, spurred by both theoretical and methodological advances and the new issues created by the interweaving of natural, technological, and social hazards in contemporary society. As human interventions in physical space have produced more complex sociospatial relations, risks have been transformed from localized events into phenomena that have roots in the very essence of contemporary life, in what sociologists have termed *risk society.* The multidimensionality of contemporary hazards has made such hybrid hazards a challenge for today's researchers. Natural hazards, traditionally studied as earthquakes, droughts, floods, or intense rains, took on a new dimension, inserted in societal dynamics and the more encompassing perspective of environment. Natural hazards became environmental hazards.

In this context, vulnerability has emerged as a key concept, revealing the other side of the event—the conditions and the resources available for response. Hazards came to be studied not only in terms of risk factors and damage, but above all in their relational, circumstantial, and spatial dimension; each place, society, and individual, exposed to the same hazards, may be affected differently.

Vulnerability Concept and Its Elements

Vulnerability is a set of conditions and processes resulting from physical, social, economic, and environmental factors that increase the susceptibility of a community to the impact of hazards: (1) physical or material vulnerability, which includes geophysical, climatic, environmental, and infrastructural elements; (2) social or organizational vulnerability, which includes community organization, its cohesiveness, commitment, leadership, governance, and information elements; and (3) attitudinal, motivational, or cultural vulnerability, which includes community capabilities, traditional belief systems, and degree of indigenous and historical coping mechanism elements.

It is also based on the factors of climate variation (physical), exposure (physical), sensitivity (social), and adaptive capacity (social). The *Third Assessment Report* by the IPCC concluded that vulnerability to climate change is a "function of the character, magnitude, and rate of climate variation to which a system is exposed, its sensitivity and its adaptive capacity."

$$\text{Vulnerability (V)} = \text{Exposure (E)} + \text{Sensitivity (S)} - \text{Adaptive Capacity (AC)}$$

The function can also be applied to risk assessment. Hence, the system for risk analysis consists of vulnerability (V) and probability (P), and V is a function of exposure, sensitivity, and adaptive capacity.

$$\text{Risk} = \text{Vulnerability (V)} \times \text{Probability (P)}$$

Vulnerability (V) = f (E, S, AC) from the IPCC definition, where Probability (P) is the likelihood of a climate related event occuring. Adaptive capacity (AC) is the ability of a system to adjust to climate change (including climate variability and extremes) to moderate potential damages or to cope with the consequences. Sensitivity (S) is the degree to which the human population is adversely affected by climate-related stimuli. Climate-related stimuli encompass all the elements of climate change, including mean climate characteristics, climate variability, and frequency and magnitude of extremes. Exposure (E) is the number of people affected by the climate hazards and events (e.g., excessive rainfall).

Vulnerability and Schools of Thought

Various schools of thought have been in discussion on the concept of vulnerability. The economic dimension of vulnerability acknowledges economic damage potential, which can be understood as anything concrete that affects the economy of a region and can be damaged by a hazard. The economic dimension of vulnerability represents the risk to production, distribution, and consumption.

Advanced industrial societies, especially large urban centers, are especially vulnerable because the destruction of important and extensive systems of communications and infrastructure is costly and can have vast consequences on economic stability, even on the global scale. The economic dimension offers an interesting approach to regional vulnerability, especially from the insurance company point of view of damage potential.

The social dimension of vulnerability acknowledges the vulnerability of people; the emphasis is on coping capacity. Weak and poor population groups are considered especially vulnerable. Social vulnerability has to do with the different features of human beings. The most vulnerable groups are those who find it hardest to reconstruct their livelihood after a disaster. They find that, as a rule, the poor suffer more from hazards than the rich. The time dimension is relevant since reconstruction in poor areas can take a long time, which affects the economy and livelihood of the area drastically. Further, the poorer population groups do not always have a choice of where to locate; thus they might have to live in risky areas, for example on a muddy hillside.

The ecological dimension of vulnerability acknowledges ecosystem or environmental vulnerability or fragility. In the case of ecological vulnerability, it is important to find out how different kinds of natural environments cope with and recover from different hazards. According to Llewellyn Williams and Lawrence Kapustka, ecosystem vulnerability can be seen as "the inability of an ecosystem to tolerate stressors over time and space." It can be either intrinsic or extrinsic. Intrinsic vulnerability is related to factors internal to the system (ecosystem health and resilience), whereas extrinsic vulnerability contains factors external to the system (present exposure and external hazard). Ecological vulnerability thus recognizes both ecological damage potential and coping capacity.

Vulnerability Themes

The concept of vulnerability originated in research communities examining risks and hazards, climate impacts, and resilience. The vulnerability concept emerged from the recognition by these research communities that a focus on perturbations alone was insufficient for understanding the responses of, and impacts on, ecosystems exposed to such perturbations. The definition of vulnerability also encompasses response and coping, since vulnerability refers to the different variables that make people less able to absorb the impact and recover from a hazard event. However, vulnerability is understood in different ways and can be broadly categorized into three distinct themes in vulnerability research:

1. *Vulnerability as hazard exposure*: concentrates on the distribution of some hazardous condition, on human occupancy of such an area, and on the degree of loss associated with a hazardous event. Vulnerability is a pre-existing condition.
2. *Vulnerability as social response*: concentrates on response and coping capacity, including societal resistance and resilience to hazards, as well as recovery from a hazardous event. This approach highlights the social construction of vulnerability. For example, public health,

to a large extent, depends on safe drinking water, sufficient food, secure shelter, and good social conditions. A changing climate is likely to affect all these conditions.

3. *Vulnerability of places*: a combination of hazard exposure and social response within a specific geographic area. For example, extremes in maximum and minimum temperatures are expected to increase, leading to rapid mountain glacier retreat in the Himalayas, with melt water from the Himalayan glaciers contributing a sizeable portion of river flows to the Ganges, Brahmaputra, Indus, and other river systems.

Vulnerability and Urbanization

Urban areas are not disaster prone by nature; rather the socioeconomic structural processes that accelerate rapid urbanization, population movement, and population concentrations substantially increase disaster vulnerability, particularly of low-income urban dwellers. Migrants, for example, settle in areas either originally unsafe (e.g., susceptible to floods or landslides), or create the potential of human-made disaster (e.g., environmental degradation, slum fires, or health hazards).

Urban vulnerabilities are not limited to just low-income residents—a flood or a typhoon does not distinguish between residents, affecting everyone in its path. Therefore, while urban vulnerabilities are created directly by global change, such as sea-level rise and flooding (more than 80 percent of cities are on river basins or close to a coast, or both); a number of indirect causes, such as household and hazardous/toxic wastes and pollution, are also responsible, resulting in potentially higher impacts due to concentrations of infrastructure, government, population, and economic activity.

Vulnerability in an urban context should not be considered a fixed feature of a specific society or territory, but as a process with an intensity that can be reduced through adequate policies. Thus, a changing urban environment, brought about by both human-made factors and natural factors, creates vulnerabilities in cities that need to be prevented (by controling the source), protected (by building to withstand), and controlled (through land-use planning and zoning).

Vulnerability Assessment for Urban Areas

Vulnerability assessment is a method to identify the vulnerability of an area—the people, infrastructure, and property that are likely to be affected. It includes everyone who enters the area, including employees, commuters, shoppers, tourists, and others. Inventorying the jurisdiction's assets to determine the number of buildings, their value, and population in hazard areas can also help determine vulnerability. A jurisdiction with many high-value buildings in a high-hazard zone will be extremely vulnerable to financial devastation brought on by a disaster event. Identifying hazard-prone critical facilities is vital because they are necessary during response and recovery phases.

Decision-making tools, such as multihazard mapping, can aid during a disaster for the evacuation process and in evolving a management strategy. Critical facilities necessary for the health and welfare of an area and that are essential during response to a disaster include hospitals, fire stations, police stations, and other emergency facilities; transportation systems such as highways, airways, and waterways; utilities, water treatment plants, communications systems, and power facilities; high-potential-loss facilities such as bulk fuel storage facilities; and hazardous materials sites. These can be identified with the vulnerability assessment decision-making tools.

Specific Vulnerability

Specific vulnerability, a weakness that is exposed when a hazard grows into a disaster, can be ascertained by considering vulnerable components. The components in this case can be people and assets. The vulnerability depends on the planning that has gone into the development of these components. Specific vulnerability can also be predetermined or calculated by taking into account all the susceptible elements, after factoring in the resilience of those elements to the hazard. The elements in this case can be people, economic installations of any kind, and other assets. The susceptibility also depends on the amount of planning that has gone into the formation of these elements. In the case of people, the amount of planning can be related to the amount of understanding about the hazards and to the facilities available to the people, enabling them to cope with the risks posed by such hazards.

Multihazard Vulnerability and Mapping

With multihazard vulnerability, the spread of specific vulnerabilities on the delineated area is favorable in the development of combined development and risk mitigation plans. When citing the reasoning of unpredictability and uncertainty, multihazard vulnerability assessment is the way forward.

The same reasoning of uncertainty can in turn be used in the case of hazards being modified from forces beyond immediate control, such as global warming. Then again, there will be areas that will face risks from a certain hazard far more than others. For example, states such as Rajasthan in India might face heat waves more than an earthquake or heavy rainfall. In such cases, multihazard vulnerability planning is required and carries primary importance.

Multiple hazard mapping (MHM) is an excellent tool to create awareness in mitigating multiple hazards. It is a comprehensive analytical tool for assessing vulnerability and risk, especially when combined with the mapping of critical facilities. When an area is exposed to more than one hazard, an MHM helps the planning team analyze all of them for vulnerability and risk. For example, the effects and impact of a single hazard event, as in the case of a flood or a landslide, have different severities and affect different locations. In either, MHM can play a valuable role in the planning of new development projects or the incorporation of hazard-reduction techniques into existing developments.

The main purpose of MHM is to gather together in one map the different hazard-related information for a study area to convey a composite picture of the natural hazards of varying magnitude, frequency, and area of effect. An MHM may also be referred to as a "composite," "synthesized," and "overlay" hazard map. Using individual maps to convey information on each hazard can be cumbersome and confusing for planners and decision-makers because of their number and possible differences in area covered, scales, and detail.

There are several benefits of MHM: (1) a more concise focus on the effects and impacts of natural phenomena on a particular area is possible during early planning stages; (2) many hazards and the trigger mechanism of each can be viewed at the same time. Common reduction or mitigation techniques can be recommended for the same portion of the study area. Inadequate or missing hazard information (location, severity, or frequency) can be more easily identified; (3) a study area or a subarea can be expanded, reduced, or deleted. Study areas can be divided into subareas requiring more information, additional assessments, or specific reduction techniques; (4) more realistic evaluation of risks to new development are possible; (5) appropriate hazard-reduction techniques can be more easily built into the investment project formulation; and (6) selection of appropriate land uses can become more rational.

Agriculture, especially in coastal areas, is vulnerable to weather disasters. At this Wallace, North Carolina, poultry farm, Hurricane Floyd claimed 23,000 birds at a cost of $85,000, plus cleanup costs. The owner also lost 2,500 hogs and his corn crop.

Vulnerability and Resilience

Vulnerability and resilience are closely related concepts, and there is a dialectic relationship between the two. Resilience refers generally to persistence and sustainability, while vulnerability refers overall to the capacity to withstand and adapt. However, when conditions of vulnerability are addressed and decreased, the capacity to respond and adapt improves and the persistence and sustainability (resilience) of the system is supported. This relationship is important for the development of resilient socioecological systems, as well as for the promotion of human security. Vulnerability points to the need for systems to change. When these changes include preparedness, recovery, and mitigation, geared to alleviate the effects of specific hazards, resilience increases, which in turn results in the reduction of vulnerability.

Vulnerability and Research

While there is great deal of ongoing research in identifying the key vulnerabilities of climate change and anthropogenic interference, there are significant gaps in the existing knowledge base. Bridging this gap requires an interdisciplinary and integrated approach to the current research, one that can capture both the biogeophysical and socioeconomic systems of a society. For example, new and improved estimates for sea-level rise, melting of the Himalaya, and rainfall patterns are urgently needed. Recent theory includes consideration of physical vulnerability; the (in)capacities linked to particular environmental, geographic, and other conditions; social vulnerability; and the historic, sociocultural, economic, political, and other conditions that structure adaptive and coping (in)capacity for individuals and societies. There is also a need for improving knowledge of climate sensitivity and risk management and the probable distribution of impacts of climate change. Although information is a key to vulnerability, other systems, such as better knowledge of institutional and organizational dynamics and stakeholder inputs, can help in the assessment of vulnerabilities.

Lessons From Vulnerability

In terms of their spatial distribution, natural hazards affect socioeconomic groups differentially. Some are widespread and affect all groups (e.g., snowstorms, earthquakes, droughts, and storms), whereas others occur in areas in which the primary population group exposed tends to be poorer because residence in these hazard-prone areas is tied to poverty, such as in areas prone to floods and landslides. With increasing frequency and intensity, these events are likely to affect growing numbers of persons, requiring societal—no longer merely sectoral—interventions.

Not only are poor, developing countries—with their weak institutional mechanisms for predicting and responding to natural hazards—more vulnerable, but poorer, unprotected social segments of wealthier countries, in spite of sophisticated prediction technology and elaborate civil defense systems, are also vulnerable. Institutional, political, economic, cultural, and geographical factors all contribute to vulnerability, with distinct differences among persons and places. While a more encompassing understanding of the relations between the components and dimensions of vulnerability is necessary, it is just as important to continue efforts to understand the specific causal nexus in specific places, because it is in specific places that the different dimensions of vulnerability materialize, giving clues as to the nature of such interactions. The hazards-of-place approach permits the observation of hazards at this scale, allowing us to make the transcalar connection, starting from the impacted area and moving toward greater understanding in regional and global terms.

Komalirani Yenneti
University of Birmingham

See Also: Climate Change, Effects of; Cyclones; Developing Countries; Floods; *Fourth Assessment Report*; Gender and Climate Change; Hurricanes and Typhoons; Intergovernmental Panel on Climate Change; Poverty and Climate Change; Refugees, Environmental and Climate; Sea Level, Rising; Tsunamis.

Further Readings

Blaikie, P., et al. *At Risk: Natural Hazards, People's Vulnerability and Disasters.* New York: Routledge, 1994.

Füssel, Hans-Martin and Richard Klein. "Climate Change Vulnerability Assessments: An Evolution of Conceptual Thinking." *Climatic Change,* v.75/3 (2006).

McCarthy, J. J. "Climate Change 2001: Impacts, Adaptation and Vulnerability." In *Working Group II Third Assessment Report,* edited by Intergovernmental Panel of Climate Change. Cambridge: Cambridge University Press, 2001.

Parry, L. M. "Climate Change 2001: Impacts, Adaptation and Vulnerability." In *Working*

Group II Fourth Assessment Report, edited by Intergovernmental Panel of Climate Change. Cambridge: Cambridge University Press, 2007.

Watson, T. R., M. C. Zinyowera, and R. H. Moss, eds. *The Regional Impacts of Climate Change: An Assessment of Vulnerability.* Cambridge: Cambridge University Press, 1997.

Williams, Llewellyn R. R. and Lawrence A. Kapustka. "Ecosystem Vulnerability: A Complex Interface With Technical Components." *Environmental Toxicology and Chemistry,* v.19/4 (April 2000).

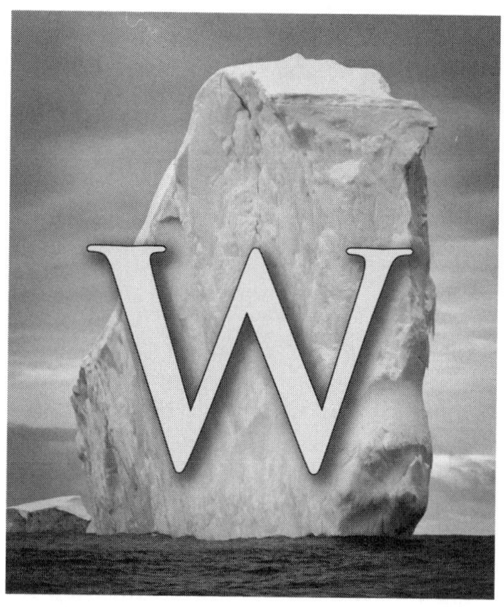

Walker, Gilbert Thomas

Gilbert Thomas Walker (1868–1958) was the British physicist and statistician who first described the phenomenon of Southern Oscillation, a coherent interannular fluctuation of atmospheric pressure over the tropical Indo-Pacific region that produces wind anomalies. This was part of Walker's project to determine the connections between the Asian monsoon and other climatic fluctuations in the global climate in an effort to predict unusual monsoon years that cause drought and famine in the Asian sector. Although he was not a meteorologist by education, Walker greatly advanced the study of global climate with his discovery.

Gilbert Thomas Walker was born on June 14, 1868, in Rochdale, Lancashire. He was the eldest son and fourth child in a large family of eight. Soon after his birth, his family relocated to Croydon, where his father became the borough's chief engineer. From 1876 on, Gilbert attended Whitgift School, and in 1881, he won a scholarship to St. Paul's School. He already excelled in mathematics from these early years and passed the London matriculation in 1884. However, he did not stay in London for his degree, opting instead for Trinity College, Cambridge, where he enrolled in 1886, thanks to a scholarship. In 1891, he was elected a Trinity Fellow. He received an M.A. in 1893, and two years later, he was appointed a lecturer in mathematics. The heavy work necessary to attain these successes eventually took their toll on Walker's health, which broke down in 1890, forcing the mathematician to spend the next three winters in Switzerland. As a result of his precarious health, Walker did not publish many significant papers in the following years, but his 1899 work, "Aberration and Some Other Problems Connected With the Electromagnetic Field," earned him the prestigious Adams Prize from Cambridge University. Walker was elected a fellow of the Royal Society in 1904, the same year that he received a Sc.D. from Cambridge University.

In the summer of 1903, Walker resigned his academic positions to become assistant to Sir John Eliot, who was the meteorological reporter to the government of India and director-general of Indian Observatories. The choice of Walker as special assistant was surprising, as Walker was not a meteorologist, but a mathematician. At the end of 1903, Eliot retired, and Walker became the sole person responsible for the Indian Meteorological Department. He continued Eliot's quest for professional individuals to become members of his staff. He made prestigious appointments, including J. H. Field, J. Patterson, and G. C. Simpson, who later became directors of meteorological services in India, Canada, and the United Kingdom, respectively. From the beginning of his appointment, Walker devoted his research to the problems of the monsoon and, in 1909, published his first meteorological papers. In 1908, Walker also gave lectures at the University of Calcutta, which were then published in 1910 by the Cambridge University Press. Walker married May Constance Carter in 1908, and the couple had a son and a daughter.

Seasonal Foreshadowing

Walker's interest in the monsoon resulted from the famine that the absence of rains had caused in India in 1899. Walker soon understood that he could not tackle meteorological problems through mathematical analysis and tried to develop more empirical techniques. He called his methodology *seasonal foreshadowing*, rather than weather forecasting, as the phrase indicated a vaguer prediction. Walker calculated statistical delay correlations between antecedent meteorological events both within and outside India and the subsequent behavior of the Indian monsoon. He was one of the first scientists to establish relationships between apparently separate events.

The sets of relationships that he established, subsequently called the Walker circulation in his honor, create a system resembling a global heat engine, influencing the world's climate. The Walker circulation works like a swing in which warm, moist air rises in the western Pacific, becomes drier at high elevation, and displaces eastward, where heavy air sinks and returns westward. The phenomenon thus creates high air currents moving from the west to the east and, at the same time, east-to-west trade winds near the ocean surface. Global warming theorists have predicted that rising temperatures will eventually slow down this mechanism.

Walker retired as director of the Indian Meteorological Department in 1924 and became a professor of meteorology at the Imperial College of Science and Technology in London. There, Walker continued his research into global weather and simultaneously carried out laboratory experiments in physics to study the convection of unstable fluids, particularly in its applications to cloud formations. He retired from Imperial College in 1934 and moved to Cambridge, where he lived until 1950. Although retired, Walker remained an active researcher and served as the editor of the *Quarterly Journal of the Royal Meteorological Society* from 1934 to 1941. He died in Coulsdon, Surrey, on November 4, 1958. Although he had only mixed success in his original goal, the prediction of monsoonal failures, Walker conceived of theories that allowed his successors to move beyond local observation and forecasting toward comprehensive models of worldwide climate.

Throughout his distinguished career, Walker was a member of prestigious societies and received numerous honors. He was elected a fellow of the Royal Meteorological Society in 1905 and served as president of the society in 1926 and 1927. He was also vice president of the society three times and was an ordinary member of council in 1925 and from 1935 to 1939. He was awarded the society's Symons Gold Medal in 1934. While working in India, Walker was the president of the Royal Society of Bengal and president of the Indian Science Congress. He became a Companion of the Order of the Star of India in 1911 and was knighted on the king's birthday in 1924. Walker was also an honorary fellow of Imperial College and a fellow of the Royal Astronomical Society.

Luca Prono
Independent Scholar

See Also: Climate Models; Southern Oscillation Index.

Further Readings

Walker, Gilbert. "On the Meteorological Evidence for Supposed Changes of Climate in India." *Indian Meteorological Memoirs*, v.21/1 (1910).

Walker, J. M. "Pen Portrait of Sir Gilbert Walker, CSI, MA, ScD, FRS." *Weather*, v.52/7 (1997).

Walker Circulation

The Walker circulation is an atmospheric system of air flow in the equatorial Pacific Ocean. The trade winds across the tropical Pacific flow from east to west: Air rises above the warm waters of the western Pacific, flows eastward at high altitudes, and descends over the eastern Pacific. A weaker Walker circulation (in the reverse direction) occurs over the Indian Ocean.

Sir Gilbert Walker assumed the post of director-general of the observatory in India following catastrophic famines in the late 1800s, that resulted from a general failure of the South Asian monsoon. In an effort to predict the monsoons, Walker undertook an investigation into the regional climate system. Over time, he recognized that the monsoonal system extended to a panoceanic scale. Walker observed that an inverse relation of atmospheric pressures at sea level generally existed between the two sides of the Pacific Ocean. A high-pressure phase in South America was usually accompanied by low pressure in the western Pacific and vice versa—the Southern Oscillation (SO). The generally accepted measure of the SO is the inverse relationship between surface air pressure at Darwin, Australia, and Tahiti (stations used by Walker); the SO is normally identified by the SO Index (SOI), that is, the difference in atmospheric pressure at sea level between these stations (Tahiti minus Darwin). The greater the SOI, the greater the intensity of the trade winds. Historically, the SO has exhibited a more or less cyclical pattern, in that it weakens or reverses every few years.

Jacob Bjerknes: Explaining the Mechanism

On the basis of data collection initiated during the International Geophysical Year in 1957 to 1958, in the 1960s, Jacob Bjerknes of the University of California described the general nature of the mechanism linking the system. He extended the horizontal picture of the SO vertically by theorizing that to complete the system of the trade winds and atmospheric air pressure, there needed to be a countercirculation of air from west to east at high altitudes, descending over the eastern Pacific.

Atmospheric circulation is intimately coupled with the movement of water in the tropical Pacific. At the surface, the trade winds initiate a westward flow of surface water across the Pacific Ocean, producing an increase of sea level in the western Pacific of approximately 40 cm. The equatorial heating of this water produces high seasurface temperatures (SSTs) in oceanic waters near Indonesia. The resulting low atmospheric pressure and evaporation fuels the pan-Pacific upper-atmospheric circulation characterized by convection (low atmospheric pressure) in the west and subsidence (high atmospheric pressure) in the east. This is termed the Walker circulation in honor of Sir Gilbert.

The Walker circulation is closely connected to oceanic upwelling off the coast of South America. Fluctuations in the circulation are closely linked to El Niño and La Niña events—together termed the El Niño/SO system (ENSO). A weakening or reversal of the Walker circulation is closely linked with the El Niño phenomenon, with warmer-than-average SSTs in the eastern Pacific as upwelling diminishes. In contrast, the opposite phase, a particularly strong Walker circulation, produces a La Niña event, with cooler SSTs caused by increased upwelling. Interannular switches in the dipole are linked to global-scale changes in patterns of weather. Several explanations for the variation in the Walker circulation have been hypothesized, but the nature of the mechanisms initiating the change in phase has not been fully identified.

Evidence suggests that the Walker circulation may have been weakening since the mid-19th century. However, there is a high degree of uncertainty concerning the potential effects of climate change on the Walker circulation. Transient warming may dominate before the ocean has had a chance to reach equilibrium. In the short term, warming may occur more quickly in the western Pacific, thereby enhancing the circulation. In contrast, atmospheric models have indicated that climate change will, as part of a weakening of the entire tropical circulation, lead to a general decrease in the strength of the Walker circulation. The specific mechanisms involved are not fully known, and projections remain rather speculative and are a focus of intense research.

Michal J. Bardecki
Ryerson University

See Also: El Niño and La Niña; Pacific Ocean; Southern Oscillation Index; Trade Winds; Walker, Gilbert Thomas.

Further Readings

Ravelo, Ana Cristina. "Walker Circulation and Global Warming: Lessons From the Geologic Past." *Oceanography*, v. 19/4 (December 2006).

Vecchi, Gabriel A., Brian J. Soden, Andrew T. Wittenberg, Isaac M. Held, Ants Leetmaa, and Matthew J. Harrison "Weakening of Tropical Pacific Atmospheric Circulation Due to Anthropogenic Forcing." *Nature*, v.441 (May 4, 2006).

Washington

Washington is a state in the Pacific Northwest, sharing borders with Oregon, Idaho, and British Columbia, Canada. The state's 2,300 mi. (3,701 km) of shoreline include a Pacific Ocean coast and Puget Sound, an inlet of the Pacific Ocean and saltwater estuary fed by freshwater discharged from the Cascade Mountain and Olympic watersheds. The state has a varied climate and geography, ranging from rainforest in the west to a semiarid basin in the east (which lies in the rain shadow of the Cascade range). Washington is the leading hydroelectric power producer in the United States, and the Grand Coulee dam hydroelectric power plant is the highest-capacity electrical plant in the country. Nearly 75 percent of the state's electricity is generated by hydroelectric power, and Washington is also a leading producer of energy from other renewable sources such as wind, wood, and wood waste.

Washington has a varied economy that includes major manufacturers such as Boeing, software developers such as Microsoft, online retailers such as amazon.com, agriculture, fisheries, forest products, and tourism. About 11 percent of Washington's economy is agriculture, a $35 billion industry that employs 160,000 people. Leading products include apples (the state produces 55 percent of the apples grown in the United States), dairy, wheat, potatoes (the second-largest producer in the country), cattle, hops (the leading U.S. producer) and grapes (the second-largest U.S. producer). Washington is also the leading

Mount St. Helens in Washington is an active volcano, and had an enormous eruption at 8:32 A.M. on May 18, 1980. The debris blasted down nearly 230 sq. mi. (596 sq. km) of forest and buried much of it beneath volcanic mud deposits. The eruption created a cloud of ash 2,500 mi. long and 1,000 mi. wide. In 2004, the volcano was active again, emitting between 50 and 250 tons of sulfur dioxide per day. In 2004, all of the state's industries combined produced about 120 tons of the noxious gas per day.

U.S. producer of sweet cherries, pears, and red raspberries and the second-largest producer of premium wines. About 21 million acres of land in Washington is forested, with 18 million of those classified as timberland with the remainder as primarily parks or wilderness areas. Major owners of commercial forestland in Washington state include the federal government (about 5 million acres), the timber industry (4.6 million acres), small tree farmers and forestland owners (2.9 million acres), tribes (1.4 million acres), and state and local governments (2.2 million acres).

Average temperatures in the Pacific Northwest increased about 1.5 degrees F (0.83 degree C) over the past 100 years and are expected to increase another 3–10 degrees F (1.6–5.5 degrees C) during this century. Washington is already feeling the effects of climate change. For instance, 50 glaciers have disappeared from the Cascade Mountains, and sea levels are rising along the coast. By 2040, the snowpack is expected to shrink by 40 percent, endangering the state's supply of freshwater for agriculture, municipal uses, hydropower generation, and supporting aquatic ecosystems. The amount of precipitation falling as rain, rather than snow, has increased due to higher average temperatures October through March, endangering the current water-supply infrastructure that assumes that most water needed in the summer will be stored as snowpack.

The Puget Sound area, which is heavily populated, will be especially endangered by an expected sea-level rise of 13 in. (33 cm) by 2100. Summer moisture deficits will increase the risk of forest fires and limit forest productivity, with some wildlife species facing decline or extinction due to loss of habitat. Rising water temperatures and decreasing summer water flow will endanger the state's salmon population.

In November 2007, Washington passed the Clean Energy Initiative (I-937), making it the second state to adopt a renewable energy standard by ballot initiative and the 21st state overall with such a standard. The initiative requires the state's largest utilities to use 15 percent renewable energy sources by 2010 and to pursue energy conservation opportunities with their customers and in the communities they serve. Successful implementation of the initiative is estimated to reduce the state's carbon emissions by 4.6 million metric tons by 2025, the equivalent of removing 750,000 automobiles from the road. In 2009, the state legislature approved State Agency Climate Leadership Act SB 5560, which includes provisions to create an integrated response to climate change. This act includes four topic advisory groups: the Built Environment, Infrastructure, and Communities; Human Health and Security; Ecosystems, Species and Habitats; and Natural Resources.

In 2008, *Forbes* magazine ranked Washington among the top three U.S. states in terms of both quality of the environment and business climate, reflecting the state's determination to promote the green economy and green business practices. In 2008, Washington State University and the Washington State Employment Security Department conducted the state's first Green Jobs Survey, identifying over 47,000 jobs in the state contributing to energy efficiency, renewable energy, and the prevention, reduction, mitigation, and cleanup of pollution. In 2009, the state created a strategic framework for further growing its green economy.

Sarah Boslaugh
Kennesaw State University

See Also: Agriculture; Food Production; Forests; Green Economy; University of Washington.

Further Readings

Doughton, Sandi. "Mount St. Helens the State's No. 1 Air Polluter." *Seattle Times* (December 1, 2004). http://seattletimes.nwsource.com/html/localnews/2002105397_volcano01m.html (Accessed July 2011).

State of Washington Community, Trade and Economic Development. "Washington State's Green Economy: A Strategic Framework." http://www.ecy.wa.gov/climatechange/CTEDdocs/GreenEconomy_StrategicFramework.pdf (Accessed July 2011).

State of Washington Department of Ecology. "Climate Change: Meeting the Challenges and Seizing the Opportunities." http://www.ecy.wa.gov/climatechange/index.htm (Accessed July 2011).

Union of Concerned Scientists. "Backgrounder: Northwest." http://www.ucsusa.org/assets/documents/global_warming/us-global-climate-change-report-northwest.pdf (Accessed July 2011).

Washington, Warren

Warren M. Washington (1936–) is an African American meteorologist and atmospheric scientist whose research focuses on the development of computer models that describe and predict the Earth's climate. Washington was one of the first developers of atmospheric computer models in the early 1960s at the National Center for Atmospheric Research in Boulder, Colorado. These computer models use the basic laws of physics to predict future states of the atmosphere. Because these equations are extremely complex, it is almost impossible to solve them without a powerful computer system. Washington's book, *An Introduction to Three-Dimensional Climate Modeling* (1986), coauthored with Claire Parkinson, is a standard reference for climate modeling. In his subsequent research, Washington worked with others to incorporate ocean and sea ice physics as part of a climate model. Such models now involve atmospheric, ocean, sea ice, surface hydrology, and vegetation components.

Washington is a senior scientist, chief scientist, and former director of the Climate Change Research Section of the National Center for Atmospheric Research (NCAR) in the center's Climate and Global Dynamics Division in Boulder, Colorado. (Roger Wakimoto became director in February 2010.) He has advised the U.S. Congress and several U.S. presidents on climate-system modeling, serving on the President's National Advisory Committee on Oceans and Atmosphere from 1978 to 1984. He received a National Medal of Science from President Barack Obama in 2010, and was one of the 2007 Nobel Peace Prize winners.

Early Beginnings

Washington was born in Portland, Oregon, in 1936. His father, Edwin Washington, Jr., wanted to be a schoolteacher, but in the 1920s, it was impossible for African Americans to be hired as teachers in Portland public schools. Thus, Edwin was forced to work as a waiter in Pullman cars to support his family. His wife, Dorothy Grace Morton Washington, became a practical nurse after Warren and his four brothers grew up. Washington's interest in scientific research was apparent from an early age, and he was encouraged by his high school teachers. When Washington had to choose what to do after high school, his counselor advised him to attend a business school, rather than college. However, his ambition was to be a scientist, so he enrolled at Oregon State University, where he earned his bachelor's degree in physics in 1958. During his undergraduate years, Washington became interested in meteorology while working on a project at a weather station near the campus. The project involved using radar equipment to follow storms as they came in off the coast. Because of his growing interest in meteorology, Washington began a master's degree in meteorology at Oregon State, graduating in 1960. He then began a Ph.D. at Pennsylvania State University, graduating in 1964, thus becoming one of only four African Americans to receive a doctorate in meteorology.

Washington began working for NCAR in Boulder, Colorado, in 1963, and has remained associated with that institution throughout his career. His research at the center has attempted to describe patterns of oceanic and atmospheric circulation. Washington has contributed to the creation of complex mathematical models that include the effects of surface and air temperature, soil and atmospheric moisture, sea ice volume, various geographical traits, and other factors on past and current climates.

Washington's research has helped further understanding of the greenhouse effect. He has contributed to determining the process in which excess carbon dioxide in the Earth's atmosphere causes the retention of heat, thus leading to global warming. Washington's research also shed light on other mechanisms of global climate change. In interviews and statements, Washington has made it clear that he firmly believes that global warming is the result of human actions: "For researchers in climate science, the question of whether or not climate change is attributable to human activity was put to rest several years ago with our DOE-supported simulations showing that the only way to duplicate the sharp increase of the global average temperature observed in the late 20th century was to include human generated greenhouse gases in the simulations. When the same simulation was run without the human-generated greenhouse gas increases, the model simulations show that the Earth would be in a

Warren M. Washington is a senior scientist and chief scientist in the Climate Change Research section of the National Center for Atmospheric Research (NCAR) in the center's Climate and Global Dynamics Division in Boulder, Colorado.

slight cooling trend with the natural forcings of volcanic and solar activities. For us, that was the smoking gun for human-induced climate change." He has thus pleaded for climate science to rise in priority as a science problem for American administrations. He has also claimed that the Department of Energy has a particular responsibility to help find solutions for the global warming problem. According to Washington, "as the impacts of climate change become more apparent with increased severity of heat waves, droughts, water shortages, and more severe hurricanes, there will be more emphasis on understanding how we can better mitigate and adapt to the changes." The meteorologist has suggested that the Department of Energy study the carbon footprint and effect of various technology paths for the production of energy. He has also argued for an increased focus on what strategies to use to mitigate climate change and to find a long-term stabilization for carbon dioxide and other greenhouse gases in the atmosphere.

Throughout his career, Washington has published over 100 professional articles about atmospheric science, scientific textbooks, and an autobiographical volume. He has also served as a member and a director of prestigious institutes and commissions. Washington was appointed the director of the Climate and Global Dynamics Division at NCAR in 1987. In 1994, he was elected president of the American Meteorological Society. He is a fellow of the American Association for the Advancement of Science and a member of its board of directors, a fellow of the African Scientific Institute, a distinguished alumnus of Pennsylvania State University, a fellow of Oregon State University, and founder and president of the Black Environmental Science Trust, a nonprofit foundation that encourages African American participation in environmental research and policymaking. From 1974 to 1984, Washington served on the President's National Advisory Committee on Oceans and Atmosphere. In 1995, he was appointed by President Bill Clinton to a six-year term on the National Science Board. In 1997, he was awarded the Department of Energy Biological and Environmental Research Program Exceptional Service Award for Atmospheric Sciences in the development and application of advanced coupled atmospheric–ocean general circulation models to study the effects of anthropogenic activities on future climates.

Luca Prono
Independent Scholar

See Also: Climate Models; Clinton Administration; Department of Energy, U.S.; Education; Obama Administration.

Further Readings

SciDAC. "Advanced Computing for Understanding and Adapting to Climate Change: Interview With Warren Washington." http://www.scidacreview.org/0702/html/interview.html (Accessed March 2012).

Washington, Warren M. and Mary C. Washington. *Odyssey in Climate Modeling, Global Warming, and Advising Five Presidents*. Morrisville, NC: Lulu.com, 2008.

Waves, Gravity

Atmospheric gravity waves are generated by atmospheric disturbances such as storm fronts, strong wind shears, and flow over mountains and play a key role in coupling the lower and upper atmosphere, causing a redistribution of momentum and energy from the troposphere and lower stratosphere into the upper atmospheric regions of the middle stratosphere, mesosphere, and lower thermosphere. Gravity waves trigger convection and induce mixing and transport of atmospheric chemicals such as ozone. Gravity waves typically form within or near the back edge of a precipitation shield. The strongest upward motions of gravity waves occur just following the surface pressure trough and lead to maximum precipitation rates just ahead of the ridge.

Atmospheric Waves

Atmospheric gravity waves can occur at all altitudes in the atmosphere and are important for the transport of energy and momentum from one region of the atmosphere to another, to initiate and modulate convection and subsequent hydrological processes, and to inject energy and momentum into the flow. When the gravity wave breaks, the resulting turbulence mixes atmospheric chemicals. These wave-breaking processes occur globally and affect climate of the mesosphere and stratosphere.

These mesoscale–regional scale processes have global significance because of their accumulative effects from the global distribution of various wave sources. The primary challenges to observational, numerical, and analytical studies are how to better quantify gravity wave excitation as it is related to various tropospheric processes, the global distribution of the wave sources, their propagation and breaking, and the multiscale interactions involving gravity waves.

The difficulty in producing the observed Arctic climate change in models may be a result of not including gravity waves in the models. The most likely energy source mechanisms are latent heat release in deep convection and shear instability, in which waves can extract energy from the jet stream when vertical wind shear is sufficiently strong to reduce the Richardson number below 0.25. Alternatively, wave energy loss can be prevented by an efficient wave duct, which appears to be the most prevalent of the three mechanisms described.

Gravity waves are maintained by wave-ducting processes requiring a layer of static stability (the duct depth near the surface), no critical levels (wind moving in the direction of the wave at the same speed) in the lower stable layer that would absorb the wave's energy, and a reflecting layer above the stable layer to keep the wave from losing its energy.

Gravity waves can affect an existing cloud pattern in several ways as they propagate: Through modulating the cloud pattern, with the development of wave cloud formations, the wave and cloud can propagate in tandem with little effect on the overall cloud pattern. Convection can generate a broad spectrum of waves, ranging from short-period waves excited by the development of convective cells along a thunderstorm gust front to large wavelength disturbances from the release of latent heat in a thunderstorm complex.

The challenge of including gravity waves in global climate models stems from the resolution ability of computers; with increasing computer power, more complex equations over smaller distances can be resolved to examine gravity waves. Current models often use one of the available gravity wave drag parameters and assume a fixed gravity wave source for proper representation of turbulence on the small scale.

The vast spatial and temporal extent of gravity waves has important implications for the atmosphere, from the mesoscale to the global scale, and poses a stiff challenge to improving weather and climate predictions at all ranges.

An important part of the National Center for Atmospheric Research mission is to understand the coupling of the lower and upper atmosphere through dynamical, chemical, and radiative processes. Sudden stratospheric warming involves dynamical changes on vastly different scales from the troposphere to the lower thermosphere, and thus provides an opportunity to understand the coupling process. Further wave source sensitivity studies and observations will help to define gravity wave sources and behavior.

Lyn Michaud
Independent Scholar

See Also: Climate Models; Jet Streams; National Center for Atmospheric Research.

Further Readings

Geller, Marvin A., Hanli Liu, Jadwiga H. Richter, Dong Wu, and Fuqing Zhang, "Gravity Waves in

Weather, Climate, and Atmospheric Chemistry: Issues and Challenges for the Community." Gravity Wave Retreat, Boulder, Colorado, June 2006.

Koch, Steven, Hugh D. Cobb III, and Neil A. Stuart. "Notes on Gravity Waves—Operational Forecasting and Detection of Gravity Waves." *Weather and Forecasting* (June 1997).

NCAR Earth System Laboratory at NCAR. http://www.nesl.ucar.edu (Accessed March 2012).

Waves, Internal

An internal wave is a wave that develops below the surface of a fluid, along with changes in density. With increased depth, this change may be gradual, or it may occur abruptly at the interface. Similar to the transmission of energy by wind along the surface of the ocean, the density interface beneath the ocean surface transmits energy to produce internal waves. The greater the difference in density between the two fluids, the faster the wave will move.

There are a variety of causes for internal waves, including the tidal pull of water, wind stress, and energy put into the water by moving vessels. Year-round internal waves caused by tidal forces carry between 30 and 50 percent of their energy away from their source. Seasonal (stormy winter months) internal waves caused by wind and storms carry at least 15 to 20 percent of the energy input from their source.

Internal waves reach greater highs (above 100 m, or 328 ft.) from a smaller energy input than do the waves resulting from large energy input at the ocean's surface. This is because they move along interfaces with less density difference than between the ocean surface and the atmosphere.

Eventually, internal waves run out of energy and break, similar to surface waves. When they break in the deep ocean, they create turbulence. Where they create this turbulence, heat can be transferred from the upper ocean and stored in the deep ocean, with the exception of the Arctic Ocean, which is warmer in deeper water than on the surface. In the Arctic, turbulence transfers heat from the deep ocean to the surface.

Beneath the surface of the ocean exists a unique weather and climate resulting from the fluctuating currents driving wind in the atmosphere and from deep waves with similar patterns to those on the surface of the ocean—but unable to be seen from the surface—created by ocean movement caused by the tides.

Internal tide is an internal wave created from the back-and-forth flows of water over geographic features at lower depths. These internal waves radiate away in the form of tide current, carrying energy away from the source.

Internal waves are of special interest to climate modeling researchers because heat transfer is one

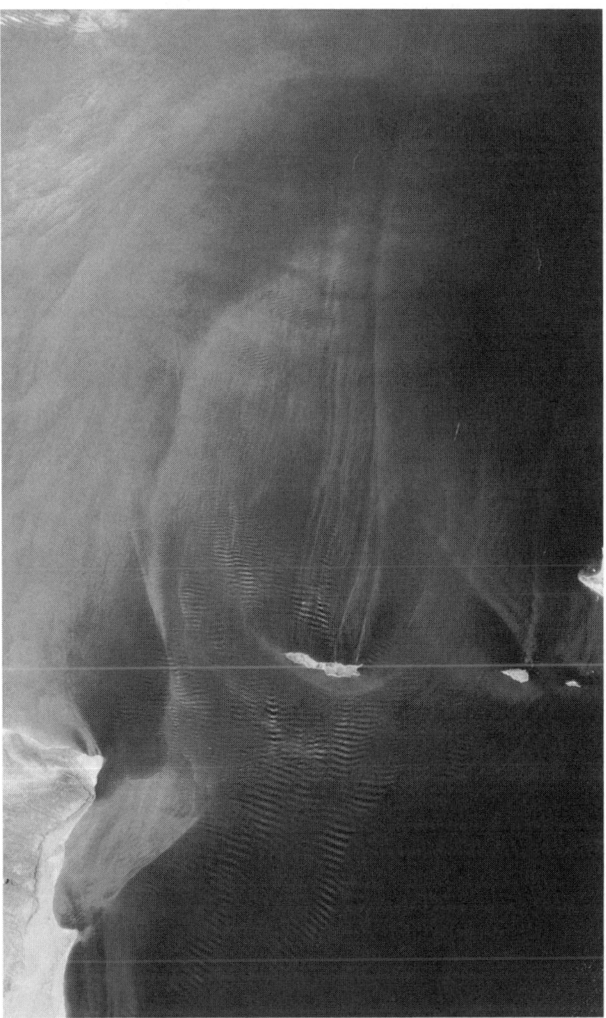

Internal waves occur beneath the water's surface. The reflection of light makes it possible to see them from high altitudes. Climate modeling researchers are interested in how these waves affect heat transfer, which plays a role in regulating the climate.

of the roles of the ocean in regulating and changing climate. For example, the amount of heat transferred from the deep ocean in the Arctic will affect the amount of floating sea ice above. In the same way that heat from the sun at the equator is transferred to the atmosphere, the upwelling of water enhances the global ocean circulation; otherwise, the depths of the ocean would be much colder and retain nutrients deeper down and away from supporting life or the ability for continuance of the carbon dioxide cycle absorption and release.

Internal waves mix and redistribute heat, salt, and nutrients in the oceans; mixing is accomplished more easily in water having uniform density. Most of this mixing occurs where internal waves break, overturning the density stratification of the ocean and creating patches of turbulence. Scientists have observed that the internal wave rates of dissipation and the redistribution of heat and nutrients is 10 percent less near the equator than at midlatitudes. This ocean dynamic would have to be accounted for in climate models.

Researchers are studying internal waves, as determining where internal wave energy might be high and where it might be low will help researchers distinguish between fluctuations in the data record originating from ocean currents and fluctuations resulting from internal waves/tides. Thus, oceanographic instruments can be deployed in oceanographically interesting locations where scientists can quantify the vertical redistribution of heat or assess the potential contribution to climate change and variability.

Lyn Michaud
Independent Scholar

See Also: Ocean Component of Models; Oceanic Changes; Oceanography.

Further Readings

Grue, John and Karsten Trulsen. *Waves in Geophysical Fluids: Tsunamis, Rogue Waves, Internal Waves and Internal Tides.* New York: Springer, 2007.

Hallberg, Robert. "Reply." *Journal of Physical Oceanography*, v.31 (July 2001).

Hines, Sandra. "Internal Waves Appear to Have the Muscle to Pump up Mid-Lats." *Science* (June 2003).

Waves, Kelvin

Kelvin waves, affecting weather and climate, occur in both the oceans and the atmosphere. These low-frequency, gravity-driven waves propagate vertically and parallel to boundaries (e.g., equator, coastline, air masses, and topography). Kelvin waves are nondispersive and carry energy from one point to another. The height or amplitude of a Kelvin wave is highest near the boundary where it propagates; the wave height decreases as the wave moves farther away from the boundary. In the Northern Hemisphere, the waves propagate anticlockwise; and in the Southern Hemisphere, the waves propagate clockwise. The flow of the Kelvin wave balances pressure perpendicular to the boundary by the forces of gravity and the Coriolis effect.

Coastal or Equatorial Waves

Kelvin waves in the ocean form as coastal waves or equatorial waves, both of which are caused by external forces—often a shift in the trade winds or resulting from temperature variations—and the water inside the Kelvin waves is usually a few degrees warmer than the surrounding water. Kelvin waves may be called external (or barotropic) if the ocean is homogenous, and internal (or baroclinic) if the ocean is stratified.

In the Northern Hemisphere, the equatorial waves propagate parallel to the equator and to the east, and the coastal Kelvin waves propagate in a counterclockwise direction, using the coastline for direction. These waves can be between 5 and 10 cm high and hundreds of km wide. Kelvin waves tend to move quickly, with a typical speed of approximately 250 km (155 mi.)/day and can cross the Pacific in approximately two months. The tidal cycles can cause Kelvin waves by the mechanism of a progressive tide wave, moving from open ocean into and out of a narrowed body of water. Because of the Earth's rotation, resulting in an anticlockwise direction of current flow inside the channel, flood tides will be greater on the right side of the channel.

The effect on climate results from the Kelvin waves causing a variation in the depth of the oceanic thermocline (the boundary between warm waters in the upper ocean and cold waters in the deep ocean). Because of this variation, Kelvin waves can be used to predict and monitor El Niño activity. In comparison with Rossby waves, which carry water back toward the western Pacific and take as long as a decade to move from the eastern Pacific to the western Pacific, the faster-moving Kelvin waves carry warm water eastward in approximately two months. In the atmosphere near the equator, Kelvin waves travel eastward and may propagate upward to higher altitudes. The formation of Kelvin waves is triggered by mountains, thunderstorm updrafts, and anything that interrupts the normal flow of stable air. The trigger forces the air upward, and the stable air sinks by gravity instead of just returning to normal; when air sinks farther, it causes the wave motion. Kelvin waves cannot happen in unstable air because the motion would just allow the movement of air to continue upward to higher altitudes.

Kelvin waves may propagate in the lower and upper stratosphere and mesosphere. In the lower stratosphere, the eastward-moving Kelvin wave is associated with periods of 10 to 30 days, and in the upper stratosphere, the Kelvin wave is associated with periods of 5 to 7 days. In the mesosphere, the Kelvin wave is associated with periods of 3 to 4 days. These Kelvin waves transport energy and eastward momentum upward and contribute to the maintenance of the eastward flow.

For predicting weather and future climate change, climate models, including the COMMA-LIM model, can reproduce the Kelvin waves to correlate with how the Kelvin wave acts in the atmosphere. The wave action can also interact with other wave types and the flow of air masses.

Lyn Michaud
Independent Scholar

See Also: Climate Models; El Niño and La Niña; Waves, Gravity; Waves, Rossby.

Further Readings

Hayashi, Y., D. G. Golder, and J. D. Mahlman. "Stratospheric and Mesospheric Kelvin Waves Simulated by the GFDL 'SKYHI' General Circulation Model." *Journal of Atmospheric Sciences*, v.41 (1984).

Phillips, T. "A Curios Pacific Wave." http://science.nasa.gov/science-news/science-at-nasa/2002/05mar_kelvinwave (Accessed June 2008).

Zhou, Cheng. *Linear and Nonlinear Kelvin Waves/Tropical Instability Waves in the Shallow-Water System*. Ann Arbor, MI: UMI Dissertation Publishing, 2011.

Waves, Planetary

Planetary-scale waves have their origin relating to the Earth's shape and rotation; the waves are so large that some of them wrap around the whole Earth and can be observed in the atmosphere through the meandering of the jet stream. A long wave or planetary wave is a weather system that circles the world, with one to three waves forming a looping path around the Earth at any given time and displacing air north and south. Planetary waves have ridges (high points) and troughs (low points). Warmer upper air is associated with an increasing number of waves or stronger waves.

Planetary waves form in the lowest part of the atmosphere, called the troposphere, and propagate upward, transferring energy into the stratosphere and heating polar air between 9 and 18 degrees F (5 and 10 degrees C). Because of a larger landmass, with the majority of the highest mountains and land–sea boundaries in the Northern Hemisphere, planetary waves form more strongly in the Northern Hemisphere. Once the wave dissipates, the polar air begins to cool. In the Southern Hemisphere, landforms also produce planetary waves, although they are weaker there because there are fewer tall mountain ranges and vast open ocean surrounding Antarctica. The warming of the Arctic stratosphere suppresses ozone destruction. Ozone exists in the lower level of the stratosphere and is caused by sunlight splitting the oxygen molecules at cooler temperatures, with less ozone destruction at warmer temperatures.

The Himalayas and other land features create the planetary atmospheric waves that serve to decrease the formation of an ozone hole at the northern pole and therefore limit solar ultraviolet radiation exposure in the Arctic. Climate change could open ozone holes in the Arctic; in the spring of 1997, weak planetary waves created conditions that formed a small ozone hole over the Arctic. The chemistry of ozone destruction requires very cold air temperatures in the stratosphere, and because of planetary wave action, the Arctic stratosphere stays warmer than the Antarctic stratosphere.

In contrast, researchers announced in 1992 that El Niño weather changes and a large number of planetary waves in the atmosphere had caused shrinking of the Antarctic ozone hole, with the ozone hole in September 2002 at half the size it was in 2000. Large-scale weather patterns (similar to a semipermanent area of high pressure) generate more frequent and stronger planetary waves. If the waves are more frequent and stronger as they move from the surface to the upper atmosphere, they warm the upper air. Because ozone breaks down more easily with colder temperatures, the warmer the upper air around the "polar vortex," or rotating column of winds that reach into the upper atmosphere where the protective ozone layer is, the less ozone is depleted.

Researchers working in Esrange, Sweden, studied the main features of planetary waves and variability of the semidiurnal tide, with planetary wave periods observed by meteor radar. They focused observation on 5, 8–10, 16, and 23 day planetary waves by meteor radar measurements in the mesosphere and lower thermosphere. In the winter, when the planetary waves are significantly amplified, a very strong periodic variability of the semidiurnal tide is also observed. This result indicates that the most probable mechanism responsible for the periodic tidal variability during winter is in situ nonlinear coupling between tides and planetary waves. They established a correlation between the planetary wave and semidiurnal tide and secondary waves with frequency, phase, and vertical wavenumber (wavelength) correlation.

The influence of planetary waves on global system dynamics with airflow and temperature distribution include the indirect effect of upper air patterns on lower air patterns through feedback, linking all layers of the atmosphere. Planetary waves (also called Rossby waves) form in the ocean and affect ocean circulation over longer periods of time—from one to 10 years.

Lyn Michaud
Independent Scholar

See Also: Arctic and Arctic Ocean; Atmospheric Boundary Layer; Atmospheric General Circulation Models; El Niño and La Niña; Jet Streams; Sweden; Waves, Gravity; Waves, Rossby.

Further Readings

Barry, Patrick L. and Tony Phillips. "Planetary Waves Break Ozone Holes: Huge Planet-Girdling Atmospheric Waves Suppress Ozone Holes Over Earth's Northern Hemisphere." http://science.nasa.gov/science-news/science-at-nasa/2001/ast11oct_1/ (Accessed June 2008).

Pancheva, Dora V. and Nicholas J. Mitchell. "Planetary Waves and Variability of the Semidiurnal Tide in the Mesosphere and Lower Thermosphere Over Esrange (68 degrees N, 21 degrees E) During Winter." *Journal of Geophysical Research*, v.109 (2004).

Waves, Rossby

NASA researchers at the Goddard Institute for Space Studies describe Rossby waves as "slow-moving waves in the ocean or atmosphere, driven from west to east by the force of Earth spinning." These are naturally occurring phenomena first recognized in 1939 by Swedish American meteorologist Carl-Gustav Rossby. These waves, which are found in both the atmosphere and the oceans, are important mechanisms for the redistribution of energy around the globe.

Atmospheric and Oceanic Waves

This phenomenon was first identified as atmospheric oscillations that occurred in the mid-latitudes in the Northern and Southern Hemispheres. In Europe and North America, people typically experience a Rossby wave as a large cold front plunging southward.

The jet stream, guised as a tongue of cold air, dips southward as a large tropical air mass moves northward. The interaction between these air masses, affected by the Coriolis affect that intensifies at lower latitudes, generates changing weather on a day-to-day and week-to-week basis. During televised weather reports, North American Rossby waves appear as large-scale oscillations of clouds moving from west to east across a continent.

Scientists later identified a similar phenomenon at work in the water of all ocean basins. Researchers discovered that oceanic Rossby waves represent a mechanism by which the ocean responds to significant atmospheric "forcing" or wind-related disruption. Rossby waves disperse the atmospheric energy across ocean basins and can be measured through satellite imagery. Because of the impact of the Earth's axial rotation and the Coriolis effect, the oceanic Rossby waves tend to spiral away from the equator in both the Northern and Southern Hemispheres.

Scientists are increasingly interested in Rossby waves because of the possible connection between these atmospheric and oceanic waves and global warming. An understanding of atmospheric Rossby waves enabled researchers to effectively study long term temperature fluctuations and provide concrete evidence that global warming was occurring.

Rossby waves can affect entire ocean basins. They also tend to move from the eastern part of the Pacific and Atlantic oceans toward the west on either side of the equator. A complementary Kelvin wave moves in the opposite direction from west to east along the equator. The multiple axes along which the ocean moves has the ability to disrupt oceanic circulation. Researchers propose that global warming will generate stronger weather events with greater frequency. Because oceanic Rossby waves transmit atmospheric disruptions, the theory is that as storms occur more frequently, the wavelength and frequency of Rossby waves will also change—with a potentially disruptive impact on ocean currents such as the Gulf Stream. A disrupted Gulf Stream could cause cooling at higher latitudes in the North Atlantic.

Researchers also see a connection between changes in Rossby waves and the intensity of El Niño and La Niña ocean surface water temperature fluctuations in the Pacific Ocean. Also known as the El Niño-Southern Oscillation (ENSO), sea surface temperature increases during El Niño events, and decreases in La Niña events can alter or intensify the monsoonal rainfall and hurricane patterns in North and South America. The challenge for climatologists and oceanographers is to understand how disrupted Rossby waves are a cause and consequence of global warming. The broader point is that global climate change is a complex process that involves the interplay between large scale atmospheric and oceanic processes operating at multiple scales.

Christopher D. Merrett
Western Illinois University

See Also: Coriolis Force; El Niño and La Niña; Jet Streams; Rossby, Carl-Gustav; Waves, Gravity; Waves, Kelvin.

Further Readings

Hansen, James and Sergej Lebedeff. "Global Trends of Measured Surface Air Temperature." *Journal of Geophysical Research*, v.92 (1987).

Hansen, James, et al. "GISS Surface Temperature Analysis. Global Temperature Trends: 2005 Summation." Goddard Institute for Space Studies. http://data.giss.nasa.gov/gistemp/2005 (Accessed March 2012).

Herring, David. "Earth's Temperature Tracker." Goddard Institute for Space Studies. http://www.giss.nasa.gov/research/features/200711_temptracker (Accessed March 2012).

Quartly, Graham, et al. "Rossby Waves: Synergy in Action." *Philosophical Transactions: Mathematical, Physical and Engineering Sciences*, v.361/1802 (2003).

Saenko, Oleg. "Influence of Global Warming on Baroclinic Rossby Radius in the Ocean: A Model Intercomparison." *Journal of Climate*, v.17 (2005).

Speth, James. *Red Sky at Morning: America and the Crisis of the Global Environment*. New Haven, CT: Yale University Press, 2005.

Vecchi, Gabriel, et al. "Weakening of Tropical Pacific Atmospheric Circulation Due to Anthropogenic Forcing." *Nature*, v.441 (2006).

Weather

Weather is the physical condition or state of the atmosphere at any given time. It is what is happening in the atmosphere at any time or over any short period of time. If there were no atmosphere,

Weather features like storms and wind develop because of the movement of heat in the atmosphere. Winds form part of larger weather systems, the most powerful of which is the hurricane. The Earth's water cycle also plays an important role in the development of many weather features, such as dew, fog, clouds, and rain. Weather, which changes each day, is not synonymous with climate, which is more stable and is generally defined as the composite of weather conditions over a considerable period of time.

there would be no weather. The principal elements of weather are temperature, pressure, wind, moisture, and precipitation. Thus, weather of any place is the sum total of its temperature, pressure, wind, moisture, and precipitation conditions for a short period of a day or a week. Temperature expresses intensity of heat. Unequal distribution of temperature over the Earth's surface causes differences in atmospheric pressure, which causes wind. Moisture is present in the atmosphere as water vapor, often condensed into clouds. It may be precipitated in the form of rain, hail, sleet, or snow. The capacity of air to gather and retain water vapor is largely dependent on its temperature. The higher the temperature, the greater the capacity of air to hold moisture. On cooling, the air is not able to retain all the moisture it gathers while warm. This leads to condensation and precipitation.

Weather, which changes daily, is not synonymous with climate, which is something more stable and is commonly defined as the average weather. Climate is the composite weather conditions over a considerable period of time. Weather conditions can change suddenly. Today may be warm and sunny, tomorrow may be cool and cloudy. Weather conditions include clouds, rain, snow, sleet, hail, fog, mist, sunshine, wind, temperature, and thunderstorms. Weather is driven by the heat stored in the Earth's atmosphere, which comes from solar energy. When heat is moved around the Earth's surface and in the atmosphere because of differences in temperature between places, this makes wind. Wind forms part of larger weather systems, the most powerful of which is the hurricane. Other weather features, like thunderstorms, also develop because of the movement of heat in the atmosphere. Some thunderstorms result in tornadoes. The Earth's water cycle plays an important role in the development of many weather features, such as dew, fog, clouds, and rain.

Weather can be described using terms such as wet or fine, warm or cold, windy or calm. The science of studying weather is called meteorology. Meteorologists measure temperature, rainfall, air pressure, humidity, sunshine, and cloudiness, and they make predictions and forecasts about what the weather will do in the future. This is important for giving people advance notice of severe weather, such as floods and hurricanes. Temperature is measured with a mercury thermometer in degrees Celsius.

Temperature is the hotness or coldness of an object. Rainfall is usually measured by collecting what falls in rain gauges and is expressed as a depth of water that has fallen, in millimeters. Wind can be observed with a weather vane, but to measure its speed, more technical equipment is needed. Alternatively, the Beaufort scale can be used to make a judgment of the strength of the wind by observing how it affects objects outdoors, such as trees. The relative humidity and dew point temperature of the air can be determined by making measurements with a hygrometer and reading a table of numbers.

The purpose of a weather map is to give a graphical or pictorial image of weather to a meteorologist. As a forecasting tool, weather maps allow a meteorologist to see what is happening in the atmosphere at virtually any location on Earth. Complex three-dimensional models of weather systems can be made by collecting weather data at multiple levels in the atmosphere. Computers then compile that information to produce the pictures that weather scientists analyze. In the early days of meteorology, these pictures were drawn by hand.

Weather Prediction and Forecasting

Weather affects virtually everyone daily. Human beings live largely at the mercy of the weather. It influences our daily lives and choices and has an enormous effect on corporate revenues and earnings. Weather can be predicted to some degree by observing the state of the sky and the wind. Weather is measured, and forecasts are usually released and used to make important decisions about travels and timing. Weather forecasts can save lives, reduce damage to property, reduce damage to crops, and tell the public and the global community about expected weather conditions. Forecasts basically predict how the present state of the atmosphere will change with time. This involves plotting weather information on special charts. Weather radar and satellites are now also used to help predict the weather. Weather forecasters measure the weather so that they can forecast it. Temperature, rainfall, wind, clouds, sunshine, and air pressure are measured all over the country, and the information is plotted on special charts. These weather charts have been used by forecasters for many years to predict what the weather will do over the next few days. Weather presenters often show simplified charts on television. Sophisticated equipment is used to help forecast the weather: Weather radar can help to show where it is raining over a country, whereas satellites are used to reveal cloud cover and the development of large weather systems. Procedures for collecting and taking the observations are determined by the World Meteorological Organization.

There are a variety of forecasting techniques. The easiest of the techniques is called persistence. In this technique, tomorrow's weather is said to be same as today's weather. Local factors that should be considered when forecasting include clouds and snow. Clouds during the day will decrease the maximum temperature expected. Without clouds, higher maximum temperatures would be obtained. With snow, the surface stays colder during the day, as less short-wave radiation is absorbed. During the night, radiational cooling effectively cools the surface. If there is some wind, the wind could create mechanical turbulence, which will mix down warmer air to the surface. Hence, winds keep the minimum nighttime temperature warmer than in calm conditions. By looking at cloud movement at different levels, one can infer the type of temperature advection that may be occurring, and therefore atmospheric stability.

Synoptic weather analysis and forecasting began after the telegraph made instantaneous long-distance communications possible after 1850. The invention of radio allowed the extension of the weather observation net over the oceans, and this was one of its first important uses. Every ship became a weather station, radioing reports at regular intervals to a central office. This great advance occurred by 1915, and weather prediction was of great service in World War I. Significant theoretical advances were made during the middle of the 20th century, especially in the properties of the upper atmosphere, elucidated by such indirect methods as sound propagation and meteor trails.

Sun, Air, and Weather

The state of the air is important in weather studies. The elements of weather are air temperature, precipitation, cloud cover and sunshine, wind speed and direction, and air pressure. The air in the atmosphere is a mixture of gases consisting of nitrogen, oxygen, water vapor, carbon (IV) oxide, and some rare gases. The amount of water vapor is very important, as it creates clouds and rain. Clouds are seen in the sky every day. They come in all shapes and sizes and bring with them all sorts of weather. A cloud is simply a visible mass of tiny water droplets that have formed because the air has become too cold at that height to store all its water as invisible vapor. This usually happens when warmer air near the ground is cooled down by rising higher in the atmosphere.

Different types of clouds can be described as they are viewed from the ground, using different terms. A cloud's name generally reflects the height at which it forms, as well as its general shape. The three main types of clouds are cirrus, cumulus, and stratus. Cirrus clouds are wispy in appearance and resemble horsetails. They are formed almost entirely of tiny ice crystals. Cumulus clouds look like fluffy balls of cotton wool. These can sometimes grow much larger, becoming cumulonimbus clouds, which bring heavy bursts of rain during thunderstorms. Cumulus clouds are clouds of vertical development and may grow upward dramatically under certain circumstances. Another type of cumulus cloud is the altocumulus cloud, which sometimes resembles fish scales. They sometimes have dark, shadowed undersides. Stratus clouds are layer clouds that form near the ground and make the weather very grey and dreary, and sometimes rainy. Stratus clouds form flat layers or uniform sheets. Only a fine drizzle can form from stratus clouds because there is no vertical development.

Air usually contains some water in the form of moisture called dew. The water is hardly seen, as it is like a gas. Humid air contains more moisture than dry air, but when the temperature of air falls, its ability to hold moisture decreases. If the temperature drops low enough, air can become saturated, even if it was originally much drier at the warmer temperature. At this point, excess moisture begins to condense, forming small water droplets on the ground called dew. As an air mass cools, it can hold less and less water vapor. If it cools down enough, it reaches a point at which the water vapor present in the air mass represents the amount needed to saturate an air mass at the lower temperature. The temperature at which saturation occurs is the dew point. The dew point depends on the amount of water vapor in the air. In winter, the temperature may fall below freezing. If dew has formed on the ground, it will freeze, forming white crystals called frost.

When the temperature of air close to the ground falls low enough, dew will form. If a larger layer of air is cooled, the condensation of excess moisture in the air forms a mist of tiny water droplets known as fog. Fog is common in the autumn and winter and usually forms when there is little or no wind to disturb it. It is also more common in hollows and valleys, where the air tends to be a little colder because it is heavier and sinks down into these places. Fog is least likely on hilltops, unless low clouds have descended to cover them.

The major driving force behind the weather is the sun. Energy from the sun is stored in the atmosphere as heat. When this heat is moved around, it makes the weather. The equator is much hotter than the poles because it receives much stronger sunlight. There is more heat stored in the atmosphere nearer the equator. Heat, however, likes to flow from warm to cold temperatures. More heat is also stored nearer the ground. This makes the air lighter than that above it, and it rises. When air rises, its temperature falls, which makes weather features such as clouds and rain.

Severe Weather

The greater part of Europe often experiences bad weather, particularly in winter. Bad weather usually comes in from the Atlantic Ocean with weather systems called depressions—regions of low pressure, strong winds, and rain. In the tropics, bad weather is much less common, but when it strikes, it can be devastating. Large tropical storms, which usually develop toward the end of summer, are called hurricanes. The most common place for hurricanes to form is in the Caribbean. Here, seawater temperature is high because the sunlight is strong, and a lot of heat is stored there. Under the right conditions, a storm will develop, which, with sufficient energy, will become a hurricane. Viewed from a satellite, a hurricane appears almost circular, with clouds spiralling out from a small center. On the ground, the weather in the center may be fairly calm, with clear skies, but as the hurricane moves over, the weather can become very nasty, with winds of over 100 mi. (161 km) per hour that are strong enough to tear roofs off houses.

Heavy rain, dark black clouds, and lightning are evidence of a thunderstorm. Thunderstorms are not nearly as strong as hurricanes, but they can be damaging, particularly if large hailstones fall out from their clouds. Thunderclouds are known scientifically as cumulonimbus clouds. Thunderstorms are more common in summer because they need a lot of energy to form. The energy comes from the heating of the ground and the surface air by the sun. If this heating is strong enough, air heated near the ground will rise up a long way into the atmosphere because it is lighter than the air around it. Warmer air is lighter than colder air. As the air rises up, it becomes colder. Moisture in the air begins to condense out as clouds, in the same way as fog forms on a calm, cool night. In thunderclouds, however, the energy is much greater, and the currents of air are strong enough to split apart the raindrops that are forming. This builds up an electric charge, which, when released, is seen as lightning. The sound of thunder is the effect of the lightning strike on the surrounding air. When rain or hail begins to fall from a thundercloud, it is usually very heavy, but it generally lasts no more than 30 minutes. Sometimes, however, the death of one thunderstorm may lead to the development of another, and the bad weather may continue for several hours.

A rapidly spiraling column of air is called a tornado. Large thunderstorms develop because there is an awful lot of energy stored in the atmosphere. Some large thunderstorms give birth to tornadoes. The tornado is usually very small in comparison to the thunderstorm, but it can wreak terrible havoc across the small area over which it moves. At a distance, a tornado is seen as a rapidly rotating funnel or spout of air, usually colored grey because of clouds and earth debris caught up inside it. The strongest tornadoes can have wind speeds of over 250 mi. (402 km) per hour—and sometimes over 300 mi. (483 km) per hour.

Most water on Earth is in the oceans. A little, however, is contained by air in the atmosphere. The water in the atmosphere is usually not seen except when it rains, as it is in the form of moisture or vapor. Water enters the atmosphere by evaporating from the surface of the oceans, lakes, and other liquid water bodies. At higher levels in the atmosphere, the air is colder, and moisture begins to condense out as fine droplets, which are seen as clouds. When conditions are right on the ground, the same process forms fog. Eventually, water in clouds forms rain, hail, sleet, or snow, which falls back to the ground. This movement of water from the Earth to the atmosphere and back again is called the water cycle. The water cycle is responsible for much of the world's weather.

Akan Bassey Williams
Covenant University

See Also: Atmospheric Composition; Clouds, Cirrus; Clouds, Cumulus; Clouds, Stratus; Rain; Thunderstorms.

Further Readings

Gore, P. *Basic Introduction to Weather.* Clarkston: Georgia Perimeter College, 2005.

Rubin, E. S. and C. I. Davidson. *Introduction to Engineering & the Environment.* New York: McGraw-Hill, 2001.

Weather World 2010 Project

In 1995, when the Department of Atmospheric Sciences (DAS) at the University of Illinois Urbana-Champaign (UIUC) began to develop the Weather World 2010 Project (WW2010), an online framework for integrating current and archived weather data with multimedia instructional resources, it used new and innovative technologies. However, the fast-evolving field of the new technologies now makes the Website look dated due to the lack of funds to adapt it to the state-of-the-art ICT tools. In spite of this, the project contains useful materials on different manifestations of global warming such as hurricanes, clouds, precipitation, and El Niño. The accuracy of the instructional resources on the Website has been reviewed by professors and scientists from the DAS at the UIUC and at the Illinois State Water Survey.

Weather Machine

Weather World 2010 is the result of a long process dating back to January 1993, when Dr. Mohan Ramamurthy and programmer John Kemp created the Weather Machine. This resource allowed users to view weather images through a Gopher server. At the same time, Steve Hall began to devise instructional modules in HyperCard for use by the educationally motivated Collaborative Visualization (CoVis) Project, which aimed to promote project-based science learning. Weather World 2010 was greatly influenced by the creation of the first Web browser Mosaic. Because this new medium represented a powerful opportunity for the exchange of information on the Internet, both weather and educational resources were made available to a larger audience in HTML format. This conversion transformed the original Weather Machine products into the Daily Planet, which was created in 1994 and represented the first attempt to realize a World Wide Web product. It soon became a popular reference for many users.

Meanwhile, Mythili Sridhar and Steve Hall made the HyperCard-based educational modules accessible on the Web in HTML format. The project team soon became aware of the necessity of integrating weather data with explicatory and didactic material. This led to the creation of the CoVis-sponsored Electronic Textbook, later named the Online Guide to Meteorology, in 1995. This server became an extremely useful source of information of archival weather data. CoVis also provided the Geosciences Web Server, a useful resource that linked users to a wide range of weather information, as well as educational material.

In the summer of 1995, the team of researchers released the Weather Visualizer, which became one of the most successful attempts to integrate real-time weather and instructional material. Its functions enabled users to create their own weather maps and to get explanations about difficult or technical terms. Seeing the rising popularity of HTML and newer technologies such as Java, the fall of 1995 witnessed the creation of the Image Animator and the Interactive Weather Report—a couple of the first Java weather tools on the Web. Thanks to real-time instant access to weather data using Java applets, these tools allowed an increased level of educational interactivity.

The first discussions of a dynamic framework for hypermedia and CD-Web interactivity started to take place in the early months of 1996, as CoVis teachers reported difficulties in accessing the large volume of information within the educational modules. Often, their connections were too slow to effectively use the resources in their classrooms. The desire to improve and redesign the existing resources, including developing better graphical interfaces and navigation systems, prompted Steve Hall and Dave Wojtowicz to begin the construction of the early form of Weather World 2010 in May 1996. This project soon grew into a much larger task than was initially conceived. Rather than a simple improvement, this new project was to provide a new framework to support the preexisting modules and to allow users access to weather data and information. In October 1996, the project was presented to the students, faculty, and staff of the Department of Atmospheric Sciences at the University of Illinois. The project drew on the different expertise of the many volunteers that it managed to involve, and in return, it provided them with a chance to learn HTML.

Initially, the team of WW2010 spent considerable time browsing the Internet, finding the most appealing features of available Websites and incorporating some of these features into the server. Although Java and other new technologies are opening up exciting possibilities, the researchers of WW2010 are still concerned about how to allow easy access to large sets of data and instructional materials and efficient navigation. In addition, their attempts to provide user interfaces suitable for varying levels of network connectivity and to upgrade the Web server have run against the lack of financial resources and the largely voluntary character of the work done for the project. Although its fate will probably be to move in the archaeology of the Internet, WW2010 remains an example of how new and innovative technologies can be used as powerful teaching and educational tools to visualize some of the effects of climate change and global warming on peoples' lives. Its model can still be an inspiration to researchers, educators, and Web designers to find ways of cooperating to make scientific information more accessible and appealing to a larger audience.

Luca Prono
Independent Scholar

See Also: Climate Change, Effects of; Global Warming; Media, Internet; Technology; Weather.

Further Readings

University of Illinois. "WW2010 (the Weather World 2010 Project)." http://www.geo.uni-bremen.de/geomod/staff/gerrit/svetlana/WW2010/files/wwhlpr/ww2010_g.htm (Accessed December 2011).

Weather World 2010 Project. http://ww2010.atmos.uiuc.edu/(Gh)/home.rxml (Accessed June 2011).

West Virginia

Located in the eastern United States, the state of West Virginia is known as the Mountain State due to its diverse topography and mountainous terrain. It encompasses approximately 24,230 sq. mi. (62,755 sq. km) and has a population of 1,859,815 (2010 U.S. census). West Virginia is situated in the central Appalachian Mountains, with an approximate mean elevation of 1,500 ft. (458 m), and is characterized by a humid subtropical climate in lower elevations and humid continental climate in higher elevations.

Temperature and precipitation patterns in the state are strongly influenced by the local Alleghany Mountains and the drainage basins of the Ohio River and Potomac/Chesapeake Bay watersheds. The state capitol, Charleston, is representative of statewide climate averages, with temperatures ranging from a mean of 33.4 degrees F (0.78 degree C) in January to 73.9 degrees F (23.2 degrees C) in July. Precipitation is uniform throughout the year, with a slight increase during the summer months; the average annual precipitation is 44.05 in. (111.89 cm). However, higher elevation regions may average 5–10 degrees F (3–6 degrees C) cooler and receive 50 percent or more precipitation, especially in the form of snow.

According to projections from the Intergovernmental Panel on Climate Change (IPCC) and the United Kingdom Hadley Centre's climate model (HadCM2), an average annual temperature increase of 3 to 4 degrees F (1.6 to 2.2 degrees C) and a 20 percent increase in precipitation could occur by 2100 across West Virginia. Since 1970, the average annual temperature in the region has already increased by 2 degrees F (1.1 degrees C). The increase has resulted in changes such as a longer growing season, increased heavy precipitation, reduced high-elevation snowpack, and an earlier winter ice thaw on lakes and rivers.

Severe Impacts Predicted

These climate change scenarios point toward several potential socioeconomic and environmental impacts across West Virginia. Hay, which is the major agricultural crop produced in the state, could see yield increases of 25 to 30 percent with an increase in temperature and precipitation. With 97 percent of West Virginia covered by forest, climate change could alter the composition and range of tree species found in the state. Eastern hardwoods may eventually be replaced by pine and scrub oaks, tree species with less commercial value. Higher elevation forests, such as old-growth and red spruce stands, would also likely significantly diminish. Warmer temperatures also could mean earlier spring snowmelts, increasing stream flows and the likelihood of flooding and flash flooding. Population centers are most concentrated along valley floors and low-elevation regions, areas most susceptible to flash flooding. Human health would also be adversely affected. Increased temperatures, precipitation, and standing water would likely increase water and vector-borne diseases such as cholera and Lyme disease, respectively. Extreme heat-related morbidity and mortality may become more common in a region where heat waves are presently rare.

The single most important economic resource (and greatest anthropogenic influence on climate) of the state is coal, which when burned as a fossil fuel releases carbon dioxide (CO_2), a major greenhouse gas (GHG). The 2003 GHG emissions report by the West Virginia Department of

Environmental Policy found that CO_2 accounted for the majority of state emissions, followed by methane (CH_4) and nitrous oxide (N_2O); all three gases originate primarily from coal combustion for electricity production and coal-mining operations. West Virginia's per-capita GHG emissions were 16.1 million metric tons of carbon equivalent in 1999, nearly three times the national average, and new estimates show an increasing GHG emissions trend.

With tens of thousands of jobs and over $3.5 billion in annual gross state product directly related to coal, the state legislature has a majority in favor of continued coal production and use. West Virginia representatives in the U.S. Congress have shown a history of voting for energy resolutions that would benefit their state's economy, while opposing federal legislation on curbing GHG emissions. A formal action plan or commission for addressing climate change initiatives at the state level, including reducing GHG emissions, had not been put forth for West Virginia (or 20 other states) as of 2011.

Anthony D. Phillips
Jill S. M. Coleman
Ball State University

See Also: Coal; Intergovernmental Panel on Climate Change; Kentucky; Renewable Energy, Bioenergy.

Further Readings

Karl, Thomas R., Jerry M. Melillo, and Thomas C. Peterson, eds. *Global Climate Change Impacts in the United States.* New York: Cambridge University Press, 2009.

U.S. Environmental Protection Agency. "Climate Change and West Virginia." http://www.as.wvu.edu/~bio105/pdf/Climate%20Change%20in%20WV.pdf (Accessed June 2011).

Western Boundary Currents

Western boundary currents are intense jet currents at the western periphery of large-scale oceanic gyres in the World Ocean. As was shown in the pioneer paper of Henry Stommel in 1948, they are the result of two following causes:

1. The β-*effect*, a term that has arisen from the traditional representation of the Coriolis force in the following formula: $f = f_0 + \beta y$, where f_0 is a Coriolis parameter at a definite latitude; in other words, the β-effect is due to spherical form of the Earth turning on its axis.
2. Low conservation of absolute vortex for oceanic motions.

Oceanic gyres are forced by horizontally inhomogeneous, large-scale wind fields (or wind vorticity). For instance, in the North Atlantic Ocean, the anticyclonic subtropical gyre is situated under the northeastern trade wind and midlatitude westerly wind as a result of clockwise wind vorticity, whereas the north tropical cyclonic gyre is a result of anticlockwise wind vorticity between the Intertropical Convergence Zone and the northeastern trade wind. Currents in the western part of each gyre are more intense than in the eastern part because they are dictated by the law of conservation of absolute (relative plus planetary) vortex. Each particle moving northward (southward) gets an additional (loses) planetary vorticity as a result of spherical form of the rotating Earth. In the clockwise gyre, this should be compensated by the increase of relative negative vorticity; that is, by the intensification of clockwise rotatation. In the counterclockwise gyre, this should be compensated by the increase of relative positive vorticity; that is, by the intensification of counterclockwise rotatation. In both cases, this leads to intensification of currents in the western periphery of the basin. In the eastern part of a gyre, all particles move in the opposite direction in comparison with the western part. It leads to the weakening of circulation in the eastern gyre's end.

The β-effect may be also understood in terms of Rossby waves. Long, nondispersive Rossby waves carry (kinetic) energy from the east to the west within each gyre. After their reflection from the western boundary of the basin, the short, dispersive Rossby waves are generated and move to the east. However, the short Rossby waves are dissipated in the relatively narrow vicinity of the near-

A conductivity, temperature, and depth package is deployed east of Abaco Island, Bahamas, as part of a NOAA climate variability study that takes snapshot sections and time series moorings to monitor the transport in the deep Western Boundary Current.

coastal zone just as a result of their shortness and dispersive properties, which leads to more affective realization of dissipative processes. Thus, the kinetic energy of the planetary Rossby waves is accumulated in the vicinity of the western periphery of the gyres.

In fact, the western boundary currents (especially in the Atlantic Ocean) are also controlled by thermohaline factors. The β-effect impacts the thermohaline circulation and causes the intensification of the thermohaline currents in the western part of the basin. Deep thermohaline currents in the North Atlantic Ocean (generating in the region of the sinking of deep Atlantic Ocean water and spreading at depths between 1.5 and 2.5 mi., or 2.5 and 4 km) are southward, while compensative thermohaline currents in the upper baroclinic layer (between the surface and 0.6 to 1.2 mi., or 1 to 2 km) are northward. As a result of superposition of the meridional thermohaline circulation, the wind-driven, northward western boundary currents in the clockwise gyres of the North Atlantic Ocean intensify, while southward currents in the counterclockwise gyres weaken.

The most intense western boundary currents in the Northern Hemisphere are the Gulf Stream, Labrador current, North Brazilian current (Atlantic Ocean), Kuroshio current (Pacific Ocean), and Somali current (Indian Ocean). The velocity in these currents' axes reaches or even exceeds 6.5 ft. (2 m) per second. Detailed analysis of the structure and origins of western boundary currents (such as the Gulf Stream) was conducted by Henry Stommel in 1958 and 1966.

Western boundary currents in the North Atlantic Ocean carry ~100 Sv (1 Sverdrup = 10^6 cu. m per second) of water in the upper baroclinic layer. The wind vorticity accounts for about 30 to 60 Sv (30-60 multiplied by 10^6 cu. m per second). The average power of the source of deep Atlantic Ocean water is about 20 Sv (20 multiplied by 10^6 cu. m per second). Therefore, the joint effect of wind vorticity and meridional thermohaline circulation can explain up to 80 percent of observed transport of the western boundary currents in the North Atlantic Ocean.

The remaining (at least) 20 percent of total transport is a result of the mesoscale eddies. In fact, the western boundary currents, which look like meandered jets, generate the intense mesoscale eddies, the "rings." The typical horizontal size of rings is about 60 mi. (~100 km), and orbital velocity is 3.3 to 6.5 ft. (1 to 2 m) per second. Rings trap the water in their central part and carry it with a typical speed of about a few centimeters per second. The lifetime of the rings may reach four years, after which time most of them are recirculated and feed the western boundary currents. Thus, the mesoscale eddies account for a significant portion of volume transport of the western boundary currents. Recirculation of the Gulf Stream is one of the integral manifestations of mesoscale effects.

Alexander Boris Polonsky
Marine Hydrophysical Institute, Sevastopol

See Also: Coriolis Force; Gulf Stream; Intertropical Convergence Zone; Kuroshio Current; Stommel, Henry; Thermohaline Circulation; Waves, Planetary; Waves, Rossby; Wind-Driven Circulation.

Further Readings

NOAA Ocean Service Education. "Boundary Currents." http://oceanservice.noaa.gov/education/tutorial_currents/04currents3.htm (Accessed December 2011).

Stommel, Henry. *The Gulfstream. A Physical and Dynamical Description*. Berkeley: University of California Press, 1958.
Stommel, Henry. "The Westward Intensification of the Wind-Driven Ocean Currents." *Transactions, American Geophysical Union*, v.29/2 (1948).

Western Regional Climate Center

The Western Regional Climate Center (WRCC), based in Reno, Nevada, and inaugurated in 1986, is part of the network of six regional climate centers in the United States (the other five are the High Plains, midwestern, southern, southeast, and northeast climate centers). The WRCC is administered by the National Oceanic and Atmospheric Administration. Specific supervision is provided by the National Climatic Data Center of the National Environmental Satellite, Data, and Information Service.

The mission of the WRCC is to make reliable and high-standard climate data and information concerning the western region of the United States available to the general public and policymakers. As part of its initiatives to promote more informed decisions by policymakers on weather-related issues, the center carries out applied research related to climate and coordinates climate-related activities at the state, regional, and national levels. The center receives queries from a wide variety of users for different purposes: lawyers, media, insurance companies, different businesses, teachers and students, contractors, the Forest Service, state and local government, and individuals interested in weather observation.

Climate Change Focus in the West

Climate change is one of the center's main areas of research, and it strategically places the effects of climate variability within the geographical context of the western United States. The center investigates the relationship of El Niño/Southern Oscillation to western climate, as well as the climatic trends and fluctuations in the west. It is also the home of the regional climatologist.

The data collected by the center include daily climate observations for a digital period of record (from 6,781 stations, about 2,608 that are now active); summarized monthly climate data (5,240 stations); hourly precipitation data (1,937 stations); upper-air soundings recorded twice a day (about 50 stations); and surface airway hourly observations (over 1,800 stations nationwide). In addition, the center provides access to these databases: Remote Automatic Weather Station, historic lightning data though 1996, Access to Natural Resources Conservation Service SNOTEL, and other western databases.

The WRCC coordinates the work of federal resource management agencies and western committees and commissions. It liaises with other centers and programs, such as the National Climate Data Center in Asheville, North Carolina; regional climate centers; state climatologists and state climate programs; Climate Analysis Center, Washington, D.C.; and National Weather Service.

Differing Views

Although the WRCC is concerned with some of the most evident phenomena of global warming, including the effect of El Niño on western climate, the climatologists at the center do not agree on a single definition of global warming. For example, in July 2007, Jim Ashby, one of the WRCC climatologists, challenged the credibility of weather data collected by weather stations, as they are often moved. Even moving the weather station just few hundred yards could make a difference in temperature and moisture, Ashby maintained, creating a situation in which data belonging to different areas are compared. This comparison of inhomogeneous data would be, according to Ashby, the same as comparing apples and oranges. Ashby has also stressed that even stations that remain in the same place can have changing circumstances that alter weather readings. Atmospheric readings can be altered by the surroundings (the presence of trees favors cooler temperatures, whereas new buildings and more roads and rooftops retain more heat). Ashby uses Reno, where the WRCC is situated, as a classic example of a place where weather is being changed by urbanization. Average low temperatures in the city have risen about 10 degrees F (5.5 degrees C) in the last 20 years. "Most of the warming is due to the fact Reno is growing like a weed," Ashby argued. "The weather station used to be out in a field somewhere. Now it's surrounded by asphalt." This line of reasoning is echoed by global warming skeptics who claim that these moves and changes at weather stations have affected the validity of the climate record.

Another climatologist at the center, Kelly Redmond, has instead pointed out the visible effects of global warming on the vegetation of the region. Western states are heavily dependent on snowpack for their flora. Melting snow provides three fourths of the water in streams. Over the past 35 years, temperatures across the region have risen from 1 to 3 degrees F (0.5 to 1.6 degrees C), causing the snow to melt as much as three weeks earlier and leading to widespread droughts.

In spite of these differences in assessing the extent and impact of the phenomenon, the future of the WRCC, like the other five regional centers, relies on its cooperation with the Regional Integrated Science and Assessments and the state climate offices to provide crucial information on climate change and adaptability to the population. The teamwork among these entities has already produced the WestMap application, which is linked to the WRCC's Website. This tool allows users to create weather maps for western states, taking into account the variability over time of several factors, such as precipitation and temperatures.

Luca Prono
Independent Scholar

See Also: Climate Change, Effects, of; Desert Research Institute; Nevada; Oregon Climate Service; Weather.

Further Readings

Associated Press. "West Getting Burned by Global Warming." http://www.heatisonline.org/contentserver/objecthandlers/index.cfm?id=4659&method=full (Accessed June 2011).

Krier, Robert. "Weather Station Moves Add Degree of Difficulty." http://www.signonsandiego.com/news/nation/20070709-9999-1n9station.html (Accessed June 2011).

Western Regional Climate Center. http://www.wrcc.dri.edu/wrccmssn.html (Accessed June 2011).

Wind-Driven Circulation

Wind-driven circulation (WDC) refers to ocean currents initiated and propelled by winds blowing across the surface of the ocean. Wind-driven circulation is part of the complex system of energy and heat redistribution that helps to draw tropical heat away from equatorial regions to moderate the Earth's climate.

Wind Generation and Oceanic Circulation

Wind is caused by the uneven solar heating of the Earth's surface, which creates areas of high and low atmospheric pressure. Pressure differentials cause air to circulate from areas of high to low pressure. Solar heating is greatest near the equator, causing a region of low pressure. Rising air moves northward as cooler air from high pressure, mid-latitude regions moves toward the equator. The Coriolis effect, the rotational action of the Earth spinning on its access, deflects the winds toward the east, north of the equator, and to the west, south of the equator. This explains why prevailing winds in the mid-latitude Northern Hemisphere come from the west.

An understanding of wind generation helps to explain the operation of ocean currents. There are two general types of ocean circulation and both result from solar heating: WDC and thermohaline circulation (THC). Differential solar heating generates wind. Wind blowing across the surface of the ocean causes the water to move. The direction of the ocean current is determined in part by the direction of the prevailing winds at a given latitude. WDC is responsible for currents near the ocean's surface.

THC drives deep-water ocean currents and vertical movement of water. THCs occur as water density increases, either through cooling as water moves toward the poles, or because of increased salinity. In both cases, the THC is responsible for carrying warmer surface water to the ocean depths. There is an interplay, then, between the WDCs and THCs that is exemplified in the Gulf Stream that moves warm water from the Caribbean in a northeasterly direction, along the eastern seaboard of North America, toward northern Europe. As WDCs carry water away from the tropics, the water cools and becomes more saline. It consequently becomes more dense, which causes it to sink along a THC. This mechanism, whereby warm surface water is transported toward the poles, descends as it becomes cooler, to be re-circulated southward as cool deepwater currents, is sometimes referred to as the North Atlantic conveyor.

Interplay of WDCs and THCs

The Gulf Stream starts as a WDC because it gets its initial energy from the trade winds. Global warming could affect the Gulf Stream in two ways. The first, and more widely reported mechanism occurs because global warming could melt Arctic ice, decreasing sea salinity. This will in turn hinder the ability of the water to descend as part of the THC. The end result is the cooling of northern latitudes.

Global warming could also affect the initial forcing of the Gulf Stream. Some researchers suggest that global warming will disrupt and weaken the trade winds. Diminished trade winds means less energy driving the Gulf Stream northward, also contributing to a cooler climate at high latitudes. This could cause dramatic cooling in northern Europe, which needs the North Atlantic Conveyor to moderate its climate. The paradox is that global warming could ultimately cause regional cooling in some parts of the world by affecting both the THCs and WDCs.

Christopher D. Merrett
Western Illinois University

See Also: Coriolis Force; El Niño and La Niña; Jet Streams; Trade Winds; Winds, Easterlies; Winds, Westerlies.

Further Readings

Saenko, Oleg, et al. "On the Response of the Oceanic Wind-Driven Circulation to Atmospheric CO_2 Increase." *Climate Dynamics*, v.25 (2005).

Toggweiler, J. R. and J. Russell. "Ocean Circulation in a Warming Climate." *Nature*, v.451 (2008).

Winds, Easterlies

Winds are defined by their origins. The "easterly" descriptor refers to winds with an easterly zonal component (coming from the east). These include northeasterly and southeasterly winds. Easterly winds occur at all atmospheric scales, including local and synoptic. However, the term *easterly winds* generally refers to large-scale belts of winds operating within the global circulation of the atmosphere.

In the atmosphere's general circulation, two distinct bands of easterly winds exist: the trade winds and the polar easterlies. They are found at low and high latitudes, respectively, and arise from the dynamics of air flow among pressure systems.

Trade Winds

Among the most consistent winds on Earth, the trade winds (or trades) are part of the Hadley cell circulation found from approximately 0 degrees to 30 degrees north and south of the equator. Air rises near the equator as a result of a combination of convection spurred by intense solar radiation and the low-level convergence of wind in a circumpolar zone called the Intertropical Convergence Zone. As the rising air approaches the tropopause, it turns. At approximately 30 degrees, the air subsides, or sinks, resulting in the belt of persistent subtropical highs. Air diverges out of these anticyclones and flows toward the equatorial low. This outflow of air gives rise to the trades. As the air moves toward the equatorial low, it is deflected as a consequence of the Coriolis force (or effect). This deflection results in northeasterly trade winds in the Northern Hemisphere and southeasterly trades in the Southern Hemisphere. The trade winds are also referred to as tropical easterlies, particularly when the associated vertical wind shear is large.

Polar Easterlies

The polar easterlies, a belt of winds found from approximately 60 to 90 degrees in each hemisphere, constitute another component of the general circulation. These winds originate from dynamics similar to those of the Hadley cell. Thermally driven high pressure exists at the poles; similarly, a circumpolar zone of low pressure is found near 60 degrees. As air flows from the polar high to the subpolar low (as a result of the presence of a pressure gradient), it is deflected. This deflection leads to a belt of easterly winds in both Northern and Southern Hemispheres.

Rising air over the western tropical Pacific, with sinking air over the eastern tropical Pacific, comprises the Walker air current. The trades, as part of the Walker circulation, push warm surface waters toward the west. Any weakening of the trades would disrupt this oceanic transport of water. This dynamic is similar to an El Niño, in which the trade winds slow and even break down. Recent research suggests that climate changes, specifically global warming, would weaken or slow the trade winds. The Walker circulation has already diminished by 3.5 percent since the 1800s. This slowing is projected to continue, and much of this change is attributed to anthropogenic activity.

The culprit in the slowing trade winds is the balance between evaporation and precipitation. To maintain a balance, the atmospheric absorption of moisture must be balanced by its release by precipitation. Water vapor, transported west by the trade winds, is precipitated out over the western Pacific. Warmer temperatures, then, spur the absorption of additional water vapor by enhancing evaporation rates. However, precipitation rates increase more slowly. To balance these processes, wind flow must decrease.

Weakened trade winds would result in numerous consequences, including the disruption of normal weather and climate patterns, as well as suppressed oceanic upwelling and potentially reduced biological productivity. The latter would have important economic ramifications, particularly for the Pacific fishing industries.

Petra A. Zimmermann
Ball State University

See Also: El Niño and La Niña; Walker Circulation; Winds, Westerlies.

Further Readings

Ahrens, C. Donald. *Meteorology Today*. 8th ed. Belmont, CA: Thomson Brooks/Cole, 2007.

Robinson, Peter J. and Ann Henderson–Sellers. *Contemporary Climatology*. 2nd ed. New York: Prentice Hall, 1999.

Vecchi, Gabriel A., Brian J. Soden, Andrew T. Wittenberg, Isaac M. Held, Ants Leetma, and Matthew J. Harrison. "Weakening of Tropical Pacific Circulation as a Result of Anthropogenic Forcing." *Nature*, v.441/4 (May 2006).

Winds, Westerlies

The westerlies are the prevailing winds in the middle latitudes blowing from the subtropical high pressure toward the poles. The westerlies originate due to pressure differences between the subtropical high-pressure zone and the subpolar low-pressure zone. The westerlies curve to the east due to the Coriolis effect caused by the Earth's rotation. In the Northern Hemisphere, the westerlies blow predominantly from the southwest, while in the Southern Hemisphere, they blow predominantly from the northwest. The equatorward boundary is fairly well defined by the subtropical high-pressure belts, while the poleward boundary is more variable. The westerlies can be quite strong, particularly in the Southern Hemisphere, where less land causes friction to slow them down. The strongest westerly winds typically occur between 40 and 50 degrees latitude.

Winds transport heat from warmer areas to cooler areas and help the Earth maintain equilibrium of its thermal environment. In the mid-latitude, the westerlies play a large role in the weather and atmospheric circulation in the middle latitudes. They transport warm, moist air to polar fronts and are also responsible for the formation of extratropical cyclones. In winter, they collect warm, moist air from over the oceans, move it to the cooler continents, and bring heavy rainfall to areas like the northwest coast of the United States. In summer, they collect hot, drier air from over the continents and move it to the oceans.

Global Warming and Westerlies

Does global warming influence westerlies? A recent study of westerlies in the Southern Hemisphere shows that the westerlies are shifting southward toward Antarctica. No conclusion has been made yet, however. Some scientists believe that recent observations are related to global warming, while others believe they are a part of natural variations. The North Atlantic Oscillation (NAO) is one indicator that shows the relationship between global warming and westerlies. NAO is calculated by the difference in pressure between the permanent low-pressure system located over Iceland (Icelandic Low) and the permanent high-pressure system located over the Azores (Azores High). Global warming can reduce the difference in pressure between two places. At a high NAO index, a large pressure difference between the two places induces stronger westerly flows. The storm track advances northward and Europe experiences milder winters but more frequent rainfall in central Europe and nearby. At a low NAO index with suppressed westerlies, the storm track moves more toward the Mediterranean and results in colder winters in Europe and southern Europe, and North Africa receives more storms and higher rainfall.

El Niño-Southern Oscillation (ENSO) is another indicator. In winters of El Niño years, the polar jet stream in the Northern Hemisphere moves further poleward and brings warmer winter weather to the northeastern part of the United States. In the winter of 2006–07, the warming induced was about 9 degrees F (5 degrees C), which was as much as five times the air temperature increase compared to warming in a typical El Niño year. Changes in both surface and upper-level westerlies due to El Niño patterns can also influence the development, intensity, and track of hurricanes over the tropical Atlantic Ocean. In the fall of 2006, El Niño strengthened the upper-level westerlies, increased wind shear, and discouraged tropical cyclogenesis over the tropical Atlantic. Whether or not global warming is behind these stronger El Niño patterns is still being researched. A 2007 climate model by Joellen L. Russell and colleagues indicates that westerlies influence the temperature of the Southern Ocean. According to the model, the southward movement of the Southern Hemisphere westerlies in recent years transfers more heat and carbon dioxide into the deeper waters of the Southern Ocean. This poleward shift of the westerlies has intensified the strength of the westerlies near Antarctica. The pattern could slow down global warming somewhat, but induce ocean levels to rise in Antarctica.

How global warming influences the westerlies still remains in question. The recent observation, however, suggests that global warming brings noticeable change in the westerlies.

Jongnam Choi
Western Illinois University

See Also: El Niño and La Niña; Indian Ocean; Monsoons; Southern Ocean; Southern Oscillation Index.

Further Readings

Chen, G., J. Lu, and D. M. W. Frierson. "Phase Speed Spectra and the Latitude of Surface Westerlies: Interannual Variability and Global Warming Trend." *Journal of Climate*, v.21 (2008).

Toggweiler, J. R. "Shifting Westerlies." *Science*, v.323/5920 (March 13, 2009).

Weisse, Ralf and H. V. Storch. *Marine Climate and Climate Change: Storms, Wind Waves and Storm Surges*. Berlin: Springer/Praxis, 2009.

Wisconsin

Known as America's Dairyland, Wisconsin lies in the northcentral United States between Lake Superior, Upper Michigan, Lake Michigan, and the Mississippi and Saint Croix rivers. The state extends about 295 mi. (475 km) east to west and about 320 mi. (515 km) north to south. The total area of the state is 56,153 sq. mi. (145,436 sq. km), of which 54,426 sq. mi. (140,963 sq. km) is land and 1,727 sq. mi. (4,473 sq. km) is inland water. There are approximately 15,000 lakes. The Wisconsin mainland has at least 574 mi. (925 km) of lakeshore and holds jurisdiction over 10,062 sq. mi. (26,061 sq. km) of lake waters. Wisconsin's landscape can be divided into four main geographical regions: the Superior Upland; the Driftless Area; a large, crescent-shaped plain in central Wisconsin; and a large, glaciated lowland plain in the east and southeast along Lake Michigan. The population was estimated at 5.69 in 2010.

Wisconsin was 63 to 86 percent forested before European settlement. In 2006, Wisconsin's forests covered 46 percent of the state's land area. The most heavily forested region is in the north. Hardwood forest is the most abundant forest type. In 2006, lumber production totaled 583 million board ft. The Chequamegon–Nicolet National Forest, the state's only national forest, is located in northern Wisconsin, covering 616,150 hectares. The 10 state forests cover 190,741 hectares. The economy of Wisconsin is driven by both manufacturing and agriculture. Although Wisconsin is often perceived as a farming state, manufacturing accounts for a far greater part of the state's income.

Changing Climate and Impacts

The Wisconsin climate is typically humid continental, with some modification by Lakes Michigan and Superior. The winters are cold and snowy and the summers are warm and humid. The average annual temperature varies from 39 degrees F (21.6 degrees C) in the north to about 50 degrees F (27.7 degrees C) in the south. The long-term mean annual precipitation ranges from 30 to 34 in. over most of the western uplands and northern highlands, then diminishes to about 28 in. along most of the Wisconsin central plains and Lake Superior coastal area. About two-thirds of the annual precipitation falls during the growing season. It is normally adequate for vegetation, although drought is occasionally reported.

Wisconsin's climate has been warming. Weather station records from 1950 to 2006 show that the average temperature has warmed 1.1 degrees F (0.61 degree C), with a 2.5 degrees F (1.39 degrees C) increase reported in northern Wisconsin. The greatest warming has been in winter, with 2.5 degrees F (1.39 degrees C), and spring, with 1.7 degrees F (0.94 degree C). Annual precipitation has also increased in the past 60 years by about 3.1 in. per year, especially in autumn. The key changes projected in Wisconsin include significant warming, particularly during winter and spring, with more warming during nighttime than daytime; a longer growing season, including an earlier onset of spring; more hot days and fewer cold nights; shorter frost and snow seasons; greater annual precipitation, particularly in winter and spring; and prolonged droughts, with rainfall more concentrated into fewer, more intense downpours.

Changing climate in Wisconsin will affect natural habitats for forests, wildlife, and plant communities. The ranges for some plant and animal species are expanding northward. Forest biomass will increase. Boreal trees, such as black spruce and balsam fir, may no longer grow in the state by the end of this century. Animals like the American marten, spruce grouse, and snowshoe hare may disappear from Wisconsin. Warming temperature in the spring and fall would help boost agricultural production, while increased warming in the summer months could reduce yields of corn and soybeans. Climate change influences water resources, society, and the built environment. In many places in Wisconsin, warming conditions increase the risks of stormwater management and drinking water systems. The impacts on coastal regions include increased shoreline erosion and increased vulnerability of shoreline infrastructure. Coastal wetlands face increased sedimentation from runoff and flooding and greater threats from invasive species. In addition, warming may bring new and more public health problems as heat waves become more frequent and climatic conditions boost air pollutants. Waterborne diseases may multiply as a result of heavy rainfall.

Canadian geese seek refuge in the comparatively warm water temperature during minus 20°F weather in Mequon, Wisconsin, near Lake Michigan. Key climate changes projected for the state include warming during winter and spring and an earlier spring.

Adaptations to a Changing Climate

Over the past several years, Wisconsin has taken important steps to limit global warming emissions from power plants and to boost energy efficiency. Wisconsin has already committed to several actions that will curb the growth of carbon dioxide (CO_2) emissions by 2020. Energy-efficiency programs have helped reduce natural gas and electricity consumption. Wisconsin has also adapted several measures to promote renewable energy and biofuels. In 2006, Wisconsin passed a Renewable Energy and Energy Efficiency Bill that requires 10 percent of the electricity sold in the state in 2015 to come from renewable sources.

In 2007, Wisconsin created a task force on global warming to study measures to reduce a variety of the state's greenhouse gas (GHG) emissions. Wisconsin joined the Climate Registry in 2008. Also in 2008, the task force released Wisconsin's Strategy for Reducing Global Warming. In 2010, the Wisconsin state legislature introduced a new bill, the Clean Energy Jobs Act, to increase renewable energy use, decrease GHG emissions, and meet energy conservation goals.

The Wisconsin Department of Natural Resources is following planned or proposed federal initiatives, including potential federal climate legislation and regulation of GHGs under the Clean Air Act. In 2008, the Wisconsin Environment Research and Policy Center published the *Blueprint for Action: Policy Options to Reduce Wisconsin's Contribution to Global Warming.*

In 2011, the Wisconsin Initiative on Climate Change Impacts (WICCI) released its first adaptation report, which outlines how Wisconsin's climate has been changing and how climate changes will impact Wisconsin's future. These data are informing policymakers, natural resource managers, and stakeholders on possible strategies for lessening the contributions to global warming (mitigation) and for adjusting to changes in the future (adaptation). Many other programs, initiatives, and actions aimed at addressing climate change and reducing Wisconsin's dependence on fossil fuel have been or are being put in place.

Weimin Xi
University of Wisconsin, Madison

See Also: Agriculture; Forests; Midwestern Regional Climate Center.

Further Readings

Governor's Task Force on Global Warming. *Wisconsin's Strategy for Reducing Global Warming.* Madison: Public Service Commission of Wisconsin, 2008.

Waller, Donald M. and Thomas P. Ronney. *The Vanishing Present.* Chicago: University of Chicago Press, 2008.

Wisconsin Department of Natural Resources. http://dnr.wi.gov (Accessed June 2011).

Wisconsin Environment Research and Policy Center. http://www.wisconsinenvironment.org/center (Accessed July 2010).

Wisconsin Initiative on Climate Change Impacts. http://www.wicci.wisc.edu (Accessed June 2011).

Woods Hole Oceanographic Institution

The Woods Hole Oceanographic Institution (WHOI), a world-renowned private, nonprofit organization founded in 1930, is the largest independent oceanographic research institution in the United States. WHOI is committed to multidisciplinary and collaborative higher-education and scientific research that furthers the understanding of the world's oceans and their role within the Earth's ecosystem as a whole. In addition, WHOI disseminates its research findings and information to the public and policymakers to foster understanding and decision making for the greater good of society. At any given moment, projects are in progress around the world, representing a wide range of scientific inquiries in which collaboration and creativity are highly valued and encouraged. Most recently, WHOI released findings that ocean acidification, exacerbated by greenhouse gas (GHG) pollution, is potentially so extensive that shellfish populations are at risk.

WHOI recruits qualified scientists, engineers, staff, and students and provides an interdisciplinary and flexible setting in which students, as future scientists and engineers, can thrive. In 2009, WHOI awarded 99 fellowships and scholarships, as well as 275 stipends. The research activities of WHOI are centered in five departments: (1) applied ocean physics and engineering, encompassing ocean acoustics, observation systems, and immersible vehicles; (2) biology, studying the spatial and temporal distribution of marine organisms; (3) marine chemistry and geochemistry, focusing on chemical analyses and modeling of ocean processes; (4) geology and geophysics, including the study of the oceans' role in past climate change and tectonics; and (5) physical oceanography, including the examination of the geography and physics of the ocean currents and water properties. In December 2010, WHOI celebrated the 50th anniversary of the launch of a buoy into waters off Bermuda. The buoy was in place for 79 days and in turn marked the launch of the WHOI Buoy Group and a new stage in physical oceanography.

WHOI's research efforts are enhanced by four ocean institutes, established in 2000, which address the concerns of members of the general public and policymaking bodies and make research findings available as quickly as possible. The Coastal Ocean Institute encourages pioneering, interdisciplinary experiments, and field missions to increase knowledge and understanding about basic ocean processes. The Coastal Ocean Institute provides access to publications, including its newsletter and annual report, on WHOI's Website. The Ocean Life Institute sponsors studies of the oceans' organisms and processes to understand the evolution of life and adaptability of species to their natural surroundings.

The Ocean and Climate Change Institute (OCCI), among other undertakings, supports research on the effect of GHGs on the ocean and the effect of ocean dynamics that may cause large, sudden climate shifts. OCCI supported 13 research grants for WHOI investigations in 2010, including Arctic Warming and Destabilization of Gas Hydrates (Chris German, Richard Camilli, and Dana Yoerger), Bioenergetic Assessment of Climate Change Impacts on Arctic Copepods (Rubao Ji and Carin Ashijian), and Examining the Effects of Arctic Warming on Coastal Landforms and Estuarine Ecosystems (Jeffrey Donnelly, Joan Bernhard, Liviu Giosan, Andrew Ashton, Kris Karnauskas, and Andrea Hawkes).

WHOI's fourth institute, the Deep Ocean Exploration Institute, encourages multidisciplinary endeavors throughout WHOI and advocates the development of deep-sea technology, including vehicles such as *Alvin*, the deep sea submergence vehicle owned by the U.S. Navy and operated by WHOI. The National Deep Submergence Facility (NDSF), funded by the federal government and located at WHOI, oversees the operation of the remotely operated vehicle *Jason/Medea* and the robotic underwater vehicle *Sentry*, which recently replaced the Autonomous Benthic Explorer (ABE). ABE was lost at sea in early 2010. *Sentry*, like its predecessor, can be used as a stand-alone vehicle or in tandem with another, but can dive to greater depths and has longer deployment capability than ABE. *Alvin* has remained a state-of-the-art vehicle because it has been disassembled, inspected, and reassembled every 3 to 5 years. Every part of *Alvin* has been replaced at least once in the vehicle's lifetime. *Alvin* was slated for retirement in 2008, but instead will be upgraded with new components and a larger personnel sphere, enabling it to eventually dive to over 21,000 ft. (6,500 m).

High Technology

WHOI promotes the use of advanced instrumentation and systems in the laboratory and at sea. WHOI's research fleet, which in addition to its submergence vehicles includes one of the United States' newest research vessels, the navy-owned *Atlantis*, provides students in the joint program offered by WHOI and the Massachusetts Institute of Technology (MIT) incomparable opportunities for research and learning. In addition to the *Atlantis*, WHOI operates two additional ships, the navy-owned *Knorr* and the *Oceanus*—owned by the National Science Foundation—as part of the University–National Oceanographic Laboratory System (UNOLS). WHOI also solely owns the *Tiaga*, which is designed for work close to shore. The *Knorr* is expected to be replaced by a new ship in 2014. WHOI's ship operations are focused in Central America, the Caribbean, North America, South America, and Europe, including Iceland and Greenland.

The MIT/WHOI Joint Program in Oceanography ranks among the leading graduate marine science programs in the world. The program grew out of a memorandum of understanding between MIT and WHOI in 1968. In addition, WHOI's Geophysical Fluids Dynamic (GFD) Program, which was founded in 1959, offers eight to 10 new graduate fellows each year the unique experience of participating in an intensive, 10-week interdisciplinary research program. The students present a lecture and prepare a written report for inclusion in the GFD proceeding's volumes; the newsletters summarizing highlights of the program are available online.

Students who have fulfilled the requirements of a Ph.D. program may be awarded one of several postdoctoral appointments: as a scholar, fellow, or investigator. Postdoctoral scholarships are awarded for 18 months in the fields of oceanography, biology, chemistry, engineering, geology, physics, and biology. In addition, professionals in law, social sciences, or natural sciences may apply for Marine Policy Fellowships, which focus on the examination of maritime conflicts. WHOI also appoints postdoctoral investigators to positions that fall within the parameters of existing research contracts or grants.

In addition to its premier graduate and postdoctoral programs, WHOI provides opportunities for undergraduate students to gain experience through its Summer Student and Minority Fellowship programs, working in partnership with scientists and engineers on a wide range of scientific topics. WHOI also grants a limited number of undergraduate students and certain advanced high school students the chance to participate in WHOI's education programs as guest students for up to one year.

Middle school teachers benefit from participation in professional development workshops presented by WHOI scientists and engineers at the WHOI Exhibit Center, which highlights the institution's research programs and vessels. WHOI also participates in a wide variety of special programs to increase citizens' understanding of the oceans, including The Woods Hole Science and Technology Education Partnership (WHSTEP); the New England Center of Ocean Science Education Excellence (COSEE-NE), one of several centers around the United States funded by the National Science Foundation; and the WHOI Sea Grant program.

In partnership with the Marine Biological Laboratory (MBL), an international research center for biology, biomedicine, and ecology, WHOI makes a critical contribution to the everyday operation of the MBLWHOI Library. The library, acclaimed for having one of the world's largest collections—both print and electronic—in the areas of biomedicine, oceanography, marine biology, and ecology, meets the daily needs of WHOI and MBL scientists and associated researchers and students. The library encourages access to the collection by other institutions and researchers, but its services are restricted to those compatible with the needs of WHOI and MBL.

The MBLWHOI Library houses the Data Library and Archives, which, through a wide-ranging collection of administrative records, oral papers and histories, personal papers and diaries, photographs, films, videos, instruments, and other documents, makes the history of WHOI accessible to WHOI's scientific community. The MBLWHOI Library, committed to preserving and digitizing significant works, offers a large collection of online resources, including the *Alvin* dive log database, *Alvin* ocean floor photos, digital photos from the WHOI archives, and a searchable database of nearly 4,000 films and videotapes in the library's collection. The MBLWHOI Library is one of 12 natural history and botanical libraries that comprise the Biodiversity Heritage Library (BHL), a cooperative undertaking to digitize and make available for open access the literature of biodiversity held in their individual collections. The BHL contributes to the Encyclopedia of Life (EOL), a global partnership dedicated to making knowledge about the world's organisms freely available online.

WHOI receives funding from private contributions and endowment income, but the majority of its funding is generated by peer-reviewed grants and contracts from government agencies, including the National Science Foundation. WHOI has a written donor privacy policy and makes its governing documents, conflict of interest policy, and financial statements available through its Website. Charity Navigator, an independent evaluator of charities in the United States, assigned its highest ranking of four stars to WHOI. Evaluations are based on annual financial information each charity provides to the U.S. Internal Revenue Service.

Robin K. Dillow
Oakton Community College

See Also: Navy, U.S.; Oceanography; Technology.

Further Readings

Charity Navigator. "Marine Biological Laboratory: Biological Discovery in Woods Hole." http://www.charitynavigator.org/index.cfm?bay=search.summary&orgid=4046 (Accessed July 2011).

GuideStar. "Woods Hole Oceanographic Institution." http://www2.guidestar.org/organizations/04-2105850/woods-hole-oceanographic-institution.aspx (Accessed June 2011).

University-National Oceanographic Laboratory System. http://www.unols.org (Accessed June 2011).

Woods Hole Oceanographic Institution. http://www.whoi.edu (Accessed June 2011).

World Bank

The World Bank was established on December 27, 1945, following the ratification of the Bretton Woods agreement. The World Bank was conceived of in July 1944 at the United Nations (UN) Monetary and Financial Conference to provide development assistance to facilitate the reconstruction of Europe following World War II. Since then, the World Bank has provided financial assistance to developing countries following natural disasters and humanitarian emergencies to facilitate postconflict rehabilitation and economic liberalization and development.

The organization of the World Bank consists of two agencies of the five that make up the World Bank Group (WBG). The two agencies are the International Bank for Reconstruction and Development (IBRD) and the International Development Association (IDA). The World Bank consists of 187 member countries, all of whom are shareholders, represented by a board of governors—the ultimate policymakers of the bank. The board of governors consists of member countries' ministers of finance or development, and it meets annually. The governors delegate specific duties to 24 on-site executive directors. The five largest shareholders, France, Germany, Japan, the United Kingdom, and the United States, each appoint one executive director, and the remaining member countries are represented by 19 other executive directors, thus making up the 24. The president of the World Bank, currently Robert B. Zoellick, who serves until 2012, is responsible for chairing the meetings of the board of directors and for overall management of the bank. The president of the bank serves a renewable five-year term. The president is, by tradition, a U.S. citizen nominated by the president of the United States—the bank's largest shareholder. The presidential nominee is confirmed by the board of governors.

Reducing Global Poverty

The World Bank's activities are focused on the reduction of global poverty. It focuses on achieving the Millennium Development Goals (MDGs) that call for the elimination of poverty and the implementation of sustainable development. The constituent parts of the bank, the IBRD and the IDA, achieve their objectives through the provision of low- or no-interest loans and grants to countries with little or no access to international credit markets. The bank is a market-based, nonprofit organization, using its high credit rating to make up for the lack of low-interest rate loans.

The bank's mission is to aid developing countries and their inhabitants achieve the MDGs through the alleviation of poverty by developing an environment for investment, jobs, and sustainable growth. The bank aims to promote economic growth, and through investment in and empowerment of the poor, enables them to participate in development. The bank focuses on four key factors necessary for economic growth and the creation of a business environment: (1) capacity building (strengthening governments and educating government officials); (2) infrastructure creation (the implementation of legal and judicial systems to encourage business, including protecting individual and property rights and honoring contracts); (3) development of financial systems; and (4) combating corruption (eradicating corruption to ensure optimal effect of actions).

The bank obtains funding for its operations primarily through the IBRD's sale of AAA-rated bonds in the world's financial markets. Although this generates some profit, the majority of the IBRD's income is generated from lending its own capital. The IDA obtains the majority of its funds from 40 donor countries that replenish the bank's funds every three years, and from loan repayments, which then become available for relending.

The bank offers two basic types of loans: investment loans and development policy loans. The former support economic and social development projects, whereas the latter provide quick financial dispersal to support policy and institutional reforms in various countries. Although the IBRD provides loans with a low interest rate (between 0.5 and 1 percent for a standard bank loan), the IDA's loans are interest free. The borrowers' project proposals are evaluated for their economical, financial, social, and environmental aspects to ensure that they are viable before any amount of money that is distributed.

One recent controversial loan was the April 2010 loan of $3 billion to the South African state utility company, Eskom, for the construction of a 4,800 megawatt, coal-fired power plant in the Limpopo region of the country. The loan was not

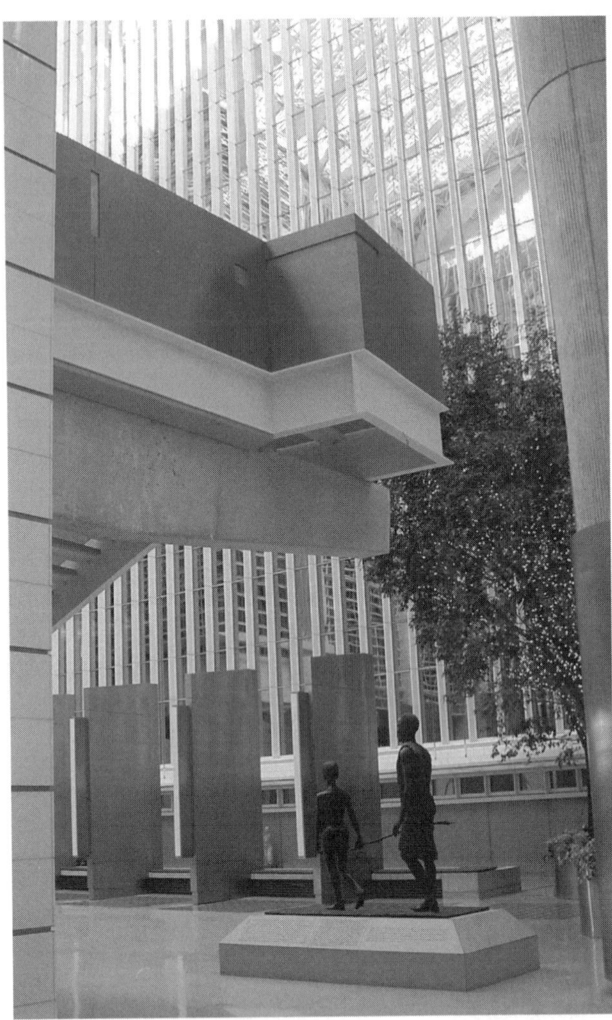

The atrium of the World Bank in Washington, D.C. It has funded a number of climate-related programs, with activities focused on achieving the Millennium Development Goals that call for sustainable development and the elimination of poverty.

supported by the United States, Britain, or the Netherlands due to environmental concerns about the project. The bank argued that the plant would provide much-needed electrical energy, while also spurring economic growth, not only in South Africa, but also within the subregion. The bank approved $750 million in financing for renewable and energy-efficiency projects. Opposition to the project was also questioned, noting that both the United States and Britain allow for the development of coal-powered plants in their countries. The South African plant would use the same technology that the United States and other developing countries use to lower carbon emissions.

The bank also distributes grants to facilitate development projects by encouraging innovation, cooperation among organizations, and participation of local stakeholders in projects. IDA grants are predominantly used for debt relief in the most indebted and poverty-stricken countries, creation of initiatives for the reduction of greenhouse gases (GHGs), amelioration of sanitation and water supply issues, support of vaccination and immunization programs, and support for civil society organizations.

Partnerships and Programs

With respect to its work on issues of global warming, the World Bank has created and funded a number of climate-related partnerships and programs with other agencies and national governments, including the UN Framework Convention on Climate Change (UNFCCC), Global Environment Facility (GEF), Carbon Finance, Energy Sector Management Assistance Program (ESMAP), the Asia Alternative Energy Program (ASTAE), Global Facility for Disaster Reduction and Recovery (GFDRR), Vulnerability and Adaptation Resources Group (VARG), and Global Gas Flaring Reduction partnership (GGFR).

The UNFCCC is an international treaty through which countries consider ways to reduce global warming and cope with inevitable temperature increases. The World Bank is an observer to the UNFCCC and takes part in a number of technical discussions conducted by the UNFCCC Secretariat, such as by the Subsidiary Body for Implementation and the Subsidiary Body for Scientific and Technological Advice.

The GEF is the financing mechanism for the UNFCCC, as well as other key international environmental agreements. As a GEF-implementing agency, the World Bank helps identify, prepare, and implement projects that reduce poverty and benefit the local and global environment. Climate change was the second most active focal area of the GEF's active portfolio at the end of fiscal year 2006.

The World Bank's Carbon Finance Unit offers a means to leverage new private and public investment in projects that mitigate climate change by reducing GHG emissions while promoting sustainable development. Projects relate to rural electrification, renewable energy, energy efficiency, urban infrastructure, waste management, pollution abatement, forestry, and water resources management.

ESMAP, cosponsored by the World Bank and the UN Development Programme, is a global technical assistance program that provides policy advice on sustainable development issues to the governments of developing countries and economies in transition. ESMAP also contributes to technology and knowledge transfer in energy sector management and, since its creation in 1983, has operated in 100 different countries through some 450 different activities. Recently, a new window in ESMAP has opened to support the goals of the Clean Energy Investment Framework.

The GGFR partnership, led by the World Bank, facilitates and supports national efforts to use flared gas by promoting effective regulatory frameworks and tackling the constraints on gas utilization, such as insufficient infrastructure and poor access to local and international energy markets, particularly in developing countries. Launched in 2002, the GGFR partnership brings together representatives of the governments of oil-producing countries, state-owned companies, and major international oil companies so they can overcome the barriers to reducing gas flaring by sharing global best practices and implementing country-specific programs.

Another World Bank initiative is the Sustainable Development Network, a part of which includes a climate change team within the Environment Department of the World Bank. The climate change team provides resources and expertise for the World Bank's participation in international climate change negotiations under the UNFCCC. The team also provides technical advice to the World Bank's GEF Program on the preparation of GEF climate change mitigation projects in energy efficiency, renewable energy, and the development of strategic initiatives with the GEF. The team also is leading the bank's efforts related to climate change vulnerability and adaptation issues for its client countries.

The bank recognizes that achieving objectives related to climate change is a long-term process, requiring the integration of GHG mitigation and vulnerability and adaptation agendas into mainstream operational work. These instruments include planning, policy dialogue, generation and dissemination of knowledge, and investment lending, all of which are primarily aimed at promoting national development priorities. The bank provides support to clients to help them develop better techniques to manage climate change. The support focuses on three key areas: mitigation of GHG emissions, reduction of vulnerability and adaptation to climate change, and capacity building. In the area of GHG mitigation, the bank promotes policy and regulations, as these tend to have large and sustainable effects on improving the efficiency of resource use and, consequently, reducing GHG emissions. In the context of these reforms, the bank mobilizes resources from the GEF and Prototype Carbon Fund to support GHG abatement measures that simultaneously address poverty reduction and sustainable development goals.

In areas of vulnerability and adaptation where the decision on UNFCCC support is pending, the bank will mobilize donor financing for a Vulnerability and Adaptation Facility to better prepare for climate change. Over the medium term, the bank will focus on improving the understanding of the potential effects of climate change and on identifying and implementing no-regrets measures to reduce vulnerability to the current climate and to future climate change. Finally, the bank will assist clients in building the capacity needed to deal with GHG abatement and with vulnerability and adaptation.

Resource and Training Materials

As part of its work in climate change, the World Bank has developed a variety of resource and training materials addressing the fundamental issues underlying climate change. These include examples of successfully integrating climate change concerns into project work or underlying analysis; and basic tools for accurately identifying the climate change effects of projects, baselines, and alternatives. The climate-risk screening tool kit is referred to as ADAPT (Assessment and Design for Adaptation to Climate Change), which is a prototype tool that will screen proposed development projects for potential risks posed by climate change and variability. The bank has also developed a variety of tools and examples to help its staff and clients more readily address the methodological, technical, and economic issues underlying the incorporation of GHG issues into project development and economic analysis. In the area of renewable energy, the Renewable Energy Toolkit comprises a range of tools to help bank staff and country counterparts improve the design and implementation of renewable energy projects. It aims to provide practical implementation needs at each stage of the project cycle and helps project staff determine sustainable business models, financing mechanisms, and regulatory approaches.

Michael Joseph Simsik
U.S. Peace Corps

See Also: Climate Change, Effects of; Economics, Cost of Affecting Climate Change; Economics, Impact From Climate Change; Emissions; Trading; Global Environment Facility; South Africa.

Further Readings

Burton, Ian and Maarten K. Van Aalst. *Come Hell or High Water: Integrating Climate Change Vulnerability and Adaptation into Bank Work.* Washington, DC: World Bank, 1999.

Mathur, Ajay and Maarten K. Van Aalst. *An Adaptation Mosaic: A Sample of the Emerging World Bank Work in Climate Change Adaptation.* Washington, DC: World Bank Global Climate Change Team, 2004.

World Bank. *Contributions from the National Strategy Studies Program to COP 6 Negotiations Regarding CDM and JI.* Washington, DC: World Bank, 2001.

World Bank. *Look Before You Leap: A Risk Management Approach for Incorporating Climate Change Adaptation in World Bank Operations.* Washington, DC: World Bank, 2004.

World Bank. *Making Sustainable Commitments: Environment Strategy for the World Bank.* Washington, DC: World Bank, 2001.

World Bank. *Poverty and Climate Change: Reducing the Vulnerability of the Poor through Adaptation.* Washington, DC: World Bank, 2003.

World Bank. *Sustainable Development and Global Environment: The Experience of the World Bank Group-Global Environment Facility Program.* Washington, DC: World Bank, October 2002.

"World Bank Approves Loan for Coal-Fired Power Plant in South Africa." Reuters (April 9, 2010).

World Business Council for Sustainable Development

The World Business Council for Sustainable Development (WBCSD) was founded by Swiss industrialist Stephan Schmidheiny in 1992 and contains approximately 200 member businesses from over 30 countries and a variety of industrial sectors. The WBCSD maintains a global network of partner organizations, primarily located in developing nations and known as Business Councils for Sustainable Development. The WBCSD also works with governments, intergovernmental organizations, and nongovernmental organizations (NGOs) at the national and international level. Its guiding interest is the relationship between business and sustainable development, advocating the leadership role of business. Energy and climate represent one of the council's four core areas of commitment. Recent emphasis has focused on the partnerships between government, business, and society.

Schmidheiny founded the WBCSD prior to the 1992 Rio Earth Summit, seeking to ensure that business would have an active voice in the Summit, as well as future sustainable development debates and initiatives. His work with Summit Secretary General Maurice Strong resulted in the publication of the book *Changing Course: A Global Business Perspective on Development and the Environment.* After the summit, Schmidheiny and his partners decided to continue their work, merging the WBCSD with the World Industry Council. It is headquartered in Geneva, Switzerland, with a second office in Washington, D.C. Bjorn Stigson assumed the presidency in 1995.

Business and Development: Hand in Hand

The broad mission of the WBCSD centers on the belief that business should provide a leadership role in the drive toward sustainable development and to ensure that sustainable development and business growth go hand in hand. Its stated objectives are to demonstrate past and current business efforts toward sustainable development, share best practices and new technologies among members, advocate for business on sustainable development issues, actively engage in policy development with the goal of creating a suitable framework to allow business to effectively contribute to sustainable development, develop and promote a business platform for sustainable development, and work toward sustainability in developing and transitioning nations.

The WBCSD has undertaken a variety of programs and initiatives in the field of climate change. Project areas have included energy efficiency in buildings, ecopatent commons, and urban infrastructure, while industry areas have included cement, electric utilities, forest products, mining and minerals, and tires. The WBCSD also produces a variety of publications related to climate change and global warming, as well as numerous other sustainable development issues.

With the International Chamber of Commerce, the WBCSD hosts an annual global business day coordinated with the United Framework Convention on Climate Change (UNFCCC) climate change conference to emphasize the role of business within global climate change mitigation efforts. The first such day was held in Bali in 2007, followed by Poznan (2008), Copenhagen (2009), and Cancun (2010). The WBCSD aids the Climate Investment Funds in supporting and financing sustainable development in the area of climate, and informs businesses of the funds through the publication of *Business Guide to the Climate Investment Funds*. The council also helps the self-selection of private sector observers and participates in the Partnership Forum of Funds stakeholders. The WBCSD has served as a partner organization on projects such as the Support to Regulatory Activities for Carbon Capture and Storage (STRACO2) project to develop a regulatory framework in China.

Business and Public Policy

The Council also plays an active role in the interaction between business and public policy, such as the reduction of greenhouse gas (GHG) emissions among signatory nations of the 1995 UNFCCC and the 1997 Kyoto Protocol. The council worked with Ecofys and Climate Focus in the 2010 study report titled "Options for Institutional Engagement in the UNFCCC Process," which focused on private-sector engagement in international climate change policy development. The council also participated in other climate change dialogues, including the Major Economies Forum on Energy and Climate, the G8 and G20, and the Mexican Dialogues of 2010. The council worked with the World Economic Forum to publish the 2008 *CEO Climate Policy Recommendations to G8 Leaders*.

The WBCSD seeks to provide information and tools to private-sector businesses to aid in that sector's adoption of climate mitigation programs and requirements. One of the council's biggest achievements has been the development of the Greenhouse Gas Protocol (GHG Protocol). The WBCSD partnered with the World Resources Institute, as well as large corporations and environmental groups, to create the GHG Protocol, a standardized international tool for the measurement and reduction of GHG emissions. The project began in 1998 when climate change policy became a key international focus as a way to provide corporations and governments with an effective tool to standardize the measurement of GHG emissions. The first edition, titled *The Greenhouse Gas Protocol: A Corporate Accounting and Reporting Standard*, was published in 2001.

The Corporate GHG Accounting and Reporting Standard or Corporate Module offers advice and standard guidelines for the preparation of GHG emissions inventories. The GHG Protocol for Project Accounting or Project Module offers a standard protocol for the quantification of GHG emissions reductions of programs aimed at climate change mitigation. The GHG Protocol is one of the most widely used accounting tools among governments, businesses, developing countries, environmental groups, and standards and programs across the globe.

Marcella Bush Trevino
Barry University

See Also: Greenhouse Gas Emissions; Policy, International; Sustainability.

Further Readings

Greenhouse Gas Protocol Initiative. http://www.ghgprotocol.org (Accessed July 2011).

World Business Council for Sustainable Development. http://www.wbcsd.org/home.aspx (Accessed July 2011).

World Climate Research Programme

The World Climate Research Programme (WCRP) is sponsored by the International Council for Science (ICSU), World Meteorological Organization (WMO), and Intergovernmental Oceanographic Commission (IOC) of the United Nations Educational, Scientific and Cultural Organization (UNESCO). The program brings together the intellectual and structural potentialities related to climate and climate change of more than 185 countries. The program thus aims to work as an international forum to share scientific discoveries and facilitates the understanding of the phenomena that influence climate. The two underlying objectives of the WCRP are to determine the predictability of climate and to assess the effect of human activities on it.

These two objectives stem from the needs identified by the UN Framework Convention on Climate Change. To achieve its objectives, the WCRP adopts a multidisciplinary approach, organizes extensive observational and modeling projects, and encourages research on aspects of climate too large and complex to be addressed by any one nation or single scientific discipline. The WCRP is not open exclusively to scientists. On the contrary, it aims to involve different groups such as policymakers, information end-users, and sponsors in a scientifically accurate debate on climate change and variability.

Climate Data, Research, and Policy

The WCRP was established in 1980. It was initially joint-sponsored by the ICSU and the WMO. Since 1993, the IOC of UNESCO has also become a sponsor of the program. Since its establishment, the WCRP has contributed to the advancement of climate science. Thanks to WCRP researchers, climate scientists can monitor, simulate, and project global climate with improved accuracy so that climate information can be used for governance, in decision making, and in support of a wide range of practical applications. In 2005, after 25 years of serving science and society, the WCRP launched its Strategy Framework 2005–2015, which expresses the program's commitment to working efficiently and effectively toward strengthening knowledge and increasing capabilities with regard to climate variability and change. Titled the Coordinated Observation and Prediction of the Earth System, the framework aims "to facilitate analysis and prediction of Earth system variability and change for use in an increasing range of practical applications of direct relevance, benefit and value to society." The WCRP is thus devoted to providing a larger series of products and services to an ever-increasing group of users. The WCRP intends to reach this goal through the integration of observations and models to generate new understanding and improve climate predictions.

Today, the WCRP covers studies of the different parts of the Earth's climate system: global atmosphere, oceans, sea and land ice, the biosphere, and the land surface. The major core projects, diverse working groups, various cross-cutting activities, and many cosponsored activities of the WCRP all aim to improve scientific understanding of processes that can enable better forecasts.

The WCRP also lays the scientific foundation for meeting the research challenges posed in Agenda 21, a plan of action by the United Nations in every area where humans impact the environment, and provides the international framework for scientific cooperation in the study of global climate change. Through an annual series of reports, WCRP research addresses the many outstanding issues of scientific uncertainty in the Earth's climate system, and WCRP scientists contribute significantly to the collection and improvement of climate observations, model development, and understanding of the climate system necessary for the detection and attribution of past climate change, and the provision of climate information.

The Global Energy and Water Cycle Experiment (GEWEX) project studies the dynamics and thermodynamics of the atmosphere, the atmosphere's interactions with the Earth's surface (especially over land), and the global water cycle. GEWEX uses suitable models to represent and forecast the variations of the global hydrological regime and its effect on atmospheric and surface dynamics. GEWEX also focuses on variations in regional hydrological processes and water resources and their response to changes in the environment, such as the increase in greenhouse gases. GEWEX projects are divided into three focus areas corresponding to the key elements in the global energy and water cycle: radiation, hydrometeorology, and modeling and prediction.

The Climate Variability and Predictability (CLIVAR) project, set up in 1995, specifically targets climate variability. Its mission is to observe, simulate, and predict the Earth's climate system, with a focus on ocean–atmosphere interactions. CLIVAR seeks to develop predictions of climate variations on seasonal to centennial timescales and to refine the estimates of anthropogenic climate change. CLIVAR also includes a Working Group on Seasonal to Interannual Prediction, which oversees development of improved models, assimilation systems, and observing system requirements for seasonal prediction.

The section of WCRP dealing with the Stratospheric Processes and Their Role in Climate (SPARC), founded in 1993, carries out research on the chemistry of the climate system. In particular, it focuses on the interaction of dynamic, radiative, and chemical processes. SPARC's projects include the construction of stratospheric reference climatology and the improvement of understanding of trends in temperature, ozone, and water vapor in the stratosphere. SPARC also studies gravity wave processes, their role in stratospheric dynamics, and how these may be represented in models.

The Climate and Cryosphere (CliC) project, founded in 2000, measures the effects of climatic variability and change on components of the cryosphere and their consequences for the climate system. CliC is also charged with the task of improving the management of data and information relating to the cryosphere and climate, and with making data more readily available for use by the broader scientific community. To this end, CliC has established a Web-based Data and Information Service for CliC.

The Surface Ocean–Lower Atmosphere Study aims to quantify the key biogeochemical–physical interactions and feedbacks between the ocean and the atmosphere. WCRP cosponsors the project jointly with the Commission on Atmospheric Chemistry and Global Pollution, the International Geosphere-Biosphere Programme, and the Scientific Committee on Oceanic Research. The project investigates biogeochemical interactions and feedbacks between ocean and atmosphere, exchange processes at the air–sea interface and the role of transport and transformation in the atmospheric and oceanic boundary layers, and air–sea fluxes of carbon dioxide and other long-lived radiatively active gases.

The Working Group on Surface Fluxes was established in 2007 to review the requirements of the different WCRP schemes for surface sea fluxes, including biogeochemical fluxes, to coordinate the various related research initiatives and to encourage research and facilitate operational activities on surface fluxes.

Titles of major, ongoing research initiatives include the Atmospheric Model Intercomparison Project and the Global Energy and Water Cycle Experiment.

In August 2011, the WCRP held its First Open Science Conference, Climate Science in Service to Society, in Denver, Colorado, a gathering of the international scientific research community focused on climate variability and change.

Luca Prono
Independent Scholar

See Also: Carbon Dioxide; Climate; Climate Models; Policy, International.

Further Readings

Slaymaker, Olav and Richard Kelly. *Cryosphere and Global Environmental Change*. Hoboken, NJ: Wiley & Sons, 2007.

World Climate Research Programme. http://www.wcrp-climate.org (Accessed March 2012).

World Health Organization

The World Health Organization (WHO) is a specialized agency of the United Nations that acts as a coordinating authority on international public health. Established on April 7, 1948, with headquarters in Geneva, Switzerland, the agency inherited the mandate and resources of its predecessor, the Health Organization, which was an agency of the League of Nations. WHO's constitution states that its objective "is the attainment by all people of the highest possible level of health."

History, Structure, and Leadership

WHO is one of the original agencies of the United Nations, its constitution formally coming into force on April 7, 1948. The remaining activities of the League of Nations Health Organization were transferred to the newly formed WHO. Additionally, the epidemiological service of the French Office International d'Hygiène Publique was incorporated into the World Health Organization.

WHO has 193 member states, which includes the 191 members of the United Nations, the Cook Islands, and Niue. Nonstate territories of member states may join as associate members, which means they receive full information but limited participation and voting rights; Puerto Rico and Tokelau are associate members. The third category of membership is observer status, which has been granted to Palestine, the Holy See, and the Republic of China. Countries that have diplomatic recognition, but that are nonmembers, are Liechtenstein and the rest of the nonmember states. In addition to the observer states and diplomatic entities, the observer organizations of the International Committee of the Red Cross and International Federation of Red Cross and Red Crescent Societies have entered into "official relations" with WHO and are invited as observers.

Governance, Financing, and Activities

WHO member states appoint delegations to the World Health Assembly, WHO's supreme decision-making body. The World Health Assembly generally meets once a year, in May. The primary functions of the assembly are appointing a director-general for a five-year term, evaluating the organization's financial policies, considering the financial policies of the organization, and reviewing and approving the proposed program budget. The assembly elects 34 members, technically qualified in the field of health, to the executive board for three-year terms. The main functions of the board are to carry out the decisions and policies of the assembly, advise it, and facilitate its work in general.

WHO is financed by contributions from member states and donors. In recent years, WHO's work has involved increasing collaboration with external bodies; there are currently around 80 partnerships with nongovernmental organizations (NGOs), such as the Bill and Melina Gates Foundation and the pharmaceutical industry. By 2007, voluntary contributions to WHO from national and local governments, NGOs, and the private sector were more than double the level of dues from member nations.

WHO coordinates internationally to control outbreaks of infectious diseases such as malaria, tuberculosis, and influenza, and sponsors programs to prevent, treat, and eradicate such diseases. The greatest success of this program was the elimination of smallpox in 1980—the first disease in history to be eliminated by human effort.

A second initiative of WHO is to develop and promote the use of evidence-based tools to inform health policy options. It oversees the implementation of international health regulations and publishes the *International Statistical Classification of Diseases and Related Health Problems*. The organization has published tools for monitoring the capacity of national health systems and international workforces and the ability of nations to meet primary healthcare goals. The major international health policy frameworks in which the organization participates are the code of International Health Regulations (annual), the Framework Convention on Tobacco Control (2003), and the Global Code of Practice on the International Recruitment of Health Personnel (2010).

The third initiative of the organization is carrying out health-related campaigns and advocating for global health. The most visible campaign each year is World Health Day, which is focused on a different health topic each year. Recent examples include maternal health and water safety.

The final activity of WHO is conducting research in a variety of health-related areas, such as communicable diseases, reproductive health, neglected tropical diseases, health policy, and health systems. To successfully conduct research, the organization relies on the expertise of many world-renowned scientists, such as the Expert Committee on Biological Standardization. The organization advocates the activities of member states to use the research produced to address their national health needs and promote knowledge transfer.

Each year, the World Health Organization produces a series of publications. The most famous is the *World Health Report*, which is a series of global health policy reports. The next major publication is the *Bulletin of the World Health Organization*. The organization also co-publishes, with BioMed Central, a scientific journal called *Human Resources for Health*.

Stance on Climate Change

The World Health Organization acknowledges global climate change and dedicated World Health Day 2008 to health risks caused by climate changes. Its beliefs are affirmed by this statement:

> The core concern is succinctly stated: climate change endangers human health. The warming of the planet will be gradual, but the effects of extreme weather events—more storms, floods, droughts, and heat waves—will be abrupt and acutely felt. Both trends can affect some of the most fundamental determinants of health: air, water, food, shelter, and freedom from disease.

The organization has noted that global warming occurring since the 1970s caused over 140,000 excess deaths annually by 2004. Many of the major killers, such as diarrhea, malnutrition, malaria, and dengue fever, are highly climate sensitive; WHO expects these to worsen as the climate changes. Areas with weak health infrastructure, primarily in developing countries, will be the most vulnerable without assistance to prepare themselves for climate change and respond to its impacts. Besides the benefits to the climate,

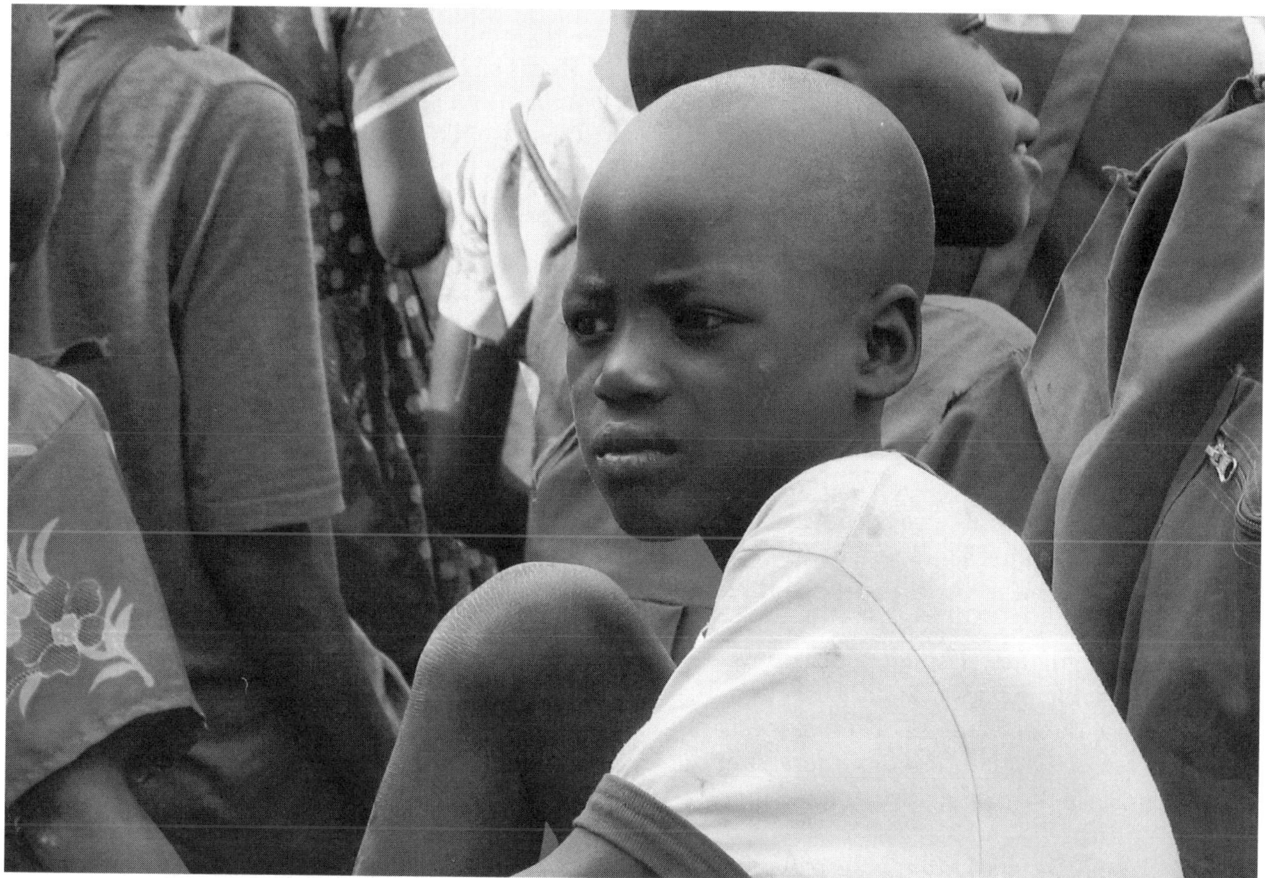

Children in Uganda: The World Health Organization (WHO) coordinates reviews of scientific evidence on the links between climate change and health. Increased temperatures, rising sea levels, and changing weather patterns would affect food, water, and shelter. The official stance of WHO is that although global warming may bring some localized benefits, such as fewer winter deaths and increased food production in some areas, the overall health impacts are likely to be overwhelmingly negative.

reducing greenhouse gas emissions through better transportation, food, and energy-use choices can result in improved health.

WHO is concerned about many climate-change related risks on health around the world:

- *Extreme heat:* High heat contributes to cardiovascular and respiratory disease. In the European heat wave during the summer of 2003, more than 70,000 excess deaths were recorded. Pollen counts, which trigger asthma, are also higher in extreme heat.
- *Natural disasters*: Since the 1960s, weather-related natural disasters have reportedly more than tripled worldwide. These disasters result in over 60,000 deaths annually, mainly in developing countries.
- *Rising sea levels*: Rising seas and extreme weather events will destroy homes, medical facilities and other essential services. More than half of the world's population lives within 60 km of the sea. Climate change refugees are more at risk of a range of negative health effects.
- *Variable rainfall patterns*: Fresh water supplies are likely to be affected by a change in precipitation, and the lack of safe water can increase the risk of poor hygiene and diarrheal disease. Extreme water scarcity can lead to drought and famine. The WHO estimates that by the 2090s, climate change will likely increase the area, frequency, and average duration of drought.
- *Flooding*: Floods, which contaminate freshwater and increase the risk of waterborne diseases, are becoming more frequent and intense. Floods also cause drownings and physical injuries and disrupt health services.
- *Food production*: Rising temperatures and variable precipitation are likely to decrease the production of staple foods in many of the poorest regions, increasing the occurrence of malnutrition-related illness and deaths.
- *Disease*: Climatic conditions strongly affect the transmission season and geographic range of diseases carried by insects, snails or other cold blooded animals, and affect their transmission seasons of important vector-borne diseases and to alter their geographic range.

Caitlin M. Augustin
University of Miami

See Also: Climate Change, Effects of; Diseases; Health; Poverty and Climate Change.

Further Readings

Iriye, Akira. *Global Community: The Role of International Organizations in the Making of the Contemporary World.* Berkeley: University of California Press, 2002.

Siddiqi, Javed. *World Health and World Politics: The World Health Organization and the UN System.* Columbia: University of South Carolina Press, 1995.

World Health Organization. "Constitution of the World Health Organization." http://www.who.int/entity/governance/eb/who_constitution_en.pdf (Accessed July 2007).

World Health Organization. *Monitoring the Building Blocks of Health Systems: A Handbook of Indicators and Their Measurement Strategies.* Geneva: WHO, 2010. http://www.who.int/healthinfo/systems/monitoring/en/index.html (Accessed July 2007).

World Meteorological Organization

The World Meteorological Organization (WMO) is a specialized agency of the United Nations (UN). It articulates UN policy on the state and behavior of the Earth's atmosphere, its interaction with the oceans, the climate it produces, and the resulting distribution of water resources. Because of this emphasis, WMO increasingly regards global warming as one of its major concerns. The organization supports intergovernmental legal agreements on major global environmental concerns, such as ozone-layer depletion, climate change, desertification, and biodiversity. These include the UN Framework Convention on Climate Change (UNFCCC), UN Convention to Combat Desertification, and Vienna Convention on the Protection of the Ozone Layer. In particular, WMO supports the UNFCCC through the Global Atmosphere Watch (GAW) program, which contributes to implementing the Global Climate Observing System. In recognition of its role in the study of climate change and in campaigns to contain its most negative consequence, the Intergovernmental Panel on Climate Change—jointly established by the WMO and UN Environment Programme in 1988—was the recipient with former Vice President Al Gore of the 2007 Nobel Peace Prize.

WMO has a membership of 188 countries and territories. It originated from the International Meteorological Organization, which was founded in 1873. Established in 1950, the following year, WMO was assigned the role of specialized agency of the UN for meteorology (weather and climate), operational hydrology, and related geophysical sciences. It is based in Geneva, Switzerland.

International Cooperation and Mission

WMO seeks to provide the framework for an international cooperation regarding climate matters. The organization points out that weather, climate, and the water cycle do not belong to single nations, so international cooperation at a global scale is essential for the development of meteorology and operational hydrology. Since its establishment, WMO has exercised its leadership in international programs and services for the preservation of the environment and the welfare of humanity as a whole. The National Meteorological and Hydrological Services, for example, are designed to prevent human losses and damages to infrastructures due to natural disasters, to preserve the environment, and to ensure that all sectors of society have access to water resources, food security, and transportation.

WMO also encourages the free exchange of data and information, products, and services in real- or near-real-time on these topics. In the specific case of disasters related to weather, climate, and water, which represent nearly 90 percent of all natural catastrophes, WMO issues advance warnings with the goal of saving lives and reducing damages to property and the environment. In addition, the organization works to reduce the effects of human-induced disasters, such as those associated with chemical and nuclear accidents, forest fire, and volcanic ash. WMO's central role in international efforts to monitor and protect the environment lies in its support of the implementation of a number of environmental conventions and in advising governments on environmental issues.

WMO supports the establishment of national and international networks for meteorological, climatological, hydrological, and geophysical observations, as well as for the exchange, processing, and standardization of related data. It also provides technology transfer, training, and research, and fosters collaboration between the National Meteorological and Hydrological Services and its members. The organization conceives of meteorology as supplying public weather services to the different economic sectors of society, such as agriculture, tourism, aviation, and shipping. Thus, WMO contributes to policymaking in these areas at national and international levels. The organization claims that its activities contribute toward ensuring the sustainable development and well-being of nations.

In application of its mission to global warming related issues, WMO coordinates the international collection of data to assess atmospheric–ocean processes and interactions, such as El Niño/La Niña, and water resources availability. Most significantly for global warming, the WMO lists among its programs the Global Atmosphere Watch (GAW), which includes a coordinated global network of 80 countries that host observing stations for the collection of climate data.

WMO's interest in a program of atmospheric chemistry and the meteorological aspects of air pollution dates back to the 1950s. This included assuming responsibility for standard procedures for uniform ozone observations and establishing the Global Ozone Observing System during the 1957 International Geophysical Year. In the late 1960s, the Background Air Pollution Monitoring Network was set up and was subsequently consolidated with the Global Ozone Observing System into the current GAW in 1989. GAW provides data to scientifically explain the changes in the chemical composition and related physical characteristics of the atmosphere that may negatively affect the environment. The priorities of the scheme have been identified in greenhouse gases (GHGs) for possible climate change, ozone and ultraviolet radiation for both climate and biological concerns, and certain reactive gases and the chemistry of precipitation. GAW is intended to provide accessible, high-quality atmospheric data to the scientific community. These components include measurement stations, calibration and data-quality centers, data centers, and external scientific groups for program guidance. To meet the future challenges of global warming, GAW intends to focus more closely on understanding and monitoring human actions on the atmosphere. In addition to the traditional surface-based network, GAW has increasingly been developing airborne and space-based observations through aircraft and satellite measurements. These data are particularly important to understanding the composition of the troposphere and the concentration of GHGs in it.

Luca Prono
Independent Scholar

See Also: American Meteorological Society; Antarctic Meteorology Research Center; Climate; Global Warming; Intergovernmental Panel on Climate Change; Royal Meteorological Society.

Further Readings

Global Atmosphere Watch. http://www.wmo.ch/pages/prog/arep/gaw/gaw_home_en.html (Accessed July 2011).

World Meteorological Organization. http://www.wmo.ch/pages/index_en.html (Accessed July 2011).

World Resources Institute

The World Resources Institute (WRI) is a nonpartisan and nonprofit environmental organization working to find practical applications for theoretical research on the protection of the Earth and the improvement of people's lives. WRI supplies information and proposals for policies that promote sustainable development, both in environmental and social terms. It focuses on environmental preservation and sustainable development, climate protection from damages caused by greenhouse gases (GHGs), and providing guidelines on how to adapt to those aspects of climate change that appear inevitable.

WRI is based in Washington, D.C., and has a staff of more than 100 scientists, economists, policy experts, business analysts, statistical analysts, mapmakers, and communicators. The institute is convinced that the shift to low-carbon technology will only occur if business owners and shareholders can be persuaded of its profitability. To WRI, business investors are instrumental in solving the climate crisis. WRI produces a highly respected biennial publication, the *World Resources Report*, which supplies data and in-depth analysis on current environmental issues, such as the importance of efficient ecosystem management for rural poverty relief. The report is a collaborative product of WRI with the World Bank, the United Nations (UN) Environment Programme, and UN Development Programme (UNDP). The *World Resources Report* was launched in 1986 to bridge the gap in information about the conditions of the world's natural resources.

History

WRI was founded in June 1982 by American lawyer and environmental activist James Gustave Speth, a former chairman of the U.S. Council on Environmental Quality and later professor of law at Georgetown University. He also acted as WRI's first director. He held this position until January 1993, when he became director of the UNDP and was succeeded by Jonathan Lash, senior staff attorney at the Natural Resources Defense Council from 1978 to 1985. The John D. and Catherine T. MacArthur Foundation of Chicago provided $15 million to help finance WRI's first five years of operation. The institute was organized as a nonprofit Delaware corporation that could receive tax-deductible contributions. WRI's current mission is not to promote environmental activism and militancy, but to work as an independent institution that should carry out scientifically sound research and suggest viable policies.

In 1985, WRI was one of the first research centers to organize an international meeting on the rising emissions of carbon dioxide (CO_2) and other GHGs into the atmosphere. During its over two decades of activity, WRI has attracted other centers that have chosen to merge with the institute, including the North American office of the International Office for Environment and Development and the Management Institute for Environment and Business. In 1990, the UNDP commissioned WRI with a study that eventually resulted in the creation of the Global Environment Facility. Throughout the 1990s, WRI played a key role in initiatives aimed to contain the phenomenon of global warming. In 1992, the institute made important contributions to the development of the Convention on Biological Diversity, which was then signed at the Earth Summit in Rio de Janeiro. In 1999, WRI committed to stopping its emissions of CO_2. The institute has designed a specific project, U.S. Climate Action, designed to support President Obama's pledge to reduce U.S. greenhouse emissions by 17 percent below 2005 levels by 2020.

Projects and Programs

The activities of the institute are structured around four key areas and goals. In the areas of people and ecosystems, the institute aims to counter the fast decline of ecosystems and guarantee that they continue to provide humans with vital goods and services. In its governance and access program areas, WRI works to improve public knowledge about decisions on natural resources and the environment. In this way, the organization attempts to make local governments and international bodies more accountable to people in their environmental and social policies.

In the area of climate protection, energy, and transport, which is more directly linked to climate change, WRI promotes awareness among citizens and policymakers on the need for economic and energy measures to reduce the degree of climate change induced by humans. The institute is also committed to helping humanity and the natural world adapt to the climate change that is already taking place. As for the areas of markets and enterprise, WRI seeks to promote economic development that may increase social opportunities and protect the environment. This area is also particularly relevant to global warming, as the institute works with nations to show that reductions in greenhouse emissions do not necessarily limit economic and industrial development.

Several WRI projects are devoted to show how growth can be powered by clean energy sources: Low-Carbon Development in Emerging Economies, Low-Carbon Energy Technology, and Business and Climate. WRI has worked with the private sector to find profitable solutions that bring both economic and environmental benefits. WRI also created the GHG accounting system, which companies all over the world use to account for their emissions.

The institute has also created several important networks, such as the Global Forest Watch, which monitors the conditions of forests; the Access Initiative, a global forum of civil society organizations committed to improving citizen access to information and favor their participation in decisions that affect the environment; and the Green Power Market Development Group, a partnership of Fortune 500 companies devoted to establishing corporate markets for renewable energy. WRI has also been responsible, in partnership with Mexico City, for the creation of the Bus Rapid Transit Corridor, a transport system designed to reduce environmental damages. The institute is collaborating with metropolises such as Shanghai, Hanoi, and Istanbul for the creation of similar systems. In 2011, WRI was one of the co-sponsors of the Sixth Asia Clean Energy Forum.

Luca Prono
Independent Scholar

See Also: Climate Change, Effects of; United Nations Development Programme; United Nations Environment Programme.

Further Readings

Charity Navigator. "World Resources Institute." http://www.charitynavigator.org/index.cfm?bay=search.summary&orgid=4766 (Accessed December 2011).

Greenhouse Gas Protocol. "The Sustainability Consortium members Vote to Adopt the Greenhouse Gas Protocol Product Standard." http://www.ghgprotocol.org/feature/sustainability-consortium-members-vote-adopt-greenhouse-gas-protocol-product-standard (Accessed December 2011).

World Resources Institute. http://www.wri.org (Accessed July 2011).

World Systems Theory

An underexamined aspect of global warming is the geographic separation between nations that generate greenhouse gases and the people who are most likely to be hurt by global warming. World Systems Theory (WST) attempts to explain how climate change is both a cause and a consequence of existing social and regional inequalities, as well as how geopolitical structures in the global economy ensure that the benefits and costs of burning fossil fuels are not shared equally.

WST evolved to counter free-market economists such as Walter Rostow, who argued that countries were poor because of deficiencies within those countries. Less developed countries could advance by emulating wealthy countries, which had moved through several stages of development.

Dependency Theory

A chorus of critics challenged this view because, from their perspective, adherents to the stages of development view blamed poor countries for their poverty. In response, critics argued that the lack of development in places such as South America was a result of the economic and political structures imposed by wealthy countries through colonial and lingering postcolonial relationships. Foremost among these scholars was Andre Gunder Frank.

His dependency theory argued that as capitalism diffused outward from the economic core of western Europe over the past 500 years, it transformed the places at the periphery that were drawn into the growing capitalist world system. These new territories in Africa and Latin America were thrust into a subordinate role, and their economies were restructured to suit the needs of the colonial powers in Europe. Frank summarized this subordination as the "development of underdevelopment." Capitalist penetration into new regions underdevelops or undermines the economic potential of these places. Frank wanted to shift the blame for poverty away from the processes within poor countries and to place responsibility for poverty on the structural relationships imposed by colonial, and later, neocolonial, core powers.

Building on dependency theory, scholars such as Immanuel Wallerstein proposed WST to provide more historical context to the processes leading to the development of inequality. WST sorts the countries of the world into three broad categories, according to the type of economic processes that predominate in those countries. Countries in the periphery are largely dependent on agriculture and other forms of natural resource extraction. Social inequality is high because most jobs are typically low skill and low wage. Core countries have diverse economies based on manufacturing, services, and information technology. An intermediate category called the semiperiphery includes countries that are more economically diverse than peripheral countries, but that have invested less in manufacturing and other tertiary sectors. Hence, the level of development is less than that of core countries.

WST improves on the rigid core-periphery dichotomy of dependency theory by adding the semiperipheral category. This modification helps to explain the historical reality of peripheral countries such as Korea ascending to core status, or how the global economy can relegate former colonial powers such as Portugal to semiperipheral status. WST claims that the international division of labor creates a system of unequal exchange and income inequality between core, semiperipheral, and peripheral countries. Peripheral countries extract raw materials and ship them largely unprocessed to the semiperipheral and core countries for processing. The finished goods are then shipped back to the peripheral countries at much higher prices. This system of unequal exchange is perpetuated by protectionist international trade agreements; copyright and patent laws, which limit the diffusion of processing technology to the periphery; multinational corporations, which repatriate profits to the core that were earned by extracting resources in the periphery; and wealthy local elites in poor countries, who benefit from status quo relationships between the core and periphery.

The creation and perpetuation of global economic and political inequality has direct implications for understanding the effects of global climate change. Countries in the periphery are far more reliant on agriculture than countries in the core. Countries in the core are much more invested in manufacturing. As a consequence, they produce disproportionately large amounts of greenhouse gases. That means that there is a geographic separation between the largest producers of greenhouse gases and the regions that will suffer the most harm from global warming. Climate changes will have a disproportionately negative impact on the economies and communities in the periphery. The fact that countries at the periphery have the fewest resources to adapt to climate change makes matters even worse. WST is a useful counterbalance to traditional neoclassical discussions of market-based solutions to greenhouse gas reductions because it brings a geopolitical perspective to the understanding of climate change.

Christopher D. Merrett
Western Illinois University

See Also: Developing Countries; Economics, Cost of Affecting Climate Change.

Further Readings

Frank, Andre Gunder. *Capitalism and Underdevelopment in Latin America*. New York: Monthly Review Press, 1967.

O'Brien, Karen and Robin M. Leichenko. "Winners and Losers in the Context of Global Change." *Annals of the Association of American Geographers*, v.93/1 (2003).

Rice, James. "Ecological Unequal Exchange: Consumption, Equity, and Unsustainable Structural Relationships Within the Global Economy." *International Journal of Comparative Sociology*, v.48/1 (2007).

Roberts, Timmons. "Global Inequality and Climate Change." *Society and Natural Resources*, v.14/6 (2001).

Rostow, Walter. *The Stages of Economic Growth: A Non-Communist Manifesto*. Cambridge: Cambridge University Press, 1960.

Wallerstein, Immanuel. *The Capitalist World-Economy*. Cambridge: Cambridge University Press, 1979.

World Trade Organization

The World Trade Organization (WTO), headquartered in Geneva, Switzerland, is the global body that deals with the global rules of trade in order to facilitate trade between nations. The WTO was established in 1995 after the completion of the Uruguay Round Talks and is the successor to the General Agreement on Tariffs and Trade (GATT). Only nations can be members in the WTO. The various WTO agreements involve the areas of goods and services, intellectual property rights, investment-related measures, and trade in agricultural products. There is a dispute settlement mechanism under WTO that functions as a mechanism to deal with trade disputes between member countries.

Expansion in Global Trade

Global trade has expanded in volume 32 times between 1950 and 2007. With global economic integration and the end of the cold war, more countries have been participating in international trade under WTO rules than ever before. The relationship between increased production, higher trade volume, and emissions is unclear and under debate. As increased levels of economic activity and global trade require greater material and energy use, an increase in greenhouse gas (GHG) emissions is inevitable; however, because of greater energy efficiency, an increase in productivity, and technological advances, the increase in emissions may not be directly proportional to the expansion of the economy, the increase in production of goods, and the provision of services. Further, while more open trade can facilitate increases in the availability and consumption of climate-friendly goods and services and diffusion of environmentally sound technologies (ESTS), studies indicate that more open trade is more likely to result in increases in CO_2 emissions, despite an increase in energy efficiency and a decrease in energy intensity in production.

In the trade-environment debate, environmental advocates and nongovernmental organizations argue that uncontrolled trade expansion is harmful to the environment and will result in more emissions, but many economists dismiss such views as too simplistic and point out that the overall impact of trade on global emissions cannot be easily determined.

The Development Doha Round (DDA), the trade-negotiation round begun in November 2001 under the auspices of the WTO, is in progress. One of the objectives of the DDA is to liberalize trade in the area of environmental goods and services. Countries have made proposals for liberalization by reduction in tariffs and other measures, but at this stage, it is not clear whether the DDA negotiations will conclude soon, with binding commitments in environmental goods and services.

Liberalized Trade and Industry

While liberalized trade in environmental goods and services may be helpful for the environment, countries are less keen to opt for such liberalization if it runs counter to the interests of local industry, which may not be able to compete against cheap imports. Within the WTO Committee on Trade and Environment, issues such as carbon border adjustments and carbon footprint schemes are being discussed. In most of these issues, the key question is whether such measures will be used as disguised protectionism, and whether such measures are compatible with rules under WTO agreements.

The issue is complex because not all WTO member states are parties to the Kyoto Protocol, and may not be parties to any future agreement on climate change or a successor to the Kyoto Protocol, and vice versa. Thus, a country that has not signed the climate change agreement or the Kyoto Protocol can still use the WTO dispute settlement mechanism to challenge measures taken by another country or group of countries for being inconsistent with WTO Agreements—if both are members of WTO—under the protocol or climate change agreement or any unilateral trade measure that imposes some conditions to protect environment or to reduce emissions.

Article XX of GATT permits the use of product-related measures and trade restrictive measures under some circumstances, if they are "necessary to protect human, animal, or plant life or health," or "relate to conservation of exhaustible natural resources," but the introductory paragraph of Article XX, also known as "chapeau," prohibits measures that are means for "arbitrary or unjustifiable discrimination between countries" or measures that are "disguised restrictions on international trade."

The interpretation of Article XX by the panels and appellate bodies set up under the dispute settlement mechanism indicates that WTO rules permit use of trade restrictions under some circumstances, but as the decisions are given on a case-by-case basis, in some circumstances trade restrictions citing protection of environment may not be accepted as compatible with WTO rules. Generally, a necessity test has been incorporated in WTO agreements (such as GATT's Technical Barriers to Trade agreement).

The necessity test determines if the trade restrictive measures are really necessary to fulfill the objective for the measures being imposed. The WTO panels examining the issue take into account whether another, less trade-restrictive measure could have achieved the same objective. If such a measure is available, the application of the original measure can be challenged for consistency with WTO rules. The panels consider the contribution made by the measure to the objective, the impact of the measure on trade, and the importance of the objective in protecting natural resources and the environment in assessing its compatibility with WTO rules.

Scholars consider that there are many unresolved issues in the trade–environment interface under WTO rules, and this is equally applicable to measures that may be taken to limit emissions through carbon taxes, border adjustment norms, mandatory standards, or schemes such as mandatory ecolabeling. Only a case-by-case inquiry can determine whether each situation is compatible with WTO rules.

Krishna Ravi Srinivas
Research Information System for Developing Countries

See Also: Emissions, Trading; Kyoto Protocol; Technology Transfer; United Nations Conference on Trade and Development; World Bank.

Further Readings

Center for International Environmental Law and Friends of the Earth. *Is World Trade Law a Barrier to Saving Our Climate?* Geneva: CIEL, 2009.

Hufbauer, G. C., S. Charnovitz, and J. Kim. *Global Warming and World Trading System.* Washington, DC: Peterson Institute for International Economics, 2009.

World Trade Organiation and United Nations Environment Programme. *Trade and Climate Change.* Geneva: WTO, 2009.

World Weather Watch

The World Weather Watch is the central program of the World Meteorological Organization (WMO), the United Nations' agency for cooperation among national weather bureaus, founded in 1950. The Fourth World Meteorological Congress approved the idea of the program in 1963, and the WMO, which has 188 member countries and territories, subsequently established the World Weather Watch to make an integrated worldwide weather-forecasting system available.

The World Weather Watch includes the Tropical Cyclone Program, the Antarctic Activities Program, an Emergency Response Activities Program for environmental emergencies, and the Instruments and Methods of Observation Program to ensure the quality of the observations that are vital for weather forecasting and climate monitoring.

Through the World Weather Watch, a system is in place for countries around the world to obtain daily weather forecasts. The core components of the World Weather Watch—the Global Observatory System (GOS), the Global Telecommunications System (GTS), and the Global Data-Processing and Forecasting System (GDPFS)—enable the World Weather Watch to provide basic meteorological data to the WMO and other related international organizations.

The GOS allows for observing, documenting, and communicating data about the weather and climate for the creation of forecast and warning services. Monitoring the climate and environment is a priority of the WMO, and the GOS is critical to the effective and efficient operations of the WMO. Long-term objectives of the GOS include the standardization of observation practices and the optimization of global observation systems.

The GTS consists of land and satellite telecommunication links that connect meteorological telecommunication centers. The GTS provides efficient and reliable communication service among the three World Meteorological Centers in Melbourne, Moscow, and Washington, and the 15 Regional Telecommunication Hubs that make up the Main Telecommunication Network. The six Regional Meteorological Telecommunication Networks, covering Africa; Asia; South America, North America, Central America, and the Caribbean; the south–west Pacific; and Europe, ensure the collection and distribution of data to members of the WMO. The National Meteorological Telecommunication Networks make it possible for the National Meteorological Centers to collect data and to receive and disseminate weather information on a national level.

The primary aim of the GDPFS is to prepare and provide meteorological analyses to members in the most cost-effective manner possible. The GDPFS is organized to implement functions at international, regional, and national levels through the World Meteorological Centers, Regional Specialized Meteorological Centers, and National Meteorological Centers. Real-time functions include preprocessing and postprocessing of data and the preparation of forecast products. Non-real-time functions include long-term storage of data and the preparation of products for climate-related analysis.

Increasingly, the World Weather Watch provides support for developing international programs related to global climate and other environmental issues, and sustainable development. The entire continent of Africa has only 1,150 World Weather Watch stations—one per 26,000 sq. km (10,038 sq. mi.)—even though the continent's land mass is as large as North America, Europe, Australia, and Japan put together. This represents coverage eight times lower than the WMO's recommended minimum level. The changing climate of Africa necessitates greater capacity building on the part of institutions prepared to address the likely crises that lie ahead. The World Weather Watch is vital in developing that capacity.

The World Weather Watch and its parent organization, the WMO, through the development of a permanent global weather data network, have proven critical to defining global warming as a given. As a consequence, the political and policymaking debates about climate change and its very real consequences, such as those facing Africa, have moved to a new arena. Although the World Weather Watch cannot compel individual governments to act on its findings, it has framed the issue of climate change on a truly global scale.

Robin K. Dillow
Oakton Community College

See Also: Climate; United Nations Environment Programme; World Meteorological Organization.

Further Readings

Edwards, Paul N. "Meteorology as Infrastructural Globalism." *OSIRIS*, v.21 (2006).

United Nations Environment Programme. "The Environment in the News." http://www.unep.org/cpi/briefs/2006Nov20.doc (Accessed March 2012).

"The View from Space." *Weatherwise*, v.48/3 (1995).

World Meteorological Association. http://www.wmo.int (Accessed March 2012).

World Wildlife Fund

Although the World Wildlife Fund has changed its name to the World Wide Fund for Nature in most countries, the original name remains the official moniker in the United States and Canada. In fact, when it changed its name in 1986, it still kept its initials (WWF) around the world. As an international nongovernmental organization, it was founded in 1961 in Switzerland to help with the conservation, research, and restoration of the natural environment.

Although over many years the WWF became famous for its protection of endangered fauna—its symbol remains a panda bear—it has also been keen to preserve natural environments, seeing its role as helping endangered flora as much as fauna, with the name change reflecting this. The WWF now recognizes that the single biggest threat to the environment today comes from global warming, and as a result, it has campaigned for companies and individuals to reduce their greenhouse gas (GHG) emissions. As an organization, it has been involved in campaigning over climate change and global warming since at least 1990, although individuals within the WWF were active even earlier. The stated aim of the WWF is to reduce GHG emissions to 80 percent of 1990 levels by 2050. It has campaigned for the U.S. government and Congress to act far more decisively.

The major area where the WWF initially concentrated its energies was in reducing deforestation, especially in Brazil, Central Africa, and the Russian Far East. U.S. experts from the WWF–U.S. have taken part in many projects in these regions, as well as other parts of the world. They have been involved in recording levels of deforestation—in many cases illegal logging—and notifying the relevant governments, as well as bringing extreme levels of deforestation to world attention. This campaign has been relatively successful at alerting governments and populations of countries to widespread, often illegal, deforestation. Meanwhile, the organization has also been involved in detailing fauna that is threat-

In October 2008, a study released by the World Wildlife Fund claimed that half to three-quarters of major Antarctic penguin colonies could be damaged or wiped out if global temperatures are allowed to climb by more than 3.6 degrees F (2 degrees C).

ened by the cutting of forests and highlighting the problems faced by some animals. In other cases, the WWF has highlighted the plight of animals that would be affected more generally by global warming and climate change. This is particularly true of Arctic and Antarctic animals; their landscape has transformed with the melting of parts of the polar regions. The plight of polar bears has been particularly highlighted.

Community, Education, and Business

Traditionally, the WWF has organized throughout the United States at the city, town, and village level, with the education of young people at the forefront of its approach. This means that the WWF has devoted much of its time and energy to encouraging students to gain a greater interest in the environment and the threat of global warming through the provision of resource kits, booklets, and lectures. Many of these items have been available free of charge, or heavily subsidized, with many schoolchildren becoming interested in the world of the WWF through television documentaries and other media sources such as the Internet.

This has resulted in the development of educational problems to teach more students how to plan ways of reducing carbon dioxide (CO_2) emissions. Among the children who have been involved in WWF projects have been those displaced by Hurricane Katrina, who have been better able to understand the problems leading to the hurricane. To that end, the WWF hosts the Southeast Climate Witness Program, which allows students to attend a Climate Camp in June 2008 and take part in the Youth Summit in Washington, D.C., the following month. Many schools around the United States also raise money for the WWF that is used for the campaign against climate change.

Although it has long been a community movement, the WWF has also started working heavily with businesses. This change has seen the WWF and some of its partners collaborating with 12 prominent companies, including the Collins Companies, IBM, Johnson & Johnson, Nike, Polaroid, and Sony. These large corporations have agreed to work toward reducing their CO_2 emissions by over 10 million tons each year, which has led to many smaller companies becoming aware of the effects of their greenhouse gas emissions and working to reduce them. The WWF has also tried to push for, with less success, energy utility companies to reduce the emissions of their operations. The organization has made some gradual progress in this area, but at a much slower pace than the WWF and other climate change activists would like.

Justin Corfield
Geelong Grammar School

See Also: Charismatic Megafauna; Climate Change, Effects of; Deforestation; Economics, Cost of Affecting Climate Change; Greenpeace International; Marine Mammals; Movements, Environmental; Penguins; Polar Bears.

Further Readings

Flippen, J. Brooks. *Conservative Conservationist: Russell E. Train and the Emergence of American Environmentalism*. Baton Rouge: Louisiana State University Press, 2006.

World Wildlife Fund. "Climate Overview." http://www.worldwildlife.org/climate (Accessed July 2011).

Worldwatch Institute

The Worldwatch Institute has stated that it is "dedicated to fostering the evolution of an environmentally sustainable and socially just society, where human needs are met in ways that do not threaten the health of the natural environment or the prospects of future generations." It describes itself as "an independent, globally focused environmental and social policy research organization" with a "unique blend of interdisciplinary research and accessible writing." Worldwatch is essentially a think tank, with its closest environmental movement analogues being Resources for the Future, the World Resources Institute, and the Earth Policy Institute. The latter is headed by Lester Brown, who founded Worldwatch in 1974, and served as its president through 2000. Worldwatch's current executive director is Robert Engelman, a former environmental reporter and a founder of the Society of Environmental Journalists.

Worldwatch prides itself on its accessible writing style and its fact-based analysis of critical global issues. It focuses on the underlying causes of these issues and seeks, through education and dissemination of information, to inspire people to act in positive ways. A search of its Website shows large numbers of publications related to climate change, which it has addressed in its publications since at least 1984. *Worldwatch Reports* (formerly *Worldwatch Papers*), one of its signature publications, has sought to educate the public regarding "pressing economic, environmental, and social issues" since 1975. Since 1984, Worldwatch has published *State of the World*, a widely read and influential annual report; the 2009 volume, "Into a Warming World," focused on climate change. Although Worldwatch does not lobby Congress directly, this comprehensive report is read by legislators as well as world leaders, students, and ordinary individuals and has been translated into 25 languages.

In 1992, *Vital Signs: The Trends That Are Shaping Our Future* was first published. It was designed to be an accessible, annual series with brief entries on more than 50 issues that affect the world each year. From 1988 to 2010, the group published a bimonthly magazine, *World Watch*. Many of those articles, along with summaries of current Worldwatch research and blogs, and a subscription service to Vital Signs Online are available on the Institute's Website.

More Than One Issue

Worldwatch is not a one-issue organization, having written about a very wide range of environmental issues including energy, water pollution and water availability, soil erosion and other agricultural concerns, population, biodiversity, materials recycling and conservation, forests, and toxic materials. However, it seeks to foster recognition that these issues are inextricably tied to issues of social justice and peace. It began paying consistent attention to the relationship between social and environmental issues, particularly in international settings, much earlier than most environmental organizations. It began calling attention to the need for a sustainable society in at least 1982, five years before sustainability began to gain widespread attention with the publication of the Brundtland Commission report, *Our Common Future*.

The institute's three major program areas are food and agriculture, environment and society, and climate and energy. Projects as of 2011 included providing research and policy advice in the following areas: low-carbon, economic development strategies; environmental and climate impacts of new natural gas reserves and extraction methods; status reports on renewable energy and energy efficiency; security implications of transitioning to low-carbon technologies; climate and energy challenges in India and China; connecting green industry and sustainability advocates with policymakers via its New Economy Council; and ReVolt, a blog on international climate and energy policy. Current research publications include findings from the *World Nuclear Industry Status Report*, focusing on the effects of the nuclear industry on climate change, and a report on the interrelation and correlation between climate change, population, and women's lives.

A desire to inspire change in societal attitudes and actions from a grassroots perspective is a hallmark of this organization. It seeks to effect change, not by force from the top, but by educating the public and thereby inspiring them to demand change. It carries on this vision with 26 staff members.

Gordon P. Rands
Pamela J. Rands
Western Illinois University

See Also: Developing Countries; Resources for the Future; Sustainability; World Resources Institute.

Further Readings

State of the World 2011: Innovations that Nourish the Planet. Washington, DC: Worldwatch Institute, 2011.

Worldwatch Institute. http://www.worldwatch.org (Accessed May 2011).

Wyoming

Wyoming is the least populous U.S. state, although the 10th-largest by area, with a population of 544,270 in 2009. The climate is continental, with wide temperature extremes; most of the state receives less than 10 in. (25 cm) of rain annually.

The major industries are ranching and mining, with the latter providing about one-quarter of the gross state product. The eastern part of the state is part of the Great Plains regions while the western part is mountainous. Almost half of the area of Wyoming is owned by the U.S. government, including numerous national forests and two national parks, Yellowstone and Grand Teton.

Rich in Coal

Wyoming produced 14.9 percent of all U.S. energy in 2008, including 2.6 percent of crude oil, 10.8 percent of natural gas, and 40.2 percent of coal. Most of the Powder River Basin, the largest coal-producing region in the United States, is in Wyoming. Wyoming also has major reserves of energy, including 583 million barrels of oil (2.8 percent of the U.S. total), 35,283 billion cu. ft. of dry natural gas (12.9 percent of the U.S. total), 1,010 million barrels of natural gas plant liquids (11.8 percent of the U.S. total), and 6,917 million short tons of recoverable coal (39.6 percent of the U.S. total). Almost all (97 percent) of electricity generated in Wyoming is produced by coal-fired power plants, with most of the remainder produced by hydroelectric plants or renewable sources. In 2006, Wyoming produced 779,986 kilowatt hours of electricity from wind power, ranking 11th among U.S. states.

The Intergovernmental Panel on Climate Change estimates that if global warming continues on its present pace, the average temperature in Wyoming will rise by 6.75 degrees F (3.75 degrees C) by 2100. Wyoming's glaciers are already melting at a rapid pace, and if the predicted warming takes place, many or all of them could disappear. Global warming is also expected to bring with it increased wildfires, more periods of drought, and more extreme rain events, which could result in increased flash flooding. Warmer winter temperatures will result in less snowpack and earlier snowmelt in the mountains, leading to more winter runoff and reduced water flow in the summer. Warmer winters are expected to result in increased populations of destructive insects such as the pine bark beetle, which attacks conifer forests. Warmer, drier conditions are not favorable to some of Wyoming's common tree species, and it is predicted that the area covered by the state's whitebark pine forests will be reduced by as much 90 percent over the next 50 years if temperatures rise as expected.

Wyoming is home to a rich diversity of wildlife, including 295 birds, 109 mammals, 22 reptiles, 56 fish, and 12 amphibians. Hunting, fishing, and wildlife tourism are important contributors to the state's economy, producing revenues of over $904 million in 2006 and supporting over 16,000 jobs. Global warming could destroy the habitat of many of these species and sharply reduce their numbers: For instance, half the trout stream habitats in the Rocky Mountain region are expected to become nonviable by the end of the century if expected temperature increases occur. Reduced snowfall would also seriously damage the state's skiing industry and require resorts to increasingly rely on artificial snow.

Natural resource extraction has brought prosperity to Wyoming, which has one of the lowest unemployment rates in the country, but has also brought pollution. In 2011, ozone levels near the state's gas fields were measured at 124 parts per billion, far above the Environmental Protection Agency's (EPA) maximum healthy limit of 75 parts per billion and higher than the worst measured level in Los Angeles in 2010. Wyoming has a history of resisting regulation of greenhouse gases (GHGs), and a provision in the state's 1999 Environmental Quality Act, known as the anti-Kyoto Protocol law, specifically prohibits the state from participating in the regulation of GHGs. State governor Dave Freudenthal claimed in September 2010 that this provision of state law exempts Wyoming from complying with the EPA's "tailoring rule," which requires facilities that emit 25,000 tons or more per year of GHGs to seek special permits. In February 2011, newly elected governor Matt Mead filed a legal challenge to these same EPA regulations, arguing that the state did not have enough time to meet the new requirements.

Sarah Boslaugh
Kennesaw State University

See Also: Coal; Forests; Kyoto Protocol; Oil, Production of; Natural Gas; Renewable Energy, Geothermal; Renewable Energy, Wind; Species Extinction; Tourism.

Further Readings

Bleizeffer, Dustin. "Governor: Wyoming Can't Control Greenhouse Gases." *Wyoming Star Tribute* (September 11, 2010).

Koch, Wendy. "Wyoming's Smog Exceeds Los Angeles' Due to Gas Drilling." *USA Today* (March 9, 2011).

National Wildlife Federation. "Global Warming and Wyoming" (February 2, 2009). http://www.nwf.org/globalwarming/pdfs/Wyoming.pdf (Accessed July 15, 2011).

Nelson, Gabriel. "Wyoming Joins Texas in Suing EPA Over Rollout of Greenhouse Gas Regulations." *New York Times* (February 16, 2011).

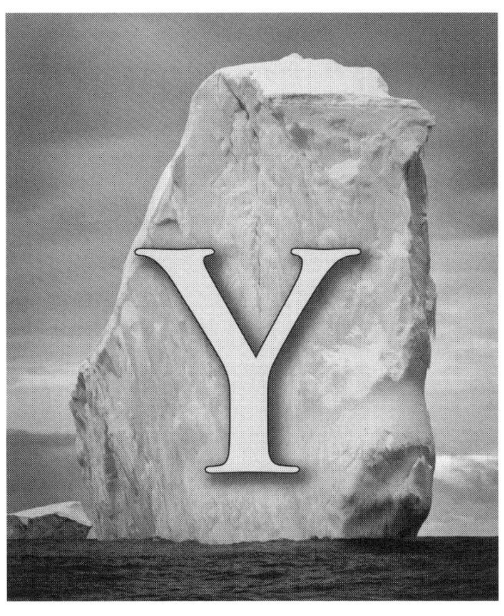

Yemen

Yemen is located on the Arabian Peninsula, with an area of 203,849 sq. mi. (527,968 sq. km), sharing borders with Oman and Saudi Arabia. It has 1,184 mi. (1,906 km) of coastline on the Red Sea and the Gulf of Aden (opposite Somalia, Djibouti, and Eritrea), which includes the Bab al Mandab, a narrow passage between the Red Sea and the Gulf of Aden, a strategic oil shipping route; in 2010, an estimated 3.5 million barrels of oil passed through this route. Yemen includes the island of Jabal Al-Tair in the Red Sea, which has seen volcanic activity as recently as 2007.

The terrain on the mainland is a coastal plain backed by hills and mountains, with central desert plains sloping into the interior desert of the Arabian peninsula. The climate is primarily hot and dry, with the western coast having a hot and humid climate and the western mountains having a temperate climate affected by seasonal monsoons. Less than 3 percent of the land is arable, and Yemen has limited freshwater resources and groundwater is rapidly being depleted. Yemen's population of just over 24 million (estimated as of July 2011) has a median age of 18.1 years and is growing at a rate of 2.6 percent annually (the 21st highest in the world). The gross domestic product in 2010 was $27,000, among the lowest in the Middle East, with an unemployment rate of 35 percent and 45 percent of families living below the poverty line.

Although over 90 percent Yemen's exports and 75 percent of its government revenues come from hydrocarbons, oil revenues are declining. However, it is hoped that further development of its natural gas reserves (Yemen began exporting liquefied natural gas in 2009) will offset falling income from oil. In addition, the government has begun efforts to diversity the economy into sectors such as fisheries, manufacturing, services, and finance. Yemen has crude oil reserves of 3 billion barrels and in 2010 produced an average of 260,000 barrels day, down from its peak production of 440,000 barrels per day in 2001. Increasing domestic consumption (156,000 barrels per day in 2010) has also contributed to a decline in exports. Yemen has 16.9 trillion cu. ft. of proven natural gas reserves and produced 509 billion cu. ft. of gross natural gas in 2009. Generation and consumption of electricity have risen steadily over the past 30 years: in 1980, 0.471 billion kilowatt hours (kWh) of electricity were generated and the same amount was consumed, while in 2008, 6.152 billion kWh were generated and 4.646 billion kWh were consumed. Carbon dioxide emissions from consumption of fossil fuels has also risen over the past 30 years, from 6.43 million metric tons in 1980 to 22.84 million metric tons in 2008 (the 79th highest in the world).

Global warming poses a serious threat to Yemen's stability and well-being. Water shortages are already occurring in some areas, and annual mean temperatures have been increasing: For instance, in Aden (Yemen's capital, a port on the Red Sea), warming of about 2.5 degrees F (1.4 degrees C) has been observed in the past century, while on average across the country, temperatures have increased about 0.9 degree F (0.5 degree C) and summer precipitation declined in the highlands. Mean annual temperatures are expected to increase 2.1–5.9 degrees F (1.2–3.3 degrees C) by 2060, with greater increases in the interior regions of the country. It is predicted that runoff will increase by 13 percent, the number of hot days will increase, and the number of cold days will decrease. Hazards posed by these changes include floods, droughts, sea-level rise, landslides, and increasing desertification. Yemen submitted a National Adaptation Program of Action to the World Bank in 2009, which lays the groundwork for a national response to climate change. Proposals include efforts to develop and implement coastal management programs, water conservation through improved irrigation techniques and reuse of treated water, rainwater harvesting, sustainable management of fisheries, and planting and replanting of palms and mangroves in response to anticipated sea-level rise.

Sarah Boslaugh
Kennesaw State University

See Also: Desertification; Natural Gas; Oil, Production of; Sea Level, Rising.

Further Readings

Global Facility for Disaster Reduction and Recovery. "Climate Risk and Adaptation Country Profile: Yemen" (April 2011)." http://sdwebx.worldbank.org/climateportal/doc/GFDRRCountryProfiles/wb_gfdrr_climate_change_country_profile_for_YEM.pdf (Accessed July 2011).

U.S. Energy Information Administration. "Country Analysis Brief: Yemen" (February 2011). http://www.eia.gov.countries.cab.cfm?fips=YM (Accessed July 2011).

World Bank. "Climate Change: Yemen." http://go.worldbank.org/URGQ7RJ2F0 (Accessed July 2011).

Younger Dryas

Marking the boundary between the Holocene and Pleistocene epochs, the Younger Dryas, a period of glacial conditions between 12,900 and 11,500 years ago, is named for *Dryas octopetala*, a flower that is adapted to the cold. *Dryas*'s pollen is found in abundance in strata of this age. *Dryas*'s pollen is also found in older strata, necessitating the term *Younger Dryas* to distinguish this time from older periods in which *Dryas*' pollen is abundant. Locked in an ice age, Earth had finally warmed and the glaciers had begun to retreat 15,000 years ago. Counteracting this warming trend, the Younger Dryas reduced temperatures 50 degrees F in only a decade. Glaciers once more advanced in North America and Europe. Rainfall diminished, and frigid winds carried dust from central Asia throughout Europe.

The Younger Dryas was part of the Cenozoic ice age, which locked the world in glaciers 100,000 years ago. The climate was particularly cold as recently as 18,000 years ago. From these frigid conditions, the climate gradually warmed, until 15,000 years ago, the glaciers began to retreat. The Younger Dryas interrupted this warming trend, restoring glacial conditions to Earth.

Three Possible Causes

Climatologists have advanced three causes of the Younger Dryas, though it is uncertain whether all three operated at the same time. The fact that the Southern Hemisphere cooled before the Northern Hemisphere suggests that some mechanism cooled the south, whereas no mechanism was then operating in the north. The rapid change in climate the Younger Dryas may have caused the extinction of large mammals in North America and the collapse of the first Native American culture. In western Asia, the Younger Dryas may have prompted humans to invent agriculture. The end of the Younger Dryas ushered in the modern climate.

The leading explanation focuses on ocean currents. The Gulf Stream brings warm water from the tropics to the North Atlantic Ocean, warming the Atlantic coasts of North America and Europe. The Gulf Stream remained undisturbed as the North American glacier began to retreat north 15,000 years ago. In the initial centuries of retreat, the ice sheet emptied its water down the

Mississippi River and into the Gulf of Mexico. By 12,900 years ago, however, the North American glacier had retreated to the Great Lakes. Melted water no longer flowed south down the Mississippi River, but now went east along the St. Lawrence River to the Atlantic Ocean. This cold water shut down the Gulf Stream, robbing North America and Europe of its warmth and returning the climate to glacial conditions.

Climatologists have identified a second cause in the impact of an asteroid near the Great Lakes 12,900 years ago. Upon impact, the asteroid ejected enormous amounts of debris, dust, and ash into the atmosphere, blocking out sunlight and cooling the Earth. A third cause might have been the sudden, and unexplained, cessation of El Niño. Every two to seven years, warm water from the western Pacific Ocean and Indian Ocean flows east, warming the west coasts of South and North America. Without El Niño, these continents cooled, returning them to glacial conditions. Possibly more than one cause initiated the Younger Dryas.

The Younger Dryas ended as abruptly as it had begun, when temperatures rose 50 degrees F (28 degrees C) in just 10 years. Glaciers retreated to Antarctica, Greenland, and the North Pole, and rainfall again became abundant. The Cenozoic ice age, having cooled the Earth for 100,000 years, finally ended with the close of the Younger Dryas. Forests returned to Scandinavia, Germany, and North America. The return of warmth and rainfall, along with the invention of agriculture, allowed humans to settle in communities. With some exceptions, humans were no longer nomads. The end of the Younger Dryas initiated the modern climate. Although temperatures have fluctuated in modernity, the retreat of glaciers has so far been permanent. Perhaps glaciers will return one day, though there is no evidence that they will come back soon.

Christopher Cumo
Independent Scholar

See Also: Cenozoic Era; Climate; Cretaceous Period; Greenhouse Effect; Greenhouse Gas Emissions.

Further Readings

Cox, John D. *Climate Crash: Abrupt Climate Change and What It Means for Our Future.* Washington, DC: Joseph Henry Press, 2005.

Rothschild, Lynn J. and Adrian M. Lister, eds. *Evolution on Planet Earth: The Impact of the Physical Environment.* San Diego, CA: Academic, 2003.

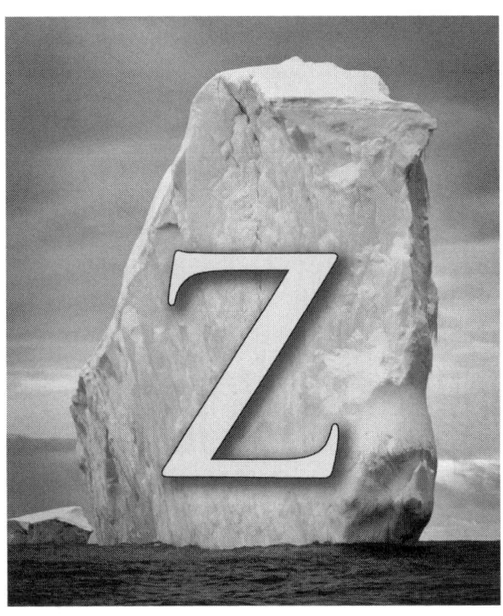

Zambia

The perception of climate change or urban health hazards of different communities and its association with empirical evidence should form an important geographical approach in applied research.

In the context of Zambia, Cuthbert Makondo wrote his monograph titled *Micro Climate Change in Zambia* with a focus on the Chikuni area. As elsewhere in Zambia, forest encroachment and deforestation have increased considerably in many parts of Zambia. In the southern province, as Makondo uses as an example, it is strikingly evident, although there are several marked spatial anomalies and differences that can be attributed to the underlying local factors, such as population size, cultivation methods, and types and sources of energy.

According to Makondo's study, many deforested areas experience microclimate change, such as rising temperatures and rainfall reduction, resulting in less growth of vegetation regeneration and crop failure. Thus, the Chikuni area, as one of the most highly deforested areas with frequent droughts in the southern province, has triggered great concern and interest to investigate and account for its local climate from 1950 to 2000 to discern whether there is any significant microclimate change (see Figures 1 and 2).

According to these findings, there is a general temperature rise, as well as rainfall reduction in the Chikuni area. This is evident in both climatic models and graphic displays of both rainfall and temperature. Further, comparisons of significant tests have indicated that there was climatic change in the Chikuni area in the period under investigation, although changes have been more notable after 1975. In addition, rainfall anomaly has become a serious challenge in Zambia's development process. According to M. Mudenda, Zambia has begun to experience heavy rains at a time when the number of dwellers in unplanned settlements is increasing at a fast rate. Resilience of many ecosystems is likely to be exceeded this century by an unprecedented combination of climate change; associated disturbances such as flooding, drought, wildfire, and insects; and other global change drivers, such as land-use change, pollution, fragmentation of natural resources, and overexploitation of resources.

Cuthbert Munyati argued in a study published in 1997 that climate change, especially the occurrence of drought, is related to the El-Niño/Southern Oscillation. There is a direct link between drought and El Niño events in the entire region, but with local variations related to land-use patterns. In Zambia, the research analysis by Munyati has indicated rainfall fluctuations from year to year. However, during the period from 1978

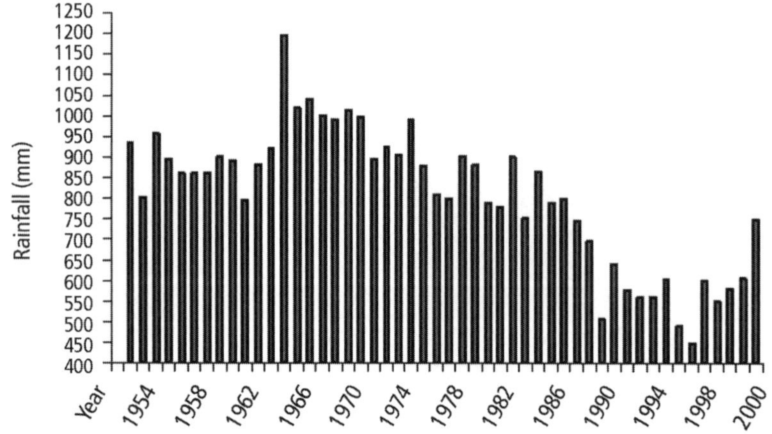

Figure 1 Rainfall total variations for the Chikuni area (1950–2000)

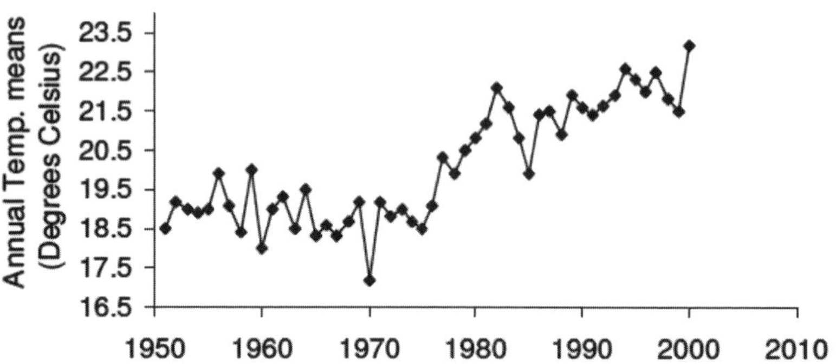

Figure 2 Annual temperature means for the Chikuni area (1950–2000)

Source: "Micro-Climate Change in Zambia (Chikuni Area)" study by Cuthbert Makondo (BA-Geography) 2002.

to 1996, the rainy season appears to be a time of reducing rainfall compared to the period before 1978 (1950 to 1977).

Significant Effect of Climate Change

The impact of climate change on agriculture in Zambia has also been tremendous. According to studies, Zambia has experienced an increase in drought frequency and intensity in the last 20 years. The droughts from 1991 to 1992, 1994 to 1995, and 1997 to 1998 worsened the quality of life for vulnerable groups such as subsistence farmers. According to a 2006 policy report from the Centre for Environmental Economics and Policy in Africa (CEEPA), the results indicate that an increase in mean temperature in November and December and a reduction in mean precipitation in January and February have negative impacts on net farm revenue, whereas an increase in mean temperature in January and February and an increase in mean annual runoff have positive impacts on net farm revenue.

Globally, a rise in temperature has resulted in outbreaks of malaria and cholera. In Zambia as well, a study conducted by researchers from the Madrid Carlos III Institute of Health highlighted the association between an increase in cholera cases and climatic determinants. The analysis of data confirm that the increase in environmental temperature six weeks before the rainy season increases the number of people affected by this sickness by 4.9 percent. Miguel Ángel Luque, one of the study's authors and a researcher from the Madrid Carlos III Institute of Health, explains, "This is the first time that it has become evident in the sub-Saharan region that the increase in environmental temperature is related to the increase in cholera cases." The research project, carried out in Lusaka (Zambia) between 2003 and 2006, analyzes data from three cholera epidemics that occurred in a consecutive fashion. The results show that climatic variables (rain and environmental temperature) are related to the increase in cholera cases during the epidemic period. The study maintains that a 1.8 degrees F (1 degree C) increase in temperature six weeks before the beginning of the outbreak explains the 5.2 percent increase in cholera cases during an epidemic.

Rais Akhtar
Jawaharlal Nehru University

See Also: Botswana; Burundi; Deforestation; Diseases; Namibia; Small Farmers.

Further Readings

de Wit, M. "Climate Change and African Agriculture." CEEPA Policy Note No. 27 (August 2006). http://www.ceepa.co.za/docs/ POLICY%20 NOTE%2027.pdf (Accessed July 2011).

Makondo, Cuthbert. *Micro-Climate Change in Zambia (Chikuni Area)*. Lusaka: University of Zambia, 2002.

Mudenda, M. M. "Climatic Change in Zambia: Ignore, Mitigate, or Adapt? TS 2E–Climate Change and Environmental Threats." Paper presented at FIG Congress 2010: Facing the Challenge—Building the Capacity, Sydney, Australia, April 11–16, 2010.

Munyati, C. "Extremes in Climatic Parameters as Indicators of Change in Kafue Basin, Zambia." *University of Zambia Journal of Science and Technology*, v.1/2 (1997).

Zimbabwe

The Republic of Zimbabwe is a middle-income developing country of approximately 12 million people in southern Africa; it gained its independence from Great Britain in 1980 after a protracted civil war. Originally among the continent's most prosperous countries, its economy has declined significantly since 2000, when the government began a program of land reform resulting in the seizure of large numbers of commercial farms. Because of these economic and related political difficulties, there have been relatively few efforts to address the potential impacts of climate change or participate in international mitigation and adaptation programs. Zimbabwe is a geographically diverse country with a mixture of ecozones ranging from semiarid savannah or bushveld through subtropical, tropical, and montane forest. The main distinguishing feature of the country is its altitude variation, ranging from the central plateau (high veld) with altitudes between 3,900 and 5,250 ft. (1,200 and 1,600 m), and the lowveld, which comprises about 20 percent of the country and is generally under 2,900 ft. (900 m). Rainfall is extremely varied and seasonal.

Zimbabwe's environmental diversity is mirrored in its economic diversity. Until recently, the country had a relatively complex economy, with a strong agricultural sector based on tobacco and maize, a related agroindustrial sector, and a robust mining sector generating significant revenue from platinum, chromium, and gold. After initial contraction in the early 2000s, the mining sector is now strong and even growing, but the agricultural and industrial sectors have yet to recover.

Overall, per-capita income has plummeted from $2,400 in 2000 to $500 in 2010. Until 2009, when a temporary power-sharing government was formed after intervention by regional authorities, inflation had risen from an annual rate of 32 percent in 1998 to an official estimated high of 11,200,000 percent in August 2008. The Zimbabwe dollar, periodically reissued because of inflation, was at the time denominated in the trillions. Since the power-sharing government took control, the country has abandoned its own currency and all transactions are now in U.S. dollars, bringing inflation down to approximately 5 percent in 2010.

Climate Change Record

Zimbabwe's record with regard to climate change policy and actions is mixed. The country has signed and ratified both the Climate Change Convention and the Kyoto Protocol and submitted its first National Communication in 1998. There have been no significant policy developments since that time, although the Ministry of Environment and Natural Resources Management held a National Climate Change Conference in November 2010 and indicated at the time that it was preparing a national strategy document. The absence of a national strategy has seriously impeded access to various United Nations (UN) programs and funding sources, including climate-specific funds accessed via the Global Environment Facility. Because Zimbabwe is not listed as a Least Developed Country (LDC) by the UN (despite its economy being ranked 225th out of 228 countries), it has not been required to submit a National Adaptation Program of Action (NAPA)—but neither does it have access to special funding for this purpose.

The country's private and nongovernmental organization sectors are nevertheless quite active and have completed a substantial amount of research on potential Clean Development Mechanism (CDM) projects and the impacts of climate change on forestry and agriculture. During the earlier trial period of CDM (referred to as activities implemented jointly, or AIJ), Zimbabwe consultants completed a draft National Strategy that identified the following areas for potential CDM development:

- use of coalbed methane for ammonia generation;
- investment in a mini-hydroelectric project to supply electric power to rural and semi-urban consumers;
- increasing boiler efficiency in industry;
- improving energy efficiency in tobacco curing; and
- generation of power from methane produced at a sewage plant.

All of these projects were believed to have a high potential for replication in the country.

In 2009, the Zimbabwe government launched a public relations campaign titled Our Climate, Our Future. This program sought to raise awareness in the country about the climate change issue, and to bring about a significant change in agricultural practices.

Zimbabwe established a designated national authority (DNA) for CDM in 2007, and is in the process of recruiting a national program coordinator for climate change.

Geoff Stiles
Independent Scholar

See Also: Botswana; Climate Change Knowledge Network; Developing Countries; Poverty and Climate Change.

Further Readings

Chagutah, Tigere. "Climate Change Vulnerability and Adaptation Preparedness in Southern Africa: Zimbabwe Country Report." Nairobi, Kenya: Heinrich Böll Stiftung Southern Africa, 2010.

Frost, Peter G. H. *Zimbabwe and the United Nations Framework Convention on Climate Change*. Mount Pleasant, Harare: University of Zimbabwe, 2001.

Ministry of Mines, Environment and Tourism. "Zimbabwe's Initial National Communication on Climate Change" (1998). http://unfccc.int/resource/docs/natc/zimnc1.pdf (Accessed July 2011).

Zooplankton

The term *zooplankton* is derived from the Greek words *zoo* and *plagktos*, which mean "animal" and "wandering," respectively. Zooplankton are aquatic animals that possess little or no swimming capabilities and are found in marine and freshwater ecosystems. They are generally unable to swim against the flow and thus drift with the water current. Zooplankton usually implies the complete community of such animals found drifting within water bodies such as lakes, rivers, and oceans. The great diversity of organisms comprising zooplankton includes species sensitive to changes in the biological, chemical, and physical aspects of the environment, as well as others less sensitive to environmental factors. This variable sensitivity to environmental factors makes it possible to use zooplankton as indicators of environmental quality and climate change.

Most zooplankton are microscopic in size and include representatives of the phylum Rotifera—the "wheel-bearing" or ciliated animals, such as *Keratella quadrata*, as well as many crustaceans, including the strikingly iridescent epipelagic copepods belonging to the genera *Sapphirina* and *Copilia*. Also included are single-celled protozoans, some annelids, and certain much larger coelenterates, such as the jellyfish *Aurelia aurita*. Crustaceans in the Order Cladocera, such as waterfleas in the genus *Daphnia*, are often the most abundant zooplankton in freshwater, while genera in the Class Copepoda, such as *Calamus*, *Diaptomus*, and *Cyclops*, are abundant in marine waters. In coastal waters, insects can be a significant component of the plankton community.

Links Between Producers and Consumers

In marine and freshwater food webs, zooplankton are one of the most significant links between the producers and higher consumers. Most zooplankton are primary consumers—herbivores that graze on plants—although zooplankton may also be carnivorous, feeding on other zooplankton species. Some may be parasitic for part of their life or detrivores feeding on organic matter suspended in the water column. A water-filtering mechanism helps zooplankton extract and feed on the chlorophyll-containing aquatic organisms known as phytoplankton. Consisting mostly of

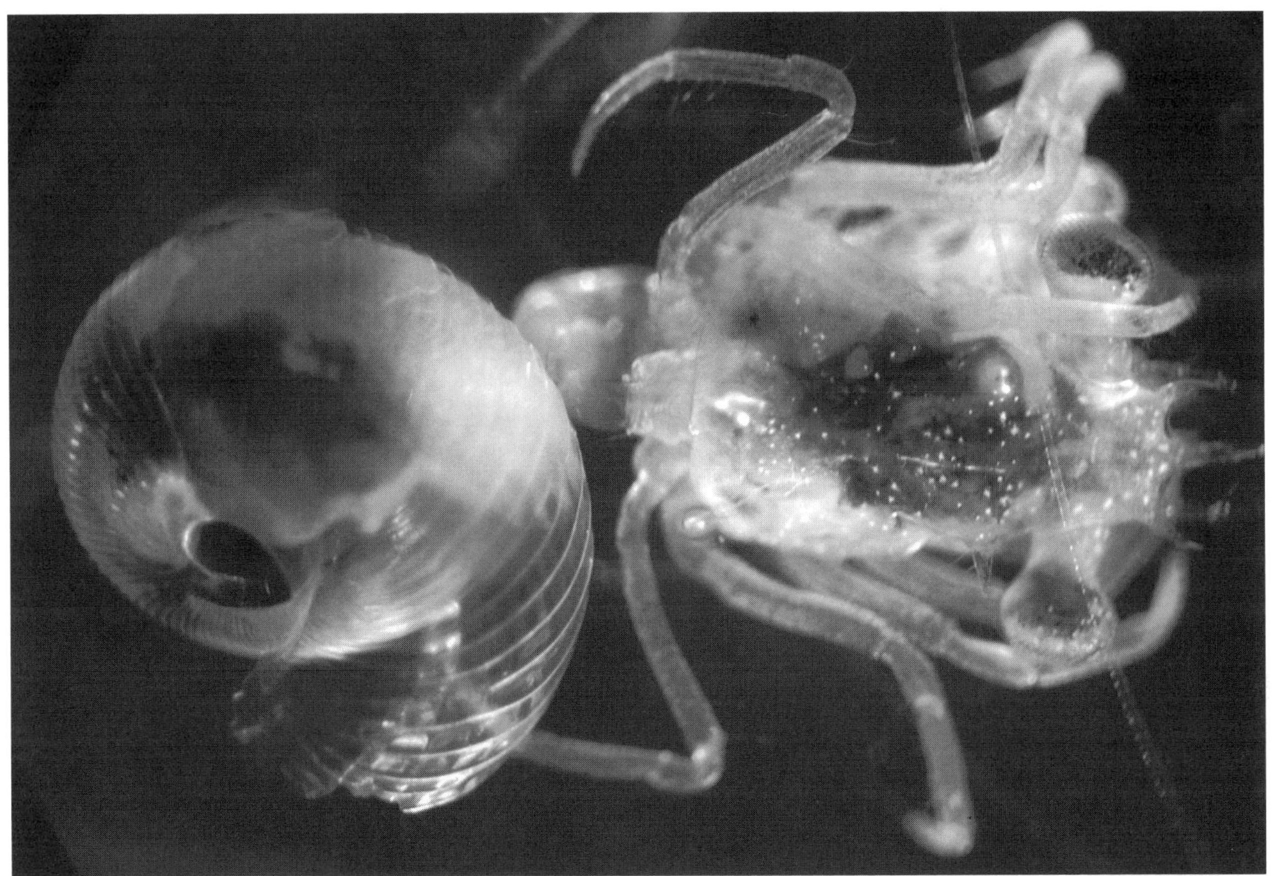

Crustacean larva and a pteropod. Data collected from a decadal series of ocean sampling in the Pacific Ocean from the Southern to Northern Hemispheres confirms that the oceans are becoming more acidic from absorbing vast amounts of carbon dioxide, backing up earlier modeling predictions, as well as other ocean field studies. An increase in dissolved CO_2 in ocean water, and thus acidity, will have a detrimental effect on marine life such as corals and plankton by reducing skeletal growth rates of calcium-secreting organisms.

unicellular or colonial algae, the photosynthesizing phytoplankton are the foundation of the aquatic food web. Phytoplankton abundance increases seasonally in response to greater sunlight availability and is followed shortly by an increase in zooplankton abundance. This cycle continues year after year, and the productivity of zooplankton is thus directly dependent on phytoplankton productivity. Zooplankton are an important dietary component of secondary consumers such as planktivorous fish larvae, herring, and salmon, which in turn are consumed by tertiary consumers and top predators in the food chain such as harbor seals, sharks, and killer whales.

Compilation of biogeographical, ecological, taxonomic, and genetic information on the world's zooplankton is a huge task being conducted at universities and institutes around the world. Protocols have been established for sampling zooplankton found at the water surface and at different depths in the water. In addition to traditionally used nets and trawls, advanced technologies developed for submersible vehicles, remote sensing, optical sensor systems, and molecular genetics are critical to understanding the global patterns of zooplankton distribution. Zooplankton production varies with available solar energy, nutrients such as nitrogen and phosphorous, water temperature, and various other physical and biological characteristics of the aquatic environment. Since the majority of these characteristics vary with latitude and altitude, both are important determinants of plankton production and biomass. The tropical zone demonstrates low productivity in general as a result of the relatively rapid uptake of nutrients by phytoplankton, while sunlight is the main limiting factor in the otherwise nutrient-rich Arctic region.

Climate warming is fragmenting and depleting the sea ice cover, which functions as a heat insulator and unique habitat for a range of organisms, including zooplankton. Climate warming has also been associated with changes in water circulation patterns between the Arctic, Pacific, and Atlantic oceans, with implications for both flora and fauna. Species of boreal plankton from the Pacific and Arctic oceans are known to have begun traveling into the North Atlantic, resulting in changes in species assemblages. Studies have linked the surface warming of southern California's Pacific waters by 2.7 degrees F (1.5 degrees C) at some places since the 1950s to an 80 percent decline in the biomass of the 20–200 mm sized macrozooplankton during the same period. Carbon dioxide, a greenhouse gas that contributes to climate warming, is absorbed at the ocean surface, and the carbon component is taken up by phytoplankton. Zooplankton contribute to the ocean's greenhouse gas buffering ability by grazing on carbon-fed algae and then descending and releasing the carbon in deeper ocean waters.

Rahul J. Shrivastava
Florida International University

See Also: Arctic and Arctic Ocean; Atlantic Ocean; Charismatic Megafauna; Climatic Data, Lake Records; Climatic Data, Sea Floor Records; Oceanic Changes; Pacific Ocean; Phytoplankton.

Further Readings

Greene, Charles H. and Andrew J. Pershing. "Climate Drives Sea Change." *Science*, v.315/23 (February 2007).

Johnson, William S. and M. Allen Dennis. *Zooplankton of the Atlantic and Gulf Coasts: A Guide to Their Identification and Ecology.* Baltimore, MD: John Hopkins University Press, 2005.

Lalli, Carol M. and Timothy Parsons. *Biological Oceanography: An Introduction.* 2nd ed. Oxford: Butterworth-Heinemann, 1997.

Roemmich, Dean and John McGowan. "Climatic Warming and the Decline of Zooplankton in the California Current." *Science*, v.267/3 (March 1995).

Travers, Bridget E., ed. *The Gale Encyclopedia of Science.* Vol. 6. Detroit, MI: Gale Research, 1996.

Waller, Geoffrey, ed. *Sealife: A Complete Guide to the Marine Environment.* Washington, DC: Smithsonian Institutions Press, 1996.

Glossary

Abrupt Climate Change
A change in the climate that takes place over a few decades or less, persists for at least a few decades, and causes substantial disruptions in human and natural systems.

Acid Rain
Also called acid precipitation or acid deposition, acid rain is precipitation containing harmful amounts of nitric and sulfuric acids formed primarily by sulfur dioxide and nitrogen oxides released into the atmosphere when fossil fuels are burned. It can be wet precipitation (rain, snow, or fog) or dry precipitation (absorbed gaseous and particulate matter, aerosol particles, or dust). Acid rain has a pH below 5.6. Normal rain has a pH of about 5.6, which is slightly acidic. The term pH is a measure of acidity or alkalinity and ranges from 0 to 14. A pH measurement of 7 is regarded as neutral. Measurements below 7 indicate increased acidity, while those above 7 indicate increased alkalinity.

Adaptation
Adjustment in natural or human systems to a new or changing environment. Adaptation to climate change refers to adjustment in natural or human systems in response to actual or expected climatic stimuli or their effects, which moderates harm or exploits beneficial opportunities. Various types of adaptation can be distinguished, including anticipatory and reactive adaptation, private and public adaptation, and autonomous and planned adaptation.

Aerosol
A collection of airborne solid or liquid particles, with a typical size between 0.01 and 10 micrometers (μm) and residing in the atmosphere for at least several hours. Aerosols may be of either natural or anthropogenic origin. Aerosols may influence climate in two ways: directly through scattering and absorbing radiation, and indirectly through acting as condensation nuclei for cloud formation or modifying the optical properties and lifetime of clouds.

Afforestation
The practice of restoring and re-creating of non-forest land to a new forest, or restoring a forest that was deforested many years ago from human activities, such as agriculture or habitation.

Albedo

The fraction of solar radiation reflected by a surface or object, often expressed as a percentage. Most snow-covered surfaces have a high albedo; the albedo of soils ranges from high to low; vegetation-covered surfaces and oceans have a low albedo. The Earth's albedo varies mainly through varying cloudiness, snow, ice, leaf area, and land cover changes.

Alleroed

A village in Denmark; its name is used for a warm period at the end of the last glacial period.

Alliance of Small Island States (AOSIS)

The group of Pacific and Caribbean nations that call for relatively fast action by developed nations to reduce greenhouse gas emissions. The AOSIS countries fear the effects of rising sea levels and increased storm activity predicted to accompany global warming. Its plan is to hold Annex I Parties to a 20 percent reduction in carbon dioxide emissions by 2005.

Alternative Energy

Energy derived from nontraditional sources (e.g., compressed natural gas, solar, hydroelectric, and wind).

Annex I Parties

Industrialized countries that, as parties to the Framework Convention on Climate Change, have pledged to reduce their greenhouse gas emissions by 2000 to 1990 levels. Annex I Parties consist of countries belonging to the Organisation for Economic Co-operation and Development (OECD) and countries designated as economies in transition.

Anthropogenic

Made by people or resulting from human activities. Usually used in the context of emissions that are produced as a result of human activities.

Atmosphere

The gaseous envelope surrounding the Earth. The dry atmosphere consists almost entirely of nitrogen (78.1 percent volume mixing ratio) and oxygen (20.9 percent volume mixing ratio), together with a number of trace gases, such as argon (0.93 percent volume mixing ratio), helium, radiatively active greenhouse gases such as carbon dioxide (0.035 percent volume mixing ratio), and ozone. In addition, the atmosphere contains water vapor, the amount of which is highly variable but typically 1 percent volume mixing ratio. The atmosphere also contains clouds and aerosols. The atmosphere can be divided into a number of layers according to its mixing or chemical characteristics, generally determined by its thermal properties (temperature). The layer nearest the Earth is the troposphere, which reaches up to an altitude of about about 5 mi. (8 km) in the polar regions and up to nearly 11 mi. (17 km) above the equator. The stratosphere, which reaches to an altitude of about 31 mi. (50 km), lies atop the troposphere. The mesosphere, which extends up to 50–56 mi. (80–90 km), is atop the stratosphere, and finally the thermosphere, or ionosphere, gradually diminishes and forms a fuzzy border with outer space.

Atmospheric Lifetime

The lifetime of a greenhouse gas refers to the approximate amount of time it would take for the anthropogenic increment to an atmospheric pollutant concentration to return to its natural level (assuming emissions cease) as a result of either being converted to another chemical compound or being taken out of the atmosphere via a sink. This time depends on the pollutant's sources and sinks as well as its reactivity. The lifetime of a pollutant is often considered in conjunction with the mixing of pollutants in the atmosphere; a long lifetime will allow the pollutant to mix throughout the atmosphere. Average lifetimes can vary from about a week (sulfate aerosols) to more than a century (chlorofluorocarbons [CFCs] and carbon dioxide).

Baseline Emissions

The emissions that would occur without policy intervention (in a business-as-usual scenario). Baseline estimates are needed to determine the effectiveness of emissions reduction programs.

Basket of Gases

The basket of gases includes six greenhouse gases, which are: carbon dioxide (CO_2), nitrous oxide (N_2O), perfluorocarbons (PFCs), methane (CH_4), hydrofluorocarbons (HFCs), and sulfur hexafluoride (SF_6). These six greenhouse gases are regulated under the Kyoto Protocol.

Berlin Mandate
A ruling negotiated at the first Conference of the Parties (COP 1), which took place in March, 1995, concluding that the present commitments under the Framework Convention on Climate Change are not adequate. Under the Framework Convention, developed countries pledged to take measures aimed at returning their greenhouse gas emissions to 1990 levels by 2000.

Biogeochemical Cycle
The chemical interactions that take place among the key chemical constituents essential to life, such as carbon, nitrogen, oxygen, and phosphorus.

Biomass
Organic non-fossil materials that are biological in origin, including organic material (both living and dead) from above and below ground (e.g., trees, plants, crops, roots, animals, and animal waste).

Biomass Energy
Energy produced by combusting renewable biomass materials such as wood. The carbon dioxide emitted from burning biomass will not increase total atmospheric carbon dioxide if this consumption is done on a sustainable basis (i.e., if in a given period of time, regrowth of biomass takes up as much carbon dioxide as is released from biomass combustion). Biomass energy is often suggested as a replacement for fossil fuel combustion, which produces large greenhouse gas emissions.

Biosphere
The part of the Earth system comprising all ecosystems and living organisms in the atmosphere, on land (terrestrial biosphere), or in the oceans (marine biosphere), including derived dead organic matter such as litter, soil organic matter, and oceanic detritus.

Black Carbon
Operationally defined species based on measurement of light absorption and chemical reactivity and/or thermal stability; consists of soot, charcoal, and/or possible light-absorbing refractory organic matter.

Borehole
Any exploratory hole drilled into the Earth or ice to gather geophysical data. Climate researchers often take ice core samples, a type of borehole, to predict atmospheric composition in earlier years.

Capital Stocks
The accumulation of machines and structures that are available to an economy at any point in time to prune goods or render services. These activities usually require a quantity of energy that is determined largely by the rate at which that machine or structure is used.

Carbon Cycle
All parts (reservoirs) and fluxes of carbon. The cycle is usually thought of as four main reservoirs of carbon interconnected by pathways of exchange. The reservoirs are the atmosphere, terrestrial biosphere (usually includes freshwater systems), oceans, and sediments (includes fossil fuels). The annual movements of carbon, the carbon exchanges between reservoirs, occur because of various chemical, physical, geological, and biological processes. The ocean contains the largest pool of carbon near the surface of the Earth, but most of that pool is not involved with rapid exchange with the atmosphere.

Carbon Dioxide (CO_2)
The concentration of this greenhouse gas is affected directly by human activities. CO_2 also serves as the reference to compare all other greenhouse gases. The major source of CO_2 emissions is fossil fuel combustion. CO_2 emissions are also a product of forest clearing, biomass burning, and non-energy production processes such as cement production. Atmospheric concentrations of CO_2 have been increasing at a rate of about 0.5 percent per year and are now about 30 percent above preindustrial levels.

Carbon Dioxide Equivalent
A metric measure used to compare the emissions of the different greenhouse gases based upon their global warming potential (GWP). Carbon dioxide equivalents are commonly expressed as "million

metric tons of carbon dioxide equivalents (MMT-CO_2Eq)." The carbon dioxide equivalent for a gas is derived by multiplying the tons of the gas by the associated GWP. The use of carbon equivalents (MMTCE) is declining.

$$\text{MMTCO}_2\text{Eq} = (\text{million metric tons of a gas})^* \\ (\text{GWP of the gas})$$

Carbon Dioxide Fertilization
The enhancement of the growth of plants as a result of increased atmospheric CO_2 concentration. Depending on their mechanism of photosynthesis, certain types of plants are more sensitive to changes in atmospheric CO_2 concentration. In particular, C_3 plants generally show a larger response to CO_2 than C_4 plants.

Carbon Sequestration
The uptake and storage of carbon. Trees and plants, for example, absorb carbon dioxide, release the oxygen, and store the carbon. Fossil fuels were at one time biomass and continue to store the carbon until burned.

Carbon Sinks
Carbon reservoirs and conditions that take in and store more carbon (carbon sequestration) than they release. Carbon sinks can serve to partially offset greenhouse gas emissions. Forests and oceans are common carbon sinks.

Chlorofluorocarbons and Related Compounds
This family of anthropogenic compounds includes chlorofluorocarbons (CFCs), bromofluorocarbons (halons), methyl chloroform, carbon tetrachloride, methyl bromide, and hydrochlorofluorocarbons (HCFCs). These compounds have been shown to deplete stratospheric ozone, and therefore are typically referred to as ozone-depleting substances. The most ozone-depleting of these compounds are being phased out under the Montreal Protocol.

Climate
Climate in a narrow sense is usually defined as the "average weather," or more rigorously, as the statistical description in terms of the mean and variability of relevant quantities over a period of time ranging from months to thousands of years. The classical period is three decades, as defined by the World Meteorological Organization (WMO). These quantities are most often surface variables such as temperature, precipitation, and wind. Climate in a wider sense is the state, including a statistical description, of the climate system.

Climate Change
Climate change refers to any significant change in measures of climate (such as temperature, precipitation, or wind) lasting for an extended period (decades or longer). Climate change may result from natural factors, such as changes in the sun's intensity or slow changes in the Earth's orbit around the sun; natural processes within the climate system (e.g., changes in ocean circulation); human activities that change the atmosphere's composition (e.g., through burning fossil fuels); and the land surface (e.g., deforestation, reforestation, urbanization, and desertification).

Climate Change Action Plan (CCAP)
Unveiled in October 1993 by President Clinton, the CCAP is the U.S. plan for meeting its pledge to reduce greenhouse gas emissions under the terms of the Framework Convention on Climate Change (FCCC). The goal of the plan was to reduce U.S. emissions of greenhouse gases to 1990 levels by 2000.

Climate Feedback
An interaction mechanism between processes in the climate system is called a climate feedback, when the result of an initial process triggers changes in a second process that in turn influences the initial one. A positive feedback intensifies the original process, and a negative feedback reduces it.

Climate Lag
The delay that occurs in climate change as a result of some factor that changes only very slowly.

Climate Model
A quantitative way of representing the interactions of the atmosphere, oceans, land surface, and ice.

Climate Modeling
The simulation of the climate using computer-based models.

Climate Sensitivity

The equilibrium response of the climate to a change in radiative forcing; for example, a doubling of the carbon dioxide concentration.

Climate System (or Earth System)

The five physical components (atmosphere, hydrosphere, cryosphere, lithosphere, and biosphere) that are responsible for the climate and its variations.

Cloud Condensation Nuclei

Cloud condensation nuclei are more commonly known as cloud seeds. These airborne particles are transformed from gas state to liquid state through the process of condensation, and potentially form cloud droplets.

Coalbed Methane (CBM)

Coalbed methane is methane contained in coal seams, and is often referred to as virgin coalbed methane, or coal seam gas.

Coal Mine Methane

Coal mine methane is the subset of CBM that is released from coal seams during the process of coal mining.

Conference of the Parties (COP)

The supreme body of the United Nations Framework Convention on Climate Change (UNFCCC). It comprises more than 180 nations that have ratified the convention. Its first session was held in Berlin, Germany, in 1995, and it is expected to continue meeting on a yearly basis. The COP's role is to promote and review the implementation of the convention.

Coral Bleaching

The process that takes place when corals lose the microscopic organisms called algae that live within their tissues. These algae provide the coral with nutrients, and they're responsible for the color of the coral. If a disturbance such as rising water temperature causes the algae to leave, corals will appear white and could eventually die.

Cryosphere

One of the interrelated components of the Earth's system, the cryosphere is frozen water in the form of snow, permanently frozen ground (permafrost), floating ice, and glaciers. Fluctuations in the volume of the cryosphere cause changes in ocean sea level, which directly impact the biosphere.

Deforestation

Those practices or processes that result in the change of forested lands to nonforest uses. This is often cited as one of the major causes of the enhanced greenhouse effect for two reasons: (1) the burning or decomposition of the wood releases carbon dioxide; and (2) trees that once removed carbon dioxide from the atmosphere in the process of photosynthesis are no longer present and contributing to carbon storage.

Desertification

Land degradation in arid, semi-arid, and dry subhumid areas resulting from various factors, including climatic variations and human activities.

Diurnal Temperature Range (DTR)

The diurnal temperature range is determined by calculating the difference between the minimum and maximum temperatures during a 24-hour period.

Economic Potential

The portion of the technical potential for greenhouse gas emissions reductions or energy-efficiency improvements that could be achieved cost-effectively in the absence of market barriers. The achievement of the economic potential requires additional policies and measures to break down market barriers.

Ecosystem

A natural community of plants, animals, and other living organisms and the physical environment in which they live and interact.

El Niño

A climatic phenomenon occurring irregularly, but generally every three to five years. El Niños often first become evident during the Christmas season (El Niño means Christ-child) in the surface oceans of the eastern tropical Pacific Ocean. The phenomenon involves seasonal changes in the direction of the tropical winds over the Pacific and abnormally warm surface ocean temperatures.

The changes in the tropics are most intense in the Pacific region; these changes can disrupt weather patterns throughout the tropics and can extend to higher latitudes.

Emissions
The release of a substance (usually a gas when referring to climate change) into the atmosphere.

Energy Vampire
An appliance or device that uses electricity even when it is turned off.

Enhanced Greenhouse Effect
The natural greenhouse effect has been enhanced by anthropogenic emissions of greenhouse gases. Increased concentrations of carbon dioxide, methane, nitrous oxide, chlorofluorocarbons (CFCs), hydrochlorofluorocarbons (HFCs), perfluorocarbons (PFCs), sulfur hexafluoride (SF6), nitrogen trifluoride (NF3), and other photochemically important gases caused by human activities such as fossil fuel consumption trap more infrared radiation, thereby exerting a warming influence on the climate.

Evapotranspiration
The sum of evaporation and plant transpiration. Potential evapotranspiration is the amount of water that could be evaporated or transpired at a given temperature and humidity, if there was water available. Actual evapotranspiration can not be any greater than precipitation, and will usually be less because some water will run off in rivers and flow to the oceans. If potential evapotranspiration is greater than actual precipitation, then soils are extremely dry during at least a major part of the year.

Fluorinated Gas
A group of powerful greenhouse gases that can stay in the atmosphere for hundreds to thousands of years. Fluorinated gases are manmade; they do not occur naturally. They are used in refrigeration and air-conditioning systems, fire extinguishers, and foam products.

Fluorocarbons
Carbon-fluorine compounds that often contain other elements such as hydrogen, chlorine, or bromine. Common fluorocarbons include chlorofluorocarbons (CFCs), hydrochlorofluorocarbons (HCFCs), hydrofluorocarbons (HFCs), and perfluorocarbons (PFCs).

Forcing Mechanism
A process that alters the energy balance of the climate system, that is, changes the relative balance between incoming solar radiation and outgoing infrared radiation from Earth. Such mechanisms include changes in solar irradiance, volcanic eruptions, and enhancement of the natural greenhouse effect by emissions of greenhouse gases.

Fossil Fuel
A type of fuel that forms deep within the Earth. Examples of fossil fuels include coal, oil, and natural gas. Fossil fuels are created over millions of years as dead plant and animal material becomes trapped and buried in layers of rock, and heat and pressure transform this material into a fuel. All fossil fuels contain carbon, and when people burn these fuels to produce energy, they create carbon dioxide.

General Circulation Model (GCM)
A global, three-dimensional computer model of the climate system that can be used to simulate human-induced climate change. GCMs are highly complex and they represent the effects of such factors as reflective and absorptive properties of atmospheric water vapor, greenhouse gas concentrations, clouds, annual and daily solar heating, ocean temperatures, and ice boundaries. The most recent GCMs include global representations of the atmosphere, oceans, and land surface.

Glacier
A multi-year surplus accumulation of snowfall in excess of snowmelt on land and resulting in a mass of ice at least 0.1 km^2 in area that shows some evidence of movement in response to gravity. A glacier may terminate on land or in water. Glaciers are on every continent except Australia.

Global Warming
Global warming is an average increase in the temperature of the atmosphere near the Earth's surface and in the troposphere, which can contribute to changes in global climate patterns. Global

warming can occur from many causes, both natural and human induced.

Global Warming Potential (GWP)
Global warming potential is the cumulative radiative forcing effects of a gas over a specified time horizon resulting from the emission of a unit mass of gas relative to a reference gas. The GWP-weighted emissions of direct greenhouse gases in the U.S. inventory are presented in terms of equivalent emissions of carbon dioxide (CO_2), using units of teragrams of carbon dioxide equivalents (Tg CO_2 Eq.).

$$\text{Conversion: Tg} = 10^9 \text{ kg} = 10^6 \text{ metric tons} = 1 \text{ million metric tons}$$

The molecular weight of carbon is 12, and the molecular weight of oxygen is 16; therefore, the molecular weight of CO_2 is 44 (i.e., 12+[16 x 2]), as compared to 12 for carbon alone. Thus, carbon comprises 12/44ths of carbon dioxide by weight.

Greenhouse Effect
Trapping and buildup of heat in the atmosphere (troposphere) near the Earth's surface. Some of the heat flowing back into space from the Earth's surface is absorbed by water vapor, carbon dioxide, ozone, and several other gases in the atmosphere and then reradiated back toward the Earth's surface. If the atmospheric concentrations of these greenhouse gases rise, the average temperature of the lower atmosphere will gradually increase.

Greenhouse Gas (GHG)
Any gas that absorbs infrared radiation in the atmosphere. Greenhouse gases include, but are not limited to, water vapor, carbon dioxide (CO_2), methane (CH_4), nitrous oxide (N_2O), chlorofluorocarbons (CFCs), hydrochlorofluorocarbons (HCFCs), ozone (O_3), hydrofluorocarbons (HFCs), perfluorocarbons (PFCs), and sulfur hexafluoride (SF_6).

Halocarbons
Compounds containing either chlorine, bromine, or fluorine and carbon. Such compounds can act as powerful greenhouse gases in the atmosphere. The chlorine and bromine containing halocarbons are also involved in the depletion of the ozone layer.

Heterotrophic Respiration
Organic matter that is converted to CO_2 by organisms other than plants.

Hockey Stick
A plot of the past millennium's temperature that shows the drastic influence of humans in the 20th century. Specifically, temperature remains essentially flat until about 1900, then shoots up, like the upturned handle of a hockey stick.

Hydrocarbons
Hydrocarbons are substances containing only hydrogen and carbon. Fossil fuels are made up of hydrocarbons.

Hydrochlorofluorocarbons (HCFCs)
Compounds containing hydrogen, fluorine, chlorine, and carbon atoms. Although ozone-depleting substances, they are less potent at destroying stratospheric ozone than chlorofluorocarbons (CFCs). They have been introduced as temporary replacements for CFCs and are also greenhouse gases.

Hydrologic Cycle
The process of evaporation, vertical and horizontal transport of vapor, condensation, precipitation, and the flow of water from continents to oceans. It is a major factor in determining climate through its influence on surface vegetation, the clouds, snow and ice, and soil moisture. The hydrologic cycle is responsible for 25 to 30 percent of the midlatitudes' heat transport from the equatorial to polar regions.

Hydrosphere
The component of the climate system comprising liquid surface and subterranean water, such as oceans, seas, rivers, freshwater lakes, and underground water.

Ice Cap
A dome-shaped accumulation of glacier ice and perennial snow that completely covers a mountainous area or island, so that no peaks or *nunataks* poke through.

Ice Core
A cylindrical section of ice removed from a glacier or an ice sheet in order to study climate patterns of the past. By performing chemical analyses on the air trapped in the ice, scientists can estimate the percentage of carbon dioxide and other trace gases in the atmosphere at a given time.

Ice Sheet
A thick, subcontinental- to continental-scale accumulation of glacier ice and perennial snow that spreads from a center of accumulation, typically in all directions. Also called a *continental glacier*.

Ice Shelf
The floating terminus of a glacier, typically formed when a terrestrial glacier flows into a deep water basin, such as in Antarctica and the Canadian Arctic.

Infrared Radiation
Radiation emitted by the Earth's surface, the atmosphere, and clouds. It is also known as terrestrial or longwave radiation. Infrared radiation has a distinctive range of wavelengths longer than the wavelength of the red color in the visible part of the spectrum of infrared radiation. It is practically distinct from that of solar or short-wave radiation because of the difference in temperature between the sun and the Earth-atmosphere system.

Intergovernmental Panel on Climate Change (IPCC)
The IPCC was established jointly by the United Nations Environment Programme and the World Meteorological Organization in 1988. The purpose of the IPCC is to assess information in the scientific and technical literature related to all significant components of the issue of climate change. The IPCC draws upon hundreds of the world's expert scientists as authors and thousands as expert reviewers. Leading experts on climate change and environmental, social, and economic sciences from some 60 nations have helped the IPCC to prepare periodic assessments of the scientific underpinnings for understanding global climate change and its consequences. With its capacity for reporting on climate change, its consequences, and the viability of adaptation and mitigation measures, the IPCC is also looked to as the official advisory body to the world's governments on the state of the science of the climate change issue. For example, the IPCC organized the development of internationally accepted methods for conducting national greenhouse gas emission inventories.

Landfill
Land waste disposal site in which waste is generally spread in thin layers, compacted, and covered with a fresh layer of soil each day.

Level of Scientific Understanding (LOSU)
Level of Scientific Understanding is a subjective, four-point scale ranging from very low, to low, to medium, to high. The LOSU is intended to differentiate the level of scientific understanding of the radiative forcing agents that influence climate change.

Longwave Radiation
The radiation emitted in the spectral wavelength greater than 4 micrometers corresponding to the radiation emitted from the Earth and atmosphere.

Mauna Loa Record
The Mauna Loa Observatory in Mauna Loa, Hawai'i, has been collecting atmospheric CO_2 concentration data since 1958 and this data is called the Mauna Loa record. This record shows that the average yearly atmospheric CO_2 concentrations have been steadily increasing.

Methane (CH_4)
A hydrocarbon that is a greenhouse gas with a global warming potential most recently estimated at 23 times that of carbon dioxide (CO_2). Methane is produced through anaerobic (without oxygen) decomposition of waste in landfills, animal digestion, decomposition of animal wastes, production and distribution of natural gas and petroleum, coal production, and incomplete fossil fuel combustion.

Metric Ton
Common international measurement for the quantity of greenhouse gas emissions. A metric ton is equal to 2,205 lbs. or 1.1 short tons.

Montreal Protocol
The Montreal Protocol on Substances that Deplete the Ozone Layer (i.e., a protocol developed at the Vienna Convention for the Protection of the Ozone Layer), or commonly referred to as the Montreal Protocol. The Montreal Protocol is an international treaty that is structured around phasing out the production of several groups of halogenated hydrocarbons alleged to be responsible for ozone depletion, such as chlorofluorocarbons (CFCs) and other substances that contain chlorine and bromine.

Mount Pinatubo
An active volcano located in the Philippine Islands that erupted in 1991. The eruption of Mount Pinatubo ejected enough particulate and sulfate aerosol matter into the upper atmosphere to block some of the incoming solar radiation from reaching Earth's atmosphere. This effectively cooled the planet from 1992 to 1994, masking the warming that had been occurring for most of the 1980s and 1990s.

Natural Gas
Underground deposits of gases consisting of 50 to 90 percent methane (CH_4) and small amounts of heavier gaseous hydrocarbon compounds such as propane (C_3H_8) and butane (C_4H_{10}).

Nitrogen Oxides (NOx)
Gases consisting of one molecule of nitrogen and varying numbers of oxygen molecules. Nitrogen oxides are produced in the emissions of vehicle exhausts and from power stations. In the atmosphere, nitrogen oxides can contribute to formation of smog, can impair visibility, and have health consequences.

Nitrous Oxide (N_2O)
A powerful greenhouse gas with a global warming potential of 296 times that of carbon dioxide (CO_2). Major sources of nitrous oxide include soil cultivation practices, especially the use of commercial and organic fertilizers, fossil fuel combustion, nitric acid production, and biomass burning.

Non-Annex B Parties
These are countries not listed in the Kyoto Treaty under Annex B.

Non-Annex I Parties
These are countries not listed in the UNFCCC under Annex I.

North Atlantic Oscillation (NAO)
The NAO is a large-scale fluctuation in atmospheric pressure between the subtropical high pressure system located near the Azores in the Atlantic Ocean and the subpolar low pressure system near Iceland and is quantified in the NAO Index. Surface pressure drives surface winds and wintertime storms from west to east across the North Atlantic, affecting climate from New England to western Europe as far eastward as central Siberia and the eastern Mediterranean and southward to West Africa.

Oxidize
To chemically transform a substance by combining it with oxygen.

Ozone (O_3)
Ozone, the triatomic form of oxygen (O_3), is a gaseous atmospheric constituent. In the troposphere, it is created both naturally and by photochemical reactions involving gases resulting from human activities (photochemical smog). In high concentrations, tropospheric ozone can be harmful to a wide range of living organisms. Tropospheric ozone acts as a greenhouse gas. In the stratosphere, ozone is created by the interaction between solar ultraviolet radiation and molecular oxygen (O_2). Stratospheric ozone plays a decisive role in the stratospheric radiative balance. Depletion of stratospheric ozone, due to chemical reactions that may be enhanced by climate change, results in an increased ground-level flux of ultraviolet (UV-) B radiation.

Ozone-Depleting Substance (ODS)
A family of manmade compounds including chlorofluorocarbons, bromofluorocarbons (halons), methyl chloroform, carbon tetrachloride, methyl bromide, and hydrochlorofluorocarbons (HCFCs). These compounds, typically referred to as ODSs, have been shown to deplete stratospheric ozone.

Ozone Layer
The layer of ozone that begins approximately 9 mi. (15 km) above Earth and thins to an almost

negligible amount at about 31 mi. (50 km), shielding the Earth from harmful ultraviolet radiation from the sun.

Ozone Precursors
Chemical compounds such as carbon monoxide, methane, nonmethane hydrocarbons, and nitrogen oxides, which in the presence of solar radiation react with other chemical compounds to form ozone, mainly in the troposphere.

Particulate Matter (PM)
Very small pieces of solid or liquid matter such as particles of soot, dust, fumes, mists, or aerosols. The physical characteristics of particles, and how they combine with other particles, are part of the feedback mechanisms of the atmosphere.

Parts per billion (ppb)
Number of parts of a chemical found in one billion parts of a gas, liquid, or solid mixture.

Parts per million (ppm)
Number of parts of a chemical found in 1 million parts of a particular gas, liquid, or solid.

Perfluorocarbons (PFCs)
A group of human-made chemicals composed of carbon and fluorine only. These chemicals were introduced as alternatives, along with hydrofluorocarbons, to the ozone-depleting substances.

Photosynthesis
The process by which plants take in CO_2 from the atmosphere (or bicarbonate in water) to build carbohydrates, releasing O_2 in the process. There are several pathways of photosynthesis, with different responses to atmospheric CO_2 concentrations.

Precession
The comparatively slow torquing of the orbital planes of all satellites with respect to the Earth's axis, due to the bulge of the Earth at the equator, which distorts the Earth's gravitational field.

Radiation
Energy transfer in the form of electromagnetic waves or particles that release energy when absorbed by an object.

Radiative Forcing
Radiative forcing is the change in the net vertical irradiance (expressed in Watts per square meter: Wm^2) at the tropopause due to an internal change or a change in the external forcing of the climate system, such as a change in the concentration of carbon dioxide or the output of the sun. Usually radiative forcing is computed after allowing for stratospheric temperatures to readjust to radiative equilibrium, but with all tropospheric properties fixed at their unperturbed values.

Recycling
Collecting and reprocessing a resource so it can be used again, for example, collecting aluminum cans, melting them down, and using the aluminum to make new cans or other aluminum products.

Reforestation
Reforestation is the natural and/or artificial restocking and regeneration of trees in recently depleted forests and woodlands as a result of natural or man-made activities, such as fires, storms, flooding, landslides, insect infestations, volcanic eruptions, slash-and-burn clearing, logging, or clear-cutting.

Residence Time
The average time spent in a reservoir by an individual atom or molecule. With respect to greenhouse gases, residence time usually refers to how long a particular molecule remains in the atmosphere.

Respiration
The biological process whereby living organisms convert organic matter to CO_2, releasing energy and consuming O_2.

Short Ton
Measurement for a ton in the United States. A short ton is equal to 2,000 lbs. or 0.907 metric ton.

Sink
Any process, activity, or mechanism that removes a greenhouse gas, aerosol, or precursor of a greenhouse gas or aerosol from the atmosphere.

Soil Carbon
A major component of the terrestrial biosphere pool in the carbon cycle. The amount of carbon

in the soil is a function of the historical vegetative cover and productivity, which in turn is dependent in part upon climatic variables.

Solar Radiation
Radiation emitted by the sun. It is also referred to as short-wave radiation. Solar radiation has a distinctive range of wavelengths (spectrum) determined by the temperature of the sun.

Storm Surge
An abnormal rise in sea level accompanying a hurricane or other intense storm, the height of which is the difference between the observed level of the sea surface and the level that would have occurred in the absence of the cyclone. Storm surge is usually estimated by subtracting the normal or astronomic tide from the observed storm tide.

Stratosphere
Region of the atmosphere between the troposphere and mesosphere, having a lower boundary of approximately 5 mi. (8 km) at the poles to 9 mi. (15 km) at the equator and an upper boundary of approximately 31 mi. (50 km). Depending upon latitude and season, the temperature in the lower stratosphere can increase, be isothermal, or even decrease with altitude, but the temperature in the upper stratosphere generally increases with height due to absorption of solar radiation by ozone.

Streamflow
The volume of water that moves over a designated point over a fixed period of time. It is often expressed as cubic feet per second ($ft.^3/sec.$).

Sulfate Aerosols
Particulate matter that consists of compounds of sulfur formed by the interaction of sulfur dioxide and sulfur trioxide with other compounds in the atmosphere. Sulfate aerosols are injected into the atmosphere from the combustion of fossil fuels and the eruption of volcanoes.

Sulfur Hexafluoride (SF_6)
A colorless gas, soluble in alcohol and ether, slightly soluble in water. A very powerful greenhouse gas used primarily in electrical transmission and distribution systems and as a dielectric in electronics.

Thermohaline Circulation (THC)
Large-scale, density-driven circulation in the ocean, caused by differences in temperature and salinity. In the North Atlantic, the thermohaline circulation consists of warm surface water flowing northward and cold deep water flowing southward, resulting in a net poleward transport of heat.

Trace Gas
Any one of the less common gases found in the Earth's atmosphere. Nitrogen, oxygen, and argon make up more than 99 percent of the Earth's atmosphere. Other gases, such as carbon dioxide, water vapor, methane, oxides of nitrogen, ozone, and ammonia are considered trace gases.

Tropopause
The tropopause is located between the troposphere and the stratosphere.

Troposphere
The lowest part of the Earth's atmosphere from the surface to about 6 mi. (10 km) in altitude in mid-latitudes (ranging from 5.5 mi. [9 km] in high latitudes to 10 mi. [16 km] in the tropics on average) where clouds and weather phenomena occur.

Ultraviolet Radiation (UV)
The energy range just beyond the violet end of the visible spectrum. Although ultraviolet radiation constitutes only about 5 percent of the total energy emitted from the sun, it is the major energy source for the stratosphere and mesosphere, playing a dominant role in both energy balance and chemical composition.

United Nations Framework Convention on Climate Change (UNFCCC)
The Convention on Climate Change sets an overall framework for intergovernmental efforts to tackle the challenge posed by climate change. It recognizes that the climate system is a shared resource, and its stability can be affected by industrial and other emissions of carbon dioxide and other greenhouse gases.

Wastewater
Water that has been used and contains dissolved or suspended waste materials.

Water Vapor
The most abundant greenhouse gas, it is the water present in the atmosphere in gaseous form. Water vapor is a part of the natural greenhouse effect.

Weather
Atmospheric condition at any given time or place. It is measured in terms of such things as wind, temperature, humidity, atmospheric pressure, cloudiness, and precipitation.

Sources: U.S. Environmental Protection Agency, U.S. Forest Service, Northwest Power and Conservation Council, U.S. Geological Survey, and National Oceanic and Atmospheric Administration.

Compiled by Andrew Hund
Independent Scholar

Resource Guide

Books

Abrahamson, D. E., ed. *The Challenge of Global Warming*. Washington, DC: Island Press, 1989.

Adger, N., et al. *Climate Change 2007: Impacts, Adaptation and Vulnerability: Contribution of Working Group II to the Fourth Assessment Report of the Intergovernmental Panel on Climate Change*. Cambridge: Cambridge University Press, 2007.

Aguado, E. and James J. Burt. *Understanding Weather and Climate*. Upper Saddle River, NJ: Prentice Hall, 2006.

Ahrens, C. Donald. *Meteorology Today*. Belmont, CA: Thomson Brooks/Cole, 2007.

Alley, R. *The Two-Mile Time Machine: Ice Cores, Abrupt Climate Change and Our Future*. Princeton, NJ: Princeton University Press, 2005.

Archer, D. *Global Warming: Understanding the Forecast*. Malden, MA: Blackwell, 2007.

Archer, D. *The Long Thaw: How Humans Are Changing the Next 100,000 Years of Earth's Climate*. Princeton, NJ: Princeton University Press, 2008.

Baumert, K., et al. *Climate Data: Insight and Observations*. Arlington, VA: Pew Center on Global Climate Change, 2004.

Baxter, W. *Today's Revolution in Weather*. New York: International Economic Research Bureau, 1953.

Bernaerts, A. *Climate Change and Naval War: A Scientific Assessment*. Bloomington, IN: Trafford Publishing, 2006.

Bickel, J. and L. Lane. *An Analysis of Climate Engineering as a Response to Climate Change*. Copenhagen, Denmark: Copenhagen Consensus Center, 2009.

Blair, T. *Climatology, General and Regional*. New York: Prentice Hall, 1942.

Bolin, B. *A History of the Science and Politics of Climate Change: The Role of the Intergovernmental Panel on Climate Change*. Cambridge: Cambridge University Press, 2007.

Bostrom, N. and M. Cirkovic, eds. *Global Catastrophic Risks*. Oxford: Oxford University Press, 2008.

Bowen, M. *Thin Ice: Unlocking the Secrets of Climate in the World's Highest Mountains*. New York: Henry Holt, 2005.

Boykoff, M. *Who Speaks for Climate? Making Sense of Media Reporting on Climate Change.* Cambridge: Cambridge University Press, 2010.

Broecker, W. *The Great Ocean Conveyor: Discovering the Trigger for Abrupt Climate Change.* Princeton, NJ: Princeton University Press, 2010.

Broecker, W. and R. Kunzig. *Fixing Climate: What Past Climate Changes Reveal About the Current Threat—and How to Counter It.* New York: Hill and Wang, 2008.

Brooks, C. E. P. *Climate Through the Ages: A Study of the Climatic Factors and Their Variations.* London: Benn, 1949.

Brown, N. *History and Climate Change: A Eurocentric Perspective.* London: Routledge, 2001.

Burroughs, W. *Climate Change in Prehistory: The End of the Reign of Chaos.* Cambridge: Cambridge University Press, 2005.

Burton, I., N. Diringer, and J. Smith. *Adaptation to Climate Change: International Policy Options.* Alexandria, VA: Pew Center on Global Climate Change, 2006.

Campbell, K., et al. *The Age of Consequences: The Foreign Policy and National Security Implications of Global Climate Change.* Washington, DC: Center for Strategic & International Studies, 2007.

Clark, P., et al. *Abrupt Climate Change. A Report by the U.S. Climate Change Science Program and the Subcommittee on Global Change Research. Synthesis and Assessment Product Sap 3.4.* Washington, DC: U.S. Geological Survey, 2008.

Climate: Long-Range Investigation, Mapping and Prediction (CLIMAP) Project. *Seasonal Reconstruction of the Earth's Surface at the Last Glacial Maximum.* Boulder, CO: Geological Society of America Map and Chart Series MC-36, 1981.

CNA Corporation Military Advisory Board. (Gen. Gordon R. Sullivan, Chair.) *National Security and the Threat of Climate Change.* Alexandria, VA: CNA Corporation, 2007.

Consultative Group on International Agricultural Research (CGIAR). *Global Climate Change: Can Agriculture Cope?* Washington, DC: CGIAR, 2009.

Coward, Harold and Thomas Hurka, eds. *Ethics and Climate Change: The Greenhouse Effect.* Waterloo, Canada: Wilfred Laurier University Press, 1993.

Cracknell, Basil. *Outrageous Waves: Global Warming and Coastal Change in Britain Through Two Thousand Years.* Chichester, UK: Phillimore & Co., 2005.

Davoudi, S., J. Crawford, and A. Mehmood, eds. *Planning for Climate Change: Strategies for Mitigation and Adaptation for Spatial Planners.* London: Earthscan, 2009.

DiMento, Joseph and Pamela Doughman, eds. *Climate Change: What It Means for Our Children and Our Grandchildren.* Cambridge, MA: MIT Press, 2007.

Douglass, A. *Climatic Cycles and Tree-Growth: A Study of the Annual Rings of Trees in Relation to Climate and Solar Activity.* Washington, DC: Carnegie Institution of Washington, 1936.

Emanuel, Kerry. *What We Know About Climate Change.* Cambridge, MA: MIT Press, 2007.

Energy Information Administration. *Impact of the Kyoto Protocol on U.S. Energy Markets and Economic Activity.* Washington, DC: U.S. Department of Energy, 1998.

Fagan, Brian. *The Little Ice Age: How Climate Made History, 1300–1850.* New York: Basic Books, 2000.

Fagan, Brian. *The Long Summer: How Climate Changed Civilization.* London: Granta Books, 2005.

Feenstra, J., I. Burton, J. B. Smith, and R. Tol, eds. *Handbook on Methods for Climate Change Impact Assessment and Adaptation Strategies.* Amsterdam and Nairobi: United Nations Environment Programme and Institute for Environmental Studies, 1998.

Fleming, James R. *Fixing the Sky: The Checkered History of Weather and Climate Control.* New York: Columbia University Press, 2010.

Fleming, James R. *Historical Perspectives on Climate Change.* Oxford: Oxford University Press, 2004.

Frakes, Lawrence. *Climate Throughout Geologic Time.* Amsterdam: Elsevier/North-Holland, 1979.

Frederick, Kenneth D. and Peter H. Gleick. *Water and Global Climate Change: Potential Impacts on U.S. Water Resources*. Alexandria, VA: Pew Center on Global Climate Change, 1999.

Friel, H. *The Lomborg Deception: Setting the Record Straight About Global Warming*. New Haven, CT: Yale University Press, 2010.

Fry, Carolyn. *The Impact of Climate Change: The World's Greatest Challenge in the Twenty-First Century*. London: New Holland, 2008.

Glover, Leigh. *Postmodern Climate Change*. London: Routledge, 2006.

Graedel, Thomas. *Atmosphere, Climate and Change*. New York: W. H. Freeman, 1995.

Greenpeace. *Dealing in Doubt: The Climate Denial Industry and Climate Science*. Amsterdam: Greenpeace International, 2010.

Hann, J. *Handbook of Climatology*. Translation of *Handbuch der Klimatologie*. 2nd ed. (1897). New York: Macmillan, 1903.

Hansen, James. *Storms of My Grandchildren: The Truth About the Coming Climate Catastrophe and Our Last Chance to Save Humanity*. London: Bloomsbury, 2009.

Hartmann, Dennis L. *Global Physical Climatology*. San Diego, CA: Academic Press, 1994.

Harvey, H. *The Chemistry and Fertility of Sea Water*. Cambridge: Cambridge University Press, 1955.

Harvey, L. D. Danny. *Climate and Global Environmental Change*. Upper Saddle River, NJ: Prentice Hall, 2000.

Henson, Robert. *The Rough Guide to Climate Change: The Symptoms. The Science. The Solutions*. London: Rough Guides, 2006.

Hoggan, J. and R. Littlemore. *Climate Cover-Up. The Crusade to Deny Global Warming*. Vancouver, Canada: Greystone, 2009.

Hulme, M. *Why We Disagree About Climate Change*. Cambridge: Cambridge University Press, 2009.

Intergovernmental Panel on Climate Change. *Climate Change: Contribution of Working Group I to the Third Assessment Report*. Cambridge: Cambridge University Press, 2001.

Intergovernmental Panel on Climate Change. *Climate Change 1994: Radiative Forcing of Climate Change*. Cambridge: Cambridge University Press, 1995.

Intergovernmental Panel on Climate Change. *Climate Change 1995: The Science of Climate Change*. Cambridge: Cambridge University Press,1996.

Intergovernmental Panel on Climate Change. *Climate Change 2001: The Scientific Basis*. Cambridge: Cambridge University Press, 2001.

Intergovernmental Panel on Climate Change. *Climate Change 2007: The Physical Basis: Contribution of Working Group I to the Fourth Assessment Report of the Intergovernmental Panel on Climate Change*. Cambridge: Cambridge University Press, 2007.

Kendrew, Wilfrid. *The Climates of the Continents*. Oxford: Clarendon, 1961.

Lamb, H. *The Changing Climate: Selected Papers*. London: Methuen, 1966.

Lichter, S. *Climate Scientists Agree on Warming*. STATS Articles 2008. Washington, DC: Statistical Assessment Service, 2008.

Linacre, Edward and Bart Geerts. *Climates and Weather Explained*. London: Routledge, 1997.

Linden, E. *The Winds of Change: Climate, Weather, and the Destruction of Civilizations*. New York: Simon & Schuster, 2006.

Maroto, M. and M. Valer, eds. *Environmental Challenges and Greenhouse Gas Control for Fossil Fuel Utilization in the 21st Century*. New York: KluwerAcademic/Plenum Publishers, 2002.

Oxfam. *Adapting to Climate Change, What's Needed in Poor Countries and Who Should Pay*. Oxfam. Washington, DC: Oxfam, 2007.

Petty, G. W. *A First Course in Atmospheric Radiation*. Madison, WI: Sundog Publishing, 2004.

Philander, George S. *Is the Temperature Rising? The Uncertain Science of Global Warming*. Princeton, NJ: Princeton University Press, 2000.

Philander, George S. *Our Affair With El Niño: How We Transformed an Enchanting Peruvian Current Into a Global Climate*

Hazard. Princeton, NJ: Princeton University Press, 2005.

Rampino, Michael R. *Climate: History, Periodicity, and Predictability*. New York: Van Nostrand Reinhold, 1987.

Raupach, M. R., G. Marland, P. Ciais, J. C. Le Quéré, G. Canadell, G. Klepper, and C. B. Field. *Global and Regional Drivers of Accelerating CO_2 Emissions*. Washington, DC: Proceedings of the National Academy of Sciences, 2007.

Robinson, Peter and Ann Henderson-Sellers. *Contemporary Climatology*. Upper Saddle River, NJ: Prentice Hall, 1999.

Rosenzweig, C. and D. Hillel. *Climate Change and the Global Harvest: Potential Effects of the Greenhouse Effect on Agriculture*. Oxford: Oxford University Press, 1998.

Shanley, Robert A. *Presidential Influence and Climate Change*. Westport, CT: Greenwood Press, 1992.

Siedler, Gerold, John Church, and John Gould, eds. *Ocean Circulation and Climate: Observing and Modelling the Global Ocean*. London: Academic Press, 2001.

Singer, S. Fred and Dennis T. Avery. *Unstoppable Global Warming: Every 1,500 Years*. Lanham, MD: Rowman & Littlefield, 2007.

Singh, Ram Babu. *Urban Sustainability in the Context of Global Change: Towards Promoting Healthy and Green Cities*. Enfield, NH: Science Publishers, 2001.

Spray, Sharon. *Global Climate Change*. Lanham, MD: Rowman & Littlefield, 2002.

Stern, Nicholas. *The Economics of Climate Change: The Stern Review*. Cambridge: Cambridge University Press, 2007.

Stevens, W. *The Change in the Weather: People, Weather and the Science of Climate*. New York: Delacorte Press, 1999.

United Nations Framework Convention on Climate Change. *An Introduction to the Kyoto Protocol Compliance Mechanism*. New York: United Nations, 2006.

Weart, S. *The Discovery of Global Warming*. 2nd ed. Cambridge, MA: Harvard University Press, 2008.

Williams, Mary E. *Global Warming: An Opposing Viewpoints Guide*. Farmington Hills, MI: Greenhaven Press, 2006.

Yamin, Farhana and Joanna Depledge. *The International Climate Change Regime: A Guide to Rules, Institutions and Procedures*. Cambridge: Cambridge University Press, 2004.

Yoshino, Masatoshi. *Climates and Societies: A Climatological Perspective*. Dordrecht, Netherlands: Kluwer Academic Publishers, 1997.

Articles

Allen, M., et al. "Warming Caused by Cumulative Carbon Emissions: Towards the Trillionth Tonne." *Nature*, v.458 (April 30, 2009).

Bates, Diane. "Environmental Refugees? Classifying Human Migrations Caused by Environmental Change." *Population and Environment*, v. 23/5 (May 2002).

Bindschadler, Robert A. "Hitting the Ice Sheets Where It Hurts." *Science*, v.311 (2006).

Blanchon, Paul, et al. "Rapid Sea-Level Rise and Reef Back-Stepping at the Close of the Last Interglacial Highstand." *Nature*, v.458 (2009).

Bolin, Bert. "The Carbon Cycle." *Scientific American* (September 1970).

Bray, J. "An Analysis of the Possible Recent Change in Atmospheric Carbon Dioxide Concentration." *Tellus*, v.11 (1959).

Broecker, Wallace S. "The Great Ocean Conveyor." *Oceanography*, v.4 (1991).

Brooks, C. E. P. "Selective Annotated Bibliography on Climatic Changes." *Meteorological Abstracts and Bibliography*, v.1/4 (1950).

Bryden, Harry L., Hannah R. Longworth, and Stuart A. Cunningham. "Slowing of the Atlantic Meridional Overturning Circulation at 25° N." *Nature*, v.438 (December 2005).

Bryson, Reid A. "A Perspective on Climatic Change." *Science*, v.184 (1974).

Burkett, V. R., et al. "Nonlinear Dynamics in Ecosystem Response to Climatic Change: Case Studies and Policy Implications." *Ecological Complexity*, v.2/4 (2005).

Calanca, P. "Climate Change and Drought Occurrence in the Alpine Region: How Severe Are Becoming the Extremes? *Global and Planetary Change*, v.57/1–2 (2007).

Caldeira, Ken and Michael E. Wickett. "Anthropogenic Carbon and Ocean pH." *Nature*, v.425/6956 (September 2003).

Callendar, G. S. "Variations in the Amount of Carbon Dioxide in Different Air Currents." *Quarterly Journal of the Royal Meteorological Society*, v.66 (1940).

Camill, P. and. J. S. Clark. "Long-Term Perspectives on Lagged Ecosystem Responses to Climate Change: Permafrost in Boreal Peatlands and Grassland/Woodland Boundary." *Ecosystems*, v.3/6 (2000).

Charlson, Robert J. "A Stone Age Greenhouse." *Nature*, v.438 (2005).

Church, John A. and Neil J. White. "20th Century Acceleration in Global Sea-Level Rise." *Geophysical Research Letters*, v.33 (2006).

Climate: Long-Range Investigation, Mapping and Prediction. "The Last Interglacial Ocean." *Quaternary Research*, v.21 (1984).

Climate: Long-Range Investigation, Mapping and Prediction and A. McIntyre, et al. "The Surface of the Ice-Age Earth." *Science*, v.191 (1976).

Coughlin, Steven S. "Educational Intervention Approaches to Ameliorate Adverse Public Health and Environmental Effects From Global Warming." *Ethics in Science and Environmental Politics* v. 2006/13–14 (2006).

Cox, Peter M., et al. "Increasing Risk of Amazonian Drought Due to Decreasing Aerosol Pollution." *Nature*, v.453 (2008).

Crary, A. P., et al. "Evidences of Climate Change From Ice Island Studies." *Science*, v.122 (1955).

Crowley, T. J. "Causes of Climate Change Over the Past 1000 Years." *Science*, v.289 (2000).

Crutzen, Paul J., et al. "N_2O Release From Agro-Biofuel Production Negates Global Warming Reduction by Replacing Fossil Fuels." *Atmospheric Chemistry and Physics*, v.8 (2008).

Dansgaard, W., et al. "Evidence for General Instability of Climate From a 250-Kyr Ice-Core Record." *Nature*, v.364 (1993).

Dessler, Andrew E. "A Determination of the Cloud Feedback From Climate Variations Over the Past Decade." *Science*, v.330 (2010).

Doherty, Sarah J. "Initiative to Improve Process Representation in Chemistry–Climate Models." *Eos: Transactions of the American Geophysical Union*, v.90 (2009).

Doran, P. T. and M. Kendall Zimmerman. "Examining the Scientific Consensus on Climate Change." *Eos: Transactions of the American Geophysical Union*, v.90 (2009).

Douglass, David H., et al. "A Comparison of Tropical Temperature Trends With Model Predictions." *International Journal of Climatology*, v.28 (2008).

Dybas, Cheryl. "Increase in Rainfall Variability Related to Global Climate Change." *Earth Observatory, NASA* (December 12, 2002).

Ekholm, Nils. "On the Variations of the Climate of the Geological and Historical Past and Their Causes." *Quarterly Journal of the Royal Meteorological Society*, v.27 (1901).

Elsner, James B., et al. "The Increasing Intensity of the Strongest Tropical Cyclones." *Nature*, v.455 (2008).

Etkins, Robert and Edward S. Epstein. "The Rise of Global Mean Sea Level as an Indication of Climate Change." *Science*, v.215 (1982).

European Project for Ice Coring in Antarctica (EPICA). "One-to-One Coupling of Glacial Climate Variability in Greenland and Antarctica." *Nature*, v.444 (2006).

Eyring, V., et al. "Assessment of Temperature, Trace Species, and Ozone in Chemistry–Climate Model Simulations of the Recent Past." *Journal of Geophysical Research*, v.111 (2006).

Fauria, M. Macias, et al. "Unprecedented Low Twentieth Century Winter Sea Ice Extent in the Western Nordic Seas Since A.D. 1200." *Climate Dynamics*, v.34 (2009).

Ganopolski, A. and S. Rahmstorf. "Rapid Changes of Glacial Climate Simulated in a Coupled Climate Model." *Nature*, v.409 (2001).

Gordon, Arnold L. "Inter-Ocean Exchange of Thermocline Water." *Journal of Geophysical Research*, v.91 (1986).

Hansen, J., et al. "Global Temperature Change." *Proceedings of the National Academy of Sciences of the US (PNAS)*, v.103/3 (2006).

Haque, C. and I. Burton. "Adaptation Options Strategies for Hazards and Vulnerability Mitigation: An International Perspective." *Mitigation and Adaptation Strategies for Global Change*, v.10 (2005).

Hay, J. and N. Mimura. "Sea Level Rise: Implications for Water Resources Management." *Mitigation and Adaptation Strategies for Global Change*, v.10 (2005).

Hayhoe, K., et al. "Past and Future Changes in Climate and Hydrological Indicators in the U.S. Northeast." *Climate Dynamics*, v.28/4 (2007).

Holland, Marika M., et al. "Future Abrupt Reductions in the Summer Arctic Sea Ice." *Geophysical Research Letters,* v.33 (2006).

Howat, Ian M., et al. "Rapid Changes in Ice Discharge From Greenland Outlet Glaciers." *Science,* v.315 (2007).

Huybers, P. "Comment on 'Hockey Sticks, Principal Components, and Spurious Significance' by S. Mcintyre and R. McKitrick." *Geophysical Research Letters,* v.32 (2005).

Jones, P. D., et al. "Urbanization Effects in Large-Scale Temperature Records, With an Emphasis on China." *Journal of Geophysical Research,* v.113 (2008).

Kaufman, Darrell S., et al. "Recent Warming Reverses Long-Term Arctic Cooling." *Science,* v.324 (2009).

Keeling, Charles D. "Is Carbon Dioxide From Fossil Fuel Changing Man's Environment?" *Proceedings of the American Philosophical Society,* v.114 (1970).

Kintisch, Eli. "Projections of Climate Change Go From Bad to Worse, Scientists Report." *Science,* v.323 (2009).

Lenton, T., et al. "Tipping Elements in the Earth's Climate System." *Proceedings of the National Academy of Sciences of the U.S. (PNAS)*, v.105/6 (2008).

Lozier, M. Susan. "Deconstructing the Conveyor Belt." *Science,* v.328 (2010).

Madden, Roland A. and V. Ramanathan. "Detecting Climate Change Due to Increasing Carbon Dioxide." *Science,* v.209 (1980).

Meier, Mark F., Mark Dyurgerov, Rick Ursula, Shad O'Neel, W. Tad Pfeffer, Robert Anderson, Suzanne Anderson, and Andrey Glazovsky. "Glaciers Dominate Eustatic Sea-Level Rise in the 21st Century." *Science*, v.298/5602 (December 2002).

Meinshausen, M., et al. "Greenhouse-Gas Emission Targets for Limiting Global Warming to 2°C." *Nature*, v.458 (April 30, 2009).

Mirza, M. M. Q., R. A. Warrick, and N. J. Ericksen. "The Implications of Climate Change on Floods of the Ganges, Brahmaputra and Meghna Rivers in Bangladesh." *Climate Change*, v.57/3 (2003).

Monastersky, Richard. "Climate Science on Trial." *Chronicle of Higher Education,* v.53/3 (September 8, 2006).

Myhre, M., E. J. Highwood, K. P. Shine, and F. Stordal. "New Estimates of Radiative Forcing Due to Well Mixed Greenhouse Gases." *Geophysical Research Letters*, v.25/14 (1998).

Nierenberg, Nicolas, et al. "Early Climate Change Consensus at the National Academy: The Origins and Making of 'Changing Climate.'" *Historical Studies in the Natural Sciences,* v.40 (2010).

Ormerod, W. G., P. Ferund, and A. Smith. "Ocean Storage of CO_2." Cheltenham, UK: International Energy Agency Greenhouse Gas R&D Program, 2002.

Parmesan, C. "Climate and Species Range." *Nature*, v.382/6594 (1996).

Rahmstorf, S. "A Semi-Empirical Approach to Projecting Future Sea-Level Rise." *Science* (January 19, 2007).

Ramanathan, V. and Y. Feng. "On Avoiding Dangerous Anthropogenic Interference With the Climate System: Formidable Challenges Ahead." *Proceedings of the National Academy of Sciences of the U.S. (PNAS)*, v.105/38 (September 23, 2008).

Randerson, James. "Should Governments Play Politics With Science?" *New Scientist*, v.184/2468 (October 9, 2004).

Roots, E. "Climate Change: High Latitude Regions." *Climatic Change*, v.15/1–2 (1989).

Rosenzweig, C., et al. "Climate Change and Extreme Weather Events: Implications for Food Production, Plant Diseases, and Pests." *Global Change and Human Health*, v.2/2 (2001).

Sagarin, R., et al. "Climate-Related Change in an Intertidal Community Over Short and Long Time Scales." *Ecological Monographs*, v.69/4 (1999).

Scheffer, Marten, et al. "Positive Feedback Between Global Warming and Atmospheric CO_2 Concentration Inferred From Past Climate Change." *Geophysical Research Letters,* v.3 (2006).

Schubert, C. "Global Warming Debate Gets Hotter." *Science News*, v.159/24 (June 16, 2001).

Simmonds, Mark P. and Steven J. Isaac. "The Impacts of Climate Change on Marine Mammals: Early Signs of Significant Problems." *Oryx*, v.41 (2007).

Sokolov, A. P., et al. "Probabilistic Forecast for 21st Century Climate Based on Uncertainties in Emissions (Without Policy) and Climate Parameters." *Journal of Climate* (2009).

Solomon, S., et al. "Irreversible Climate Change Due to Carbon Dioxide Emissions." *Proceedings of the National Academy of Sciences of the U.S.* (*PNAS*), v.106/6 (2009).

Supreme Court of the United States. *Opinion of the Court: Commonwealth of Massachusetts, et al., v. U.S. Environmental Protection Agency, et al.* (549 U.S. No. 05.1120) (2007).

UNESCO. "Changes in Climate." *Arid Zone Research*, No. 20 (1963).

Periodicals

Annals of Glaciology
International Glaciological Society

Climate Research: Interactions of Climate With Organisms, Ecocsystems, and Human Societies
Inter-Research Science Center

Energy and Environment
Multi-Science Publishing

Environmental Ethics
Center for Environmental Philosophy

Environmental Justice: Issues, Policies, and Solutions Environmental Law
Oxford University Press

Environmental Science and Technology
Center for Environment and Energy Research and Studies

Eos, Transactions of the American Geophysical Union
American Geophysical Union

EPA Journal
Environmental Protection Agency

Geophysical Research Letters
American Geophysical Union

Global Environment Politics
MIT Press

Global Environmental Change
Royal Society of Chemistry

Hazardous Waste
BPI News

Journal of Applied Meteorology and Climatology (formerly *Journal of Applied Meteorology*)
American Meteorological Society

Journal of Climate
American Meteorological Society

Journal of Environment and Development
Sage Publications

Journal of Environmental Economics and Management
Academic Press

Journal of Environmental Education
Heldref Education

Journal of Environmental Management
Academic Press

Journal of Forestry
Oxford University Press

Journal of Geochemical Exploration
Elsevier Science

Journal of Geophysical Research
American Geophysical Union

Journal of the Atmospheric Sciences (until 1960 titled *Journal of Meteorology*)
American Meteorological Society

National Geographic
National Geographic Society

Nature
Palgrave Macmillan

Oryx
Fauna and Flora International

Paleoceanography
American Geophysical Union

Proceedings of the American Philosophical Society
American Philosophical Society

Quaternary Science Reviews
Elsevier

Science Trends in Ecology and Evolution
Oxford University Press

Internet

ACCESS: Africa Centre for Climate and Earth Systems Science
http://africaclimatescience.org
Amity Institute of Global Warming and Ecological Studies
http://www.amity.edu/aigwes
Asia-Pacific Network for Global Change Research
http://www.apn-gcr.org/newAPN/indexe.htm
Center for Climate and Energy Solutions (C2ES)
http://www.pewclimate.org
Center for Integrated Study of the Human Dimensions of Global Change
http://hdgc.epp.cmu.edu
Center for International Climate and Environmental Research (CICERO)
http://www.cicero.uio.no/home/index_e.aspx
Climate Action Network
http://www.climatenetwork.org
Climate Analysis Indicators Tool (CAIT)
http://cait.wri.org
Climate Ark
http://www.climateark.org
Climate Change at the National Academies
http://dels-old.nas.edu/climatechange
Climate Change Education.Org (Global Warming Education Climate Change Science Education)
http://www.climatechangeeducation.org
Climate Change Policy—The Europe Center
http://europe.stanford.edu/research/climate
Climate Justice Initiative
http://www.corpwatch.org/section.php?id=100
Climate Institute
http://www.climate.org
Climate Science Forum
http://climate-science.org
Environmental Change Network
http://www.ecn.ac.uk
Environmental Protection Agency (EPA) Global Warming Page
http://www.epa.gov/climatechange/index.html
Green Teacher
http://www.greenteacher.com
ICLEI Europe—Climate Change Adaptation
http://www.iclei-europe.org/topics/climate-change-adaptation
Intergovernmental Panel on Climate Change
http://www.ipcc.ch
M.I.T. Joint Program on the Science and Policy of Global Change
http://globalchange.mit.edu
National Aeronautics and Space Administration (NASA) Climate Change Solutions
http://climate.nasa.gov/solutions
National Aeronautics and Space Administration Global Change Data Center (GCDC)
http://science.gsfc.nasa.gov/sed/index.cfm?fuseAction=home.main&&navOrgCode=610.2)
Natural Environment Research Council (NERC)
http://www.nerc.ac.uk
PBS *NOVA/Frontline*: What's Up with the Weather?
http://www.pbs.org/wgbh/warming/index.html
South East European Virtual Climate Change Center (SEEVCCC)
http://www.seevccc.rs/?p=668
True North: Adapting Infrastructure to Climate Change in Northern Canada
http://nrtee-trnee.ca/climate/true-north
Union of Concerned Scientists: Global Warming
http://www.ucsusa.org/global_warming
U.S. Forest Service—Climate Change Resource Center (CCRC)
http://www.fs.fed.us/ccrc
U.S. Global Change Research Program
http://www.globalchange.gov
Pew Center on Global Climate Change
http://www.pewclimate.org
Woods Hole Research Center: Global Warming
http://www.whrc.org

Compiled by Andrew Hund
Independent Scholar

Appendix

Graphic Plots and Text Prepared by
Robert A. Rohde

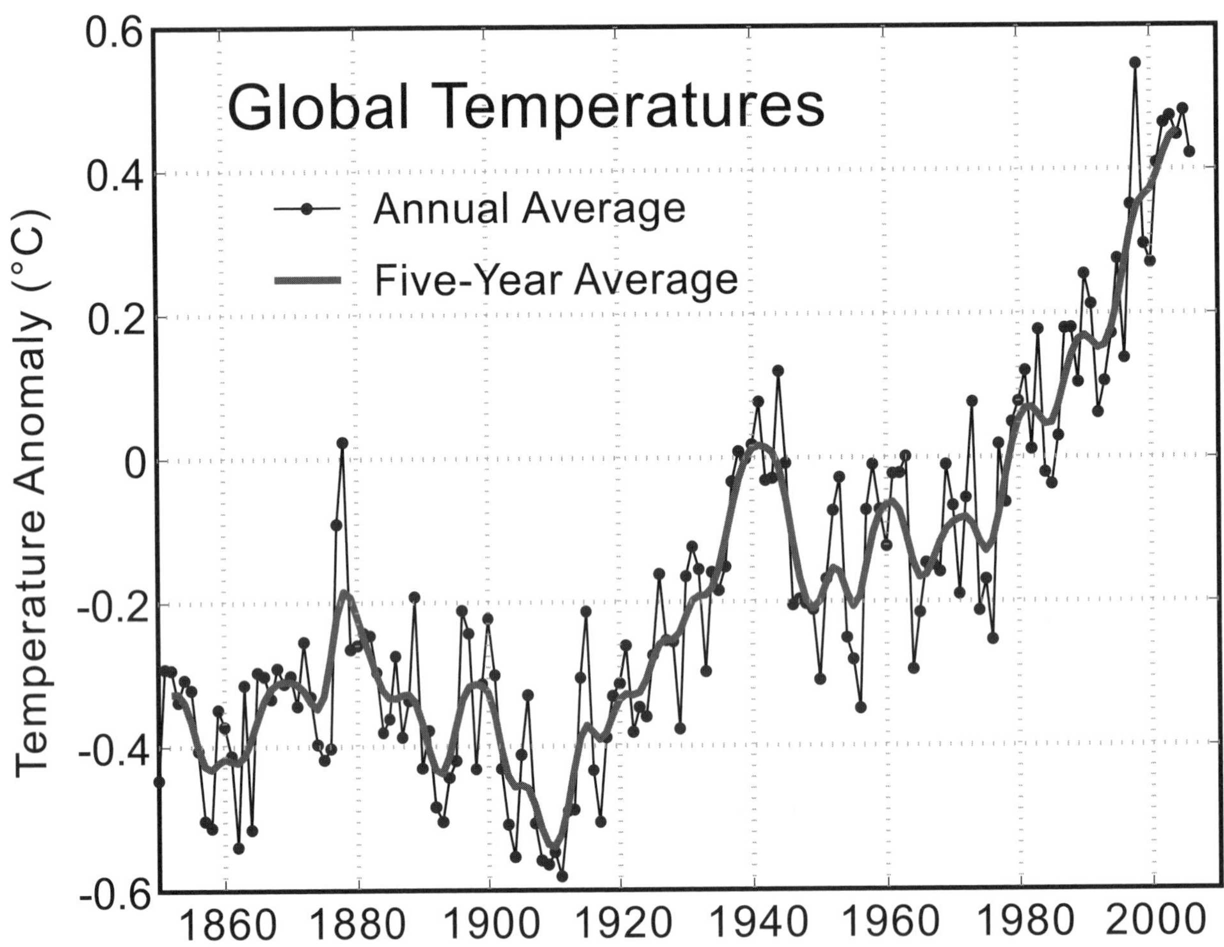

This image shows the instrumental record of global average temperatures as compiled by the Climatic Research Unit of the University of East Anglia and the Hadley Centre of the UK Meteorological Office. Data set HadCRUT3 was used, which follows the methodology outlined by Brohan et al. (2006). Following the common practice of the Intergovernmental Panel on Climate Change (IPCC), the zero on this figure is the mean temperature from 1961 to 1990.

The uncertainty in the analysis techniques leading to these measurements is discussed in Foland et al. (2001) and Brohan et al. (2006). They estimate that global averages since ~1950 are within ~0.05 degree C of their reported value with 95 percent confidence. In the recent period, these uncertainties are driven primarily by considering the potential impact of regions where no temperature record is available. For averages prior to ~1890, the uncertainty reaches ~0.15 degree C, driven primarily by limited sampling and the effects of changes in sea surface measurement techniques. Uncertainties between 1880 and 1890 are intermediate between these values.

Incorporating these uncertainties, Foland et al. (2001) estimated the global temperature change from 1901 to 2000 as 0.57 ± 0.17 degree C, which contributed to the 0.6 ± 0.2 degree C estimate reported by the Intergovernmental Panel on Climate Change (IPCC 2001a, [1]). Both estimates are 95 percent confidence intervals.

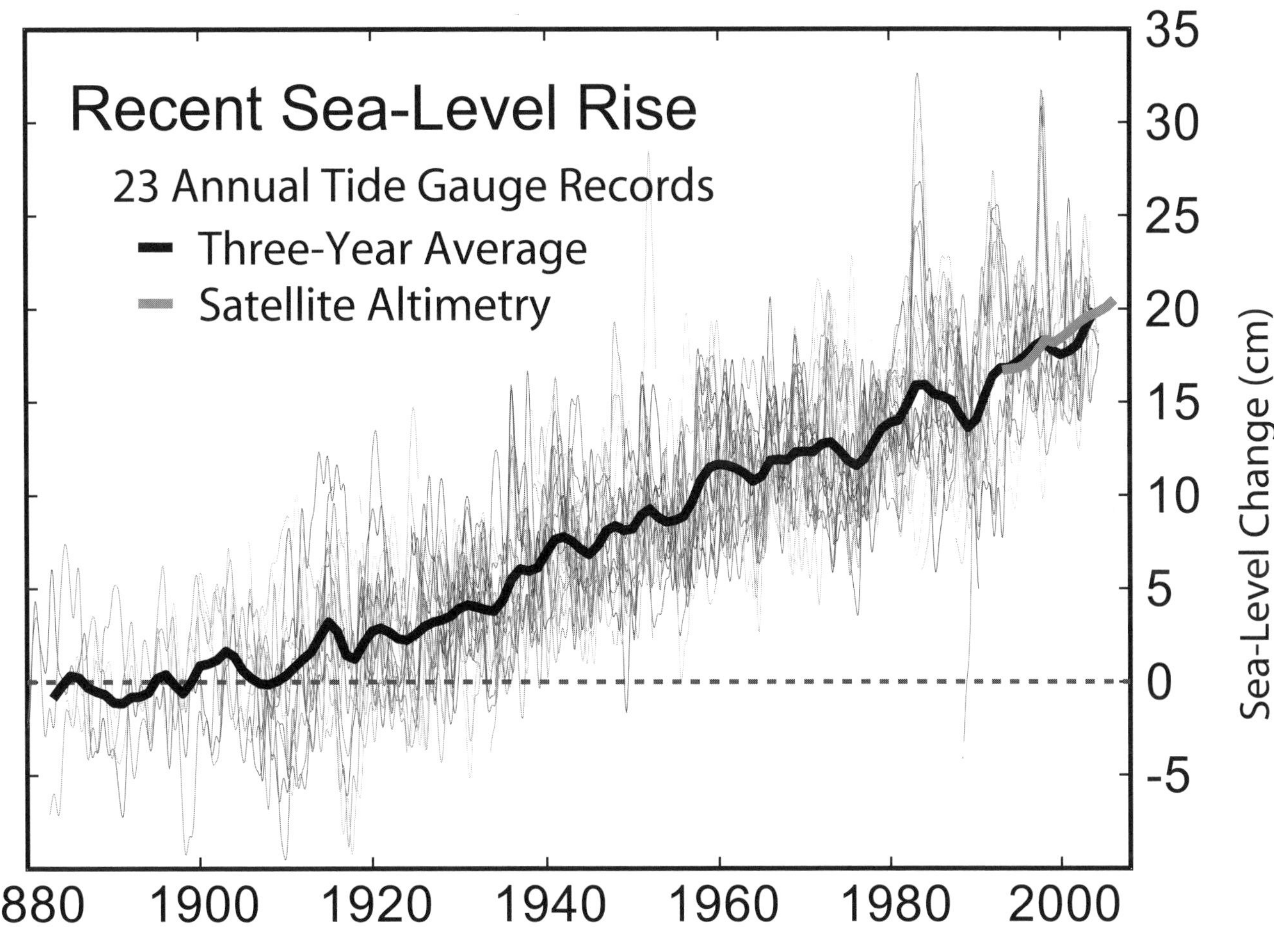

This figure shows the change in annually averaged sea level at 23 geologically stable tide gauge sites with long-term records as selected by Douglas (1997). The thick, dark line is a three-year moving average of the instrumental records. These data indicate a sea-level rise of ~18.5 cm from 1900 to 2000. Because of the limited geographic coverage of these records, it is not obvious whether the apparent decadal fluctuations represent true variations in global sea level, or merely variations across regions that are not resolved.

For comparison, the recent annually averaged satellite altimetry data from TOPEX/Poseidon are shown in the thick gray line. These data indicate a somewhat higher rate of increase than tide gauge data, however the source of this discrepancy is not obvious. It may represent systematic error in the satellite record and/or incomplete geographic sampling in the tide gauge record. The month-to-month scatter on the satellite measurements is roughly the thickness of the plotted gray curve.

Much of recent sea-level rise has been attributed to global warming.

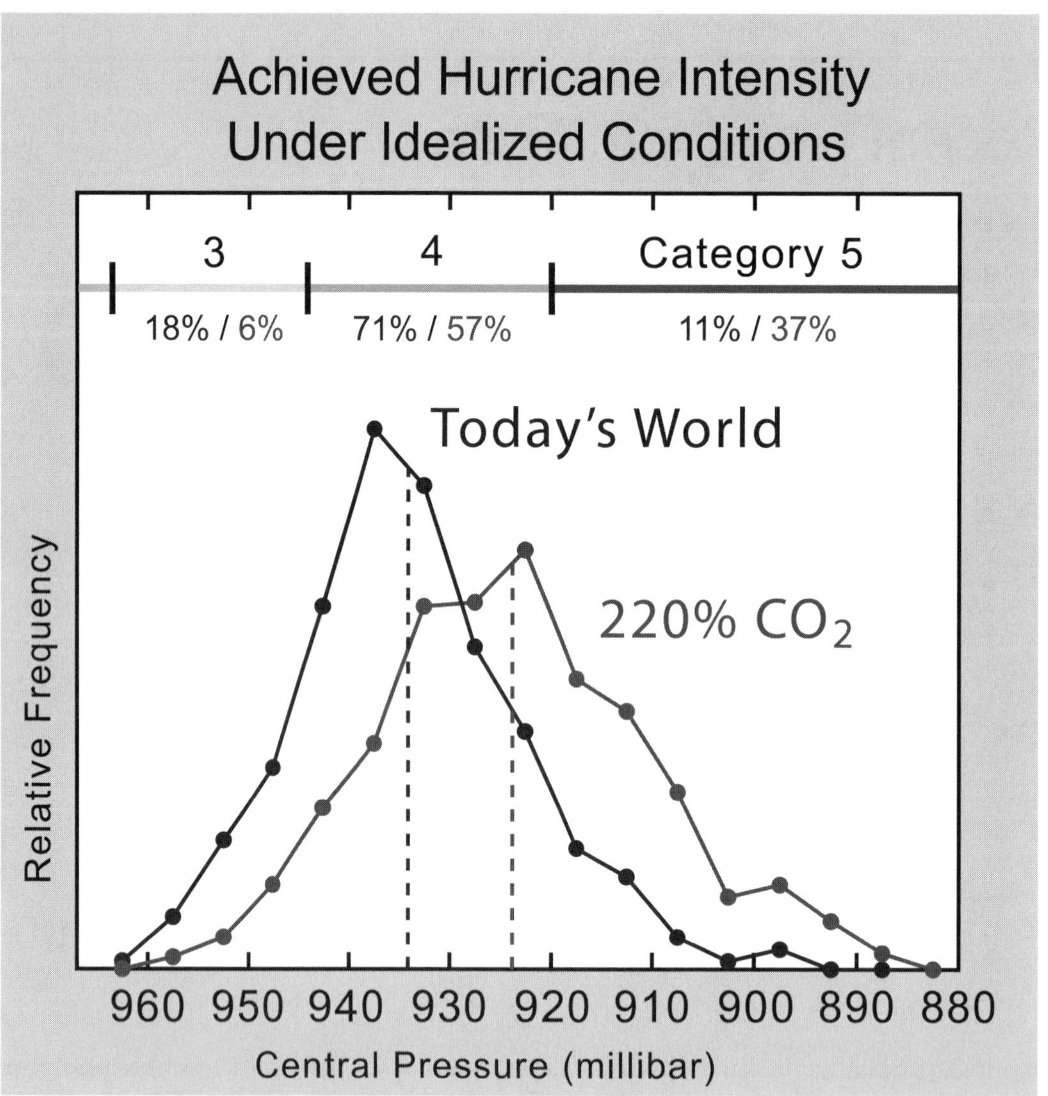

This figure, which reproduces one of the key conclusions of Knutson and Tuleya (2004), shows a prediction for how hurricanes and other tropical cyclones may intensify as a result of global warming. Specifically, Knutson and Tuleya performed an experiment using climate models to estimate the strength achieved by cyclones allowed to intensify over either a modern summer ocean or over an ocean warmed by carbon dioxide concentrations 220 percent higher than the present day. A number of different climate models were considered, as well as conditions over all the major cyclone-forming ocean basins. Depending on site and model, the ocean warming involved ranged from 0.8 to 2.4 degrees C. Results, which were found to be robust across different models, showed that storms intensified by about one-half category (on the Saffir-Simpson Hurricane Scale) as a result of the warmer oceans. This is accomplished with a ~6 percent increase in wind speed or equivalently a ~20 percent increase in energy (for a storm of fixed size). Most significantly, these results suggest that global warming may lead to a gradual increase in the probability of highly destructive Category 5 hurricanes. This work does not provide any information about future frequency of tropical storms. Also, since it considers only the development of storms under nearly ideal conditions for promoting their formation, this work is primarily a prediction for how the maximum achievable storm intensity will change. Hence, this does not directly bear on the growth or development of storms under otherwise weak or marginal conditions for storm development (such as high upper-level wind shear). However, it is plausible that warmer oceans will somewhat extend the regions and seasons under which hurricanes may develop.

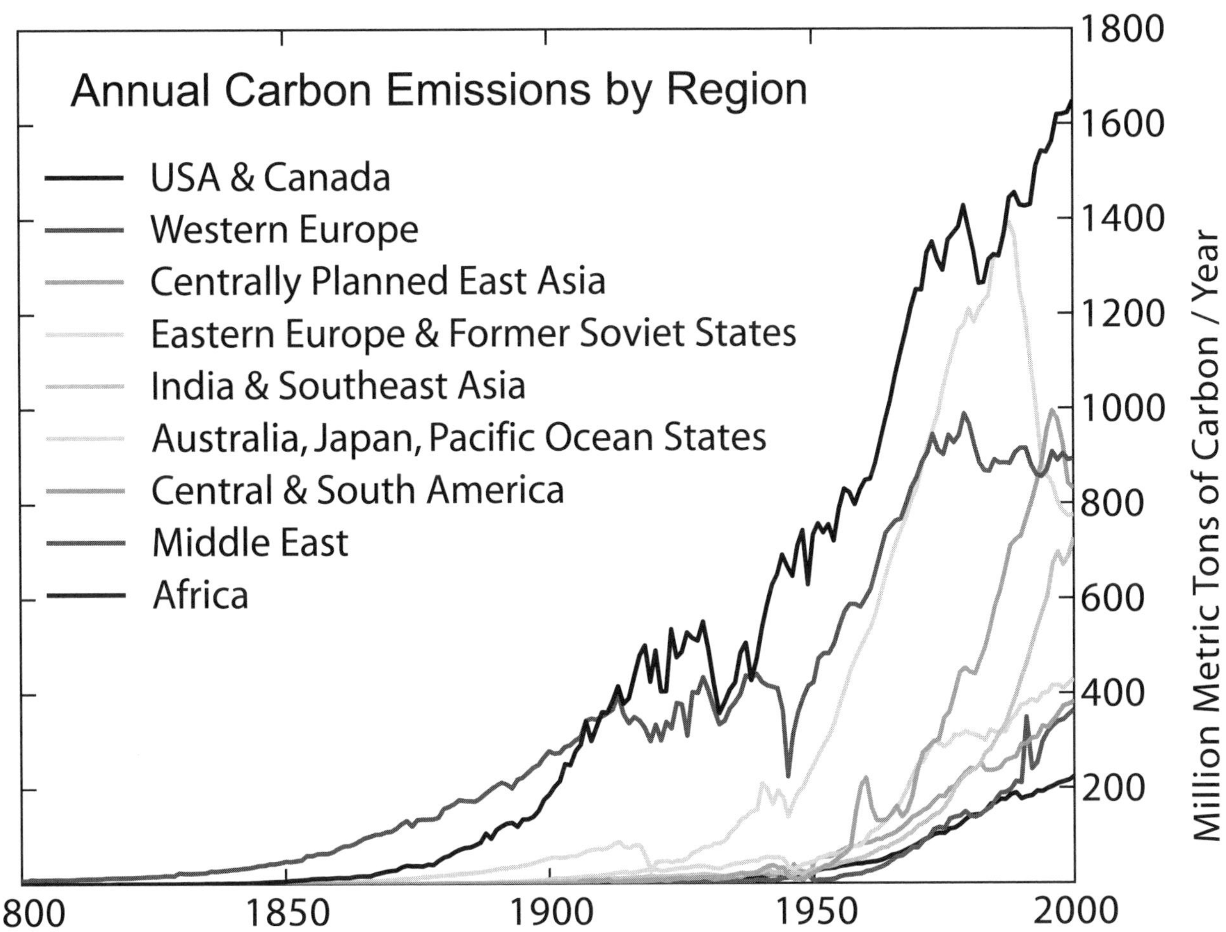

This figure shows the annual fossil fuel carbon dioxide emissions, in million metric tons of carbon, for a variety of non-overlapping regions covering the Earth. Data source: Carbon Dioxide Information Analysis Center.

Regions are sorted from largest emitter (as of 2000) to the smallest:

United States and Canada
Western Europe (plus Germany)
Communist East Asia (China, North Korea, Mongolia, etc.)
Eastern Europe, Russia, and Former Soviet States
India and Southeast Asia (plus South Korea)
Australia, Japan, and other Pacific Island States
Central and South America (includes Mexico and the Caribbean)
Middle East
Africa

Editor's Note: According to Reuters news agency, China acknowledged in November 2010 that it was the world's largest emitter of greenhouse gases, surpassing the United States, the world's top emitter for the 20th century.

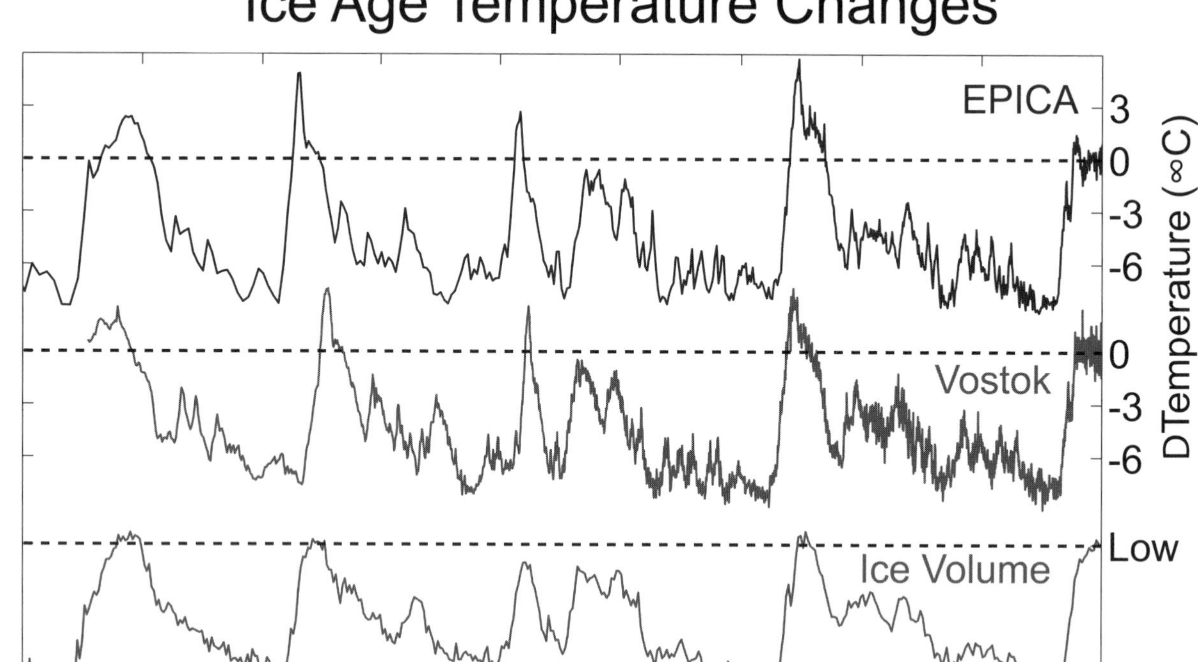

This figure shows the Antarctic temperature changes during the last several glacial/interglacial cycles of the present ice age and a comparison to changes in global ice volume. The present day is on the right.

The first two curves show local changes in temperature at two sites in Antarctica as derived from deuterium isotopic measurements (δD) on ice cores (EPICA Community Members 2004, Petit et al. 1999). The final plot shows a reconstruction of global ice volume based on δ18O measurements on benthic foraminifera from a composite of globally distributed sediment cores and is scaled to match the scale of fluctuations in Antarctic temperature (Lisiecki and Raymo 2005). Note that changes in global ice volume and changes in Antarctic temperature are highly correlated, so one is a good estimate of the other, but differences in the sediment record do not necessarily reflect differences in paleotemperature. Horizontal lines indicate modern temperatures and ice volume. Differences in the alignment of various features reflect dating uncertainty and do not indicate different timing at different sites.

The Antarctic temperature records indicate that the present interglacial is relatively cool compared to previous interglacials, at least at these sites. It is believed that the interglacials themselves are triggered by changes in Earth's orbit known as Milankovitch cycles, and that the variations in individual interglacials can be partially explained by differences within this process. For example, Overpeck et al. (2006) argues that the previous interglacial was warmer because of increased solar radiation at high latitudes. The Liesecki and Raymo (2005) sediment reconstruction does not indicate significant differences between modern ice volume and previous interglacials, though some other studies do report slightly lower ice volumes/higher sea levels during the 120 ka and 400 ka interglacials (Karner et al. 2001, Hearty and Kaufman 2000). Temperature changes at the typical equatorial site are believed to have been significantly less than the changes observed at high latitude.

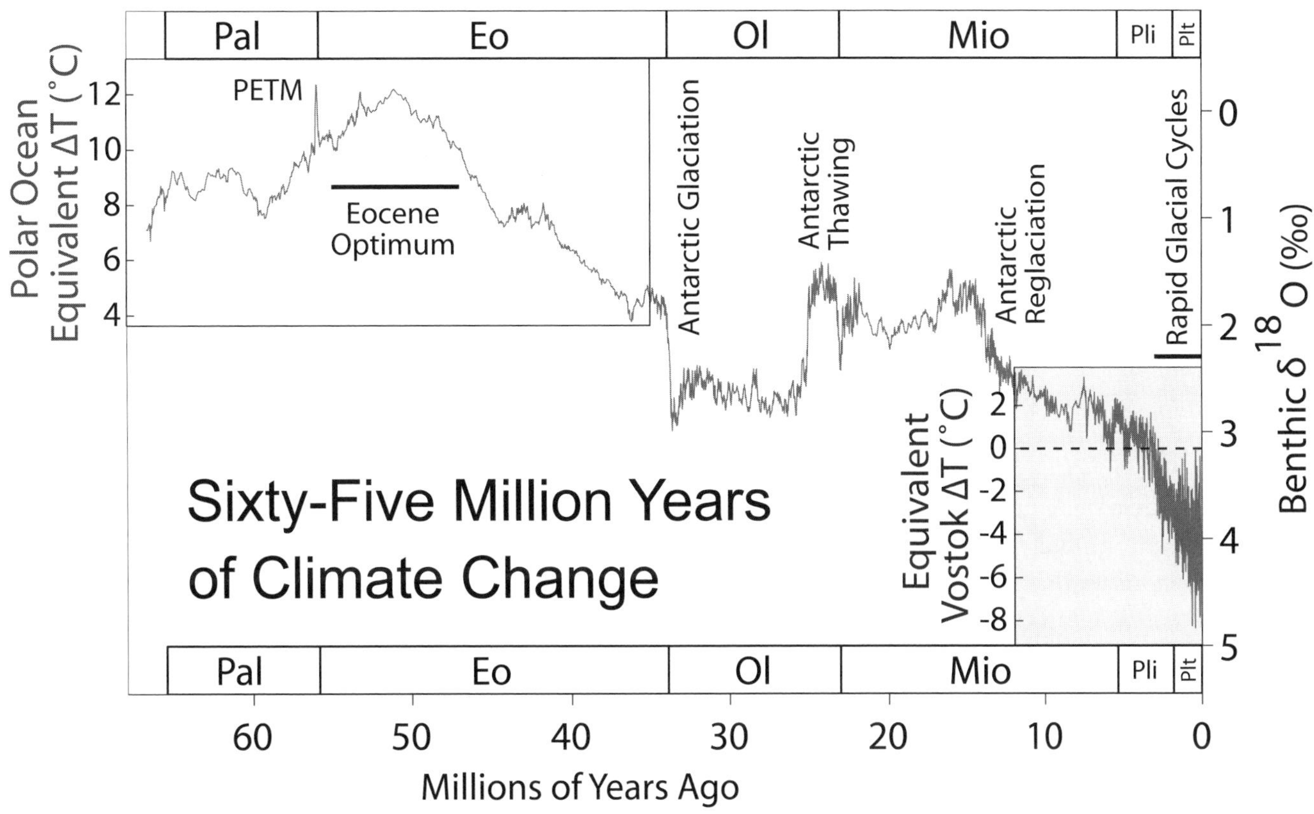

This figure shows climate change over the last 65 million years. The data is based on a compilation of oxygen isotope measurements (δ18O) on benthic foraminifera by Zachos et al. (2001), which reflect a combination of local temperature changes in their environment and changes in the isotopic composition of seawater associated with the growth and retreat of continental ice sheets.

Because it is related to both factors, it is not possible to uniquely tie these measurements to temperature without additional constraints. For the most recent data, an approximate relationship to temperature can be made by observing that the oxygen isotope measurements of Lisiecki and Raymo (2005) are tightly correlated to temperature changes at Vostok, Antarctica, as established by Petit et al. (1999). Present day is indicated as 0. For the oldest part of the record, when temperatures were much warmer than today, it is possible to estimate temperature changes in the polar oceans (where these measurements were made) based on the observation that no significant ice sheets existed and hence all fluctuation (in δ18O) must result from local temperature changes (as reported by Zachos et al.).

The intermediate portion of the record is dominated by large fluctuations in the mass of the Antarctic ice sheet, which first nucleates approximately 34 million years ago, then partially dissipates around 25 million years ago, before re-expanding toward its present state 13 million years ago. These fluctuations make it impossible to constrain temperature changes without additional controls. Significant growth of ice sheets did not begin in Greenland and North America until approximately 3 million years ago, following the formation of the Isthmus of Panama by continental drift. This ushered in an era of rapidly cycling glacials and interglacials (upper right). Also appearing on this graph are the Eocene Climatic Optimum, an extended period of very warm temperatures, and the Paleocene-Eocene Thermal Maximum (labeled PETM). Due to the coarse sampling and averaging involved in this record, it is likely that the full magnitude of the PETM is underestimated by a factor of 2 to 4 times its apparent height.

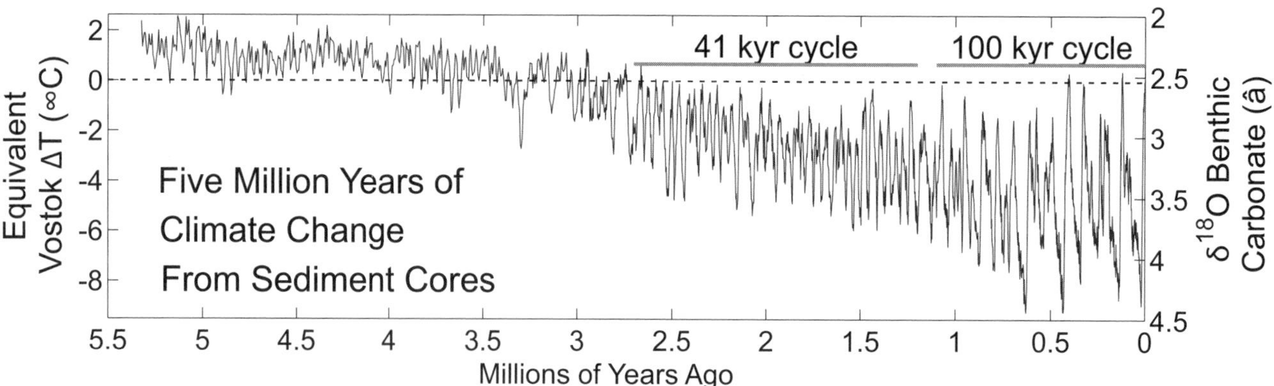

This figure shows the climate record of Lisiecki and Raymo (2005) constructed by combining measurements from 57 globally distributed deep-sea sediment cores. The measured quantity is oxygen isotope fractionation in benthic foraminifera, which serves as a proxy for the total global mass of glacial ice sheets.

Lisiecki and Raymo constructed this record by first applying a computer-aided process of adjusting individual "wiggles" in each sediment core to have the same alignment (i.e., wiggle matching). Then the resulting stacked record is orbitally tuned by adjusting the positions of peaks and valleys to fall at times consistent with an orbitally driven ice model (see Milankovitch Cycles). Both sets of these adjustments are constrained to be within known uncertainties on sedimentation rates and consistent with independently dated tie points (if any). Constructions of this kind are common, however, they presume that ice sheets are orbitally driven, and hence data such as this cannot be used in establishing the existence of such a relationship.

The observed isotope variations are very similar in shape to the temperature variations recorded at Vostok, Antarctica, during the 420 kyr for which that record exists. Hence, the right-hand scale of the figure was established by fitting the reported temperature variations at Vostok (Petit et al. 1999) to the observed isotope variations. As a result, this temperature scale should be regarded as approximate and its magnitude is only representative of Vostok changes. In particular, temperature changes at polar sites, such as Vostok, frequently exceed the changes observed in the tropics or in the global average. A horizontal line at 0 degrees C indicates modern temperatures (ca. 1950).

Labels are added to indicate regions where 100 kyr and 41 kyr cyclicity is observed. These periodicities match periodic changes in Earth's orbital eccentricity and obliquity, respectively, and have been previously established by other studies (not relying on orbital tuning).

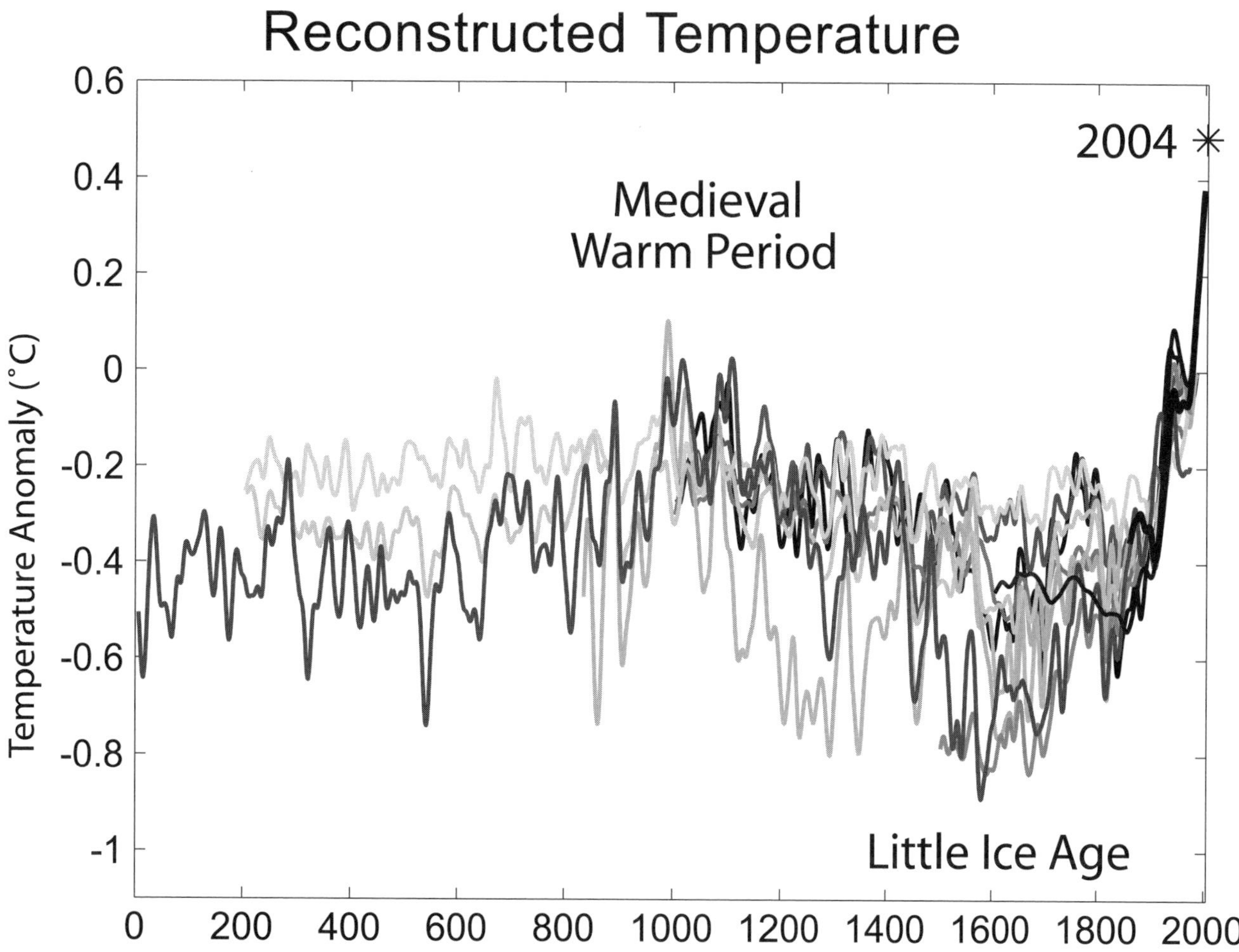

This image is a comparison of 10 different published reconstructions of mean temperature changes during the last 2,000 years. More recent reconstructions are plotted toward the front and in redder colors, older reconstructions appear toward the back and in bluer colors. An instrumental history of temperature is also shown in black. The medieval warm period and Little Ice Age are labeled at roughly the times when they are historically believed to occur, though it is still disputed whether these were truly global, or only regional events. The single, unsmoothed annual value for 2004 is also shown for comparison.

It is unknown which, if any, of these reconstructions is an accurate representation of climate history; however, these curves are a fair representation of the range of results appearing in the published scientific literature. Hence, it is likely that such reconstructions, accurate or not, will play a significant role in the ongoing discussions of global climate change and global warming.

For each reconstruction, the raw data has been decadally smoothed with a $\delta = 5$ yr. Gaussian weighted moving average. Also, each reconstruction was adjusted so that its mean matched the mean of the instrumental record during the period of overlap.

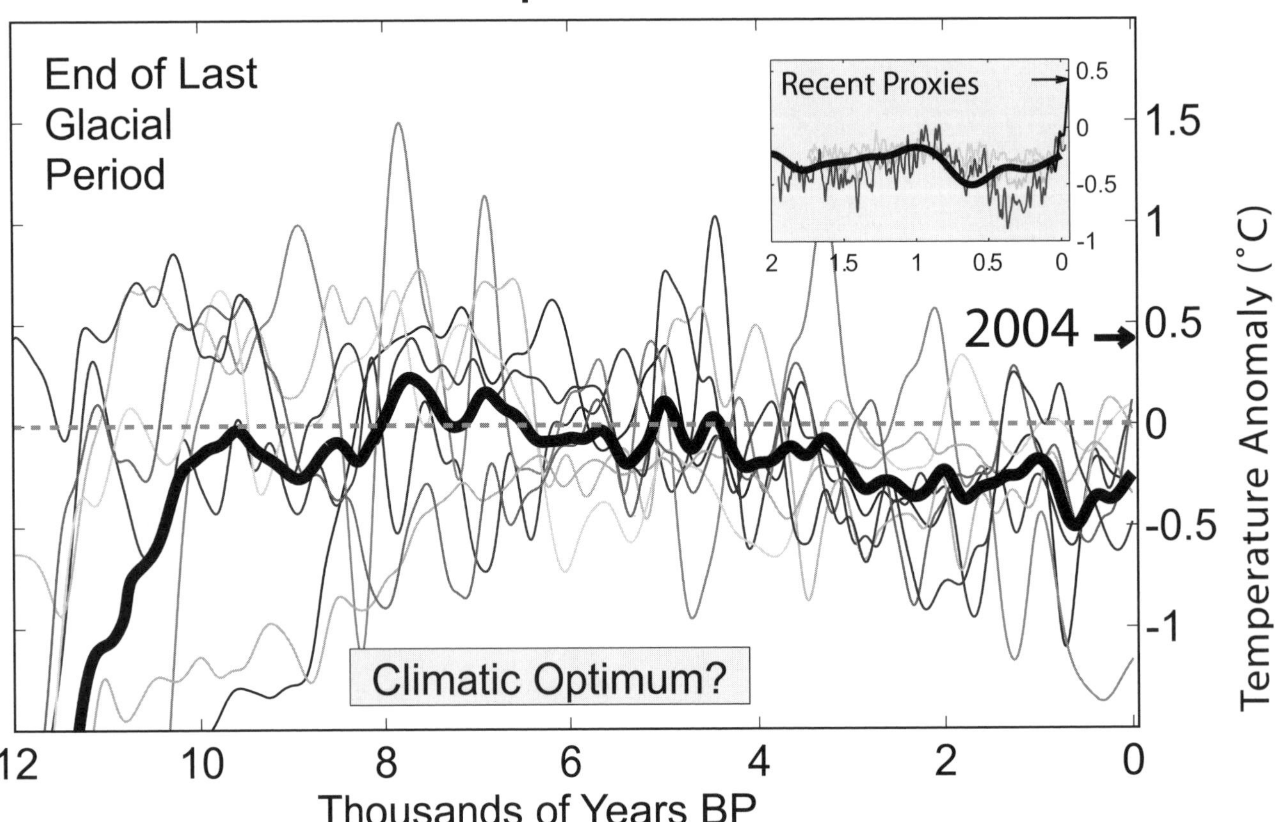

The main figure shows eight records of local temperature variability on multi-centennial scales throughout the course of the Holocene, and an average of these (thick dark line). The records are plotted with respect to the mid-20th-century average temperatures, and the global average temperature in 2004 is indicated. The inset plot compares the most recent two millennia of the average to other high-resolution reconstructions of this period. At the far left of the main plot, the climate emerges from the last glacial period of the current ice age into the relative stability of the current interglacial. There is general scientific agreement that during the Holocene, temperatures have been quite stable compared to the fluctuations during the preceding glacial period. The average curve above supports this belief. However, there is a slightly warmer period in the middle, which might be identified with the proposed Holocene climatic optimum. The magnitude and nature of this warm event is disputed, and it may have been largely limited to summer months and/or high northern latitudes.

Because of the limitations of data sampling, each curve in the main plot was smoothed, and consequently, this figure cannot resolve temperature fluctuations faster than approximately 300 years. Further, while 2004 appears warmer than any other time in the long-term average, an observation that might be a sign of global warming, the 2004 measurement is from a single year. It is impossible to know whether similarly large short-term temperature fluctuations may have occurred at other times, but are unresolved by the resolution available in this figure. The next 150 years will determine whether the long-term average centered on the present appears anomalous with respect to this plot. Since there is no scientific consensus on how to reconstruct global temperature variations during the Holocene, the average shown here should be understood as only a rough, quasi-global approximation to the temperature history of the Holocene. In particular, higher resolution data and better spatial coverage could significantly alter the apparent long-term behavior.

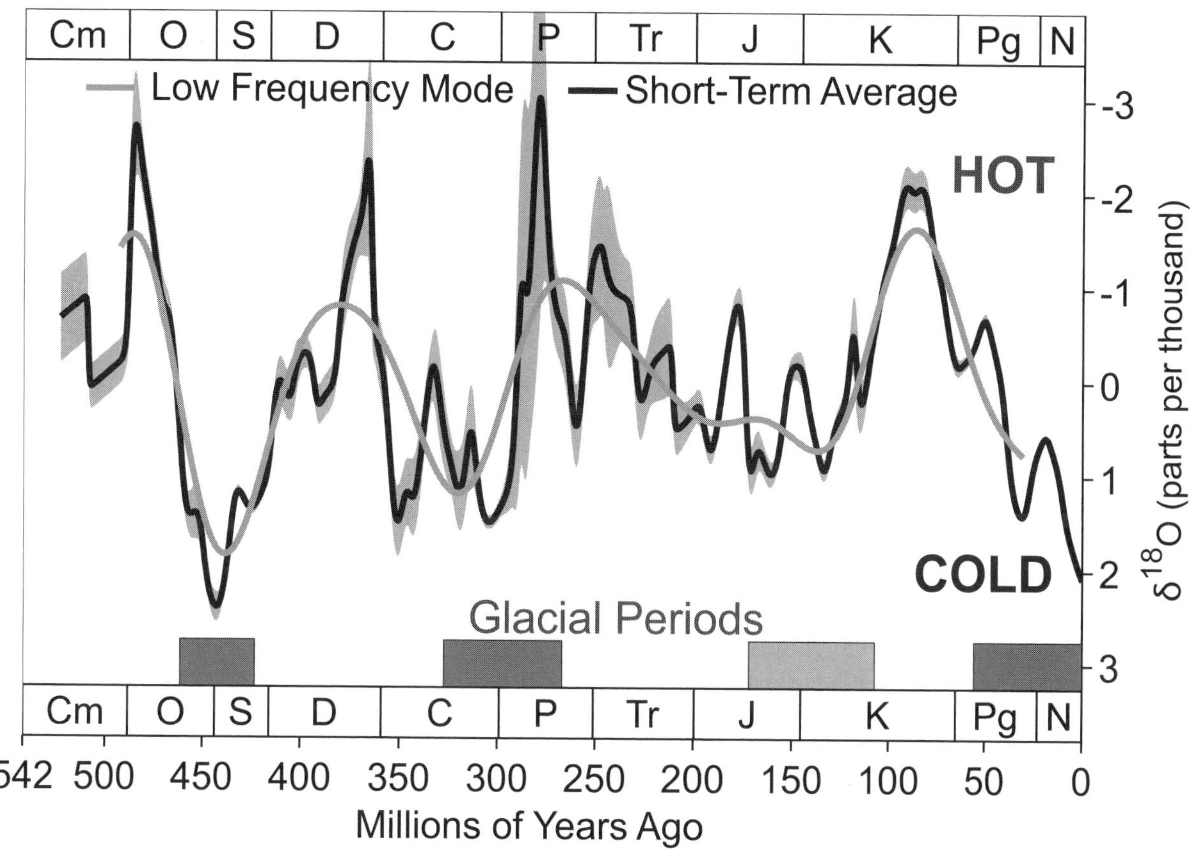

This figure shows the long-term evolution of oxygen isotope ratios during the Phanerozoic eon as measured in fossils, reported by Veizer et al. (1999), and updated online in 2004 [1]. Such ratios reflect both the local temperature at the site of deposition and global changes associated with the extent of continental glaciation. As such, relative changes in oxygen isotope ratios can be interpreted as rough changes in climate. Quantitative conversion between this data and direct temperature changes is a complicated process subject to many systematic uncertainties; however, it is estimated that each 1 part 1,000 change in δ18O represents roughly a 1.5–2 degrees C change in tropical sea surface temperatures (Veizer et al. 2000). Also shown on this figure are blue bars showing periods when geological criteria (Frakes et al. 1992) indicate cold temperatures and glaciation as reported by Veizer et al. (2000). All data presented here have been adjusted to the 2004 ICS geologic timescale. The "short-term average" was constructed by applying a δ = 3 Myr Gaussian weighted moving average to the original 16,692 reported measurements. The gray bar is the associated 95 percent statistical uncertainty in the moving average. The "low frequency mode" is determined by applying a band-pass filter to the short-term averages in order to select fluctuations on timescales of 60 Myr or greater.

On geologic timescales, the largest shift in oxygen isotope ratios is due to the slow radiogenic evolution of the mantle. It is not possible to draw any conclusion about very long-term (>200 Myr) changes in temperatures from this data alone. However, it is usually believed that temperatures during the present cold period and during the Cretaceous thermal maximum are not greatly different from cold and hot periods during most of the rest the Phanerozoic. Some recent work has disputed this (Royer et al. 2004), suggesting instead that the highs and lows in the early part of the Phanerozoic were both significantly warmer than their recent counterparts. Common symbols for geologic periods are plotted at the top and bottom of the figure for reference.

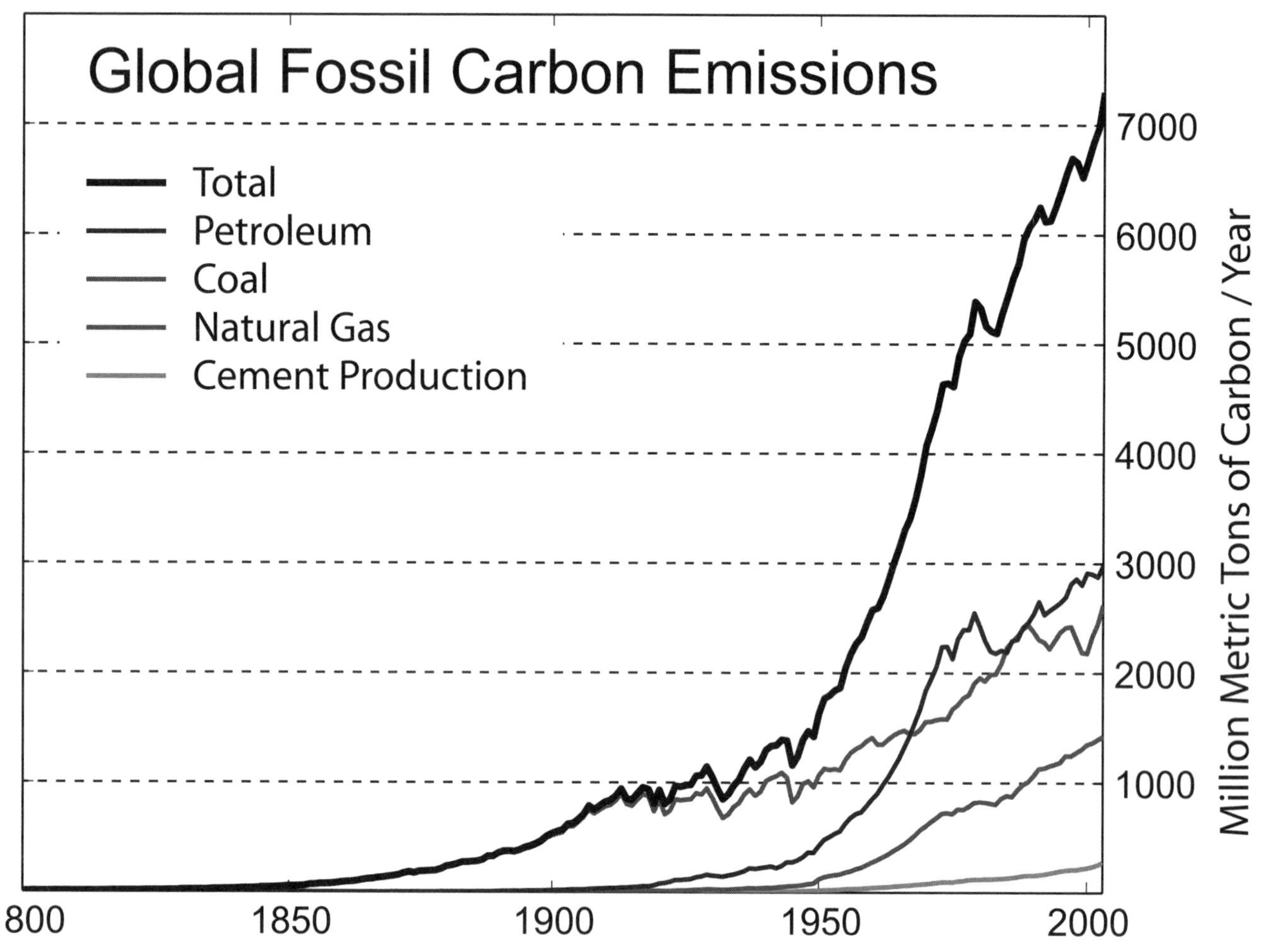

Global annual fossil fuel carbon dioxide emissions, in million metric tons of carbon, as reported by the Carbon Dioxide Information Analysis Center.

Original data: [full text] Marland, G., T. A. Boden, and R. J. Andres (2003). "Global, Regional, and National CO2 Emissions" in *Trends: A Compendium of Data on Global Change*. Oak Ridge, Tenn., U.S.A.: Carbon Dioxide Information Analysis Center, Oak Ridge National Laboratory, U.S. Department of Energy.

The data are originally presented in terms of solid (e.g., coal), liquid (e.g., petroleum), gas (i.e., natural gas) fuels, and separate terms for cement production and gas flaring (i.e., natural gas lost during oil and gas mining). In the plotted figure, gas flaring (the smallest of all categories) was added to the total for natural gas. Note that the carbon dioxide releases from cement production result from the thermal decomposition of limestone into lime, and so technically are not a fossil fuel source.

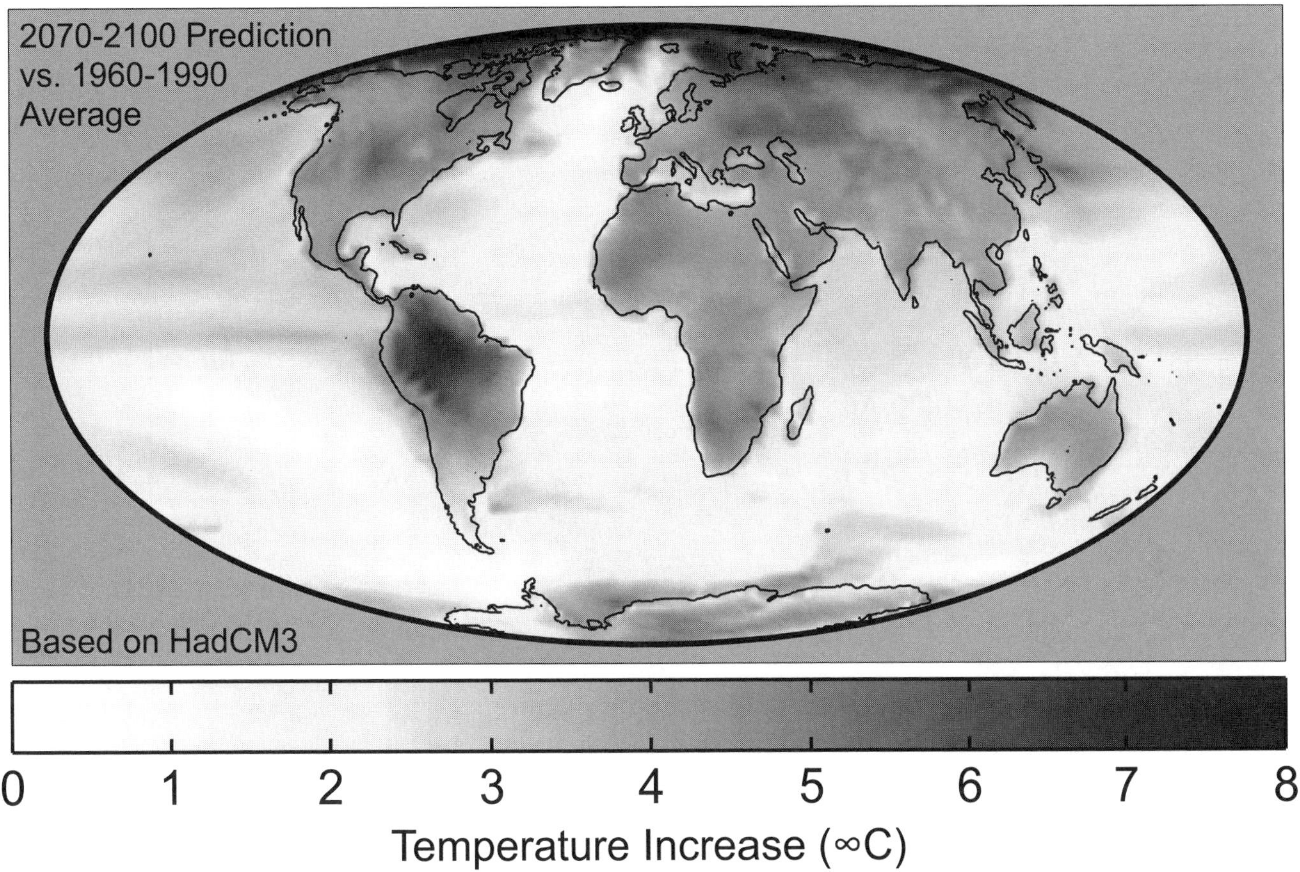

This figure shows the predicted distribution of temperature change due to global warming from the Hadley Centre HadCM3 climate model. These changes are based on the IS92a ("business as usual") projections of carbon dioxide and other greenhouse gas emissions during the next century, and assume normal levels of economic growth and that no significant steps are taken to combat global greenhouse gas emissions.

The plotted gray tints show predicted surface temperature changes expressed as the average prediction for 2070–2100 relative to the model's baseline temperatures in 1960–90. The average change is 3.0 degrees C, placing this model on the lower half of the Intergovernmental Panel on Climate Change's 1.4-5.8 degrees C predicted climate change from 1990 to 2100. As can be expected from their lower specific heat, continents are expected to warm more rapidly than oceans, with an average of 4.2 degrees C and 2.5 degrees C in this model, respectively. The lowest predicted warming is 0.55 degree C south of South America and the highest is 9.2 degrees C in the Arctic Ocean (points exceeding 8 degrees C are plotted as black).

This model is fairly homogeneous, except for strong warming around the Arctic Ocean related to melting sea ice and strong warming in South America related to predicted changes in the El Niño cycle and Brazillian rainforest. This pattern is not a universal feature of models, as other models can produce large variations in other regions (e.g., Africa and India) and less extreme changes in places like South America.

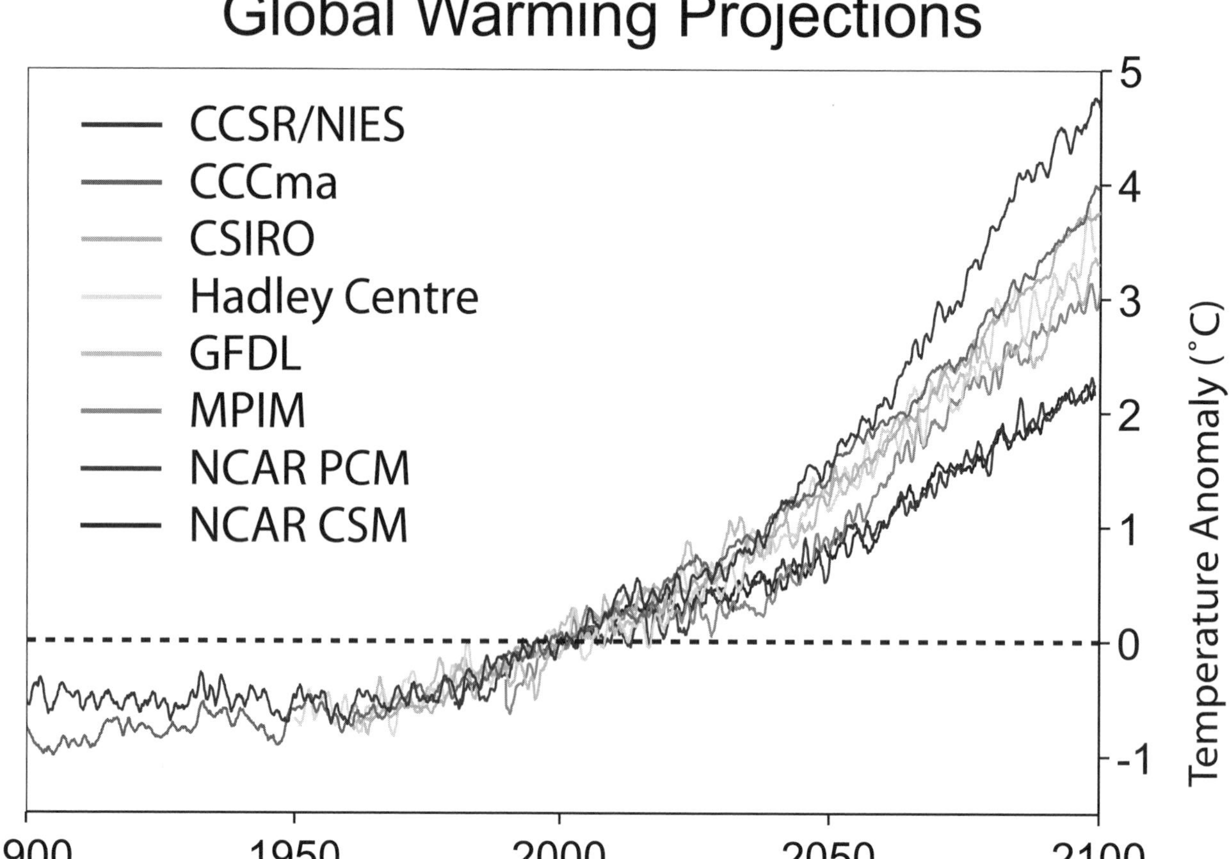

This figure shows climate model predictions for global warming under the SRES A2 emissions scenario relative to global average temperatures in 2000. The A2 scenario is characterized by a politically and socially diverse world that exhibits sustained economic growth, but does not address the inequities between rich and poor nations, and takes no special actions to combat global warming or environmental change issues. This world in 2100 is characterized by large population (15 billion), high total energy use, and moderate levels of fossil fuel dependency (mostly coal). At the time of the IPCC *Third Assessment Report*, the A2 scenario was the most well-studied of the SRES scenarios.

The IPCC predicts global temperature change of 1.4–5.8 degrees C due to global warming from 1990 to 2100 (IPCC 2001a). As evidenced above (a range of 2.5 degrees C in 2100), much of this uncertainty results from disagreement among climate models, though additional uncertainty comes from different emissions scenarios.

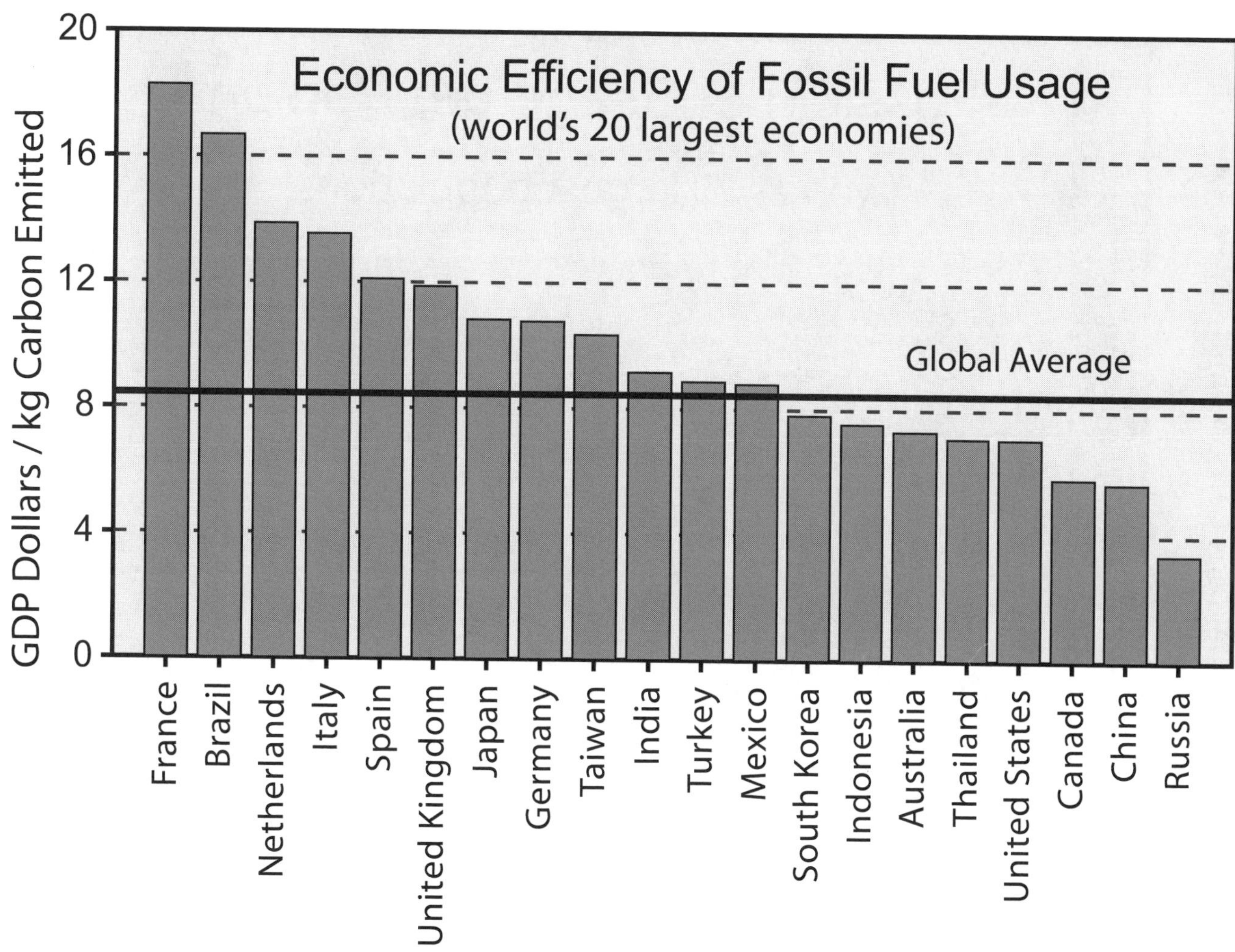

This figure shows an estimate of how efficiently the world's 20 largest economies convert fossil fuel usage into wealth as expressed by the ratio of their gross domestic product (or GDP, calculated by the method of purchasing power parity in U.S. dollars) over the number of kilograms of fossil fuel carbon released into the atmosphere each year. The relatively narrow range of variation between most countries in this figure suggests that the pursuit of wealth in the present world is strongly tied to the availability of fossil fuel energy sources.

As countries may be reluctant to combat fossil fuel emissions in ways that cause economic decline, this figure serves to suggest the degree to which different large economies can decrease emissions through short-term improvements in efficiency and alternative fuel programs.

The two countries that produce the highest GDP per kilogram carbon, Brazil and France, are heavily reliant on alternative energy sources, hydroelectric and nuclear power, respectively.

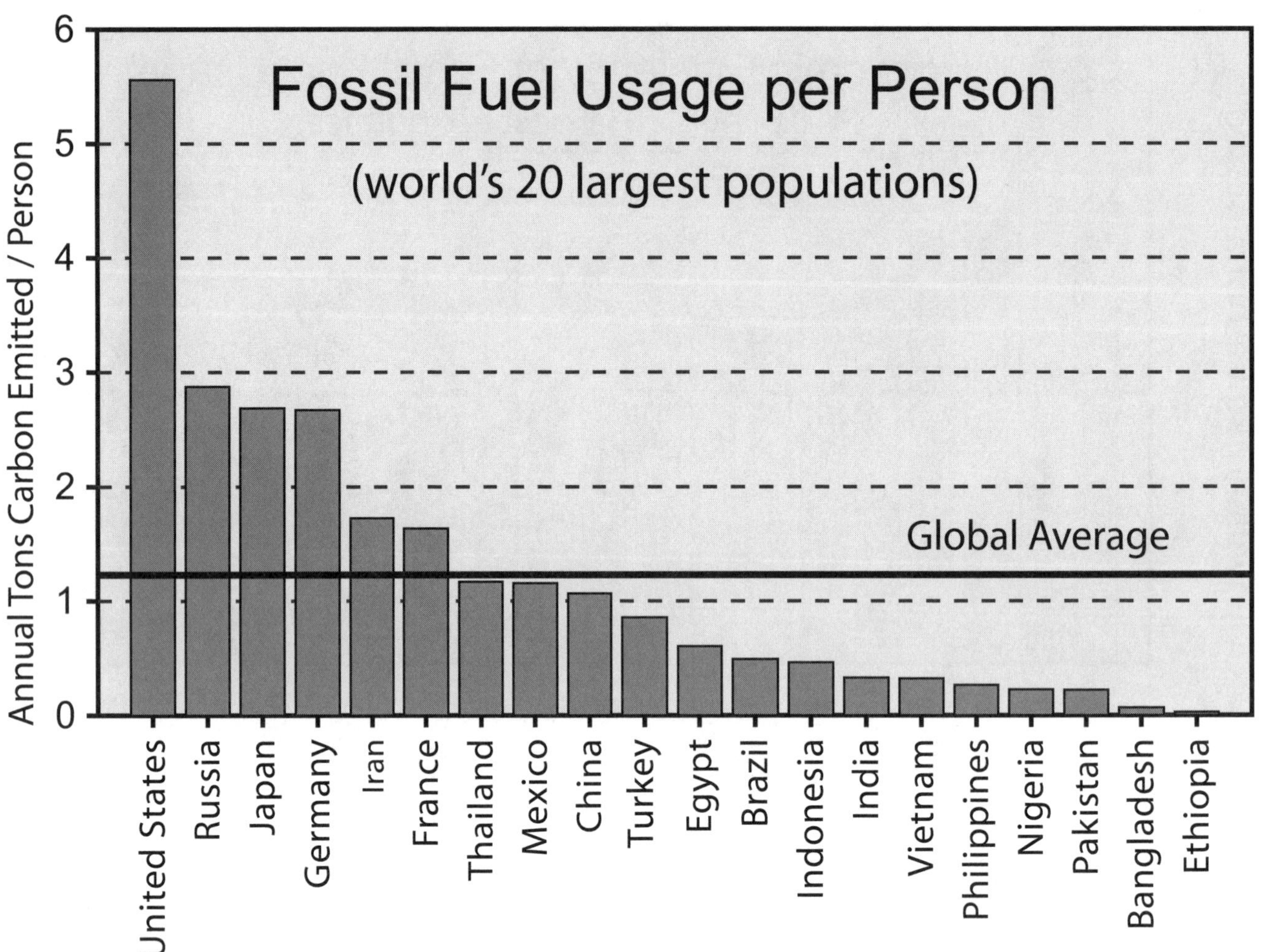

This figure shows the disparity in fossil fuel consumption per capita for the countries with the 20 largest populations. The large range of variation is indicative of the separation between the rich, industrialized nations and the poor/developing nations. The global average is also shown.

As most countries desire wealth and aim to develop that wealth through the development of industry, this figure suggests the degree to which poor nations may strive to increase their emissions in the course of trying to match the industrial capacity of the developed world. Managing such increases and dealing with the apparent social inequality of the present system will be one of the challenges involved in confronting global warming.

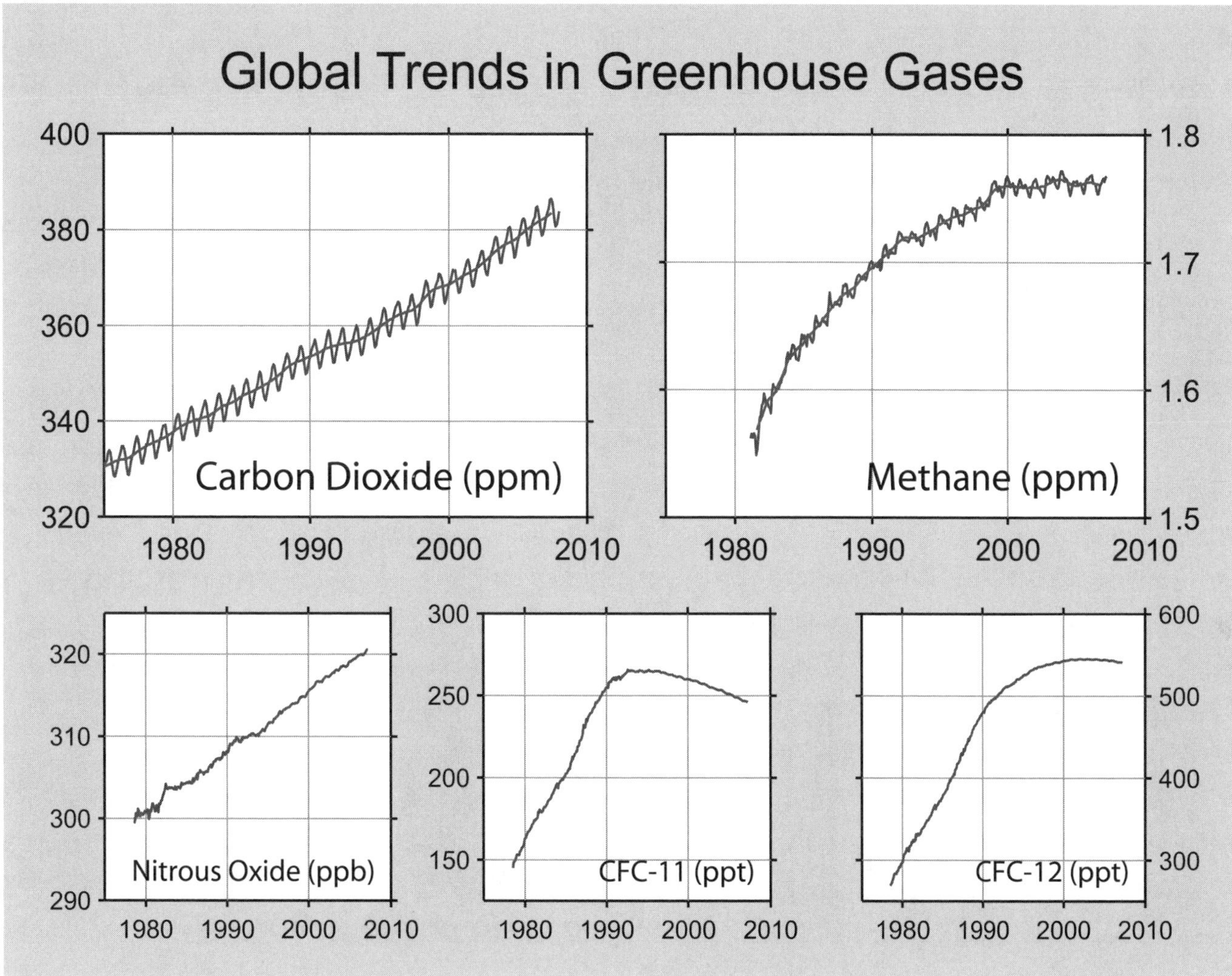

Global trends in major greenhouse gas concentrations. The rise of greenhouse gases, and their resulting impact on the greenhouse effect, are believed to be responsible for most of the increase in global average temperatures during the last 50 years. This change, known as global warming, has provoked calls to limit the emissions of these greenhouse gases (e.g., the Kyoto Protocol). Notably, the chlorofluorocarbons CFC-11 and CFC-12 shown above have undergone substantial improvement since the Montreal Protocol severely limited their release due to the damage they were causing to the ozone layer.

At present, approximately 99 percent of the 100-year global warming potential for all new emissions can be ascribed to just three gases: carbon dioxide, methane, and nitrous oxide.

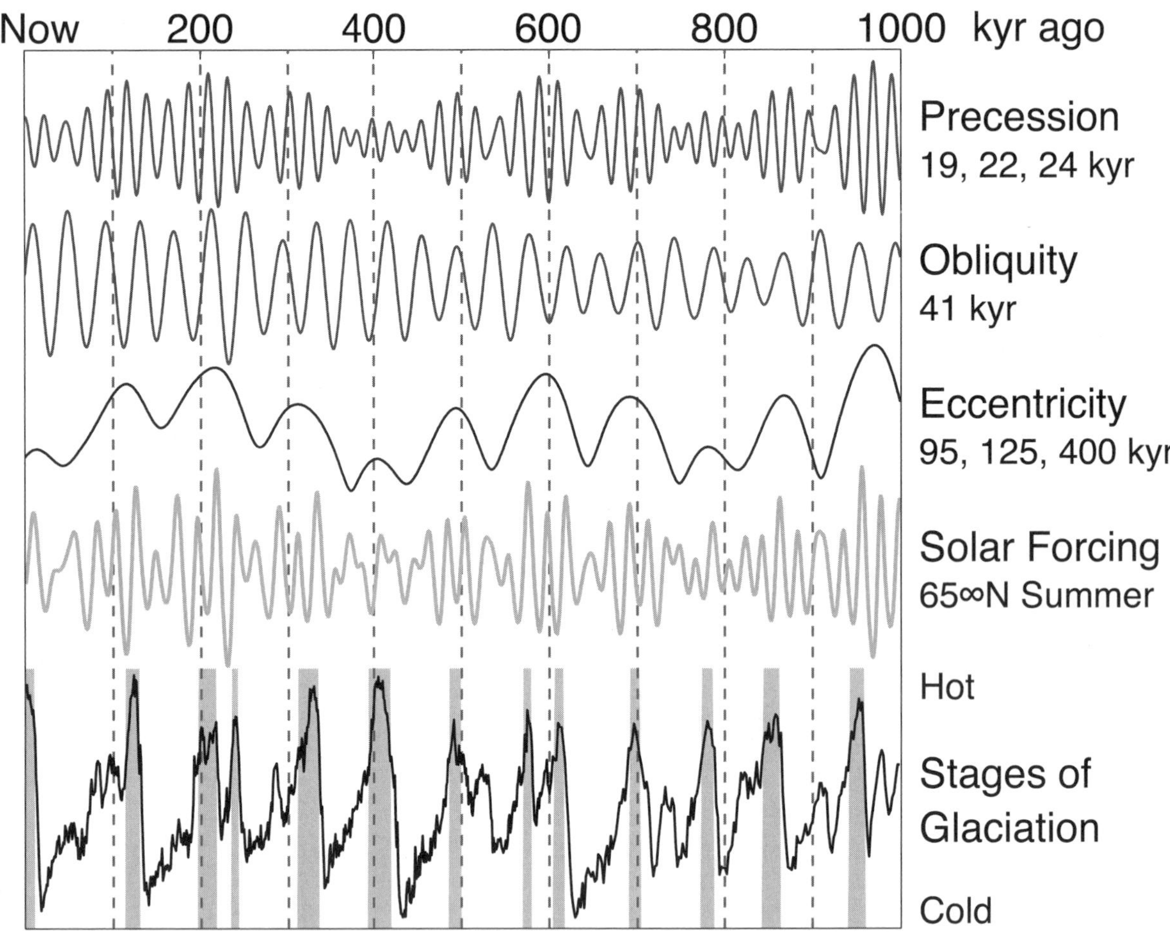

Milankovitch Cycles

The Earth's orbit around the sun is slightly elliptical. Over time, the gravitational pull of the moon and other planets causes the Earth's orbit to change following a predictable pattern of natural rhythms, known as Milankovitch cycles. Over a ~100,000-year cycle, the Earth migrates from an orbit with near-zero eccentricity (a perfect circle) to one with approximately 6 percent eccentricity (a slight ellipse). In addition, the tilt of the Earth axis, known as its obliquity, varies from 21.5 to 24.5 degrees, with a 41,000-year rhythm. And last, the orientation of the Earth's axis rotates with a ~20,000-year cycle relative to the orientation of the Earth's orbit. This cycle, known as "precession," affects the intensity of the seasons.

The figure shows the pattern of changes in each of the three modes of orbital variability: eccentricity, obliquity, and precession. These changes in the Earth's orbit lead to a complex series of changes in the amount of sunlight that a given location on Earth can expect to receive during a given season. An example is shown for summer sunlight near the Arctic circle. Sunlight at this location is believed to influence the growth and decay of ice sheets during ice ages. The last line shows measured changes in climate during the last million years, with warm interglacials highlighted in gray bands. As can be seen, such interglacials appear to preferentially occur near maxima in eccentricity and slightly following times of maximum summer sunlight.

Index

Index note: Volume numbers are in **boldface**. Article titles and their page numbers are in **boldface**.

B

J

P

S

U

X

Y

Z

Photo Credits

VOLUME 1: Library of Congress: 232, 381; National Aeronautics and Space Administration: 49, 116, 125, 152, 168, 279, 284, 397, 418, 542; National Oceanic and Atmospheric Administration: 21, 80, 236, 335; National Science Foundation: 56; Photos.com: 2,5,13,17,40, 35, 71, 85, 90, 101, 130, 134, 148, 177, 180, 186, 195, 209, 212, 221, 240, 245, 250, 264, 269, 272, 304, 310, 314, 318, 340, 346, 348, 355, 372, 406, 411,423, 427, 441, 456, 459, 462, 469, 501, 521, 526; Sandia National Laboratories: 510; United Nations Educational, Scientific and Cultural Organization: 470; U.S. Archives and Records: 391; USAID: 202, 259, 434, 448; U.S. Coast Guard: 324; U.S. Department of Agriculture Agricultural Research Service: 18, 141, 477; U.S. Department of Energy: 293, 391; U.S. Embassy: 161; U.S. Environmental Protection Agency: 62, 491; U.S. Fish and Wildlife Service: 408; U.S. Geological Survey: 26; U.S. Navy: 535; Wikimedia Commons: 108 (left and right), 364, 482.

VOLUME 2: Argonne National Laboratory: 859; Federal Emergency Management Agency: 798; Lawrence Livermore National Laboratory: 1036; Morguefile: 929; National Aeronautics and Space Administration: 594 (left and right), 704, 725, 735, 869, 976; National Oceanic and Atmospheric Administration: 563, 969, 983; National Science Foundation: 668, 946; Photos.com: 553, 556, 568, 575, 580, 587, 606, 625, 632, 650, 657, 685, 695, 713, 716, 742, 760, 787, 805, 812, 819, 822, 831, 843, 876, 885, 900, 908, 920, 936, 952, 1009, 1012, 1051, 1060, 1073; StockExhng.com: 999; Taylor Shellfish Farms: 1044; USAID/Kendra Helmer: 1020; U.S. Department of Agriculture Agricultural Research Service: 849, 960; U.S. Department of Defense: 728, 767, 913; U.S. Department of Energy: 643; U.S. Department of Housing and Urban Development: 792; U.S. Department of the Interior: 891; U.S. Fish and Wildlife Service: 1068; U.S. Geological Survey: 616, 779, 941; U.S. Navy: 992; White House: 662; Wikimedia Commons: 864, 1078.

VOLUME 3: Celestia Motherlode/Neethis; 1263; Climate Prediction Center: 1281; Federal Emergency Management Agency: 1447; Getty Images: 1464; Andy Green: 1378, 1482, 1489; International Institute for Sustainable Development: 1238; Morguefile: 1477; National Aeronautics and Space Administration: 1147, 1163, 1204, 1235, 1269, 1299, 1304, 1307, 1352, 1418, 1426, 1459; National Aeronautics and Space Administration/Boeing: 1181; National Oceanic and Atmospheric Administration: 1157, 1167, 1205, 1215, 1232, 1243, 1297, 1310, 1409, 1471, 1511; National Park Service: 1357; National Science Foundation: 1121; Nellis Air Force Base: 1193; Oxfam East Africa: 1027; Photos.com: 1084, 1087, 1094, 1114, 1126, 1157, 1253, 1286, 1324, 1333, 1336, 1341, 1349, 1375, 1431, 1437, 1440, 1454, 1498; Pieternella Pieterse/Concern Worldwide/USAID: 1260; Sebastian Schulze: 1196; StockExhng.com: 1214, 1224, 1248, 1274, 1321; United Nations/Evan Schneider: 1387; U.S. Department of Agriculture Agricultural Research Service: 1107; U.S. Department of Energy: 1188; U.S. Geological Survey: 1142; U.S. House of Representatives: 1392; U.S. Navy: 1362; White House: 1174; Wikimedia Commons: 1098, 1139, 770, 1314, 1370, 1406; Wikimedia Commons:/John Southern: 1344; Wikimedia Commons:/Yoo Chul Chung: 836; J. Young, Natural History Museum, London/U.S. Department of Energy: 1403.